Introduction to
Internal Combustion Engines

Other SAE books of interest

Engine Combustion: Pressure Measurement and Analysis
by David R. Rogers
(Product Code: R-388)

An Introduction to Engine Testing and Development
by Richard D. Atkins
(Product Code: R-344)

Internal Combustion Engine Handbook
by Richard Van Basshuysen, Fred Schaefer
(Product Code: R-345)

For more information or to order a book, contact SAE International at
400 Commonwealth Drive, Warrendale, PA 15096-0001 USA;
phone 877-606-7323 (U.S. and Canada only) or 724-776-4970
(outside U.S. and Canada);
fax 724-776-0790;
email CustomerService@sae.org;
website: http://books.sae.org

Introduction to
Internal Combustion Engines

By Richard Stone

Department of Engineering Science
University of Oxford

Fourth Edition

Warrendale, Pennsylvania
USA

400 Commonwealth Drive
Warrendale, PA 15096-0001 USA

E-mail: CustomerService@sae.org
Phone: 877-606-7323 (inside USA and Canada)
         724-776-4970 (outside USA)
Fax: 724-776-1615

ISBN 978-0-7680-2084-7
Library of Congress Catalog Number 2012943931
SAE Order No. R-391

Information contained in this work has been obtained by SAE International from sources believed to be reliable. However, neither SAE International nor its authors guarantee the accuracy or completeness of any information published herein and neither SAE International nor its authors shall be responsible for any errors, omissions, or damages arising out of use of this information. This work is published with the understanding that SAE International and its authors are supplying information, but are not attempting to render engineering or other professional services. If such services are required, the assistance of an appropriate professional should be sought.

To purchase bulk quantities, please contact:
SAE Customer Service
E-mail: CustomerService@sae.org
Phone: 877-606-7323 *(inside USA and Canada)*
         724-776-4970 *(outside USA)*
Fax: 724-776-1615

Visit the SAE Bookstore at
http://store.sae.org

Short contents

Contents

Preface to the Fourth Edition

This book is intended to provide engineers and students with the background that is pre-supposed in many articles, papers and advanced texts. Since the book is primarily aimed at students, it has sometimes been necessary to give only outline or simplified explanations. However, numerous references have been made to sources of further information.

Internal combustion engines form part of most thermodynamics courses at Colleges and Universities. This book should be useful to students who are following specialist options in internal combustion engines, and also to students at earlier stages in their courses – especially with regard to laboratory work.

Practising engineers should also find the book useful when they need an overview of the subject, or when they are working on particular aspects of internal combustion engines that are new to them.

The subject of internal combustion engines draws on many areas of engineering: thermodynamics and combustion, fluid mechanics and heat transfer mechanics, stress analysis, materials science, electronics and computing. However, internal combustion engines are not just subject to thermodynamic or engineering considerations – the commercial (marketing, sales etc.) and economic aspects are also important, and these are discussed as they arise.

In order to make this book self-contained, Appendix A contains thermodynamic tables for combustion calculations; this should also ensure consistency in the answers if these tables are used in any calculations. The tables contain equilibrium constants, and molar thermodynamic property data (internal energy, enthalpy, Gibbs energy and entropy) for the species to be found in the reactants and products. The datum for enthalpy adopted here is a datum of *zero enthalpy for elements when they are in their standard state at a temperature of 25°C*. The enthalpy of any molecule at 25°C will thus correspond to its enthalpy of formation, ΔH_f^0. This choice of datum (although not used to my knowledge in any other tables) will be seen to facilitate energy balances in combustion. Unlike tables that use an identical datum for reactant and product species, there is no need to include enthalpies or internal energies of reaction. Chapter 3 contains material to explain and illustrate the use of these tables. The relevant examples have been solved using the old and new approaches for two reasons. Firstly, to demonstrate that combustion problems can be analysed more simply using data presented in the format of Appendix A. Secondly, because some students may have to use the format in existing tables for their examinations.

Since the first publication of *Introduction to Internal Combustion Engines* the preferred practice for denoting molar quantities is to use a lower case letter (to emphasise that it is a specific quantity) with a tilde above, for example ñ. However, for consistency within the book (and in common with many other publications) molar quantities will be denoted here by upper case letters. This requires the reader to decide whether the symbol refers to a molar specific quantity or a property value of a complete system.

Preparing the fourth edition has also provided the opportunity to include further worked examples and many problems (with numerical answers). There is a Solutions Manual available online for free download for lecturers adopting this book at www.palgrave.com/engineering/stone which contains the questions, answers and the thermodynamic data that can be found in Appendix A. This provides an opportunity for some additional discussion to be included.

The first seven chapters still provide an introduction to internal combustion engines, and these can be read in sequence.

Chapter 1 provides an introduction, with definitions of engine types and operating principles. The essential thermodynamics is provided in chapter 2, while chapter 3 provides the background in combustion and fuel chemistry. The differing needs of spark ignition engines, direct injection spark ignition engines (a new chapter) and compression ignition engines are discussed in chapters 4, 5 and 6 respectively. Chapter 7 describes how the induction and exhaust processes are controlled.

Chapter 12 provides a discussion of some topics in the mechanical design of engines, and explains some of the criteria that influence material selection. Chapter 14 provides an overview of the experimental equipment that is an essential part of engine development. This chapter has substantial additions to describe gas analysis equipment and its use, and an introduction to computer-based combustion analysis.

The remaining chapters pre-suppose a knowledge of the material in the earlier chapters. Chapter 8 contains a treatment of two-stroke engines, a topic that has gained widened interest since the first edition; both spark ignition and compression ignition engines are discussed.

Chapter 9 is mainly concerned with turbulence: how it is measured, how it is defined, and how it affects combustion. Such knowledge is pre-supposed in technical papers and the contributions in specialist texts.

Chapter 10 discusses the application of turbochargers to spark ignition and compression ignition engines, and there is a new section on supercharging. There is further material on turbocharging in chapter 11 (Engine modelling) as a turbocharged engine is used as an example.

Chapter 12 covers aspects of mechanical design, and this theme continues with chapter 13. The treatment of engine cooling in chapter 13 has been provided since it is a subject that is often overlooked, yet it is an area that offers scope for reductions in both fuel consumption and emissions.

The material in the book has been used by the author for teaching final year students, contributing to MSc courses, and short courses for industry. These experiences have been invaluable in preparing the fourth edition, as have the written comments from readers. The material within the book has come from numerous sources. The published sources have been acknowledged, but of greater importance have been the conversations and discussions with colleagues and researchers in the area of internal combustion engines, especially when they have provided me with copies of relevant publications. If you have a paper that contains material that might be included in a later edition, do please email me.

I would again welcome any criticisms or comments on the book, either concerning the detail or the overall concept.

Autumn 2011

RICHARD STONE
richard.stone@eng.ox.ac.uk

Acknowledgements

The author and publisher wish to thank the following, who have kindly given permission for the use of copyright material:

The American Society of Mechanical Engineers, New York, for figures from technical publications.

Dr W. J. D. Annand (University of Manchester) for figures from W. J. D. Annand and G. E. Roe, *Gas Flow in the Internal Combustion Engine*, published by Foulis, Yeovil, 1974.

Atlantic Research Associates, Tunbridge Wells, and Martin H. Howarth for three figures from M. H. Howarth, *The Design of High Speed Diesel Engines*, published by Constable, London, 1966.

Blackie and Son Ltd, Glasgow, for three figures from H. R. Ricardo and J. G. G. Hempson, *The High Speed Internal Combustion Engine*, 1980.

Butterworth–Heinemann, Oxford, for figures from technical publications.

Butterworths, Guildford, for five figures from K. Newton, W. Steeds and T. K. Garrett, *The Motor Vehicle*, 10th edn, 1983.

Cummins Turbo Technologies Ltd for two figures from technical publications.

The Engineer, London, for a figure from a technical publication.

Eureka Engineering Materials & Design, Horton Kirby, Kent, for a figure from a technical publication.

Ford of Europe, Inc., Brentwood, for eleven figures from Ford technical publications.

Froude Hofmann Ltd, Worcester, for two figures from *Publication No. 526/2*.

GKN Engine Parts Division, Maidenhead, for a figure from *Publication No. EPD 82100*.

HMSO. Crown Copyright is reproduced with the permission of the Controller of Her Majesty's Stationery Office.

Hutchinson Publishing Group Ltd, London, for a figure from A. Baker, *The Component Contribution* (co-sponsored by the AE Group), 1979.

IMechE and their publishers, Professional Engineering Publications.

Johnson Matthey Chemicals Ltd, Royston, for three figures from *Catalyst Systems for Exhaust Emission Control from Motor Vehicles*.

Longmans Group Ltd, Harlow, for four figures from G. F. C. Rogers and Y. R. Mayhew, *Engineering Thermodynamics*, 3rd edn, 1980; two figures from H. Cohen, G. F. C. Rogers and H. I. H. Saravanamuttoo, *Gas Turbine Theory*, 2nd edn, 1972.

Lucas CAV Ltd, London, for eight figures from *CAV Publications 586, 728, 730, 773* and *C2127E*, and a press release photograph.

Lucas Electrical Ltd, Birmingham, for two figures from publications *PLT 6339* (Electronic Fuel Injection) and *PLT 6176* (Ignition).

Mechanical Engineering Publications, Bury St Edmunds, for figures from technical publications.

MIT Press, Cambridge, Massachusetts, and Professor C. F. Taylor for a figure from C. F. Taylor, *The Internal Combustion Engine in Theory and Practice*, 1966/8.

Orbital Engine Corporation Ltd, Perth, Australia, for figures from technical publications.

Oxford University Press for two figures from Singer's *History of Technology*.

Patrick Stephens Ltd, Cambridge, for three figures from A. Allard, *Turbocharging and Supercharging*, 1982.

Pergamon Press Ltd, Oxford, for a figure from R. S. Benson and N. D. Whitehouse, *Internal Combustion Engines*, 1979; three figures from H. Daneshyar, *One-Dimensional Compressible Flow*, 1976.

Plenum Press, New York, and Professor John B. Heywood (MIT) for a figure from J. B. Heywood, *Combustion Modelling in Reciprocating Engines*, 1980.

Society of Automotive Engineers, Warrendale, Pennsylvania, for ten figures and copy from *SAE 790291, SAE 820167, SAE 821578* and *Prepr. No. 61A* (1968).

Verein Deutscher Ingenieure, Dusseldorf, for figures from technical publications.

Wärtsilä for figures from the *Sulzer Technical Review*, Nos 1 and 3 (1982).

Material is acknowledged individually throughout the text of the book.

Every effort has been made to trace all copyright holders but, if any have been inadvertently overlooked, the publisher will be pleased to make the necessary arrangements at the first opportunity.

Notation

abdc	after bottom dead centre
atdc	after top dead centre
A	piston area (m^2)
A_c	curtain area for poppet valve (m^2)
A_e	effective flow area (m^2)
A_f	flame front area (m^2)
A_o	orifice area (m^2)
A/F (or AFR)	air/fuel ratio
bbdc	before bottom dead centre
bdc	bottom dead centre
bmep	brake mean effective pressure (N/m^2)
bsac	break specific air consumption
bsfc	brake specific fuel consumption
btdc	before top dead centre
B	bore diameter (m)
BHP	brake horse power
c	sonic velocity (m/s)
c_p	specific heat capacity at constant pressure (kJ/kg K)
c_v	specific heat capacity at constant volume (kJ/kg K)
ca (or CA)	crank angle
C_D	discharge coefficient
C_o	orifice discharge coefficient
C_p	molar heat capacity at constant pressure (kJ/kmol K)
C_v	molar heat capacity at constant volume (kJ/kmol K)
CI	compression ignition
CoV	coefficient of variation
CV	calorific value (kJ/kg)
dc	direct current
dohc	double overhead camshaft
D_v	valve diameter (m)
DI	direct injection (compression ignition engine)
DISI	direct injection spark ignition
DPF	diesel particulate filter
EGR	exhaust gas recirculation
f	fraction of exhaust gas residuals
ff	turbulent flame factor
fmep	frictional mean effective pressure
fwd	front wheel drive
g	gravitational acceleration (m/s^2)
G	Gibbs energy (kJ/kg)
GDI	gasoline direct injection
h	specific enthalpy (kJ/kg); manometer height (m)
h_d	mean height of indicator diagram (m)
H	enthalpy (kJ)

HCV	higher calorific value (J/kg)
imep	indicated mean effective pressure (N/m^2)
isac	indicated specific air consumption (kg/MJ)
isfc	indicated specific fuel consumption (kg/MJ)
I	current (A)
IDI	indirect injection (compression ignition engine)
jv	just visible (exhaust smoke)
k	constant
K	equilibrium constant
KLSA	knock limited spark advance
l	length; connecting-rod length (m)
l_b	effective dynamometer lever arm length (m)
l_d	indicator diagram length (m)
L	stroke length (m); inductance (H)
L_D	duct length (m)
L_v	valve lift (m)
LBT	lean best torque
LCV	lower calorific value (J/kg)
LDA	laser Doppler anemometer
LDV	laser Doppler velocimeter
m	mass (kg)
$\dot{m}_a$	air mass flow rate (kg/s)
$\dot{m}_f$	fuel mass flow rate (kg/s)
m_r	reciprocating mass (kg)
mfb	mass fraction burn (equation 14.51)
M	mutual inductance (H); molar mass (kg); moment of momentum flux (N m)
MBT	minimum (ignition) advance for best torque (degrees)
MPI	multipoint injection
n	number of moles, cylinders
N^*	rev/s for two-stroke, rev/2s for four-stroke engines
N'	total number of firing strokes/s ($\equiv n.N^*$)
NMOG	non-methane organic gases
Nu	Nusselt number (dimensionless heat transfer coefficient)
ohc	overhead camshaft
ohv	overhead valve
p	pressure (N/m^2)
p'	partial pressure (N/m^2)
$\bar{p}$	mean effective pressure (N/m^2)
Pm	particulate mass
Pn	particulate number
Pr	Prandtl number (ratio of momentum and thermal diffusivities)
PFI	port fuel injection
PM	particulate matter
Q	heat flow (kJ)
r	crankshaft throw ($\equiv 1/2$ engine stroke) (m)
r_p	pressure ratio
r_v	volumetric compression ratio
R	specific gas constant (kJ/kg K)
R_o	molar (or universal) gas constant (kJ/kmol K)
Re	Reynolds number
s	specific entropy (kJ/kg K)
sac	specific air consumption (kg/MJ); this usually means brake specific air consumption
sfc	specific fuel consumption (kg/MJ); this usually means brake specific fuel consumption

S_L	laminar flame speed
S_V	laminar burning velocity
SGDI	spray guided direct injection
SI	spark ignition
SMD	Sauter mean diameter (equation 6.13)
t	time (s)
tdc	top dead centre
T	absolute temperature (K); torque (N m)
T_0	absolute temperature of the environment (K)
u_l	laminar flame front velocity (m/s)
u_t	turbulent flame front velocity (m/s)
U	internal energy (J)
v	velocity (m/s)
$\bar{v}_p$	mean piston velocity (m/s)
V	volume (m^3)
$\dot{V}_a$	volumetric flow rate of air (m^3/s)
V_s	engine swept volume (m^3)
VCO	valve covered orifice
wmmp	weakest mixture for maximum power
W	work (kJ)
$\dot{W}$	power (kW)
W_b	brake work (kJ)
W_c	compressor work (kJ)
W_f	friction work (kJ)
W_i	indicated work (kJ)
W_{REV}	work output from a thermodynamically reversible process (kJ)
W_t	turbine work (kJ)
WOT	wide-open throttle
x	a length (m); mass fraction
Z	Mach index
α	cut-off (or load) ratio
γ	ratio of gas heat capacities, c_p/c_v or C_p/C_v
ΔH_0	enthalpy of reaction (combustion) $\equiv -CV$
Δp	pressure difference (N/m^2)
$\Delta\theta_b$	combustion duration (crank angle, degrees)
ε	heat exchanger effectiveness = (actual heat transfer)/(max. possible heat transfer)
η	efficiency
η_b	brake thermal efficiency $\equiv \eta_o$
η_c	isentropic compressor efficiency
η_{Diesel}	ideal air standard Diesel cycle efficiency
η_{FA}	fuel/air cycle efficiency
η_i	indicated (arbitrary overall) efficiency
η_m	mechanical efficiency
η_o	arbitrary overall efficiency
η_{Otto}	ideal air standard Otto cycle efficiency
η_R	rational efficiency, W/W_{REV}
η_t	isentropic turbine efficiency
η_v	volumetric efficiency
$\lambda\,(1/\phi)$	actual AFR/stoichiometric AFR
θ	crank angle (degrees)
θ_0	crank angle at the start of combustion (degrees)
μ	dynamic viscosity (N s/m^2)
ρ	density (kg/m^3)
ρ_u	density of the unburnt gas (kg/m^3)

σ	surface tension (N/m)
ϕ	equivalence ratio = (stoichiometric air/fuel ratio)/(actual air/fuel ratio)
	(*note that* sometimes the reciprocal definition is used in other publications)
ω	specific humidity (kg water/kg dry air); angular velocity (rad/s)

Suffixes

a	air
b	brake
c	compressor
f	friction
g	gas
i	indicated, chemical species
l	liquid
m	mechanical
t	turbine
v	volumetric

CHAPTER 1

Introduction

1.1 | Fundamental operating principles

The reciprocating internal combustion engine must be by far the most common form of engine or prime mover. As with most engines, the usual aim is to achieve a high work output with a high efficiency; the means to these ends are developed throughout this book. The term 'internal combustion engine' should also include open circuit gas turbine plant where fuel is burnt in a combustion chamber. However, it is normal practice to omit the prefix 'reciprocating'; none the less this is the key principle that applies to both engines of different types and those utilising different operating principles. The divisions between engine types and between operating principles can be explained more clearly if direct injection spark ignition (DISI) and Wankel-type engines are ignored initially; hence these are not discussed until section 1.4.

The two main types of internal combustion engine are: spark ignition (SI) engines, where the fuel is ignited by a spark; and compression ignition (CI) engines, where the rise in temperature and pressure during compression is sufficient to cause spontaneous ignition of the fuel. The spark ignition engine is also referred to as the petrol, gasoline or gas engine from its typical fuels, and the Otto engine, after the inventor. The compression ignition engine is also referred to as the Diesel or oil engine; the fuel is also named after the inventor.

During each crankshaft revolution there are two strokes of the piston, and both types of engine can be designed to operate in either four strokes or two strokes of the piston. The four-stroke operating cycle can be explained by reference to figure 1.1:

1 *The induction stroke.* The inlet valve is open, and the piston travels down the cylinder, drawing in a charge of air. In the case of a spark ignition engine the fuel is usually pre-mixed with the air.

2 *The compression stroke.* Both valves are closed, and the piston travels up the cylinder. As the piston approaches top dead centre (tdc), ignition occurs. In the case of compression ignition engines, the fuel is injected towards the end of the compression stroke.

3 *The expansion, power or working stroke.* Combustion propagates throughout the charge, raising the pressure and temperature, and forcing the piston down. At the end of the power stroke the exhaust valve opens, and the irreversible expansion of the exhaust gases is termed 'blow-down'.

4 *The exhaust stroke.* The exhaust valve remains open, and as the piston travels up the cylinder the remaining gases are expelled. At the end of the exhaust stroke, when the exhaust valve closes some exhaust gas residuals will be left; these will dilute the next charge.

The four-stroke cycle is sometimes summarised as 'suck, squeeze, bang and blow'. Since the cycle is completed only

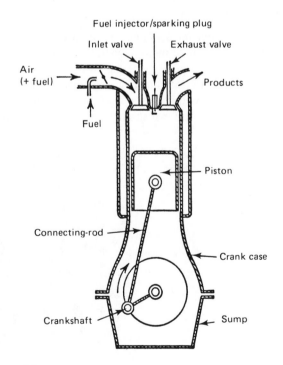

Figure 1.1
A four-stroke engine (reproduced with permission from Rogers and Mayhew, 1980a).

once every two revolutions the valve gear (and ignition or fuel injection equipment) have to be driven by mechanisms operating at half engine speed. Some of the power from the expansion stroke is stored in a flywheel, to provide the energy for the other three strokes.

The two-stroke cycle eliminates the separate induction and exhaust strokes; and the operation can be explained with reference to figure 1.2:

1 *The compression stroke* (figure 1.2a). The piston travels up the cylinder, so compressing the trapped charge. If the fuel is not pre-mixed, the fuel is injected towards the end of the compression stroke; ignition should again occur before top dead centre. Simultaneously, the underside of the piston is drawing in a charge through a spring-loaded non-return inlet valve.

2 *The power stroke.* The burning mixture raises the temperature and pressure in the cylinder, and forces the piston down. The downward motion of the piston also compresses the charge in the crankcase. As the piston approaches the end of its stroke the exhaust port is uncovered (figure 1.2b) and blowdown occurs. When the piston is at bottom dead centre (figure 1.2c) the transfer port is also uncovered, and the compressed charge in the crankcase expands into the cylinder. Some of the remaining exhaust gases are displaced by the fresh charge; because of the flow mechanism this is called 'loop scavenging'. As the piston travels up the cylinder, first the transfer port is closed by the piston, and then the exhaust port is closed.

For a given size engine operating at a particular speed, the two-stroke engine will be more powerful than a four-stroke engine since the two-stroke engine has twice as many power strokes per unit time. Unfortunately the efficiency of a two-stroke engine is likely to be lower than that of a four-stroke engine. The problem with two-stroke engines is ensuring that the induction and exhaust processes occur efficiently, without suffering charge dilution by the exhaust gas residuals. The spark ignition engine is particularly troublesome, since at part throttle operation the crankcase pressure can be less than atmospheric pressure. This leads to poor scavenging of the exhaust gases, and a rich air/fuel mixture becomes necessary for all conditions, with an ensuing low efficiency (see chapter 4, section 4.1).

These problems can be overcome in two-stroke direct injection engines by supercharging, so that the air pressure at inlet to the crankcase is greater than the exhaust back-pressure. This ensures that when the transfer port is opened, efficient scavenging occurs; if some air passes straight through the engine, it does not lower the efficiency since no fuel has so far been injected.

Originally engines were lubricated by total loss systems with oil baths around the main bearings or splash lubrication from oil in the sump. As engine outputs increased a circulating high-pressure oil system became necessary; this also assisted the heat transfer. In two-stroke spark ignition engines a simple system can be used in which oil is pre-mixed with the fuel; this removes the need for an oil pump and filter system.

An example of an automotive four-stroke compression ignition engine is shown in figure 1.3, and a two-stroke spark ignition motor cycle engine is shown in figure 1.4.

The size range of internal combustion engines is very large, especially for compression ignition engines. Two-stroke compression ignition engines vary from engines for models with swept volumes of about 1 cm^3 and a fraction of a kilowatt output, to large marine engines with a cylinder bore of about 1 m, up to 12 cylinders in-line, and outputs of up to 50 MW.

An example of a large two-stroke engine is the Sulzer RTA engine (see figure 1.5) described by Wolf (1982). The efficiency increases with size because the effects of clearances and cooling losses diminish. As size increases the operating speed reduces and this leads to more efficient combustion;

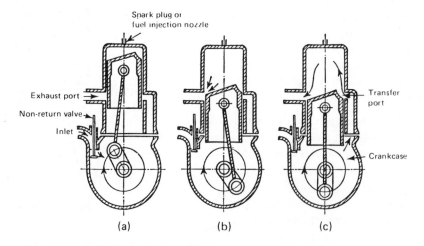

Figure 1.2
A two-stroke engine (reproduced with permission from Rogers and Mayhew, 1980a).

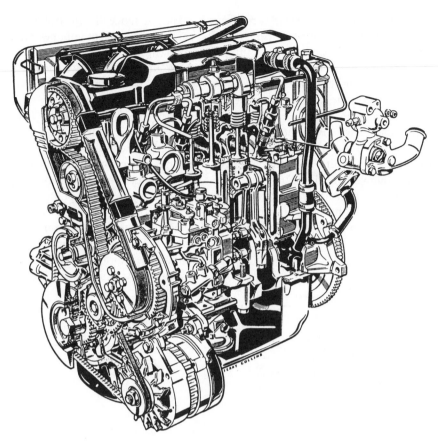

Figure 1.3
Ford 1.6-litre diesel engine (courtesy of Ford).

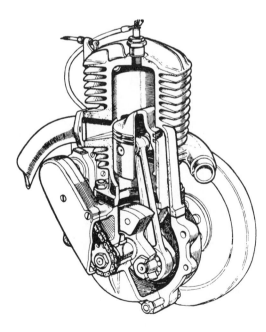

Figure 1.4
Two-stroke spark ignition engine.

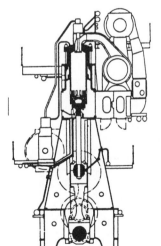

Power/cylinder	2960 kW
Swept volume/cylinder	1.32 m^3
Speed	87 rpm
Peak cylinder pressure	76 bar
Mean effective pressure	8.65 bar

Figure 1.5
Sulzer RTA two-stroke compression ignition engine (courtesy of Wärtsilä).

also the specific power demand from the auxiliaries reduces. The efficiency of such an engine can exceed 50 per cent, and it is also capable of burning low-quality residual fuels. A further advantage of a low-speed marine diesel engine is that it can be coupled directly to the propeller shaft. To run at low speeds an engine needs a long stroke, yet a stroke/bore ratio greater than 2 leads to poor loop scavenging.

The Sulzer RTA engine has a stroke/bore ratio of about 3 and uses uniflow scavenging. This requires the additional complication of an exhaust valve. Figure 1.6 shows arrangements for loop, cross and uniflow scavenging. The four-stroke compression ignition engines have a smaller size range, from about 400 cm³ per cylinder to 60 litres per cylinder with an output of 600 kW per cylinder at about 600 rpm with an efficiency of over 45 per cent.

The size range of two-stroke spark ignition engines is small, with the total swept volumes rarely being greater than 1000 cm³. The common applications are in off-highway applications such as chainsaws, where the high output, simplicity and low weight are more important than their poor fuel economy and high emissions.

Since the technology of two-stroke engines is rather different from that of four-stroke engines, chapter 8 is devoted to two-stroke engines. The interest in two-stroke engines is because of the potential for higher specific power outputs and a more frequent firing interval. More detail can be found in the books published by Blair (1990, 1996).

Automotive four-stroke spark ignition engines usually have cylinder volumes in the range 50–500 cm³, with the total swept volume rarely being greater than 5000 cm³. Engine outputs are typically 50 kW/litre, a value that can be increased seven-fold by tuning and turbocharging.

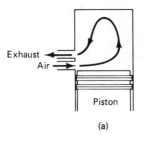

(a)

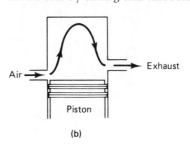

(b)

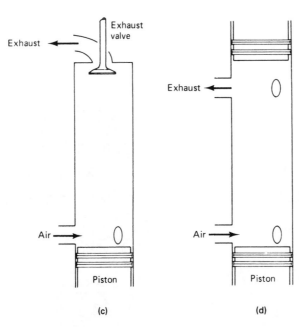

(c)

(d)

Figure 1.6

Two-stroke scavenging systems: (a) loop scavenging; (b) cross scavenging;
(c) uniflow scavenging with exhaust valve;
(d) uniflow scavenging with opposed pistons.

The largest spark ignition engines are gas engines; these are usually converted from large medium-speed compression ignition engines. High-output spark ignition engines have been developed for racing and in particular for aero-engines. A famous example of an aero-engine is the Rolls-Royce Merlin V12 engine. This engine had a swept volume of 27 litres and a maximum output of 1.48 MW at 3000 rpm; the specific power output was 1.89 kW/kg.

1.2 | Early internal combustion engine development

As early as 1680 Huygens proposed to use gunpowder for providing motive power. In 1688 Papin described the engine to the Royal Society of London, and conducted further experiments. Surprising as it may seem, these engines did not use the expansive force of the explosion directly to drive a piston down a cylinder. Instead, the scheme was to explode a small quantity of gunpowder in a cylinder, and to use this effect to expel the air from the cylinder. On cooling, a partial vacuum would form, and this could be used to draw a piston down a cylinder – the so-called 'atmospheric' principle.

Papin soon found that it was much more satisfactory to admit steam and condense it in a cylinder. This concept was used by Newcomen who constructed his first atmospheric steam engine in 1712. The subsequent development of atmospheric steam engines, and the later high-pressure steam engines (in which the steam was also used expansively), overshadowed the development of internal combustion engines for almost two centuries. When internal combustion engines were ultimately produced, the technology was based heavily on that of steam engines.

Throughout the late 18th and early 19th century there were numerous proposals and patents for internal combustion engines; only engines that had some commercial success will be mentioned here.

The first engine to come into general use was built by Lenoir in 1860; an example of the type is shown in figure 1.7. The engine resembled a single-cylinder, double-acting horizontal steam engine, with two power strokes per revolution. Induction of the air/gas charge and exhaust of the burnt mixture were controlled by slide valves; the ignition was obtained by an electric spark. Combustion occurred on both sides of the piston, but considering just one combustion chamber the sequence was as follows:

1 In the first part of the stroke, gas and air were drawn in. At about half stroke, the slide valves closed and the mixture was ignited; the explosion then drove the piston to the bottom of the stroke.

2 In the second stroke, the exhaust gases were expelled while combustion occurred on the other side of the piston.

Question 2.12, in the next chapter, analyses the Lenoir engine by modelling it with an air standard cycle. This provides an upper bound on its efficiency, which can be seen to be about one-third of that of a typical spark ignition engine with compression.

The next significant step was the Otto and Langen atmospheric or free-piston engine of 1866; the fuel consumption was about half that of the Lenoir engine. The main features of the engine were a long vertical cylinder, a heavy piston and a racked piston rod (figure 1.8). The racked piston rod was engaged with a pinion connected to the output shaft by a ratchet. The ratchet was arranged to free-wheel on the upward stroke, but to engage on the downward stroke. Starting with the piston at the bottom of the stroke the operating sequence was as follows:

1 During the first tenth or so of the stroke, a charge of gas and air was drawn into the cylinder. The charge was ignited by a flame transferred through a slide valve, and the piston was forced to the top of its stroke without delivering any work, the work being stored as potential energy in the heavy piston.

2 As the cylinder contents cooled, the partial vacuum so formed, and the weight of the piston, transferred the work to the output shaft on the downward stroke. Exhaust occurred at the end of this stroke.

The piston had to weigh about 70 kg per kW of output, and by its nature the engine size was limited to outputs of a few kilowatts; none the less some 10 000 engines were produced within five years.

At the same time commercial exploitation of oil wells in the USA was occurring, as a result of the pioneer drilling by Drake in 1859. This led to the availability of liquid fuels that were

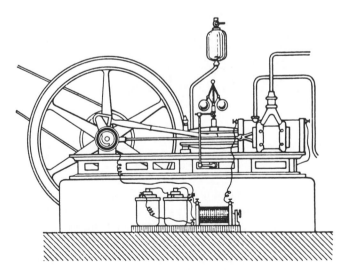

Figure 1.7
Lenoir gas engine of 1860 (reproduced with permission from Singer, *History of Technology*, Oxford University Press, 1978).

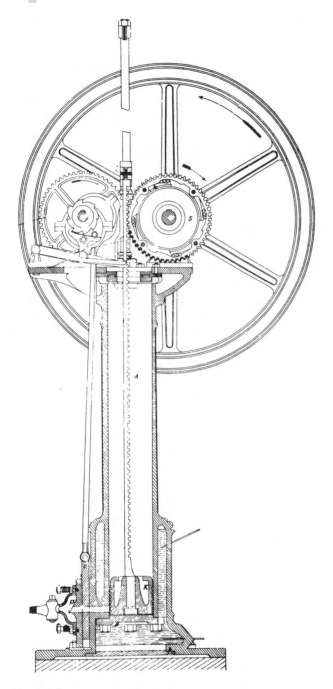

Figure 1.8
Otto and Langen free-piston engine.

much more convenient to use than gaseous fuels, since these often needed a dedicated gas-producing plant. Liquid fuels without doubt accelerated the development of internal combustion engines, and certainly increased the number of different types available, with oil products providing both the lubricant and the fuel. For the remainder of the 19th century any engine using a gaseous fuel was called a gas engine, and any engine using a liquid was called an oil engine; no

reference was necessarily made to the mode of ignition or the different operational principles.

In 1876 the Otto silent engine using the four-stroke cycle was patented and produced. As well as being much quieter than the free-piston engine, the silent engine was about three times as efficient. Otto attributed the improved efficiency to a conjectured stratification of the charge. This erroneous idea was criticised by Sir Dugald Clerk, who appreciated that the improved efficiency was a result of the charge being compressed before ignition. Clerk subsequently provided the first analysis of the Otto cycle (see chapter 2, section 2.2.1).

The concept of compression before ignition can be traced back to Schmidt in 1861, but perhaps more remarkable is the work of Beau de Rochas. As well as advocating the four-stroke cycle, Beau de Rochas included the following points in 1862:

1 There should be a high volume-to-surface ratio.
2 The maximum expansion of the gases should be achieved.
3 The highest possible mixture pressure should occur before ignition.

Beau de Rochas also pointed out that ignition could be achieved by sufficient compression of the charge.

Immediately following the Otto silent engine, two-stroke engines were developed. Patents by Robson in 1877 and 1879 describe the two-stroke cycle with under-piston scavenging, while patents of 1878 and 1881 by Clerk describe the two-stroke cycle with a separate pumping or scavenge cylinder.

The quest for self-propelled vehicles needed engines with better power-to-weight ratios. Daimler was the first person to realise that a light high-speed engine was needed, which would produce greater power by virtue of its higher speed of rotation, 500–1000 rpm. Daimler's patents date from 1884, but his twin cylinder 'V' engine of 1889 was the first to be produced in quantity. By the end of the 19th century the petrol engine was in a form that would be currently recognisable, but there was still much scope for development and refinement.

The modern compression ignition engine developed from the work of two people, Akroyd Stuart and Rudolf Diesel. Akroyd Stuart's engine, patented in 1890 and first produced in 1892, was a four-stroke compression ignition engine with a compression ratio of about 3 – too low to provide spontaneous ignition of the fuel. Instead, this engine had a large uncooled pre-chamber or vaporiser connected to the main cylinder by a short narrow passage. Initially the vaporiser was heated externally, and the fuel then ignited after it had been sprayed into the vaporiser at the end of the compression stroke. The turbulence generated by the throat to the vaporiser ensured rapid combustion. Once the engine had been started the external heat source could be removed. The fuel was typically a light petroleum distillate such as kerosene or fuel oil; the efficiency of about 15 per cent was comparable with that of the Otto silent engine. The key innovations with the Akroyd Stuart engine were the induction, being solely of air, and the injection of fuel into the combustion chamber.

Diesel's concept of compressing air to such an extent that the fuel would spontaneously ignite after injection was published in 1890, patented in 1892 and achieved in 1893; an early example is shown in figure 1.9. Some of Diesel's aims were unattainable, such as a compression pressure of 240 bar, the use of pulverised coal and an uncooled cylinder. None the less, the prototype ran with an efficiency of 26 per cent, about twice the efficiency of any contemporary power plant and a figure that steam power plant achieved only in the 1930s.

Diesel injected the fuel by means of a high-pressure (70 bar) air blast, since a liquid pump for 'solid' or airless injection was not devised until 1910 by McKechnie. Air-blast injection necessitated a costly high-pressure air pump and storage vessel; this restricted the use of diesel engines to large stationary and marine applications. Smaller high-speed compression ignition engines were not used for automotive applications until the 1920s. The development depended on experience gained from automotive spark ignition engines, the development of airless quantity-controlled fuel injection pumps by Bosch (chapter 6, section 6.5.2), and the development of suitable combustion systems by people such as Ricardo.

1.3 | Characteristics of internal combustion engines

The purpose of this section is to discuss one particular engine – the Ford 'Dover' direct injection in-line six-cylinder truck engine (figure 1.10). This turbocharged engine has a swept volume of 6 litres, weighs 488 kg and produces a power output of 114 kW at 2400 rpm.

In chapter 2 the criteria for judging engine efficiency are derived, along with other performance parameters, such as mechanical efficiency and volumetric efficiency, which provide insight into why a particular engine may or may not be efficient. Thermodynamic cycle analysis also indicates that, regardless of engine type, the cycle efficiency should improve with higher compression ratios. Furthermore, it is shown that for a given compression ratio, cycles that compress a fuel/air mixture have a lower efficiency than cycles that compress pure air. Cycle analysis also indicates the work that can be extracted by an exhaust gas turbine.

The type of fuel required for this engine and the mode of combustion are discussed in chapter 3. Since the fuel is injected into the engine towards the end of the compression stroke, the combustion is not pre-mixed but controlled by diffusion processes – the diffusion of the fuel into the air, the diffusion of the air into the fuel, and the diffusion of the combustion products away from the reaction zone. Turbu-

Figure 1.9
Early (1898) diesel engine (output: 45 kW at 180 rpm) (reproduced with permission from Singer, *History of Technology*, Oxford University Press, 1978).

lence is essential if these processes are to occur in the small time available. The main fuel requirement is that the fuel should readily self-ignite; this is the exact opposite to the requirements for the spark ignition engine. Fuel chemistry and combustion are discussed in chapter 3, along with additives, principally those that either inhibit or promote self-ignition. One of the factors limiting the output of this, and any other, diesel engine is the amount of fuel that can be injected before unburnt fuel leaves the engine as smoke (formed by agglomerated carbon particles). These and other engine emissions are discussed in sections 3.8 and 6.6.

A key factor in designing a successful compression ignition engine is the design of the combustion chamber, and the correct matching of the fuel injection to the in-cylinder air motion. These factors are discussed in chapter 6, the counterpart to chapter 4 which discusses spark ignition engines. The design and manufacture of fuel injection

Figure 1.10
Ford Dover compression ignition engine (courtesy of Ford).

equipment are undertaken by specialist manufacturers; however, the final matching of the fuel injection equipment to the engine still has to be done experimentally. This turbocharged engine is likely to have a lower compression ratio than its naturally aspirated counterpart, in order to limit the peak pressures during combustion. One effect of this is that starting is made more difficult; the compression ratio is often chosen to be the minimum that will give reliable starting. None the less, cold weather or poor fuel quality can lead to starting difficulties, and methods to improve starting are discussed in section 6.4.

The induction and exhaust processes are controlled by poppet valves in the cylinder head. The timing of events derives from the camshaft, but is modified by the clearances and the elastic properties of the valve gear. The valve timing for the turbocharged engine will keep the valves open for longer periods than in the naturally aspirated version, since appropriate turbocharger matching can cause the inlet pressure to be greater than the exhaust. These aspects, and the nature of the flow in the inlet and exhaust passages, are discussed in chapter 7. In the turbocharged engine the inlet and exhaust manifold volumes will have been minimised to help reduce the turbocharger response time. In addition, care will have to be taken to ensure that the pressure pulses from each cylinder do not interfere with the exhaust system. The design of naturally aspirated induction and exhaust systems can be very involved if the engine performance is to be

optimised. Pressure pulses can be reflected as rarefaction waves, and these can be used to improve the induction and exhaust processes. Since these are resonance effects, the engine speed at which maximum benefit occurs depends on the system design and the valve timing, see sections 7.4, 7.5 and 7.6.

Chapter 9 discusses the in-cylinder motion of the air in some detail. It provides definitions of turbulence and swirl, and explains how both can be measured. Chapter 6 should be read before chapter 9, as this is where the significance of swirl and turbulence is explained.

The turbocharger is another component that is designed and manufactured by specialists. Matching the turbocharger to the engine is difficult, since the flow characteristics of each machine are fundamentally different. The engine is a slow-running (but large) positive displacement machine, while the turbocharger is a high-speed (but small) non-positive displacement machine, which relies on dynamic flow effects. Turbochargers inevitably introduce a lag when speed or load is increased, since the flow rate can only increase as the rotor speed increases. These aspects, and the design of the compressor and turbine, are covered in chapter 10, along with applications to spark ignition engines. Turbochargers increase the efficiency of compression ignition engines since the power output of the engine is increased more than the mechanical losses.

Chapter 11 describes how computer modelling can be applied to internal combustion engines. Since the interaction of dynamic flow devices and positive displacement devices is complex, the chapter on computer modelling ends with an example on the computer modelling of a turbocharged diesel engine.

Some of the mechanical design considerations are dealt with in chapter 12. A six-cylinder in-line engine will have even firing intervals that produce a smooth torque output. In addition, there will be complete balance of all primary and secondary forces and moments that are generated by the reciprocating elements. The increased cycle temperatures in the turbocharged engine make the design and materials selection for the exhaust valve and the piston assembly particularly important. The increased pressures also raise the bearing loads and the role of the lubricant as a coolant will be more important. Computers are increasingly important in design work, as component weights are being reduced and engine outputs are being raised.

Chapter 13 is devoted to engine cooling systems. These are often taken for granted, but chapter 13 explains the principles, with particular emphasis on liquid cooled engines. There is also a description of how cooling systems can be designed to accelerate engine warm-up, and to provide higher efficiencies at steady-state operating points.

The use of computers is increasing in all aspects of engine work: modelling the engine to estimate performance, matching of the fuel injection equipment, selection and performance prediction of the turbocharger, estimation of vehicle performance (speed, fuel consumption etc.) for different vehicles, transmission and usage combinations.

None the less, when an engine such as the Dover diesel is being developed there is still a need to test engines. The designers will want to know about its performance at all loads and speeds. Some measurements, such as the fuel efficiency, are comparatively simple to make, and these techniques are introduced at the start of chapter 14. Chapter 14 also describes how exhaust emissions are measured, and the procedures needed for in-cylinder combustion analysis. The chapter ends with a description of computer-controlled test facilities.

1.4 Additional types of internal combustion engine

Two types of engine that fall outside the simple classification of reciprocating spark ignition or compression ignition engine are the Wankel engine and the direct injection spark ignition engine.

1.4.1 The Wankel engine

The Wankel engine is a rotary combustion engine, developed from the work of Felix Wankel. The mode of operation is best explained with reference to figure 1.11. The triangular rotor has a centrally placed internal gear that meshes with a sun gear that is part of the engine casing. An eccentric that is an integral part of the output shaft constrains the rotor to follow a planetary motion about the output shaft. The gear ratios are such that the output shaft rotates at three times the speed of the rotor, and the tips of the rotor trace out the two-lobe epitrochoidal shape of the casing. The compression ratio is dictated geometrically by the eccentricity of the rotor and the shape of its curved surfaces. The convex surfaces shown in the diagram minimise the sealed volumes, to give the highest compression ratio and optimum gas exchange. A recess in the combustion chamber provides a better-shaped combustion chamber.

The sequence of events that produces the four-stroke cycle is as follows. In figure 1.11a with the rotor turning in a clockwise direction a charge is drawn into space 1, the preceding charge is at maximum compression in space 2, and the combustion products are being expelled from space 3. When the rotor turns to the position shown in figure 1.11b, space 1 occupied by the charge is at a maximum, and further rotation will cause compression of the charge. The gases in space 2 have been ignited and their expansion provides the power stroke. Space 3 has been reduced in volume, and the exhaust products have been expelled. As in the two-stroke loop or cross scavenge engine there are no valves, and here the gas flow through the inlet and exhaust is controlled by the position of the rotor apex.

For effective operation the Wankel engine requires efficient seals between the sides of the rotor and its casing, and the more demanding requirement of seals at the rotor tips. Additional problems to be solved were cooling of the rotor, the casing around the spark plug and the exhaust passages. Unlike a reciprocating engine, only a small part of the Wankel engine is cooled by the incoming charge. Furthermore, the spark plug has to operate reliably under much hotter conditions. Not until the early 1970s were the sealing problems sufficiently solved for the engine to enter production. The advantages of the Wankel are its compactness, the apparent simplicity, the ease of balance and the potential for high outputs by running at high speeds.

The major disadvantages of the Wankel engine are its low efficiency (caused by limited compression ratios) and the high exhaust emissions resulting from the poor combustion chamber shape. By the mid 1970s concern over firstly engine emissions, and secondly fuel economy led to the demise of the Wankel engine. Experiments with other types of rotary combustion engine have not led to commercial development.

1.4.2 Direct injection spark ignition engines

Direct injection spark ignition engines, as the name implies, have the fuel injected directly into the cylinder. They can operate in two different regimes:

1 At full load, injection occurs during induction, so that

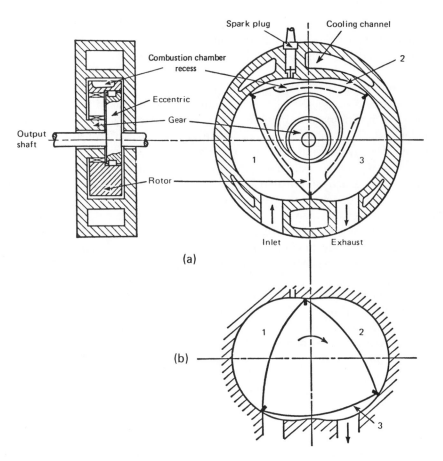

Figure 1.11
The Wankel engine (reproduced with permission from Rogers and Mayhew, 1980a).

there is time for a homogeneous air/fuel mixture to be formed. This early injection also facilitates evaporative cooling of the air and combustion chamber, so that a high mass of air is trapped in the cylinder. This enables the maximum power output to be comparable with a conventional spark ignition engine.

2 At part load, injection occurs during the compression stroke, and it has to be timed so that there is a stratified charge, with a flammable mixture formed at the spark plug, at the time that ignition is required. This gives the potential of a higher efficiency, since the throttling losses can be reduced or eliminated. There is thus the potential for the fuel economy to approach that of diesel engines.

The principle behind stratified charge engines is to have a readily ignitable mixture in the vicinity of the spark plug, and a weaker (possibly non-ignitable) mixture in the remainder of the combustion chamber. The purpose of this arrangement is to control the power output of the engine by varying only the fuel supply without throttling the air, thereby eliminating the throttling pressure-drop losses. The stratification of the charge was initially obtained by division of the combustion chamber

to produce a pre-chamber that contains the spark plug. Typically fuel would also be injected into the pre-chamber, so that charge stratification is controlled by the timing and rate of fuel injection. Thus fuel supply is controlled in the same manner as compression ignition engines, yet the ignition timing of the spark controls the start of combustion.

Another means of preparing a stratified charge was to provide an extra valve to the pre-chamber, which controlled a separate air/fuel mixture. This was the method used in the Honda CVCC engine (figure 1.12), the first stratified charge engine in regular production.

However, of great significance is the potential for direct injection spark ignition (DISI) engines to achieve the specific output of gasoline engines, yet with fuel economy that is comparable to diesel engines. Mitsubishi launched such a DISI engine, and this has been described by Kume *et al.* (1996) and Ando (1997). The Mitsubishi engine used a combustion chamber formed by a carefully shaped piston, which controls the interaction between the air motion and fuel injection. More recently, much higher fuel injection pressures have been used (150 bar) so that a much more finely atomised fuel spray is formed, and the intention is to form a flammable

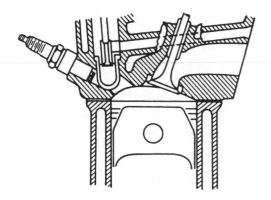

Figure 1.12
Honda CVCC engine, the first in regular production (from Campbell, 1978).

mixture in the region of the spark plug without fuel wetting the piston. When liquid fuel contacts the piston it leads to a 'pool fire' – a diffusion flame that is a source of particulate matter emissions.

The operation of direct injection spark ignition engine is discussed further in chapter 5. The design of stratified charge engines is complex, and the operating envelope for stratified charge operation is also restricted by emissions legislation. With weak mixtures, the conventional three-way catalyst will not reduce NO_x emissions, therefore a stratified charge engine has to operate with a very weak mixture, so that the engine-out exhaust emissions will satisfy the NO_x emissions requirements.

A proper discussion of stratified charge engines requires a knowledge of chapters 2–6.

1.5 | Prospects for internal combustion engines

The future of internal combustion engines will be influenced by two factors: the future cost and availability of suitable fuels, and the development of alternative power plants.

Liquid fuels are by far the most convenient energy source for internal combustion engines, and the majority (over 99 per cent) of such fuels come from crude oil. The oil price is largely governed by political and taxation policy, and there is no reason to suppose that these areas of control will change.

In 2008 the world consumption of oil was about 84 million barrels per day. This figure has increased by 15 per cent in the last decade, with over half of the increase attributable to Asia. Current known oil reserves would then imply a supply of crude oil for another 40 years or so. However, exploration for oil continues and new reserves are being found; it must also be remembered that oil companies cannot justify expensive exploration work to demonstrate reserves for, say, the next 100 years. The ratio of oil reserves to the rate of production is shown in figure 1.13. This suggests a continuing equilibrium between supply and demand.

Internal combustion engines can also be fuelled from renewable energy sources. Spark ignition engines run satisfactorily on alcohol-based fuels, and compression ignition engines can operate on vegetable oils. Countries such as Brazil, with no oil reserves but plentiful sources of vegetation, are already operating an alcohol-fuelled policy.

The other major source of hydrocarbons is coal, and even conservative estimates show a 200-year supply from known reserves. One approach is to introduce a suspension of coal particles into the heavy fuel oil used by large compression ignition engines. A more generally applicable alternative is the preparation of fuels by 'liquefaction' or 'gasification' of coal. A

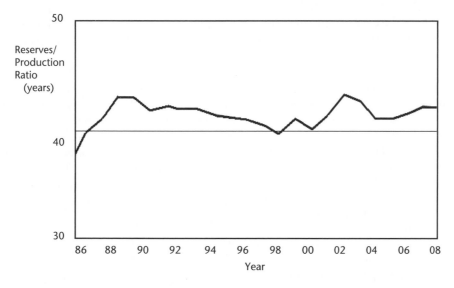

Figure 1.13
Worldwide reserves/production ratio (data derived from the *BP Statistical Review of World Energy*, 2009).

comprehensive review of alternative processes is given by Davies (1983), along with the yields of different fuels and their characteristics.

The preceding remarks indicate that the future fuel supply is assured for internal combustion engines, but that other types of power plant may supersede them; the following is a discussion of some of the possibilities.

Steam engines have been used in the past, and would have the advantages of external combustion of any fuel, with readily controlled emissions. However, if it is possible to overcome the low efficiency and other disadvantages it is probable that this would have already been achieved and they would already have been adopted; their future use is thus unlikely.

Stirling engines have been developed, with some units for automotive applications built by United Stirling of Sweden. The fuel economy at full and part load is comparable to compression ignition engines, but the cost of building the complex engine is about 50 per cent greater. As the Stirling engine has external combustion it too has the capability of using a wide range of fuels with readily controlled emissions.

Gas turbines present another alternative: conventionally they use internal combustion, but external combustion is also possible. For efficient operation a gas turbine would require a high efficiency compressor and turbine, high pressure ratio, high combustion temperatures and an effective heat exchanger. These problems have been solved for large aero and industrial gas turbines, but scaling down to even truck engine size changes the design philosophy. The smaller size would dictate the less efficient radial flow compressor, with perhaps a pressure ratio of 5:1 and a regenerative heat exchanger to preserve the efficiency. All these problems could be solved, along with reductions in manufacturing cost, by the development of ceramic materials. However, part load efficiency is likely to remain poor, although this is of less significance in truck applications. The application to private passenger vehicles is even more remote because of the importance of part load efficiency, and the reductions in efficiency that would follow from the smaller size. Marine application has been limited because of the low efficiency. In addition, the efficiency can deteriorate rapidly if the turbine blades corrode as a result of the combustion of salt-laden air.

Electric vehicles present an interesting possibility which is currently restricted by the lack of a suitable battery. Performance figures have been improved to give a typical speed of about 100 km per hour, and a range of 120 km. Battery technology is being improved, and lithium-ion and lithium-polymer battery systems are being developed, with two to three times the energy density of lead/acid batteries. However, batteries have less than 1 per cent of the energy density of hydrocarbon fuels. Electric vehicles could meet the majority of personal transport needs, and are viable for local delivery vehicles. However, electric vehicles are expensive, and their use has to be either subsidised or encouraged by fiscal or other means.

This leads to the possibility of a hybrid vehicle that has both an internal combustion engine and an electric motor. This is obviously an expensive solution but one that is versatile and efficient by using the motor and/or engine.

The simplest hybrid vehicles are those with parallel or series configurations. In the parallel configuration, the vehicle can be propelled by the electric motor and engine, either independently, or together. In the series configuration, the engine is connected to a generator, and the wheels are only connected to the electric motor.

Ingenious systems have been developed by Volkswagen, in which the electric motor/generator is integrated with the flywheel; see Walzer (1990). However, the greatest potential is offered by dual or split hybrid systems which combine elements of both the series and hybrid systems (Yamaguchi, 1997). During normal driving, some of the engine power can be sent via the generator and electric motor to a differential type gearbox, so that adjusting the electric motor speed effectively controls the overall gearing ratio from the engine to the wheels.

Perhaps the greatest challenge to internal combustion engines will come from fuel cells – devices that allow the direct conversion of fuels to electricity. These are of sufficient importance to be the subject of the next section.

1.6 | Fuel cells

Fuel cells are electrochemical devices that permit the direct conversion of chemical energy to electricity, with the potential of a very high efficiency. As fuel cells are open-circuit devices, their maximum efficiency is not limited by their operating temperature and the Carnot cycle efficiency. This will become clearer in chapter 2, section 2.1, after the necessary thermodynamics has been developed. It will be seen shortly that fuel cells have quite specific fuel requirements, and an appreciation of how these can be obtained from conventional fuel cells will become clear by studying chapter 3.

The first fuel cell was built by Sir William Grove in 1839, who had recognised the occurrence of 'reverse electrolysis'. The fuel cell used oxygen and hydrogen with platinum electrodes. The requirement for pure oxygen and pure hydrogen is a major inconvenience, and the most notable applications of fuel cells have been in space, where the fuels are available for propulsion. There are currently four basic fuel cell types with significant commercial potential:

SPFC   Solid Polymer Fuel Cells
PAFC   Phosphoric Acid Fuel Cells
MCFC   Molten Carbonate Fuel Cells, and
SOFC   Solid Oxide Fuel Cells.

Their operation and operating temperature ranges are summarised in figure 1.14.

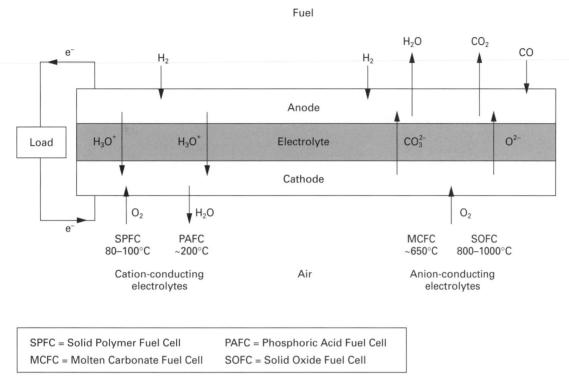

Figure 1.14
Basic fuel cell types (Gardner, 1997). ©IMechE/Professional Engineering Publishing Limited.

The electrolyte is a thin gas-tight ion conducting membrane with porous electrodes on each side. The membrane/electrolyte assembly (MEA) separates the fuel and the oxidant, and allows their chemical interaction. The name of the fuel cell type indicates the nature of the electrolyte, and figure 1.14 indicates a further subdivision into:

1 cation (positively charged ion) conducting electrolytes (SPFC and PAFC), in which the products appear on the air side, and

2 anion (negatively charged ion) conducting electrolytes (MCFC and SOFC), in which the products appear on the fuel side.

The cathode-seeking cation is a hydrated proton (a positively charged hydrogen atom H^+, attached to one or more water molecules). The anode-seeking anion is a carbonate ion (CO_3^{2-}) in the MCFC, and an oxygen ion (O^{2-}) in the SOFC. The SPFC and PAFC both require hydrogen (H_2) as their fuel, while the MCFC and SOFC both have the advantage that carbon monoxide (CO) can be used as well as hydrogen. In general, the reformation of hydrocarbon fuels yields both carbon monoxide and hydrogen, so it is a thermodynamic disadvantage if only the hydrogen can be utilised. Each fuel will produce a different open-circuit voltage, and it will be the lower voltage that will govern. Fortunately, in the case of carbon monoxide and hydrogen, these theoretical voltages are very close in the range of 800–900°C, so the SOFC is able to minimise this loss of electrical energy.

Electrical energy is extracted from the fuel cell, when the free electrons pass through an external load circuit, and not through the electrolyte. The electrolyte has to permit the passage of the ions (driven by the chemical potential arising from the oxidation of the fuel), but prevent the passage of the electrons.

The stationary and non-stationary applications of fuel cells are likely to employ different technologies. For vehicular applications there is limited scope for utilising 'waste heat', so the low-temperature solid polymer fuel cell (SPFC) is likely to be used. An efficiency (based on the lower calorific value of a hydrocarbon fuel) of about 30–40 per cent is likely. Although the fuel cell is at a comparatively low temperature (80–100°C), if on-board reformation of a hydrocarbon or oxygenate fuel is to be used to supply the hydrogen, then temperatures in the region of 200–300°C are required for the reformation of methanol, and temperatures in the region of 800°C would be needed for the reformation of hydrocarbons.

For stationary applications the 'waste heat' can be used, and this makes the solid oxide fuel cell (SOFC) the most attractive, since it operates at the highest temperature. Another advantage of high-temperature operation is that there is scope for internally reforming the fuel, so that (for example) methane (CH_4) can be the feedstock. In the SOFC the

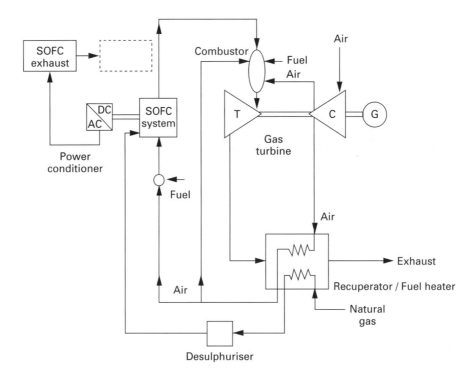

Figure 1.15
An integrated solid oxide fuel cell (SOFC)/gas turbine system (adapted from Bevc, 1997).

electrolyte is a ceramic, so great care is needed in the design to allow for the ceramic's brittle nature, and it is also necessary to avoid thermal shocks.

The power density of all fuel cells can be increased by operation at elevated pressures, and in the case of high-temperature fuel cells this can lead to direct integration with a gas turbine system. Figure 1.15 shows such a system that is discussed in detail by Bevc (1997). The waste heat from the turbine is used to heat both the fuel and the air, and some of this air goes directly to the gas turbine combustion chamber. The SOFC is designed to operate at pressures of up to 10 bar. The exhaust from the SOFC will contain some unoxidised fuel (CO and H_2), and some unreformed fuel (CH_4), but this oxidation can be completed in the gas turbine combustion chamber. The overall efficiency of such a SOFC/gas turbine system is predicted to be 60–70 per cent, but it will only be an economic proposition if the capital costs of the fuel cell are low enough.

1.7 | Concluding remarks

The main application of internal combustion engines in terms of numbers is in automotive transport, and it is very difficult to make fair comparisons between conventional vehicles, hybrid vehicles and those with fuel cells. The results depend on the assumptions about the vehicle, its drive cycle, the type of fuel, and how it has been refined, transported and stored. There have been many 'well-to-wheel' studies, but very comprehensive studies are undertaken under the auspices of the European Commission Joint Research Centre and led by CONCAWE.

Figure 1.16 compares different vehicle systems (for which the different internal combustion engine technologies are reviewed in subsequent chapters):

PISI – Port Injection Spark Ignition (a conventional gasoline engine)

DISI – Direct Injection Spark Ignition (an alternative gasoline engine)

DICI – Direct Injection Compression Ignition (a conventional diesel engine)

DPF – Diesel Particulate Filter (the pressure drop across the filter and the need to regenerate the filter impose a fuel consumption penalty)

CNG – Compressed Natural Gas

C-H_2 – Compressed Hydrogen

L-H_2 – Liquefied Hydrogen

Figure 1.16a shows that diesel engines are more efficient than gasoline, and that vehicle energy consumption is reduced by hybridisation. The well-to-tank energy requirement is very low compared to the calorific value of the fuel, because of the ease of extracting, handling and refining liquid

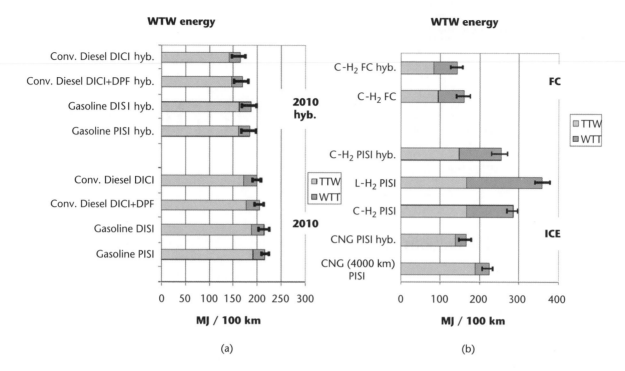

Figure 1.16
Well-to-wheel (WTW) performance of different vehicle systems, with the energy consumption apportioned between fuel production and distribution to the vehicle (WTT – well-to-tank) and vehicle usage (TTW – tank-to-wheel), adapted from CONCAWE (2007). (a) Conventional and hybrid vehicles with liquid fuels. (b) Gaseous fuelled vehicles, with internal combustion engines (ICE) and fuel cells (FC).

fuels. Figure 1.16b shows that Compressed Natural Gas (CNG) has a slightly lower energy consumption in a spark ignition engine than gasoline, but that the energy requirement for the fuel distribution is greater, leading to an overall lower well-to-wheel efficiency. The use of hydrogen with internal combustion engines leads to an overall increase in energy consumption because of the high energy cost in its production and storage. Liquid hydrogen (L-H_2) has a particularly high-energy requirement because at ambient pressure its boiling point is 20 K. Only a compressed hydrogen (C-H_2) fuel cell vehicle has the potential to achieve a lower energy consumption than that for a diesel vehicle. The CONCAWE modelling assumed that hydrogen was being produced from methane (CH_4, the major constituent of Natural Gas). When hydrogen is manufactured from methane (or any hydrocarbon) it is necessary to remove the carbon dioxide, and this could then be stored ('sequestered'). The disadvantage of fuel cell vehicles would be their very high cost using current technology. The batteries on purely electric vehicles make them very expensive, and with hybrid vehicles the extra cost is divided between the batteries, electric motor(s)/generator(s) and the power electronics.

1.8 | Question

1.1 What are the key technological developments for spark ignition engines in the first half of the 20th century?

CHAPTER 2

Thermodynamic principles

2.1 | Introduction and definitions of efficiency

This chapter provides criteria by which to judge the performance of internal combustion engines. Most important are the thermodynamic cycles based on ideal gases undergoing ideal processes. However, internal combustion engines follow a mechanical cycle, not a thermodynamic cycle. The start and end points are mechanically the same in the cycle for an internal combustion engine, whether it is a two-stroke or four-stroke mechanical cycle.

The internal combustion engine is a non-cyclic, open-circuit, quasi steady-flow, work-producing device. None the less it is very convenient to compare internal combustion engines with the ideal air standard cycles, as they are a simple basis for comparison. This can be justified by arguing that the main constituent of the working fluid, nitrogen, remains virtually unchanged in the processes. The internal combustion engine is usually treated as a steady-flow device since most engines are multi-cylindered, with the flow pulsations smoothed at inlet by air filters and at exhaust by silencers.

Air standard cycles have limitations as air and, in particular, air/fuel mixtures do not behave as ideal gases. These effects are discussed at the end of this chapter where computer modelling is introduced. Despite this, the simple air standard cycles are very useful, as they indicate trends. Most important is the trend that as compression ratio increases cycle efficiency should also increase.

At this stage it is necessary to define engine efficiency. This is perhaps the most important parameter to an engineer, although it is often very carelessly defined.

The fuel and air (that is, the reactants) enter the power plant at the temperature T_0 and pressure p_0 of the environment (that is, under ambient conditions). The discharge is usually at p_0, but in general the exhaust products from an internal combustion engine are at a temperature in excess of T_0, the environment temperature. This represents a potential for producing further work if the exhaust products are used as the heat source for an additional cyclic heat power plant. For maximum work production all processes must be reversible and the products must leave the plant at T_0, as well as p_0.

Availability studies show that the work output from an ideal, reversible, non-cyclic, steady-flow, work-producing device is given by

$$W_{REV} = B_{in} - B_{out} \tag{2.1}$$

where the steady-flow availability function $B \equiv H - T_0 S$.

This can be derived by considering a reversible steady-flow single-stream process, with initial state 1, and a final state equal to the ground state 0. The steady-flow energy equation on a specific basis (that is, per kg) can be written

$$gZ_1 + h_1 + \frac{v_1^2}{2} + [q]_1^0 = gZ_0 + h_0 + \frac{v_2^2}{2} + [W_x]_1^0 \tag{2.2}$$

Rearranging gives

$$[W_x]_1^0 = [q]_1^0 + g(Z_1 - Z_0) + (h_1 - h_0) + \frac{1}{2}(v_1^2 - v_0^2) \tag{2.3}$$

For a reversible process, the Second Law of Thermodynamics can be written as

$$\frac{[q]_1^0}{T_0} = S_0 - S_1 = \Delta S \tag{2.4}$$

Combining and rearranging the First and Second Law equations gives

$$[w_x]_{1 \atop Rev}^0 = T_0(S_0 - S_1) + g(Z_1 - Z_0) + (h_1 - h_0) + \frac{1}{2}\left(v_1^2 - v_0^2\right) \tag{2.5}$$

If we neglect changes in the potential energy and kinetic energy, and defining

$$b \equiv h - T_0 s \tag{2.6}$$

then

$$[w_x]_{1 \atop Rev}^0 = (h_1 - T_0 S_1) - (h_0 - T_0 S_0) = b_1 - b_0 = \Delta b \tag{2.7}$$

Table 2.1

Fuel	Reaction	$-\Delta G_0$	$-\Delta H_0$ (calorific value)
		MJ/kg of fuel	
C	$C + O_2 \rightarrow CO_2$	32.84	32.77
CO	$CO + \frac{1}{2}O_2 \rightarrow CO_2$	9.19	10.11
H_2	$H_2 + \frac{1}{2}O_2 \rightarrow H_2O$ liq.	117.6	142.0
	$H_2 + \frac{1}{2}O_2 \rightarrow H_2O$ vap.	113.4	120.0
CH_4	$CH_4 + 2O_2 \rightarrow 2H_2O + CO_2$ vap.	57.6	50.0

For a change from state 1 to 2, we can treat this as a change from $1 \rightarrow 0$, followed by a change from $0 \rightarrow 2$. For a system (of unit mass), equation (2.7) becomes

$$[w_x]_1^2 = b_1 - b_2 = (h_1 - T_0 s_1) - (h_2 - T_0 s_2) \quad (2.8)$$
$$\text{Rev}$$

In the case of an internal combustion engine the ideal case would be when both reactants and products enter and leave the power plant at the temperature and pressure of the environment, albeit with different composition. Thus

$$W_{REV} = B_{R0} - B_{P0} = (H - T_0 S)_{R0} - (H - T_0 S)_{P0} \quad (2.9)$$
$$= (H_{R0} - T_0 S_{R0}) - (H_{P0} - T_0 S_{P0}) \quad (2.10)$$

Recalling the definition of the Gibbs energy ($G = H - TS$) enables us to write

$$W_{REV} = G_{R0} - G_{P0} = -\Delta G_0 \quad (2.11)$$

where
$G_{R0} = G$ of the reactants at p_0, T_0
$G_{P0} = G$ of the products at p_0, T_0

Equation (2.1) defines the maximum amount of work (W_{REV}) that can be obtained from a given chemical reaction. As such it can be used as a basis for comparing the actual output of an internal combustion plant or any open-circuit system, including fuel cells. This leads to a definition of rational efficiency, η_R:

$$\eta_R = \frac{W}{W_{REV}} \quad (2.12)$$

where W is the actual work output.

It is misleading to refer to the rational efficiency as a thermal efficiency, as this term should be reserved for the cycle efficiency of a cyclic device. The upper limit to the rational efficiency can be seen to be 100 per cent, unlike the thermal efficiency of a cyclic device.

This definition is not widely used, since G_0 cannot be determined from simple experiments. Instead the calorific value (CV) of a fuel is used

$$CV \equiv (H_{R0} - H_{P0}) \equiv -\Delta H_0 \quad (2.13)$$

where $H_{R0} \equiv H$ of the reactants at p_0, T_0
$H_{P0} \equiv H$ of the products at p_0, T_0
and $\Delta H_0 \equiv H_{P0} - H_{R0}$

The difference in the enthalpies of the products and reactants can be readily found from the heat transfer in a steady-flow combustion calorimeter. This leads to a convenient, but arbitrary, definition of efficiency as

$$\eta_b \equiv \eta_0 \equiv \frac{W}{CV} \equiv \frac{W}{-\Delta H_0} \quad (2.14)$$

where η_0 = arbitrary overall efficiency
η_b = brake (thermal) efficiency
and W = work output.

The terms 'brake efficiency' and 'brake power output' derive from using friction to 'brake' or load the engine using a brake dynamometer (see section 14.2.1). While the arbitrary overall efficiency is often of the same order as the thermal efficiency of cyclic plant, it is misleading to refer to it as a thermal efficiency.

Table 2.1, derived from Haywood (1980), shows the difference between $-\Delta G_0$ and $-\Delta H_0$.

The differences in table 2.1 are not particularly significant, especially as the arbitrary overall efficiency is typically about 30 per cent.

In practice, engineers are more concerned with the fuel consumption of an engine for a given output rather than with its efficiency. This leads to the use of specific fuel consumption (sfc), the rate of fuel consumption per unit power output:

$$sfc = \frac{\dot{m}_f}{\dot{W}} \text{ kg/J} \quad (2.15)$$

where $\dot{m}_f$ = mass flow rate of fuel
$\dot{W}$ = power output.

It can be seen that this is inversely proportional to the arbitrary overall efficiency, and is related by the calorific value of the fuel:

$$sfc = \frac{1}{-\Delta H_0 \eta_0} \quad (2.16)$$

It is important also to specify the fuel used. The specific fuel consumption should be quoted in SI units (kg/MJ), although it is often quoted in metric units such as (kg/kW h) or in British Units such as (lbs/BHP h).

Sometimes a volumetric basis is used, but this should be avoided as there is a much greater variation in fuel density than calorific value.

2.2 | Ideal air standard cycles

Whether an internal combustion engine operates on a two-stroke or four-stroke cycle and whether it uses spark ignition or compression ignition, it follows a mechanical cycle not a thermodynamic cycle. However, the overall efficiency of such an engine is assessed by comparison with the thermal efficiency of air standard cycles, because of the similarity between the engine indicator diagram and the state diagram of the corresponding hypothetical cycle. The engine indicator diagram is the record of pressure against cylinder volume, recorded from an actual engine. Pressure/volume diagrams are very useful, as the enclosed area equates to the work in the cycle.

2.2.1 The ideal air standard Otto cycle

The Otto cycle is usually used as a basis of comparison for spark ignition and high-speed compression ignition engines. The cycle consists of four non-flow processes, as shown in figure 2.1. The compression and expansion processes are assumed to be adiabatic (no heat transfer) and reversible, and thus isentropic. The processes are as follows:

1–2   isentropic compression of air through a volume ratio V_1/V_2, the compression ratio r_v
2–3   addition of heat Q_{23} at constant volume
3–4   isentropic expansion of air to the original volume, $V_4/V_3 = r_v$
4–1   rejection of heat Q_{41} at constant volume to complete the cycle.

The efficiency of the Otto cycle, η_{Otto} is

$$\eta_{\text{Otto}} = \frac{W}{Q_{23}} = \frac{Q_{23} - Q_{41}}{Q_{23}} = 1 - \frac{Q_{41}}{Q_{23}} \qquad (2.17)$$

By considering air as a perfect gas we have constant specific heat capacities, and for mass m of air the heat transfers are

$$\begin{aligned} Q_{23} &= mc_v(T_3 - T_2) \\ Q_{41} &= mc_v(T_4 - T_1) \end{aligned} \qquad (2.18)$$

thus

$$\eta_{\text{Otto}} = 1 - \frac{T_4 - T_1}{T_3 - T_2} \qquad (2.19)$$

For the two isentropic processes 1–2 and 3–4, $TV^{\gamma-1}$ is a constant. Thus

$$\frac{T_2}{T_1} = \frac{T_3}{T_4} = r_v^{\gamma-1} \qquad (2.20)$$

where γ is the ratio of gas specific heat capacities, c_p/c_v. Thus

$$T_3 = T_4 r_v^{\gamma-1} \quad \text{and} \quad T_2 = T_1 r_v^{\gamma-1} \qquad (2.21)$$

and substituting into equation (2.19) gives

$$\eta_{\text{Otto}} = 1 - \frac{T_4 - T_1}{r_v^{\gamma-1}(T_4 - T_1)} = 1 - \frac{1}{r_v^{\gamma-1}} \qquad (2.22)$$

The value of η_{Otto} depends on the compression ratio, r_v, and not the temperatures in the cycle. To make a comparison with a real engine, only the compression ratio needs to be specified. The variation in η_{Otto} with compression ratio is shown in figure 2.2 along with that of η_{Diesel}.

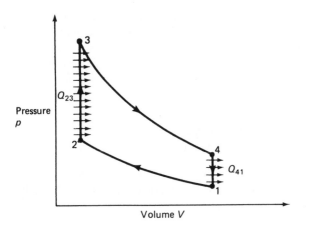

Figure 2.1
Ideal air standard Otto cycle.

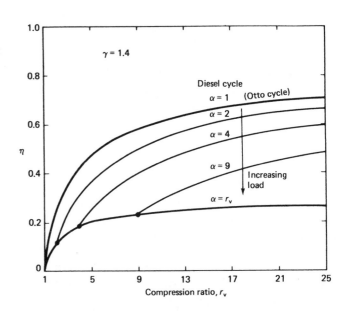

Figure 2.2
Diesel cycle efficiency for different load ratios α.

2.2.2 The ideal air standard Diesel cycle

The Diesel cycle has heat addition at constant pressure, instead of heat addition at constant volume as in the Otto cycle. With the combination of high compression ratio, to cause self-ignition of the fuel, and constant-volume combustion the peak pressures can be very high. In large compression ignition engines, such as marine engines, fuel injection is arranged so that combustion occurs at approximately constant pressure in order to limit the peak pressures.

The four non-flow processes constituting the cycle are shown in the state diagram (figure 2.3). Again, the best way to calculate the cycle efficiency is to calculate the temperatures around the cycle. To do this it is necessary to specify the cut-off ratio or load ratio:

$$\alpha \equiv V_3/V_2 \tag{2.23}$$

The processes are all reversible, and are as follows:

1–2 isentropic compression of air through a volume ratio V_1/V_2, the compression ratio r_v
2–3 addition of heat Q_{23} at constant pressure while the volume expands through a ratio V_3/V_2, the load or cut-off ratio α
3–4 isentropic expansion of air to the original volume
4–1 rejection of heat Q_{41} at constant volume to complete the cycle.

The efficiency of the Diesel cycle, η_{Diesel}, is

$$\eta_{\text{Diesel}} = \frac{W}{Q_{23}} = \frac{Q_{23} - Q_{41}}{Q_{23}} = 1 - \frac{Q_{41}}{Q_{23}} \tag{2.24}$$

By treating air as a perfect gas we have constant specific heat capacities, and for mass m of air the heat transfers are

$$Q_{23} = mc_p(T_3 - T_2)$$
$$Q_{41} = mc_v(T_4 - T_1) \tag{2.25}$$

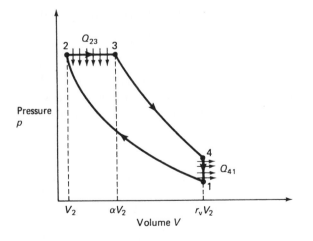

Figure 2.3
Ideal air standard Diesel cycle.

Note that the process 2–3 is at constant pressure, thus

$$\eta_{\text{Diesel}} = 1 - \frac{1}{\gamma} \frac{T_4 - T_1}{T_3 - T_2} \tag{2.26}$$

For the isentropic process 1–2, $TV^{\gamma-1}$ is a constant:

$$T_2 = T_1 r_v^{\gamma-1} \tag{2.27}$$

For the constant pressure process 2–3

$$\frac{T_3}{T_2} = \frac{V_3}{V_2} = \alpha \quad \text{thus} \quad T_3 = \alpha r_v^{\gamma-1} T_1 \tag{2.28}$$

For the isentropic process 3–4, $TV^{\gamma-1}$ is a constant:

$$\frac{T_4}{T_3} = \left(\frac{V_3}{V_4}\right)^{\gamma-1} = \left(\frac{\alpha}{r_v}\right)^{\gamma-1} \tag{2.29}$$

thus

$$T_4 = \left(\frac{\alpha}{r_v}\right)^{\gamma-1} T_3 = \alpha r_v^{\gamma-1} \left(\frac{\alpha}{r_v}\right)^{\gamma-1} T_1 = \alpha^\gamma T_1 \tag{2.30}$$

Substituting for all the temperatures in equation (2.26) in terms of T_1 gives

$$\eta_{\text{Diesel}} = 1 - \frac{1}{\gamma} \cdot \frac{\alpha^\gamma - 1}{\alpha r_v^{\gamma-1} - r_v^{\gamma-1}} = 1 - \frac{1}{r_v^{\gamma-1}} \left[\frac{\alpha^\gamma - 1}{\gamma(\alpha - 1)}\right] \tag{2.31}$$

At this stage it is worth making a comparison between the air standard Otto cycle efficiency (equation 2.22) and the air standard Diesel cycle efficiency (equation 2.31).

The Diesel cycle efficiency is less convenient; it is not solely dependent on compression ratio, r_v, but is also dependent on the load ratio, α. The two expressions are the same, except for the term in square brackets

$$\left[\frac{\alpha^\gamma - 1}{\gamma(\alpha - 1)}\right]$$

The load ratio lies in the range $1 < \alpha < r_v$, and is thus always greater than unity. Consequently the expression in square brackets is always greater than unity, and the Diesel cycle efficiency is less than the Otto cycle efficiency *for the same compression ratio*. This is shown in figure 2.2 where efficiencies have been calculated for a variety of compression ratios and load ratios. There are two limiting cases. The first is, as $\alpha \to 1$, then $\eta_{\text{Diesel}} \to \eta_{\text{Otto}}$. This can be shown by rewriting the term in square brackets, and using a binomial expansion:

$$\left[\frac{\alpha^\gamma - 1}{\gamma(\alpha - 1)}\right] = \left[\frac{\{1 + (\alpha - 1)\}^\gamma - 1}{\gamma(\alpha - 1)}\right]$$
$$= \left[\frac{1}{\gamma(\alpha - 1)}\left\{1 + [\gamma(\alpha - 1)] + \frac{\gamma(\gamma - 1)}{2!}(\alpha - 1)^2 + \ldots - 1\right\}\right] \tag{2.32}$$

As $\alpha \to 1$, then $(\alpha - 1) \to 0$ and the $(\alpha - 1)^2$ terms and higher can be neglected, thus the term in square brackets tends to unity.

The second limiting case is when $\alpha \to r_v$ and point $3 \to 4$ in the cycle, and the expansion is wholly at constant pressure; this corresponds to maximum work output in the cycle. Figure 2.2 also shows that as load increases, with a fixed compression ratio the efficiency reduces. The compression ratio of a compression ignition engine is usually greater than for a spark ignition engine, so the diesel engine is usually more efficient.

2.2.3 The ideal air standard Dual cycle

In practice, combustion occurs neither at constant volume nor at constant pressure. This leads to the Dual, Limited Pressure, or Mixed cycle which has heat addition in two stages, firstly at constant volume, and secondly at constant pressure. The state diagram is shown in figure 2.4; again all processes are assumed to be reversible. As might be expected, the efficiency lies between that of the Otto cycle and the Diesel cycle. An analysis can be found in several sources such as Heywood (1988) or Benson and Whitehouse (1979), but is merely quoted here as the extra complication does not give results significantly closer to reality:

$$\eta = 1 - \frac{1}{r_v^{\gamma-1}}\left[\frac{r_p\alpha^\gamma - 1}{(r_p - 1) + \gamma r_p(\alpha - 1)}\right] \qquad (2.33)$$

where r_p is the pressure ratio during constant-volume heat addition; $r_p = p_3/p_2$. As before, $\alpha = V_4/V_3$. This cycle analysis is the subject of question 2.6.

2.2.4 The ideal air standard Atkinson cycle

This is commonly used to describe any cycle in which the expansion stroke is greater than the compression stroke; the analysis is presented in example 2.3. Figure 2.5 shows the

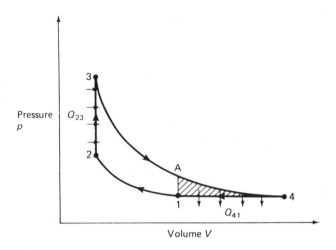

Figure 2.5
Ideal air standard Atkinson cycle.

limiting case for the Atkinson cycle in which expansion is down to pressure p_1. All processes are reversible, and processes 1–2 and 3–4 are also adiabatic.

The hatched area (1A4) represents the increased work (or reduced heat rejection) when the Atkinson cycle is compared to the Otto cycle. The mechanical difficulties of arranging unequal compression and expansion strokes have prevented the development of engines working on the Atkinson cycle. However, if the inlet valve closure is very early (or very late), then the effective compression ratio becomes lower than the geometric compression ratio, and this system is sometimes known as a Miller cycle. This also reduces the throttling loss in spark ignition engines operating at part load; there is further discussion in section 7.4.2 (Variable valve timing). The expansion A4 can be arranged in a separate process, for example an exhaust turbine. This subject is treated more fully in chapter 10. The Atkinson cycle analysis is illustrated by example 2.3.

2.3 | Comparison between thermodynamic and mechanical cycles

Internal combustion engines operate on a mechanical cycle, not a thermodynamic cycle. Although it is an arbitrary procedure it is very convenient to compare the performance of real non-cyclic engines with thermal efficiencies of hypothetical cycles. This approach arises from the similarity of the engine indicator diagram (figure 2.6) and the state diagram of a hypothetical cycle (figure 2.1 or figure 2.3).

Figure 2.6 is a stylised indicator diagram for a high-speed four-stroke engine, with exaggerated pressure difference between the induction and exhaust strokes. As before, r_v is the compression ratio, and V_c is the clearance volume with the piston at top dead centre (tdc). The diagram could be from either a spark ignition or a compression ignition engine, as in

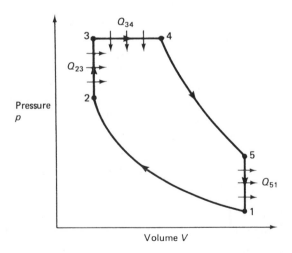

Figure 2.4
Ideal air standard Dual cycle.

Plate 1

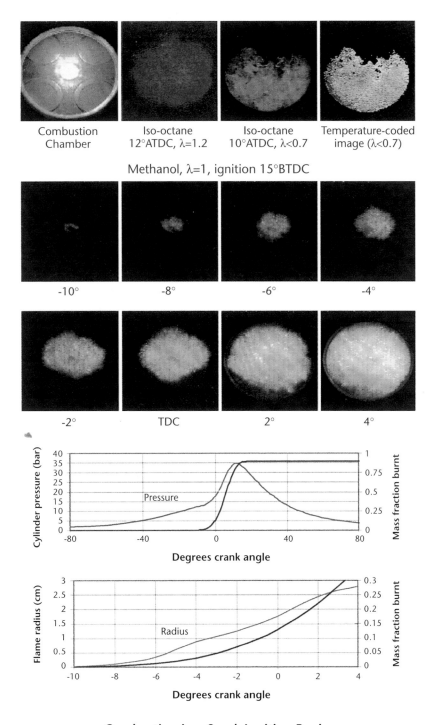

Combustion Chamber · Iso-octane 12°ATDC, λ=1.2 · Iso-octane 10°ATDC, λ<0.7 · Temperature-coded image (λ<0.7)

Methanol, λ=1, ignition 15°BTDC

-10° · -8° · -6° · -4°

-2° · TDC · 2° · 4°

Combustion in a Spark Ignition Engine

Plate 2

Photographs of combustion in a high-speed direct injection Diesel engine

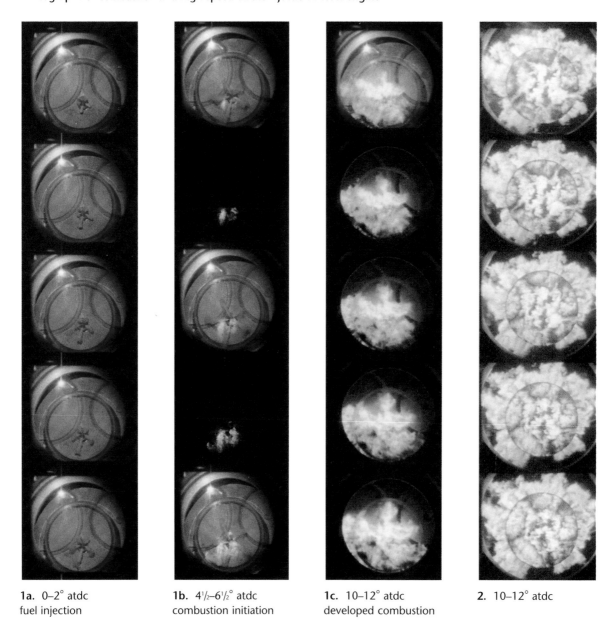

1a. 0–2° atdc
fuel injection

1b. 4½–6½° atdc
combustion initiation

1c. 10–12° atdc
developed combustion

2. 10–12° atdc

Sequence 1 is at low load, with copper vapour laser illumination of every frame in sequence 1a, and alternate frames in sequence 1b (Zambare, 1998).

Sequence 2 is at high load showing the whole of the piston crown, unpublished work.

Photographs reproduced by permission of D. E. Winterbone, E. Clough and V. V. Zambare.

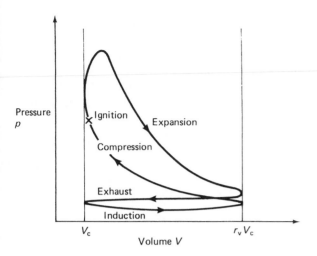

Figure 2.6
Stylised indicator diagram for four-stroke engine.

both cases combustion occurs neither at constant pressure nor at constant volume. For simplicity it can be idealised as constant-volume combustion (figure 2.7) and then compared with the Otto cycle (figure 2.1).

In the idealised indicator diagram, induction 0–1 is assumed to occur with no pressure drop. The compression and expansion (1–2, 3–4) are not adiabatic, so neither are they isentropic. Combustion is assumed to occur instantaneously at constant volume, 2–3. Finally, when the exhaust valve opens blow-down occurs instantaneously, with the exhaust expanding into the manifold 4–1, and the exhaust stroke occurring with no pressure drop, 1–0. The idealised indicator diagram is used as a basis for the simplest computer models.

In comparison, the Otto cycle assumes that:

1 Air behaves as a perfect gas with constant specific heat capacity, and all processes are fully reversible.

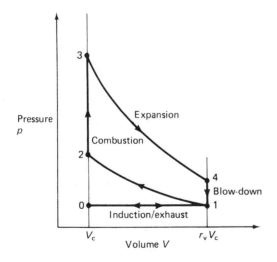

Figure 2.7
Idealised indicator diagram for four-stroke engine.

2 There is no induction or exhaust process, but a fixed quantity of air and no leakage.

3 Heat addition is from an external source, in contrast to internal combustion.

4 Heat rejection is to the environment to complete the cycle, as opposed to blow-down and the exhaust stroke.

2.4 | Additional performance parameters for internal combustion engines

The rational efficiency and arbitrary overall efficiency have already been defined in section 2.1. The specific fuel consumption which has a much greater practical significance has also been defined. The additional parameters defined here relate to the work output per unit swept volume in terms of a mean effective pressure, and the effectiveness of the induction and exhaust strokes.

There are two key types of mean effective pressure, based on either the work done by the gas on the piston or the work available as output from the engine.

(a) *Indicated mean effective pressure (imep)*
The area enclosed on the p–V trace or indicator diagram from an engine is the indicated work (W_i) done by the gas on the piston. The imep is a measure of the indicated work output per unit swept volume, in a form independent of the size and number of cylinders in the engine and engine speed.

The imep is defined as

$$\text{imep (N/m}^2) = \bar{p}_i$$
$$= \frac{\text{indicated work output (N m) per cylinder per mechanical cycle}}{\text{swept volume per cyclinder (m}^3)}$$

(2.34)

Figure 2.8 shows an indicator diagram with a hatched area, equal to the net area of the indicator diagram. In a four-stroke cycle the negative work occurring during the induction and exhaust strokes is termed the *pumping loss*, and has to be subtracted from the positive indicated work of the other two strokes. When an engine is throttled the pumping loss increases, thereby reducing the engine efficiency. In figure 2.8 the hatched area has the same volume scale as the indicator diagram, so the height of the hatched area must correspond to the imep.

The *pumping loss* or *pumping work* can also be used to define a pumping mean effective pressure (*pmep*), $\bar{p}_p$

$$\bar{p}_p = \frac{\text{pumping work (N m) per cylinder per mechanical cycle}}{\text{swept volume per cylinder (m}^3)}$$ (2.35)

It should be noted that the definition of imep used here is not universally adopted. Sometimes (most notably in the USA), the imep does not always incorporate the pumping work, and this leads to the use of the terms *gross imep* and *net imep*:

$$\text{gross imep} = \text{net imep} + \text{pmep}$$ (2.36)

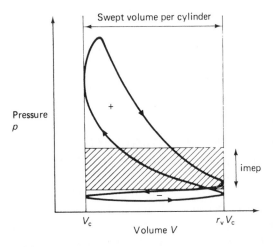

Figure 2.8
Indicated mean effective pressure.

Unfortunately, imep does not always mean net imep, so it is necessary to check the context to ensure that it is not a gross imep.

The imep bears no relation to the peak pressure in an engine, but is a characteristic of engine type. The imep in naturally aspirated four-stroke spark ignition engines will be smaller than the imep of a similar turbocharged engine. This is mainly because the turbocharged engine has greater air density at the start of compression, so more fuel can be burnt.

(b) *Brake mean effective pressure (bmep)*
The work output of an engine, as measured by a brake or dynamometer, is more important than the indicated work output. This leads to a definition of bmep, $\bar{p}_b$, which is very similar to equation (2.34):

$$\bar{p}_b \ (N/m^2) = \frac{\text{brake work output (N m) per cylinder per mechanical cycle}}{\text{swept volume per cylinder (m}^3)}$$

(2.37)

or in terms of the engine brake power

$$\text{brake power} = \bar{p}_b L A N'$$
$$= \bar{p}_b (L A n) N^* = \bar{p}_b V_s N^*$$

(2.38)

where L = piston stroke (m)
   A = piston area (m^2)
   n = number of cylinders
   V_s = engine swept volume (m^3)

and N' = number of mechanical cycles of operation per second (for all pistons)
   $N^* = N'/n$

$$= \begin{cases} \text{rev./s for two-stroke engines} \\ \dfrac{\text{rev./s}}{2} \text{ for four-stroke engines.} \end{cases}$$

The bmep is a measure of work output from an engine, and not of pressures in the engine. The name arises because its unit is that of pressure.

(c) *Mechanical efficiency, η_m, and Frictional mean effective pressure, fmep*
The difference between indicated work and brake work is accounted for by friction, and work done in driving essential items such as the lubricating oil pump. Frictional mean effective pressure (fmep), $\bar{p}_f$, is the difference between the imep and the bmep:

$$\text{fmep} = \text{imep} - \text{bmep}$$
$$\bar{p}_f = \bar{p}_i - \bar{p}_b$$

(2.39)

Mechanical efficiency is defined as

$$\eta_m = \frac{\text{brake power}}{\text{indicated power}} = \frac{\text{bmep}}{\text{imep}}$$

(2.40)

(d) *Indicated efficiency, η_i*
When comparing the performance of engines it is sometimes useful to isolate the mechanical losses. This leads to the use of indicated (arbitrary overall) efficiency as a means of examining the thermodynamic processes in an engine:

$$\eta_i = \frac{\dot{W}_i}{\dot{m}_f . CV} = \frac{\dot{W}}{\dot{m}_f . CV . \eta_m} = \frac{\eta_0}{\eta_m}$$

(2.41)

(e) *Volumetric efficiency, η_v*
Volumetric efficiency is a measure of the effectiveness of the induction and exhaust processes. Even though some engines inhale a mixture of fuel and air it is convenient, but arbitrary, to define volumetric efficiency as

$$\eta_v = \frac{\text{mass of air inhaled per cylinder per cycle}}{\text{mass of air to occupy swept volume per cylinder at 'ambient' } p \text{ and}}$$

Assuming air obeys the Gas Laws, this can be rewritten as

$$\eta_v = \frac{\text{volume of 'ambient' density air inhaled per cylinder per cycle}}{\text{cylinder swept volume}}$$
$$= \frac{\dot{V}_a}{V_s N^*}$$

(2.42)

where $\dot{V}_a$ = volumetric flow rate of air with 'ambient' density
   V_s = engine swept volume

$$N^* = \begin{cases} \text{rev./s for two-stroke engines} \\ \dfrac{\text{rev./s}}{2} \text{ for four-stroke engines.} \end{cases}$$

This now raises the question of what is meant by 'ambient'. In the case of naturally aspirated engines it will be the atmospheric conditions, but in the case of a supercharged or turbocharged engine it will refer to the inlet manifold conditions.

The output of an engine is related to its volumetric efficiency, and this can be shown by considering the mass of fuel that is burnt (m_f), its calorific value (CV), and the brake efficiency of the engine:

$$m_f = \frac{\eta_v V_s \rho_a}{\text{AFR}}$$

(2.43)

where AFR is the gravimetric (that is, by mass) air/fuel ratio and ρ_a is the ambient air density.

The amount of brake work produced per cycle (W_b) is given by

$$W_b = \bar{p}_b V_s$$
$$= m_f \eta_b CV \qquad (2.44)$$

Combining equations (2.43) and (2.44), and then rearranging gives

$$\bar{p}_b = \eta_v \eta_b \rho_a CV / AFR \qquad (2.45)$$

and similarly

$$\bar{p}_i = \eta_v \eta_i \rho_a CV / AFR \qquad (2.46)$$

In the case of supercharged engines, compressor or inter-cooler delivery conditions should be used instead of ambient conditions. Volumetric efficiency has a direct effect on power output, as the mass of air in a cylinder determines the amount of fuel that can be burnt. In a well-designed, naturally aspirated spark ignition engine the volumetric efficiency can be over 90 per cent, and over 100 per cent for a tuned induction system.

Volumetric efficiency depends on the density of the gases at the end of the induction process; this depends on the temperature and pressure of the charge. There will be pressure drops in the inlet passages and at the inlet valve owing to viscous effects. The charge temperature will be raised by heat transfer from the induction manifold, mixing with residual gases, and heat transfer from the piston, valves and cylinder. In a petrol engine, fuel evaporation can cool the charge by as much as 25 K, and alcohol fuels have much greater cooling effects; this improves the volumetric efficiency.

In an idealised process, with charge and residuals having the same specific heat capacity and molar mass, the temperature of the residual gases does not affect volumetric efficiency. This is because in the idealised process induction and exhaust occur at the same constant pressure, and when the two gases mix the contraction on cooling of the residual gases is exactly balanced by the expansion of the charge.

In practice, induction and exhaust processes do not occur at the same pressure and the effect this has on volumetric efficiency for different compression ratios is discussed by Taylor (1985a). The design factors influencing volumetric efficiency are discussed in chapter 7, section 7.3.

(f) *Specific air consumption (sac)*

As with specific fuel consumption, specific air consumption can be evaluated as a brake specific air consumption (bsac) or as an indicated specific air consumption (isac). The specific air consumption indicates how good a combustion system is at utilising the air trapped within the cylinder, and it is most likely to be encountered with diesel engines. This is because the output of diesel engines (especially when naturally aspirated) is often limited by the fuelling level that causes smoke. In other words, when the specific air consumption is

evaluated at full load, it tells the designer how well the air and fuel have been mixed.

When the gravimetric air/fuel ratio is known (AFR), the relationship with brake specific fuel consumption (bsfc) and indicated specific fuel consumption (isfc) is

$$isac = AFR \times isfc \qquad (2.47)$$
$$bsac = AFR \times bsfc \qquad (2.48)$$

Equations (2.47) and (2.48) show that the specific air consumption is also dependent on the engine efficiency. Indicated specific air consumption thus provides an insight into the combustion and thermodynamic performance; brake specific air consumption also includes the mechanical performance.

2.5 | Fuel/air cycle

The simple ideal air standard cycles overestimate the engine efficiency by a factor of about 2. A significant simplification in the air standard cycles is the assumption of constant specific heat capacities. Heat capacities of gases are strongly temperature-dependent, as shown by figure 2.9.

The molar constant-volume heat capacity will also vary, as will γ, the ratio of heat capacities, since:

$$C_p - C_v = R_0, \qquad \gamma = C_p/C_v \qquad (2.49)$$

If this is allowed for, air standard Otto cycle efficiency falls from 57 per cent to 49.4 per cent for a compression ratio of 8.

When allowance is made for the presence of fuel and combustion products, there is an even greater reduction in cycle efficiency. This leads to the concept of a fuel/air cycle which is the same as the ideal air standard Otto cycle, except that allowance is made for the real thermodynamic behaviour of the gases. The cycle assumes instantaneous complete combustion, no heat transfer, and reversible compression and expansion. Taylor (1985a) discusses these matters in detail and provides results in graphical form. Figures 2.10 and 2.11 show the variation in fuel/air cycle efficiency as a function of equivalence ratio for a range of compression ratios. Equivalence ratio ϕ is defined as the chemically correct (stoichiometric) air/fuel ratio divided by the actual air/fuel ratio. The datum conditions at the start of the compression stroke are pressure (p_1) 1.013 bar, temperature (T_1) 115°C, mass fraction of combustion residuals (f) 0.05, and specific humidity (ω) 0.02 – the mass ratio of water vapour to air.

The fuel 1-octene has the formula C_8H_{16}, and structure

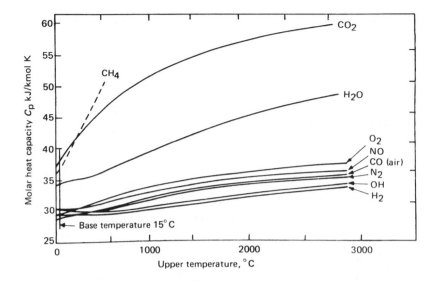

Figure 2.9
Molar heat capacity at constant pressure of gases above 15°C quoted as averages between 15°C and abscissa temperature (adapted from Taylor, 1985a).

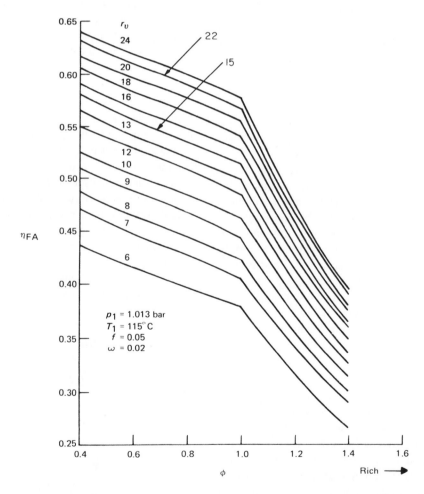

Figure 2.10
Variation of efficiency with equivalence ratio for a constant-volume fuel/air cycle with 1-octene fuel for different compression ratios (adapted from Taylor, 1985a).

Figure 2.10 shows the pronounced reduction in efficiency of the fuel/air cycle for rich mixtures. The improvement in cycle efficiency with increasing compression ratio is shown in figure 2.11, where the ideal air standard Otto cycle efficiency has been included for comparison.

In order to make allowances for the losses due to phenomena such as heat transfer and finite combustion time, it is necessary to develop computer models.

Consider a spark ignition engine with a compression ratio of 10, for which the Otto cycle efficiency predicts an efficiency of 60 per cent and the fuel/air cycle predicts an efficiency of 47 per cent. In reality such an engine might have a full throttle brake efficiency of 30 per cent, and this means there are 17 percentage points to be accounted for, perhaps as follows:

	percentage points
mechanical losses	3
finite speed of combustion:	
20 degree 10–90 per cent burn	1
40 degree 10–90 per cent burn	3
blow-by and unburnt fuel in the exhaust	1
cycle-by-cycle variations in combustion	2
exhaust blow-down and gas exchange	2
heat transfer	7

The 10–90 per cent burn represents the main part of the combustion process, and this will be treated more formally in chapter 3, section 3.5. Cycle-by-cycle variations in combustion are discussed in chapter 4, section 4.4.

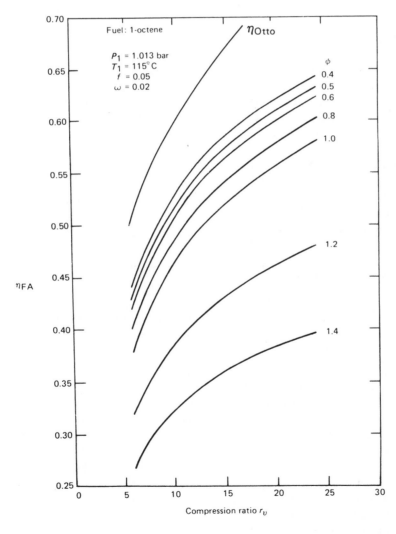

Figure 2.11
Variation of efficiency with compression ratio for a constant-volume fuel/air cycle with 1-octene fuel for different equivalence ratios (adapted from Taylor, 1985a).

'Finite piston speed losses' occur since combustion takes a finite time and cannot occur at constant volume. This leads to the rounding of the indicator diagram.

'Heat losses' are most significant between the end of the compression stroke and the beginning of the expansion stroke. However, with no heat transfer the cycle temperatures would be raised and the fuel/air cycle efficiencies would be reduced slightly because of increasing gas specific heats and dissociation.

The exhaust valve opening before the end of the expansion stroke promotes gas exchange but reduces the expansion work.

Since the fuel is injected towards the end of the compression stroke in compression ignition engines (unlike the spark ignition engine where it is pre-mixed with the air) the compression process will be closer to ideal in the compression ignition engine than in the spark ignition engine. This is another reason for the better fuel economy of the compression ignition engine.

2.6 | Computer models

In internal combustion engines the induction, compression, expansion and exhaust strokes are all influenced by the combustion process. In any engine model it is necessary to include all processes, of which combustion is the most complex. Combustion models are discussed in chapter 3, section 3.9.

Benson and Whitehouse (1979) provide a useful introduction to engine modelling by giving FORTRAN programs for spark ignition and compression ignition engine cycle calculations. The working of the programs is explained and typical results are presented. These models have now been superseded but none the less they provide a good introduction.

Ferguson (1986) presents more comprehensive FORTRAN programs for modelling spark ignition and compression ignition engines. These programs include: variable gas properties (specific heat capacity), dissociation, heat transfer, finite speed of combustion and blow-by (leakage from the cylinder). Developments of these models will be used later to illustrate engine performance.

The use of engine models is increasing as engine testing becomes more expensive and computing becomes cheaper. Additionally, once an engine model has been set up, results can be produced more quickly. However, it is still necessary to check model results against engine results to calibrate the model. Engine manufacturers and research organisations either develop their own models or buy-in the expertise from specialists.

The aspects of engine models discussed in this section are the compression and expansion processes. The main difference between spark ignition and compression ignition cycles is in the combustion process. The other significant difference is that spark ignition engines usually induct and compress a fuel/air mixture.

This section considers the compression and expansion strokes in the approach adopted by Benson and Whitehouse (1979).

The First Law of Thermodynamics expressed in differential form is

$$dQ = dU + dW \qquad (2.50)$$

The heat transfer, dQ, will be taken as zero in this simple model. Heat transfer in internal combustion engines is still very poorly understood, and there is a shortage of experimental data. A widely used correlation for heat transfer inside an engine cylinder is due to Annand (1963); this and other correlations are discussed in chapter 11, section 11.2.4.

The change in absolute energy of the cylinder contents, dU, is a complex function of temperature, which arises because of the variation of the gas specific heat capacities with temperature. Equation (2.50) cannot be solved analytically, so a numerical solution is used, which breaks each process into a series of steps. Consider the ith to the $(i + 1)^{th}$ steps, for which the values of U can be found as functions of the temperatures

$$dU = U(T_{i+1}) - U(T_i) \qquad (2.51)$$

The work term, dW, equals pdV for an infinitesimal change. If the change is sufficiently small the work term can be approximated by

$$dW = \frac{(p_i + p_{i+1})}{2}(V_{i+1} - V_i) \qquad (2.52)$$

These results can be substituted into equation (2.50) to give

$$0 = U(T_{i+1}) - U(T_i) + \frac{(p_{i+1} + p_i)}{2}(V_{i+1} - V_i) \qquad (2.53)$$

To find p_{i+1}, the state law is applied

$$pV = nR_0T$$

If the gas composition is unchanged, the number of moles, n, is constant, thus

$$P_{i+1} = \frac{(V_i)}{(V_{i+1})}\frac{(T_{i+1})}{(T_i)}p_i \qquad (2.54)$$

For each step change in volume the temperature (T_{i+1}) can be found, but because of the complex nature of equation (2.29) an iterative solution is needed. The Newton–Raphson method is used because of its rapid convergence.

For the $(n + 1)^{th}$ iteration

$$(T_{i+1})_{n+1} = (T_{i+1})_n - \frac{f(E)_n}{f'(E)_n} \qquad (2.55)$$

where

$$f'(U) = \frac{d}{dT}(f(U)) \qquad (2.56)$$

For each volume step the first estimate of T_{i+1} is made by assuming an isentropic process with constant specific heat capacities calculated at temperature T_i.

This model is equivalent to that used in section 2.5 on the fuel/air cycle; the efficiencies quoted from this model are slightly higher than those of Taylor (1985a). The differences can be attributed to Taylor considering starting conditions of 115°C with 5 per cent exhaust residuals and 2 per cent water vapour, and fuel of octene (C_8H_{16}) as opposed to octane (C_8H_{18}). As an example, figure 2.12 shows the variation of efficiency with equivalence ratio for a compression ratio of 8 using octane as fuel; this can be compared with figure 2.10.

This simple computer model would give comparable results to the fuel/air cycle as no account is taken of either heat transfer or finite combustion times. More complex computer models are discussed in chapter 11, and these make allowance for finite combustion time, heat transfer and reaction rate kinetics. Complex models have close agreement to results that might be obtained from an experimental determination of indicated arbitrary overall efficiency. This is not surprising, as the complex models will have empirical constants derived from experiments to calibrate the model.

2.7 | Conclusions

This chapter has devised criteria by which to judge the performance of actual engines, and it has also identified some of the means of increasing both efficiency and power output, namely:

    raising the compression ratio
    minimising the combustion duration (load ratio, $\alpha \rightarrow 1$)

Reciprocating internal combustion engines follow a mechanical cycle – not a thermodynamic cycle since the start and end points are thermodynamically different. For this reason their performance should be assessed using a rational (or exergetic) efficiency. Unlike the thermal efficiency of a cyclic plant, the upper limit to the rational efficiency is 100 per cent. None the less, the arbitrary overall efficiency based on calorific value (equation 2.14) is widely used because of its convenience. Furthermore, arbitrary overall efficiencies are often compared with thermal efficiencies, since both are typically in the range 30–40 per cent. The problem of how to quote efficiency is often side-stepped by quoting specific fuel consumption, in which case the fuel ought to be fully specified.

The use of the air standard cycles arises from the similarity between engine indication (p–V) diagrams and the state diagrams of the corresponding air standard cycle. These hypothetical cycles show that, as the volumetric compression ratio is increased, the efficiency also increases (equations 2.22 and 2.31). The air standard Diesel cycle also shows that, when combustion occurs over a greater fraction of the cycle, then the efficiency reduces. No allowance is made in these simple analyses for the real properties of the working fluid, or for the changes in composition of the working fluid. These shortcomings are overcome in the fuel/air cycle and in computer models.

Despite combustion occurring more slowly in compression ignition engines, they are more efficient than spark ignition engines because of several factors:

1 Their higher compression ratios.

2 Their power output is controlled by the quantity of fuel injected, not by throttling, with its associated losses.

3 During compression the behaviour of air is closer to ideal than the behaviour of a fuel/air mixture.

Additional performance parameters have also been defined: imep, bmep, fmep, pmep, mechanical efficiency, indicated efficiency and volumetric efficiency. The indicated performance parameters are particularly useful since they measure the thermodynamic performance, as distinct from brake performance which includes the associated mechanical losses.

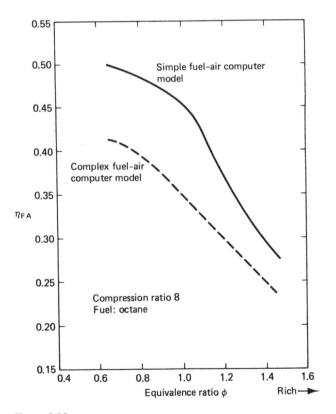

Figure 2.12
Variation of efficiency with equivalence ratio for simple fuel/air computer model and complex fuel/air model with allowance for heat transfer and combustion time (reprinted with permission from Benson and Whitehouse, 1979, © Pergamon Press Ltd).

2.8 | Examples

EXAMPLE 2.1

(i) The Rolls-Royce CV12 turbocharged four-stroke direct injection diesel engine has a displacement of 26.1 litres. The engine has a maximum output of 900 kW at 2300 rpm and is derated to 397.5 kW at 1800 rpm for industrial use. What is the bmep for each of these types?

(ii) The high-performance version of the CV12 has a bsfc of 0.063 kg/MJ at maximum power, and a minimum bsfc of 0.057 kg/MJ. Calculate the arbitrary overall efficiencies for both conditions and the fuel flow rate at maximum power. The calorific value of the fuel is 42 MJ/kg.

Solution:

(i) Using equation (2.38)

$$\text{brake power} = \bar{p}_b V_s N^*$$

remembering that for a four-stroke engine $N^* = (\text{rev/s})/2$.
Rearranging gives

$$\text{bmep}, \bar{p}_b = \frac{\text{brake power}}{V_s \dfrac{\text{rev/s}}{2}}$$

For the high-performance engine

$$\bar{p}_b = \frac{900 \times 10^3}{26.1 \times 10^{-3} \times (2300/120)} = 18.0 \times 10^5 \ \text{N/m}^2$$

$$= \underline{18 \ \text{bar}}$$

For the industrial engine

$$\bar{p}_b = \frac{397.5 \times 10^3}{26.1 \times 10^{-3} \times (1800/120)} = 10.15 \times 10^5 \ \text{N/m}^2$$

$$= \underline{10.15 \ \text{bar}}$$

(ii) Equation (2.16) relates specific fuel consumption to arbitrary overall efficiency:

$$\text{bsfc} = \frac{1}{-\Delta H_0 \eta_0} \quad \text{or} \quad \eta_0 = \frac{1}{\text{CV} \cdot \text{sfc}}$$

At maximum power

$$\eta_0 = \frac{1}{0.063 \times 42} = 37.8 \ \text{per cent}$$

At maximum economy

$$\eta_0 = \frac{1}{0.057 \times 42} = 41.8 \ \text{per cent}$$

Finally calculate the maximum flow rate of fuel, $\dot{m}_f$:

$$\dot{m}_f = \text{brake power} \cdot \text{sfc}$$
$$= (900 \times 10^3) \times (0.063 \times 10^{-6}) = \underline{0.0567 \ \text{kg/s}}$$

EXAMPLE 2.2

A high-performance four-stroke SI engine has a swept volume of 875 cm^3 and a compression ratio of 10:1. The indicated efficiency is 55 per cent of the corresponding ideal air standard Otto cycle. At 8000 rpm, the mechanical efficiency is 85 per cent, and the volumetric efficiency is 90 per cent. The air/fuel ratio (gravimetric, that is, by mass) is 13:1 and the calorific value of the fuel is 44 MJ/kg. The air is inducted at 20°C and 1 bar.

Calculate: (i) the arbitrary overall efficiency and the sfc

(ii) the air mass flow rate, power output and bmep.

Solution:

(i) The first step is to find the arbitrary overall efficiency, η_0:

$\eta_0 = \eta_m . \eta_i$, and the question states that $\eta_i = 0.55\,\eta_{Otto}$

$$\eta_{Otto} = 1 - \frac{1}{r_v^{\gamma-1}}$$

$$= 1 - \frac{1}{10^{(1.4-1)}} = 0.602 \tag{2.22}$$

Thus $\eta_0 = 0.85 \times 0.55 \times 0.602 = \underline{28.1 \text{ per cent}}$

From equation (2.16)

$$\text{sfc} = \frac{1}{CV . \eta_0} = \frac{1}{44 \times 0.281} = \underline{0.0809 \text{ kg/MJ}}$$

(ii) The air mass flow rate is found from the volume flow rate of air using the equation of state:

$$p\dot{V}_a = \dot{m}_a R_a T, \qquad R_a = 287 \text{ J/kg K}$$

From equation (2.42) $\dot{V}_a = V_s \eta_v N^*$, where $N^* = (\text{rev/s})/2$ for a four-stroke engine. Combining and rearranging gives

$$\dot{m}_a = \frac{pV_s\eta_v N^*}{R_a T} = \frac{10^5 \times 875 \times 10^{-6} \times 0.9 \times (8000/120)}{287 \times 293}$$

$$= 0.0624 \text{ kg/s}$$

The power output can be found for the fuel flow rate, since the efficiency (or sfc) is known. The fuel flow rate is then found for the air flow rate, from the air/fuel ratio (A/F):

$$\dot{m}_f = \dot{m}_a / A/F$$

$$\text{Brake power output} = \frac{\dot{m}_f}{\text{sfc}} = \frac{0.0624}{13.0809 \times 10^{-6}} = \underline{59.3 \text{ kW}}$$

Finally, from equation (2.38)

$$\bar{p}_b = \frac{\dot{W}_b}{V_s N^*} = \frac{59.3 \times 10^3}{875 \times 10^{-6} \times (8000/120)} = \underline{10.2 \text{ bar}}$$

EXAMPLE 2.3

Reciprocating internal combustion engines have been fitted with ingenious mechanisms that allow the expansion ratio (r_e) to be greater than the compression ratio (r_c). When such a system is modelled by an ideal gas cycle there is:

(i)   heat addition at constant volume (as in the Otto cycle)
(ii) some heat rejection at constant volume (as in the Otto cycle)
(iii) and some heat rejection at constant pressure, to complete the cycle.

The processes are shown in figure 2.13 on the T–s plane. Depict the processes on the p–V plane.
   The constant-volume temperature rise (2–3) is θT_1. Derive an expression for the cycle efficiency in terms of r_c, r_e and θ; state any assumptions.
   The limiting case is the Atkinson cycle, in which the expansion ratio is sufficient to provide full expansion to the initial pressure, and all heat rejection is then at constant pressure. Determine the limiting value of the expansion ratio (r_e) in terms of θ and r_c, and the Atkinson cycle efficiency.

Solution:
Assume (figure 2.14)

$1 \rightarrow 2$	isentropic compression
$2 \rightarrow 3$	constant-volume heat input
$3 \rightarrow 4$	isentropic expansion
$4 \rightarrow 5$	constant-volume heat rejection
$5 \rightarrow 1$	isobaric heat rejection

Perfect gas

with r_c = compression ratio
       r_e = expansion ratio
       V_c = clearance volume.

$$\eta = 1 - \frac{Q_0}{Q_i} = 1 - \frac{Q_{45} + Q_{51}}{Q_{23}}$$

$$\left. \begin{array}{l} Q_{23} = c_v m(T_3 - T_2) \\ Q_{45} = c_v m(T_4 - T_5) \\ Q_{51} = c_p m(T_5 - T_1) \end{array} \right\}$$

giving $$\eta = \frac{c_v(T_4 - T_5) + c_p(T_5 - T_1)}{c_v(T_3 - T_2)} \qquad (1)$$

It is possible to find all the temperatures in the cycle by working around the cycle, starting from state 1:

for isentropic compression

$$1 \rightarrow 2, \qquad T_2 = T_1 r_c^{\gamma-1}$$

for constant-volume heat addition

$$2 \rightarrow 3, \qquad T_3 = T_2 + \theta T_1 = T_1(r_c^{\gamma-1} + \theta)$$

Now working from state 1 in the opposite direction:

for constant-pressure heat rejection

$$5 \rightarrow 1, \qquad T_5 = (V_5/V_1) \times T_1 = T_1 \times (r_e/r_c)$$

for constant-volume heat rejection

$$4 \rightarrow 5, \qquad T_4 = T_3/r_e^{\gamma-1} = T_1\left(\frac{\theta + r_c^{\gamma-1}}{r_e^{\gamma-1}}\right)$$

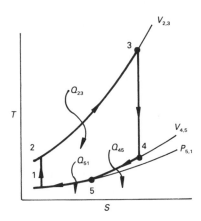

Figure 2.13
The temperature/entropy state diagram for the Atkinson cycle.

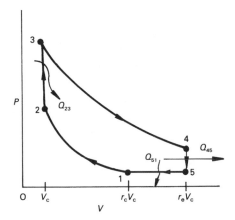

Figure 2.14
The pressure/volume state diagram for the Atkinson cycle.

Substitute into equation (1) in terms of T_1, for T_2, T_3, T_4 and T_5:

$$\eta = 1 - \frac{c_v\left(\dfrac{\theta + r_c^{\gamma-1}}{r_e^{\gamma-1}} - \dfrac{r_e}{r_c}\right) + c_p\left(\dfrac{r_e}{r_c} - 1\right)}{\theta c_v}$$

Multiply top and bottom by $\dfrac{r_c r_c^{\gamma-1}}{c_v}$:

$$\eta = 1 - \frac{\theta r_c + r_c^{\gamma} - r_e^{\gamma} + \gamma(r_e - r_c) \times r_e^{\gamma-1}}{\theta r_c r_e^{\gamma-1}}$$

or

$$\eta = 1 - \frac{(\gamma - 1)r_e^{\gamma} + r_c(\theta - \gamma r_e^{\gamma-1}) + r_c^{\gamma}}{\theta r_c r_e^{\gamma-1}} \tag{2.57}$$

Check: as $r_c \to r_e$, $r_c = r_e = r_v$, $\eta \to \eta_{\text{Otto}} = 1 - \dfrac{1}{r_v^{\gamma-1}}$

The limiting case (the Atkinson cycle) is when full expansion occurs, that is, state 4 merges with state 5. Thus $T_4 = T_5$, and

$$T_1\left(\frac{\theta + r_c^{\gamma-1}}{r_e^{\gamma-1}}\right) = T_1\left(\frac{r_e}{r_c}\right)$$

Rearranging to give an expression for r_e:

$$r_e^{\gamma} = \theta r_c + r_c^{\gamma}, \qquad r_e = \sqrt[\gamma]{(\theta r_c + r_c^{\gamma})}$$

$$\eta_{\text{Atkinson}} = 1 - \frac{Q_{51}}{Q_{23}} = 1 - \frac{c_p(r_e/r_c - 1)T_1}{c_v \theta T_1}$$

$$\eta_{\text{Atkinson}} = 1 - \frac{\gamma(r_e - r_c)}{\theta r_c} = 1 - \frac{\gamma(r_e - r_c)}{r_e^{\gamma} - r_c^{\gamma}}$$

The limiting compression ratio of an SI engine is around 10; suppose $T_1 = 300$ K.
Consider the combustion of a stoichiometric mixture, with a gravimetric air/fuel ratio of 15, and a calorific value (CV) of 44 MJ/kg of fuel:

$$\theta = \frac{T_{23}}{T_1} = \frac{CV/(AFR + 1)}{C_v/T_1} = \frac{44 \times 10^3/16}{0.75 \times 300}$$

$$\theta \sim 12$$

This enables us to find the expansion ratio corresponding to a compression ratio of 10:

$$\begin{aligned}
r_e &= \sqrt[\gamma]{(\theta r_c + r_c^{\gamma})} \\
&= \sqrt[1.4]{(12 \times 10 + 10^{1.4})} \\
&= 35
\end{aligned}$$

An expansion ratio 3.5 times the compression ratio will be very difficult to obtain:

$$\begin{aligned}
\eta_{\text{Atkinson}} &= 1 - \frac{\gamma(r_e - r_c)}{\theta r_c} = 1 - \frac{1.4(35 - 10)}{12 \times 10} \\
&= 70.8 \text{ per cent}
\end{aligned}$$

$$\eta_{\text{Otto}} = 1 - \frac{1}{r_v^{\gamma-1}} = 60.2 \text{ per cent}$$

Thus, at most, η_{Atkinson} gives a 16 per cent improvement in cycle efficiency and work output. However, in practice, this is unlikely to be obtained, because of the increased frictional losses. Also, the engine would be very bulky, about three times the size of a conventional engine that has the same output.

2.9 | Questions

Unless otherwise stated in these questions, the ratio of heat capacities (γ) should be taken as 1.4, and the specific heat capacity of air can be assumed constant at 1.01 kJ/kg K.

2.1 For the ideal air standard Diesel cycle with a volumetric compression ratio of 17:1 calculate the efficiencies for cut-off ratios of 1, 2, 4, 9. Take $\gamma = 1.4$. The answers can be checked with figure 2.2.

2.2 Outline the shortcomings of the simple ideal cycles, and explain how the fuel/air cycle and computer models overcome these problems.

2.3 A 2-litre four-stroke indirect injection diesel engine is designed to run at 4500 rpm with a power output of 45 kW; the volumetric efficiency is found to be 80 per cent. The bsfc is 0.071 kg/MJ and the fuel has a calorific value of 42 MJ/kg. The ambient conditions for the test were 20°C and 1 bar. Calculate the bmep, the arbitrary overall efficiency and the air/fuel ratio.

2.4 A twin-cylinder two-stroke engine has a swept volume of 150 cm^3. The maximum power output is 19 kW at 11 000 rpm. At this condition the bsfc is 0.11 kg/MJ, and the gravimetric air/fuel ratio is 12:1. If ambient test conditions were 10°C and 1.03 bar, and the fuel has a calorific value of 44 MJ/kg, calculate the bmep, the arbitrary overall efficiency and the volumetric efficiency.

2.5 A four-stroke 3-litre V6 spark ignition petrol engine has a maximum power output of 100 kW at 5500 rpm, and a maximum torque of 236 Nm at 3000 rpm. The minimum bsfc is 0.090 kg/MJ at 3000 rpm, and the air flow rate is 0.068 m^3/s. The compression ratio is 8.9:1 and the mechanical efficiency is 90 per cent. The engine was tested under ambient conditions of 20°C and 1 bar; take the calorific value of the fuel to be 44 MJ/kg.

(a) Calculate the power output at 3000 rpm and the torque output at 5500 rpm.
(b) Calculate for both speeds the bmep and the imep.
(c) How does the arbitrary overall efficiency at 3000 rpm compare with the corresponding air standard Otto cycle efficiency?
(d) What is the volumetric efficiency and air/fuel ratio at 3000 rpm?

2.6 Show that the air standard Dual cycle efficiency is given by

$$\eta = 1 - \frac{1}{r_v^{\gamma-1}} \left[\frac{r_p \alpha^\gamma - 1}{(r_p - 1) + \gamma r_p (\alpha - 1)} \right] \tag{2.33}$$

where r_v = volumetric compression ratio
  r_p = pressure ratio during constant-volume heat addition
  α = volumetric expansion ratio during constant-pressure heat addition.

   Make reference to figure 2.4.

2.7 Explain why air standard cycles are used to represent the performance of real internal combustion engines.

2.8 A four-cylinder, four-stroke petrol engine is to develop 40 kW at 40 rev/s when designed for a volumetric compression ratio of 10.0:1. The ambient air conditions are 1 bar and 18°C, and the calorific value of the fuel is 44 MJ/kg.

(a) Calculate the specific fuel consumption in kg per MJ of brake work if the indicated overall efficiency is 50 per cent of the corresponding air standard Otto cycle, and the mechanical efficiency is 90 per cent. The specific heat capacity ratio for air is 1.4.

(b) The required gravimetric air/fuel ratio is 15.4 and the volumetric efficiency is 92 per cent. Estimate the required total swept volume, and the cylinder bore if the bore is to be equal to the stroke. Calculate also the brake mean effective pressure.

2.9 A compression ignition engine has a volumetric compression ratio of 15. Find the thermal efficiency of the following air standard cycles having the same volumetric compression ratio as the engine. The specific heat capacity ratio for air is 1.4:

(a) An Otto cycle.
(b) A Diesel cycle in which the temperature at the beginning of compression is 18°C, and in which the heat supplied per unit mass of air is equal to the energy supplied by the fuel (in terms of its calorific value). The gravimetric air/fuel ratio is 28:1; the calorific value of the fuel is 44 MJ/kg; assume the specific heat of air at constant pressure is 1.01 kJ/kg K and is independent of temperature.

2.10 A petrol engine of volumetric ratio 9:1 takes in a mixture of air and fuel in the ratio 17:1 by weight; the calorific value of the fuel is 44 MJ/kg. At the start of compression the temperature of the charge is 50°C. Assume that compression and expansion are reversible with pv^n constant, and $n = 1.325$ and 1.240 respectively, and that combustion occurs instantaneously at minimum volume. Combustion can be regarded as adding heat equal to the calorific value to the charge.

However, there is a finite combustion efficiency, and heat transfer from the combustion chamber. Combustion is thus equivalent to a net heat input that corresponds to 75 per cent of the calorific value of the fuel being burnt.

Calculate the temperatures after compression, and at the start and end of expansion. Calculate the net work produced by the cycle and thus evaluate the indicated efficiency of the engine. Why is it inappropriate to calculate the indicated efficiency in terms of the heat flows?

Use the following thermodynamic data:

	Molar mass, kg	Specific heat capacity at constant volume, kJ/kg K
Air/fuel mixture	30	0.95
Combustion products	28	0.95

2.11 Derive the expression for the air standard Diesel cycle efficiency, in terms of the ratio of gas specific heat capacities (γ), the volumetric compression ratio (r_v) and the load ratio (α). State clearly the assumptions in your analysis.

Show that the indicated mean effective pressure ($\bar{p}_i$) of the air standard Diesel cycle is given by

$$\bar{p}_i = \frac{P_1}{\gamma - 1} \times \frac{r_v}{r_v - 1}\left[\gamma r_v^{\gamma-1}(\alpha - 1) - (\alpha^\gamma - 1)\right]$$

where P_1 is the pressure at the start of compression.

2.12 The Lenoir engine operates without compression before ignition, and it can be modelled by the following processes in an air standard cycle:

$1 \to 2$ heat addition at constant volume, to increase the temperature to ϕT_1,
$2 \to 3$ reversible and adiabatic expansion through a volumetric expansion ratio r_e,
$3 \to 4$ heat rejection at constant volume, to decrease the pressure to p_1, and
$4 \to 1$ heat rejection at constant pressure, to decrease the temperature to T_1.

Draw the processes on a p–V diagram, and show that the air standard cycle efficiency is given by

$$\eta = 1 - \frac{(\gamma - 1)r_e^\gamma + \phi - \gamma r_e^{\gamma-1}}{(\phi - 1)r_e^{\gamma-1}}$$

where r_e = volumetric expansion ratio, and ϕ = temperature ratio during constant-volume heat addition.

Show that the Atkinson cycle efficiency (see example 2.3) reduces to this result when the compression ratio is set to unity.

What is the air standard cycle efficiency when the expansion ratio (r_e) is 2, and the temperature ratio during constant-volume heat addition (ϕ) is 6?

2.13 In an Otto cycle, the temperature rise during constant-volume heat addition is ϕT_1. If the conditions at the start of compression are defined as state 1, and the volumetric compression ratio is r_v, find expressions for the compression work, the expansion work, the net work, and the work ratio (net work divided by the expansion work).

CHAPTER 3

Combustion and fuels

3.1 | Introduction

The fundamental difference between spark ignition (SI) and compression ignition (CI) engines lies in the type of combustion that occurs, and not in whether the process is idealised as an Otto cycle or a Diesel cycle. The combustion process occurs at neither constant volume (Otto cycle), nor constant pressure (Diesel cycle). The difference between the two combustion processes is that spark ignition engines usually have pre-mixed flames while compression ignition engines have diffusion flames. With pre-mixed combustion the fuel/air mixture must always be close to stoichiometric (chemically correct) for reliable ignition and combustion. To control the power output a conventional spark ignition engine is throttled, thus reducing the mass of fuel and air in the combustion chamber; this reduces the cycle efficiency. In contrast, for compression ignition engines with fuel injection the mixture is close to stoichiometric only at the flame front. The output of compression ignition engines can thus be varied by controlling the amount of fuel injected; this partly accounts for their superior part load fuel economy.

With pre-mixed reactants the flame moves relative to the reactants, so separating the reactants and products. An example of pre-mixed combustion is with oxy-acetylene equipment; for welding, the flame is fuel-rich to prevent oxidation of the metal, while, for metal cutting, the flame is oxygen-rich in order to burn as well as to melt the metal.

With diffusion flames, the flame occurs at the interface between fuel and oxidant. The products of combustion diffuse into the oxidant, and the oxidant diffuses through the products. Similar processes occur on the fuel side of the flame, and the burning rate is controlled by diffusion. A common example of a diffusion flame is the candle. The fuel is melted and evaporated by radiation from the flame, and then oxidised by air; the process is evidently one governed by diffusion as the reactants are not pre-mixed.

The Bunsen burner, shown in figure 3.1, has both a pre-mixed flame and a diffusion flame. The air entrained at the base of the burner is not sufficient for complete combustion

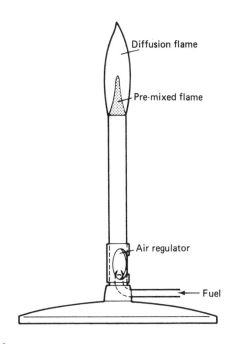

Figure 3.1
Bunsen burner with pre-mixed and diffusion flames.

with a single pre-mixed flame. Consequently, a second flame front is established at the interface where the air is diffusing into the unburnt fuel.

The physics and chemistry of combustion are covered in some detail by both Gaydon and Wolfhard (1979) and Lewis and von Elbe (1961), but neither book devotes much attention to combustion in internal combustion engines. Hydrocarbon/air mixtures have an ambient laminar burning velocity of up to 0.5 m/s, a notable exception being acetylene/air with a value of 1.58 m/s. An order of magnitude calculation for the burning time in a cylinder of 100 mm diameter with central ignition gives about 100 ms. However, for an engine running at 3000 rpm the combustion time can only be about 10 ms. This shows the importance of turbulence in speeding combustion by at least an order of magnitude.

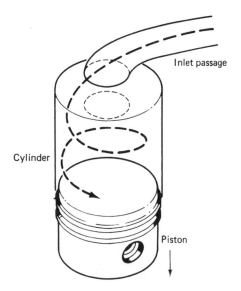

Figure 3.2
Swirl generation from tangential inlet passage.

Turbulence is generated as a result of the induction and compression processes, and the geometry of the combustion chamber. In addition there can be an ordered air motion such as swirl which is particularly important in diesel engines. This is obtained from the tangential component of velocity during induction (figure 3.2).

For pre-mixed combustion the effect of turbulence is to break up, or wrinkle the flame front. There can be pockets of burnt gas in the unburnt gas and vice versa. This increases the flame front area and speeds up combustion. Figure 3.3 shows a comparison of laminar and turbulent flame fronts.

For diffusion-controlled combustion the turbulence again enhances the burning velocity. Fuel is injected as a fine spray into air which is hot enough to vaporise and ignite the fuel. The ordered air motion is also important because it sweeps away the vaporised fuel and combustion products from the fuel droplets; this is shown in figure 3.4.

Finally, it should be noted that there are compression ignition engines with pre-mixed combustion, for example 'model diesel' engines that use a carburetted ether-based fuel. Conversely, there are spark ignition engines such as some stratified charge engines that have diffusion processes; neither of these exceptions will be considered separately.

Before discussing combustion in spark ignition engines and compression ignition engines in any greater detail, it is necessary to study combustion chemistry and dissociation.

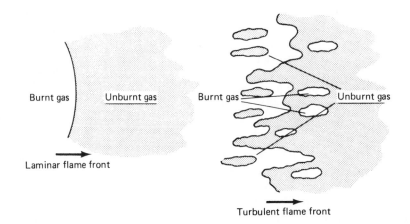

Figure 3.3
Comparison between laminar and turbulent flame fronts for pre-mixed combustion.

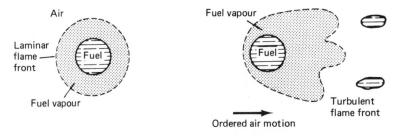

Figure 3.4 Comparison between laminar flame front in stagnant air with turbulent flame front and ordered air motion for diffusion-controlled combustion.

3.2 | Combustion chemistry and fuel chemistry

Only an outline of combustion chemistry will be given here as the subject is treated in general thermodynamics books such as Rogers and Mayhew (1980a) or more specialised works like Goodger (1979). However, neither of these books emphasises the difference between rational efficiency (η_R) and arbitrary overall efficiency (η_0).

For reacting mixtures the use of molar quantities is invaluable since reactions occur between integral numbers of molecules, and the mole is a unit quantity of molecules. The mole is the amount of substance in a system that contains as many elementary entities as there are atoms in 0.012 kg of carbon 12. The normal SI unit is the kilomole (kmol), and the molar number (Avogadro constant) is 6.023×10^{26} entities/kmol.

Consider the reaction between two molecules of carbon monoxide (CO) and one molecule of oxygen (O_2) to produce two molecules of carbon dioxide (CO_2):

$$2CO + O_2 \rightarrow 2CO_2$$

There is conservation of mass, and conservation of the number of atoms.

It is often convenient to consider a unit quantity of fuel, for instance a kilomole, so the above reaction can be written in terms of kilomoles as

$$CO + \tfrac{1}{2}O_2 \rightarrow CO_2$$

A further advantage of using kilomoles is that, for the same conditions of temperature and pressure, equal volumes of gas contain the same number of moles. Thus the volumetric composition of a gas mixture is also the molar composition of the gas mixture. This is obviously not the case if liquid or solids are also present.

As well as the forward reaction of carbon monoxide oxidising to carbon dioxide, there will be a reverse reaction of carbon dioxide dissociating to carbon monoxide and oxygen:

$$CO + \tfrac{1}{2}O_2 \rightleftharpoons CO_2$$

When equilibrium is attained the mixture will contain all possible species from the reaction. In addition the oxygen can also dissociate:

$$O_2 \rightleftharpoons 2O$$

With internal combustion engines, dissociation is important. Furthermore, there is not always sufficient time for equilibrium to be attained. Initially, complete combustion will be assumed.

Fuels are usually burnt with air, which has the following composition:

| Molar | 21.0 per cent O_2 | 79 per cent N_2^* |
| Gravimetric | 23.2 per cent O_2 | 76.8 per cent N_2^* |

Atmospheric nitrogen, N_2^*, represents all the constituents of air except oxygen and water vapour. Its thermodynamic properties are usually taken to be those of pure nitrogen. The molar masses (that is, the masses containing a kilomole of molecules) for these substances are:

O_2 31.999 kg/kmol; N_2 28.013 kg/kmol
N_2^* 28.15 kg/kmol; air 28.96 kg/kmol

When carbon monoxide is burnt with air the reaction (in kmols) is

$$CO + \tfrac{1}{2}\left(O_2 + \frac{79}{21}N_2^*\right) \rightarrow CO_2 + \tfrac{1}{2}\frac{79}{21}N_2^*$$

The nitrogen must be kept in the equation even though it does not take part in the reaction; it affects the volumetric composition of the products and the combustion temperature. The molar or volumetric air/fuel ratio is

$$1 : \tfrac{1}{2}\left(1 + \frac{79}{21}\right)$$

or   1 : 2.38

The gravimetric air/fuel ratio is found by multiplying the number of moles by the respective molar masses — (12 + 16) kg/kmol for carbon monoxide, and 29 kg/kmol for air:

$$1.(12 + 16) : 2.38(29)$$

or   1 : 2.47

So far the reacting mixtures have been assumed to be chemically correct for complete combustion (that is, stoichiometric). In general, reacting mixtures are non-stoichiometric; these mixtures can be defined in terms of the excess air, theoretical air or equivalence ratio. Consider the same reaction as before with 25 per cent excess air, or 125 per cent theoretical air:

$$CO + \frac{1.25}{2}\left(O_2 + \frac{79}{21}N_2^*\right) \rightarrow CO_2 + \frac{0.25}{2}O_2 + \frac{1.25}{2}\cdot\frac{79}{21}N_2^*$$

The equivalence ratio

$$\phi = \frac{\text{stoichiometric air/fuel ratio}}{\text{actual air/fuel ratio}} = \frac{1}{1.25} = 0.8$$

The air/fuel ratio can be either gravimetric or molar; the usual form is gravimetric and this is often implicit.

Fuels are often mixtures of hydrocarbons, with bonds between carbon atoms, and between hydrogen and carbon atoms. During combustion these bonds are broken, and new bonds are formed with oxygen atoms, accompanied by a release of chemical energy. The principal products are carbon dioxide and water.

As combustion does not pass through a succession of equilibrium states it is irreversible, and the equilibrium position will be such that entropy is a maximum. The different compounds in fuels are classified according to the number of carbon atoms in the molecules. The size and geometry of the molecule have a profound effect on the chemical properties. Each carbon atom requires four bonds; these can be single bonds or combinations of single, double and triple bonds. Hydrogen atoms require a single bond.

An important family of compounds in petroleum (that is, petrol or diesel fuel) are the alkanes, formerly called the paraffins. Table 3.1 lists some of the alkanes; the different prefixes indicate the number of carbon atoms.

The alkanes have a general formula C_nH_{2n+2}, where n is the number of carbon atoms. Inspection shows that all the carbon bonds are single bonds, so the alkanes are termed 'saturated'. For example, propane has the structural formula

An alternative description of the alkanes is RH, where R represents a radical. The alkanes are based on the alkyl radical (C_nH_{2n+1}), with CH_3 being the methyl radical, C_2H_5 the ethyl radical and so on.

When four or more carbon atoms are in a chain molecule it is possible to form isomers. Isomers have the same chemical formula but different structures, which often leads to very different chemical properties. Iso-octane is of particular significance for spark ignition engines; although it should be called 2,2,4-trimethylpentane, the isomer implied in petroleum technology is

(Hydrogen atoms not shown)

The 'backbone' of five carbon atoms is numbered from left to right, and the numbers 2, 2, 4 designate the carbon atoms to which the methyl (CH_3) groups are attached. This isomer of octane might also be called 4,4,2-trimethylpentane, but the convention is to use the lowest numbers possible.

Compounds that have straight chains with a single double bond are termed alkenes (formerly olefins); the general formula is C_nH_{2n}. An example is propylene, C_3H_6:

Table 3.1 *Alkane family of compounds*

Formula	Name	Comments	
CH_4	methane	'Natural gas'	LPG
C_2H_6	ethane	⎫	Liquefied
C_3H_8	propane ⎫	⎬	Petroleum
C_4H_{10}	butane ⎭	'Bottled gas' ⎭	Gases
C_5H_{12}	pentane ⎫		
C_6H_{14}	hexane	Liquids at room	
C_7H_{16}	heptane	temperature	
C_8H_{18}	octane		
•			
•			
•			
$C_{16}H_{34}$	cetane ⎭		
•			
•			
etc.			

Such compounds are termed 'unsaturated' as the double bond can be split and extra hydrogen atoms added, a process termed 'hydrogenation'.

With butene, there are two straight chain isomers, depending on where the double bond is located:

1-Butene   2-Butene

The number denotes the number of the first double-bonded carbon atom from the end closest to the double bond. Side-chain isomers also occur, and since there cannot be rotation about a double bond, there is a second isomer of 2-butene. Hydrocarbons with two carbon double bonds are called dienes, and with three carbon double bonds are called trienes:

1,3-Butadiene   1,4-Pentadiene

Butadiene is of particular concern as an exhaust emission, because of its reactivity. The dienes have a general formula C_nH_{2n-2}, which is the same as the alkynes (acetylenes), but the alkynes have a carbon triple bond. The simplest member of this family is acetylene: $H-C\equiv C-H$.

Most of the alkene content in fuels comes from catalytic cracking. In this process the less volatile alkanes are passed under pressure through catalysts such as silica or alumina at about 500°C. The large molecules are decomposed, or cracked, to form smaller more volatile molecules. A hypothetical example might be

$C_{20}H_{42} \rightarrow C_4H_8 + C_5H_{10} + C_4H_{10} + C_6H_{14} +$ C
alkane   alkenes   alkanes   carbon

A disadvantage of alkenes is that they can oxidise when the fuel is stored in contact with air. The oxidation products reduce the quality of the fuel and leave gum deposits in the engine. Oxidation can be inhibited by the addition of alkylphenol (see page 63), typically 100 ppm (parts per million by weight).

Hydrocarbons can also form ring structures, which can be saturated or unsaturated. Cyclo-alkanes are saturated and have a general formula C_nH_{2n}; in petroleum technology they are called naphthenes. The smallest example is cyclopropane.

Aromatic compounds are unsaturated and based on the benzene molecule, C_6H_6. This has an unsaturated ring best represented as

The inner circle signifies two groups of three electrons, which form molecular orbitals each side of the carbon atom plane. The structure accounts for the distinct properties of the aromatic compounds. Benzene and its derivatives occur in many crude oils but they also come from the distillation of coal.

The hydrogen atoms in benzene can be substituted by methyl groups. Substitution of one hydrogen atom gives methylbenzene $(C_6H_5.CH_3)$, which is widely known as toluene. Substitution of two hydrogen atoms gives dimethylbenzene, $C_6H_5.(CH_3)_2$, which is traditionally known as xylene. Xylene can have three isomers, when the methyl groups are adjacent, separated by one hydrogen atom, or separated by two hydrogen atoms. Toluene can be considered as a combination of phenyl (C_6H_5) and methyl (CH_3) radicals; somewhat confusingly the benzyl radical is $C_6H_5.CH_2$. Only when the side chain is complex is the substance given the prefix phenyl; toluene is not called phenylmethane.

Two or three benzene rings can be joined together to form polycyclic (or polynuclear) aromatic hydrocarbons. The simplest case is naphthalene $(C_{10}H_8)$ with two benzene rings sharing a pair of carbon atoms, and as with the benzene ring, hydrogen atoms can be interchanged with methyl groups. With three benzene rings sharing two or more pairs of carbon atoms, there are three possible isomers:

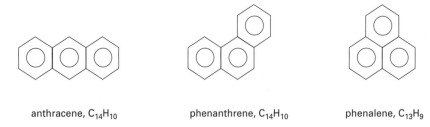

anthracene, $C_{14}H_{10}$ phenanthrene, $C_{14}H_{10}$ phenalene, $C_{13}H_9$

Polycyclic aromatic hydrocarbons are significant in both diesel fuel, and in the exhaust products.

The final class of fuels that have significance for internal combustion engines are the alcohols, in particular methanol (CH_3OH) and ethanol (C_2H_5OH):

$$H-\underset{\underset{H}{|}}{\overset{\overset{H}{|}}{C}}-O-H \qquad\qquad H-\underset{\underset{H}{|}}{\overset{\overset{H}{|}}{C}}-\underset{\underset{H}{|}}{\overset{\overset{H}{|}}{C}}-O-H$$

The resurgent interest in alcohols is due to their manufacture from renewable energy sources, such as the destructive distillation of wood and fermentation of sugar respectively.

Simple alcohols contain a single oxygen atom, and this requires two bonds. The generic formula is $C_nH_{2n+1}OH$, and with three or more carbon atoms then isomers can be formed:

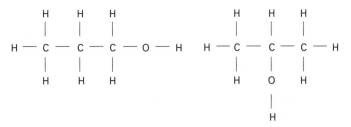

n-propanol, or 1-propanol   iso-propanol, or 2-propanol

The number indicates the location of the OH group, in terms of the carbon atom closest to the end of the molecule. When the hydrogen atom in a benzene ring is replaced by the OH group, phenol is formed: C_6H_5OH. If another hydrogen atom is replaced with an alkyl group, then an alkyl phenol is formed:

OH

C_nH_{2n+1}

Dihydric alcohols (called glycols or diols) contain two OH groups. The simplest is ethylene glycol, which is widely used as an antifreeze agent in engine cooling systems:

$$H-O-\underset{\underset{H}{|}}{\overset{\overset{H}{|}}{C}}-\underset{\underset{H}{|}}{\overset{\overset{H}{|}}{C}}-O-H$$

Other oxygenated compounds are important as fuel additives, intermediate compounds during combustion or as exhaust emissions:

aldehydes   ketones   ethers

where R and R' denote radicals that can be the same. The aldehydes and ketones are clearly related, since the aldehyde has a hydrogen atom rather than a radical. The simplest aldehyde is formaldehyde (CH_2O), and its acrid smell is often detectable from spark ignition engines operating on propane. The ethers are ignition improvers for diesel engines, and dimethylether (DME), $(CH_3)_2O$, has been proposed as a diesel fuel. Methyl tertiary butyl ether (MTBE), $(CH_3)_3COCH_3$, is an important oxygenate for improving the octane rating of gasoline, but its addition is usually limited (to about 10 or 15 per cent) by legislation.

The calculations for the combustion of fuel mixtures are not complicated, but are best shown by worked solutions; see examples 3.1 and 3.2. At this stage it is sufficient to note that the mean composition of molecules will be close to C_xH_{2x} and that most of the bonds will be single carbon–carbon or carbon–hydrogen. This implies that the stoichiometric (gravimetric) air/fuel ratio is always close to 14.8:1 (see question 3.1). Furthermore, because of the similarity in bond structure, the calorific value will vary only slightly for a range of fuels; this is shown in figure 3.5 using data from Blackmore and Thomas (1977) and Taylor (1985a).

More detailed information on the energy content of different fuels is presented in figure 3.6. Obviously the molar calorific value of fuels increases with the molar mass, so it is

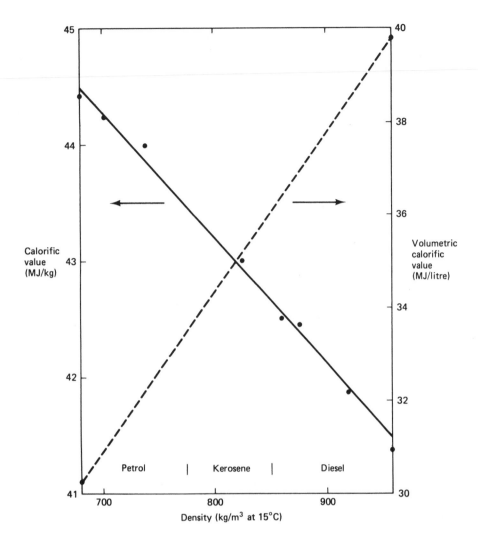

Figure 3.5
Variation in calorific value for different fuel mixture (with acknowledgement to Blackmore and Thomas, 1977).

more instructive to show the trends on a gravimetric basis: both in terms of a kg of fuel, and a kg of stoichiometric mixture. There is a general trend that the fuel calorific value decreases with increasing numbers of hydrogen atoms, with only slight differences between alkanes (paraffins), olefins and acetylenes. The aromatics have slightly lower calorific values, but the same trend. However, alcohols have gravimetric calorific values significantly lower than those of pure hydrocarbons, but they increase with the number of carbon atoms. One way of looking at this is to say that the oxygen already represents a partial oxidation of the fuel, and the effect of one oxygen atom becomes less significant as more CH_2 groups are added.

When the calorific values are considered on the basis of one kilogram of stoichiometric mixture, then the differences are more significant. The stoichiometric air/fuel ratios for the alkanes are the highest, and this then leads to them having the lowest mixture calorific values. The aromatics are significant in automotive fuels, and these can lead to higher engine power

outputs. Alcohols have very low gravimetric stoichiometric air/fuel ratios, and this can more than compensate for their low calorific value. Although engine performance may not be much affected by using alcohols, it must be remembered that the fuel will be bought on a volumetric or mass basis, and that there would be a higher fuel consumption rate.

3.3 | Combustion thermodynamics

Only combustion under the simple conditions of constant volume or constant pressure will be considered here. However, the energy analysis for combustion with varying volume has already been presented in chapter 2, section 2.6, and such analyses require the calculation of the internal energy of reacting mixtures, and this corresponds to constant-volume combustion. The gases, both reactants and products,

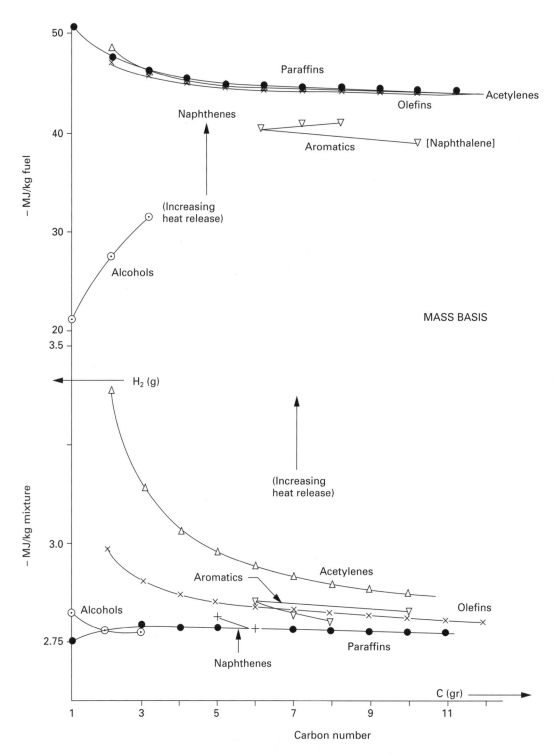

Figure 3.6
Lower calorific values for fuels in the vapour state, using both the fuel and a stoichiometric mixture as the gravimetric basis (adapted from Goodger, 1975).

will be considered as ideal (a gas that obeys the equation of state $pV = RT$, but has specific heat capacities that vary with temperature but not pressure).

It will be seen shortly that combustion calculations require

tabulations of thermodynamic data, such as internal energy. Two alternative approaches are to be adopted here, and they will be designated a and b. Section 3.3a follows a conventional approach using widely available tabulations of thermodynamic data. However, it will be seen that section 3.3b (using thermodynamic data tabulated in Appendix A) leads to more direct solutions. The relevant examples in section 3.11 are solved using both approaches, using the same designation (a and b).

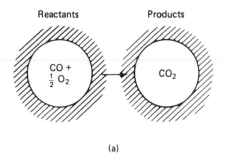

(a)

3.3a Use of conventional thermodynamic tabulations

There are many tabulations of molar enthalpy for the species encountered in combustion, and two common ones are:

(a) Haywood (1972), who uses a datum of zero enthalpy at 0 K, and

(b) Rogers and Mayhew (1988), who use a datum of zero enthalpy at 298.15 K.

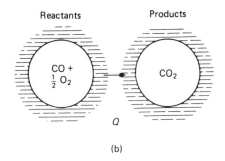

(b)

Each set of tables will be self-consistent, but obviously the data should not be mixed without datum corrections. If only molar enthalpy is tabulated (Haywood, 1972), molar internal energy has to be deduced:

$$H \equiv U + pV \quad \text{or} \quad H \equiv U + R_0 T$$

The datum temperatures for these properties are arbitrary, but 25°C is convenient as it is the reference temperature for enthalpies of reaction.

Consider a rigid vessel containing a stoichiometric mixture of carbon monoxide and oxygen; the reactants (R) react completely to form carbon dioxide, the products (P). The first case (figure 3.7a) is when the container is insulated, so the process is adiabatic.

Apply the First Law of Thermodynamics:

$$\Delta U = Q - W$$

but $Q = 0$ since the process is adiabatic
and $W = 0$ since the container is rigid.
Thus $\Delta U = 0$ since there is no change in potential or kinetic energy
and $U_R = U_P$.

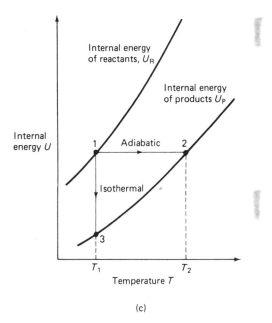

(c)

Figure 3.7
Constant-volume combustion: (a) rigid insulated bomb calorimeter; (b) rigid isothermal bomb calorimeter; (c) internal energy of products and reactants.

This process is represented by the line 1–2 on figure 3.7c, a plot that shows how the internal energy of the reactants and products vary with temperature. Temperature T_2 is the adiabatic combustion temperature.

The second case to consider (figure 3.7b) is when the container is contrived to be isothermal, perhaps by using a water bath. Again apply the First Law of Thermodynamics:

$$\Delta U = Q - W$$

Again $W = 0$ since the container is rigid
but $Q \neq 0$ since the process is isothermal.
Thus

$$\Delta U = Q = U_P - U_R \qquad (3.1)$$

Normally energy is released in a chemical reaction and Q is negative (exothermic reaction); if Q is positive the reaction is said to be endothermic. This method is used to find the constant-volume or isochoric calorific value of a fuel in a combustion bomb.

$$\text{Isochoric calorific value} = (CV)_{T,V}$$
$$= -(\Delta U)_{T,V} = (U_R - U_P)_{T,V}$$

The process is shown by the line 1–3 in figure 3.7c.

Combustion at constant pressure can be analysed by considering a rigid cylinder closed by a free-moving perfectly sealed piston (figure 3.8a). As before, the combustion can be contrived to be adiabatic or isothermal.

Again apply the First Law of Thermodynamics:

$$Q = \Delta U + W$$

In this case W is the displacement work. For constant pressure:

$$W = p\Delta V = \Delta pV$$

Thus

$$Q = \Delta U + \Delta pV = \Delta H = H_P - H_R \qquad (3.2)$$

Adiabatic and isothermal combustion processes are shown on figure 3.8b. Calorific value normally refers to the isobaric isothermal calorific value, $CV = -\Delta H_{Tp} = (H_R - H_P)_{T,p}$. The datum temperature T_0 for tabulating calorific value is usually 298.15 K (25°C); this is denoted by $-\Delta H_0$. To evaluate calorific values at other temperatures, say T_1, recall that the calorific value at this temperature, $-\Delta H_{T_1,p}$, will be independent of the path taken.

Referring to figure 3.8b, 1R–0R–0P–1P will be equivalent to 1R–1P.

Thus

$$\Delta H_{T_1,p} = (H_{0P} - H_{1P}) + \Delta H_{T_0,p} + (H_{1R} - H_{0R}) \qquad (3.3)$$

This principle is demonstrated by example 3.3. These calorific values can be found using steady-flow combustion calorimeters.

Since most combustion occurs at constant pressure (boilers, furnaces and gas turbines) isobaric calorific value ($-\Delta H_0$) is tabulated instead of isochoric value (ΔU_0); consequently there is a need to be able to convert from isobaric to isochoric calorific value.

Consider the reactants at the same temperature, pressure and volume (T, p, V_R). The difference between the calorific values is given by

$$(CV)_{p,T} - (CV)_{V,T} = (H_R - H_P)_{T,p} - (U_R - U_P)_{T,V_R}$$

Recalling that $H \equiv U + pV$

$$(CV)_{p,T} - (CV)_{V,T} = (U_R - U_P)_{T,p} + p(V_R - V_P)_{T,p} - (U_R - U_P)_{T,V_R}$$

and assuming that internal energy is a function only of temperature and not of pressure, that is

$$(U_P)_{p,T} = (U_P)_{V_R,T}$$

thus

$$(CV)_{p,T} - (CV)_{V,T} = p(V_R - V_P)_{p,T}$$

Neglect the volumes of any liquids or solids, and assume ideal gas behaviour, $pV = nR_0T$; so

$$(CV)_{p,T} - (CV)_{V,T} = (n_{GR} - n_{GP})R_0T \qquad (3.4)$$

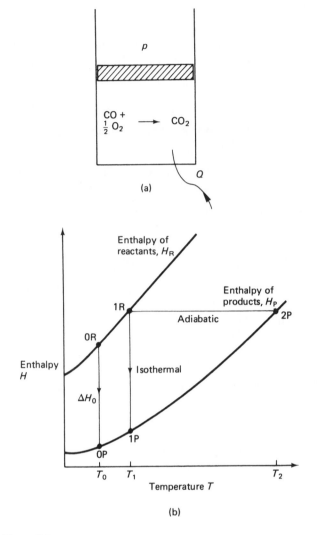

Figure 3.8
Constant-pressure combustion:
(a) constant-pressure calorimeter;
(b) enthalpy of reactants and products.

where n_{GR} number of moles of gaseous reactants/unit of fuel
$\quad\quad n_{GP}$ number of moles of gaseous products/unit of fuel.

The difference between the calorific values is usually very small.

3.3b Use of thermodynamic tabulations in Appendix A

The alternative approach (advocated here) is to use thermodynamic tables that incorporate the enthalpy of formation. This means that there is no need for separate tabulations of calorific values (or enthalpies or internal energies of reaction), and it will be found that energy balances for combustion calculations are greatly simplified. The enthalpy of formation

(ΔH_f^0) is perhaps more familiar to physical chemists than to engineers:

The *enthalpy of formation* (ΔH_f^0) of a substance is the standard reaction enthalpy for the formation of the compound from its elements in their reference state.

The *reference state* of an element is its most stable state (for example carbon *atoms*, but oxygen *molecules*) at a specified temperature and pressure, usually 298.15 K and a pressure of 1 bar. In the case of atoms that can exist in different forms, it is necessary to specify their form – for example, carbon is as graphite, not diamond.

In section 3.3a, since all species used the same datum, it was necessary to incorporate the enthalpy (or internal energy) of combustion, and a reaction would be analysed in three steps (for example, in figure 3.8, 1R–0R, 0R–0P and 0P–2P). The thermodynamic data tables in Appendix A incorporate the enthalpies of formation, and are thus designated 'absolute'. Combustion calculations are now a single step, 1R–2P.

In the case of the isothermal isobaric combustion process (1R–1P) the temperature (T) is obviously constant, and the difference in absolute enthalpy corresponds to the isobaric calorific value $(-\Delta H_T^0)$.

$$\Delta H_T^0 = H_{R,T} - H_{P,T} \qquad (A.2)$$

As a simple example, consider a stoichiometric mixture of hydrogen and oxygen at 25°C and burnt at constant pressure, and it is assumed that the products are solely water vapour. Estimate the adiabatic flame temperature.

Solution

$$H_2 + \tfrac{1}{2}O_2 \rightarrow H_2O$$

At 25°C, the enthalpy of the reactants (H_R) is

$$H_R = 0 + 0 = 0$$

Since the flame is adiabatic, then

$$H_P = H_R = 0$$

Inspection of the Absolute Enthalpy Tables (table A.4 in Appendix A) for water vapour shows that the flame temperature will be just below 5000 K (linear interpolation indicates 4998 K, but this is somewhat pointless when we know that dissociation is going to lead to a much lower temperature, in fact 3077 K).

3.3.1 The effect of the state of the reactants and products on calorific values

In most combustion problems any water produced by the reaction will be in the vapour state. If the water were condensed, the calorific value would be increased and then be referred to as the higher calorific value (HCV). The relationship between higher and lower calorific value (LCV) is

gravimetric basis: $HCV - LCV = yh_{fg}$

molar basis: $HCV - LCV = n_{H_2O}H_{fg}$ $\qquad$ (3.5)

where y is the mass of water per unit quantity of fuel and h_{fg} (or H_{fg}) is the enthalpy of evaporation of water at the temperature under consideration.

For hydrocarbon fuels the difference is significant, but lower calorific value is invariably used or implied. See example 3.3 for the use of calorific values. Similarly the state of the fuel must be specified, particularly if it could be liquid or gas. However, the enthalpy of vaporisation for fuels is usually small compared to their calorific value. For example, for iso-octane at 298.15 K (table A.3):

$$H_{fg} = 35.1 \text{ MJ/kmol}$$
$$-\Delta H_0 = 5100.5 \text{ MJ/kmol}$$

3.4 | Dissociation

Dissociation has already been introduced in this chapter by discussing the equilibrium of carbon dioxide, oxygen and carbon dioxide:

$$CO + \tfrac{1}{2}O_2 \rightleftharpoons CO_2$$

According to Le Châtelier's Principle, an equilibrium will always be displaced in such a way as to minimise any changes imposed from outside the system. This equilibrium can be affected in three ways: a change in concentration of a constituent, a change in system pressure, or a change in temperature. Considering one change at a time:

(i)   Suppose excess oxygen were added to the system.
The reaction would move in the forwards direction, as this would reduce the concentration of oxygen.

(ii) Suppose the system pressure were increased.
Again the reaction would move in the forwards direction, as this reduces the total number of moles (n), and the pressure reduces since

$$pV = nR_0T$$

(iii) Suppose the temperature of the system were raised.
The reaction will move in a direction that absorbs heat; for this particular reaction, that will be the reverse direction.

Care must be taken in defining the forwards and reverse directions as these are not always self-evident. For example, take the water–gas reaction

$$CO_2 + H_2 \rightleftharpoons CO + H_2O \quad \text{or} \quad CO + H_2O \rightleftharpoons CO_2 + H_2$$

To study this matter more rigorously it is necessary to introduce the concept of equilibrium constants, K_p; these are also called dissociation constants.

Consider a general stoichiometric reaction of

a kmols A + b kmols B $\rightleftharpoons$ c kmols C + d kmols D

It can be shown by use of a hypothetical device, the Van t'Hoff equilibrium box, that

$$K_p = \Pi p_i^{*\nu_i} = \left(p_C^{*c} \times p_D^{*d} \right) / \left(p_A^{*a} \times p_B^{*b} \right) \tag{3.6}$$

where p_i^* is the numerical value of the partial pressure (p_i') of component i, when the pressure is expressed in units of *bar*, and ν_i is the stoichiometric coefficient of species i.

This derivation can be found in thermodynamics texts such as Rogers and Mayhew (1980a), while a more generalised derivation is included here in Appendix A. This derivation also shows how the equilibrium constant can be determined from the change in Gibbs energy of the reaction at the relevant temperature:

$$\ln K_p = -\Delta G^0 / R_0 T$$

where $\Delta G^0 = \Sigma \nu_i G_i^0$, and G^0 is the absolute molar Gibbs energy at 1 bar. K_p is a function only of temperature and will have units of pressure to the power $(c + d - a - b)$. p', the partial pressure of a component, is defined as

$$p_a' = \frac{a'}{a' + b' + c' + d'} p$$

where p is the system pressure, and a', b', c' and d' are the actual numbers of kmol present of each species.

To solve problems involving dissociation introduce a variable x

$$CO + \tfrac{1}{2}O_2 \rightarrow (1-x)CO_2 + xCO + \frac{x}{2}O_2$$

$$K_p = \frac{p'_{CO_2}}{(p'_{CO})(p'_{O_2})^{\frac{1}{2}}} \quad \text{where } p'_{CO_2} = \frac{1-x}{(1-x) + x + \frac{x}{2}} p \tag{3.7}$$

$$= \frac{1-x}{1 + \frac{x}{2}} p, \text{ etc.}$$

so

$$K_p = \frac{\left(\frac{1-x}{1+\frac{x}{2}}\right)p}{\left(\frac{x}{1+\frac{x}{2}}\right)p \sqrt{\frac{x/2}{(1+\frac{x}{2})} p}}, \quad \text{or} \quad K_p = \frac{1-x}{x(x/2)^{\frac{1}{2}}} \left(\frac{1 + \frac{x}{2}}{p} \right)^{\frac{1}{2}}$$

As the equilibrium constant varies strongly with temperature, it is most convenient to tabulate log values; the numerical value will also depend on the pressure units adopted (Rogers and Mayhew, 1988 or Haywood, 1972). Again, the use of equilibrium constants is best shown by a worked problem, example 3.5. With internal combustion engines there will be several dissociation mechanisms occurring, and the simultaneous solution is best performed by computer.

However, the most significant dissociation reactions are

$$CO + \tfrac{1}{2}O_2 \rightleftharpoons CO_2$$

and the water–gas reaction

$$CO_2 + H_2 \rightleftharpoons CO + H_2O$$

The dissociation of carbon dioxide means that carbon monoxide will be present, even for the combustion of weak mixtures. Of course, carbon monoxide is most significant when rich mixtures are being burned, and there is insufficient oxygen for full oxidation of the fuel. When carbon monoxide is present from the combustion of hydrocarbons, then there will also be water vapour present. The water–gas reaction implies that the water vapour can be dissociated to produce hydrogen, and this will be seen later in figure 3.19. The water–gas equilibrium can be assumed to be the sole determinant of the burned gas composition, and this is illustrated by example 3.6.

Somewhat counter-intuitively, the effect of dissociation is greatest for a stoichiometric mixture, and this follows from Le Châtelier's Principle. With a weak mixture, the excess air displaces the equilibrium in such a way as to minimise the amount of oxygen. In contrast, for a rich mixture, the excess fuel displaces the equilibrium in such a way as to minimise the amount of partially burnt fuel (CO, H_2). For a stoichiometric mixture of methane and air at ambient conditions, dissociation leads to about 1 per cent CO, and a 100 K lower combustion temperature. If the final pressure is higher then dissociation is reduced, while if the initial temperature is higher dissociation will be increased.

3.4.1 Calculation of the equilibrium combustion temperature and pressure

In many combustion processes, the products leave the system at a comparatively low temperature (say below 1500°C). At these temperatures the reaction rates are comparatively low, and the composition of the products will not change much with further cooling. For example, the exhaust from internal combustion engines operating rich of stoichiometric will have both carbon monoxide and hydrogen present. The relative proportions correspond to the equilibrium of the water–gas reaction at a temperature of about 1800 K:

$$CO_2 + H_2 \rightleftharpoons CO + H_2O$$

This is a useful way of estimating the exhaust gas composition, perhaps as the first guess in a more comprehensive equilibrium analysis.

Consider now the combustion of an arbitrary fuel ($C_\alpha H_\beta O_\gamma N_\delta$) with air. If the products of combustion are limited to CO_2, CO, H_2O, H_2, O_2 and N_2, then the generalised combustion equation with air can be written as

$$C_\alpha H_\beta O_\gamma N_\delta + \lambda(\alpha + \beta/4 - \gamma/2)\left\{ O_2 + \frac{79.05}{20.95} N_2 \right\} \rightarrow$$

$$n_1 CO_2 + n_2 CO + n_3 H_2O + n_4 H_2 + n_5 O_2 + n_6 N_2 \tag{3.8}$$

where the excess air ratio (λ) is unity for stoichiometric reactions, and greater than unity for weak mixtures.

Four *atomic* balances can be written:

C balance : $\qquad \alpha = n_1 + n_2$ $\qquad\qquad\qquad$ (3.9)

H balance : $\qquad \beta = 2n_3 + 2n_4$ $\qquad\qquad\quad$ (3.10)

O balance : $\qquad \gamma + \lambda(\alpha + \beta/4 - \gamma/2)2 = 2n_1 + n_2 + n_3 + 2n_5$

$\qquad\qquad\qquad\qquad\qquad\qquad\qquad$ (3.11)

N balance : $\qquad \delta + \lambda(\alpha + \beta/4 - \gamma/2)2 \times \dfrac{79.05}{20.95} = 2n_6$

$\qquad\qquad\qquad\qquad\qquad\qquad\qquad$ (3.12)

With six unknowns and only four simultaneous equations, then two further equations are needed. A convenient simplification is to assume no oxygen in the products of rich combustion, and no hydrogen or carbon monoxide in the products of weak combustion. In other words:

rich mixtures ($\lambda < 1$): $\qquad\quad n_5 = 0$ $\qquad$ (3.13)

weak mixtures ($\lambda > 1$): $\qquad n_2 = n_4 = 0$ $\quad$ (3.14)

stoichiometric mixtures ($\lambda = 1$): $\quad n_2 = n_4 = n_5 = 0$ (3.15)

For rich mixtures a further equation is still required, and this is provided by the water–gas equilibrium:

$$CO_2 + H_2 \rightleftharpoons CO + H_2O \qquad\qquad (3.16)$$

Equation (3.6) has to be applied:

$$K_p = \Pi p_i^{*v_i} = (p_C^{*c} \times p_D^{*d})/(p_A^{*a} \times p_B^{*b})$$

where $\quad$ A represents species 1, CO_2,
$\qquad\quad$ B represents species 4, H_2,
$\qquad\quad$ C represents species 2, CO,
$\qquad\quad$ D represents species 3, H_2O.

Equation (3.16) can be written more formally as

$$CO + H_2O - CO_2 - H_2 = 0$$

so the stoichiometric coefficients are: $b = c = 1$, and $a = d = -1$.

Noting the definitions of partial pressures, $p_i = (n_i/\Sigma n)p = n_i(p/\Sigma n)$, so substitution into equation (3.6) gives

$$K_p = \frac{n_2(p/\Sigma n)n_3(p/\Sigma n)}{n_1(p/\Sigma n)n_4(p/\Sigma n)}$$

in which the $(p/\Sigma n)$ terms cancel because the stoichiometric coefficients sum to zero to give

$$K_p = \frac{n_2 n_3}{n_1 n_4} \qquad\qquad\qquad\qquad (3.17)$$

The simplification when the stoichiometric coefficients sum to zero is very significant since:

1 the presence of other species does not matter, and

2 the system pressure does not matter.

Simultaneous solution of equations (3.9) to (3.15) and (3.17) yields the results that are summarised in table 3.2.

For the rich products of combustion in table 3.2 the compositions have been written with variable n_2 included; this can now be eliminated by use of the equilibrium defined in equation (3.17) to give a quadratic equation in which the solution required is the one that makes all molar quantities positive:

$$n_2 = \frac{-b \pm \sqrt{b^2 - 4ac}}{2a} \qquad\qquad (3.18)$$

where $\quad a = K_p - 1$
$\qquad b = (-3K_p + 2\lambda K_p - 2\lambda + 2)\alpha + (-K_p + \lambda K_p - \lambda)\beta/2$
$\qquad\qquad + (1 - \lambda)(K_p - 1)\gamma$
$\qquad c = \alpha K_p(1 - \lambda)(2\alpha + \beta/2 - \gamma).$

Solution of equation (3.18) requires a knowledge of the equilibrium constant (K_p), and this can be found in Appendix A.

For a more realistic combustion calculation, the temperature is found by the simultaneous solution of two sets of non-linear equations, so as to satisfy:

1 the energy balance (the internal energy and enthalpy are non-linear functions of temperature for the high temperatures encountered with combustion), and the

2 equilibrium (the equilibrium composition is a highly non-linear function of temperature).

In the case of constant-volume combustion it is also necessary to calculate the pressure, and the pressure and temperature also have to satisfy the equation of state ($pV = nR_0T$).

Thus constant-pressure combustion has two interrelated sets of equations (for temperature and composition), while constant-volume combustion has three interrelated sets of equations (for temperature, composition and the equation of state).

Table 3.2 *Simplified products of combustion*

Species	i	Weak ($\lambda > 1$)	Rich ($\lambda < 1$)
CO_2	1	α	$\alpha - n_2$
CO	2	0	n_2
H_2O	3	$\beta/2$	$\gamma + \lambda(\alpha + \beta/4 - \gamma/2)2 - 2\alpha + n_2$
H_2	4	0	$\beta/2 - \gamma - \lambda(\alpha + \beta/4 - \gamma/2)2 + 2\alpha - n_2$
O_2	5	$(\lambda - 1)(\alpha + \beta/4 - \gamma/2)$	0
N_2	6	$\lambda(\alpha + \beta/4 - \gamma/2)79.05/20.95 + \delta/2$	$\lambda(\alpha + \beta/4 - \gamma/2)79.05/20.95 + \delta/2$

The simple equilibrium at the start of this section had four atomic species and six species present in the products. Only one equilibrium equation was needed since with rich mixtures there was assumed to be no oxygen, and for weak mixtures no hydrogen or carbon monoxide. This simplification could be relaxed by including the carbon monoxide equilibrium:

$$CO + \tfrac{1}{2}O_2 \rightleftharpoons CO$$

So there would be six unknown species and six equations (four atomic balances and two equilibrium equations). Each time another product species is included another independent equilibrium equation is needed, so that clearly only the simplest equilibrium combustion temperature problems can be solved manually. Instead, extensive use is made of computer routines and packages. More complex equilibria need to be considered in combustion, because although species such as H, OH, O etc. have very little effect on the computed equilibrium combustion temperature, they have a profound effect on reaction rates and the formation of emissions.

Two alternative approaches to equilibrium combustion calculations are:

1 solution using the equilibrium constant formulation, and
2 minimisation of the Gibbs energy.

The two approaches are of course equivalent, since the definition of the equilibrium constant

$$\ln K_p = -\frac{\Delta G^0}{R_0 T} \qquad (A.30)$$

was obtained by minimising the Gibbs energy.

A very comprehensive equilibrium calculation program for equilibrium combustion in FORTRAN IV was published by NASA (Gordon and McBride, 1971, 1976). A simpler program by Olikara and Borman (1975), considering only 12 species in the products, is also in the public domain, but uses imperial units. Ferguson (1986) adopts an approach similar to Olikara and Borman for 10 product species (omitting argon and atomic nitrogen) and employs SI units. The Ferguson subroutines are useful for incorporation into user-written programs, such as the cycle simulation of internal combustion engines.

Disadvantages of the equilibrium constant formulation are the difficulties in dealing with: condensed species, non-ideal equations of state and numerical problems as $K_p \to 0$. The Gibbs energy minimisation approach was used by Gordon and McBride (1976) in the current version of the NASA program and by Reynolds (1992) in the STANJAN package that dates from 1981. STANJAN is freely available for educational use, and has been used for devising and checking the problems in this book. Also freely available is GASEQ from http://www.gaseq.co.uk/.

3.5 | Pre-mixed combustion in spark ignition engines

Combustion either occurs normally – with ignition from a spark and the flame front propagating steadily throughout the mixture – or abnormally. Abnormal combustion can take several forms, principally pre-ignition and self-ignition. Pre-ignition is when the fuel is ignited by a hot spot, such as the exhaust valve or incandescent carbon combustion deposits. Self-ignition is when the pressure and temperature of the fuel/air mixture are such that the remaining unburnt gas ignites spontaneously. Pre-ignition can lead to self-ignition and vice versa; these processes will be discussed in more detail after normal combustion has been considered.

3.5.1 Normal combustion

When the piston approaches the end of the compression stroke, a spark is discharged between the sparking plug electrodes. The spark leaves a small nucleus of flame that propagates into the unburnt gas. Until the nucleus is of the same order of size as the turbulence scale, the flame propagation cannot be enhanced by the turbulence.

This early burn period comprises the initial laminar combustion, and the transition to fully turbulent combustion, and is sometimes referred to as the 'delay period'. The delay period is of approximately constant time duration. Figure 3.9 compares the pressure diagrams for the cases when a mixture is ignited and when it is not ignited. The point at which the pressure traces diverge is ill-defined, but it is used to denote the end of the delay period. The delay period is typically of 1–2 ms duration, and this corresponds to 15–30° of crank angle at 2500 rpm. A more rigorous approach is to calculate the mass fraction burn (mfb) from the pressure history (see

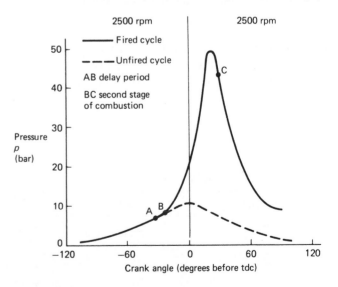

Figure 3.9
Hypothetical pressure diagram for a spark ignition engine.

chapter 14, section 14.5.2), but there are of course uncertainties with both these calculations, and the choice of a mass fraction burn that corresponds to the delay period – 1 per cent, 5 per cent and 10 per cent mass fraction burn durations are all used. The 1 per cent mfb duration corresponds to laminar combustion, but its calculation is very susceptible to experimental and computational uncertainties. The 5 per cent and 10 per cent mass fraction burn durations encompass the laminar burning phase and the transition to fully turbulent combustion, and both are widely used and are often referred to as the early burn period or phase. The early burn period depends on the temperature, pressure and composition of the fuel/air mixture, but it is a minimum for slightly richer than stoichiometric mixtures, in other words, when the laminar flame speed is highest. The main combustion period is dominated by turbulent combustion, and can be defined as the 10–90 per cent or 5–95 per cent mass fraction burn duration.

The end of the second stage of combustion is also ill-defined on the pressure diagram, but occurs shortly after the peak pressure. The second stage of combustion is affected in the same way as the early burn period, and also by the turbulence. This is very fortunate since turbulence increases as the engine speed increases, and the time for the second stage of combustion reduces almost in proportion. In other words, the second stage of combustion occupies an approximately constant number of crank angle degrees. In practice, the maximum cylinder pressure usually occurs 5–20° after top dead centre (Benson and Whitehouse, 1979).

The final stage of combustion is one in which the flame front is contacting more of the combustion chamber, with a reduced flame front area in contact with the unburned mixture, the remaining unburned mixture in the combustion chamber being burnt more slowly. The cylinder pressure should also be falling, so unburned mixture will be leaving crevices, and some of the fuel previously absorbed into the oil films on the cylinder wall will be desorbed. This final stage of combustion is very slow, and will not be complete by the time the exhaust valve opens. Even if there is some further oxidation in the exhaust port, combustion will still not be complete, as evidenced by the emissions of unburnt hydrocarbons. The unburnt hydrocarbons should be no more than about 1 per cent of the original fuel.

Since combustion takes a finite time, the mixture is ignited before top dead centre (btdc), at the end of the compression stroke. This means that there is a pressure rise associated with combustion before the end of the compression stroke, and an increase in the compression (negative) work. Advancing the ignition timing causes both the pressure to rise before top dead centre and also the compression work to increase. In addition the higher pressure at top dead centre leads to higher pressures during the expansion stroke, and to an increase in the expansion (positive) work. Obviously there is a trade-off between these two effects, and this leads to an optimum ignition timing. Since the maximum is fairly insensitive to ignition timing, the minimum ignition advance is used; this is referred to as 'minimum (ignition) advance for best torque' (MBT). This is discussed further in chapter 4, section 4.1.

By using the minimum advance, the peak pressures and temperatures in the cylinder are reduced; this helps to restrict heat transfer, engine noise, emissions and susceptibility to abnormal combustion. Similar arguments apply to compression ignition engines.

During the early stages of combustion, while the flame nucleus is still small, it can be displaced from the sparking plug region by large-scale flows in the cylinder. This can occur in a random way, and can have a significant effect on the subsequent propagation of combustion. This is readily shown by the non-repeatability of consecutive indicator diagrams from an engine, and is called variously 'cyclic dispersion', 'cycle-by-cycle variation' or 'cyclic variation'. (This is illustrated later by figure 4.25 where there is further discussion of cyclic dispersion.)

Plate 1 illustrates combustion in a spark ignition engine, using a through-the-piston optical access arrangement described by figure 14.1. The combustion images were taken in the Rover-Oxford K4 engine (a single-cylinder version of the Rover K16 engine, which is the case study in chapter 15, section 15.2). The first picture shows an image of the combustion chamber, illuminated by a tungsten filament bulb, placed at the location of the spark plug. The piston window is 75 per cent of the bore diameter, and it appears slightly elliptical because the CCD camera pixels are slightly rectangular. The exhaust valves are at the top, and the inlet valves are at the bottom; since the combustion chamber has a pent-roof (see figure 4.11) these valves appear elliptical.

All of the combustion images were taken at 1000 rpm, full throttle, with ignition 15°btdc. The first two images are of iso-octane combustion with a weak mixture (excess air ratio, $\lambda = 1.2$) and a rich mixture (excess air ratio, $\lambda = 0.7$). Further details of all of the experimental techniques used in producing this plate can be found in Simonini (1998). The characteristic blue colour of a weak mixture is attributable to emissions from the C_2 and CH radicals. With the rich mixture, these faint emissions are masked by very strong yellowish radiation from glowing soot particles. The soot is emitting thermal radiation, and by measuring the intensity at two known wavelengths it is possible to estimate the soot temperature; see for example Gaydon and Wolfhard (1979). The temperature-coded image of the rich iso-octane combustion image has been produced in an adaptation of this 'two-colour' method.

A threshold can be applied to the image intensities, from which the enflamed area can be measured. This is a line-of-sight image of a three-dimensional flame, but none the less the enflamed area can be converted to the radius of an equivalent circle. The sequence of images from methanol combustion was obtained from separate cycles, with several images taken at each crank angle. Methanol oxidation emits negligible radiation in the visible spectrum, so a small quantity of a sodium salt was added to the methanol. The yellow emissions are thus attributable to the sodium ions. Simultaneous pressure measurements enabled a representative selection of images to be assembled, and the pressure measurements were also analysed to give the average pressure history and the mass fraction burnt (using the Rassweiler and Withrow method described in chapter 14, section 14.5.2). These results have been plotted 80° either side of tdc and, although ignition was 15°btdc, there was no detectable flame image until 10°btdc, at which point the pressure history shows no evidence of combustion. The 'broken' nature of the flame front is evidence of turbulent combustion.

The piston window is 6 cm in diameter, and by 4°atdc the image has filled the field of view; this is why the gradient of the radius against crank angle plot decreases at 2°atdc. At this point over half the combustion chamber is enflamed, but only about 20 per cent of the fuel has been burnt – this is of course because the density of the unburnt mixture is about 5 times the density of the burnt mixture. Combustion photography is particularly useful for studying the very early stages of combustion, when there has been negligible effect from combustion on the pressure history.

3.5.2 Abnormal combustion

Surface ignition is caused by the mixture igniting as a result of contact with a hot surface, such as an exhaust valve. Surface ignition is often characterised by 'running-on'; that is, the engine continues to fire after the ignition has been switched off. If the engine is operating with the correct mixture strength, ignition timing and adequate cooling, yet there is surface ignition, the usual explanation is a build-up of combustion deposits, or 'coke'. If the surface ignition occurs in advance of the spark, then it is called pre-ignition. Pre-ignition causes an increase in the compression work and this causes a reduction in power. In a multi-cylinder engine, with pre-ignition in just one cylinder, the consequences can be particularly serious as the other cylinders continue to operate normally. Pre-ignition leads to higher peak pressures, and this in turn can cause self-ignition.

Self-ignition occurs when the pressure and temperature of the unburnt gas are such as to cause spontaneous ignition (figure 3.10). The flame front propagates away from the spark plug, and the unburnt (or 'end') gas is compressed as a result of the combustion process, and this rise in pressure increases the temperature. If spontaneous ignition of the unburnt gas occurs, there is a rapid pressure rise which can be characterised by a 'knocking'. The 'knock' is audible, caused by resonances of the combustion chamber walls. As a result of knocking, the thermal boundary layer at the combustion chamber walls can be destroyed. This causes increased heat transfer which might then lead to certain surfaces causing pre-ignition.

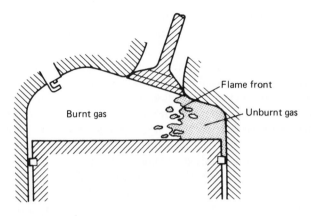

Figure 3.10
Combustion in a spark ignition engine.

If the pre-ignition occurs earlier in the compression stroke, the spontaneous ignition will occur earlier and lead to more severe knock. This in turn makes the pre-ignition even earlier, and leads to a situation known as run-away knock, since turning the ignition off will have no effect. Run-away knock causes overheating of the piston (the melting point of typical aluminium alloys is quite low, and the piston material is comparatively soft at about 250°C). Engine failure can be through holes being formed in the piston and or piston material being deposited on the cylinder liner leading to seizure.

For auto-ignition (self-ignition) to occur, the temperature in the end-gas has to be high enough for sufficient time, t_{ai}. The simplest model for describing auto-ignition of the end-gas is the Livengood–Wu integral (Livengood and Wu, 1955):

$$\int_0^{t_{ai}} \frac{1}{\tau} dt = 1 \tag{3.19}$$

where τ is the induction time – the time at a particular temperature and pressure that would lead to auto-ignition. The integral is needed since the temperature and pressure in the end-gas are varying, and the concept is similar to that of cumulative fatigue damage in materials. The induction time can be modelled by an Arrhenius-type expression:

$$\tau = A p^{-n} \exp\left(\frac{B}{T}\right) \tag{3.20}$$

in which A, B and n are constants for different fuels. The constants are typically evaluated in rapid compression machines, or by kinetic modelling of the decomposition and oxidation reactions. Strictly speaking the induction time will also depend on the composition, and for methane Law (2006) reports:

$$\tau = 2.5 \times 10^{-15} \exp(26700/T)[CH_4]^{0.32}[O_2]^{-1.02}$$

where $[CH_4]$ and $[O_2]$ are the molar concentrations (mole/cm^3) of the methane and oxygen.

For a mixture (by volume) of 62 per cent iso-octane, 29 per cent toluene and 9 per cent n-heptane (with a RON of 98 and a MON of 91) Kalghatgi *et al.* (2009) suggest:

$$\tau = 83.1 \times 10^{-6} \exp(6626/T)p^{-1.21} \text{ with } \tau\,(\text{s}), p\,(\text{bar}), T\,(\text{K})$$

Further discussion of the mechanisms behind knock is given in chapter 5, section 5.6.1, in the context of gasoline direct injection engines.

In chapter 2 it was shown how increasing the compression ratio should improve engine performance. Unfortunately, raising the compression ratio also increases the susceptibility to knocking. For this reason, much research has centred on the fundamental processes occurring with knock. These mechanisms are discussed in section 3.7.

3.6 | Combustion in compression ignition engines

Near the end of the compression stroke, liquid fuel is injected as one or more jets. The injector receives fuel at very high pressures in order to produce rapid injection, with high velocity jet(s) of small cross-sectional area; in all but the largest engines there is a single injector. The fuel jets entrain air and break up into droplets; this provides rapid mixing which is essential if the combustion is to occur sufficiently fast. Sometimes the fuel jet is designed to impinge onto the combustion chamber wall; this can help to vaporise the fuel and break up the jet. There will be large variations in fuel/air mixtures on both a large and small scale within the combustion chamber.

Plate 2 shows combustion in a high-speed direct injection diesel engine, using through-the-piston optical access. A description of the experimental equipment and the techniques can be found in Rao *et al.* (1992). The combustion system in these images is very similar to the Ford HSDI diesel engine, which is the case study in chapter 15, section 15.3. There are two valves per cylinder, and the piston bowl is centred around the 5 hole injector nozzle, which is offset 2.3 mm from the cylinder axis. The piston bowl minimum diameter is 43.0 mm, which is 45.9 per cent of the bore diameter. Sequence 1 is for a medium-load case with the engine operating at 1200 rpm. Injection started 0.5°btdc with an injection pressure of about 450 bar, and combustion commenced at about 4.6°atdc, corresponding to an ignition delay of 0.7 ms. The fuel injection spray in sequence 1a is illuminated by a copper vapour laser, which is synchronised to fire at the camera framing rate. The light is directed into the combustion chamber by a fibre optic light guide. The spray rapidly penetrates the combustion bowl, and the effect of the clockwise swirl is to deflect the fuel spray slightly.

In sequence 1b the laser only illuminates alternate frames, so as to make it easier to detect the start of combustion. The fuel sprays can be seen impinging onto the sides of the piston bowl. Some of this fuel will form a liquid film, while some of it will be atomised by the impact. Combustion is initiated at a single site, but because the mixture elsewhere is at the point of igniting, ignition occurs shortly later at numerous other sites. Sequence 1c shows that combustion spreads rapidly throughout the combustion chamber, and it also spreads slightly into the region above the piston bowl. Sequence 2 shows the engine operating at high load, with a transparent piston crown that gives optical access to the entire bore. This high load case shows the substantial amount of combustion that occurs in the small gap between the piston crown and the cylinder head.

Figure 3.11 shows the pressure diagram for a compression ignition engine; when compared to figure 3.9 (spark ignition engine) it can be seen immediately that the pressures are higher, especially for the unfired cycle. Referring to figure 3.11, there are several stages of combustion, not distinctly separated:

(i) *Ignition delay*, AB. After injection there is initially no apparent deviation from the unfired cycle. During this period the fuel is breaking up into droplets being vaporised, and mixing with air. Chemical reactions will be starting, albeit slowly.

(ii) *Rapid or uncontrolled combustion*, BC. A very rapid rise in pressure caused by ignition of the fuel/air mixture prepared during the ignition delay period.

(iii) *Controlled combustion*, CD. Combustion occurs at a rate determined by the preparation of fresh air/fuel mixture.

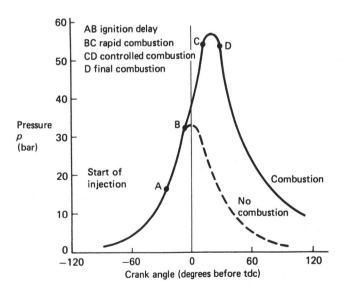

Figure 3.11

Hypothetical pressure diagram for a compression ignition engine.

(iv) *Final combustion*, D. As with controlled combustion the rate of combustion is governed by diffusion until all the fuel or air is utilised.

As with spark ignition engines the initial period is independent of speed, while the subsequent combustion occupies an approximately constant number of crank angle degrees. In order to avoid too large a rapid combustion period, the initial fuel injection should be carefully controlled. The 'rapid' combustion period can produce the characteristic 'diesel knock'. Again this is caused by a sudden pressure rise, but is due to self-ignition occurring too slowly. Its cure is the exact opposite to that used in spark ignition engines; fuels in compression ignition engines should self-ignite readily. For a given fuel and engine, diesel knock can be reduced by avoiding injection of too much fuel too quickly. Some systems inject a small quantity of fuel before the main injection, a system known as pilot injection. Alternatively, the engine can be modified to operate with a higher compression ratio. This increases the temperature and pressure during the compression stroke, and this will reduce the ignition delay period. The modelling of the ignition delay and the subsequent combustion are treated in section 11.2.

To obtain maximum output the peak pressure should occur about 10–20° after top dead centre. Sometimes the injection is later, in order to retard and to reduce the peak pressure. The fuel ignites spontaneously at many sites, and produces an intense flame. There is significant radiation from the flame front, and this is important for vaporising the fuel. Since the combustion occurs from many sites, compression ignition engines are not susceptible to cyclic variation or cyclic dispersion.

Compression ignition engines can operate over a wide range of air/fuel mixtures with equivalence ratios in the range 0.14–0.90. The power output of the engine is controlled by the amount of fuel injected, as in the combustion region the mixture is always approximately stoichiometric. This ensures good part load fuel economy as there are no throttling losses. The fuel/air mixture is always weaker than stoichiometric, as it is not possible to utilise all the air. At a given speed the power output of an engine is usually limited by the amount of fuel that causes the exhaust to become smoky.

3.7 | Fuels and additives

The performance and in particular fuel economy of internal combustion engines should not be considered in isolation, but also in the context of the oil refinery. Oil refining and distribution has an overall efficiency of about 88 per cent (Francis and Woollacott, 1981). The more recent data for the well-to-tank and tank-to-wheel energy usage in figure 1.16 indicates an efficiency of 87 per cent. This efficiency would be changed if the demand for the current balance of products was changed. Oil refining can be compared to the work of a butcher; each process has a raw material (a barrel of oil or a carcass) that has to be cut in such a way that all the products can be sold at a competitive price. Just as there are different animals, there are also different types of crude oil, depending on the source. However, the oil refinery can change the type of product in additional ways, such as cracking, although there is an energy cost associated with this.

The energy content of a typical petrol is 44 MJ/kg or 31.8 MJ/litre and of a typical diesel fuel 42 MJ/kg or 38.15 MJ/litre; associated with these is an energy content at the refinery of typically 2.7 MJ/kg and 1.6 MJ/kg (Francis and Woollacott, 1981). This gives an effective primary energy of 46.7 MJ/kg or 35.5 MJ/litre for petrol and 43.65 MJ/kg or 39.95 MJ/litre for diesel fuel. There may be circumstances in which it is more appropriate to use primary energy density than to use calorific value. These figures also highlight the difference in energy content of unit volume of fuel. This is often overlooked when comparing the fuel economy of vehicles on a volumetric basis.

3.7.1 Characteristics of petrol

The properties of petrol are discussed thoroughly by Blackmore and Thomas (1977). The two most important characteristics of petrol are its volatility and octane number (its resistance to self-ignition).

Volatility is expressed in terms of the volume percentage that is distilled at or below fixed temperatures. If a petrol is too volatile, when it is used at high ambient temperatures the petrol is liable to vaporise in the fuel lines of carburetted engines and form vapour locks. This problem is most pronounced in vehicles that are being restarted, since under these conditions the engine compartment is hottest. If the fuel is not sufficiently volatile the engine will be difficult to start, especially at low ambient temperatures. The volatility also influences the cold start fuel economy. Spark ignition engines are started on very rich mixtures, and in part this is to ensure adequate vaporisation of fuel. Increasing the volatility of the petrol at low temperatures will evidently improve the fuel economy during and after starting. Blackmore and Thomas (1977) point out that in the USA as much as 50 per cent of all petrol is consumed on trips of 10 miles or less. Short journeys have a profound effect on vehicle fuel economy, yet fuel consumption figures are invariably quoted for steady-state conditions.

The European Standards Organisation (CEN) has defined a common standard for its member countries, and for unleaded gasoline this standard has the designation EN 228: 1993, and this includes eight different volatility specifications. Members of CEN can then use their national standards (BS EN 228 in the case of the UK), to specify particular volatilities for different seasons. The volatility is also expressed in terms of the Reid Vapour Pressure (RVP). The Reid apparatus is shown in figure 3.12, and it consists of a fuel chamber connected to

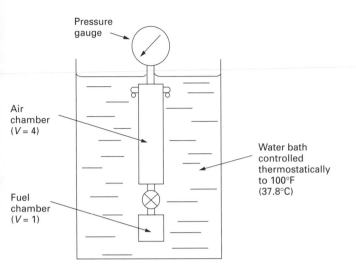

Figure 3.12
Reid vapour pressure apparatus (from Goodger, 1975).

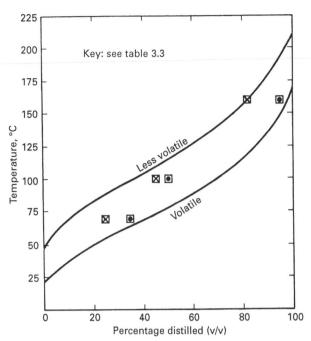

Figure 3.13 Distillation curves for petrol (with acknowledgement to Blackmore and Thomas, 1977).

an air chamber of four times its volume. The fuel chamber is filled with chilled fuel, connected to the air chamber, and then immersed in a water bath at 37.8°C and shaken. The method is defined in ASTM Procedure D323. The final pressure has to be corrected for atmospheric pressure and the effect of heating the air in the air chamber, and the result is an absolute value.

The fuel volatility specified in BS EN 228: 1993 is compared with typical fuel specifications from Blackmore and Thomas (1977) in table 3.3. This is plotted with further data in figure 3.13.

Table 3.3 shows how the specification of petrol varies to suit climatic conditions. Petrol stored for a long time in vented

tanks is said to go stale; this refers to the loss of the more volatile components that are necessary for easy engine starting.

The octane number of a fuel is a measure of its anti-knock performance. A scale of 0–100 is devised by assigning a value of 0 to n-heptane (a fuel prone to knock), and a value of 100 to iso-octane (in fact 2,2,4-trimethylpentane, a fuel resistant to knock). A 95 octane fuel has the performance equivalent to

Table 3.3 *Volatility of different petrol blends*

	BS EN 228		Less volatile	Volatile	North-west Europe		Central Africa
	Grade 4 Min. Max.	Grade 8 Min. Max			Summer	Winter	
Reid vapour pressure (bar)	0.45–0.80	0.65–1.00					
Distillate evaporated at 70°C (per cent v/v)	15–45	20–50	10	42	25	35	10
Distillate evaporated at 100°C (per cent v/v)	40–60	43–70	38	70	45	50	38
Distillate evaporated at 160°C (per cent v/v)			80	98	80	95	80
Distillate evaporated at 180°C (per cent v/v)	>85	>85					
Final boiling point °C	215	215					
Residue (per cent v/v)	2	2					
Symbol used in figure 3.13	–	–	–	–	⊠	⊡	

16 April to 15 October Grade 4 Minima
1 June to 31 August Grade 4 Maxima
16 October to 15 April Grade 8 Minima
1 September to 31 May Grade 8 Maxima

that of a mixture of 95 per cent iso-octane and 5 per cent n-heptane by volume. The octane requirement of an engine varies with compression ratio, geometrical and mechanical considerations, and also its operating conditions. There are two commonly used octane scales, research octane number (RON) and motor octane number (MON) covered by British Standards EN 25164: 1994 and EN 25163: 1994 respectively. Both standards refer to the *Annual Book of ASTM* (American Society for Testing and Materials) *Standards Volume 05.04 – Test Methods for Rating Motor, Diesel and Aviation Fuels.*

The tests for determining octane number are performed using the ASTM-CFR (Cooperative Fuel Research) engine; this is a variable compression ratio engine similar to the Ricardo E6 engine. In a test the compression ratio of the engine is varied to obtain standard knock intensity. With the same compression ratio two reference fuel blends are found whose knock intensities bracket that of the sample. The octane rating of the sample can then be found by interpolation. The different test conditions for RON and MON are quoted in *ASTM Standards Volume 05.04*, and are summarised in table 3.4.

Table 3.4 shows that the motor octane number has more severe test conditions since the mixture temperature is greater and the ignition occurs earlier. There is not necessarily any correlation between MON and RON as the way fuel components of different volatility contribute to the octane rating will vary. Furthermore, when an engine has a transient increase in load, excess fuel is supplied. Under these conditions it is the octane rating of the more volatile components that determines whether or not knock occurs.

A worldwide summary of octane ratings used to be published by the Associated Octel Co. Ltd, London. In North America the octane rating quoted is an average of the RON and the MON, and it called the Anti-Knock Index, AKI = 0.5(RON + MON). Elsewhere, when the octane rating is quoted it will be the RON, as this is higher (see table 3.6). The difference between the RON and MON values is known as the sensitivity, and this is discussed further is section 3.7.2.

As the RON and MON are not good at describing the anti-knock quality of a practical fuel, an alternative is to use an Octane Index (OI), defined by

$$OI = (1 - K)RON + KMON = RON - KS$$

where RON and MON are the Research and Motor Octane Numbers, K is a constant depending only on the pressure and temperature variation in the engine, and S is the sensitivity (RON – MON).

So, an engine for which the RON test conditions were representative would have $K = 0$, while an engine which operated more like the MON test conditions (which have a higher unburned gas temperature for a given pressure) would have $K = 1$. Thus K decreases as the compression temperature in the unburned gas at a given pressure in the engine decreases and K can be negative if this temperature is lower than in the RON test. As spark ignition (SI) engine designers seek higher efficiency, knock becomes more likely. Also, developments such as direct injection, higher compression ratios and downsizing with turbocharging all reduce the unburned gas temperature for a given pressure and push the value of K downwards.

Table 3.4 *Summary of RON and MON test conditions*

Test conditions	Research octane number	Motor octane number
Engine speed, rpm	600 ± 6	900 ± 9
Crankcase oil, SAE grade	30	30
Oil pressure at operating temperature, psi	25–30	25–30
Crankcase oil temperature	135 ± 15°F (57 ± 8.5°C)	135 ± 15°F (57 ± 8.5°C)
Coolant temperature		
Range	212 ± 3°F (100 ± 1.5°C)	212 ± 3°F (100 ± 1.5°C)
Constant within	± 1°F (0.5°C)	± 1°F (0.5°C)
Intake air humidity, grains of water per lb of dry air	25–50	25–50
Intake air temperature	See ASTM Standards Volume 05.04	100 ± 5°F (38 ± 2.8°C)
Mixture temperature		300 ± 2°F (149 ± 1.1°C)
Spark advance, deg. btdc	13	14–26 depending on compression ratio
Spark plug gap, in.	0.020 ± 0.005	0.020 ± 0.005
Breaker point, gap, in.	0.020	0.020
Valve clearances, in.		
Intake	0.008	0.008
Exhaust	0.008	0.008
Fuel/air ratio	Adjusted for maximum knock	

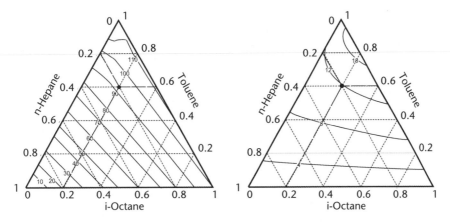

Figure 3.14 Ternary plots for iso-octane, n-heptane and toluene of the Research Octane Number (RON) and Sensitivity (S = RON − MON), adapted from Smallbone *et al.* (2010).

Kalghatgi (2005) reports that in Europe there is evidence of a monotonic decrease in the average K value from 1987 to 1992. Tests with 37 different Japanese and European cars (34 models) equipped with knock sensors had K values that were negative in most cases. Thus for a given RON, a fuel of lower MON has higher OI and will give better anti-knock performance. So a vehicle fitted with a knock sensor will be able to operate with more advanced ignition timing and give improved acceleration and more power.

The requirement that the MON should be as low as possible (for a given RON value) is counter-intuitive, but the important point is that this is for a given RON value, and in general fuels that have a low MON will tend to have a low RON. Legislation invariably specifies a minimum MON value and Kalghatgi (2005) argues that this can be inappropriate.

Mixtures of the Primary Reference Fuels (PRFs, n-heptane and iso-octane) by definition will give the same RON and MON values, but the use of a third component enables sensitivity (S) to be obtained. Commercial gasoline contains aromatics and toluene is used in tri-component blends (comprising iso-octane, n-heptane and toluene) that are referred to as Toluene Reference Fuels (TRFs). Data on the RON and MON performance of these blends has been reported by Smallbone *et al.* (2010). These and other data have then been surface-fitted to produce the ternary plots shown in figure 3.14.

As an example, a mixture of 20 per cent iso-octane, 20 per cent n-heptane and 60 per cent toluene has a RON of 92 and a Sensitivity of 12. The Research Octane Number (RON) and Sensitivity (S = RON − MON) do not blend linearly, in other words, the contours of constant RON or 'sensitivity' are curved. The curvature is greatest in the fuel Sensitivity diagram but also at the highest RON values, and this is the region that is most relevant to contemporary gasoline. This indicates that the usual linear mixing models are prone to significant errors.

Modelling is an alternative to testing, and Smallbone *et al.* (2010) have used a Stochastic Reactor Model (SRM), which is derived from a more general probability density transport model. Ensembles of particles are modelled, in order to track the history of species, pressure and temperature and their interactions. The chemical kinetics were based on a toluene/n-heptane/iso-octane oxidation mechanism developed by Andrae *et al.* (2008) that contained 137 species and had previously been optimised and validated against ignition delay times from shock tubes and rapid compression machines, observed flame speeds from counterflow flames and HCCI engine ignition timings. As well as modelling the octane rating tests, such models can also be used to assess the fuel requirements of production engines.

The attraction of high octane fuels is that they enable high compression ratios to be used. Higher compression ratios give increased power output and improved economy. This is shown in figure 3.15 using data from Blackmore and Thomas (1977). The octane number requirements for a given compression ratio vary widely, but typically a compression ratio of 7.5 requires 85 octane fuel, while a compression ratio of 10.0 would require 100 octane fuel. There are even wide variations in octane number requirements between supposedly identical engines.

Of the various fuel additives, those that increase octane numbers have greatest significance. In 1922 Midgely and Boyd discovered that lead-based compounds improved the octane rating of fuels. By adding 0.5 gram of lead per litre, the octane rating of the fuel is increased by about 5 units.

The lead additives take the form of lead alkyls, either tetramethyl lead $(CH_3)_4Pb$, or tetraethyl lead $(C_2H_5)_4Pb$. Since the active ingredient is lead, the concentration of the additives is expressed in terms of the lead content.

Thus

$$0.5 \text{ g Pb/litre} = 0.645 \text{ g } (CH_3)_4Pb/\text{litre}$$

and

$$0.5 \text{ g Pb/litre} = 0.780 \text{ g } (C_2H_5)_4Pb/\text{litre}$$

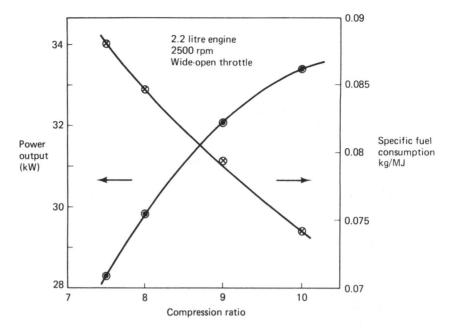

Figure 3.15
Effect of changing compression ratio on engine power output and fuel economy (with acknowledgement to Blackmore and Thomas, 1977).

Most countries now have restrictions on the use of lead in automotive fuels for environmental reasons. However, the use of lead additives in aviation gasoline is still very significant as in low lead fuel there is up to 0.56 g Pb/litre, giving a RON of about 100. As well as the possible dangers of lead pollution, catalysts for the conversion of other engine pollutants are made inactive by lead. There are very strict controls for type approval of aviation engines and their fuels, so change is inhibited. The key issue with engines designed for leaded fuel is their susceptibility for valve seat recession (see the end of this section) if they are run on unleaded fuel. ASTM D910–07a *Standard Specification for Aviation Gasolines* defines the octane requirements and volatility. The test procedures differ from the automotive ones (octane ratings are evaluated for both lean and rich mixtures), and the volatility of the fuel is lower, so as to avoid vapour lock at high altitudes.

To understand how lead alkyls can inhibit knocking, the chemical mechanism involved in knocking must be considered in more detail. Two possible causes of knocking are cool flames or low-temperature auto-ignition, and high-temperature auto-ignition. Cool flames can occur in many hydrocarbon fuels and are studied by experiments with fuel/oxygen mixtures in heated vessels. If some mixtures are left for a sufficient time, a flame is observed at temperatures that are below those for normal self-ignition; the flames are characterised by the presence of peroxide and aldehyde species. Engine experiments have been conducted in which the concentrations of peroxide and aldehyde species have been measured, giving results that imply the presence of a cool flame. However, knock was obtained only with higher compression ratios, which implied that it is a subsequent

high-temperature auto-ignition that causes the rapid pressure rise and knock. Cool flames have not been observed with methane and benzene, so when knock occurs with these fuels it is a single-stage high-temperature auto-ignition effect.

Downs and Wheeler (1951–52) and Downs *et al.* (1961) discuss the possible chemical mechanisms of knock and how tetraethyl lead might inhibit knock. In the combustion of, say, heptane, it is unlikely that the complete reaction

$$C_7H_{16} + 11O_2 \rightarrow 7CO_2 + 8H_2O$$

can occur in one step. A gradual degradation through collisions with oxygen molecules is much more likely, finally ending up with CO_2 and H_2O. This is a chain reaction where oxygenated hydrocarbons such as aldehydes and peroxides will be among the possible intermediate compounds. A possible scheme for propane starts with the propyl radical $(C_3H_7)^-$, involves peroxide and aldehyde intermediate compounds, and finally produces a propyl radical so that the chain reaction can then repeat:

$$\underset{\text{propyl radical}}{CH_3CH^-CH_3} + O_2 \rightarrow \underset{\text{hydroperoxide}}{CH_3CH(OO)CH_3}$$

$$\rightarrow \underset{\text{aldehyde}}{CH_3CHO + CH_3O} \xrightarrow{+C_3H_8} \underset{\text{alcohol}}{CH_3OH} + CH_3CH^-CH_3 + CH_3CHO$$

or

$$\rightarrow C_2H_5CHO + \underset{\substack{\text{hydroxyl} \\ \text{radical}}}{OH}$$

There are many similar chain reactions that can occur and some of the possibilities are discussed by Lewis and von Elbe (1961) in greater detail.

Table 3.5 *Typical gasoline volumetric composition (per cent), with identification of some individual components*

n-alkanes	9	n-butane	4.3
		n-pentane	1.9
		n-hexane	1.5
		other n-alkanes	1.3
iso-alkanes	43	2-methylpentane	14.3
		isopentane	5.8
		3-methylpentane	4.1
		iso-octane	3.5
		2-methylhexane + 2,3-dimethylpentane	3.5
		2,3-dimethylbutane	3.3
		other iso-alkanes	8.5
aromatics	34	xylene	10.2
		toluene	8.0
		1,2,4-trimethylbenzene	2.6
		ethylbenzene	2.2
		2-ethyltoluene	1.7
		other aromatics	9.3
alkenes	11	2-methylbut-2-ene	1.9
		pent-2-ene	1.5
		other alkenes	7.6
cyclo-alkanes	3		

Tetraethyl lead improves the octane rating of the fuel by modifying the chain reactions. During the compression stroke, the lead alkyl decomposes and reacts with oxygenated intermediary compounds to form lead oxide, thereby combining with radicals that might otherwise cause knock.

The suggested mechanism is as follows:

$$PbO + OH \rightarrow PbO(OH)$$
$$PbO(OH) + OH \rightarrow PbO_2 + H_2O$$
$$PbO_2 + R \rightarrow PbO + RO$$

where R is a radical such as the propyl radical, C_3H_7.

One disadvantage with the lead alkyls is that lead compounds are deposited in the combustion chamber. These can be converted to more volatile lead halides by alkyl halide additives such as dichloroethane ($C_2Cl_2H_4$) or dibromoethane ($C_2Br_2H_4$). However, some lead halides remain in the combustion chamber and these deposits can impair the insulation of spark plugs, and thus lead to misfiring. By adding aryl phosphates to petrol, lead halides are converted to phosphates, which have greater electrical resistivity. A second benefit is that lead phosphates are less prone to cause pre-ignition by surface ignition. These and other additives are discussed more fully by Blackmore and Thomas (1977).

In the EC leaded fuel was withdrawn in the year 2000. The composition of a typical unleaded gasoline (with a RON of 95) is shown in table 3.5. Adding 0.15 g Pb/litre would raise the octane rating by about 2 units, since its effectiveness is greatest at low concentrations.

The composition of the fuel depends in part on the source of the crude oil. Crude oil from the Far East tends to be high in aromatics, while oil from the North Sea is high in alkanes. When a crude oil is distilled it produces what is known as Straight Run Gasoline (SRG). A SRG from North Sea Oil would be high in alkanes, and have an unacceptably low octane rating (a RON of about 75). When such a SRG is cracked (section 3.2) its alkene content will be increased, thereby increasing its octane rating.

The influence of the chemical structure on the Research Octane Number (RON) is illustrated by figure 3.16. This shows how the octane number:

1 decreases with increasing chain length;

2 increases with increasing number of side chains for the same number of carbon atoms;

3 increases with ring structures (cyclo-alkanes and aromatics).

This can also be seen by studying table 3.6, which summarises the octane rating (RON and MON) for a wide range of fuel components.

Table 3.6 clearly shows how the alkenes and ring compounds (both the cyclo-alkanes and aromatics) can be used to produce fuels with a high octane rating.

Even prior to the legislation restricting and banning the use of lead additives, there was much research into alternative additives for increasing the octane rating; Taylor (1985b) lists 45 additives. Organic additives are much less effective than organo-metallic additives. This is unfortunate, since organo-metallic compounds tend to be both toxic, and incompatible with catalysts, since they leave deposits which are likely to poison or block the catalyst. Table 3.7 indicates the relative effectiveness of some organo-metallic additives.

Research
Octane Number

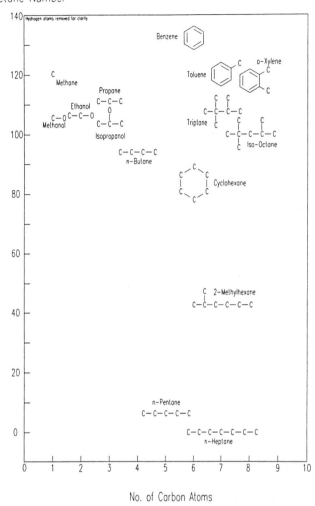

Figure 3.16
The influence of the chemical structure on the Research Octane Number (RON) (Pashley, 1994).

Table 3.7 needs to be used with extreme caution, because the effectiveness of an additive depends on many factors, including:

1 the concentration that is being used (the effectiveness reduces with increasing concentrations), and

2 the composition of the gasoline; additives that are effective with alkanes can be less effective with aromatics.

These and other factors are discussed in detail by Owen and Coley (1995). MMT has been widely used in North America, and one attraction was its synergistic performance with lead additives. Iron additives have also been used in the past, but they can lead to excessive engine wear. Even when there are no restrictions on the use of organo-metallic compounds, their use can be precluded on the basis of cost or stability.

Aniline ($C_6H_5NH_2$) is one of the most effective organic anti-knock additives, but this is only about 3 per cent as effective as tetraethyl lead on a mass basis. Thus large quantities would have to be used, and it is more economical to achieve high octane ratings in unleaded gasoline by refining and blending, and the use of oxygenate compounds (these are discussed in the next section).

Valve seat recession

One consequence of the banning of leaded fuel is that some vehicles with older engines may be susceptible to valve seat recession. Lead additives give rise to deposits that inhibit wear of the valve and its seat. The mechanism of exhaust valve seat wear is well established, and comprises oxidation and material transfer, to form a hard and rough surface that then abrades the valve seat.

An authoritative review on valve seat recession and the use of unleaded gasoline has been written by Vincent (1991). This review establishes that valve recession was well known in the 1920s and 1930s, especially for the exhaust valves of engines subject to prolonged high load and speed operation, such as bus and truck operation on highways. There was speculation that lead anti-knock additives would reduce valve seat recession, but leaded fuel was not used exclusively, so at the same time valve seat inserts were also introduced. However, with the wide-scale use of leaded fuel from the mid 1940s, engines without valve seat inserts did not suffer valve seat recession.

In the absence of suitable fuel additives to prevent valve seat recession, some engines will need to be fitted with valve seat inserts. It is possible for valve seat recession to be avoided by not subjecting an engine to high speed or high load operation. However, it would be prudent to monitor both the engine usage pattern and the clearances in the valve train.

3.7.2 In-vehicle performance of fuels, and the potential of alcohols

Alcohols have certain advantages as fuels, particularly in countries without oil resources, or where there are sources of the renewable raw materials for producing methanol (CH_3OH) or ethanol (C_2H_5OH). Car manufacturers have extensive programmes for developing alcohol-fuelled vehicles. Alcohols can also be blended with oil-derived fuels and this improves the octane ratings. Both alcohols have high octane ratings (ethanol has a RON of 106) and high enthalpy of vaporisation; this improves the volumetric efficiency but can cause starting problems. For cold ambient conditions it may be necessary to start engines with petrol. The other main disadvantages are the lower energy densities (about half that of petrol for methanol and two-thirds for ethanol), and the miscibility with water.

In Western Europe there is common legislation for unleaded gasoline, identified as BS EN 228: 1993 in the UK,

Table 3.6 *Octane rating of hydrocarbon fuel components (octane ratings above 100 are obtained from comparisons with leaded iso-octane) – Lovell (1948) and Obert (1973)*

Fuel	Formula	Structure	RON	MON
Alkanes				
methane	CH_4	CH_4	120	120
ethane	C_2H_6	$(CH_3)_2$	115	99
propane	C_3H_8	$CH_3(CH_2)CH_3$	112	97
n-butane	C_4H_{10}	$CH_3(CH_2)_2CH_3$	94	89
i-butane	C_4H_{10}	$CH(CH_3)_3$	102	98
n-pentane	C_5H_{12}	$CH_3(CH_2)_3CH_3$	62	63
i-pentane (2-methylbutane)	C_5H_{12}	$CH_3CH(CH_3)CH_2CH_3$	93	90
n-hexane	C_6H_{14}	$CH_3(CH_2)_4CH_3$	62	63
i-hexane (2-methylpentane)	C_6H_{14}	$CH_3CH(CH_3)(CH_2)_2CH_3$	92	93
i-hexane (2,2-dimethylbutane)	C_6H_{14}	$CH_3C(CH_3)_2CH_2CH_3$	92	93
n-heptane	C_7H_{16}	$CH_3(CH_2)_5CH_3$	0	0
i-heptane (2,2-dimethylpentane)	C_7H_{16}	$CH_3C(CH_3)_2(CH_2)_2CH_3$	93	96
i-heptane (2,4-dimethylpentane)	C_7H_{16}	$CH_3CH(CH_3)CH_2CH(CH_3)CH_3$	83	84
i-heptane (2,2,3-trimethylbutane) [triptane]	C_7H_{16}	$CH_3C(CH_3)_2CH(CH_3)CH_3$	112	100
n-octane	C_8H_{18}	$CH_3(CH_2)_6CH_3$	−20	−17
i-octane (2,2,4-trimethylpentane)	C_8H_{18}	$CH_3C(CH_3)_2CH_2CH(CH_3)CH_3$	100	100
Alkenes				
1-butene	C_4H_8	$CH_2CHCH_2CH_3$	98	80
2-butene	C_4H_8	$CH_3CHCHCH_3$	100	84
1-pentene	C_5H_{10}	$CH_2CH(CH_2)_2CH_3$	91	77
2-methyl-2-butene	C_5H_{10}	$CH_3C(CH_3)CHCH_3$	97	85
Cycloalkanes				
cyclopentane	C_5H_{10}	$(CH_2)_5$	101	95
cyclohexane	C_6H_{12}	$(CH_2)_6$	83	77
cycloheptane	C_7H_{14}	$(CH_2)_7$	39	41
cyclo-octane	C_8H_{16}	$(CH_2)_8$	71	58
Aromatics				
benzene	C_6H_6		–	115
toluene (methylbenzene)	$C_6H_5CH_3$		120	109
xylene (dimethylbenzene)	$C_6H_4(CH_3)_2$		118	115

and part of this specification is in table 3.8. In the UK (and some other countries) the oil companies have introduced an unleaded fuel with a higher octane rating. This is defined in the UK by BS 7800, and it has a RON of 98 and a MON of 87, so it has an octane rating equivalent to that of a premium leaded fuel. In the European Union the sulphur level limit was reduced from 150 ppm to 50 ppm (by mass) in 2003 and to 10 ppm in 2009.

The improvement in the anti-knock rating of the unleaded fuel (and leaded fuels with reduced lead content) has been achieved by adding fuel components that have high octane ratings. Notable in this respect are oxygenates and benzene. Since oxygenate fuels have a lower calorific value, a different density and a different stoichiometric air/fuel ratio, they can lead to engines operating away from design conditions. Furthermore, oxygenate fuels can absorb moisture, and this can lead to corrosion and fuel separation problems. BS EN 228 limits the total amount of oxygenates by limiting the oxygen content of the fuel to 2.5 per cent by mass. There are also limits on the individual oxygenate components and on benzene (for reasons of carcinogenity). Table 3.9 summarises these limits, and indicates their octane rating.

Table 3.7 *Effectiveness of organo-metallic additives relative to tetraethyl lead, based on the mass of the metal; derived from Owen and Coley (1995)*

Compound	Formula	Effectiveness
MMT	$CH_3C_5H_4Mn(CO)_3$	1.65
Tetraethyl lead	$Pb(C_2H_5)_4$	1.00
Iron pentacarbonyl	$Fe(CO)_5$	0.88
Nickel carbonyl	$Ni(CO)_4$	0.50
Tetraethyl tin	$Sn(C_2H_5)_4$	0.13

MMT – Methylcyclopentadienyl manganese tricarbonyl.

Table 3.8 *Octane number requirements of different fuel grades in Europe*

Description	RON	MON
Regular unleaded	91	82.5
Premium unleaded	95	85
Super Plus	99	88

The improvement that oxygenate fuels give to the resulting fuel blend depends on the oxygenate fuel and the fuel that it is being blended with. By definition, when mixtures of iso-octane and n-heptane are prepared, there is a linear relationship between composition and the octane rating of the blend. However, many fuel components exhibit non-linear mixing properties when evaluated on a volumetric basis, and this is why a blending value has been indicated in table 3.9. For a given component, the blending value will depend on what it is being mixed with, but it is convenient to assume a linear mixing relation and a blending value for oxygenates, to estimate the octane rating of a fuel blend.

However, Anderson *et al.* (2010) have carefully reviewed the octane rating data for alcohol/gasoline blends and the use of blending octane numbers (bON, which account for the non-linear mixing behaviour when alcohols and gasoline are blended). Unfortunately the value of the bON depends on the volume fraction of alcohol that is being added. Anderson *et al.* (2010) argue that since the knock mechanisms will depend on the partial pressures of the components in the vaporised fuel, then the octane rating of alcohol/gasoline blends should be calculated by using a molar weighted average. This remarkably simple procedure then means that the octane rating of an alcohol/gasoline blend can be calculated using the pure alcohol RON and MON values. It is necessary to know the density and mean molar mass of the gasoline, but Anderson *et al.* (2010) only needed to use assumed values of 750 kg/m^3 and 110 kg/kmol. They also provide a comprehensive review

of the octane rating data for methanol and ethanol, and highlight the reasons for the discrepancies in the data (in part due to changes in test procedures).

Most fuels have different values of RON and MON (the difference is known as the sensitivity), and this illustrates that the anti-knock performance of the fuel depends on the test conditions. Furthermore, there are several limitations associated with the standard tests in the CFR engine. Firstly, both MON and RON are evaluated at low engine speeds, and secondly the air/fuel ratio is adjusted to give the maximum knock. Thus it is not surprising that fuels behave rather differently in multi-cylinder engines, whether they are installed on dynamometers or in vehicles.

Clearly it is impractical to vary the compression ratio of a production engine. Instead the ignition timing is modified until the ignition timing is sufficiently advanced just to cause knock; this defines the knock limited spark advance (KLSA). Palmer and Smith (1985) report that a unity increase in RON will increase the KLSA by $1\frac{1}{2}°$ to $2°$ crank angle. Fuels can thus be tested in a steady-state test at full throttle, in which the KLSA is obtained as a function of speed for each fuel. The KLSA can then be compared with the manufacturer's ignition advance characteristics to establish the knock margin.

Transients are also of great importance for engines that do not have direct injection. During an increase in load, the throttle is opened and the pressure in the inlet manifold rises. At a low throttle setting there will be a high proportion of fuel vaporised, as the partial pressure of the fuel represents a larger fraction of the manifold pressure. When the throttle opens, and the manifold pressure rises, the partial pressure of the fuel cannot rise, so that some previously vaporised fuel condenses. Furthermore, when the throttle is opened, the air flow increases almost instantaneously (within a cycle or so), but the larger fuel droplets and the fuel film on the manifold will lag behind the air flow. For these two reasons, extra fuel is supplied during a throttle opening transient, to ensure that the mixture supplied to the engine is flammable. During an

Table 3.9 *The limits imposed by BS EN 228 and BS 7800 on oxygenates and benzene (BV denotes Blending Value)*

Component	Limit per cent (v/v)	RON	MON	BV (RON)
Benzene C$_6$H$_6$	1.0		115	
Methanol CH$_3$OH	3.0	106	92	112
Ethanol C$_2$H$_5$OH	5.0	107	89	110
Isopropyl alcohol (propan-2-ol) CH$_3$CHOHCH$_3$	5.0	–	–	118
Tertiary butyl alcohol C$_3$H$_7$CHOH	7.0	–	–	107
Ethers containing five or more carbon atoms	10.0	–	–	–
Other oxygenates specified by BS EN 228 and BS 7800	7.0	–	–	–

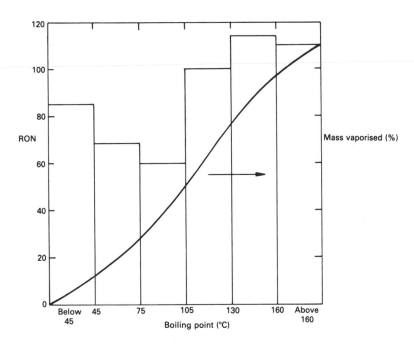

Figure 3.17
The RON and mass per cent of the fractions in a catalytic reformate fuel of 91 RON (data from Palmer and Smith, 1985).

increase in load it will be the more volatile components of the fuel (the front end) that will vaporise preferentially and enter the engine. The different fractions within a fuel will have different octane ratings, and this is illustrated by figure 3.6. The overall RON of the catalytic reformate fuel of figure 3.17 is 91, yet the RON of the fractions lies with the range of 58.5 to 114.

Such a fuel is unlikely to give an acceptable engine performance, as the fractions that boil in the temperature range of 45–105°C have a low RON. A useful concept here is the Delta octane number (ΔON). This is the difference between the knock rating of the whole gasoline, and the knock rating of the gasoline boiling below 100°C. The lower the ΔON, then the better the transient performance of the gasoline in avoiding knock.

However, it is the in-vehicle performance that is important, and this can be evaluated by an acceleration test. Suppose the vehicle is subject to a series of full throttle acceleration tests in a fixed gear (say direct drive) between specified speeds (for example, 50–100 km/h). In each test, the ignition timing can be advanced incrementally away from the vehicle manufacturer's setting until knock is detected.

Since a wide range of spark ignition engines are used in vehicles, fuel companies have to undertake extensive test programmes to ensure that their fuel will meet the market requirements. The test programmes have to cover a wide range of engines, and include supposedly identical engines in order to allow for variations in manufacture.

When methanol and ethanol are readily available, their properties make them attractive substitutes for petrol. Since alcohols burn more rapidly than petrol, the ignition timings in the tests for RON and MON are over-advanced, and this leads to an underestimate of the anti-knock performance. However, methanol in particular is susceptible to pre-ignition, and this is more likely to restrict the compression ratio (so as to limit surface temperatures in the combustion chamber).

The evaporative and combustion properties of alcohols are given in table 3.10. Even though smaller percentages of the

Table 3.10   *The evaporative and combustion properties of alcohols*

Fuel	Boiling point (°C)	Stoichiometric gravimetric air/fuel ratio	Enthalpy of evaporation		Enthalpy of combustion (LCV)	
			(kJ/kg fuel)	(kJ/kg stoichiometric mixture)	(MJ/kg fuel)	(MJ/kg stoichiometric mixture)
Methanol	65	6.5	1191	159	20.0	2.68
Ethanol	78.5	9.0	922	92	26.9	2.69
Petrol	25–175	14.5	310	20	42.0	2.71

alcohols evaporate during mixture preparation, the greater evaporative cooling effect is such that the alcohols produce lower mixture temperatures, with consequential improvements to the volumetric efficiency of the engine. The lower air/fuel ratios for alcohols mean that the chemical energy released per kg of stoichiometric mixture burnt during combustion is very similar to that for petrol, despite the lower specific enthalpies of combustion. Table 3.10 lists the enthalpy of combustion on the basis of 1 kg of a stoichiometric mixture. The improved volumetric efficiency and the similar combustion energy increase the output of the engine, and thus reduce the significance of mechanical losses, thereby improving the overall efficiency. Goodger (1975) reports the comparisons with hydrocarbon fuels made by Ricardo at a fixed compression ratio, in which there was a 5 per cent improvement in efficiency using ethanol, and a 10 per cent improvement when methanol was used. In racing applications, methanol is particularly attractive as the power output increases more with a richer mixture than it does for petrol. Also, the more rapid combustion of alcohols, and their greater ratio of moles of products to moles of reactants, both improve the cycle efficiency. Finally, alcohols can operate with leaner mixtures, and this leads to lower emissions.

The disadvantages associated with alcohols are their lower volatility and energy density (table 3.10), their miscibility with water, and a tendency towards pre-ignition.

Alcohols are attractive as additives (or more correctly extenders) to petrol, since the higher octane rating of the alcohols raises the octane rating of the fuel. Goodger (1975) reports that up to 25 per cent ethanol yields a linear increase in the octane rating (RON) of 8 with 92 octane fuel and an increase of 4 with 97 octane fuel.

In Europe, blends of petrol with 5 per cent alcohol have been commonly used to improve the octane rating, and Germany has sponsored a programme to introduce 15 per cent of methanol into petrol.

Palmer (1986) reports on vehicle tests, which showed that all oxygenate blends gave a better anti-knock performance during low speed acceleration than hydrocarbon fuels of the same octane rating. Furthermore, there is a tendency for the anti-knock benefits of oxygenate fuels to improve in unleaded fuels. However, care is needed with ethanol blends to avoid possible problems of high-speed knock. Fuel consumption on a volumetric basis is almost constant as the percentage of oxygenate fuel in the blend increases. The lower calorific value of the oxygenates is in part compensated for by the increase in fuel density. None the less, Palmer (1986) shows that energy consumption per kilometre falls as the oxygenate content in the fuel increases.

Ethanol is entirely miscible with petrol, while methanol is only partially miscible. The miscibility of both alcohols in petrol reduces with the presence of water and lower temperatures. To avoid phase separation, as moisture becomes absorbed in the fuel, chemicals such as benzene, acetone or the higher alcohols can be added to improve the miscibility. Palmer (1986) also reports on materials compatibility, hot and cold weather driveability, altitude effects, and the exhaust emissions from oxygenate blends.

The major disadvantages of methanol and ethanol as alternative fuels concern their production and low volatility. Methanol can be produced from either coal or natural gas, but there is an associated energy cost and impact on the environment. Monaghan (1990) points out that when methanol is derived from natural gas, then the contribution to greenhouse gases (in terms of carbon dioxide equivalent) is less than from a petrol fuelled vehicle. However, for methanol derived from coal, there is a greater contribution of greenhouse gases. The greenhouse effect is discussed in chapter 4, section 4.3.1.

The low volatility of methanol means that with conventional engine technology, priming agents are needed for cold starting below 10°C (Beckwith et al., 1986). Priming agents are hydrocarbons added to methanol, to improve the low-temperature vapour pressure and the flammability. Beckwith et al. report tests on a range of priming agents for methanol, with concentrations of up to 18 per cent by volume. They point out that the priming agents all led to increased vapour pressure at high temperatures which might lead to vapour lock problems. They concluded that straight run gasoline (with cut points up to 150°C) gave the best compromise between cold and hot weather performance.

An ethanol blend known as E85 (meaning 85 per cent by volume ethanol) is available in several countries.

Ethanol is mostly simply made from the fermentation of sugars, but competition with food use means that second-generation or cellulosic-ethanol needs to be exploited. Ethanol has been produced from cellulose for over 100 years, but there is now a rapid increase in the commercialisation of the process. Lignocellulosic materials (such as wood) have to be finely divided and then hydrolysed (by an acid) to release the cellulose. Enzymes then break the cellulose into simple sugars such as glucose which can be fermented to produce ethanol which is then separated by distillation.

3.7.3 Petrol additives

The use of additives to increase the octane rating has already been discussed in section 3.7.1, along with the scavenging additives associated with lead additives. Mention has already been made of anti-oxidation additives in section 3.2. The other important additives in petrol are detergents, corrosion inhibitors, metal deactivators and biocides. Sometimes anti-valve-seat-recession, anti-static and lubricity additives are used. Owen and Coley (1995) provide much useful information on additives.

Detergents are used to inhibit the formation of deposits. In the case of carburettors or port fuel injection systems, deposits can form in the induction system. With port fuel injection systems, fuel is sprayed towards the back of the inlet valve, and if no detergent is present then carbonaceous deposits form. Fuel injectors and deposits formed in the nozzles will have an adverse effect on the spray formation and subsequent mixture preparation. Gasoline direct injection engines are the most demanding because the fuel injector will be hotter as it is mounted in the cylinder head. Tests would consist of alternately running the engine at a medium to high load (say 10–15 minutes) and then turning the engine off (for say 30–45 minutes) so that, with the absence of cooling flows, component temperatures rise.

Anti-oxidation additives are needed because the unsaturated hydrocarbons (olefins) are prone to oxidation. The products of oxidation are gums and lacquers that can change the size of orifices that control flow in carburettors and fuel injectors. Alkylphenols are used in the range of 5–100 ppm, and they react with free radicals that are produced during hydrocarbon oxidation.

Metal deactivators prevent materials such as copper catalysing oxidation reactions; they work by reacting with the metal to form an inert layer. **Biocides** prevent bacterial growth that might occur at the interface with any water that has come from air by condensation.

Anti-valve-seat-recession additives can be added by users, and are needed because some engines that were designed in the era of leaded fuel suffer wear of the exhaust valve (valve-seat-recession), as discussed at the end of section 3.7.1. The additives are based on phosphorus, manganese and sodium compounds, but their effectiveness varies between engines and with operating regimes.

3.7.4 Characteristics of diesel fuel

The most important characteristic of diesel fuel is the cetane number, as this indicates how readily the fuel self-ignites. Viscosity is also important, especially for the lower-grade fuels used in the larger engines; sometimes it is necessary to have heated fuel lines. Another problem with diesel fuels is that, at low temperatures, the high molecular mass components can precipitate to form a waxy deposit. This is defined in terms of the cold filter plugging point.

The European Standards Organisation (CEN) common standard for diesel fuel is EN 590: 1993, and this includes six different cold filter plugging point temperature specifications in the range $-20°C$ to $5°C$. The UK specification is BS EN 590: 1997, and part of this is quoted in table 3.11.

The flashpoint is the temperature to which the liquid has to be heated for the vapour to form a combustible mixture with air at atmospheric pressure. Since the flashpoint of diesel fuel is at least $55°C$, this makes it a safer fuel to store than either petrol or kerosene. The flashpoints of petrol and kerosene are about $-40°C$ and $30°C$ respectively.

If an engine runs on a fuel with too low a cetane number, there will be diesel knock. Diesel knock is caused by too rapid combustion and is the result of a long ignition delay period, since during this period fuel is injected and mixes with air to form a combustible mixture. Ignition occurs only after the

Table 3.11 *Specifications for diesel fuel*

Property		BS EN 590
Viscosity at 40°C, mm²/s		2.0–4.5
Density at 15°C, kg/m³		820–845
Cetane number	min.	51
Carbon residue, Ramsbottom per cent by mass on 10 per cent residue	max.	0.3 per cent
Distillation, recovery at 350°C, per cent by volume	min.	85
Flashpoint, closed Pensky–Martins °C	min.	55
Water content, mg/kg	max.	200
Sediment, mg/litre	max.	24
Ash, per cent by mass	max.	0.01
Sulphur	max.	10 ppm
Copper corrosion test	max.	1
Cold filter plugging point (°C) max.	Summer (16 Mar–21 Oct)	−5
	Winter (22 Oct–15 Mar)	−15
Fatty Acid Methyl Ester (FAME) per cent by volume	max.	7.0

pressure and temperature have been above certain limits for sufficient time, and fuels with high cetane numbers are those that self-ignite readily.

As with octane numbers, a scale of 0–100 is constructed; originally a value of 0 was assigned, to α-methylnaphthalene ($C_{10}H_7CH_3$, a naphthenic compound with poor self-ignition qualities), and a value of 100 was assigned to n-cetane ($C_{16}H_{34}$, a straight-chain alkane with good self-ignition qualities). A 65 cetane fuel would have ignition delay performance equivalent to that of a blend of 65 per cent n-cetane and 35 per cent α-methylnaphthalene by volume. An isocetane, heptamethylnonane (HMN), is now used to define the bottom of the scale with a cetane number of 15.

The tests for determining cetane number in BS 5580: 1978 refer to the *Annual Book of ASTM Standards Volume 05.04*. The tests are performed with an ASTM-CFR engine equipped with a special instrument to measure ignition delay. With standard operating conditions the compression ratio of the engine is adjusted to give a standard delay period with the fuel being tested. The process is repeated with reference fuel blends to find the compression ratios for the same delay period. When the compression ratio of the fuel being tested is bracketed by the reference fuels, the cetane number of the test fuel is found by interpolation.

As would be expected, fuels with high cetane numbers have low octane numbers and vice versa. This relationship is shown in figure 3.18 using data from Taylor (1985b); surprisingly there is a single line, thus showing independence of fuel composition.

Sometimes a cetane index is used, as the only information needed is fuel viscosity and density with no need for engine tests. The cetane index can be used only for straight petroleum distillates without additives.

Table 3.12 presents the cetane ratings of some pure fuel components. This study by Hurn and Smith (1951) was prompted by the need to use catalytically cracked components in diesel fuel and other components with lower cetane ratings. The study also included combustion bomb studies of the ignition delay (at high pressure and various temperatures).

Throughout the 1980s and 1990s there was a steady relative reduction in the use of fuel oil (the heavy fractions from crude oil that are used for heating and some large diesel engines), while at the same time there was a corresponding increase in the demand for the middle distillates (which

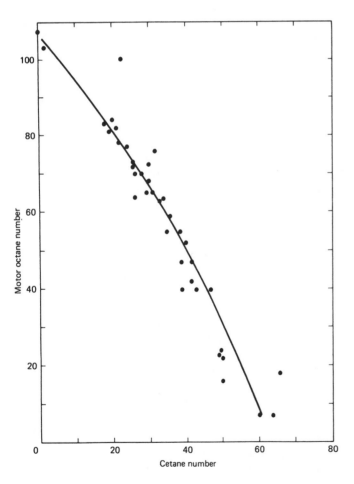

Figure 3.18
Relationship between cetane number and octane number for fuels distilled (not cracked) from crude oil (adapted from Taylor, 1985b).

Table 3.12 *Cetane ratings of some pure fuel components – Hurn and Smith (1951)*

Fuel	Formula	Cetane number
Alkanes		
n-heptane	C_7H_{16}	56
n-octane	C_8H_{18}	64
n-decane	$C_{10}H_{22}$	77
n-dodecane	$C_{12}H_{26}$	88
n-tetradecane	$C_{14}H_{30}$	96
n-hexadecane (cetane)	$C_{16}H_{34}$	100
heptamethylnonane	$C_{16}H_{34}$	15
octadecane	$C_{18}H_{38}$	103
Alkenes		
1-octene	C_8H_{16}	41
1-decene	$C_{10}H_{20}$	60
1-dodecene	$C_{12}H_{24}$	71
1-tetradecene	$C_{14}H_{28}$	83
1-hexadecene	$C_{16}H_{32}$	84
1-octadecene	$C_{18}H_{36}$	90
Ring structures		
α-methylnaphthalene	$C_{10}H_7CH_3$	0
methylcyclohexane	$C_6H_{11}CH_3$	20
decahydronaphthalene (decalin, two 'cyclohexanes' side-by-side)	$C_{10}H_{18}$	42
dicyclohexyl (two 'cyclohexanes' end-to-end)	$C_{12}H_{22}$	47

Table 3.13 *Comparison of the change in world (excluding the former Soviet Union) crude oil consumption (million tonnes/year) – from the BP Statistical Review of World Energy, 1998 and 2009*

Year	Gasoline (per cent)	Middle distillates (per cent)	Fuel oil (per cent)	Others (per cent)	Total (million tonnes)
1987	28	34	20	17	2527
1997	28	36	17	18	3197
2008	31	37	11	21	4192

includes kerosenes and diesel). Table 3.13 summarises some of these trends.

A consequence of these changes in demand is that the middle distillates and heavier fractions have to be converted to fuel oil and gasoline. Cracking of the heavier fractions produces a diesel fuel with a low cetane rating, so cracking has to be combined with hydrogenation (hydrocracking). Hydrocracking converts polycyclic aromatic compounds to form monocyclic compounds with side chains; with further hydrogenation these side chains are detached and become alkanes. The higher molecular mass alkanes are split to generate alkenes and alkanes, and the alkenes are hydrogenated to form alkanes. An alternative method for increasing the cetane rating is to use additives, as discussed in the next section.

The other main refining requirement with diesel fuel is sulphur removal. In 1985 sulphur levels were typically in the range 0.2–0.5 per cent by mass. In the US the sulphur level was reduced to 0.05 per cent by mass in 1993, and Europe adopted this level in 1996. In the European Union the sulphur-level limit was reduced from 350 ppm in 2000 to 10 ppm in 2009 for highway use, and to 1000 ppm in 2008 for off-highway use. In the US, the sulphur content of diesel for off-highway use was reduced to 500 ppm in 2007 with a further reduction to 15 ppm in 2010 – the level set for highway use in 2006. Sulphur levels have been reduced because the sulphur forms either:

1 sulphur dioxide (SO_2) which oxidises to SO_3 and combines with water to form sulphuric acid (H_2SO_4), a component of acid rain, or

2 sulphates (SO_4^-) which contribute to the particulate emissions.

Sulphur is removed by hydrodesulfurisation, in which the fuel is reacted with hydrogen in the presence of a cobalt and molybdenum catalyst. The hydrogen sulphide (H_2S) which this process generates then has to be converted into elemental sulphur. Low sulphur diesel fuels have lower lubricity, lower electrical conductivity and reduced stability, but additives can be used to compensate (Merchant *et al.*, 1997) and these are discussed in the next section.

3.7.5 Diesel fuel additives

Additives to improve the cetane number will be discussed first, then additives to lower the cold filter plugging point temperature, followed by additives that are used with low sulphur fuels, and finally other additives.

Additives in diesel fuel to improve the cetane number are also referred to as ignition accelerators. Typically an improvement of 6 on the cetane scale is obtained by adding 1 per cent by volume of amyl nitrate, $C_5H_{11}ONO_2$. Other effective substances are ethyl nitrate, $C_2H_5ONO_2$ and ethyl nitrite, C_2H_5ONO. However, the most widely used additive is currently 2-ethyl hexyl nitrate (2EHN), because of its good response in a wide range of fuels and comparatively low cost (Thompson *et al.*, 1997). Adding 1000 ppm of 2EHN will increase the cetane rating by about 5 units. In some parts of the world there is legislation limiting the nitrogen content of diesel fuels; this is because, although the mass of nitrogen is negligible to that available from the air, fuel-bound nitrogen contributes disproportionately to nitric oxide formation. Under these circumstances peroxides can be used, such as ditertiary butyl peroxide (Nandi and Jacobs, 1995).

Ignition delay is most pronounced at slow speeds because of the reduced temperature and pressure during compression. Cold-starting can be a problem, and is usually remedied by providing a facility on the injector pump to inject excess fuel. Under severe conditions, additional starting aids such as heaters may be needed, or volatile fuels with high cetane numbers, such as ether, can be added to the intake air.

Diesel fuel contains molecules with about 12–22 carbon atoms, and many of the higher molecular mass components (such as cetane, $C_{16}H_{34}$, melting point 18°C) would be solid at room temperature if they were not mixed with other hydrocarbons. Thus, when diesel fuel is cooled, a point will be reached at which the higher molecular mass components will start to solidify and form a waxy precipitate. As little as 2 per cent wax out of the solution can be enough to gel the remaining 98 per cent. This will affect the pouring properties and (more seriously at a slightly higher temperature) block the filter in the fuel injection system. These and other related low-temperature issues are discussed comprehensively by

Owen and Coley (1995), who point out that as much as 20 per cent of the diesel fuel can consist of higher molecular mass alkanes. It would be undesirable to remove these alkanes, since they have higher cetane ratings than many of the other components. Instead, use is made of anti-waxing additives that modify the shape of the wax crystals.

Wax crystals tend to form as thin 'plates' which can overlap and interlock. Anti-waxing additives do not prevent wax forming; they work by modifying the wax crystal shape to a dendritic (needle-like) form, and this reduces the tendency for the wax crystals to interlock. The crystals are still collected on the outside of the filter, but they do not block the passage of the liquid fuel. The anti-waxing additives in commercial use are copolymers of ethylene and vinyl acetate, or other alkene-ester copolymers. The performance of these additives varies with different fuels, and the improvement decreases as the dosage rate is increased. It is possible for 200 ppm of additive to reduce the cold filter plugging point (CFPP) temperature by about 10 K.

As noted previously, additives can be used with low sulphur diesel fuel to compensate for their lower lubricity, lower electrical conductivity and reduced stability. To restore the lubricity of a low sulphur fuel to that of a fuel with 0.2 per cent sulphur by mass, a dosage of order 100 mg/litre is needed. Care is required in the selection of the additive, if it is not to interact unfavourably with other additives (Batt *et al.*, 1996).

Electrical conductivity is not normally subject to legislation, but if fuels have a very low conductivity then there is the risk of a static electrical charge being built up. If a road tanker, previously filled with gasoline, is being filled with diesel, then there is the possibility of a flammable mixture being formed. The conductivity of untreated low sulphur diesel fuels can be less than 5 pS/m (Merchant *et al.*, 1997). Conductivities over 100 pS/m can be obtained by adding a few ppm of a chromium-based static dispersant additive. Low sulphur fuels and fuels that have been hydro-treated to reduced the aromatic content are also prone to the formation of hydroperoxides. These are known to degrade neoprene and nitrile rubbers, but this can be prevented by using antioxidants such as phenylenediamines (only suitable in low sulphur fuels) or hindered phenols (Owen and Coley, 1995).

Other additives used in diesel fuels are:

Detergents (for example, amines and amides) are used to inhibit the formation of combustion deposits. Most significant are deposits around the injector nozzles that interfere with the spray formation. This can then lead to poor air/fuel mixing and particulate emissions. A typical dosage level is 100–200 ppm.

Anti-ices (for example, alcohols or glycols) have a high affinity for water, and are soluble in diesel fuel. Water is present through contamination, and as a consequence of humid air above the fuel in vented tanks being cooled below its dew point temperature. If ice was formed, then it could block both fuel pipes and filters.

Biocides Anaerobic bacteria can form growths at the water/diesel interface in storage tanks, and these are capable of blocking fuel filters.

Anti-foamants (10–20 ppm silicone-based compounds) facilitate the rapid and complete filling of vehicle fuel tanks.

3.7.6 Alternative diesel fuels

A radical alternative to diesel fuel which has scope for very clean combustion is dimethyl ether (DME) – (Fleisch and Meurer, 1995). Dimethyl ether (CH_3OCH_3) is currently produced by the dehydration of methanol, but it can also be produced directly from natural gas. It is thus a route for using natural gas in diesel engines. DME has a cetane number above 55, but since its normal boiling point is –25°C, it has to be pressurised to about 5 bar for storage at ambient temperatures. The storage requirements are comparable to LPG used in spark ignition engines, and its energy density (excluding storage overheads) is about 60 per cent of that of diesel fuel. The small molecular size and the presence of the oxygen atom lead to a complete absence of particulates – the fuelling level can thus be increased, giving scope for about a 10 per cent increase in output. The lower combustion temperatures lead to about a halving of NO_x, and the higher cetane rating reduces the ignition delay period to give quieter combustion. DME is less expensive to produce than liquefied natural gas (LNG), and is easier to transport, so there is the possibility of DME being produced at remote natural gas fields, and being transported in preference to LNG which has to be stored at –162°C. Although no major engine modifications are needed, the fuelling system elastomers need to be compatible with DME, and the fuel injection system might dictate the use of additives to compensate for the low lubricity of DME.

Alternative fuels can be popular in countries that have no indigenous crude oil and a surplus of an alternative fuel. These circumstances can help justify the higher intrinsic cost of alternative fuels. More recently alternative fuels have become important because of the so-called 'zero option'. This is when the carbon dioxide emissions from burning a biomass fuel are balanced by the carbon dioxide that the crop will have absorbed during its growth.

The literature on alternative fuels demonstrates that a diesel engine can be made to run on virtually any liquid fuel, and Onion and Bodo (1983) provide a useful summary of alternative diesel fuels. Mittelbach and Remschmidt (2006) provide a very comprehensive treatment of bio-diesel, from the raw materials, bio-diesel manufacture, composition and fuel properties, to exhaust emissions and after-treatment, and environmental impact. As diesel engines are invariably developed to run on conventional fuels of crude oil origin, then alternative fuels can lead to operational difficulties. The different physical properties of alternative fuels lead to a different ignition delay and combustion performance. These differences can be reduced by running a diesel engine on

Table 3.14 *Physical properties of sunflower oil and diesel fuel*

Property	Sunflower oil	Diesel fuel	Ref.
Density at 15°C (kg/m³)	922	846	a
Distillation (°C)			
initial boiling point	477	189	a
10 per cent vaporised at	548	215	a
50 per cent vaporised at	588	261	a
90 per cent vaporised at	–	300	a
final boiling point	–	324	a
Enthalpy of vaporisation (kJ/kg)	837	251	b
Flash point (°C)	314	90	b
Self-ignition temperature (°C)	357	330	b
Cetane number	36	50	a
Viscosity at 40°C (cSt)	34	2.4	a
Pour point (°C)	–11	–18	a
Gravimetric elemental analysis (per cent)			
Carbon	77.1	86.7	a
Hydrogen	11.6	13.0	a
Oxygen	11.2	–	a
Sulphur		0.3	a
Stoichiometric gravimetric air/fuel ratio	12.4	14.4	b
Lower calorific value at 25°C			
(MJ/kg)	37.0	42.6	a
(MJ/l)	34.1	35.0	
Lower calorific value of stoichiometric reactants (MJ/kg of mixture)	2.76	2.77	

a: Needham and Doyle (1985).
b: Araya and Tsunematsu (1987).

mixtures of an alternative fuel with diesel, but this does not necessarily eliminate the durability problems.

A typical vegetable oil is sunflower oil, and table 3.14 compares its physical properties with a typical diesel fuel. Sunflower oil consists of unsaturated fatty acids with 16–22 carbon atoms (Ziejewski, 1983). Sunflower oil is produced by first sieving the seeds to remove the chaff. It is then crushed in a press or between rollers, leaving about 5 per cent of the oil unrecovered. This residue can be solvent extracted, leaving only 1 per cent of the oil behind.

Inspection of table 3.14 shows that sunflower oil has a lower calorific value, but its higher density means that the calorific value of sunflower on a volumetric basis approaches the volumetric calorific value of diesel fuel. However, it is essential to compare the fuel economy of the two fuels using brake efficiency and not on a specific fuel consumption basis. When this is done, see for example Stone and Ladommatos (1992a), there is no significant difference in the brake efficiency when sunflower oil is substituted for diesel fuel. An obvious difference between the two fuels is the comparatively low value of the cetane number for sunflower oil (36), despite only a slightly higher self-ignition temperature. The low cetane number is also a consequence of the low volatility of the sunflower oil.

Engine tests show that sunflower oil has a longer ignition delay period than diesel fuel, but the difference in delay period reduces as either load or speed is increased. Indeed at high loads and speeds the sunflower oil can have a shorter ignition delay period than the diesel fuel, since these conditions lead to greater vaporisation of the sunflower oil. Although sunflower oil tends to have a longer ignition delay period, this does not necessarily lead to a higher maximum rate of pressure rise and combustion noise or heat release rate. This is because the volatility of sunflower oil at low temperatures is low (when the ignition delay is long), so that only a limited amount of flammable air/fuel mixture is formed during the ignition delay period (Stone and Ladommatos, 1992a). However, sunflower oil leads to more combustion deposits; there is also the possibility of the sunflower oil forming gum deposits when it mixes with the lubricant.

The combustion characteristics of vegetable oils can be improved by esterification. In the case of sunflower oil, this raises the cetane number to 60, and lowers the boiling point range to 339–411°C (Needham and Doyle, 1985). Transesterification, in which the oil is reacted with an alcohol, reduces the molecular size, thereby reducing the viscosity and the boiling point. Combustion performance and durability are improved, but transesterification can double the cost of the fuel.

Esterified vegetable oil is often referred to as FAME – Fatty Acid Methyl Ester. A fatty acid is R_1COOH with its methyl ester being R_1COOCH_3, and its general form being R_1COOR_2, with R_1 and R_2 representing any hydrocarbon radical (including those with one or more double bonds). Vegetable oil (or fat) and animal fats are triglycerides (more properly known as triacylglyceride, TAG), which are esters of fatty acids with glycerol (a trihydric alcohol, commonly known as glycerine). Transesterification is the transformation from one type of ester to another, but it is also colloquially (and confusingly) referred to as esterification.

triglyceride   + 3 methanol →   glycerol   +   3 methyl esters

R_1, R_2, and R_3 represent long radicals that can be different from one another, and in the case of sunflower oil they are unsaturated (that is, they have double bonds) with 16 to 22 carbon atoms. If ethanol is used, then ethyl esters will be produced. The reaction is very slow, but can be catalysed by a strong base, such as sodium hydroxide (NaOH) or potassium hydroxide (KOH) which is dissolved in methanol. The strong base removes the hydrogen atom (deprotonates) the alcohol to form a RO radical which associates with the carbon atom that has the double bond to form an intermediate that leads to the formation of the methyl esters. For methanol, the reaction is kept at about 70°C, since every 10 K increase in temperature will about double the reaction rate. The reactants have to be dry, or otherwise salts of the fatty acids (soap) will be formed, a process known as saponification. Purification is comparatively simple, since the glycerol is denser than the methyl ester. Excess methanol will have been used, and this is recovered by distillation, and soap present can be removed by washing with warm water.

Schönborn et al. (2008) have conducted a detailed study on the combustion and emissions characteristics of individual fatty acid esters (six methyl esters with alkyl radicals ranging from C_{12} to C_{22} and ethyl oleate (C18 radical)). They present a

detailed summary of the physical properties of the different esters. As expected, the longer the alkyl radical then the shorter the ignition delay period, and if there is no injection timing compensation, then the earlier ignition leads to higher NO_x emissions. Similarly, for a given radical length (C18), an increased number of double bonds (in other words being less saturated) increased the ignition delay period. The emission of total particulate mass was not significantly affected by an increase in fatty acid chain length from 12 to 18 carbon atoms, so long as the ignition delay was held constant. However, an increase in the number of double bonds caused a significant increase in NO_x emissions and particulate mass emissions even when the ignition delay and heat release histories were the same.

The engine lubricant is contaminated by blow-by gases, unburned fuel and soot. FAME is less volatile than diesel fuel so there can be higher levels of lubricant dilution by fuel. This can lead to a reduced oxidation stability of the crankcase lubricant, which can reduce the engine durability. Szybist et al. (2007) conclude that FAME can reduce the amount of particulate matter produced compared with conventional diesel fuel and that 100 per cent FAME led to a 30 K lower temperature to initiate regeneration in a Diesel Particulate Filter (DPF, see also chapter 6, section 6.6.3). The regeneration was linked to the concentration of NO_2 (nitrogen dioxide) which is produced from NO (nitric oxide) in the oxidation catalyst that precedes the DPF. The higher NO emissions can be caused by the earlier ignition of the fuel (see section 3.7.5), as FAME has a higher cetane number. Boehman et al. (2005) note that soot trapped in a DPF has a higher organic content with bio-diesel, and that this is another reason for the lower regeneration initiation temperature.

There are also suggested limits on the percentage of unsaturated esters, since these are particularly susceptible to oxidation and polymerisation. The resulting products are prone to blocking filters and fuel pumps. There are also concerns that FAME leads to degradation of polymeric seals in fuelling systems. In Europe EN 14214 gives a specification for bio-diesel, and the oxidation stability and ash limits are of particular importance. Fuel-injection equipment manufacturers, and engine manufacturers, frequently only allow 5 per cent by volume (designated as B5) of FAME if it meets the EN 14214 bio-diesel specification. The European Union specification for diesel fuel, EN 590, allows up to 7 per cent FAME, but with a more rigorous oxidation stability limit. FAME can also contain high levels of water (about 1000 ppm), which provides a medium for bacteria, and the presence of oxygen in the molecules can then support aerobic bacteria that decompose the fuel. Such decomposition is an advantage if there is a fuel spillage, but obviously a disadvantage for engine operation. FAME does have good lubricity and this is an advantage compared with low sulphur conventional (mineral) diesel fuel.

Vegetable oils are (in most cases) food crops, but a notable exception is jatropha which can be grown in marginal arable conditions not suited to food crops. Jatropha has a yield of about 2 m^3/ha/yr, which is about double that of sunflower or rapeseed oil (Schenk *et al.*, 2008). Of renewed interest is the production of bio-diesel from algae. In the US a 20-year project concluded in 1998 that bio-diesel from algae was not economically feasible (Sheehan *et al.*, 1998). However, the economics have changed and there is increased concern with climate change, so this has led to great interest in bio-diesel from algae, and this forms part of a comprehensive review by Schenk *et al.* (2008). Since algae have evolved to grow under low light conditions, they have a much higher efficiency of photosynthesis than do plants. This leads to a higher yield of about 12 m^3/ha/yr, with projections for 100 m^3/ha/yr. Algae can utilise salt and waste water streams, thereby greatly reducing freshwater use.

A problem with fatty acid methyl esters is that contamination from manufacture (for example, if there is alkali or soap present) leads to high levels of ash. Ash is incombustible material, so it is a measure of the amount of metal or other inorganic contaminants in the fuel. In vehicles fitted with a Diesel Particulate Filter (DPF or 'particle trap'), ash leads to the accumulation of material on these filters, leading to more frequent filter maintenance and an increased back-pressure in the exhaust system that will reduce vehicle fuel economy. EN 14214 limits the sulphated ash to 0.02 per cent by mass, and alkali metals to 5 ppm by mass. The alkali metals come from the sodium hydroxide (NaOH) or potassium hydroxide (KOH) used as catalysts in the transesterification reaction.

Gas to liquid (GTL)

As already mentioned, natural gas has a boiling point of about $-162°C$, so if it could be converted to a liquid hydrocarbon by a gas to liquid (GTL) process it would be readily transportable. Natural gas is partially oxidised to form a mixture of CO, CO_2 and H_2, known as synthesis gas (syngas). The carbon dioxide is removed and the Fischer–Tropsch process is used to make high molecular mass hydrocarbons.

The Fischer–Tropsch process comprises many competing reactions, but the choice of catalyst enables selected products to be formed. Of particular importance for diesel engines is the formation of alkanes:

$$(2n + 1)H_2 + nCO \rightarrow C_nH_{2n+2} + nH_2O$$

The reactions take place at temperatures in the range of 150–300°C to increase the rate of reaction, and at high pressures to favour the forward reaction. The water can also be a very useful by-product in desert regions. The alkanes are mostly straight chained (n-alkanes), so this forms a diesel fuel with a high cetane number. Also, the absence of aromatic compounds reduces the particulate matter emissions.

BioMass to liquid (BTL)

Syngas can also be made from biomass, and again it is a partial oxidation process, but in this context it is usually termed pyrolysis and such gasification is a critical part of the process. There are essentially two types of biomass raw material – woody and herbaceous. Currently, woody material accounts for about 50 per cent of total world bioenergy potential, and another 20 per cent is straw-like feedstock, obtained as a by-product from agriculture. In Europe, willow is harvested in 2–4 year intervals when it has grown to about 6 m; the willow stumps then grow and another crop can be harvested in 2–4 years (coppicing). The harvesting is usually done in winter to reduce the moisture content. This short-rotation forestry yields 10–15 tonnes per hectare per year, compared with ordinary forestry yielding 5–10 tonnes per hectare per year (Kavalov and Peteves, 2005).

Kavalov and Peteves (2005) also present a comprehensive review of the different gasification technologies. They note that the partial oxidation is best done with oxygen, since it is easier to separate nitrogen from air than from the syngas products. Pyrolysis/gasification is at a temperature of about 500°C, and the conversion of syngas to a liquid fuel is by the Fischer–Tropsch process, as with gas to liquid, described above.

3.8 | Engine emissions and hydrocarbon oxidation

3.8.1 Introduction

This section provides an overview of the chemistry of emissions formation, the engine factors that influence emissions and the means of their control. Reference is also made later to more detailed information on emissions formation and control in chapters 4 and 6 on spark ignition and compression ignition engines. The remaining parts of section 3.8 cover the kinetics of nitric oxide (NO) formation (section 3.8.2) and hydrocarbon oxidation (section 3.8.3), which includes the kinetics of hydrocarbon oxidation (the 'flame') and hydrocarbon and carbon monoxide emission formation. These sections (3.8.2 and 3.8.3) need some knowledge of reaction kinetics, and this background can be found in Appendix A, section A.2.3. A very comprehensive treatment of emissions, their methods of control, and their influence on the environment has been produced by Sher (1998b).

Sections 3.8.2 and 3.8.3 can be omitted at a first reading, but one important point that needs clarification here is the terminology for NO_x emissions. For practical purposes during combustion, nitric oxide (NO) is the only nitrogen oxide that is formed. However, as the NO cools it can be oxidised to nitrogen dioxide (NO_2). NO_x refers to the mixture of the NO and NO_2, and typically over 90 per cent of the NO_x from an engine is NO. However, the NO will subsequently oxidise to

NO_2 in the environment, and it is the NO_2 that can react with unburnt non-methane hydrocarbons in the presence of ultraviolet light to form a photochemical smog. On a volumetric basis the proportion of the NO or NO_2 in NO_x makes no difference. This is not true on a gravimetric basis, and since all the NO can oxidise to NO_2 (and it is the nitrogen dioxide that forms smog), then the NO_x is assigned the molar mass of NO_2 (46 kg/kmol).

The term 'engine emissions' refers primarily to pollutants in the engine exhaust. Examples of pollutants are carbon monoxide (CO), various oxides of nitrogen (NO_x), unburnt hydrocarbons (HC) and particulate matter (PM). The gaseous emissions are worse from the spark ignition engine than from the compression ignition engine. Historically, the emission of greatest concern from compression ignition engines was soot – particulate matter that was visible as grey or black smoke. However, modern automotive compression ignition engines have much lower levels of particulate matter, and since the vast majority is much smaller than about 1 micron (10^{-6} m), then it will not be visible (visible light has wavelengths in the range 0.4–0.7 micron). Concern has been expressed about possible carcinogens in the exhaust but it is not clear if these come from the diesel fuel or from the combustion process.

Concern about emissions developed in the 1960s, particularly in Los Angeles where atmospheric conditions led to the formation of a photochemical smog from NO_x and HC. Exhaust emission legislation is historically and geographically too involved for discussion here, but is dealt with by Blackmore and Thomas (1977). Strictest controls are in the USA and Japan but European legislation is comparable.

The concentrations of CO and NO_x are greater than those predicted by equilibrium thermodynamics. The rate of the forward reaction is different from the backward reaction, and there is insufficient time for equilibrium to be attained. The chemical kinetics involved are complex, for example to try and predict exhaust emissions, but work is needed (Hochgreb, 1998).

Emissions of CO, NO_x and HC vary between different engines and are dependent on such variables as ignition timing, load, speed and, in particular, fuel/air ratio. Figure 3.19 shows typical variations of emissions with fuel/air ratio for a spark ignition engine.

Carbon monoxide (CO) is most concentrated with fuel-rich mixtures, as there will be incomplete combustion. With lean mixtures, CO is always present owing to dissociation and incomplete combustion, but the concentration reduces with reducing combustion temperatures. Hydrocarbon (HC) emissions are reduced by excess air (fuel-lean mixtures) until the reduced flammability of the mixtures causes a net increase in HC emissions. These emissions originate from the flame quench layer (where the flame is extinguished by cold boundaries), crevices (regions such as piston ring grooves can be particularly important) and from the oil film. The outer edge of the quench layer can also contribute to the CO and aldehyde emissions.

The formation of NO_x is more complex since it is dependent on a series of reactions such as the Zeldovich mechanism:

$$O_2 \rightleftharpoons 2O$$
$$O + N_2 \rightleftharpoons NO + N \qquad (3.21)$$
$$N + O_2 \rightleftharpoons NO + O$$

Chemical kinetics show that the formation of NO and other oxides of nitrogen increase very strongly with increasing flame

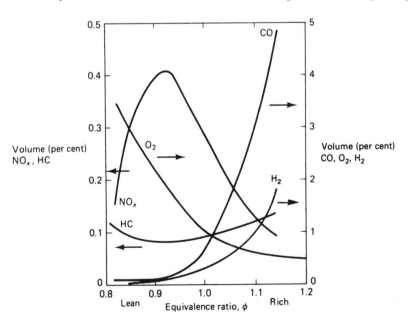

Figure 3.19
Spark ignition engine emissions for different fuel/air equivalence ratios (courtesy of Johnson Matthey).

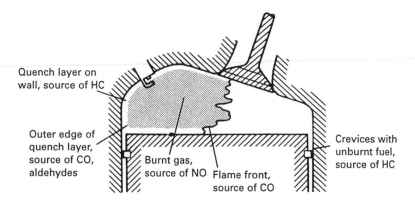

Figure 3.20
Source of emissions in spark ignition engine
(from Mattavi and Amann, 1980).

temperature. This would imply that the highest concentration of NO_x should be for slightly rich mixtures, those that have the highest flame temperature. However, oxygen is also needed for the formation of NO, so the maximum NO emissions occur just weak of stoichiometric. NO_x formation will also be influenced by the flame speed. Lower flame speeds with lean mixtures provide a longer time for NO_x to form. Similarly NO_x emissions increase with reduced engine speed. The sources of different emissions are shown in figure 3.20.

Emissions of HC and CO can be reduced by operating with lean mixtures; this has the disadvantage of reducing the engine power output. It is also difficult to ensure uniform mixture distribution to each cylinder in multi-cylinder engines. Alternatively, exhaust gas catalytic reactors or thermal reactors can complete the oxidation process; if necessary extra air can be admitted.

The ways of reducing NO_x emissions are more varied. If the flame temperature is reduced, the NO_x emissions will also be reduced. Retarding the ignition is very effective as this reduces the peak pressure and temperature, but it has an adverse effect on power output and economy. Another approach is to increase the concentration of residuals in the cylinder by exhaust gas recirculation (EGR). EGR lowers the flame temperature and gives significant reductions in NO_x. Between 5 and 10 per cent EGR is likely to halve NO_x emissions. However, EGR can lower the efficiency at full load and reduces the lean combustion limit. Catalysts can be used to reduce the NO_x to oxygen and nitrogen but this is difficult to arrange if CO and HC are being oxidised. Such systems have complex arrangements and require very close to stoichiometric mixtures of fuels with no lead-based additives; they are discussed in chapter 4, section 4.3.

Compression ignition engines have fewer gaseous emissions than spark ignition engines, but compression ignition engines have greater particulate emissions. The equivalence ratio in a diesel engine is always less than unity (fuel lean), and this accounts for the low CO emissions, about 0.1 per cent by volume. Hydrocarbon emission (unburnt fuel) is also less,

but rises towards the emission level of spark ignition engines as the engine load (bmep) rises.

The emissions of NO_x are about half those for spark ignition engines. This result might, at first, seem to contradict the pattern in spark ignition engines, for which NO_x emissions are worst for an equivalence ratio of about 0.95. In diffusion flames, fuel is diffusing towards the oxidant, and oxidant diffuses towards the fuel. The equivalence ratio varies continuously, from high values at the fuel droplet to values less than unity in the surrounding gases. The flame position can be defined for mathematical purposes as where the equivalence ratio is unity. However, the reaction zone will extend each side of the stoichiometric region to wherever the mixture is within the flammability limits. This will have an averaging effect on NO_x production. In addition, radiation from the reaction zone is significant, and NO_x production is strongly temperature-dependent. A common method to reduce NO_x emissions is to retard the injection timing, but this has adverse effects on fuel consumption and smoke emissions. Retarding the injection timing may be beneficial because this reduces the delay period and consequently the uncontrolled combustion period.

The most serious emission from compression ignition engines is particulate matter, which at very high levels has the characteristic grey or black of soot (carbon) particles. In this discussion, smoke (particulate matter) does not include the bluish smoke that signifies lubricating oil is being burnt, or the white smoke that is characteristic of unburnt fuel. These types of smoke occur only with malfunctioning engines, both compression and spark ignition.

Particulate matter from compression ignition engines originates mostly from carbon particles formed by cracking (splitting) of large hydrocarbon molecules on the fuel-rich side of the reaction zone. The carbon particles can grow by agglomeration until they reach the fuel-lean zone, where they can be oxidised. The final rate of soot release depends on the difference between the rate of formation and the rate of

Table 3.15 *Rate coefficients for the thermal NO mechanism*

	Reaction 1 $N_2 + O \rightarrow NO + O$			Reaction 2 $N + O_2 \rightarrow NO + O$			Reaction 3 $N + OH \rightarrow NO + H$		
	A	β	E/R_0	A	β	E/R_0	A	β	E/R_0
Heywood (1988)	1.6E + 13	0	0	6.4E + 09	1	3150	4.1E + 13	0	0
Miller & Bowman (1989)	3.3E + 12	0.3	0	6.4E + 09	1	3160	3.8E + 13	0	0
Hanson & Salimian (1984)	3.8E + 13	0	370	1.6E + 10	1	4470	5.4E + 13	0	1720

Notes: (1) Rate coefficients are presented in the form $k = A \times T^\beta \times \exp(-E/R_0 T)$.
(2) k_1^+ denotes the forwards reaction rate coefficient for reaction 1, and so on.
(3) Units are moles, cubic centimetres, seconds, Kelvin and calories/mole.

oxidation. The maximum fuel injected (and consequently power output) is limited so that the exhaust smoke is not visible. Particulate matter output can be reduced by advancing the injection timing or by injecting a finer fuel spray, the latter being obtained by higher injection pressures and finer nozzles. Smoke from a compression ignition engine implies a poorly calibrated injector pump or faulty injectors. Engine emissions are discussed further in chapter 4, section 4.3 for spark ignition engines, and chapter 6, section 6.6 for compression ignition engines.

3.8.2 Nitric oxide formation

The mixture of nitric oxide (NO) and nitrogen dioxide (NO$_2$) is referred to as NO$_x$. Nitric oxide is usually by far the most dominant nitrogen oxide formed during combustion. However, as already explained, subsequent further oxidation leads to nitrogen dioxide in the environment, and it is the nitrogen dioxide that reacts with the non-methane hydrocarbons in the presence of ultra-violet light, to form a photochemical smog. Thus although the major part of NO$_x$ will be NO, when emissions are calculated on a specific basis (g/MJ, g/kWh, etc.) then it is assumed that all the nitric oxide is oxidised to nitrogen dioxide.

Nitric oxide is formed in flames by three mechanisms: thermal, prompt and nitrous oxide. The thermal mechanism is based on the extended Zeldovich mechanism:

$$O + N_2 \rightleftharpoons NO + N \qquad (3.22)$$

$$N + O_2 \rightleftharpoons NO + O \qquad (3.23)$$

$$N + OH \rightleftharpoons NO + H \qquad (3.24)$$

Equations (3.22) and (3.23) were identified by Zeldovich (1946) and equation (3.24) was added by Lavoie *et al.* (1970) as it contributes significantly – this was identified by spectroscopic means. The rate constants for the thermal mechanism are very slow compared to those for combustion, and NO formation is only significant when there is a high enough temperature (say above 1800 K) and sufficient time. Thus the thermal NO mechanism is assumed to occur in the hot combustion gases, in which it can be taken that all the other species are in equilibrium apart from the NO. Measurements of the NO concentration in the burnt gases do not extrapolate

to zero at the flame. This implies that NO is formed in the flame, by the so-called prompt mechanism. The prompt mechanism is significant when there is fuel-bound nitrogen, or when the combustion temperatures are so low as to make the thermal mechanism negligible. The prompt mechanism is governed by:

$$CH + N_2 \rightarrow HCN + N \qquad (3.25)$$

The nitrous oxide mechanism is important at low temperatures and depends on a termolecular reaction:

$$N_2 + O + M \rightarrow N_2O + M \qquad (3.26)$$

with the subsequent decomposition to nitric oxide. Correa (1992) notes that the nitrous oxide sub-mechanism is significant with lean ($\lambda > 1.6$) pre-mixed laminar flames.

The kinetics of the C/H/N system have been the subject of a comprehensive review by Miller and Bowman (1989), along with a discussion of the three NO mechanisms. The kinetic values recommended by Miller and Bowman and some earlier workers for the extended Zeldovich scheme are presented in table 3.15. The differences in the table show the difficulty in obtaining rate coefficients, and the uncertainties that might be associated with predicting NO emissions. Indeed, for the predictions made later in figures 3.22 to 3.24, then different rate data can change the predictions by a factor of more than 2. It should also be appreciated that the temperatures in engine combustion can be much higher than those for which the data are validated, and that the pressure too will be much higher.

The equations for rates of reaction are developed in Appendix A, section A.2.3. The rate of NO formation by the extended Zeldovich mechanism (equations 3.22–3.24) is given by

$$\frac{d[NO]}{dt} = k_1^+[O][N_2] + k_2^+[N][O_2] + k_3^+[N][OH] - k_1^-[NO][N]$$
$$- k_2^-[NO][O] - k_3^-[NO][H] \qquad (3.27)$$

As the concentration of atomic nitrogen is low it is not necessarily computed in the evaluation of the equilibrium combustion products (see, for example, Ferguson, 1986), so instead it is assumed that the atomic nitrogen concentration is invariant. This assumption is also made in the analysis of hydrocarbon decomposition, which is presented in Appendix A, section A.2.4.

$$\frac{d[N]}{dt} = 0 = k_1^+[O][N_2] - k_2^+[N][O_2] - k_3^+[N][OH] - k_1^-[NO][N]$$
$$+ k_2^-[NO][O] + k_3^-[NO][H] \qquad (3.28)$$

Equation (3.28) is used to evaluate the concentration of atomic nitrogen, [N], and when this is substituted into equation (3.27)

$$\frac{d[NO]}{dt} = 2k_1^+[O][N_2] \frac{1 - [NO]^2/(K[O_2][N_2])}{1 + k_1^-[NO]/(k_2^+[O_2] + k_3^+[OH])} \qquad (3.29)$$

where $\quad K = k_1^+ \times k_2^+/(k_1^- \times k_2^-)$

In table 3.15 the rate coefficients have only been identified for one direction. However, the forward and reverse reactions can be related by the equilibrium concentrations ([ ]$_e$). Introducing three reaction rate variables (R_1, R_2, R_3) gives

$$R_1 = k_1^+[O]_e[N_2]_e = k_1^-[NO]_e[N]_e \qquad (3.30)$$
$$R_2 = k_2^+[N]_e[O_2]_e = k_2^-[NO]_e[O]_e \qquad (3.31)$$
$$R_3 = k_3^+[N]_e[OH]_e = k_3^-[NO]_e[H]_e \qquad (3.32)$$

After the substitution of equations (3.30)–(3.32) into equation (3.29) and much manipulation, it can be shown that

$$\frac{d[NO]}{dt} = \frac{2R_1\left\{1 - \left([NO]/[NO]_e\right)^2\right\}}{1 + \left([NO]/[NO]_e\right)R_1/(R_2 + R_3)} \qquad (3.33)$$

This differential equation can then be solved simultaneously with any other equations that are defining the combustion process.

In spark ignition engines there is normally negligible fuel-bound nitrogen, and it is usual to consider only the thermal NO mechanism. The combustion process is modelled by dividing the chamber contents into a minimum of two regions: the unburnt and the burnt gas, separated by a thin region. It is assumed that the combustion process occurs instantaneously and completely. Subsequently the burned gas is assumed to remain in thermodynamic equilibrium for the following species: CO, CO_2, H_2O, H_2, OH, O, N_2 and O_2. Also computed is the equilibrium value of the nitric oxide (denoted here as NO_e). The kinetically controlled value of the nitric oxide level (denoted here by NO_x) can then be computed by solving equation (3.33). In effect the NO kinetics have been decoupled from the combustion reaction, and this is acceptable since the hydrocarbon oxidation kinetics are orders of magnitude faster than the NO formation kinetics; table 3.16 gives an indication of various reaction timescales to support this approach.

Figure 3.21 shows the output from a spark ignition engine simulation, in which it has been assumed there is a single burned gas zone. The pressure rise (initially solely due to the piston motion) leads to isentropic compression of the unburned gas since no heat transfer has been allowed. Once combustion starts the burned gas temperature is plotted, and the change in burned and unburned gas temperatures both reflect the change in pressure. The equilibrium NO (NO_e) is present immediately after the start of combustion, and its strong temperature dependence means that it rises rapidly to a maximum (corresponding to p_{max} and $T_{b,max}$), and falls rapidly as the pressure and temperature fall. The kinetically controlled value of the nitric oxide (NO_x) starts at zero, and always lags behind the equilibrium value (NO_e). Once the burned gas temperature has fallen to a temperature of about 2000 K, then the NO kinetics are so slow that the kinetically calculated value of the nitric oxide (NO_x) remains almost constant (the composition is said to 'freeze'). As the burned gas temperature falls so the equilibrium value of the nitric oxide decreases rapidly to a level that is much below that of the kinetically predicted value.

This single burned gas zone model has in effect assumed that the gases behind the flame front are perfectly mixed. Alternatively, the burned gas can be divided into a series of zones, between which there is neither mixing nor heat transfer. The results presented here are from a model reported by Raine *et al.* (1995), which can incorporate heat transfer, blow-by and other losses. With a multi-zone model the earlier a zone burns, then the higher its temperature. This is because the temperature rise associated with combustion is almost constant, and the first burned gas is compressed by the subsequent combustion. If a given mass of gas is burned at constant volume and then compressed through a given pressure ratio, it will end with a higher temperature than if it was first compressed through that pressure ratio and then burned; this is because the compression process has a magnifying effect on temperature, while combustion has an additive effect on the temperature – illustrated by question 3.27.

Since the first burned gas has a higher temperature it will have a higher equilibrium nitric oxide level (NO_e), and this is shown in figure 3.22. Also shown are the unburnt gas

Table 3.16 *Selected chemical timescales, τ (Correa, 1992)*

Reaction	τ (s)	Reaction	τ (s)
1. $H + O_2 \rightarrow OH + O$	2×10^{-7}	2. $O + H_2 \rightarrow OH + H$	2×10^{-5}
3. $H + OH + M \rightarrow H_2O + M$	7×10^{-5}	4. $N_2 + O \rightarrow NO + N$	4×10^{-2}
5. $CH_4 + M \rightarrow CH_3 + H + M$	5×10^{-4}	6. $CH_4 + OH \rightarrow CH_3 + H_2O$	2×10^{-7}
7. $CH_3 + OH \rightarrow CH_2 + H_2O$	2×10^{-6}	8. $CO + OH \rightarrow CO_2 + H$	2×10^{-5}
9. $CH + N_2 \rightarrow HCN + N$	3×10^{-6}	10. $CH_2 + N_2 \rightarrow HCN + NH$	0.20

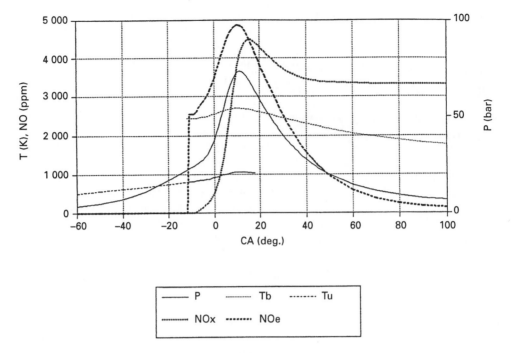

Figure 3.21
Simulation of spark ignition engine combustion with a single burned zone, showing: pressure (P), burnt gas temperature (Tb), unburnt gas temperature (Tu), and the equilibrium (NOe) and kinetic (NOx) predictions of the nitric oxide concentrations as a function of the crank angle.

temperature and pressure histories which have a negligible difference when compared with the single zone cases. However, figure 3.23 shows that the burned gas temperature of the first zone (of the five) has a higher temperature than in the single zone case (figure 3.21), and this in turn leads to higher values of the equilibrium (NO$_e$) and kinetically controlled (NO$_x$) nitric oxide levels. The final figure in this sequence (figure 3.23) shows the burned gas temperatures of zones 1, 3 and 5 and the corresponding kinetically calculated values of the nitric oxide emissions (NO$_x$). Figure 3.23 shows that there are slightly lower temperatures in the zones that are later to burn, and that this leads to substantially lower levels of the nitric oxide (NO$_x$).

3.8.3 Hydrocarbon oxidation

Even the oxidation of a simple compound like propane (C$_3$H$_8$) involves about 100 reactions and 30 species, and in order to model such a process fully there would be of the order of 30 simultaneous non-linear partial differential equations to be solved. Peters (1993) identifies 87 reactions (for which he tabulates the kinetic data) that are important in the oxidation of hydrocarbons up to propane. Some simple illustrations of the processes that are involved with hydrocarbon thermal decomposition are presented in Appendix A, section A.2.4.

Table 3.17 identifies some of the reactions for methane/air oxidation. 'M' refers to any molecule involved in a collision that leads to the 'splitting' of the molecule that it hits.

To fully determine the time history of the methane oxidation, it is necessary to have the reaction rate data for all the reactions, and a very powerful differential equation solver. The largest time step that can be taken in the solution is dictated by the fastest reaction, while the time period to be studied is dictated by the slowest reaction. Since there are many orders of magnitude difference between the fastest and slowest reactions, then it is necessary to solve the equations for numerous small steps. Such a complex situation needs to be simplified if the results are to be applicable to engineering problems. One such approach is to use reduced kinetic mechanisms, and these are comprehensively described by Peters and Rogg (1993).

Reduced kinetic mechanisms are possible, since it can be argued that not all reactions will be rate determining, nor do all species concentrations have to be determined accurately. In the case of methane, Mauss and Peters (1993) propose a four-step mechanism

$$\begin{aligned}
CH_4 + 2H + H_2O &\rightarrow CO + 4H_2 \\
CO + H_2O &\rightarrow CO_2 + H_2 \\
H + H + M &\rightarrow H_2 + M \\
O_2 + 3H_2 &\rightarrow 2H + 2H_2O
\end{aligned} \tag{3.34}$$

The assumptions necessary to achieve this simplification include the assumptions of constant O, OH, HO$_2$ and H$_2$O$_2$ concentrations. Similar approaches have been developed for the oxidation of more complex hydrocarbons.

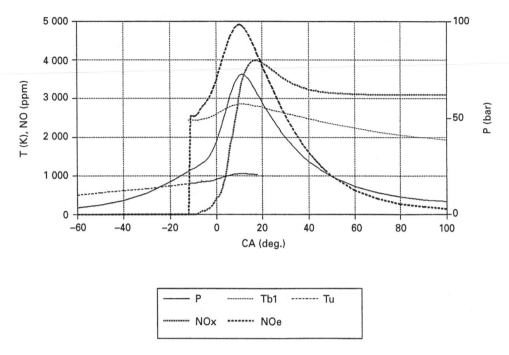

Figure 3.22
Simulation of spark ignition engine combustion with five burned zones, showing: pressure (P), burnt gas temperature of the first zone (Tb1), unburnt gas temperature (Tu), and the equilibrium (NOe) and kinetic (NOx) predictions of the nitric oxide concentrations for the first burned zone as a function of the crank angle.

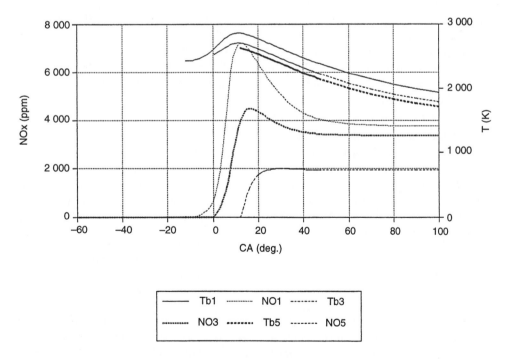

Figure 3.23
Simulation of spark ignition engine combustion with five burned zones, showing the burnt gas temperature (Tb) and the kinetic (NO) predictions of the nitric oxide concentrations for zones 1, 3 and 5 as a function of the crank angle.

Table 3.17 *Methane/air oxidation reaction scheme (Mellor, 1972)*

I.	Methane decomposition and partial oxidation reactions	$CH_4 + M \rightleftharpoons CH_3 + H + M$
		$CH_4 + O \rightleftharpoons CH_3 + OH$
		$CH_4 + H \rightleftharpoons CH_3 + H_2$
		$CH_4 + OH \rightleftharpoons CH_3 + H_2O$
		$HCO + OH \rightleftharpoons CO + H_2O$
		$CH_3 + O \rightleftharpoons H + CH_2O$
		$CH_2 + OH \rightleftharpoons HCO + H_2$
		$HCO + M \rightleftharpoons H + CO + M$
II.	Carbon monoxide oxidation	$CO + OH \rightleftharpoons CO_2 + H$
III.	O–H bimolecular propagation and chain branching reactions	$OH + H_2 \rightleftharpoons H_2O + H$
		$O_2 + H \rightleftharpoons OH + O$
		$O + H_2 \rightleftharpoons OH + H$
		$O + H_2O \rightleftharpoons OH + OH$
		$H_2O_2 \rightleftharpoons OH + OH$
IV.	O–H termolecular termination reactions	$H + H + M \rightleftharpoons H_2 + M$
		$O + O + M \rightleftharpoons O_2 + M$
		$O + H + M \rightleftharpoons OH + M$
		$H + OH + M \rightleftharpoons H_2O + M$

In spark ignition engines the unburnt hydrocarbon (HC) emissions arise from several sources:

1 Short-circuiting in which unburnt mixture flows into the exhaust during the valve overlap period.

2 Misfire (failure to ignite), or partial burn – either because combustion is too slow or started too late, or because the flame front was extinguished by high rates of strain.

3 Absorption/desorption from oil films, combustion deposits and crevices. Absorption (and flow into the crevices) occurs as the cylinder pressure rises, and desorption (and flow out of the crevices) occurs as the pressure is falling, by which time the burnt gas temperature is too low for full oxidation to occur in the available time.

4 Poor mixture preparation during transients. With a cold start an over-rich mixture has to be used to ensure a flammable air/fuel mixture, but fuel droplets inducted into the cylinder will not all be fully oxidised. As an engine warms up this tendency will be rapidly reduced, but during load changing transients there is a tendency (even with fuel injected engines) for there to be air/fuel ratio excursions from stoichiometric (this is discussed more fully in chapter 4, section 4.6.1).

Hochgreb (1998) provides an excellent overview of emissions from spark ignition engines, with a critical review of the mechanisms by which all emissions are formed, and with full details of sub-models for use in engine simulations, including the various modes of hydrocarbon absorption/desorption. Figure 3.24 indicates that about 92 per cent of the fuel is oxidised during propagation of the flame front, and that of the remaining 8 per cent (which has been absorbed into crevices etc.), about half will be oxidised after being desorbed into the burnt gases in the cylinder. This leaves about 4 per cent of the fuel entering the exhaust port, where about half will be oxidised. Thus about 2 per cent of the original fuel enters the catalyst, and with an oxidation efficiency of 95 per cent, so only about 0.1 per cent of the original fuel leaves the engine as unburnt hydrocarbons.

Crevices are the most significant source of unburnt hydrocarbons, since flames cannot propagate through narrow gaps – the so-called quench distance, which is about $\frac{1}{2}$ mm for hydrocarbon mixtures in engine-like combustion conditions. It is well established that the mass fraction of the charge stored in a crevice volume increases to

$$\frac{m_c}{m_i} = \frac{V_c T_i P_{max}}{V_i T_c p_i} \qquad (3.35)$$

where m_c is the mass trapped in the crevice
V_c is the volume of the crevice
T_c is the temperature of the crevice
p_{max} is the maximum cylinder pressure, and
suffix i denotes the conditions at inlet valve closure.

The temperature of the crevice will be approximately that of the combustion chamber (say about 450 K), so some of the unburnt mixture will be cooled as it enters the crevice, and the mass fraction trapped is greater than the volume fraction represented by the crevice. The most significant crevice is that bounded by the piston, the top ring and the cylinder. In general it will be unburnt mixture that is compressed into the crevice, since the maximum cylinder pressure is likely to be reached prior to the flame front passing over the crevice entrance. Some of the unburnt mixture will be burnt as it

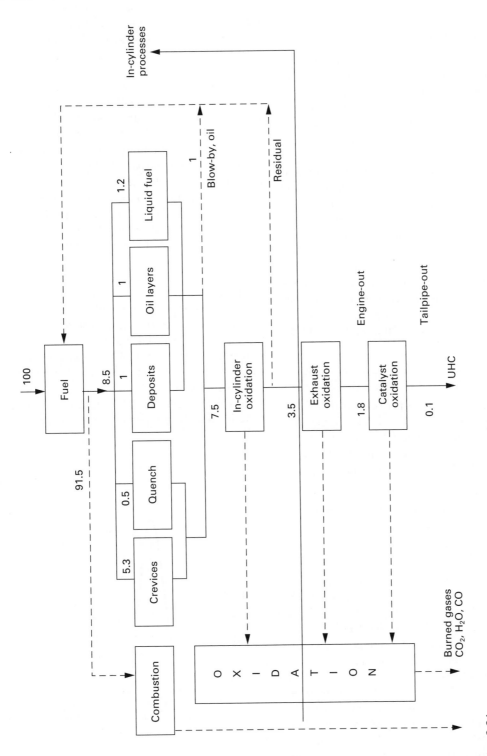

Figure 3.24
Pathways for the formation of unburnt hydrocarbons in a spark ignition engine operating at part load and mid speed (adapted from Cheng et al., 1993).

returns to the combustion chamber. The simplest way to model this is to assume that when the burnt gas is above a certain temperature (≈2000 K), then the mixture is burnt immediately. Below this threshold temperature, it is assumed to remain as unburnt mixture.

3.8.4 Carbon monoxide emissions

The carbon monoxide emissions lie between those predicted for equilibrium at peak pressure and the equilibrium values at exhaust valve opening: they tend to be closer to the maximum pressure values for rich mixtures, and closer to the exhaust valve opening values for weak mixtures. As with the nitric oxide emissions, this is a consequence of the reaction rates falling during expansion, so that a temperature is reached at which the carbon monoxide concentration appears to 'freeze'. The CO emissions can be modelled by specifying a 'freezing' temperature at which the equilibrium concentration is evaluated. More accurate models need to include the kinetics, and these are reviewed in detail by Heywood (1988). However, the main determinant of carbon monoxide emissions is the air/fuel ratio. In multi-cylinder engines operating at stoichiometric, the inter-cylinder variation in air/fuel ratio will have the biggest effect on the carbon monoxide emissions. This can be illustrated by a simple example, a two-cylinder engine operating with an overall stoichiometric mixture:

case (a) both cylinders stoichiometric, then the exhaust will contain about 0.84 per cent CO

case (b) each cylinder operating 5 per cent away from stoichiometric (one rich, the other lean), then the exhaust will contain about ([1.90 + 0.22]/2) = 1.06 per cent CO, which is comparable to the equilibrium CO level at maximum pressure

Thus for overall stoichiometric operation, inter-cylinder mixture maldistribution is likely to be more significant than correctly modelling the CO kinetics.

3.8.5 Particulate matter

Particulate matter (PM) is formed in both spark ignition and diesel engines, but because the emissions from spark ignition engines are not visible (being sub-micron) they have been ignored in the past. The particulate matter legislation for diesel engines has become increasingly stringent, and for automotive applications a diesel particulate filter (DPF) is needed. The particulate matter emissions from a compression ignition engine with a diesel particulate filter can now be lower than those from direct injection gasoline engines (see chapter 5), so it is not surprising that legislation for gasoline engines now includes particulate matter. Eastwood (2007) provides a very comprehensive review of the formation, composition, characterisation, toxicology and after-treatment of particulate matter.

Historically, the emissions legislation was just for mass, but the number and size are also significant since the deposition efficiency in the respiratory system is greatest for small particles. So it is potentially the small particles that are the most harmful, and this has led to the introduction of legislation that limits the total number of particles that are emitted. Figure 3.25 shows the particulate matter number (N) and mass (m) distributions as a function of particle size (d_p). The data are from a wall-guided gasoline direct engine (see chapter 5), and further details are in Price $et\ al.$ (2006). The mass distribution (the $dM/d\log d_p$ curve) was calculated assuming spherical particles, but the main purpose of figure 3.25 (which is plotted with size on a logarithmic scale) is to identify the nucleation mode and accumulation mode, and to show how most of the particles by number are associated with the nucleation mode, while most of the mass is associated with the accumulation mode. The size ranges assigned to the nucleation mode and accumulation mode are arbitrary, and the division might occur anywhere between 30 and 100 nm. However, the data are usually fitted well by log-normal distributions.

The effect of particulate matter on human health is highly complex and difficult to establish, since epidemiological studies are essentially the only way to link exposure to pollutants with disease. The deposition efficiency varies with the size of the particles and the location in the respiratory system. Particles larger than 1 micron are trapped in the nasal passages, and table 3.18 summarises results to be found in Eastwood (2007).

Table 3.18 shows how the smallest particles (~10 nm) are more likely to be trapped in the respiratory system than the

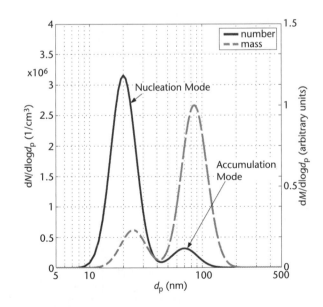

Figure 3.25 Representative particulate matter number (N) and mass (m) distributions as a function of particle size (d_p); the $dM/d\log d_p$ curve was calculated assuming spherical particles.

Table 3.18 *Deposition efficiency (per cent) of different size particles in the human respiratory system*

Location	Size (nm)				
	10	**50**	**100**	**500**	**1000**
Trachea and bronchea	20–40	5–15	4–8	4–10	4–12
Pulmonary	50–70	45–60	40–50	20–35	20–35

larger particles (say 100 nm), and that they are more likely to be trapped deep inside the lungs. Particles trapped in the nasal passages can be expelled, while those trapped inside the lungs are more likely to be absorbed. The smaller particles will have a larger surface area to mass ratio, so are more likely to have a greater proportion of organic matter.

The origins of the particles and their ensuing contribution to the particle composition are summarised in figure 3.26.

In figure 3.26 the carbonaceous fraction comes from pyrolysis of the fuel and oil, while the organic fraction is from the partial combustion of the fuel and oil. Wear products will normally be metallic and this leads to the inorganic fraction, along with any metals that have been present in the fuel or oil (frequently as additives). Any sulphur in the fuel or oil is oxidised and combines with water, leading to the production of sulphuric acid and sulphates; it is of course inorganic but of sufficient quantity to justify separate identification.

The nucleation-mode particles are generally accepted to be droplets of sulphuric acid or hydrocarbons that have condensed. Their number and size are very sensitive to the temperature of the sample being measured and how it has been diluted. The measuring instruments invariably operate at ambient temperature so the exhaust has to be cooled prior to measurements. If the exhaust is cooled without dilution, then condensation can occur. The nucleation sites for condensation are probably small (solid) inorganic particles. It is difficult to determine the composition of nucleation-mode particles because, although they are large in number, their mass is very small.

An accumulation-mode particle from a gasoline direct injection engine is shown in figure 3.27 (particles from compression ignition engines are very similar). Figure 3.27 shows how the accumulation-mode particle consists of many 'spherules' or 'primary particles', that are close to spherical in shape. These spherules are usually in the range of 30–50 nm, and they aggregate to form larger particles ('aggregates'). The image in figure 3.27 was obtained with a Transmission Electron Microscope (TEM), so the very low pressure will have evaporated any condensate on the surface – material that had either condensed directly or arrived by collision with a nucleation-mode particle. If the magnification is increased further, it is possible to determine that some of the carbon is graphitic in structure, while some of it is amorphous. X-Ray Photo Spectrometry can be used to identify the nature of the carbon–carbon bonds, and this too provides evidence of the graphitic structure.

With particulate matter, the organic fraction can be removed by a solvent (in which case it is termed the Soluble Organic Fraction – SOF), or it can be removed by heat (in which case it is termed the Volatile Organic Fraction – VOF). The sulphate fraction is the SO_4^{2-} ion that will be mostly combined with water as sulphuric acid (H_2SO_4). Sulphuric acid absorbs water from the atmosphere, so that when filters that have collected particulate matter are being weighed, it is of great importance to have kept the specimens in an environment with controlled humidity. Techniques for the measurement and characterisation of particulate matter are discussed further in chapter 14, section 14.4.

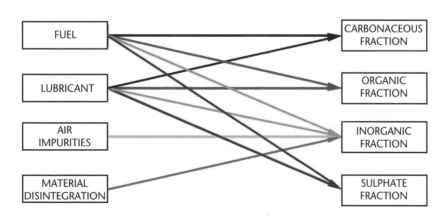

Figure 3.26 Routes for the formation of particulate matter, adapted from Eastwood (2007).

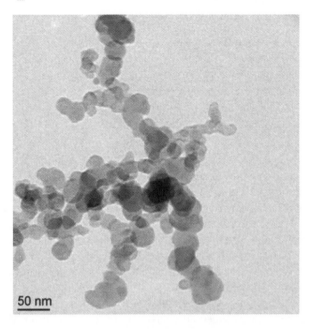

Figure 3.27 A Transmission Electron Microscope (TEM) image of an accumulation-mode particle from a gasoline direct injection engine (Price, 2009).

The mechanisms for the production of the primary particles are complex and their modelling is especially challenging, and Eastwood (2007) provides a comprehensive review. As already mentioned, the primary particles are partly graphitic in structure. Graphite has a planar structure with a hexagonal arrangement of carbon atoms – like a 'matrix' of benzene rings with the hydrogen removed and then fitted together.

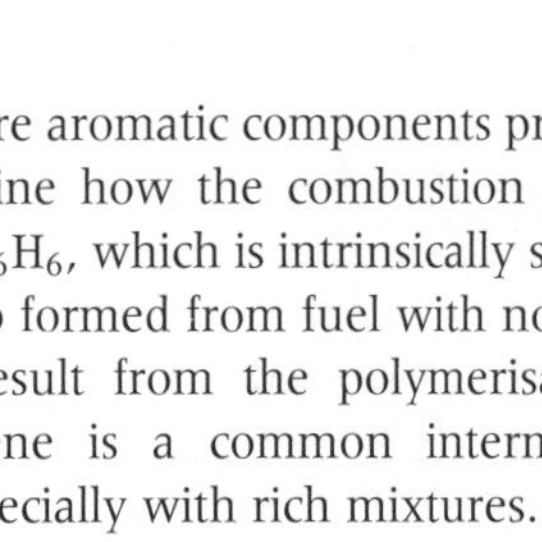

When there are aromatic components present in the fuel, it is easy to imagine how the combustion process leads to a benzene ring (C_6H_6, which is intrinsically stable), but primary particles are also formed from fuel with no aromatic content, and this can result from the polymerisation of acetylene (C_2H_2). Acetylene is a common intermediate species in combustion, especially with rich mixtures. Even if the overall air/fuel ratio is weak of stoichiometric, inhomogeneities in the mixture will lead to locally rich regions. Other species can also lead to the formation of aromatic compounds. The precursors of graphite are the polycylic aromatic hydrocarbon (PAH)

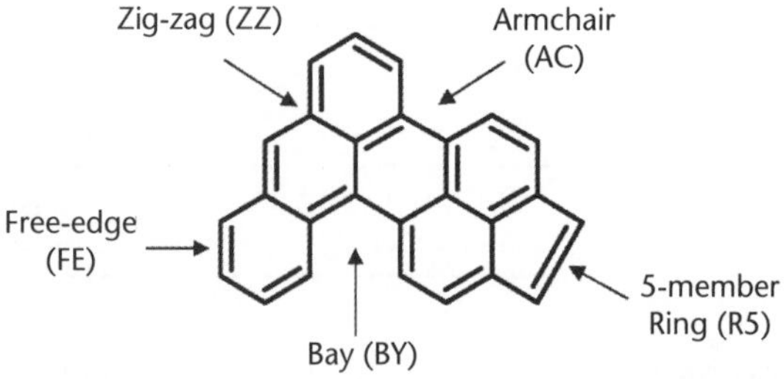

Figure 3.28 Different types of site on a PAH molecule, courtesy of Markus Kraft, University of Cambridge.

compounds, of which some examples are shown in table 3.19.

Table 3.19 shows how the same number of carbon rings can form different shapes, and that five-atom carbon rings are possible, in which case the molecule will not be planar. The level of aromaticity also varies (shown by the circle that represents molecular orbitals above and below the plane of the carbon atoms, but each containing three electrons). The position of the benzene ring can vary within some PAHs, and for a given number of rings both the number of hydrogen atoms and the number of carbon atoms can vary. The PAH molecules are mostly planar, and when they stack together a graphitic structure is formed, while if they are orientated randomly this results in an amorphous structure.

Modelling the growth of PAH molecules is computationally very demanding (Raj *et al.*, 2009), and requires identification of the different types of site on a PAH molecule, as shown in figure 3.28. PAH molecules can grow by the hydrogen-abstraction–carbon-addition (HACA) mechanism involving the addition of acetylene (C_2H_2) or other species at aromatic radical sites.

The model developed by Raj *et al.* (2009) describes the structure and growth of planar PAH molecules and is based on reactions at radical sites available in the literature, and additional reactions obtained from quantum-chemistry calculations. This provides a detailed model of soot-particle composition and the associated surface growth and oxidation reactions using quantum chemistry and Kinetic Monte Carlo (KMC) simulations. It is necessary to use a statistical representation of the PAH molecules and their functional sites because tracking every single molecule is computationally prohibitive. It is also necessary to model when the molecules are held together in a particle through physical forces (rather than chemical bonds) to form aggregates, and to include the condensation/adsorption of hydrocarbons onto the surface of the aggregates.

3.8.6 Emissions legislation

It is much easier for legislators to decree ever more stringent controls on emissions from internal combustion engines than it is to implement other strategies, such as improved traffic

Table 3.19 *Polycylic aromatic hydrocarbon (PAH) compounds*

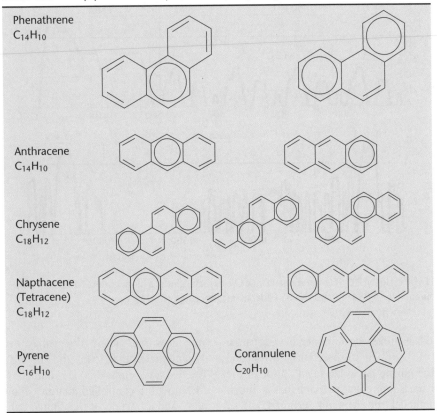

Phenathrene
$C_{14}H_{10}$

Anthracene
$C_{14}H_{10}$

Chrysene
$C_{18}H_{12}$

Napthacene
(Tetracene)
$C_{18}H_{12}$

Pyrene
$C_{16}H_{10}$

Corannulene
$C_{20}H_{10}$

management and alternatives to personal transport in congested cities. It is also much easier to impose legislation on new vehicles than it is to reduce emissions from the existing vehicle population. Concern about the environment is not likely to decrease and, as research continues, new concerns arise.

The US Federal Environmental Protection Agency (EPA) is adopting the California emissions standards as a national standard by the 2016 model year.

Table 3.20 shows the US emissions trends, and it can be seen how the Federal legislation tends to follow the Californian legislation, which now includes legislation for formaldehyde (HCHO, an irritant). A direct comparison cannot be made with the European legislation in table 3.21, since there is a different drive cycle. The drive cycles are shown in figure 3.29, and the current legislation now requires the emissions to be measured from as soon as the engine is

Table 3.20 *US exhaust emissions standards for light-duty vehicles (g/mile) up to 3856 kg (8500 pounds)*

Legislation	NMOG	CO	NO$_x$	HCHO
Federal Tier I (1994)	0.25	3.4	0.4	–
Federal Tier II (2003)	0.125	1.7	0.2	–
Tier 2 Bin 5 (2008)	0.075	3.4	0.05	0.015
(Compliance needed after				
50 000 miles)				
California				
Conventional (1993)	0.25	3.4	0.4	–
Transitional LEVs (1994)	0.125	1.7	0.2	0.015
LEVs (1997)	0.04	3.4	0.4	0.015
Ultra LEVs (1997)	0.04	3.4	0.4	0.008

LEV – low emission vehicles
HCHO – formaldehyde
NMOG – non-methane organic gases

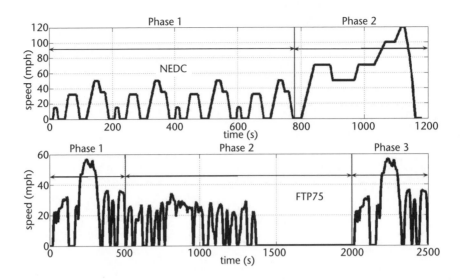

Figure 3.29 The ECE R15 Drive cycle (NEDC – New European Drive Cycle) that includes a highway test schedule (Phase 1 is of 780 s duration for 4.05 km, Phase 2 is of 400 s duration for 6.95 km), and the US Federal Test Procedure (FTP75).

started, and the HC and NO_x emissions now each have separate levels. Prior to Euro 3 (EU3: 01/2000) there was a 40-second period during which the engine could be 'idled', prior to the start of the emissions measurements. Methane has a much lower ozone-forming potential (see chapter 4, section 4.3.4), so is excluded in the US NMOG (non-methane organic gases) and EU NMHC (non-methane hydrocarbon) limits. The emissions are measured when the vehicle is tested on a chassis dynamometer that loads the driving wheels of the vehicle. The exhaust emissions are diluted (see chapter 14, figure 14.30) with a controlled quantity of air (to prevent condensation), and a fraction of the diluted exhaust is then collected in sample bags for analysis at the end of the test; a comprehensive description of these systems is given by Eggers *et al.* (2004).

Table 3.21 *EU light-duty tailpipe emissions requirements for passenger cars below 2500 kg (EU Regulation EC/715/2007 (2007))*

Positive ignition (PM and PN only applicable to Gasoline Direct Injection Engines)						
Effective timing	**CO** (g/km)	**THC** (mg/km)	**NMHC** (mg/km)	**NO$_x$** (mg/km)	**PM** (mg/km)	**PN** (1/km)
EU3: 01/2000	2.30	200	n/a	150	n/a	n/a
EU4: 01/2005	1.00	100	n/a	80	n/a	n/a
EU5: 09/2009	1.00	100	68	60	n/a	n/a
EU5b: 10/2011	1.00	100	68	60	4.5	n/a
EU6: 09/2014	1.00	100	68	60	4.5	TBD

Compression ignition					
Effective timing	**CO** (g/km)	**NO$_x$** (mg/km)	**THC + NO$_x$** (mg/km)	**PM** (mg/km)	**PN** (1/km)
EU3: 01/2000	0.64	500	560	50	n/a
EU4: 01/2005	0.50	250	300	25	n/a
EU5: 09/2009	0.50	180	230	5.0	n/a
EU5b: 10/2011	0.50	180	230	4.5	6.0×10^{11}
EU6: 09/2014	0.50	80	170	4.5	6.0×10^{11}

NMHC – non-methane hydrocarbons
TBD – to be determined
THC – total hydrocarbons
PM – particulate matter mass
PN – particulate matter number

3.9 | Combustion modelling

3.9.1 Introduction

The combustion model is one of the key elements in any computer simulation of internal combustion engine cycles. In addition, all aspects of the engine operating cycle directly influence the combustion process. Heywood (1988) provides a very good introduction to the subject and he emphasises the interdependence and complication of the combustion and engine operation. The combustion occurs in a three-dimensional, time-dependent, turbulent flow, with a fuel containing a blend of hydrocarbons, and with poorly understood combustion chemistry. The combustion chamber varies in shape, and the heat transfer is difficult to predict.

There are three approaches to combustion modelling; in order of increasing complexity, they are:

(i) *Zero-dimensional models* (or *phenomenological models.*) These use an empirical 'heat release' model, in which time is the only independent variable.

(ii) *Quasi-dimensional models*. These use a separate sub-model for turbulent combustion to derive a 'heat release' model.

(iii) *Multi-dimensional models*. These models solve numerically the equations for mass, momentum, energy and species conservation in three dimensions, in order to predict the flame propagation.

All models can be used for estimating engine efficiency, performance and emissions. The zero-dimensional and quasi-dimensional models are readily incorporated into complete engine models, but there is no explicit link with combustion chamber geometry. Consequently, these models are useful for parametric studies associated with engine development. When combustion chamber geometry is important or subject to much change, multi-dimensional models have to be used. Since the computational demands are very high, multi-dimensional models are used for combustion chamber modelling rather than complete engine modelling.

The more complex models are still subject to much research and refinement, and rely on sub-models for the turbulence effects and chemical kinetics. Review papers by Tabaczynski (1983) and by Greenhalgh (1983) illustrate the use of lasers in turbulence (laser doppler anemometry/velocimetry – LDA, LDV) and the use of lasers in chemical species measurements (spectrographic techniques), respectively. These techniques can be applied to operating engines fitted with quartz windows for optical access. All models require experimental validation with engines, and combustion films can be invaluable for checking combustion models. The ways the in-cylinder flow can be measured and defined are discussed in chapter 9, along with how turbulent combustion can be modelled in spark ignition engines.

3.9.2 Zero-dimensional models

This approach to combustion modelling is best explained by reference to a particular model, the one described by Heywood *et al.* (1979), for spark ignition engines. This combustion model makes use of three zones, two of which are burnt gas:

(i) unburnt gas

(ii) burnt gas

(iii) burnt gas adjacent to the combustion chamber – a thermal boundary layer or quench layer.

This arrangement is shown in figure 3.20, in addition to the reaction zone or flame front separating the burnt and unburnt gases. The combustion does not occur instantaneously, and can be modelled by a Wiebe function:

$$x(\theta) = 1 - \exp\left\{-a[(\theta - \theta_0)/\Delta\theta_b]^{m+1}\right\}$$

where $x(\theta)$ is the mass fraction burnt at crank angle θ

θ_0 is the crank angle at the start of combustion

and $\Delta\theta_b$ is the duration of combustion.

a and m are constants that can be varied so that a computed p–V diagram can be matched to that of a particular engine. Typically

$a = 5$ and $m = 2$

The effect of varying these parameters on the rate of combustion is shown in figure 3.30.

The heat transfer is predicted using the correlation developed by Woschni (1967); although this was developed for compression ignition engines, it is widely used for spark ignition engines. The correlation has a familiar form, in terms of Nusselt, Reynolds and Prandtl numbers:

$$Nu = a\,Re^b\,Pr^c$$

The constants a, b and c will depend on the engine geometry and speed but typical values are

$a = 0.035$, $b = 0.8$, $c = 0.333$

As well as predicting the engine efficiency, this type of model is very useful in predicting engine emissions. The concentrations of carbon–oxygen–hydrogen species in the burnt gas are calculated using equilibrium thermodynamics. Nitric oxide emissions are more difficult to predict, since they cannot be described by equilibrium thermodynamics. The Zeldovich mechanism is used as a basis for calculating the nitrogen oxide production behind the flame front. This composition is then assumed to be 'frozen' when the gas comes into the thermal boundary layer. This approach is useful in assessing (for example) the effect of exhaust gas recirculation or ignition timing on both fuel economy and nitrogen oxide emissions. Some results from the Heywood model are shown in figure 3.31.

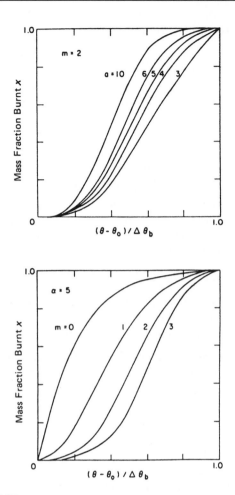

Figure 3.30
Wiebe functions (reprinted with permission from Heywood *et al.*, 1979, © Society of Automotive Engineers, Inc.).

3.9.3 Quasi-dimensional models

The simple three-zone model is self-evidently unrealistic for compression ignition engines; nor is the requirement for burn rate information very convenient – even for spark ignition engines. Quasi-dimensional models try to predict the burn rate information by assuming a spherical flame front geometry, and by using information about the turbulence as an input. For spark ignition engines this simple approach gives the rate of mass burning (dm_b/dt) as

$$\frac{dm_b}{dt} = \rho_u A_f U_t = \rho_u A_f ff\, U_1$$

where ρ_u = density of the unburnt gas
A_f = area of the flame front
U_t = turbulent flame front velocity
ff = turbulent flame factor
U_1 = laminar flame front velocity.

This approach can be made more sophisticated, in particular with regard to the turbulence. The size of the flame front compared to the turbulence scale changes, for example; this accounts for the delay period. Also, as the pressure rises during compression, the length scale of the turbulence will be reduced. Finally, allowance needs to be made for both squish and swirl. These techniques are described in chapter 9, section 9.4.

For compression ignition engines and direct injection spark ignition engines, the mixing of the air and the fuel jet is all important, and a turbulent jet-entrainment model is necessary. Since the time histories of different fuel elements will not be the same, a multi-zone combustion model is needed to trace the individual fuel elements. The prediction of NO_x emissions will provide a powerful check on any such model, since NO_x production will be very sensitive to the wide variations in both temperature and air/fuel ratio that occur in compression ignition engines.

3.9.4 Multi-dimensional models

Multi-dimensional models require the solution of the flow equations by Computational Fluid Dynamics (CFD) codes, with sub-models incorporated for processes such as ignition, combustion, flame–wall interactions and emissions modelling. Duclos *et al.* (1996) consider a number of methods for modelling combustion, and these include:

Flamelet models, which on a microscopic scale treat the flame front as laminar.

Eddy break-up models, which integrate a volumetric reaction rate, but in doing so yield a flame speed that is dependent on the flame width. This leads to an unstable flame if the flow is divergent, and in consequence the position of the flame front is unrealistic.

Probability density function models, which use an assumed turbulence intensity frequency distribution.

Coherent flame models, which use the product of a flame surface density and the local laminar burning velocity.

The very fact that different approaches are used for modelling turbulent combustion in CFD demonstrates that it is an area still subject to research. This is likely to remain the case for some while, since there is yet to be full agreement on the way to model the turbulence itself. Duclos *et al.* (1996) emphasise that flame–wall interactions are important, since a substantial amount of the reacting mixture is burnt close to the wall (due to the high density of the unburnt mixture compared to the burnt mixture), and this leads to high levels of heat transfer. The wall modifies the flow field, and a difficulty is that the mesh size is unlikely to be fine enough close to the wall for a realistic representation. Turbulence scales diminish close to a wall, so only the smaller length scales can be allowed to influence combustion.

Maly (1998) argues that CFD is able to model in-cylinder flows well, and also gaseous species, but that much work is

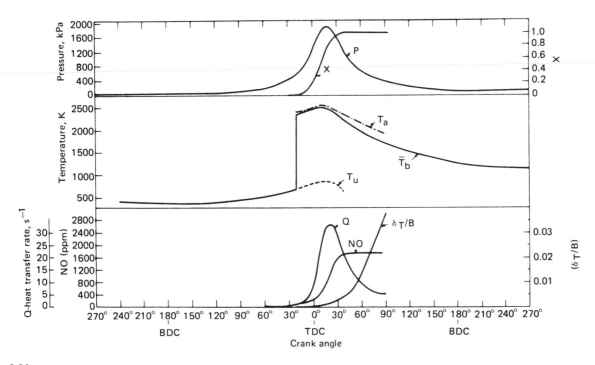

Figure 3.31
Profiles of variables predicted by the simulation throughout the four-stroke engine cycle for 5.7-litre displacement engine at base operating point. Plotted against crank angle are: mass fraction burnt X, unburnt mixture temperature T_u, mean burnt gas temperature $\bar{T}_b$, cylinder pressure P, temperature of burnt gas adiabatic core T_a, instantaneous heat transfer rate Q (normalised by the initial enthalpy of the fuel/air mixture within the cylinder), nitric oxide concentration NO, and thermal boundary layer thickness δ_T (normalised by the cylinder bore B) (reprinted with permission from Heywood *et al.*, 1979, © Society of Automotive Engineers, Inc.).

still needed on combustion modelling, and the modelling of sprays in both direct injection gasoline engines and in diesel engines. For direct injection engines there is also the need to model soot formation.

3.10 | Conclusions

A key difference between spark ignition and compression ignition engines is the difference between pre-mixed combustion in the spark ignition engine, and diffusion-controlled combustion. As a consequence, the two types of engine require different fuel properties. Spark ignition engines require volatile fuels that are resistant to self-ignition, while compression ignition engines require fuels that self-ignite readily. This also leads to the use of different fuel additives.

A proper understanding of normal and abnormal engine combustion follows from fuel and combustion chemistry. Engine emissions are also of great importance, but these are not explained entirely by predictions of dissociation from equilibrium thermodynamics. In particular, the production of nitrogen oxides involves a complex mechanism in which the different forward and reverse reaction rates are critical.

The effect of engine variables such as speed, load and ignition/injection timing on engine emissions and fuel economy is complex, and can only be explained qualitatively without the aid of engine models. In engine models, one of the critical parts is the combustion model and this in turn depends on being able to model the in-cylinder flows and reaction kinetics. Much work is still being devoted to an improved understanding of these aspects.

3.11 | Examples

EXAMPLE 3.1

A fuel has the following gravimetric composition;

hexane (C_6H_{14})	40 per cent
octane (C_8H_{18})	30 per cent
cyclohexane (C_6H_{12})	25 per cent
benzene (C_6H_6)	5 per cent

If the gravimetric air/fuel ratio is 17:1, determine the equivalence ratio.

Solution:

To calculate the equivalence ratio, first determine the stoichiometric air/fuel ratio. As the composition is given gravimetrically, this has to be converted to molar composition. For convenience, take 100 kg of fuel, with molar masses of 12 kg and 1 kg for carbon and hydrogen respectively.

Substance	Mass (m) kg	Molar mass (M) kg	No. of kmols (m/M)
C_6H_{14}	40	$(6 \times 12 + 14 \times 1)$	0.465
C_8H_{18}	30	$(8 \times 12 + 18 \times 1)$	0.263
C_6H_{12}	25	$(6 \times 12 + 12 \times 1)$	0.298
C_6H_6	5	$(6 \times 12 + 6 \times 1)$	0.064

Assuming that the combustion products contain only CO_2 and H_2O, the stoichiometric reactions in terms of kmols are found to be

$$C_6H_{14} + x\left(O_2 + \frac{79}{21}N_2^*\right) \rightarrow 6CO_2 + 7H_2O + yN_2^*$$

Variable x is found from the number of kilomoles of oxygen associated with the products, $x = 9\frac{1}{2}$, and by conservation of N_2^*, $y = (79/21)x$:

$$C_6H_{14} + 9\frac{1}{2}\left(O_2 + \frac{79}{21}N_2^*\right) \rightarrow 6CO_2 + 7H_2O + 35.74N_2^*$$

1 kmol of fuel: $9\frac{1}{2}\left(1 + \frac{79}{21}\right)$ kmols of air $= 1 : 45.23$

By inspection:

$$C_8H_{18} + 12\frac{1}{2}\left(O_2 + \frac{79}{21}N_2^*\right) \rightarrow 8CO_2 + 9H_2O + 47.0N_2^*$$

1 : 59.52 kmols

$$C_6H_{12} + 9\left(O_2 + \frac{79}{21}N_2^*\right) \rightarrow 6CO_2 + 6H_2O + 33.86N_2^*$$

1 : 42.86 kmols

$$C_6H_6 + 7\frac{1}{2}\left(O_2 + \frac{79}{21}N_2^*\right) \rightarrow 6CO_2 + 3H_2O + 28.21N_2^*$$

1 : 35.71 kmols

Fuel	kmols of fuel	Stoichiometric molar air/fuel ratio	kmols of air
C_6H_{14}	0.465	45.23:1	21.03
C_8H_{18}	0.263	59.52:1	15.65
C_6H_{12}	0.298	42.86:1	12.77
C_6H_6	0.064	35.71:1	2.29
Total	100 kg fuel		51.74

100 kg fuel: 51.74×29 kg air

$\qquad 1 : 15.00$

Equivalence ratio, $\phi = \dfrac{\text{(air/fuel ratio) stoichiometric}}{\text{(air/fuel ratio) actual}}$

$$\phi = \frac{15}{17} = 0.882$$

EXAMPLE 3.2

A fuel oil has a composition by weight of 0.865 carbon, 0.133 hydrogen and 0.002 incombustibles. Find the stoichiometric gravimetric air/fuel ratio.

When the fuel is burnt with excess air, the dry volumetric exhaust gas analysis is: CO_2 0.121, N_2^* 0.835, O_2 0.044. Determine the actual air/fuel ratio used and the wet volumetric exhaust gas analysis.

Solution:
Consider 1 kg of fuel, and convert the gravimetric data to molar data. Molar mass of carbon = 12, hydrogen = 1.

$$\frac{0.865}{12}C + \frac{0.133}{1}H + \text{Air} \rightarrow \text{Products}$$

$$0.0721C + 0.133H + a\left(O_2 + \frac{79}{21}N_2^*\right) \rightarrow 0.0721CO_2 + 0.0665H_2O + 0.396N_2^*$$

For a stoichiometric reaction, $a = 0.0721 + \frac{1}{2}(0.0665) = 0.105$.

Thus 1 kg of fuel combines with $0.105\left(1 + \dfrac{79}{21}\right)$ kmol of air

$$\text{or } 0.105\left(1 + \frac{79}{21}\right) \times 29 \text{ kg of air}$$

and the stoichiometric gravimetric air/fuel ratio is 14.5 : 1.
Note that the molar mass of the fuel is not known.

With excess air the equation can be rewritten with variable x. The incombustible material is assumed to occupy negligible volume in the products. A dry gas analysis assumes that any water vapour present in the combustion products has been removed before analysis.

$$0.0721C + 0.133H + 0.105x\, O_2 + 0.396xN_2^* \rightarrow 0.0721CO_2 + 0.0665H_2O$$
$$+ 0.396xN_2^* + 0.105(x-1)O_2$$

The variable x is unity for stoichiometric reactions; if $x > 1$ there will be $0.105(x-1)$ kmols of O_2 not taking part in the reaction with 1 kg fuel.

The dry volumetric gas analysis for each constituent in the products is in proportion to the number of moles of each dry constituent:

$$CO_2 : 0.121 = \frac{0.0721}{0.0721 + 0.396x + 0.105(x - 1)}$$

$$0.0721/0.121 = 0.0721 + (0.396 + 0.105)x - 0.105; \text{hence } \underline{x = 1.255.}$$

$$O_2 : 0.044 = \frac{0.105(x - 1)}{0.0721 + 0.396x + 0.105(x - 1)}$$

$$0.105x - 0.044(0.396 + 0.105)x = 0.044(0.0721 - 0.105) + 0.105;$$

$$\text{hence } \underline{x = 1.248.}$$

$$N_2^* : 0.835 = \frac{0.396x}{0.0721 + 0.396x + 0.105(x - 1)}$$

$$\left(0.396 + 0.105 - \frac{0.396}{0.835}\right)x = 0.105 - 0.0721; \text{hence } \underline{x = 1.230.}$$

As might be expected, each value of x is different. The carbon balance is most satisfactory since it is the largest measured constituent of the exhaust gases. The nitrogen balance is the least accurate, since it is found by difference ($0.835 = 1 - 0.044 - 0.121$). The redundancy in these equations can also be used to determine the gravimetric composition of the fuel.

Taking $x = 1.25$ the actual gravimetric air/fuel ratio is

$$14.5 \times 1.25 : 1 = \underline{18.13 : 1}$$

The combustion equation is actually

$$0.0721C + 0.0133H + 0.1313O_2 + 0.495N_2^* \rightarrow 0.0721CO_2 + 0.0665H_2O$$
$$+ 0.495N_2^* + 0.0263O_2$$

The wet volumetric analysis for each constituent in the products is in proportion to the number of moles of each constituent, including the water vapour

$$CO_2 : \frac{0.0721}{0.0721 + 0.0665 + 0.495 + 0.0263} = \frac{0.0721}{0.6599} = \underline{0.1093}$$

$$H_2O : \frac{0.0665}{0.6599} = \underline{0.1008}$$

$$N_2^* : \frac{0.495}{0.6599} = \underline{0.7501}$$

$$O_2 : \frac{0.0263}{0.6599} = \underline{0.0399}$$

Check that the sum equals unity ($0.1093 + 0.1008 + 0.7501 + 0.0399 = 1.0001$).

EXAMPLE 3.3A

Calculate the difference between the constant-pressure and constant-volume lower calorific values (LCV) for ethylene (C_2H_4) at 250°C and 1 bar. Calculate the constant-pressure calorific value at 2000 K. What is the constant-pressure higher calorific value (HCV) at 25°C? The ethylene is gaseous at 25°C, and the necessary data are in the tables by Rogers and Mayhew (1980b):

$$C_2H_4(\text{vap.}) + 3O_2 \rightarrow 2CO_2 + 2H_2O(\text{vap.}), (\Delta H_0)_{25°C} = -1\,323\,170 \text{ kJ/kmol}$$

Solution:
(i) Using equation (3.4):

$$(-\Delta H_0)_{25°C} - (-\Delta U_0)_{25°C} = (n_{GR} - n_{GP})R_0 T$$

At 25°C the partial pressure of water vapour is 0.032 bar, so it is fairly accurate to assume that all the water vapour condenses; thus

n_{GR}, number of moles of gaseous reactants $= 1 + 3$

n_{GP}, number of moles of gaseous products $= 2 + 0$

$(n_{GR} - n_{GP})R_0 T = (4 - 2) \times 8.3144 \times 298.15 = 4.958$ kJ/kmol C_2H_4

This difference in calorific values can be seen to be negligible. If the reaction occurred with excess air the result would be the same, as the atmospheric nitrogen and excess oxygen take no part in the reaction.

(ii) To calculate $(-\Delta H_0)_{2000\,K}$, use can be made of equation (3.3); this amounts to the same as the following approach:

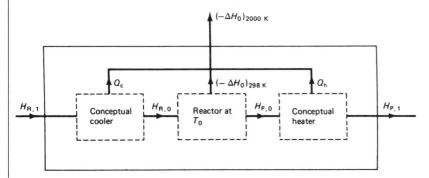

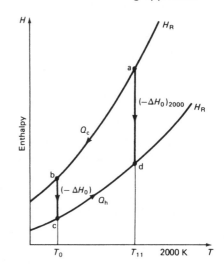

$$(-\Delta H_0)_{2000\,K} = Q_c + (-\Delta H_0)_{298\,K} + Q_h$$
$$a \rightarrow d \equiv a \rightarrow b \rightarrow c \rightarrow d$$

$$Q_c = (H_{2000\,K} - H_{298\,K})_{C_2H_4} + 3(H_{2000\,K} - H_{298\,K})_{O_2}$$
$$= (159\,390 - 0) + 3(59\,199 - 0)$$
$$= 336\,987$$

$$Q_R = 2(H_{2000\,K} - H_{298\,K})_{CO_2} + 2(H_{2000\,K} - H_{298\,K})_{H_2O}$$
$$= 2(91\,450 - 0) + 2(72\,689 - 0) = 328\,278$$

$$(HCV)_{2000\,K} = (-\Delta H_0)_{2000\,K} = 336\,987 + 1\,323\,170 - 328\,278$$
$$= \underline{1\,331\,879}\ \text{kJ/kmol}\ C_2H_4$$

Again this is not a significant difference. As before, if the reaction occurred with excess air the result would be the same, as the excess oxygen and atmospheric nitrogen take no part in the reaction, and have the same initial and final state.

(iii) As the enthalpy of reaction is for water in the vapour state, this corresponds to the lower calorific value.
Using equation (3.5)

$$HCV - LCV = yh_{fg}$$

where $y = $ mass of water vapour/kmol of fuel or

$$HCV - LCV = zh_{fg}$$

where $z = $ number of kmols of water vapour/kmol fuel.

$H_{fg} = 43\,990$ kJ/kmol of H_2O at 25°C

Thus

$$HCV = LCV + 2H_{fg}$$
$$= 1\,323\,170 + 2 \times 43\,990 = 1\,411\,150\ \text{kJ/kmol fuel}$$

EXAMPLE 3.3B

(i) Verify that the difference between the constant-pressure and constant-volume calorific values (LCV) for propane at 25°C and 1 bar are self-consistent using data from tables A.4 and A.5 (Appendix A).

(ii) Compare the constant-pressure calorific values (LCV) for propane at 25°C and 1000 K at 1 bar, and comment on the differences.

(iii) What is the constant-pressure higher calorific value (HCV) for propane at 25°C?

The propane is gaseous at 25°C, and the stoichiometric combustion equation is

$$C_3H_8 + 5O_2 \rightarrow 3CO_2 + 4H_2O$$

Solution:

(i) The lower calorific value means that the water vapour in the products of combustion is to be assumed to be in the vapour phase. For the constant-pressure calorific value:

$$(-\Delta H_0)_{25°C} = \Sigma H_R - \Sigma H_P = H_{C_3H_8} + 5H_{O_2} - (3H_{CO_2} + 4H_{H_2O})$$

Using data from tables A.4 and A.5:

$$(-\Delta H_0)_{25°C} = -104.68 + 5 \times 0 - (3 \times -393.512 + 4 \times -241.824)$$
$$= 2043.15 \text{ MJ/kmol } C_3H_8$$

(which agrees with the value in table A.3).

For the constant-volume calorific value:

$$(-\Delta U_0)_{25°C} = \Sigma U_R - \Sigma U_P = U_{C_3H_8} + 5U_{O_2} - (3U_{CO_2} + 4U_{H_2O})$$

Using data from tables A.4 and A.5:

$$(-\Delta U_0)_{25°C} = -107.159 + 5 \times -2.479 - (3 \times -395.991 + 4 \times -244.303)$$
$$= 2045.631 \text{ MJ/kmol } C_3H_8$$

Using equation (3.4):

$$(-\Delta U_0)_{25°C} - (-\Delta H_0)_{25°C} = (n_{GR} - n_{GP})R_0 T$$
$$(-\Delta U_0)_{25°C} - (-\Delta H_0)_{25°C} = 2043.152 - 2045.631 = -2.479 \text{ MJ/kmol } C_3H_8$$

and

$$(n_{GR} - n_{GP})R_0 T = (1 + 5 - [3 + 4])8.3143 \times 298.15 = -2478.9 \text{ kJ/kmol } C_3H_8$$

(ii) From part (i):

$$(-\Delta H_0)_{25°C} = 2043.15 \text{ MJ/kmol } C_3H_8$$

At 1500 K:

$$(-\Delta H_0)_{1500 K} = \Sigma H_R - \Sigma H_P = H_{C_3H_8} + 5H_{O_2} - (3H_{CO_2} + 4H_{H_2O})$$
$$(-\Delta H_0)_{1500 K} = 84.819 + 5 \times 40.571 - (3 \times -331.867 + 4 \times -193.711)$$
$$= 2058.119 \text{ MJ/kmol } C_3H_8$$

Thus the enthalpy of the reactants must have increased more with temperature than the enthalpy of the products. In other words, the reactants have a steeper gradient on the enthalpy–temperature plot (figure 3.8) than the products.

(iii) The higher calorific value (HCV) is when the water vapour in the products of combustion has been condensed to its liquid state, thus

$$HCV = LCV + (n \times H_{fg})_{H_2O} = 2043.15 + 4 \times 43.99 = 2219.11 \text{ MJ/kmol } C_3H_8$$

EXAMPLE 3.4A

In a closed combustion vessel, propane (C_3H_8) and air with an equivalence ratio of 1.11 initially at 25°C burn to produce products consisting solely of carbon dioxide (CO_2), carbon monoxide (CO), water (H_2O) and atmospheric nitrogen. If the heat rejected from the vessel is 770 MJ per kmol of fuel, show that the final temperature is about 1530°C. If the initial pressure is 1 bar, estimate the final pressure. Lower calorific value of propane $(-\Delta H_0)_a$ is 2 039 840 kJ/kmol of fuel at 25°C.

Solution:
As neither oxygen nor hydrogen is present in the combustion products, this implies that dissociation should be neglected. The stoichiometric reaction is

$$C_3H_8 + 5\left(O_2 + \frac{79}{21}N_2^*\right) \rightarrow 3CO_2 + 4H_2O + 5 \times \frac{79}{21}N_2^*$$

In this problem the equivalence ratio is 1.11, thus

$$C_3H_8 + 4.5\left(O_2 + \frac{79}{21}N_2^*\right) \rightarrow (3-x)CO_2 + xCO + 4H_2O + 16.93N_2^*$$

The temporary variable, x, can be eliminated by an atomic oxygen balance

$$4.5 \times 2 = (3-x) \times 2 + x + 4$$

Thus $x = 1$, and the actual combustion equation is

$$C_3H_8 + 4.5\left(O_2 + \frac{79}{21}N_2^*\right) \rightarrow 2CO_2 + CO + 4H_2O + 16.93N_2^*$$

Let the enthalpy of reaction be $(\Delta H_0)_b$ for this reaction.
If $(\Delta H_0)_c$ is the enthalpy of reaction for

$$CO + \tfrac{1}{2}O_2 \rightarrow CO_2$$

then $(\Delta H_0)_b = (\Delta H_0)_a - (\Delta H_0)_c$
   Using data from Rogers and Mayhew (1980b)

$$(\Delta H_0)_b = -2\,039\,840 - (-282\,990) = -1\,756\,850 \text{ kJ/kmol of fuel at 25°C}$$

Equation (3.4) converts constant-pressure to constant-volume calorific values:

where the number of kmols of gaseous reactants $n_{GR} = 1 + 4.5 \times \left(\frac{100}{21}\right)$

$$= 22.43$$

and the number of kmols of gaseous products $n_{GP} = 2 + 1 + 4 + 16.93$

$$= 23.93$$

$$\begin{aligned}(CV)_V &= (CV)_P - (n_{GR} - n_{GP})R_0 T \\ &= 1\,756\,850 - \{(22.43 - 23.93) \times 8.3144 \times 298.15\} \\ &= 1\,760\,568 \text{ kJ/kmol fuel at 25°C}\end{aligned}$$

Assume the final temperature is 1800 K and determine the heat flow. Applying the First Law of Thermodynamics

$$\begin{aligned}(CV)_V &= 2(U_{1800} - U_{298})_{CO_2} + 1(U_{1800} - U_{298})_{CO} + 4(U_{1800} - U_{298})_{H_2O} \\ &\quad + 16.93(U_{1800} - U_{298})_{N_2^*} - Q \\ 1\,760\,568 &= 2(64\,476 + 2479) + 1(34\,556 + 2479) + 4(47\,643 + 2479) \\ &\quad + 16.93(34\,016 + 2479) - Q \\ Q &= -771\,275 \text{ kJ/kmol fuel}\end{aligned}$$

The negative sign indicates a heat flow from the vessel, and the numerical value is sufficiently close to 770 MJ.

To estimate the final pressure, apply the equation of state; use a basis of 1 kmol of fuel:

$$pV = nR_0T$$

Find V for the reactants

$$V = \frac{n_R R_0 T}{P} = \frac{(1 + 4.5 + 1693)R_0 298}{10^5}$$

For the products

$$P = \frac{n_R R_0 T}{V}$$

$$= \frac{(2 + 1 + 4 + 16.93)R_0\,1800}{(1 + 4.5 + 16.93)R_0\,298} \times 10^5$$

$$= \underline{6.44\ bar}$$

EXAMPLE 3.4B

In a closed combustion vessel, propane (C_3H_8) and air with an equivalence ratio of 1.11 initially at 25°C burn to produce products consisting solely of carbon dioxide (CO_2), carbon monoxide (CO), water (H_2O) and atmospheric nitrogen. If the heat rejected from the vessel is 820 MJ per kmol of fuel, show that the final temperature is about 1800 K. If the initial pressure is 1 bar, estimate the final pressure. As neither oxygen nor hydrogen is present in the combustion products, this implies that dissociation should be neglected. The stoichiometric reaction is

$$C_3H_8 + 5\left(O_2 + \frac{79}{21}N_2^*\right) \rightarrow 3CO_2 + 4H_2O + 5 \times \frac{79}{21}N_2^*$$

Solution:
In this problem the equivalence ratio is 1.11, thus

$$C_3H_8 + 4.5\left(O_2 + \frac{79}{21}N_2^*\right) \rightarrow (3 - x)CO_2 + xCO + 4H_2O + 16.93N_2^*$$

The temporary variable, x, can be eliminated by an atomic oxygen balance

$$4.5 \times 2 = (3 - x) \times 2 + x + 4$$

Thus $x = 1$, and the actual combustion equation is

$$C_3H_8 + 4.5\left(O_2 + \frac{79}{21}N_2^*\right) \rightarrow 2CO_2 + CO + 4H_2O + 16.93N_2^*$$

We can apply the First Law of Thermodynamics to this reaction (remembering that it is occurring at constant volume):

$$\Sigma U_R + Q = \Sigma U_P$$

$$\Sigma(U_R)_{298.15\,K} = U_{C_3H_8} + 4.5U_{O_2} + 16.93U_{N_2}$$

$$= -107.159 + 4.5 \times (-2.479) + 16.93 \times (-2.480)$$

$$= -160.301\ MJ/kmol\ C_3H_8$$

$$\Sigma(U_P)_{1800\,K} = 2U_{CO_2} + U_{CO} + 4U_{H_2O} + 16.93U_{N_2}$$

$$= 2 \times (-329.117) + (-76.003) + 4 \times (-194.178) + 16.93 \times 33.983$$

$$= -935.617\ MJ/kmol\ C_3H_8$$

$$Q = \Sigma U_P - \Sigma U_R = -935.617 - (-160.301)$$

$$= -775.316\ MJ/kmol\ C_3H_8$$

The negative sign indicates a heat flow from the vessel, and the numerical value is sufficiently close to 820 MJ.

To estimate the final pressure, apply the equation of state; use a basis of 1 kmol of fuel:

$$pV = nR_0T$$

Find V for the reactants

$$V = \frac{n_R R_0 T}{p} = \frac{(1 + 4.5 + 16.93)R_0\ 298}{10^5}$$

For the products

$$p = \frac{n_R R_0 T}{V}$$
$$= \frac{(2 + 1 + 4 + 16.93)R_0\ 1800}{(1 + 4.5 + 16.93)R_0\ 298} \times 10^5$$
$$= \underline{6.44\ bar}$$

Note

The assumption of no hydrogen in the products of combustion is unrealistic and might lead to a significant error; this can be quantified by using STANJAN. This shows that there would be 0.298 kmol of hydrogen present, with equal reductions in the quantities of water vapour and carbon monoxide, and an equal increase in the quantity of carbon dioxide. This can be seen from the water–gas equilibrium equation:

$$CO + H_2O \rightleftharpoons CO_2 + H_2$$

As there is no change in the total number of moles of gas, then the final pressure will not be affected. However, the energy balance is modified, and the heat transferred would have to be increased by 8 MJ if the temperature was to remain at 1800 K.

EXAMPLE 3.5A

Compute the partial pressures of a stoichiometric equilibrium mixture of CO, O_2, CO_2 at 3000 and 3500 K when the pressure is 1 bar. Show that for a stoichiometric mixture of CO and O_2 initially at 25°C and 1 bar, the adiabatic constant-pressure combustion temperature is about 3050 K.

Solution:
Introduce a variable x into the combustion equation to account for dissociation

$$CO + \tfrac{1}{2}O_2 \rightarrow (1 - x)CO_2 + xCO + \frac{x}{2}O_2$$

The equilibrium constant K_p is defined by equation (3.7):

$$K_p = \frac{p'_{CO_2}}{(p'_{CO})\left(p'_{O_2}\right)^{\frac{1}{2}}}$$ where p'_{CO_2} is the partial pressure of CO_2 etc.

The partial pressure of a species is proportional to the number of moles of that species. Thus

$$p'_{CO_2} = \frac{1 - x}{(1 - x) + x + \dfrac{x}{2}}p, \qquad p'_{CO} = \frac{x}{1 + \dfrac{x}{2}}p, \qquad p'_{O_2} = \frac{x/2}{1 + \dfrac{x}{2}}p$$

where p is the total system pressure. Thus

$$K_p = \frac{1 - x}{x(x/2)^{\frac{1}{2}}}\left(\frac{1 + (x/2)}{p}\right)^{\frac{1}{2}}$$

Again, data are obtained from Rogers and Mayhew (1980b). At 3000 K, $\log_{10} K = 0.485$ with pressure in atmospheres:

1 atmosphere = 1.01325 bar

Thus

$$10^{0.485} = \frac{1-x}{x\sqrt{\left(\frac{x}{2}\right)}} \sqrt{\left(\frac{1+x/2}{1.01325^{-1}}\right)}$$

$$10^{0.485}(1.01325)^{-\frac{1}{2}}x^{\frac{3}{2}} = \sqrt{(2)}(1-x)\sqrt{(1+(x/2))}$$

Square all terms

$$10^{0.97}(1.01325)^{-1}x^3 = (1-x)^2(2+x)$$

$$8.211x^3 + 3x - 2 = 0$$

This cubic equation has to be solved iteratively. The Newton–Raphson method provides fast convergence; an alternative simpler method, which may not converge, is the following method (noting that $0 < x < 1$)

$$x_{n+1} = \sqrt[3]{\left(\frac{2-3x_n}{8.211}\right)} \text{ where } n \text{ is the iteration number}$$

Using a calculator without memory this readily gives

$$x = 0.436$$

At 3500 K use linear interpolation to give $\log_{10} K = -0.187$, leading to

$$x = 0.748$$

This implies greater dissociation at the higher temperature, a result predicted by Le Châtelier's Principle for an exothermic reaction.

The partial pressures are found by back-substitution, giving the results tabulated below:

T	p	x	p'_{CO_2}	p'_{O_2}	p'_{CO}	
3000 K	1	0.436	0.463	0.179	0.358	all pressures in bar
3500 K	1	0.748	0.183	0.272	0.544	

To find the adiabatic combustion temperature, assume a temperature and then find the heat transfer necessary for the temperature to be attained. For the reaction $CO + \frac{1}{2}O_2 \rightarrow CO_2$, $(\Delta H_0)_{298} = -282\,990$ kJ/kmol. Allowing for dissociation, the enthalpy for reaction is $(1-x)(\Delta H_0)_{298}$.

It is convenient to use the following hypothetical scheme for the combustion calculation

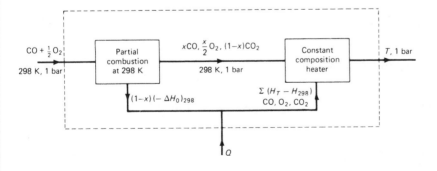

$$(1-x)(-\Delta H_0)_{298} + Q = \sum_{CO,O_2,CO_2} (H_T - H_{298})$$

If temperature T is guessed correctly then $Q = 0$. Alternatively, make two estimates and interpolate.
1st guess: $T = 3000$ K, $x = 0.436$.

$$Q = \left\{(1-x)(H_{3000})_{CO_2} + x(H_{3000})_{CO} + \frac{x}{2}(H_{3000})_{O_2}\right\} - (1-x)(-\Delta H_0)_{298}; H_{298} = 0$$

$$= \{0.564 \times 152\,860 + 0.436 \times 93\,542 + 0.218 \times 98\,098\} - 0.564 \times 282\,990$$

$$= -11\,223 \text{ kJ/kmol CO, that is, } T \text{ is too low.}$$

2nd guess: $T = 3500$ K, $x = 0.748$.

$$Q = \{0.252 \times 184\,130 + 0.748 \times 112\,230 + 0.374 \times 118\,315\} - 0.252 \times 282\,290$$

$$= 103\,462 \text{ kJ/kmol CO, that is, } T \text{ is much too high.}$$

To obtain a better estimate of T, interpolate between the 1st and 2nd guesses:

$$\frac{T_e - T_1}{Q - Q_1} = \frac{T_2 - T_1}{Q_2 - Q_1}$$

$$T_e = T_1 + (T_2 - T_1)\left(\frac{-Q_1}{Q_2 - Q_1}\right)$$

$$T_e = 3000 + 500\frac{11\,223}{103\,462 + 11\,223}$$

$$= \underline{3049 \text{ K}} \quad \text{Q.E.D.}$$

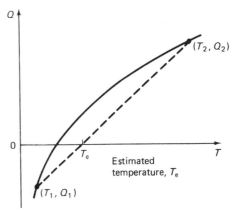

If a more accurate result is needed, recalculate Q when $T = 3050$ K. Assuming no dissociation, $T \approx 5000$ K.

EXAMPLE 3.5B

Compute the partial pressures of a stoichiometric equilibrium mixture of CO, O$_2$, CO$_2$ at 3000 and 3500 K when the pressure is 1 bar. Show that for a stoichiometric mixture of CO and O$_2$ initially at 25°C and 1 bar, the adiabatic constant-pressure combustion temperature is about 3050 K.

Solution:
Introduce a variable x into the combustion equation to account for dissociation

$$CO + \tfrac{1}{2}O_2 \rightarrow (1-x)CO_2 + xCO + \frac{x}{2}O_2$$

The equilibrium constant K is defined by equation (3.7):

$$K_p = \frac{p'_{CO_2}}{(p'_{CO})\left(p'_{O_2}\right)^{\frac{1}{2}}} \text{ where } p'_{CO_2} \text{ is the partial pressure of CO}_2 \text{ etc.}$$

The partial pressure of a species is proportional to the number of moles of that species. Thus

$$p'_{CO_2} = \frac{1-x}{(1-x)+x+\dfrac{x}{2}}p, \qquad p'_{CO} = \frac{x}{1+\dfrac{x}{2}}p, \qquad p'_{O_2} = \frac{x/2}{1+\dfrac{x}{2}}p$$

where p is the total system pressure. Thus

$$K_p = \frac{1-x}{x(x/2)^{\frac{1}{2}}}[1 + (x/2)]^{\frac{1}{2}}$$

since the system pressure was 1 bar, and the values of K_p in table A.6 are for a system pressure of 1 bar.

At 3000 K, ln $K_p = 1.108$ (reaction 6 in table A.6), thus:

$$\exp(1.108) = \frac{1-x}{x\sqrt{\left(\frac{x}{2}\right)}} - \sqrt{(1+x/2)}$$

$$x^{\frac{3}{2}} = \sqrt{(2)}(1-x)\sqrt{(1+(x/2))}$$

Square all terms

$$9.171x^3 = (1-x)^2(2+x)$$

$$8.171x^3 + 3x - 2 = 0$$

This cubic equation has to be solved iteratively. The Newton–Raphson method provides fast convergence; an alternative simpler method, which may not converge, is the following method (noting that $0 < x < 1$)

$$x_{n+1} = \sqrt[3]{\left(\frac{2-3x_n}{8.171}\right)} \text{ where } n \text{ is the iteration number}$$

Using a calculator without memory this readily gives

$$x = 0.438$$

At 3500 K, ln $K_p = -0.447$, leading to

$$x = 0.750$$

This implies greater dissociation at the higher temperature, a result predicted by Le Châtelier's Principle for an exothermic reaction.

The partial pressures are found by back-substitution, giving the results tabulated below:

T	p	x	p'_{CO_2}	p'_{O_2}	p'_{CO}	
3000 K	1	0.438	0.461	0.180	0.359	all pressures in bar
3500 K	1	0.750	0.182	0.273	0.545	

To find the adiabatic combustion temperature, assume a temperature and then find the heat transfer necessary for the temperature to be attained. As this is a constant-pressure combustion process, we need to apply the First Law of Thermodynamics in its steady-flow form for the reaction:

$$CO + \tfrac{1}{2}O_2 \rightarrow (1-x)CO_2 + xCO + x/2\,CO$$

$$\Sigma H_R + Q = \Sigma H_P$$

$$\Sigma(H_R)_{298.15\,K} = H_{CO} + 0.5H_{O_2} = -110.525 + \tfrac{1}{2}(0.0)$$
$$= -110.525 \text{ MJ/kmol CO}$$

$$\Sigma(H_P)_{3000\,K} = xH_{CO} + 0.5xH_{O_2} + (1-x)H_{CO_2}$$
$$= 0.438 \times (-16.969) + 0.219 \times (98.116) + 0.562 \times (-240.621)$$
$$= -121.174 \text{ MJ/kmol CO}$$

$$Q_1 = \Sigma H_P - \Sigma H_R = -121.174 - (-110.525) = -10.649 \text{ MJ/kmol CO}$$

As Q_1 is negative, then 3000 K is too low.

$$\Sigma(H_P)_{3000\,K} = xH_{CO} + 0.5xH_{O_2} + (1-x)H_{CO_2}$$
$$= 0.750 \times 1.706 + 0.375 \times (118.312) + 0.250 \times (-209.383)$$
$$= 6.699 \text{ MJ/kmol CO}$$

$$Q_2 = \Sigma H_P - \Sigma H_R = 6.699 - (-110.525) = 117.224 \text{ MJ/kmol CO}$$

As Q_2 is positive, then 3500 K is too high.

To obtain a better estimate of T, interpolate between the 1st and 2nd guesses:

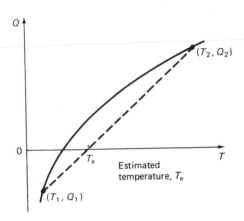

$$\frac{T_e - T_1}{Q - Q_1} = \frac{T_2 - T_1}{Q_2 - Q_1}$$

$$T_e = T_1 + (T_2 - T_1)\left(\frac{-Q_1}{Q_2 - Q_1}\right)$$

$$T_e = 3000 + 500\frac{10.649}{117.224 + 10.649}$$

$$= \underline{3042 \text{ K}} \quad \text{Q.E.D.}$$

If a more accurate result is needed, recalculate Q when $T = 3050$ K. Assuming no dissociation, $T \approx 5000$ K.

Discussion

The accuracy of this calculation can be checked using STANJAN, which yields the following results:

$T = 3042$ K, with 0.468 kmol CO

Interpolation of x between the two temperatures gives 0.464 kmol CO, which is reasonable, since the composition (unlike the enthalpy) is a non-linear function of temperature. Using STANJAN with atomic oxygen (O) included as a possible product of combustion gives

$T = 2976$ K, with 0.435 kmol CO and 0.051 kmol O

Thus the errors in using linear interpolation are much smaller than the errors caused by omitting minor species.

EXAMPLE 3.6A

In fuel-rich combustion product mixtures, equilibrium between the species CO_2, H_2O, CO and H_2 is often assumed to be the sole determinant of the burned gas composition. If methane is burnt in air with an equivalence ratio of 1.25, determine the molar fractions of the products at 1800 K and a pressure of 1 bar.

Comment on any other likely dissociation reactions, whether or not this is likely to be significant, and whether it is reasonable to assume equilibrium conditions.

Solution:
First, establish stoichiometric combustion by introducing a temporary variable x:

$$CH_4 + x\left(O_2 + \frac{79}{21}N_2\right) \rightarrow CO_2 + 2H_2O + x\frac{79}{21}N_2$$

From the molecular oxygen balance: $x = 2$.

Thus, for the combustion of a methane/air mixture with an equivalence ratio of 1.25:

$$CH_4 + \frac{2}{1.25}\left(O_2 + \frac{79}{21}N_2\right) \rightarrow aCO_2 + bCO + cH_2O + dH_2 + 6.02N_2$$

There are now four unknown variables, and the atomic balance equations can be used to give three independent equations:

carbon balance	$a + b = 1$;	$b = 1 - a$
hydrogen balance	$2c + 2d = 4$;	$d = 2 - c$
oxygen balance	$2a + b + c = 3.2$;	

Substituting for b from the carbon balance:

$2a + (1 - a) + c = 3.2; \quad a = 2.2 - c$

Or if the carbon balance is used to eliminate a:

$b = c - 1.2$

The fourth equation has to come from the water–gas equilibrium, as the nitrogen balance is not independent of the other atomic balances:

For $CO_2 + H_2 \rightleftharpoons CO + H_2O$ at 1800 K, $\quad \log_{10} K_p = 0.577$

Using equation (3.6)

$$\frac{\left(p'_{H_2O}\right)\left(p'_{CO}\right)}{\left(p'_{H_2}\right)\left(p'_{CO_2}\right)} = 3.776$$

As the stoichiometric coefficients are all unity, the system pressure is not relevant, nor is the total number of moles. Thus

$b \times c = 3.776 \times a \times d$

Substituting in terms of c:

$$(c - 1.2) \times c = 3.776(2.2 - c)(2 - c)$$
$$c^2 - 1.2c = 16.61 - 15.68c + 3.776c^2$$
$$2.776c^2 - 14.66c + 16.61 = 0$$

and solving

$$c = \frac{14.66 \pm \sqrt{(14.66^2 - 4 \times 2.776 \times 16.61)}}{2 \times 2.776}$$
$$c = 3.635 \quad \text{or} \quad c = 1.646$$

To establish which solution is valid, consider the result of the oxygen balance: as $a > 0$, then $c < 2.2$. Therefore, the required solution is $c = 1.646$. Back-substitution then gives from:

the hydrogen balance $\qquad d = 0.354$
the carbon/oxygen balance $\qquad a = 2.2 - c = 0.554$
$\qquad\qquad\qquad\qquad\qquad b = c - 1.2 = -0.446$

(As a check $(b \times c)/(d \times a) = 3.74$, which is close enough.)
   The full combustion equation is thus

$$CH_4 + 1.6\left(O_2 + \frac{79}{21}N_2\right) \rightarrow 0.554CO_2 + 0.446CO + 1.646H_2O$$
$$+ 0.354H_2 + 6.02N_2$$

At 1800 K, the water will be in its vapour state and the molar fractions are:

$\%CO_2 = a/\Sigma n_i = 0.554/9.02 = 6.14$ per cent
$\%CO = 0.446/9.02 = 4.94$ per cent
$\%H_2O = 18.25$ per cent
$\%H_2 = 3.92$ per cent
$\%N_2 = 66.7$ per cent

A possible additional equilibrium reaction is:

$$CO + \tfrac{1}{2}O_2 \rightleftharpoons CO_2$$

At 1800 K:

$$\frac{p'_{CO_2}}{p'_{CO}\sqrt{p'_{O_2}}} = 4932 \text{ atm}^{-\frac{1}{2}}$$

Assume $p = 1$ atm. p'_{O_2} is so small as not to affect p'_{CO} or p'_{CO_2}:

$$p'_{O_2} \approx \left(\frac{0.028}{4932 \times 1.916}\right)^2 = 8 \times 10^{-12} \text{ atm}$$

However, equilibrium conditions at 1800 K are unlikely to pertain, as the composition will: (a) not be in equilibrium, and (b) be frozen at its composition determined at a higher temperature before blow-down.

EXAMPLE 3.6B

In fuel-rich combustion product mixtures, equilibrium between the species CO_2, H_2O, CO and H_2 is often assumed to be the sole determinant of the burned gas composition. If methane is burnt in air with an equivalence ratio of 1.25, determine the molar fractions of the products at 1800 K and a pressure of 1 bar.

Comment on any other likely dissociation reactions, whether or not this is likely to be significant, and whether it is reasonable to assume equilibrium conditions.

Solution:
First, establish stoichiometric combustion by introducing a temporary variable x:

$$CH_4 + x\left(O_2 + \frac{79}{21}N_2\right) \rightarrow CO_2 + 2H_2O + x\frac{79}{21}N_2$$

From the molecular oxygen balance: $x = 2$.

Thus, for the combustion of a methane/air mixture with an equivalence ratio of 1.25:

$$CH_4 + \frac{2}{1.25}\left(O_2 + \frac{79}{21}N_2\right) \rightarrow aCO_2 + bCO + cH_2O + dH_2 + 6.02N_2$$

There are now four unknown variables, and the atomic balance equations can be used to give three independent equations:

carbon balance	$a + b = 1$;	$b = 1 - a$
hydrogen balance	$2c + 2d = 4$;	$d = 2 - c$
oxygen balance	$2a + b + c = 3.2$;	

Substituting for b from the carbon balance:

$$2a + (1 - a) + c = 3.2; \quad a = 2.2 - c$$

Or if the carbon balance is used to eliminate a:

$$b = c - 1.2$$

The fourth equation has to come from the water–gas equilibrium, as the nitrogen balance is not independent of the other atomic balances:

For $CO_2 + H_2 - CO - H_2O = 0$ at 1800 K, $\quad \ln K_p = -1.330$

Using equation (3.6)

$$\frac{\left(p'_{H_2}\right)\left(p'_{CO_2}\right)}{\left(p'_{H_2O}\right)\left(p'_{CO}\right)} = 0.265$$

As the stoichiometric coefficients are all unity, the system pressure is not relevant, nor is the total number of moles. Thus

$$0.265 \times b \times c = a \times d$$

Substituting in terms of c:

$$0.265(c - 1.2) \times c = (2.2 - c)(2 - c)$$
$$0.265c^2 - 0.318c = 4.4 - 4.2c + c^2$$
$$0.735c^2 - 3.882c + 4.4 = 0$$

and solving

$$c = \frac{3.882 \pm \sqrt{(3.882^2 - 4 \times 0.735 \times 4.4)}}{2 \times 0.735}$$
$$c = 3.635 \quad \text{or} \quad c = 1.647$$

To establish which solution is valid, consider the result of the oxygen balance: as $a > 0$, then $c < 2.2$. Therefore, the required solution is $c = 1.647$. Back-substitution then gives from:

the hydrogen balance $\qquad\qquad$ $d = 0.353$
the carbon/oxygen balance $\qquad$ $a = 2.2 - c = 0.553$
$\qquad\qquad\qquad\qquad\qquad\quad$ $b = c - 1.2 = -0.447$

(As a check $(d \times a)/(b \times c) = 0.266$, which is close enough.)
   The full combustion equation is thus

$$CH_4 + 1.6\left(O_2 + \frac{79}{21}N_2\right) \rightarrow 0.553CO_2 + 0.447CO + 1.647H_2O + 0.353H_2 + 6.02N_2$$

At 1800 K, the water will be in its vapour state and the molar fractions are:

$\%CO_2 = a/\Sigma n_i = 0.553/9.02 = 6.13$ per cent
$\%CO = 0.447/9.02 = 4.96$ per cent
$\%H_2O = 18.26$ per cent
$\%H_2 = 3.91$ per cent
$\%N_2 = 66.7$ per cent

A possible additional equilibrium reaction is:

$$CO - \tfrac{1}{2}O_2 + CO_2 = 0$$

At 1800 K for reaction 6 in table A.6

$\ln K_p = 8.494$
thus $\qquad$ $\dfrac{p'_{CO_2}}{p'_{CO}\sqrt{p'_{O_2}}} = 4885\dfrac{1}{\sqrt{\text{bar}}}$

Assume that p'_{O_2} is so small as not to affect p'_{CO} or p'_{CO_2}:

$$p'_{O_2} \approx \left(\frac{0.0613}{4885 \times 0.0496}\right)^2 = 6.4 \times 10^{-8} \text{ bar}$$

However, equilibrium conditions at 1800 K are unlikely to pertain, as the composition will: (a) not be in equilibrium, and (b) be frozen at its composition determined at a higher temperature before blow-down.

EXAMPLE 3.7

A gas engine operating on methane (CH_4) at 1500 rpm, full throttle, generates the following emissions measured on a dry volumetric basis:

CO_2	10.4 per cent
CO	1.1 per cent
H_2	0.6 per cent
O_2	0.9 per cent
NO	600 ppm
HC	1100 ppm (as methane)

If the specific fuel consumption is 250 g/kWh, calculate the specific emissions of carbon monoxide, nitrogen oxide and unburned hydrocarbons (that is, g/kWh). Why is this specific basis more relevant than a percentage basis?

Solution:

Consider the combustion of sufficient fuel to produce 100 kmol of dry products. Introduce the following temporary variables:

y kmol of CH_4 and x kmol of O_2 in the reactants
a kmol of H_2O and b kmol of N_2 in the products

$$yCH_4 + x\left(O_2 + \frac{79}{21}N_2\right) \rightarrow 1.1CO + 10.4CO_2 + 0.6H_2$$
$$+ 0.9O_2 + 0.11CH_4 + 0.06NO + aH_2O + bN_2$$

The quantity of N_2 is known, as the products add to 100 kmol. From the nitrogen balance:

$$b = 100 - (1.1 + 10.4 + 0.6 + 0.9 + 0.11 + 0.06) = 86.83$$

To find y, use the C balance:

$$y = 1.1 + 10.4 + 0.11 = 11.61$$

and find a from the H_2 balance:

$$2 \times 11.61 = 0.6 + 2 \times 0.11 + a$$
$$a = 22.4$$

Find x from the O_2 balance (or the N_2 balance – this will be used as a check):

$$x = \tfrac{1}{2} \times 1.1 + 10.4 + 0.9 + \tfrac{1}{2} \times 0.06 + \tfrac{1}{2} \times 22.4$$
$$x = 23.085$$

(Checking: $x \times \frac{79}{21} = 0.06/2 + 86.83$; $x = 23.089$, good enough.)
   Thus 11.61 kmol of $CH_4 \rightarrow 1.1$ kmol of CO etc., or

$$11.61 \times (12 + 4 \times 1) \text{ kg of } CH_4 \rightarrow 1.1 \times (12 + 16) \text{ kg CO}$$

However 1 kWh is produced by 250 g of fuel. Specific emissions of CO

$$= \frac{n_{CO} \times M_{CO}}{n_{CH_4} \times M_{CH_4}} \times sfc = \frac{1.1 \times (12 + 16)}{11.61 \times (12 + 4 \times 1)} \times 250$$
$$= 41.5 \text{ g CO/kWh}$$

Specific emissions of NO

$$= \frac{n_{NO} \times M_{NO}}{n_{CH_4} \times M_{CH_4}} \times sfc = (0.06 \times 30) \times 1.3458$$
$$= 2.42 \text{ g NO/kWh}$$

Specific emissions of CH_4 (with subscript P denoting products, and R denoting reactants)

$$= \frac{(n_{CH_4})_P}{(n_{CH_4})_R} \times sfc$$
$$= 2.37 \text{ g } CH_4/kWh$$

A specific basis should be used for emissions, as the purpose of the engine is to produce work (measured as kWh); the volume flow rate will be a function of the engine efficiency and the engine equivalence ratio, so the volumetric composition of the exhaust does not give an indication of the amount of pollutants.

3.12 | Questions

Table 3.22 *Composition of air, and the molar masses of its constituents*

Constituent	Mole fraction	Molar mass	Approximate molar mass kg/kmol
Nitrogen, N_2	0.7809	28.013	28
Oxygen, O_2	0.2095	31.999	32
Argon, Ar	0.0093	39.948	40
Carbon dioxide, CO_2	0.0003	44.010	44

In the following combustion calculations, values are needed for the molar masses of various species, and in the case of air we also need to define its composition – see table 3.22 above.

By multiplying the mole fraction of each species by its molar mass, and adding, we can determine the molar mass of air (M_{air}):

$$M_{air} = 28.964 \approx 29 \text{ kg/kmol}$$

In combustion calculations it is often taken that the oxygen forms 21 per cent of the air, and the remaining constituents are referred to as atmospheric nitrogen (N_2^*), as a simple way of representing the argon, carbon dioxide and other trace gases. With this assumed composition, the molar mass of the atmospheric nitrogen ($M_{N_2}^*$) can be set so as to give the correct molar mass of air:

$$M_{N_2}^* = (28.964 - 0.21 \times 32)/0.79 = 28.16 \text{ kg/kmol}$$

Molar masses are also needed for carbon and hydrogen, and are given below in table 3.23:

Table 3.23 *Molar masses of carbon and hydrogen*

Species	Molar mass, kg/kmol	Approximate molar mass, kg/kmol
Hydrogen, H_2	2.016	2
Carbon, C	12.011	12

Care must be taken with the molar masses of oxygen and hydrogen, since the values here are for molecules, *not* atoms. In the subsequent calculations the approximate values for composition and molar masses will be used. The thermodynamic data necessary for energy balance and equilibrium calculations are explained and tabulated in Appendix A.

3.1 If a fuel mixture can be represented by the general formula C_xH_{2x}, show that the stoichiometric gravimetric air/fuel ratio is 14.8:1.

3.2 A fuel has the following molecular gravimetric composition

pentane (C_5H_{12})	10 per cent
heptane (C_7H_{16})	30 per cent
octane (C_8H_{18})	35 per cent
dodecane ($C_{12}H_{26}$)	15 per cent
benzene (C_6H_6)	10 per cent

Calculate the atomic gravimetric composition of the fuel and the gravimetric air/fuel ratio for an equivalence ratio of 1.1.

3.3 The dry exhaust gas analysis from an engine burning a hydrocarbon diesel fuel is as follows: CO_2 0.121, O_2 0.037, N_2^* 0.842. Determine the gravimetric composition of the fuel, the equivalence ratio of the fuel/air mixture, and the stoichiometric air/fuel ratio.

3.4 An engine with a compression ratio of 8.9:1 draws in air at a temperature of 20°C and pressure of 1 bar. Estimate the temperature and pressure at the end of the compression stroke. Why will the temperature and pressure be less than this in practice?

The gravimetric air/fuel ratio is 12:1, and the calorific value of the fuel is 44 MJ/kg. Assume that combustion occurs instantaneously at the end of the compression stroke. Estimate the temperature and pressure immediately after combustion.

3.5 Compute the partial pressures of a stoichiometric equilibrium mixture of CO, O_2, CO_2 at (i) 3000 K and (ii) 3500 K when the pressure is 10 bar. Compare the answers with example 3.5. Are the results in accordance with Le Châtelier's Principle?

3.6 In a test to determine the cetane number of a fuel, comparison was made with two reference fuels having cetane numbers of 50 and 55. In the test the compression ratio was varied to give the same ignition delay. The compression ratios for the reference fuels were 25.4 and 23.1 respectively. If the compression ratio for the test fuel was 24.9, determine its cetane number.

3.7 Contrast combustion in compression ignition engines and spark ignition engines. What are the main differences in fuel requirements?

3.8 What is the difference between 'knock' in compression ignition and spark ignition engines? How can 'knock' be eliminated in each case?

3.9 A dual fuel engine operates by aspirating a mixture of air and fuel, which is then ignited by the spontaneous ignition of a small quantity of diesel fuel injected near the end of the compression stroke. A dual fuel engine is operating with a gravimetric air fuel ratio of 20, and the effective atomic H:C composition of the fuel is 3.1:1.

(i)   Write down the stoichiometric and actual combustion equations, and calculate the volumetric wet gas composition.
(ii) If the calorific value of the fuel is 46 MJ/kg, what percentage of this could be recovered by cooling the exhaust gas from 627°C to 77°C?
(iii) If the fuel consists of methane (CH_4) and cetane ($C_{16}H_{34}$), what is the ratio of the fuel mass flow rates? What would be the advantages and disadvantages of cooling the exhaust below the dewpoint temperature?

In a dual fuel engine, what determines the selection of the compression ratio? Explain under what circumstances emissions might be a problem.

3.10 Some combustion products at a pressure of 2 atm. are in thermal equilibrium at 1400 K, and a volumetric gas analysis yielded the following data:

CO_2	0.065
H_2O	0.092
CO	0.040

(i)   Deduce the concentration of H_2 from the equilibrium equation.
(ii) If the fuel was octane, calculate the approximate percentage of the fuel chemical energy not released as a result of the partial combustion.
(iii) Sketch how engine exhaust emissions vary with the air/fuel ratio, and discuss the shape of the nitrogen oxides and unburnt hydrocarbon curves.

3.11 The results of a dry gas analysis of an engine exhaust are as follows:

Carbon dioxide, CO_2	10.1 per cent
Atmospheric nitrogen, N_2	82.6 per cent

Stating any assumption, calculate:

(i)   the gravimetric air/fuel ratio;
(ii) the atomic and gravimetric composition of the fuel;

(iii) the stoichiometric gravimetric air/fuel ratio;
(iv) the actual equivalence ratio.

Why are the results derived from an exhaust gas analysis likely to be less accurate when the mixture is rich or significantly weak of stoichiometric? What other diesel engine emissions are significant?

3.12 A four-stroke engine is running on methane with an equivalence ratio of 0.8. The air and fuel enter the engine at 25°C, the exhaust is at 527°C, and the heat rejected to the coolant is 340 MJ/kmol fuel. Write the equations for stoichiometric combustion and the actual combustion.

(i) By calculating the enthalpy of the exhaust flow, deduce the specific work output of the engine, and thus the overall efficiency.
(ii) If the engine has a swept volume of 5 litres and a volumetric efficiency of 72 per cent (based on the air flow), calculate the power output at a speed of 1500 rpm. The ambient pressure is 1 bar.

State clearly any assumptions that you make, and comment on any problems that would occur with a rich mixture.

3.13 The results of a dry gas analysis of an engine exhaust are as follows:

Carbon dioxide, CO_2	10.7 per cent
Carbon monoxide, CO	5.8 per cent
Atmospheric nitrogen, N_2^*	83.5 per cent

Stating any assumptions, calculate:

(i) the gravimetric air/fuel ratio;
(ii) the gravimetric composition of the fuel;
(iii) the stoichiometric gravimetric air/fuel ratio.

Discuss some of the means of reducing engine exhaust emissions.

3.14 The following equation is a simple means for estimating the combustion temperature (T_c) of weak mixtures (that is, equivalence ratio, $\phi < 1.0$):

$T_c = T_m + k\phi \qquad \phi < 1.0$

where T_m is the mixture temperature prior to combustion, and k is a constant.
   By considering the combustion of an air/octane (C_8H_{18}) mixture with an equivalence ratio of 0.9 at a temperature of 25°C, calculate a value for k (neglect dissociation).
   For rich mixtures the following equation is applicable:

$$T_c = T_m + k - (\phi - 1.0) \times 1500 \qquad \phi \geq 1.0$$

Plot the combustion temperature rise for weak and rich mixtures, and explain why the slopes are different each side of stoichiometric. Sketch how these results would be modified by dissociation.

3.15 Equations are required to calculate the air/fuel ratio, the hydrogen/carbon ratio, the stoichiometric air/fuel ratio and the equivalence ratio, from the exhaust emissions readings obtained from a four gas analyser. The four gas analyser measures the following emissions on a dry gas basis:

carbon monoxide (per cent CO),
carbon dioxide (per cent CO_2),
oxygen (per cent O_2), and
unburnt hydrocarbons (as ppm of hexane, C_6H_{14})

The fuel may be assumed to be a pure hydrocarbon (that is, there are no oxygenates such as ethanol present), and it is also acceptable to ignore the presence of hydrogen in the exhaust products for the rich mixtures. Assume the following molar masses:

M_{H_2} hydrogen (H_2)	2 kg/kmol	
M_C carbon	12 kg/kmol	
M_{N_2} nitrogen (N_2)	28 kg/kmol	
M_{O_2} oxygen (O_2)	32 kg/kmol	
M_{air} (21 per cent O_2, 79 per cent N_2)	29 kg/kmol	

3.16 A rigid and well-insulated combustion vessel contains propane, oxygen and nitrogen; the initial composition is as indicated in the equation below. The initial pressure is 2 bar, and the initial temperature is 300 K. An equilibrium calculation restricted to the six principal combustion species gives the quantities of products that are identified below, and a temperature of 2718 K:

$$C_3H_8 + 5O_2 + 18.8N_2 \rightarrow 3.87H_2O + 0.13H_2 + 2.47CO_2 + 0.53CO + 0.33O_2 + 18.8N_2$$

The enthalpy of formation for propane (C_3H_8) is −104.68 MJ/kmol, and the molar specific heat capacity at constant pressure for the temperature range 200–300 K may be taken as 0.0628 MJ/kmol K.

Stating clearly any assumptions that you make:

(a) What will the final pressure be?
(b) Demonstrate that the temperature of 2718 K is consistent with the computed composition, and the conservation of energy.
(c) What other dissociation reactions are likely to be significant, and if they were to be included, what would the effect on the final temperature and pressure be?

3.17 A flow of methane at 1 bar and 298.15 K is divided into two equal flows. Flow 1 is mixed with a stoichiometric quantity of air (also at 1 bar and 298.15 K) and reacted in an engine. The hot exhaust gases from the engine (which can be assumed to be fully oxidised) are then mixed with the second flow of methane and passed into a catalytic reactor that is maintained at a temperature of 700 K, by waste heat from the engine exhaust. The outflow from the reactor is at 700 K and a pressure of 1 bar.

Write down the equation for the reactants and products entering and leaving the reactor: use a basis of 1 kmol of CH_4 entering the reactor, and y kmol of CH_4 leaving the reactor. Express the quantities of products leaving the reactor in terms of y.

For the stoichiometric equation:

$$-CH_4 - 2H_2O + CO_2 + 4H_2 = 0$$

(a) Verify that $\ln K_p = -5.973$ at 700 K.
(b) Show that if this is the only equilibrium reaction that needs to be considered, and that equilibrium is attained, then the products from the reactor will contain 0.783 kmol of CH_4.
(c) Determine the difference in the enthalpy of the reactants and products entering and leaving the reactor, and comment on how the difference can arise and its significance.
(d) State which other (if any) dissociation reaction(s) would be significant.

3.18 Octane (C_8H_{18}) is burnt with dry air in a steady-flow constant-pressure combustion system at a pressure of 15 bar and an equivalence ratio of 0.666. The fuel enters the combustion chamber as liquid at 25°C, while the air enters at a temperature of 600 K. The products leaving the combustion chamber are in equilibrium at a temperature of 1800 K, and consist solely of N_2, O_2, H_2O, CO_2 and NO.

Write down the stoichiometric and actual combustion equations, and determine the concentration of nitric oxide (NO) in the products of combustion. Calculate the heat transferred from the combustion chamber during combustion per unit mass of fuel.

Why might the concentration of nitric oxide in the exhaust differ from the equilibrium prediction?

3.19 A mixture of air and methane, initially at a temperature of 600 K and pressure of 20 bar, is burnt at a constant volume. What is the non-dimensional air/fuel ratio (lambda)?

By considering the equilibrium of the six species in the table to the right, determine the combustion temperature by considering the water–gas equilibrium, and verify that both equilibrium and the First Law of Thermodynamics are satisfied.

Species	Mols
CO	3.15222E − 01
CO_2	6.84778E − 01
H_2O	1.87628E + 00
H_2	1.23718E − 01
O_2	1.94700E − 02
N_2	6.76800E + 00

3.20 The combustion of a rich mixture of air and methane can be simplified to

$$CH_4 + 1.8O_2 + 6.77N_2 \rightarrow CO, CO_2, H_2O, H_2, N_2$$

Assume a water–gas equilibrium constant of 5.2 (the reverse of reaction 7 in table A.6 in Appendix A) to evaluate the composition of the products of combustion. By considering the energy balance, show that the flame (constant-pressure, adiabatic) temperature is close to 2200 K when the reactants are initially at 1 bar and 25°C.

3.21 For the combustion of a stoichiometric mixture of air and methane:

$$CH_4 + 2O_2 + 7.52N_2 \rightarrow CO, CO_2, H_2O, H_2, O_2, N_2$$

if the initial temperature is 25°C, show that the equilibrium adiabatic combustion temperature is 2247 K for a constant-pressure combustion process at 1 atm., by considering the energy balance, and the equilibrium of the six species.

3.22 For a fuel of arbitrary composition $C_\alpha H_\beta O_\gamma N_\delta$, show that the brake specific emissions of carbon monoxide in g CO/kWh can be evaluated from:

$$bsfc \times \frac{M_C}{M_C + (\beta/\alpha)M_H + (\gamma/\alpha)M_O + (\delta/\alpha)M_N} \times \frac{\%CO}{\%CO_2 + \%CO + ppm\ CH_4/10^4} \times \frac{M_{CO}}{M_C}$$

where   bsfc = brake specific fuel consumption (g/kWh),
        M = molar masses,
        and there is no solid carbon present in the exhaust.

Generalise the above equation for any species, and comment on the accuracy of any other atomic balance that might be used to evaluate the brake specific emissions.

3.23 A test is being conducted on a spark ignition engine which is being operated on iso-octane with an equivalence ratio (ϕ) of 1.1. The combustion products can be assumed to contain no oxygen, and the concentration of the hydrogen is half that of the carbon monoxide. Write down the combustion equation.

The temperature of the exhaust is 1000 K, and the ambient temperature is 25°C. Calculate the enthalpy of the exhaust relative to 25°C, and the calorific value of the partial products of combustion at 25°C, and express these as a proportion of the calorific value of the fuel.

If the mass flowrate of the exhaust is 0.1 kg/s, atmospheric pressure is 1 bar, and the cross-sectional area of the exhaust pipe is 10 cm^2, calculate the ratio of the exhaust kinetic energy to the fuel calorific value.

3.24 The calibration of an air/fuel ratio analyser is to be checked by using the gas mixture shown in the table to the right, which represents the dry products of combustion from a hydrocarbon fuel:

Calculate and comment on the values of:

(a) the molar hydrogen/carbon ratio of the fuel;
(b) the gravimetric air/fuel ratio (AFR) corresponding to these products;
(c) the gravimetric, stoichiometric AFR and the equivalence ratio;
(d) the water–gas constant.

Constituent	Volume
methane, CH_4	550 ppm
hydrogen, H_2	3.23 per cent
carbon monoxide, CO	6.20 per cent
carbon dioxide, CO_2	12.29 per cent
nitrogen, N_2	balance

If a fuel of this composition was to be burnt in a spark ignition engine, sketch the response of the emissions (CO, CO_2, O_2, H_2, HC, H_2O, NO_x) plotted on a wet basis to variation in the equivalence ratio. (*Hint*: It may be helpful to utilise the data from the first part of this question.)

3.25 The results of a dry gas analysis of an engine exhaust are as follows:

Carbon dioxide, CO_2	10.7 per cent
Carbon monoxide, CO	5.8 per cent
Atmospheric nitrogen, N_2^*	83.5 per cent

Stating any assumptions, calculate:

(i) the gravimetric air/fuel ratio;
(ii) the gravimetric composition of the fuel;
(iii) the stoichiometric gravimetric air/fuel ratio.

3.26 A mixture of propane (C_3H_8) and air is burnt in a constant-pressure adiabatic system at a pressure of 1 bar. The composition of the exhaust gases includes:

3.87 kmol H_2O
0.13 kmol H_2
2.49 kmol CO_2
0.51 kmol CO
0.52 kmol O_2

Assuming that there is no nitric oxide present, and stating clearly any additional assumptions:

(a) What is the equivalence ratio of the reactants?
(b) What is the temperature of the products of combustion?
(c) What is the temperature of the reactants?

3.27 An adiabatic constant-volume combustion chamber is filled with a mixture of air and methane; lambda (AFR/AFRs) = 1.3. The initial temperature of the mixture is 25°C, and its pressure is 1 bar. After laminar combustion from a single ignition source (and ignoring dissociation) the pressure would rise to 7.27 bar. Any heat transfer within the gas may also be ignored. Assume the following values of the ratio of heat capacities:

unburnt gas	1.377
burnt gas	1.237

Calculate: (a) the temperature of the gas to burn first, immediately after the start of combustion;
(b) the temperature of the gas to burn last, after the end of combustion;
(c) the temperature of the gas to burn last, immediately before the end of combustion; and
(d) the temperature of the gas to burn last, after the end of combustion.

Comment on whether dissociation is likely to be significant, and how it might affect the final pressure, the first burnt gas temperatures immediately after the start and end of combustion, and the last burnt gas temperature immediately after the end of combustion.

3.28 The Rice and Herzfeld mechanisms for thermal decomposition of hydrocarbons (in which R refers to any radical and M any molecule) are:

(1) dissociation	$M_1 \rightarrow R_1 + R_1'$	k_1
(2) propagation	$R_1 + M_1 \rightarrow R_1H + R_2$	k_2
(3) dissociation	$R_2 \rightarrow R_1 + M_2$	k_3
(4) termination	$R_1 + R_2 \rightarrow M_3$	k_4
(5) termination	$2R_1 \rightarrow M_4$	k_5
(6) termination	$2R_2 \rightarrow M_5$	k_6

where only one of the three termination or chain-termination steps (4), (5), and (6) would be expected to be important in a given reaction.

(a) If reaction (5) is the chain termination reaction, then show that the overall rate of decomposition of M_1 is to the power of three-halves:

$$-\frac{d[M_1]}{dt} \approx k_2 \left(\frac{k_1}{2k_5}\right)^{\frac{1}{2}} [M_1]^{\frac{3}{2}}$$

(b) If reaction (6) is the chain termination reaction, then show that the overall rate of decomposition of M_1 is to the power of a half:

$$-\frac{d[M_1]}{dt} \approx k_3 \sqrt{\frac{k_2}{2k_6}} [M_1]^{\frac{1}{2}}$$

CHAPTER 4

Spark ignition engines

4.1 | Introduction

This chapter considers how the combustion process is initiated and constrained in spark ignition engines. The air/fuel mixture has to be close to stoichiometric (chemically correct) for satisfactory spark ignition and flame propagation. The equivalence ratio or mixture strength of the air/fuel mixture also affects pollutant emissions, as discussed in chapter 3, and influences the susceptibility to spontaneous self-ignition (which can lead to knock). A lean air/fuel mixture (equivalence ratio less than unity) will burn more slowly and will have a lower maximum temperature than a less lean mixture. Slower combustion will lead to lower peak pressures, and both this and the lower peak temperature will reduce the tendency for knock to occur. The air/fuel mixture also affects the engine efficiency and power output. At constant engine speed with fixed throttle, it can be seen how the brake specific fuel consumption (inverse of efficiency) and power output vary. This is shown in figure 4.1 for a typical spark ignition engine at full or wide open throttle (WOT). As this is a constant-speed test, power output is proportional to torque output, and this is most conveniently expressed as bmep, since bmep is independent of engine size. Figure 4.2 is an alternative way of expressing the same data (because of their shape, the plots are often referred to as 'fish-hook' curves); additional part throttle data have also been included. At full throttle, the maximum for power output is fairly flat, so beyond a certain point a richer mixture significantly reduces efficiency without substantially increasing power output. The weakest mixture for maximum power (wmmp) will be an arbitrary point just on the lean side of the mixture for maximum power. This is also known as the mixture for Lean

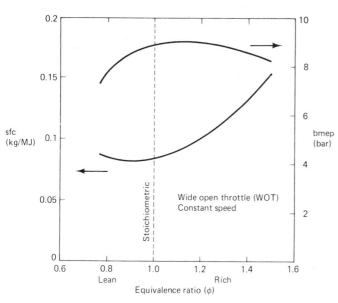

Figure 4.1
Response of specific fuel consumption and power output to changes in air/fuel ratio.

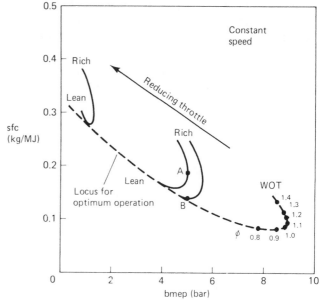

Figure 4.2
Specific fuel consumption plotted against power output for varying air/fuel ratios at different throttle settings.

Best Torque (LBT). The minimum specific fuel consumption also occurs over a fairly wide range of mixture strengths. A simplified explanation for this is as follows. With dissociation occurring, maximum power will be with a rich mixture when as much as possible of the oxygen is consumed; this implies unburnt fuel and reduced efficiency. Conversely, for maximum economy as much of the fuel should be burnt as possible, implying a weak mixture with excess oxygen present. In addition, it was shown in chapter 2 that the weaker the air/fuel mixture the higher the ideal cycle efficiency. When the air/fuel mixture becomes too weak the combustion becomes incomplete and the efficiency again falls.

The air/fuel mixture can be prepared either by a carburettor or by fuel injection (section 4.6). In a carburettor air flows through a venturi, and the pressure drop created causes fuel to flow through an orifice, the jet. There are two main types of carburettor – fixed jet and variable jet. Fixed jet carburettors have a fixed venturi, but a series of jets to allow for different engine-operating conditions from idle to full throttle. Variable jet carburettors have an accurately profiled needle in the jet. The needle position is controlled by a piston which also varies the venturi size. The pressure drop is approximately constant in a variable jet carburettor, while pressure drop varies in a fixed jet carburettor.

The alternative to carburettors is fuel injection. Early fuel injection systems were controlled mechanically, but the control is now electronic. Fuel was not traditionally injected directly into the cylinder during the induction or compression stroke. This would require high-pressure injection equipment, and it would reduce the time for preparation of an homogeneous mixture. Also, the injectors would have to withstand the high temperature and pressures during combustion and be resistant to the build-up of combustion deposits. However, these problems can be overcome, and direct injection engines have been introduced. These offer

scope for fuel economy gains, and are discussed in chapter 5. With low-pressure fuel injection systems, the fuel is usually injected close to the inlet valve of each cylinder. Alternatively a single injector can be used to inject fuel at the entrance to the inlet manifold.

The ignition timing also has to be controlled accurately, and a typical response for power output (which will also equate for efficiency, since the amount of fuel being burned is unchanged) is shown in figure 4.3. If ignition is too late, then although the work done by the piston during the compression stroke is reduced, so is the work done on the piston during the expansion stroke, since all pressures during the cycle will be reduced. Furthermore, there is a risk that combustion will be incomplete before the exhaust valve opens at the end of the expansion stroke, and this may overheat the exhaust valve. Conversely, if ignition is too early, there will be too much pressure rise before the end of the compression stroke (tdc) and power will be reduced. Thus, the increase in work during the compression stroke is greater than the increase in work done on the piston during the expansion stroke. Also, with early ignition the peak pressure and temperature may be sufficient to cause knock. Ignition timing is optimised for maximum power; as the maximum is fairly flat the ignition timing is usually arranged to occur on the late side of the maximum. This is shown by the MBT ignition timing (minimum advance for best torque) in figure 4.3. The definition of MBT timing is somewhat arbitrary. It may correspond to the timing that gives a 1 per cent fall in the peak torque. Alternatively, it might be when the operator of an engine test first detects a fall in torque, as the ignition timing is being retarded. Figure 4.3 shows that MBT ignition timing increases the knock margin. However, for some engines operating at full throttle (particularly at low speed operation) knock is likely to be encountered before the MBT ignition timing. In which case the ignition will be retarded

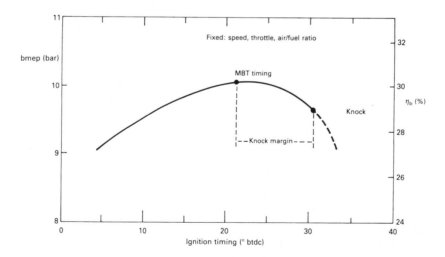

Figure 4.3
The effect of ignition timing on the output and efficiency of a spark ignition engine.

from MBT to preserve a knock margin – this is to allow for manufacturing tolerances and engine ageing making some engines more susceptible to knock. At higher engine speeds, the lower volumetric efficiency (reducing peak pressures in the cylinder), and the reduced time available for the knock mechanism to occur, mean that knock is less likely to occur. Since engines rarely operate with full throttle at low speeds, it is common for a compression ratio/octane requirement combination to be adopted, which leads to knock being encountered before MBT ignition timing at low speeds. Retarding the ignition timing in the low speed, full throttle

part of the engine operating envelope results in a local reduction in efficiency and output. However, everywhere else in the engine operating envelope, there is an efficiency and output benefit from the higher compression ratio.

The effect of the ignition timing on the pressure history can be investigated experimentally or by a model. A computer model (ISIS) has been used here, since it enables the burn rate to be kept fixed – for the cases presented here, both the 0–10 per cent and 10–90 per cent mass fraction burn durations were 20°ca. Figure 4.4a shows the effect of a 15°ca departure from the MBT ignition timing (17°ca btdc) on the pressure

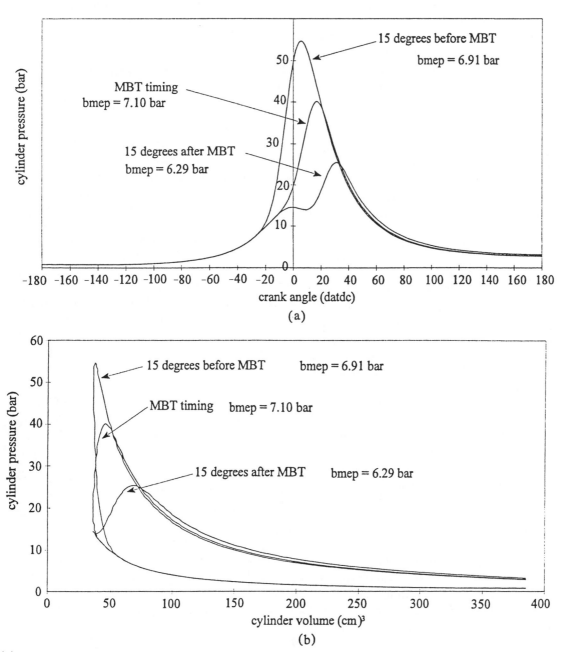

Figure 4.4
The effect of ignition timing on (a) the pressure/crank-angle history and (b) the pressure/volume diagram.

Table 4.1 *The effect of the burn durations on the MBT ignition timing and the engine performance for stoichiometric operation at 1000 rpm; based on the Rover K16 engine with a 10:1 compression ratio*

0–10 per cent burn, °ca	10–90 per cent burn, °ca	MBT timing, °ca btdc	Brake efficiency	P_{max}, bar	Angle of P_{max}, °ca atdc
20	20	17	0.266	38.5	21
20	40	23	0.250	30.2	18
40	20	37	0.266	38.3	22
Instantaneous		tdc	0.27	66	tdc

history. There are very substantial differences in the maximum cylinder pressure, and the angle at which it occurs. However, the differences in the work output (the imep) are very much less, and the reason for this can be seen in figure 4.4b. As already explained, the trade-offs between compression work and expansion work tend to cancel, and for this fast burn combustion system the combustion is always mostly complete when the piston is still very close to its top dead centre position.

The effect of different burn durations on the MBT ignition timing and the engine performance can also be studied using ISIS, and the results from such a study are shown in table 4.1. The 0–10 per cent burn duration represents the delay period and the 10–90 per cent burn duration represents the main burn period (chapter 3, section 3.5.1).

Table 4.1 shows that when the delay period (0–10 per cent burn) is lengthened from 20°ca to 40°ca, then the effect on the engine output and efficiency is unchanged, so long as the ignition timing is advanced to its new MBT value. The advance of 20°ca in the MBT ignition timing means that the phasing of the main burn period (10–90 per cent burn) is exactly the same as before, and this is why there is no change in the brake efficiency, and negligible changes in the maximum pressure and the angle of its occurrence. In contrast, when the main burn period duration is increased, then there is a lowering of the brake efficiency and the maximum pressure. The MBT igniting timing is slightly more advanced than the baseline case, but the maximum pressure is significantly lower (30 bar compared to 38 bar). If the combustion is instantaneous, then there is a very small gain in efficiency, but a substantial rise in the peak pressure to 66 bar. Some of the theoretical gains expected from instantaneous combustion are not realised because of increased dissociation and heat transfer. In summary, table 4.1 shows that it is the duration of the main burn period that influences the efficiency and output, and that a long delay (or early burn) period does not matter, so long as the combustion is correctly phased through the use of the MBT ignition timing.

At part throttle operation, the cylinder pressure and temperature are reduced and the residuals level is increased so that the flame propagation is slower; thus ignition is arranged to occur earlier at part load settings. Ignition timing can be controlled either electronically or mechanically.

The different types of combustion chamber and their characteristics are discussed in the next section.

4.2   Combustion chambers

4.2.1   Conventional combustion chambers

Initially the cylinder head was little more than a cover for the cylinder. The simplest configuration was the side valve engine (figure 4.5) with the inlet and exhaust valves together at one side of the cylinder. The most successful combustion chamber for the side valve engine was the Ricardo turbulent head, as shown in figure 4.5. This design was the result of extensive experimental studies aimed at improving combustion. The maximum compression ratio that was reasonable with this geometry was limited to about 6 : 1, but this was not a restriction since the octane rating of fuels in the 1920s and 1930s was only about 60–70.

In the Ricardo turbulent head design the clearance between part of the cylinder head and piston at the end of the

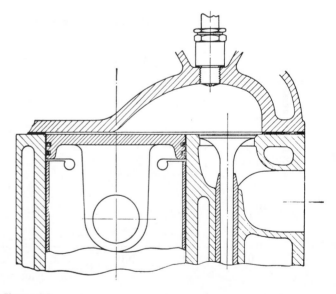

Figure 4.5
Ricardo turbulent head for side valve engines (reproduced with permission from Ricardo and Hempson, 1968).

compression stroke is very small. This forms an area of 'squish', from which gas is ejected into the main volume. The turbulence that this jet generates ensures rapid combustion. If too large a squish area is used the combustion becomes too rapid and noisy. This design also reduces the susceptibility to knock, since the gas furthest from the spark plug is in the squish area. The end gas in the squish area is less prone to knock since it will be cooler because of the close proximity of the cylinder head and piston. Excessive turbulence also causes excessive heat transfer to the combustion chamber walls, and should be avoided for this reason also.

The main considerations in combustion chamber design are:

(i) the distance travelled by the flame front should be minimised

(ii) the exhaust valve(s) and spark plug(s) should be close together

(iii) there should be sufficient turbulence

(iv) the end gas should be in a cool part of the combustion chamber.

(i) By minimising the distance between the spark plug and the end gas, combustion will be as rapid as possible. This has two effects. Firstly, it produces high engine speeds and thus higher power output. Secondly, the rapid combustion reduces the time in which the chain reactions that lead to knock can occur. This implies that, for geometrically similar engines, those with the smallest diameter cylinders will be able to use the highest compression ratios.

(ii) The exhaust valve should be as close as possible to the sparking plug. The exhaust valve is very hot (possibly incandescent) so it should be as far from the end gas as possible to avoid inducing knock or pre-ignition.

(iii) There should be sufficient turbulence to promote rapid combustion. However, too much turbulence leads to excessive heat transfer from the chamber contents and also to too rapid combustion, which is noisy. The turbulence can be generated by squish areas or shrouded inlet valves.

(iv) The small clearance between the cylinder head and piston in the squish area forms a cool region. Since the inlet valve is cooled during the induction stroke, this too can be positioned in the end gas region.

For good fuel economy all the fuel should be burnt and the quench areas where the flame is extinguished should be minimised. The combustion chamber should have a low surface-to-volume ratio to minimise heat transfer. The optimum swept volume consistent with satisfactory operating speeds is about 500 cm^3 per cylinder. For high-performance engines, smaller cylinders will enable more rapid combustion, so permitting higher operating speeds and consequently greater power output. For a given geometry, reducing the swept volume per cylinder from 500 cm^3 to 200 cm^3 might increase the maximum engine speed from about 6000 rpm to 8000 rpm.

The ratio of cylinder diameter to piston stroke is also very important. When the stroke is larger than the diameter, the engine is said to be 'under-square'. In Britain the car taxation system originally favoured under-square engines and this hindered the development of higher-performance over-square engines. In over-square engines the cylinder diameter is larger than the piston stroke, and this permits larger valves for a given swept volume. This improves the induction and exhaust processes, particularly at high engine speeds. In addition the short stroke reduces the maximum piston speed at a given engine speed, so permitting higher engine speeds. The disadvantage with over-square engines is that the combustion chamber has a poor surface-to-volume ratio, so leading to increased heat transfer. More recently there has been a return to under-square engines, as these have combustion chambers with a better surface-to-volume ratio, and so lead to better fuel economy. It will be seen later in chapter 6, section 6.2 that in general, the maximum power output of an engine is proportional to the piston area, while the maximum torque output is proportional to the swept volume.

Currently most engines have a compression ratio of about 10:1, for which a side valve geometry would be unsuitable. Overhead valve (ohv) engines have a better combustion chamber for these higher compression ratios. If the camshaft is carried in the cylinder block the valves are operated by push rods and rocker arms. A more direct alternative is to mount the camshaft in the cylinder head (ohc – overhead camshaft). The camshaft can be positioned directly over the valves, or to one side with valves operated by rocker arms. These alternatives are discussed more fully in chapter 7.

Figure 4.6a–d shows four traditional combustion chamber configurations; where only one valve is shown, the other is directly behind. Very often it will be production and economic considerations rather than thermodynamic considerations that determine the type of combustion chamber used. If combustion chambers have to be machined it will be cheapest to have a flat cylinder head and machined pistons. If the finish as cast is adequate, then the combustion chamber can be placed in the cylinder head economically.

Figure 4.6a shows a wedge combustion chamber; this is a simple chamber that produced good results. The valve drive train is easy to install, but the inlet and exhaust manifold have to be on the same side of the cylinder head. The hemispherical head, figure 4.6b, was popular for a long time in high-performance engines since it permits larger valves to be used than those with a flat cylinder head. The arrangement is inevitably expensive, with perhaps twin overhead camshafts. With the inlet and exhaust valves at opposite sides of the cylinder, it allows crossflow from inlet to exhaust. Crossflow

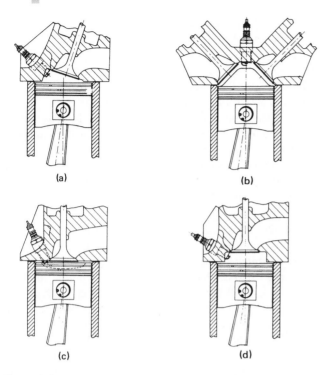

Figure 4.6
Combustion chambers for spark ignition engines:
(a) wedge chamber;
(b) hemispherical head;
(c) bowl in piston chamber;
(d) bath-tub head.

occurs at the end of the exhaust stroke and the beginning of the induction stroke when both valves are open; it is beneficial since it reduces the exhaust gas residuals. Subsequently 'pent-roof' heads with four valves per cylinder have become popular; these have a shape similar to that of a house roof. The use of four valves gives an even greater valve area than does the use of two valves in a hemispherical head. A much cheaper alternative, which also has good performance, is the bowl in piston (Heron head) combustion chamber (figure 4.6c). This arrangement was used by Jaguar for their V12 engine and during development it was only marginally inferior to a hemispherical head engine with twin overhead camshafts (Mundy, 1972). The bath-tub combustion chamber (figure 4.6d) has a very compact combustion chamber that might be expected to give economical performance; it can also be used in a crossflow engine. All these combustion chambers have:

(i)   short maximum flame travel

(ii) the spark plug close to the exhaust valve

(iii) a squish area to generate turbulence

(iv) well-cooled end gas.

The fuel economy of the spark ignition engine is particularly poor at part load; this is shown in figure 4.2. Although operating an engine on a very lean mixture can cause a reduction in efficiency, this reduction is less than if the power was controlled by throttling, with its ensuing losses. However, if a three-way catalyst is being used (section 4.3.2), the engine operation is restricted to using stoichiometric combustion. Often, engine manufacturers are concerned with performance (maximum power and fuel economy) at or close to full throttle, although in automotive applications it is unusual to use maximum power except transiently.

In chapter 3 it was stated that the maximum compression ratio for an engine is usually dictated by the incipience of knock. If the problem of knock could be avoided, either by special fuels or special combustion chambers, there would still be an upper useful limit for compression ratio. As compression ratio is raised, there is a reduction in the rate at which the ideal cycle efficiency improves, see figure 4.7. Since the mechanical efficiency will be reduced by raising the compression ratio (owing to higher pressure loadings), the overall efficiency will be a maximum for some finite compression ratio, see figure 4.7.

Some of the extensive work by Caris and Nelson (1958) is summarised in figure 4.8. This work shows an optimum compression ratio of 16:1 for maximum economy, and 17:1 for maximum power. The reduction in efficiency is also due to poor combustion chamber shape – at high compression ratios there will be a poor surface-to-volume ratio. Figures for the optimum compression ratio vary since researchers have used different engines, with different air/fuel ratios, operating points and cylinder swept volume. As the swept volume is reduced the surface-to-volume ratio deteriorates, and data

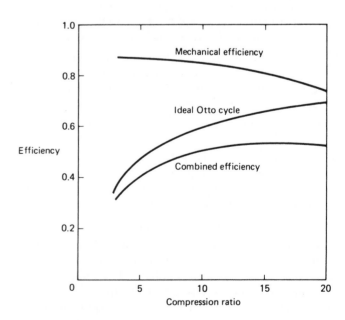

Figure 4.7
Variation in efficiency with compression ratio.

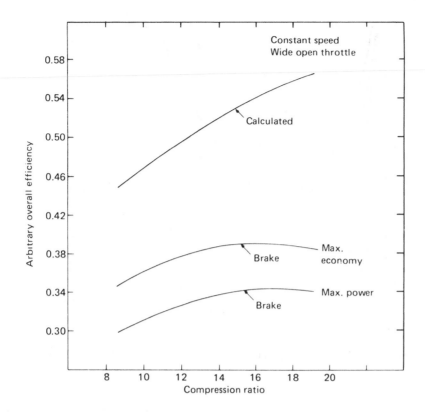

Figure 4.8
Variation in arbitrary overall efficiency with compression ratio. [Reproduced with permission, © 1958 Society of Automotive Engineers, Inc.]

assembled by Muranaka *et al.* (1987) show that the optimum compression ratio falls. With a cylinder volume of 250 cm³ the compression ratio should be around 11, while for a cylinder swept volume of 500 cm³ the optimum compression ratio is around 15. For any given engine the optimum compression ratio will also be slightly dependent on speed, since mechanical efficiency depends on speed.

4.2.2 High compression ratio combustion chambers and fast burn combustion systems

An approach that permits the use of high compression ratios with ordinary fuels is the high turbulence, compact combustion chamber engine. The concepts behind these engines and a summary of the different types are given by Ford (1982).

Increasing the turbulence allows both leaner and richer mixtures to be burned, since combustion is more rapid and can be centred around top dead centre. This is when the unburned gas temperatures will be highest, so that the combustion reactions are self-sustaining. The lean mixtures are less prone to knock, owing to the reduced combustion pressures and temperatures. Increasing the turbulence also reduces the susceptibility to knock since normal combustion occurs more rapidly. The more rapid combustion thus allows a higher compression ratio to be used, and this further widens the range of flammable mixtures. These results are sum-marised in figure 4.9. A compact combustion chamber is needed to reduce heat transfer from the gas. The chamber is concentrated around the exhaust valve, in close proximity to the spark plug; this is to enable the mixture around the exhaust valve to be burnt soon after ignition, otherwise the hot exhaust valve would make the combustion chamber prone to knock. The first design of this type was the May Fireball (May, 1979), with a flat piston and the combustion chamber in the cylinder head (figure 4.10). Subsequently another design has been developed with the combustion chamber in the piston, and a flat cylinder head. In an engine with the compression ratio raised from 9.7:1 to 14.6:1, the gain in efficiency was up to 15 per cent at full throttle, with larger gains at part throttle.

There are also disadvantages associated with these combustion chambers. Emissions of carbon monoxide should be reduced, but hydrocarbon emissions (unburnt fuel) will be increased because of the large squish areas and poor surface-to-volume ratios. Hydrocarbon emissions can be removed by a catalyst in the exhaust system. The exhaust temperatures will be low in this type of lean mixture engine, and the exhaust passages may have to be insulated to maintain sufficient temperature in the catalyst. Emissions of nitrogen oxides (NO_x) will be greater for a given air/fuel ratio, owing to the higher cylinder temperatures, but as the air/fuel mixture will be leaner overall there should be a reduction in NO_x, and

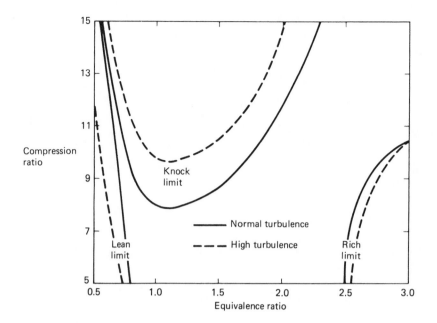

Figure 4.9
Effect of turbulence on increasing the operating envelope of spark ignition engines (adapted from Ford, 1982).

catalysts that can reduce NO_x in an oxidising environment are discussed in section 4.3.3.

More careful control is needed on mixture strength and inter-cylinder distribution, in order to stay between the lean limit for misfiring and the limit for knock. More accurate control is also needed on ignition timing. During manufacture greater care is necessary, since tolerances that are acceptable for compression ratios of 9 : 1 would be unacceptable at 14 : 1; in particular combustion chambers need more accurate manufacture. Combustion deposits also have a more significant effect with high compression ratios since they occupy a greater proportion of the clearance volume.

Three more combustion systems that have a fast burn characteristic are the Ricardo High Ratio Compact Chamber (HRCC), the Nissan NAPS-Z (or ZAPS) and the four-valve pent-roof combustion chamber. The Ricardo HRCC combustion system is similar in concept to the May Fireball (figure 4.10), but has a straight passage from the inlet valve to the combustion chamber. The four-valve pent-roof system and the Nissan NAPS-Z combustion system are shown in figure 4.11.

These three different combustion systems were the subject of extensive tests conducted by Ricardo, and reported by Collins and Stokes (1983). The main characteristic of the four-valve pent-roof combustion chamber is the large flow area provided by the valves. Consequently there is a high volumetric efficiency, even at high speeds, and this produces an almost constant bmep from mid speed upwards. The inlet tracts tend to be almost horizontal, and to converge slightly. During the induction process, barrel swirl (rotation about an axis parallel to the crankshaft) is produced in the cylinder. The reduction in volume during compression firstly causes an increase in the swirl ratio through the conservation of the moment of momentum. Subsequently, the further reduction in volume causes the swirl to break up into turbulence. This then enables weak air/fuel ratios to be burnt, thereby giving good fuel economy and low emissions (Benjamin, 1988).

The Nissan NAPS-Z combustion system has twin spark plugs, and an induction system that produces a comparatively high level of axial swirl. While the combustion initiates at the edge of the combustion chamber, the swirling flow and twin spark plugs ensure rapid combustion. With both the four-valve design and the NAPS-Z combustion chamber there is comparatively little turbulence produced by squish. In the case of the four-valve head, turbulence is also generated by

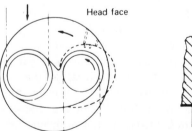
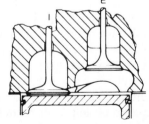

Figure 4.10
The May Fireball high-turbulence combustion chamber for high compression ratio engines burning lean mixtures.

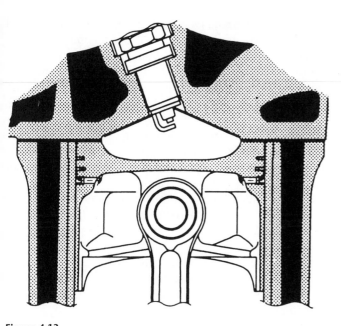

Figure 4.13
The Honda VTEC compact pent-roof combustion system (Horie and Nishizawa, 1992). © IMechE/Professional Engineering Publishing Limited.

sion stroke the number of vortices increased to three or more. The paper concludes that not only did the valve-disabled configuration result in significantly higher swirl levels than the standard configuration with both inlet valves opening, but that these vortices did not decay as fast, and hence produced higher levels of turbulence closer to tdc.

Similar results have been obtained by Horie and Nishizawa (1992), from the Honda VTEC (Variable Timing and Event Control) mechanism, which allows one of the inlet valves to be disabled. The VTEC engine combustion system is a compact pent-roof combustion system, with a bowl in the piston (figure 4.13). The bowl has a diameter of 55 mm (compared to a bore diameter of 75 mm) and a squish clearance of 0.75 mm. This compact combustion system enabled the lean-mixture limit to be extended by 2 AFR compared to a conventional flat-top piston pent-roof chamber. At a part-load operation condition of 1500 rpm and 1.6 bar bmep, Horie and Nishizawa (1992) showed that the engine would operate at an equivalence ratio of 0.66 with acceptable levels of driveability, a significantly lower brake specific fuel consumption (12 per cent less than stoichiometric) and less than 6 g/kWh of NO_x. The resulting engine management system strategy is shown in figure 4.14.

Stone *et al.* (1993) used a similar valve disablement strategy, and studied the effect on the swirl ratios, and the burn rates

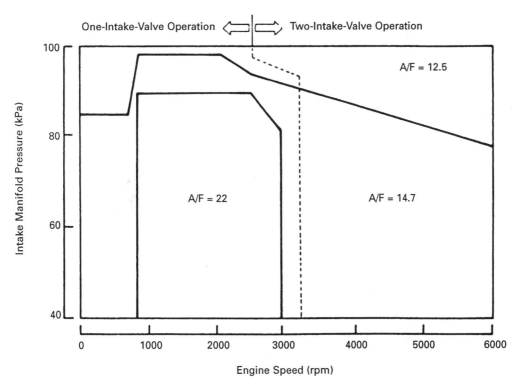

Figure 4.14
The Honda VTEC engine management system strategy, with weak-mixture operation (A/F = 22) at those part-load operating points where there was: (a) sufficient combustion stability and torque, and (b) sufficiently low NO_x emissions such that no catalytic reduction was needed. The rich-mixture operation at the higher speeds is for maximum power and to avoid catalyst overheating (Horie and Nishizawa, 1992). © IMechE/Professional Engineering Publishing Limited.

Table 4.2 *Steady-flow rig swirl measurements, (Stone et al., 1993)*

Valve operation	Ricardo swirl ratio	
	Barrel	Axial
Both	0.45	0.02
Inner	0.85	1.40
Outer	0.81	1.21

(these definitions of swirl can be found in chapter 6, section 6.2). With an inlet valve disabled, table 4.2 shows that the barrel swirl ratio is almost doubled. At a given operating point, the mass flow rate into the cylinder will be fixed, and if only one valve is being opened significantly, then the flow velocities will be almost doubled. This in turn leads to an increase in the angular velocity of the vortex, and the higher swirl ratio.

When both the inlet valves are opened, then there is negligible axial swirl. Table 4.2 shows that valve disablement is a very effective means of generating axial swirl. With inlet valve disablement, when the inner valve was opened there were consistently higher values for the axial and barrel swirl ratio, and this led to the fastest combustion. Five engine load/speed operating points were studied, and the burn durations

were typically reduced by a third; this was also accompanied by improved combustion stability. The faster and more stable combustion with an inlet valve disabled, led to an average reduction in the fuel consumption of 5.6 per cent. This reduction is of the same order as reported by Fraidl *et al.* (1990), for tests at 2000 rpm and 2 bar bmep with inlet valve deactivation. Fraidl *et al.* also investigated the greater tolerance of fast-burn combustion systems to EGR. They found that within the limit of 5 per cent CoV of imep, increasing EGR by about 10 percentage points led to an additional 2.4 per cent in the bsfc.

These and other trade-offs for stoichiometric operation can be studied by means of a computer model (ISIS). In figure 4.15 the fast burn was simulated by using 75 per cent of the 0–10 per cent and 0–90 per cent burn durations of the standard burn. For each operating point and burn rate, ignition retard and EGR were investigated, for their effects on the brake specific fuel consumption and brake specific NO_x emissions.

Consider first the effect of retarding the ignition timing with a fixed residuals level. For both the standard burn rate and the fast burn, retarding the ignition timing from MBT leads to a reduction in NO_x emissions and an increase in the fuel consumption. The NO_x emissions are reduced since the retarded ignition timing causes lower peak pressures and

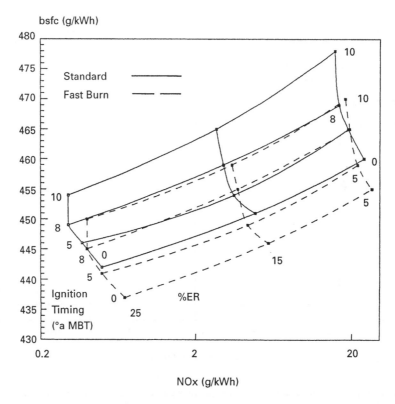

Figure 4.15
The effect of exhaust gas residuals and retarding the ignition timing from MBT on the brake specific fuel consumption and the brake specific NO_x emissions, for two different burn rates at a speed of 2000 r/min and a load of 2 bar bmep.

temperatures, and the NO_x formation is highly temperature-dependent. The fuel consumption increases with the retarded ignition timings, since the brake efficiency is lowered, and the throttle has to be opened with an increase in the air and fuel flow, in order to retain the output at the same level. The faster burn leads to higher peak pressures and temperatures and a lower brake specific fuel consumption at the MBT ignition timing. The more rapid combustion leads to higher NO_x emissions on a volumetric basis, but lower NO_x emissions when they are calculated on a brake specific basis. In other words the increase in efficiency has a greater effect on reducing the brake specific NO_x emissions than the increase in the volumetric emissions. With the fast-burn combustion system, retarding the ignition timing has a smaller effect on reducing the brake specific NO_x emissions than the slower burn combustion system. This is a consequence of the fully turbulent combustion period (the 10–90 per cent mass fraction burn) remaining closer to tdc with a fast-burn combustion system than with the standard combustion system.

Figure 4.15 also shows the strong influence that increasing exhaust gas residuals has on the reductions in both the brake specific emissions and the brake specific fuel consumption. It must be remembered that the burn rates have been assumed to be independent of the ignition timing and the exhaust residuals level. However, the two burn rates considered indicate a weak dependence on burn rate compared with changes in the ignition timing or exhaust gas residuals (ER) level. Increasing the exhaust gas residuals (ER) lowers the NO_x emissions on a volumetric and brake specific basis, because the essentially inert presence of the products of combustion lowers the combustion temperature. Increasing the exhaust gas residual level lowers the brake specific fuel consumption, as a consequence of the higher partial pressure of the residuals increasing the inlet manifold pressure, which

in turn implies a more open throttle and lower throttling losses. Figure 4.15 shows that using EGR is a better strategy for reducing NO_x emissions than retarding the ignition timing, as the EGR also gives a reduction in the brake specific fuel consumption. However, these results take no account of how the change in exhaust gas residuals level or burn rate will affect the cycle-by-cycle variations in combustion.

Most homogeneous charge spark ignition engines subject to emissions legislation have to operate at stoichiometric, so that the three-way catalyst can operate effectively (see section 4.3); the most notable exception is the Honda VTEC engine already discussed in this section. Figure 4.15 has shown how exhaust gas residuals can be used as a diluent to reduce the NO_x emissions and fuel consumption, but if it was not for the constraint of the three-way catalyst, then dilution by air would be better.

Figure 4.16 shows that dilution by air gives a greater reduction in emissions than dilution by exhaust gas recirculation (EGR), and figure 4.17 shows that dilution by air gives greater scope for reduction in the fuel consumption (Shillington, 1998). However, a lean-burn approach can only be widely adopted if catalysts for NO_x reduction in an oxidising environment become used; these NO_x catalysts are discussed in section 4.3.

4.3 | Catalysts and emissions from spark ignition engines

4.3.1 Introduction

In the late 1960s photochemical smogs were linked to pollution from vehicle engines. Smog contains hydrocarbons, peroxyacetyl nitrate (PAN, $C_2H_3NO_5$), nitrogen oxides, ozone and nitric acid (Watkins, 1991). The conditions that are

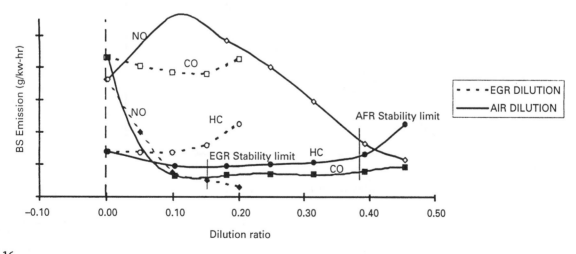

Figure 4.16
The effect of dilution ratio on the emissions from a spark ignition engine – a 0.2 dilution ratio means 20 per cent exhaust gas recirculation (EGR) with stoichiometric operation, or a lambda of 1.2 with no EGR (Shillington, 1998). © IMechE/Professional Engineering Publishing Limited.

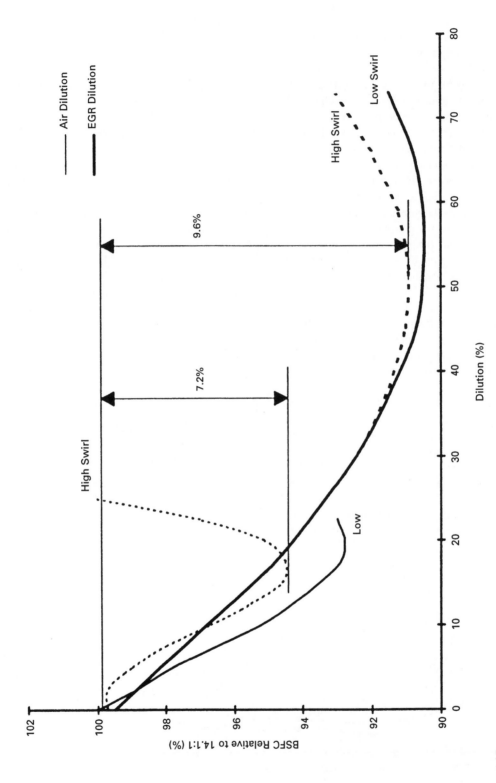

Figure 4.17

The effect of dilution ratio on the brake specific fuel consumption of a spark ignition engine – a 0.2 dilution ratio means 20 per cent exhaust gas recirculation (EGR) with stoichiometric operation, or a lambda of 1.2 with no EGR (Shillington, 1998). © IMechE/Professional Engineering Publishing Limited.

Table 4.3 *The contribution to the greenhouse effect of some greenhouse gases (excluding water vapour) – Sher (1998b)*

Gas	Current level (ppm)	Annual increase over last 20 yrs (per cent)	Relative warming effectiveness/mole	Relative contribution to the greenhouse effect (per cent)
Carbon dioxide CO_2	356	+0.4	1	50
Methane CH_4	1.74	+0.6	23	20
Nitrous oxide N_2O	0.31	+0.25	270	5
CFC-11 $CFCl_3$	0.26	+2.1	14 000	5
CFC-12 CF_2Cl_2	0.47	+2.3	19 500	10

needed for the formation of a photochemical smog are strong sunlight and no wind, so that the smog can be trapped near the ground by temperature inversion.

The smog contains oxidants (such as ozone, O_3, and nitrogen dioxide, NO_2), and since the ozone is comparatively stable it is measured as an indication of air quality. The constituents of the photochemical smog can cause many effects, including eye irritation, coughs, asthma attacks and reduced pulmonary function. In general it seems desirable for ozone levels to be below 0.1 ppm (Sher, 1998a). The mechanism for the formation of the ozone depends on the presence of nitrogen dioxide and ultra-violet (UV) light from sunlight:

$$NO_2 + UV \text{ light} \rightarrow NO + O \tag{4.1}$$

$$O + O_2 \rightarrow O_3 \tag{4.2}$$

There are additional ozone-formation mechanisms that involve unburnt hydrocarbons, and different organic compounds have different ozone-forming potentials (see table 4.6). However, since methane is so much less reactive than any other hydrocarbon (table 4.6), current US legislation excludes methane in the measurement of unburnt hydrocarbons; these are referred to as non-methane organic gases (NMOG).

The greenhouse effect causes a beneficial warming of the earth's surface of about 30 or 40 K (Waters, 1992). This is a result of selective absorption and re-radiation by the 'greenhouse' gases in the atmosphere; figure 4.18 shows the radiation balance between the earth and the sun. To understand the mechanism for these energy flows, it is necessary to consider:

(a) the way in which the intensity (I_λ) of the radiation from the earth and the sun vary with wavelength (λ), and

(b) the way in which the absorption/transmission of the radiation through the atmosphere varies with wavelength (λ).

These are both shown in figure 4.19. The incoming solar radiation is of a comparatively short wavelength, and about half reaches the earth. The other half is either reflected back to space or absorbed by the atmosphere. Since the earth is at a much lower temperature (about 300 K), the radiation that it emits is at a longer wavelength, and most of it (95 per cent) is absorbed by the atmosphere. Because the atmosphere is in thermal equilibrium, the total in-flow of heat (from the sun through the atmosphere, and the energy re-radiated by the atmosphere to the earth's surface) must equal the total outflow. Finally, since both the atmosphere and the earth are in thermal equilibrium, the incoming solar radiation (340 W/m^2) must equal that reflected, transmitted or re-radiated by the atmosphere to space. These energy flows are very large compared to the use of fossil fuels, which amounts to about 1 W/m^2 for the UK.

The greenhouse gases are those which absorb the longer wavelength radiation (say 4 μm and above). Figure 4.19 shows that there is an infra-red transmission window (that is, low absorption) in the range 7–13 μm. However, many pollutants (such as hydrocarbons, chlorofluorocarbons and nitrous oxide) can absorb radiation in this range. These molecules thus have a much greater effect on enhancing the greenhouse effect than adding more carbon dioxide or water vapour. Sher (1998b) has compiled results from several sources, and arrived at the relative significance of the different greenhouse gases that are included here in table 4.3.

Although the greenhouse effect is essential for life on earth, if its effect is increased then it can lead to global warming which is potentially damaging. Although there are no grounds for complacency, it is interesting to note that as atmospheric modelling has become more refined, predictions of global warming have been reduced. Estimates by CSTI (1992) suggested that global warming will be between 0.2 and 0.4 K/decade with no change in current fuel usage patterns. This is a level that is quite difficult to distinguish from natural changes in the earth's temperature, caused by phenomena such as variations in sun spot activity. The difficulty of interpreting observed climate changes adds importance to the need to model climate change. However, climate modelling is extraordinarily difficult for a number of reasons. Firstly, the oceans act as huge thermal stores, and they also have the ability to absorb carbon dioxide. Secondly, the most significant of the greenhouse gases is water vapour, and it is thus necessary to model the hydrological cycle – in other words, how much water is in the atmosphere, and in what form (clouds affect

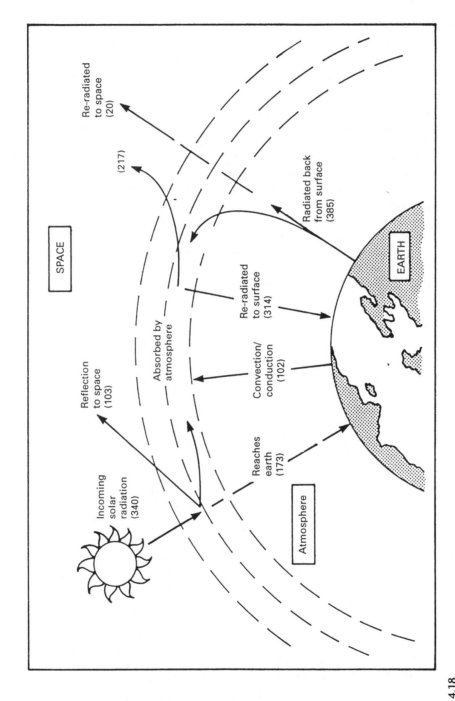

Figure 4.18
The radiation balance (W/m²) between the earth and the sun (Waters, 1992) [TRRL, Department of Transport].

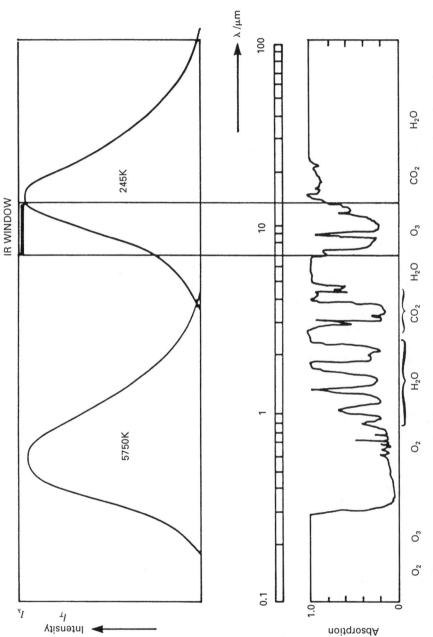

Above: Incoming short-wave radiation from the sun and outgoing long-wave radiation from the earth. Below: The pattern of absorption/transmission by the atmosphere and the principal gases involved.

Figure 4.19

The variation in intensity (I_λ) of the radiation from the earth and the sun with wavelength (λ), and the variation in the absorption/transmission of the radiation through the atmosphere (Waters, 1992) [TRRL, Department of Transport].

the incoming and outgoing infra-red radiation, and the water in clouds can be either water drops or ice crystals of varying size). Climate modelling is clearly an immensely difficult problem.

Unlike ozone close to the ground, ozone in the atmosphere at a height of 20–40 km is important in absorbing harmful UV radiation. Figure 4.19 shows that ozone can absorb UV radiation in the range of about 0.2–0.3 μm. Some UV radiation in the range 0.29–0.32 μm (known as UV-B) can be beneficial, but UV radiation in the range 0.24–0.29 μm (known as UV-C) is harmful. Ozone is formed in the stratosphere by termolecular collisions, but is consumed by reactions with ozone-depleting molecules:

$$O_2 + UV \rightarrow 2O \tag{4.3}$$

$$O_2 + O + M \rightarrow O_3 + M \quad \text{where M is another molecule} \tag{4.4}$$

$$O_3 + X \rightarrow O_2 + XO \quad \text{where X is an ozone-depleting molecule} \tag{4.5}$$

$$O_3 + XO \rightarrow O_2 + XO_2 \tag{4.6}$$

$$XO_2 \rightarrow O_2 + X \tag{4.7}$$

The ozone-depleting molecule (X) is not destroyed by its interaction with ozone, but by other processes which can have timescales of hundreds of years (DoEn, 1991). Chlorofluoro-carbons (CFCs) used in refrigeration equipment are probably the most widely known ozone-depleting substances, but nitric oxide (NO) and chlorine also cause ozone depletion. It has been reported by Phillips (1994) that chlorine released by natural phenomena such as volcanoes might be more significant in depleting ozone than are CFCs.

4.3.2 Development of three-way catalysts

Currently, the strictest emissions controls are enforced in the USA and Japan, and the legislation that led to the development of three-way catalysts is shown in table 4.4.

The US test is a simulation of urban driving from a cold start in heavy traffic. Vehicles are driven on a chassis dynamometer (rolling road), and the exhaust products are analysed using a constant-volume sampling (CVS) technique in which the exhaust is collected in plastic bags. The gas is then analysed for carbon monoxide (CO), unburnt hydrocarbons (HC) and nitrogen oxides (NO_x), using standard procedures. In 1970, three events – the passing of the American Clean Air Act, the introduction of lead-free petrol, and the adoption of cold test cycles for engine emissions – led to the development of catalyst systems.

Catalysts in the process industries usually work under carefully controlled steady-state conditions, but this is obviously not the case for engines – especially after a cold start. While catalyst systems were being developed, engine emissions were controlled by retarding the ignition and using exhaust gas recirculation (both to control NO_x) and a thermal reactor to complete oxidation of the fuel. These methods of NO_x control led to poor fuel economy and poor driveability (that is, poor transient engine response). Furthermore, the methods used to reduce NO_x emissions tend to increase CO and HC emissions and vice versa – see figure 3.19. The use of EGR and retarding the ignition also reduce the power output and fuel economy of engines.

Catalysts (Anon, 1984a) were able to overcome these disadvantages and meet the 1975 emissions requirements. The operating regimes of the different catalyst systems are shown in figure 4.20. With rich-mixture running, the catalyst promotes the reduction of NO_x by reactions involving HC and CO:

$$4HC + 10NO \rightarrow 4CO_2 + 2H_2O + 5N_2$$

and

$$2CO + 2NO \rightarrow 2CO_2 + N_2$$

Since there is insufficient oxygen for complete combustion, some HC and CO will remain. With lean-mixture conditions the catalyst promotes the complete oxidation of HC and CO:

$$4HC + 5O_2 \rightarrow 4CO_2 + 2H_2O$$

$$2CO + O_2 \rightarrow 2CO_2$$

Table 4.4 *Historical US Federal Emissions Limits (grams of pollutant per mile)*

Model year	CO	HC	NO_x	Solution
1966	87	8.8	3.6	Pre-control
1970	34	4.1	4.0	Retarded ignition, thermal reactors,
1974	28	3.0	3.1	exhaust gas recirculation (EGR)
1975	15	1.5	3.1	Oxidation catalyst
1977	15	1.5	2.0	Oxidation catalyst and improved EGR
1980	7	0.41	2.0	Improved oxidation catalysts and three-way catalysts
1981	7	0.41	1.0	Improved three-way catalysts and support materials

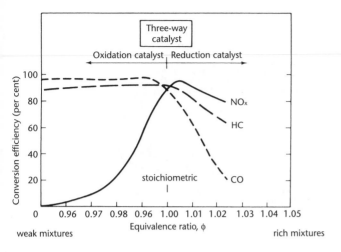

Figure 4.20
Conversion efficiencies of catalyst systems (courtesy of Johnson Matthey).

With the excess oxygen, any NO_x present would not be reduced.

Oxidation catalyst systems were the first to be introduced, but NO_x emissions still had to be controlled by exhaust gas recirculation. Excess oxygen was added to the exhaust (by an air pump), to ensure that the catalyst could always oxidise the CO and HC. The requirements of the catalyst system were:

1 High conversion of CO and HC at low operating temperatures.

2 Durability – performance to be maintained after 80 000 km (50 000 miles).

3 A low light-off temperature.

Light-off temperature is demonstrated by figure 4.21. The light-off temperature of platinum catalysts is reduced by adding rhodium, which is said to be a 'promoter'.

Dual catalyst systems control NO_x emissions without resort to exhaust gas recirculation or retarded ignition timings. A feedback system incorporating an exhaust oxygen sensor is used with a carburettor or fuel injection system to control the air/fuel ratio. The first catalyst is a reduction catalyst, and by maintaining a rich mixture the NO_x is reduced. Before entering the second catalyst, air is injected into the exhaust to enable oxidation of the CO and HC to take place in the oxidation catalyst.

Conventional reduction catalysts are liable to reduce the NO_x, but produce significant quantities of ammonia (NH_3). This would then be oxidised in the second catalyst to produce NO_x. However, by using a platinum/rhodium system the selectivity of the reduction catalyst is improved, and a negligible quantity of ammonia is produced.

Three-way catalyst systems control CO, HC and NO_x emissions as a result of developments to the platinum/rhodium catalysts. As shown by figure 4.21, very close control is needed on the air/fuel ratio. This is normally achieved by electronic fuel injection, with a lambda sensor to provide feedback by measuring the oxygen concentration in the exhaust. A typical air/fuel ratio perturbation for such a system is ±0.25 (or ±0.02ϕ).

When a three-way catalyst is used, it requires firstly, an engine management system capable of very accurate air/fuel ratio control, and secondly, a catalyst. Both these requirements add considerably to the cost of an engine. Since a three-way catalyst always has to operate with a stoichiometric air/fuel ratio, then at part load, this means that the maximum economy cannot be achieved (see figure 4.1). The fuel consumption penalty (and also the increase in carbon dioxide emissions) associated with stoichiometric operation is around 10 per cent.

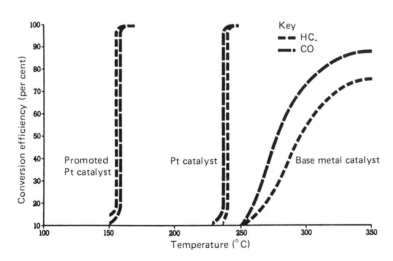

Figure 4.21
Light-off temperatures of different catalysts (courtesy of Johnson Matthey).

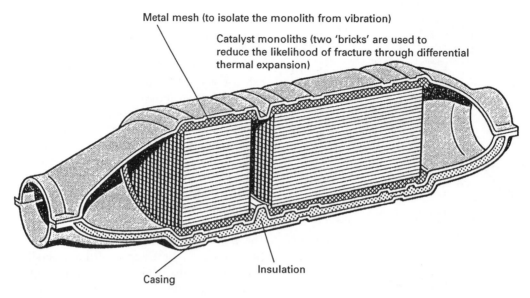

Metal mesh (to isolate the monolith from vibration)

Catalyst monoliths (two 'bricks' are used to reduce the likelihood of fracture through differential thermal expansion)

Insulation

Casing

Figure 4.22 Constructional details of a three-way catalyst, adapted from Gunther (1988).

Fortunately, the research into lean-burn combustion systems can be exploited, to improve the part-load fuel economy for engines operating with stoichiometric air/fuel ratios, by using high levels of EGR. A combustion system designed for lean mixtures can also operate satisfactorily when a stoichiometric air/fuel mixture is diluted by exhaust gas residuals. At part load up to around 30 per cent EGR can be used; this reduces the volume of flammable mixture induced, and consequently the throttle has to be opened slightly to restore the power output. With a more open throttle the depression across the throttle plate is reduced, and the pumping work (or pmep) is lower. Nakajima *et al.* (1979) showed that for a bmep of 3.24 bar with stoichiometric operation at a speed of 1400 rpm, they were able to reduce the fuel consumption by about 5 per cent through the use of 20 per cent EGR on an engine with a fast burn combustion system.

The construction of a typical three-way catalyst is shown in figure 4.22, and the technology used in the catalyst has been reviewed by Milton (1998). The catalyst is supported on either a metal or ceramic substrate (alumina Al_2O_3, aluminium oxide) which typically has an approximately square cross-sectional passage of width 1 mm. Ceramic substrates are more commonly used, and have a flow area of about 70 per cent of the cross-section, compared to about 90 per cent for metal substrates. The substrate is covered with a washcoat based on γ-Al_2O_3 (chosen for its porosity), but also containing the catalyst materials and stabilisers such as cerium oxide (CeO_2) and barium oxide (BaO). Unfortunately γ-Al_2O_3 turns to α-Al_2O_3 at 900°C, and the porosity is lost, so stabilisers are added to enable higher operating temperatures. The catalyst material is mostly platinum (about 85–95 per cent), but there are also significant quantities of rhodium.

The cerium oxide has the ability to store oxygen under weak operating conditions, and release it during rich-mixture operation. This is important for the operation of the three-way catalyst, since the oxygen sensor (section 4.7) can only determine whether a mixture is weak or rich, thus the engine management system seeks to oscillate the air/fuel ratio around stoichiometric with a frequency of about 1 Hz. This means that the ability of the catalyst to store oxygen and the pollutants is important. By installing a second lambda sensor downstream of the catalyst, and looking at the amplitude of the response compared to the upstream sensor, it is possible to assess the oxygen storage performance of the catalyst. The better the condition of the catalyst, then the greater the attenuation of the oxygen fluctuations at the catalyst exit (Rieck *et al.*, 1998).

Two important requirements of a catalyst system are:

(a) that it should be durable, with a life of say 80 000 km (50 000 miles), and

(b) that it should start working as soon as possible after the engine has started.

Durability

As a catalyst ages its performance changes in several ways. Firstly its light-off temperature rises, secondly its conversion efficiency will fall, and finally its response to different components in the exhaust will change (see section 4.3.4). The catalyst performance deteriorates through a number of mechanisms: poisoning of the catalyst, failure of the substrate, and sintering – a process by which the catalytic material agglomerates and its effective area is reduced.

Poisoning is de-activation of the catalytic material through

Table 4.5 *Federal Test Procedure – performance of two vehicles fitted with an electrically heated catalyst (EHC) – Heimrich et al. (1991)*

Configuration	NMOG (g/mi)	CO (g/mi)	NO$_x$ (g/mi)	Fuel economy (m/gal)
Veh 1, no EHC	0.15	1.36	0.18	20.2
Veh 1, with EHC[+]	0.02	0.25	0.18	19.7
Veh 2, no EHC	0.08	0.66	0.09	25.4
Veh 2, with EHC[*]	0.02	0.30	0.05	24.3

[+] With injection of 300 litres/min of air; 75 s for cold start, 30 s for hot start.
[*]With injection of 170 litres/min of air for 50 s for cold start.

deposits. Operating an engine on leaded fuel (0.125 gPb/litre) would degrade the catalyst's performance by about a quarter after 60 hours' use. Sulphur also degrades the catalyst through sulphate deposits, and European legislation limits the sulphur content to 10 ppm by mass. The catalyst will convert the sulphur dioxide to sulphur trioxide (and thence sulphuric acid) in an oxidising environment. In a rich mixture hydrogen sulphide (H$_2$S, the smell of rotten eggs) is formed, and this is often noticeable during engine warm-up, especially with the removal of nickel oxide from the catalyst on health grounds. Nickel oxide promotes the oxidation of hydrogen sulphide. A significant source of catalyst poisons can be from the additives in the lubricant (some of which is inevitably burnt), for example phosphorus is a catalyst poison.

Sintering occurs at higher temperatures over a period of time, while substrate failure occurs as a result of very short but high-temperature excursions caused by misfiring. Computational studies by Oh (1988) are supported by experimental observations. The substrate has a melting temperature of about 1700 K, and 25 per cent misfiring cycles can be tolerated, but 50 per cent misfiring leads to failure in about 10 s.

Catalyst light-off

With increasingly demanding emissions legislation, it is even more important for the catalyst to start working as soon as possible. The thermal inertia of the catalyst can be reduced by using a metal matrix, since the foil thickness is about 0.05 mm. Ceramic matrices usually have a wall thickness of about 0.3 mm, but this can be halved (Yamamoto *et al.*, 1991) to give a slight improvement in the light-off performance. Systems to promote catalyst light-off might usefully be classified as passive or active – active being when an external energy input is used. Two active systems are: electrically heated catalysts (using metal substrates), and exhaust gas ignition (EGI). EGI requires the engine to be run very rich of stoichiometric, and then adds air to the exhaust stream, so that an approximately stoichiometric mixture can then be ignited in the catalyst (Eade *et al.*, 1996). The mixture is ignited by a glow-plug situated in the chamber formed between two catalyst bricks.

Passive systems rely on thermal management. Typically a small catalyst is placed close to the engine, so that its reduced

mass and higher inlet temperatures give quicker light-off. Its small volume limits the maximum conversion efficiency, so that a second larger catalyst is then placed further downstream, under the car body. By operating the engine with very retarded ignition timing (even with ignition on the expansion stroke) then a very hot exhaust stream is produced. The late ignition means there is less work output, and the later combustion means there is less time for heat transfer. Both effects mean a higher exhaust temperature, and engines can be operated at close to full throttle, yet be at virtually no load. Proposals have also been made for storing the unburnt hydrocarbons prior to catalyst light-off, and then re-introducing them to the exhaust stream after light-off.

Electrically heated catalysts (EHCs) are placed between the close-coupled catalyst and the downstream catalyst. The electrically heated catalysts have a power input of about 5.5 kW, and are energised for 15–30 s before engine cranking, to raise their temperature to about 300°C. Once the engine is firing, the electrical power input is reduced by a controller that responds to the catalyst temperature. Results from a study of two vehicles fitted with an EHC by Heimrich *et al.* (1991) are shown in table 4.5.

It is likely that the performance of an EHC would be better than this when it is incorporated into the engine management strategy by the vehicle manufacturer. However, with EHC systems there are questions about durability, and indeed any active system is only likely to be used if it is the only solution available. A recent development from Johnson Matthey is a catalyst with a light-off temperatures in the range of 100–150°C for carbon monoxide and hydrogen. The engine is initially operated very rich (so reducing NO emissions and increasing the levels of carbon monoxide and hydrogen). Air is added after the engine to make the mixture stoichiometric, and the exothermic oxidation of the carbon monoxide and hydrogen heats up the catalyst. Initially unburnt hydrocarbons have to be stored in a trap, for release after the catalyst is fully warmed-up.

4.3.3 Lean-burn NO$_x$ reducing catalysts

It has already been seen how stoichiometric operation compromises the efficiency of engines, and that for control of NO$_x$ it is either necessary to operate at stoichiometric, or

Table 4.6 *Ozone-forming potential of some organic gases (MIR – maximum incremental reactivity)*

Fuel	Formula	Structure	MIR g O_3/g gas	Normalised reactivity
Alkanes				
methane	CH_4	CH_4	0.0148	1
ethane	C_2H_6	$(CH_3)_2$	0.25	17
propane	C_3H_8	$CH_3(CH_2)CH_3$	0.48	32
n-butane	C_4H_{10}	$CH_3(CH_2)_2CH_3$	1.02	69
n-pentane	C_5H_{12}	$CH_3(CH_2)_3CH_3$	1.045	70
i-octane (2,2,4-trimethylpentane)	C_8H_{18}	$CH_3C(CH_3)_2CH_2CH(CH_3)CH_3$	0.93	63
Alkenes				
ethene	C_2H_4	CH_2CH_2	7.29	493
propene	C_3H_6	CH_3CHCH_2	9.40	635
1-butene	C_4H_8	$CH_2CHCH_2CH_3$	8.91	602
2-butene	C_4H_8	$CH_3CHCHCH_3$	5.31	359
1,3-butadiene	C_4H_6	$CH_2CHCHCH_2$	10.89	736
Alkynes				
acetylene	C_2H_2		0.50	34
propyne	C_3H_4	$CHCCH_3$	4.10	277
1-butyne	C_4H_6	$CHCCH_2CH_3$	9.24	624
Aromatics				
benzene	C_6H_6		0.42	28
toluene (methylbenzene)	$C_6H_5CH_3$		2.73	184
ethylbenzene	$C_6H_5C_2H_5$		2.70	182
Alcohols				
methanol	CH_3OH		0.56	38
ethanol	C_2H_5OH		1.34	91
n-propanol	C_3H_7OH		2.26	153
Aldehydes				
formaldehyde	CH_2O	$HCHO$	7.16	484
acetaldehyde	C_2H_4O	CH_3CHO	5.52	373
propionaldehyde	C_3H_6O	C_2H_5CHO	6.53	441

sufficiently weak (say an equivalence ratio of 0.6) such that there is no need for NO_x reduction in the catalyst. If a system can be devised for NO_x to be reduced in an oxidising environment, then this gives scope to operate the engine at a higher efficiency.

There are a number of technologies being developed for 'DENOx'; although some technologies are more suitable for diesel engines they will none the less be discussed together here. The different systems are designated active or passive (passive being when nothing has to be added to the exhaust gases). The systems are:

(a) *SCR – Selective catalytic reduction*, a technique in which ammonia (NH_3) or urea ($CO(NH_2)_2$) is added to the exhaust stream. This is likely to be more suited to stationary engine applications. Conversion efficiencies of over 90 per cent are quoted, but the NO level needs to be known, because if too much reductant is added, then ammonia would be emitted.

(b) *Passive DENOx*. These use the hydrocarbons present in the exhaust to chemically reduce the NO. There is a narrow temperature window (in the range 160–220°C for platinum catalysts) within which the competition for HC between oxygen and nitric oxide leads to a reduction in the NO_x (Jochheim *et al.*, 1996). The temperature range is a limitation, and more suited to diesel engine operation. More recent work with copper-exchanged zeolite catalysts has shown them to be effective at higher temperatures, and by modifying the zeolite chemistry, a peak NO_x conversion efficiency of 40 per cent has been achieved at 400°C (Brogan *et al.*, 1998).

(c) *Active DENOx catalysts*. These use the injection of fuel to reduce the NO_x, and a reduction in NO_x of about 20 per cent is achievable with diesel-engined vehicles on typical drive cycles, but with a 1.5 per cent increase in the fuel consumption (Pouille *et al.*, 1998). Current systems inject fuel into the exhaust system, but there is the possibility of late in-cylinder injection with future diesel engines.

(d) *Lean NOₓ trap (LNT) catalysts*. In this technology (first developed by Toyota) a three-way catalyst is combined with a NO_x-absorbing material, to store the NO_x when the engine is operating in lean-burn mode. When the engine operates under rich conditions, the NO_x is released from the storage media, and reduced in the three-way catalyst.

NO_x trap catalysts have barium carbonate deposits between the platinum and the alumina base. During lean operation, the nitric oxide and oxygen convert the barium carbonate to barium nitrate. A rich transient (about 5 s at an equivalence ratio of 1.4) is needed about every 5 minutes so that the carbon monoxide, unburnt hydrocarbons and hydrogen regenerate the barium nitrate to barium carbonate. The NO_x that is released is then reduced by the partial products of combustion over the rhodium in the catalyst. Sulphur in the fuel causes the NO_x trap to lose its effectiveness because of the formation of barium sulphate. However, operating the engine at high load to give an inlet temperature of 600°C, with an equivalence ratio of 1.05, for 600 s can be used to remove the sulphate deposits (Brogan *et al.*, 1998).

4.3.4 Ozone-forming potential

Because different hydrocarbon emissions have differing ozone-forming potential (see table 4.6), there is an argument for taking into account the composition of the hydrocarbon emissions, not just the quantity. Exhaust gas speciation of the hydrocarbons requires the use of gas chromatography for the identification and quantification of the emissions, possibly in conjunction with mass spectrometry to identify the species. For the oxygenates in the exhaust (aldehydes and ketones), a sample has to be drawn through a silica gel cartridge containing 2,4-dinitrophenylhydrazine (DNPH), which then reacts with the aldehydes and ketones. These can then be removed from the cartridge by injecting acetonitrile, and measured by high performance liquid chromatography (HPLC).

The composition of the unburnt hydrocarbon emissions from a spark ignition engine depends on the fuel, the air/fuel ratio, the age of the catalyst and its temperature. Figure 4.23 shows the emission mass and residual ratio (the percentage of a species not converted in the catalyst) for the main hydrocarbon species when an engine is operated on baseline gasoline. The results were obtained over the Federal Test Procedure drive cycle, for vehicles fitted with an underfloor catalyst (UF), or a close-coupled catalyst and an underfloor catalyst (CC + UF) – Kojima *et al.* (1992). These results show that the addition of a close-coupled catalyst leads to a significant increase in the conversion efficiency, and that the conversion efficiency is highest for the alkenes and alkynes, and lowest for the alkanes.

Figure 4.24 shows the effect of catalyst ageing on the emissions residual ratio (the percentage of a species not converted in the catalyst) for the main hydrocarbon species

(Kojima *et al.*, 1992). The vehicle was fitted with a close-coupled and underfloor catalyst, and driven for 50 000 miles (80 000 km). These results show that the conversion efficiency falls most for the alkanes, so it is fortunate that these have the lowest ozone-forming potential. If a fuel is used that has an increased alkanes content and a decreased aromatics content, then it might be thought that this will lead to a lower ozone-forming potential for the exhaust emissions. This is not necessarily the case, since although the aromatic levels in the exhaust will decrease, the catalyst is effective at oxidising them. In contrast, increasing the alkanes in the fuel will increase the emissions of the alkenes and alkynes in the exhaust, and the catalyst is less effective at oxidising these.

It is clearly unrealistic to expect exhaust gas hydrocarbon speciation to be undertaken on a routine basis, so if ozone-forming potential is to be assessed, it is likely to be through the use of a fuel-specific Reactivity Adjustment Factor (RAF) that is multiplied by the NMOG emissions. These RAFs are not likely to vary much for gasolines, but would be higher for oxygenate fuels (because fuels like methanol will generate comparatively high levels of aldehydes). In contrast, RAFs for gaseous fuels will be low, because the dominant emission is methane.

Particulate matter emissions from conventional spark ignition engines are very low. However, they do follow the same trends as the unburned hydrocarbons, so that a retarded ignition timing gives lower particulate matter emissions. This is because there is more time for the air/fuel mixture to become fully mixed, and the temperatures in the expansion stroke will be higher, and this leads to more post-flame oxidation. The temperatures in the expansion stroke are higher because less work will have been done and there is less time for heat transfer. A rich of stoichiometric mixture leads to a marked increase in particulate matter emissions. The particulate matter emissions from gasoline direct injection engines are more significant and are subject to the legislation in table 4.7; they are discussed in chapter 5, section 5.7.

4.4 Cycle-by-cycle variations in combustion

Cycle-by-cycle variation of the combustion (or cyclic dispersion) in spark ignition engines was mentioned in chapter 3, section 3.5.1. It is illustrated here by figure 4.25, the pressure–time record for five successive cycles. Clearly not all cycles can be optimum, and Soltau (1960) suggested that if cyclic dispersion could be eliminated, there would be a 10 per cent increase in the power output for the same fuel consumption with weak mixtures. Similar conclusions were drawn by Lyon (1986), who indicated that a 6 per cent improvement in fuel economy could be achieved if all cycles burned at the optimum rate. Perhaps surprisingly, the total elimination of cyclic dispersion may not be desirable, because of engine management systems that retard the ignition when knock is

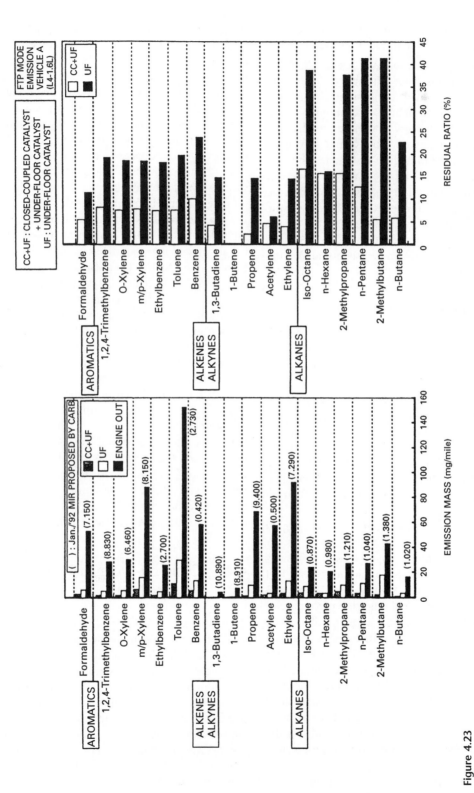

Figure 4.23

The emissions mass and residual ratio (percentage of a species not converted in the catalyst) for the main hydrocarbon species when an engine is operated on gasoline (Kojima *et al.*, 1992). © IMechE/Professional Engineering Publishing Limited.

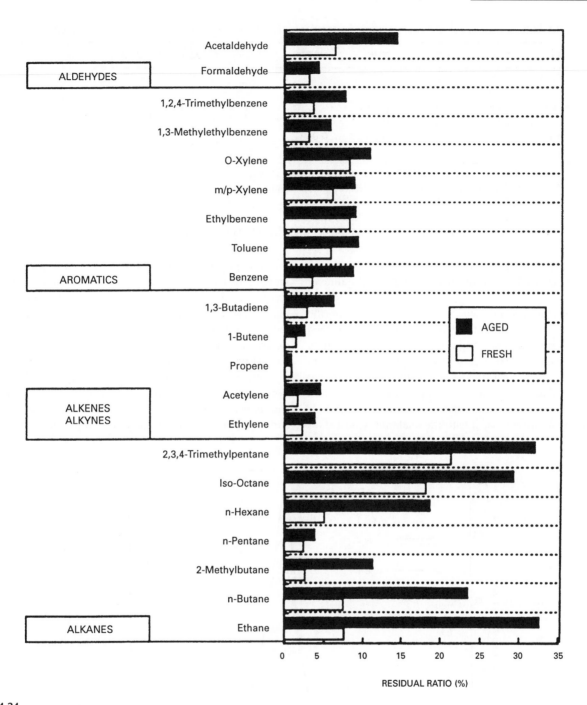

Figure 4.24
The emissions residual ratio (percentage of a species not converted in the catalyst) for the main hydrocarbon species when an engine is operated on gasoline, for a new and aged catalyst (Kojima *et al.*, 1992). © IMechE/Professional Engineering Publishing Limited.

detected. If there was no cyclic dispersion, then either none or all of the cycles would knock. It would be acceptable to the engine and driver for only a few cycles to knock, and the ignition control system can then introduce the necessary ignition retard. If all the cycles were to knock it is likely to lead to runaway knock, in which case retarding the ignition would have no effect.

Cyclic dispersion occurs because the turbulence within the cylinder varies from cycle to cycle, the air/fuel mixture is not homogeneous (there may even be droplets of fuel present) and the exhaust gas residuals will not be fully mixed with the unburned charge. It is widely accepted that the early flame development can have a profound effect on the subsequent combustion. Firstly, the formation of the flame kernel will

Table 4.7 *European exhaust emissions standards for light-duty petrol-engined vehicles (g/km); see also table 3.21*

Level	Date	CO	THC	NMHC	NO$_x$	HC + NO$_x$	Pm*	Pn*
Euro 1	1992[+]	2.72	–	–	–	0.97	–	–
Euro 2	1996[+]	2.2	–	–	–	0.5	–	–
Euro 3	2000	2.3	0.20	–	0.15	–	–	–
Euro 4	2005	1.0	0.10	–	0.08	–	–	–
Euro 5	2009	1.0	0.10	0.068	0.06	–	0.005	–
Euro 6	2014	1.0	0.10	0.068	0.060	–	0.005	##

THC – Total hydrocarbons; NMHC – non-methane hydrocarbons
Pm – Particulate (matter) mass; Pn Particulate (matter) number
– Value yet to be determined (particles/km)
* – Only applicable to Gasoline Direct Injection Engines
[+] – Excludes the first 40 s after starting and the highway part of the current drive cycle

depend on the local air/fuel ratio, the mixture motion, and the exhaust gas residuals in the spark plug gap at the time of ignition.

De Soete (1983) conducted a parametric study of the phenomena controlling the initial behaviour of spark-ignited flames. He confirmed that combustion starts as self-ignition, occurring in the volume of very hot gases (spark kernel) behind the expanding, spark-created shock wave. He also observed that spark-ignited flames pass through a non-steady propagation period before reaching a steady speed. This transient period is relatively important, compared with the total time available for combustion, in an engine cycle. In the early stages of flame growth, the flame is small compared with the turbulence length scales, and the flame can be convected away from the spark plug. If the flame nucleus is moved into the thermal-boundary layer surrounding the combustion chamber, then it will burn slowly. Furthermore, at a given flame radius, the greater contact with the wall will reduce the flame front area.

Conversely, if the flame is moved away from the combustion chamber surfaces, then it will burn more quickly. The turbulence only enhances the burn rate once the flame surface is large enough to be distorted by the turbulence, by

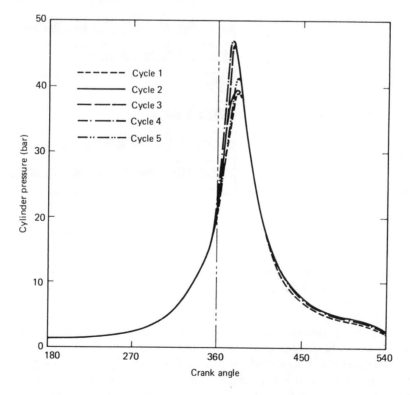

Figure 4.25
Pressure–time diagrams for five successive cycles in a Ricardo E6 engine (compression ratio 8:1, stoichiometric air/iso-octane mixture, 1000 rpm, 8.56 bar bmep) (reprinted from Stone and Green-Armytage, 1987, by permission of the Council of the Institution of Mechanical Engineers).

which time the turbulence no longer moves the flame around the combustion chamber. Lyon (1986) conducted a statistical analysis of groups of pressure distributions. This demonstrated that events early in the development of the flame kernel largely dictate the subsequent rate of combustion and pressure development. When combustion starts slowly, then it tends to continue slowly.

Figure 4.26 is a typical scatter plot that shows the relation between the 0–10 per cent mass fraction burn duration and the 10–80 per cent mass fraction burn duration. The 10–80 per cent mass fraction burn duration describes the fully turbulent combustion period, while the 0–10 per cent mass fraction burn duration covers the initial laminar combustion and the transition to fully developed turbulent combustion (this is discussed further in chapter 9, section 9.4.3). Figure 4.26 shows that a cycle that initially burns slowly will continue to burn slowly. The combustion phasing of a slow burning cycle is such that the 10–80 per cent mass fraction burn phase will occur later in the cycle, when in-cylinder temperatures will be falling in the expansion stroke, thereby resulting in a lower burn rate.

Cyclic dispersion is increased by anything that tends to slow-up the combustion process, for example: lean mixture operation, exhaust gas residuals and low load operation (in part attributable to greater exhaust gas residuals, but also attributable to lower in-cylinder pressures and temperatures). Before illustrating the variations in cyclic dispersion, it is first necessary to be able to measure it.

With modern data acquisition systems, it is possible to log the cylinder pressure from many successive cycles. It is then possible to analyse the data for each cycle, and to evaluate the maximum pressure, the maximum rate of pressure rise, the imep and the burn rate. (A burn rate analysis technique is discussed later in chapter 14, section 14.5.2.) A widely adopted way of summarising the burn rate is to note the 0–10 per cent (or sometimes 0–1 per cent), 0–50 per cent and 0–90 per cent burn durations. Simple statistical analyses yield the mean, standard deviation and coefficient of variation (CoV = standard deviation/mean), for each of the combustion parameters. It is necessary to collect a sufficiently large sample to ensure stationary values from the statistical analyses; this may require data from up to 1000 cycles when there is a high level of cyclic dispersion.

Table 4.8 presents some cyclic dispersion data from a gas engine. The table shows that each performance parameter has a different value of its coefficient of variation, which means that the way cyclic dispersion has been measured should always be defined. The peak pressure is occasionally used, since it is easy to measure. However, the imep is probably the parameter with the most relevance to the overall engine performance. At sufficiently high levels of cyclic dispersion the driver of a vehicle will become aware of fluctuations in the engine output, and this is clearly linked to the imep, because of the integrating effect of the engine flywheel. Ultimately, cyclic dispersion leads into misfire, but a driver will detect poor driveability much earlier than this. For good driveability, the coefficient of variation in the imep should be no greater than 5–10 per cent. Not only is there no direct link between the coefficients of variation of the peak pressure and the imep, but they can also respond to variables in opposite ways.

Figure 4.27 shows the variation in the mean imep, and the coefficients of variation of imep and peak pressure, when the ignition timing is varied. The imep is fairly insensitive to the changes in ignition timing (MBT ignition timing 17°C btdc), but it can be seen that the maximum imep corresponds quite closely to a minimum in the coefficient of variation of the imep, although the coefficient of variation of the peak pressure

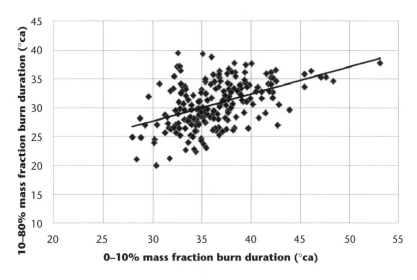

Figure 4.26 Scatter plot of the 0–10 per cent and 10–80 per cent mass fraction burn durations for a stoichiometric mixture at part throttle. Figure courtesy of Huayong Zhao.

Table 4.8 *Statistical summary of the combustion performance from a fast burn gas engine combustion system operating at 1500 rpm, full throttle with an equivalence ratio of 1.12 and MBT ignition timing*

Peak pressure	mean (bar)	59.5
	standard deviation (bar)	2.7
	coefficient of variation (per cent)	4.5
imep	mean (bar)	7.40
	standard deviation (bar)	0.14
	coefficient of variation (per cent)	1.9
0–10 per cent burn duration	mean (°ca)	10.3
	standard deviation (°ca)	1.2
	coefficient of variation (per cent)	11.6
0–50 per cent burn duration	mean (°ca)	18.4
	standard deviation (°ca)	1.54
	coefficient (per cent)	8.4
0–90 per cent burn duration	mean (°ca)	32.7
	standard deviation (°ca)	3.07
	coefficient of variation (per cent)	9.4

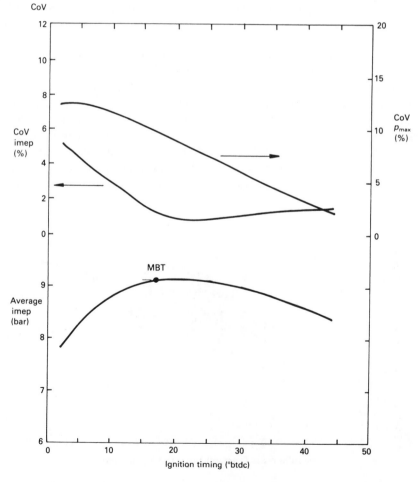

Figure 4.27
The influence of ignition on the imep, and cycle-by-cycle variation in combustion. Ricardo E6 operating at 1500 rpm with full throttle.

increases as the ignition timing is moved closer to tdc, until a maximum (7.6 per cent) occurs at about 4°btdc. In general, the coefficient of variation of imep is a minimum in the region of MBT ignition timing. This means that any uncertainty in locating MBT ignition timing has a lesser effect on the coefficient of variation of imep than the coefficient of variation of the maximum cylinder pressure. The dependence of the coefficients of variation on the ignition timing does not appear to be widely appreciated, but it has been studied in detail by Brown *et al.* (1996). When coefficients of variation are quoted it is important to identify both the ignition timing and the sensitivity of the coefficient of variation to the ignition timing.

Another useful statistical parameter is the lowest normalised value (LNV), which can be expressed as a percentage. The LNV is the lowest value of a parameter divided by the mean value. Suppose pressure data from 100 cycles were recorded, and the combustion was very stable, apart from one cycle that burnt very slowly. The standard deviation (and thus the CoV) would be very low, but this might give a misleading impression. In contrast, the LNV of the imep would be significantly less than unity, and indicate that there had been poor combustion in at least one cycle.

Figure 4.28 shows how the engine load and air/fuel ratio affect the imep, cycle-by-cycle variations in combustion and burn rate. At a fixed throttle, the richer mixture burns faster and has a higher imep and lower cycle-by-cycle variations in combustion. If combustion is phased closer to top dead centre because it is faster, the effects of the piston motion are much less significant because of the sinusoidal nature of the piston movement. At the lower load the reduced pressure in the inlet manifold will mean increased levels of residuals in the combustion chamber, and this reduces the burn rate significantly. The response of the imep to the ignition timing has already been discussed with figure 4.4, and the CoV of imep is lowest because the mean value of imep is at its maximum, and the effects of some cycles burning too quickly and others burning too slowly for the given ignition timing are at a minimum since the combustion phasing for the mean cycles is optimised.

Retarding the ignition timing tends to reduce the 0–10 per cent mass fraction burn duration as the in-cylinder temperatures will be higher, and this increases the laminar burning velocity. (The increased pressure will tend to reduce the laminar burning velocity, but the temperature effect dominates.) However, with a retarded ignition timing the phasing of the 10–80 per cent mass fraction burn phase is too late, and as with the slow-burning cycles discussed in figure 4.26, combustion is continuing too far into the expansion stroke with its falling temperatures. The 10–80 per cent mass fraction burn phase will be quickest when the piston is close to top dead centre (as this is when the unburned mixture temperature is highest), so this will be for an earlier ignition timing than the ignition timing that gives the shortest 0–10 per cent mass fraction burn duration.

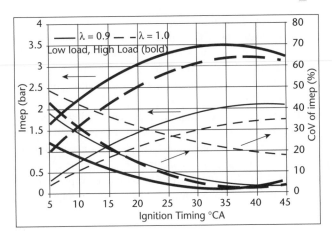

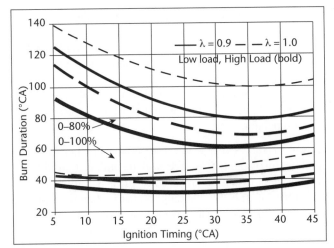

Figure 4.28 The effect of engine load and air/fuel ratio on the imep, cycle-by-cycle variations in combustion and mass fraction burn durations.

4.5 | Ignition systems

4.5.1 Ignition system overview

Most engines have a single spark plug per cylinder, a notable exception being in aircraft where the complete ignition system is duplicated to improve reliability. The spark is usually provided by a battery and coil, though for some applications a magneto is used.

For satisfactory performance, the central electrode of the spark plug should operate in the temperature range 350–700°C; if the electrode is too hot, pre-ignition will occur. On the other hand, if the temperature is too low carbon deposits will build up on the central insulator, so causing electrical breakdown. The heat flows from the central electrode through the ceramic insulator; the shape of this determines the operating temperature of the central electrode. A cool-running engine requires a 'hot' or 'soft' spark plug with a long heat flow path in the central electrode (figure 4.29a). A hot-running engine, such as a high-performance engine or a high compression ratio engine, requires a 'cool' or 'hard' spark plug. The much shorter heat flow path for a 'cool' spark plug is shown in figure 4.29b. The spark plug requires a voltage of 5–15 kV to cause electrical breakdown and form an arc; the larger the electrode gap and the higher the cylinder pressure, the greater the required voltage.

Both a traditional coil ignition system and a magneto ignition system are shown in figure 4.30. The coil is in effect a transformer with a primary or LT (Low Tension) winding of about 200 turns, and a secondary or HT (High Tension) winding of about 20 000 turns of fine wire, all wrapped round an iron core. The voltage V induced in the HT winding is

$$V = M\frac{\mathrm{d}I}{\mathrm{d}t} \tag{4.8}$$

where I is the current flowing in the LT winding
      M is the mutual inductance = $k\sqrt{(L_1 \times L_2)}$
      L_1, L_2 are the inductances of the LT and HT windings, respectively (proportional to the number of turns squared)
and k is a coupling coefficient (less than unity)
or $V = k$ (turns ratio of windings) (Low Tension Voltage).

Figure 4.31 shows that where the contact breaker closes to complete the circuit a voltage will be induced in the HT windings, but it will be small since $\mathrm{d}I/\mathrm{d}t$ is limited by the inductance of the LT winding and the battery voltage. Equation (4.9) defines the current flow in the LT winding:

$$I = \left(\frac{V_s}{R}\right)\left[1 - \exp\left(-\frac{Rt}{L_1}\right)\right] \tag{4.9}$$

where V_s = supply voltage
      R = resistance of the LT winding
      t = time after application of V_s.

When the contact breaker opens $\mathrm{d}I/\mathrm{d}t$ is much greater and sufficient voltage is generated in the HT windings to form an arc between the electrodes. A high voltage (200–300 V) is generated in the LT windings and this energy is stored in the capacitor. Without the capacitor there would be severe arcing at the contact breaker. Once the spark has ended, the capacitor discharges.

The windings of the ignition coil are connected in such a way that the voltage generated in the LT winding is added to the voltage generated in the HT winding. Once the spark has ended the capacitor discharges – the inductance and capacitance in the ignition system form a resonant circuit, so there are oscillations, but these oscillations decay because energy is dissipated by resistance in the circuit.

The energy input to the LT side (E_p) of the coil is the integration of the instantaneous current (I) and the supply voltage (V_s) over the period when the coil is switched on:

$$E_p = \int_0^{t'} I V_s \mathrm{d}t \tag{4.10}$$

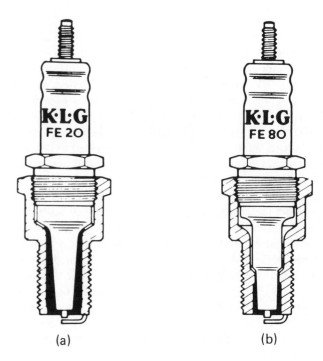

Figure 4.29
Spark plugs: (a) hot running; (b) cold running (from Campbell, 1978).

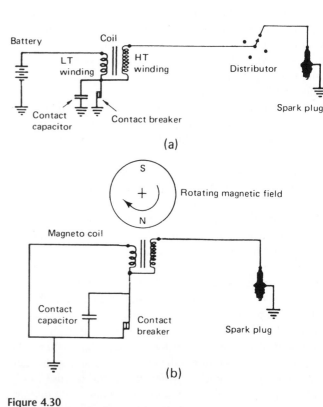

Figure 4.30
Mechanically operated ignition systems:
(a) traditional coil ignition system;
(b) magneto ignition system (adapted from Campbell, 1978).

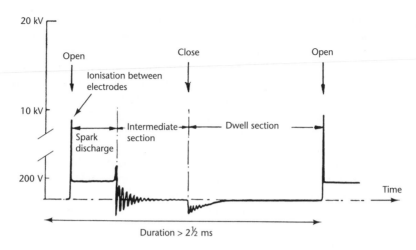

Figure 4.31
HT output from ignition coil (adapted from Campbell, 1978).

where t' is the time at which the coil is switched off, the coil-on-time.

Using equations (4.9) and (4.10) it can be shown that the energy stored in the coil (E_s) is

$$E_s = 0.5L_1I_p^2 \qquad (4.11)$$

where I_p is the LT current at the time when the coil is switched off.

The HT output is shown in figure 4.31. Initially 9 kV is needed to ionise the gas sufficiently before the spark arcs with a voltage drop of 200 V. As engine speeds increase, the dwell period becomes shorter and the spark energy will be reduced. Such a system can produce up to about 400 sparks per second; beyond this the spark energy becomes too low and this leads to misfiring. For higher spark rates, twin coil/contact breaker systems or alternatively electronic systems can be used. The current through the contact breaker can be reduced by using it to switch the base of a transistor that controls the current to the LT winding; this prolongs the life of the contact breaker. The contact breaker can be replaced by making use of opto-electronic, inductive or magnetic switching. However, all these systems use the coil in the same way as the contact breaker, but are less prone to wear and maladjustment. Without the mechanical limitations of a contact breaker, a coil ignition system can produce up to about 800 sparks per second.

In a magneto there is no need to use a battery since a current is induced in the LT winding by the changing magnetic field. Again a voltage is induced in the HT winding; as before, it is significant only when dI/dt is large at the instant when the contact breaker opens. The air gap between the rotating magnet and the iron core of the coil should be as small as possible, so that the path for the magnetic flux has as low an impedance as possible. Magneto ignition is best suited to engines that are independent of a battery.

Another type of ignition system that can be used is capacitive discharge ignition (CDI). The battery voltage is used to drive a charging circuit that raises the capacitor voltage to about 500 V. At ignition, the energy stored in the capacitor is discharged through an ignition transformer (that is, a coil with primary and secondary windings), the circuit being controlled by a thyristor. The discharge from the capacitor is such, that a short-duration (about 0.1 ms) spark is generated; the rapid discharge makes this ignition system less susceptible to spark plug fouling.

Ignition timing is usually expressed as degrees before top dead centre (°btdc), that is, before the end of the compression stroke. The ignition timing should be varied for different speeds and loads. However, for small engines, particularly those with magneto ignition, the ignition timing is fixed.

Whether the ignition is by battery and coil (positive or negative earth) or magneto, the HT windings are if possible arranged to make the central electrode of the spark plug negative. The electron flow across the electrode gap comes from the negative electrode (the cathode), and the electrons flow more readily from a hot electrode. Since the central electrode is not in direct contact with the cylinder head, this is the hotter electrode. By arranging for the hotter electrode to be the cathode the breakdown voltage is reduced.

4.5.2 Control of ignition timing

In chapter 3, section 3.5, it was explained how turbulent flame propagation occupies an approximately constant fraction of the engine cycle, since at higher speeds the increased turbulence gives a nearly corresponding increase in flame propagation rate. However, the initial period of flame growth occupies an approximately constant time (a few ms) and this corresponds to increased crank angles at increased speeds. The ignition advance was originally provided by spring-controlled centrifugal flyweights. Very often two springs of different stiffness are used to provide two stages of advance rate.

Ignition timing has to be advanced at part throttle settings since the reduced pressure and temperature and increased level of residuals in the cylinder cause slower combustion. The part throttle condition is defined by the pressure drop between atmosphere and the inlet manifold, the so-called engine 'vacuum'. An exploded view of a typical automotive distributor is shown in figure 4.32. The central shaft is driven at half engine speed (for four-stroke cycles) and the rotor arm directs the HT voltage to the appropriate spark plug via the distributor cap. For a four-cylinder engine there is a four-lobed cam that operates the contact breaker. The contact breaker and capacitor are mounted on a plate that can rotate a limited amount around the cams relative to the base plate. The position of this plate is controlled by the vacuum unit, a spring-controlled diaphragm that is connected to the inlet manifold. The cams are on a hollow shaft that can rotate around the main shaft. The relative angular position of the two shafts is controlled by the spring-regulated flyweights. The ignition timing is set by rotating the complete distributor

relative to the engine. Figure 4.33 shows typical ignition advance curves for engine speed and vacuum.

These characteristics form a series of planes, which are an approximation to the curved surface that represents the MBT ignition timing on a plot of advance against engine speed and engine vacuum. When an engine management system is used, then the MBT ignition timing can be stored as a function of load and speed; this is illustrated by figure 4.34.

Many engines that have an engine management system to control the spark, still make use of the distributor to send the HT from a single coil to the appropriate cylinder. This requires a mechanical drive, introduces a loss (due to the spark gap at the end of the rotor arm) and increases the risk of a breakdown in the HT system insulation.

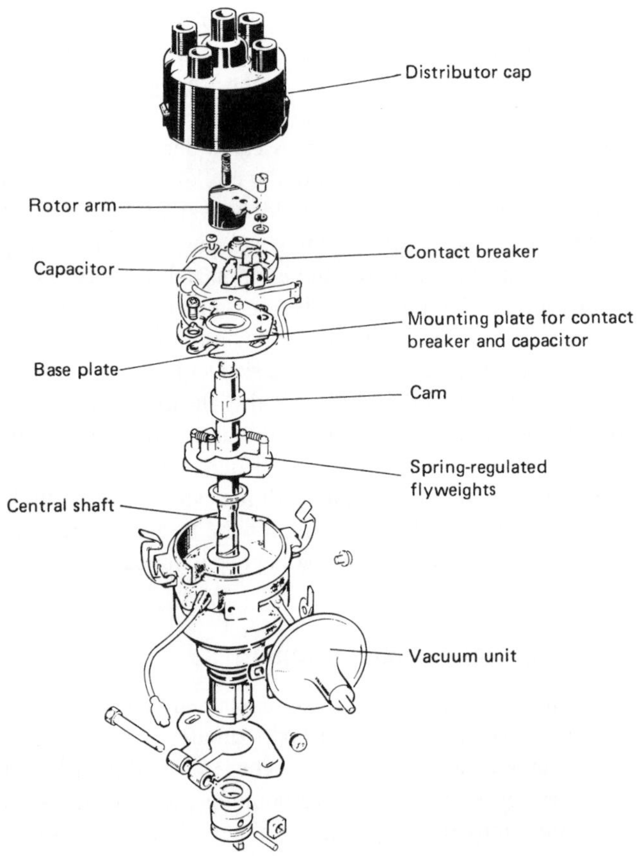

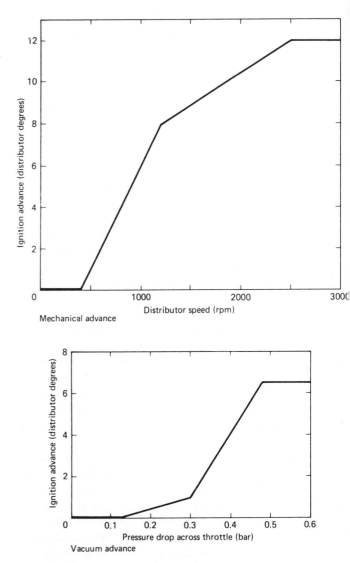

Figure 4.32
Automotive distributor (courtesy of Lucas Electrical Ltd).

Figure 4.33
Typical ignition advance curves.

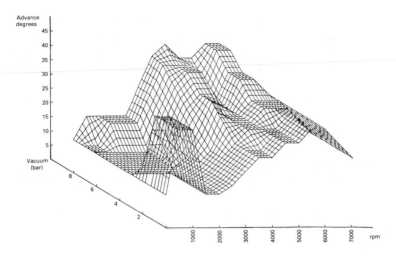

Figure 4.34
The ignition timing map (as a function of engine speed and inlet manifold pressure) as used in an engine management system (from Forlani and Ferrati, 1987).

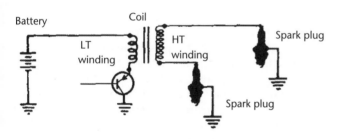

Figure 4.35 A double-ended coil ignition system; also known as a wasted spark system since a spark occurs during the exhaust stroke of the non-firing cylinder.

An alternative is to use a distributorless ignition system. In its simplest form this uses one coil per spark plug, but a more elegant solution is a double-ended coil, as shown in figure 4.35. With a double-ended coil, each end of the HT winding is connected directly to a spark plug in cylinders with a 360° phase separation. As a spark occurs at both plugs once every revolution, these systems are sometimes called wasted spark systems. The spark occurring at the end of the exhaust stroke should not have any effect, and since a spark is generated every revolution, the timing information can be collected from the flywheel (either the flywheel teeth or some other form of encoder).

4.5.3   The ignition process

The ignition process has been investigated very thoroughly by Maly and Vogel (1978) and Maly (1984). The spark that initiates combustion may be considered in the three phases shown in figure 4.36:

1 *Pre-breakdown.* Before the discharge occurs, the mixture in the cylinder is a perfect insulator. As the spark pulse occurs, the potential difference across the plug gap increases rapidly (typically 10–100 kV/ms). This causes electrons in the gap to accelerate towards the anode. With a sufficiently high electric field, the accelerated electrons may ionise the molecules they collide with. This leads to the second phase – avalanche breakdown.

2 *Breakdown.* Once enough electrons are produced by the pre-breakdown phase, an overexponential increase in the discharge current occurs. This can produce currents of the order of 100 A within a few nanoseconds. This is

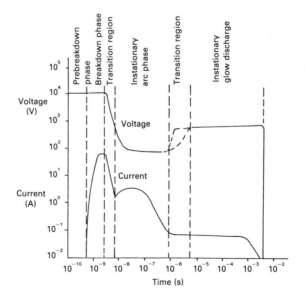

Figure 4.36
The current and voltage as a function of time during a spark discharge (adapted from Maly, 1984).

concurrent with a rapid decrease in the potential difference and electric field across the plug gap (typically to 100 V and 1 kV/cm respectively). Maly suggests that the minimum energy required to initiate breakdown at ambient conditions is about 0.3 mJ. The breakdown causes a very rapid temperature and pressure increase. Temperatures of 60 000 K give rise to pressures of several hundred bars. These high pressures cause an intense shock wave as the spark channel expands at supersonic speed. Expansion of the spark channel allows the conversion of potential energy to thermal energy, and facilitates cooling of the plasma. Prolonged high currents lead to thermionic emission from hot spots on the electrodes and the breakdown phase ends as the arc phase begins.

3   *Arc discharge*. The characteristics of the arc discharge phase are controlled by the external impedances of the ignition circuit. Typically, the burning voltage is about 100 V and the current is greater than 100 mA, and is dependent on external impedances. The arc discharge is sustained by electrons emitted from the cathode hot spots. This process causes erosion of the electrodes, with the erosion rate increasing with the plug gap. Depending on the conditions, the efficiency of the energy-transfer process from the arc discharge to the thermal energy of the mixture is typically between 10 and 50 per cent.

With currents of less than 100 mA, this phase becomes a glow discharge, which is distinguished from an arc discharge by the cold cathode. Electrons are liberated by ion impact, a less efficient process than thermionic emission. Even though arc discharges are inherently more efficient, glow discharges are more common in practice, because of the high electrode erosion rates associated with arc discharges.

The determination of the optimum spark type and duration has resulted in disagreement between researchers. Some work concludes that longer arc durations improve combustion systems, while other work indicates that short-duration (10–20 ns) high-current arcs (such as occur with capacitive discharge ignition systems) can be beneficial. These arguments have been reviewed by Stone and Steele (1989), who also report on tests in which the spark energy and the spark plug gap were varied. The spark energy was measured in a special calorimeter, and it was controlled by varying the coil-on-time and the spark plug gap. The engine performance was characterised by the bsfc and cyclic dispersion of tests at 1200 rpm with a bmep of 3.2 bar and an air/fuel ratio of 17. It was found that spark plug gap was a stronger determinant of engine performance than spark energy, and there was little to be gained by using spark plug gaps above 0.75 mm. However, for small spark plug gaps there were advantages in increasing the spark energy.

The apparent conflict between claims for long-duration and short-duration sparks can be reconciled. The short-duration spark has a better thermal conversion efficiency, and can overcome in-cylinder variations by reliable ignition and accelerated flame kernel development. In contrast, the long-duration discharge is successful, since it provides a time window long enough to mask the effects of in-cylinder variations. Similarly, a large spark plug gap is beneficial, since it increases the likelihood of there existing a favourable combination of turbulence and mixture between the electrodes.

4.6   Mixture preparation

The air/fuel mixture can be prepared by either a carburettor or a fuel injection system. Unless direct injection is used, fuel will be present in the inlet system as vapour, liquid droplets and a liquid film. Although emissions legislation has reduced the scope for using carburettors, they are still used on very small engines. There are two main types of carburettor:

(i)   fixed jet (or fixed venturi) described in section 4.6.3, or

(ii) variable jet (or variable venturi) described in section 4.6.2.

Two types of fuel injection system have been used on spark ignition engines, multi-point injection and single-point injection; both of these systems are described in section 4.6.4. The multi-point injection system employs injectors usually mounted close to the inlet port(s) of each cylinder. The single-point injection system can look very much like a carburettor, and as with the carburettor, the throttle plate and inlet manifold play an important part in mixture preparation. Even with multi-point fuel injection systems, a liquid fuel film will develop on the walls of the inlet ports.

Referring back to figure 4.2, points A and B represent the same power output but it is obviously more economical to operate the engine with a wider throttle opening and leaner mixture. When emissions legislation permits, the engine should normally receive a lean mixture and at full throttle a rich mixture. This ensures economical operation, yet maximum power at full throttle. If a lean mixture were used at full throttle, this would reduce the power output and possibly overheat the exhaust valve because of the slower combustion. When the engine is idling or operating at low load the low pressure in the inlet manifold increases the exhaust gas residuals in the cylinder, and consequently the carburettor has to provide a rich mixture. The way that the optimum air/fuel ratio changes for maximum power and maximum economy with varying power output for a particular engine at constant speed is shown in figure 4.37. The variations in the lean limit and rich limit are also shown.

4.6.1   The inlet manifold

The carburettor (or injection system) and manifold have to perform satisfactorily in both steady-state and transient condi-

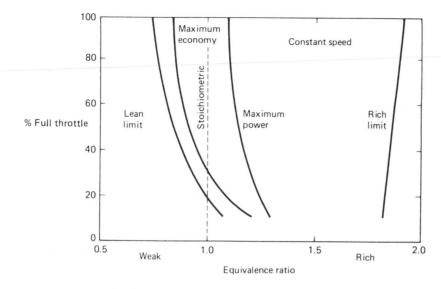

Figure 4.37
Variation in air/fuel requirement with load at constant speed.

tions. When an engine is started, extra fuel floods into the inlet manifold. Under these conditions the engine starts on a very rich mixture and the inlet manifold acts as a surface carburettor; often there will be small ribs to control the flow of liquid fuel.

In a simple branched manifold with a carburettor or single-point injection system, the intersections will often have sharp corners, although for good gas flow rounded intersections would be better. The reason is that the sharp corners help to break up the liquid film flowing on the manifold walls, see figure 4.38. In automotive applications the manifold is sometimes inclined relative to the vehicle. This is so that the fuel distribution becomes optimum when the vehicle is ascending a gradient. When carburettors are used, a way of improving the aerodynamic performance of the inlet manifold is to use multiple carburettor installations. The problem then becomes one of balancing the carburettors – that is, ensuring that the flow is equal through all carburettors, and that each carburettor is producing the same mixture strength. A cheaper alternative to twin carburettors is the twin choke carburettor. The term 'choke' is slightly misleading, as in this context it means the venturi. The saving in a twin choke carburettor is because there is only one float chamber.

When the throttle is opened, extra fuel is needed for several reasons. The air flow into the engine increases more rapidly than the fuel flow, since some fuel is in the form of droplets and some is present as a film on the manifold walls. Secondly, for maximum power a rich mixture is needed. Finally, when the throttle is opened the vaporised fuel will tend to condense. When the throttle opens the pressure in the manifold increases and the partial pressure of the fuel vapour will increase (the partial pressure of fuel vapour depends on the air/fuel ratio). If the partial pressure of the fuel rises above its saturation pressure then fuel will condense, and extra fuel is injected to compensate.

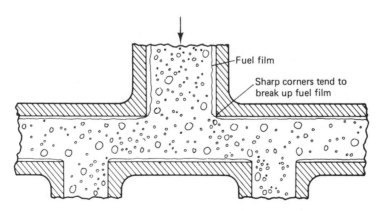

Figure 4.38
Flow of air, fuel vapour, droplets and liquid film in the inlet manifold.

Long inlet manifolds will be particularly bad in these respects because of the large volume in the manifold and the length that fuel film and droplets have to travel. In engines with horizontally opposed cylinders it is very difficult to arrange satisfactory carburation from a single carburettor.

When the throttle is suddenly closed, the reduced manifold pressure causes the fuel film to evaporate. This can provide an over-rich mixture, and so lead to emissions of unburnt hydrocarbons. The problem is overcome by a spring-loaded over-run valve on the throttle plate that by-passes air into the manifold. Sometimes heated manifolds are used to reduce the liquid film and droplets. The manifold can be heated by the engine coolant, or by conduction from the exhaust manifold. The disadvantage of a heated inlet manifold is the ensuing reduction in volumetric efficiency.

The volumetric efficiency penalty is reduced if the engine coolant is used to heat the inlet manifold, but it should be appreciated that the coolant warms-up more slowly than the exhaust manifold. However, supplementary electrical heaters can be used during warm-up. These heaters often use PTC (positive temperature coefficient) materials, so as to give automatic temperature control. The heaters employ extended surfaces (usually spines), and are located in the manifold under the carburettor or a single-point fuel injector.

Despite the careful attention paid to manifold design, it is quite usual for carburettors to give ±5 per cent variation in mixture strength between cylinders, even for steady-state operation.

4.6.2 Variable jet carburettor

A cross-section of a variable jet or variable venturi carburettor is shown in figure 4.39. The fuel is supplied to the jet ① from an integral float chamber. This has a float-operated valve that maintains a fuel level just below the level of the jet. The pressure downstream of the piston ② is in constant communication with the suction disc (the upper part of the piston) through the passage ③. If the throttle ④ is opened, the air flow through the venturi ⑤ increases. This decreases the pressure downstream of the venturi and causes the piston ② to rise. This piston will rise until the pressure on the piston is balanced by its weight and the force from the light spring ⑥. The position of the tapered needle ⑦ in the jet or orifice ① varies with piston position, thus controlling the air/fuel mixture. The damper ⑧ in the oil ⑨ stops the piston ① oscillating when there is a change in load. A valve in the damper causes a stronger damping action when the piston rises than when it falls. When the throttle is opened the piston movement is delayed by the damper, and this causes fuel enrichment of the mixture. For an incompressible fluid, the flow through an orifice or venturi is proportional to the square root of the pressure drop. Thus if both air and fuel were incompressible, the air/fuel mixture would be un-

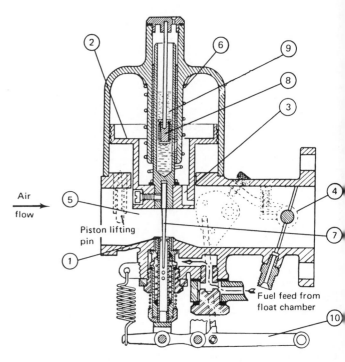

Figure 4.39
Variable jet or variable venturi carburettor (from Judge, 1970).

changed with increasing flow and a fixed piston. However, as air is compressible its pressure drop will be greater than that predicted by incompressible flow and this will cause extra fuel to flow.

For starting, extra fuel is provided by a lever ⑩ that lowers the jet. A linkage and cam also operate the throttle valve to raise the idling speed. Several modifications are possible to improve the carburettor performance. The position of the jet can be controlled by a bi-metallic strip to allow for the change in fuel properties with temperature. Fuel flow will vary with the eccentricity of the needle in the jet, with the largest flow occurring when the needle touches the side of the jet. Rather than maintain exact concentricity, the needle is lightly sprung so as always to be in contact with the jet wall; this avoids the problem of the flow coefficient being a function of the needle eccentricity.

This type of carburettor should not be confused with carburettors in which there is no separate throttle and the piston and needle are lifted directly. This simple type of carburettor is found on some small engines (such as motorcycles and outboard motors) and does not have facilities like enrichment for acceleration.

4.6.3 Fixed jet carburettor

The cross-section of a simple fixed jet or fixed venturi carburettor is shown in figure 4.40. This carburettor can only sense air flow rate without distinguishing between fully open

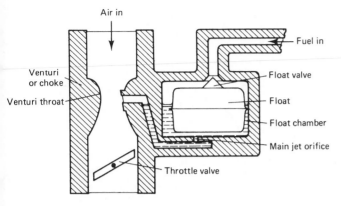

Figure 4.40
Simple fixed jet carburettor.

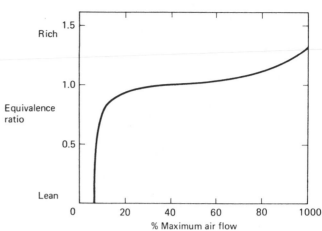

Figure 4.41
Mixture characteristics of a simple carburettor.

throttle at a slow engine speed or partially closed throttle at a higher engine speed. The fuel outlet is at the smallest cross-sectional area so that the maximum velocity promotes break-up of the liquid jet and mixing with the air; the minimum pressure also promotes fuel evaporation. The fuel outlet is a few millimetres above the fuel level in the float chamber so that fuel does not spill or siphon from the float chamber. If the air flow were reversible, there would be no pressure drop; in practice the pressure drop might be 0.05 bar at the maximum air flow rate.

The fuel/air ratio change in response to a change in air flow rate is shown in figure 4.41 for this carburettor. No fuel will flow until the pressure drop in the venturi overcomes the surface tension at the fuel outlet and the head difference from the float chamber. As the air flow increases to its maximum, the air/fuel mixture becomes richer. The maximum air flow is when the velocity at the venturi throat is supersonic. The reason for the change in air/fuel ratio is as follows.

Fuel can be treated as incompressible, and for flow through an orifice

$$\dot{m}_f = A_o C_o \sqrt{(2\rho_f \Delta p)} \qquad (4.12)$$

where $\dot{m}_f$ = mass flow rate of fuel
   A_o = orifice area
   C_o = coefficient of discharge for the orifice
   ρ_f = density of fuel
   Δp = pressure difference across the orifice.

In contrast, air is compressible, and for flow through the venturi

$$\dot{m}_a = A_t C_v \sqrt{(2\rho_a \Delta p)} \left[(r)^{1/\gamma} \sqrt{\left\{ \frac{\gamma}{\gamma - 1} \frac{1 - r^{(\gamma-1)/\gamma}}{1 - r} \right\}} \right] \qquad (4.13)$$

where $\dot{m}_a$ = mass flow rate of air
   A_t = area of venturi throat
   C_v = discharge coefficient for the venturi
   ρ_a = density of air at entry to the venturi

Δp = pressure drop between entry and the venturi throat
γ = ratio of gas specific heat capacities
$r = 1 - p/p_a$
p_a = pressure of air at entry to the venturi.

Since r is always less than unity the square bracket term in equation (4.13) will always be less than unity. This term accounts for the compressible nature of the flow. Thus, for a given mass flow rate the pressure drop will be greater than that predicted by a simple approach, assuming incompressible flow. If the pressure drop is larger than that predicted, then the fuel flow will also be larger than expected and the air/fuel ratio will be richer as well. Derivation of these formulae can be found in books on compressible flow and Taylor (1985b). A qualitative explanation of the effect is that, as the velocity increases in the venturi, the pressure drops and density also reduces. The reduction in density dictates a greater flow velocity than that predicted by incompressible theory, thus causing a greater drop in pressure. This effect becomes more pronounced as flow rates increase, until the limit is reached when the flow in the throat is at the speed of sound (Mach 1) and the venturi is said to be 'choked'.

To make allowance for the mixture becoming richer at larger flow rates a secondary flow of fuel, which reduces as the air flow rate increases, should be added. A method of achieving this is the compensating jet and emulsion tube shown in figure 4.42. The emulsion tube has a series of holes along its length, and an air bleed to the centre. At low flow rates the emulsion tube will be full of fuel. As the flow rate increases the fuel level will fall in the emulsion tube, since air is drawn in through the bleed in addition to the fuel through the compensating jet. The fuel level will be lower inside the emulsion tube than outside it, owing to the pressure drop associated with the air flowing through the emulsion tube holes. As air emerges from the emulsion tube it will evaporate

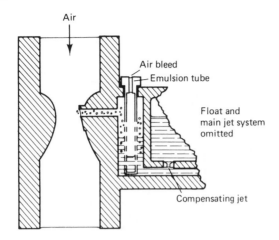

Figure 4.42
Emulsion tube and compensating jet in a fixed jet carburettor.

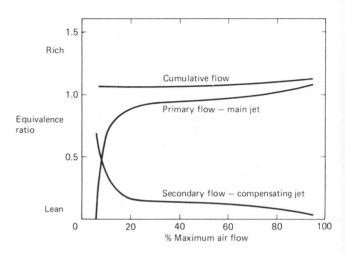

Figure 4.43
Cumulative effect of main jet and compensating jet.

the fuel and form a two-phase flow or emulsion. This secondary flow will assist the break-up of the main flow. The cumulative effects of the main and secondary flows are shown in figure 4.43.

A rich mixture for full throttle operation can be provided by a variety of means, by either sensing throttle position or manifold pressure. The mixture can be enriched by an extra jet (the 'power' jet) or the air supply to the emulsion system can be reduced. Alternatively an air bleed controlled by manifold pressure can be used to dilute a normally rich mixture. This might be a spring-loaded valve that closes when the manifold pressure approaches atmospheric pressure at full throttle.

At low air flow rates no fuel flows, so an additional system is required for idling and slow running. Under these conditions the pressure drop in the venturi is too small and advantage is taken of the pressure drop and venturi effect at the throttle valve. A typical arrangement is shown in figure 4.44. Fuel is drawn into the idling fuel line by the low-pressure region around the throttle valve. A series of ports are used to provide a smooth progression to the main jet system, and the idling mixture is adjusted by a tapered screw. When this is added to the result of the other jet systems shown in figure 4.43 the results will be as shown in figure 4.45.

With fixed jet carburettors there is no automatic mixture enrichment as the throttle is opened, instead a separate accelerator pump is linked to the throttle. For starting, a rich mixture is provided by a choke or strangler valve at entry to the carburettor. When this is closed the whole carburettor is below atmospheric pressure and fuel is drawn from the float chamber directly, and the manifold acts as a surface carburettor. The choke valve is spring loaded so that once the engine fires the choke valve is partially opened. The choke is usually linked to a cam that opens the main throttle to raise

the idling speed. A complete fixed jet carburettor is shown in figure 4.46; by changing the jet size a carburettor can be adapted for a range of engines.

4.6.4 Fuel injection

The original purpose of fuel injection was to obtain the maximum power output from an engine. The pressure drop in a carburettor impairs the volumetric efficiency of an engine and reduces its power output. The problems of balancing multiple carburettors and obtaining even distribution in the inlet manifold can also be avoided with fuel injection. Early fuel injection systems were mechanical, and complex two-dimensional cams have now been superseded by electronic systems.

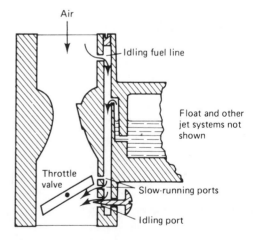

Figure 4.44
Slow-running and idling arrangement in a fixed jet carburettor.

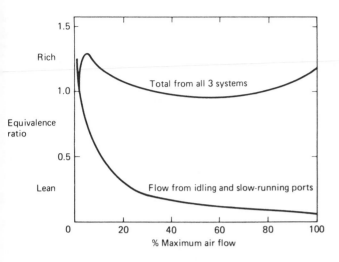

Figure 4.45
Contribution to the air/fuel mixture from the idling and slow-running ports.

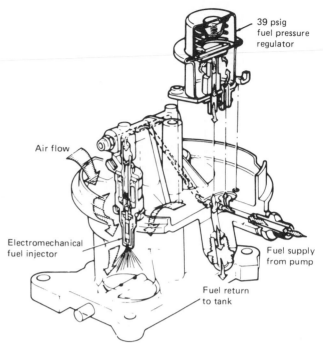

Figure 4.47
Ford single-point fuel injection system (from Ford, 1982).

As explained in section 4.6.1, there are two types of fuel injection system:

(i) single-point fuel injection shown by figure 4.47, and

(ii) multi-point fuel injection, for which an injector is shown in figure 4.48.

The single-point fuel injection system (figure 4.47) was a cheaper alternative to multi-point fuel injection; it led to about a 10 per cent lower power output than a multi-point injection system, and this helped the motor industry to maintain product differentiation. Multi-point fuel injection has the potential for a higher power output, since the

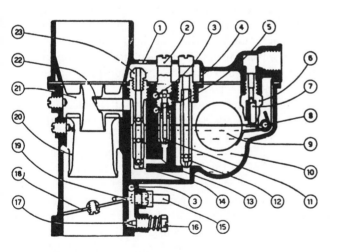

Figure 4.46
Fixed jet or fixed venturi carburettor: ① air intake; ② idling jet holder; ③ idling mixture tube; ④ air intake to the bowl; ⑤ air intake for idling mixture; ⑥ needle valve seat; ⑦ needle valve; ⑧ float fulcrum pivot; ⑨ float; ⑩ carburettor bowl; ⑪ idling jet; ⑫ main jet; ⑬ emulsioning holes; ⑭ emulsioning tube; ⑮ tube for connecting automatic spark advance; ⑯ idling mixture adjusting screw; ⑰ idling hole to the throttle chamber; ⑱ throttle butterfly; ⑲ progression hole; ⑳ choke tube; ㉑ auxiliary venturi; ㉒ discharge tube; ㉓ emulsioning tube air bleed screw.

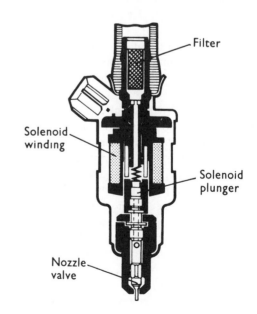

Figure 4.48
Solenoid-operated fuel injection valve (courtesy of Lucas Electrical Ltd).

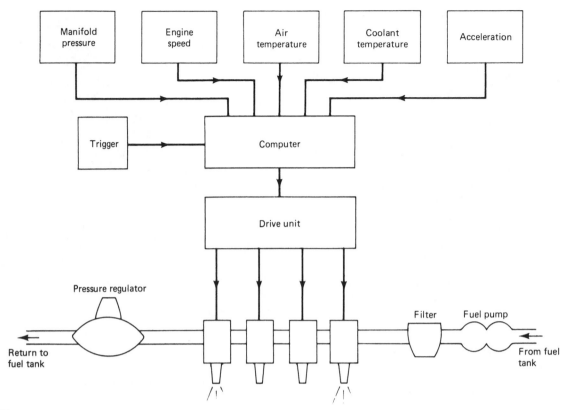

Figure 4.49
Electronic fuel injection system.

manifold can be designed for optimum air flow, and perhaps also include some induction tuning features. The fuelling level is controlled by the fuel supply pressure, and the duration of the injection pulses. As with carburettors, the single-point fuel injection system leads to fuel transport delays in the inlet manifold.

The control system for a fuel injection system is shown in figure 4.49 for a port fuel injection system, but the same principles applied to the control of the single-point injection system.

The fuel flow rate through an injector is controlled by the differential pressure across the injector. In the case of a single-point injection system, the injector sprays fuel into a region at atmospheric pressure, so a constant gauge pressure is maintained by the fuel pressure regulator. In contrast, the fuel injectors for a multi-point injection system inject into the reduced pressure of the inlet manifold. Thus, to make the fuel flow rate a function only of injection duration, the fuel pressure regulator senses the inlet manifold pressure, so as to maintain a constant differential pressure (typically 2 bar) across the injector. The control of the injectors will be discussed later in section 4.7.

In multi-point injection systems the pulse duration is typically in the range of 2–8 ms. The ratio of maximum to minimum fuel flow rate in a spark ignition engine can be 50

or so. This is a consequence of the speed range and the use of throttling. At maximum power, the injectors will be open almost continuously, while at light load, the injection duration will be an order of magnitude less. The range of pulse durations is much lower since a finite time is required to open the injector (about 2 ms).

In order to simplify the injector drive electronics, it has been common practice to fire the injectors in two groups. (If all the injectors fired together then there would be a very uneven fuel demand.) Clearly, when the fuel injection pulse starts is not going to affect the maximum power performance, since the injection is more or less continuous under these conditions. However, for part-load operation, it has been found that if injection occurs when the inlet valve is closed, then this leads to lower emissions of NO_x and unburnt hydrocarbons. The explanation for this is that fuel sprayed on to a closed inlet valve is vaporised by the hot valve (and in doing so helps to cool the valve), and there is also more time for heat transfer to the fuel from the inlet port. Thus a more homogeneous mixture is formed when the injection pulse is phased with regard to inlet valve events; these systems are known variously as time, sequential or timed sequential injection.

Although fuel transport delays are reduced with multi-point injection systems they are not eliminated, and multi-

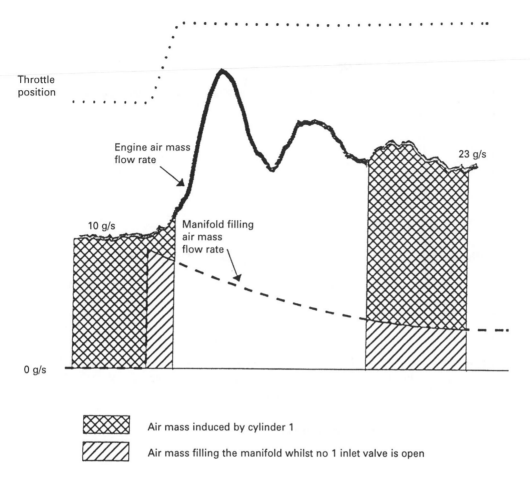

Throttle position

Engine air mass flow rate

Manifold filling air mass flow rate

10 g/s

23 g/s

0 g/s

☒☒☒☒ Air mass induced by cylinder 1

▨▨▨ Air mass filling the manifold whilst no 1 inlet valve is open

Figure 4.50
The air flow response to a sudden throttle opening transient at 2000 rpm, adapted from Rose *et al.* (1994).

point fuel injection systems can also be subject to other complications during load increase transients. In particular, when the throttle is opened rapidly, the inlet manifold pressure will rise more quickly than the fuel supply pressure (because of the lag inherent in the pressure regulator). This means that during a load increase transient the differential pressure across the injector falls. A similar problem can occur as a result of pressure pulsations in the fuel supply rail and in the inlet manifold. These pressure pulsations are an inevitable consequence of the unsteady flow, and they mean that the instantaneous pressure difference across the fuel injectors will vary, and will not correspond to the mean pressure differential being controlled by the pressure regulator.

Control of the mixture preparation during throttle opening and closing transients needs to take into account the behaviour of the fuel stored in the inlet manifold and inlet port. In measurements of the engine response to a sudden throttle opening, there is an immediate increase in the engine air flow rate. Figure 4.50 shows how this can be apportioned between filling the inlet manifold at a higher pressure, and an increase in the air flow to a cylinder. When simultaneous

measurements are made of the in-cylinder air/fuel ratio and the injector pulse width (in order to deduce the fuel flow rate), it is possible to estimate the change in the fuel mass stored in the inlet system; this is illustrated by figure 4.51. Combustion analysis, from simultaneous measurements of the in-cylinder pressure, can be used to analyse the combustion response to the air/fuel ratio excursions. A similar approach can be taken to throttle-closing transients, which lead to a reduction in the fuel stored in the inlet system. Also, by turning the injector off, and measuring the subsequent in-cylinder air/fuel ratio, then the quantity of fuel that had been stored in the inlet manifold can be determined.

4.6.5 Mixture preparation

It might be assumed that multi-point fuel injection systems give the best mixture preparation and most uniform fuel distribution; this is not necessarily the case. At very light loads the fuel injectors will only open for a very short period of time, and under these conditions differences in the response time become significant. This is because the duration of the

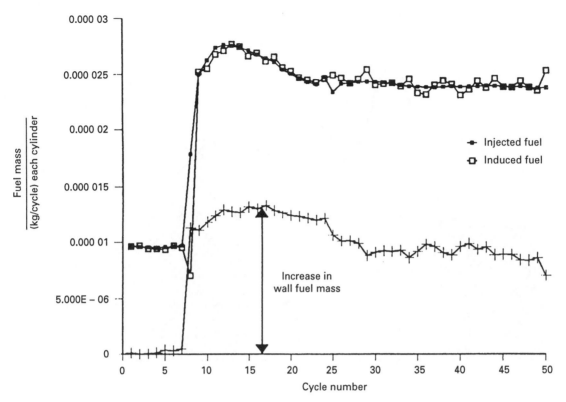

Figure 4.51
The fuel response to a sudden throttle opening transient corresponding to figure 4.50 (Rose *et al.*, 1994).

spray is small compared with the activation time of the injector. An exhaust gas oxygen sensor (described in section 4.7) can control the overall air/fuel ratio, but a single sensor cannot compensate for inter-cylinder variations. However, the exhaust gas oxygen sensor can allow for uniform ageing effects, such as the formation of deposits in injectors. The formation of gum deposits in injectors was a problem mostly associated with the USA in the late 1980s, and was linked with high levels of unsaturated hydrocarbons (notably alkyls). These problems led to detail design changes in the fuel injector and improved detergent additives in the fuel.

The above arguments have shown that the mixture maldistribution will be worst from a multi-point fuel injection system at light loads. In contrast, under light load conditions, the mixture maldistribution will be least from a carburettor or single-point fuel injection system. The mixture maldistribution arises through unequal flows of liquid fuel (from the wall film and droplets) entering the cylinders. At light loads, these problems are ameliorated for two reasons.

Firstly, at light loads the low manifold pressure ensures that a higher proportion of the fuel evaporates, and enters the cylinder as vapour. Secondly, when the pressure ratio across the throttle plate (or butterfly valve) exceeds 1.9, then the flow will become sonic in the throat between the throttle plate and the throttle body. The fuel is sprayed towards the

throttle plate and impinges on it, to form a liquid fluid. The fuel film travels to the edge of the throttle plate, and the very high air velocity in this region causes the liquid film to be broken into very small droplets. The small droplets evaporate more readily, and are also more likely to move with the air flow – as opposed to being deposited on the walls of the inlet manifold.

The spray quality of different mixture preparations systems has been systematically investigated by Fraidl (1987). Fraidl illustrated the change in droplet size distribution from a carburettor as the load increased (the droplet size distribution moved towards the larger droplet sizes), and the spatial distribution along the axis of the throttle plate spindle. Fraidl also provides a comparison between the droplet size distributions for a carburettor, single-point injector and multi-point injector at full load and part load (figure 4.52).

At full load, figure 4.52 shows that the multi-point fuel injection system gives the distribution with the smallest droplet sizes. Under these conditions the throttle plate is fully open, and so cannot contribute any improvements to the atomisation of the liquid fuel. At part load, the situation is reversed, and the ordinary multi-point fuel injection system has the droplet size distribution with the largest droplets. However, the fuel atomisation can be improved by the use of air-containment, in which an annular flow of high-velocity

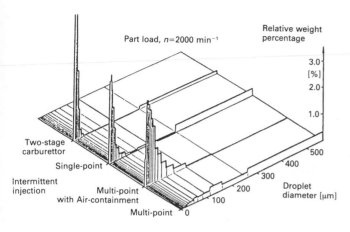

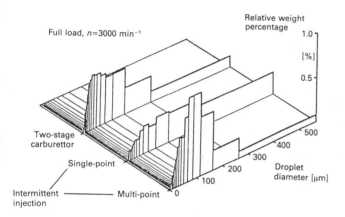

Figure 4.52
Representative droplet size spectra of mixture preparation systems at part load and full load, from Fraidl (1987) with permission of EAEC.

air encircles the fuel spray. A liquid spray is broken into droplets by the shear forces between the liquid and the surrounding gas. The higher the relative velocity between the liquid and gas, then the greater the shear force, and the smaller the size of the droplets that are formed. A high relative velocity between the liquid and gas can be provided by one of two ways. Firstly, a high velocity jet (v_j) can be produced by using a high differential pressure (Δp_i) across the injector; to a first order

$$v_j \propto \sqrt{(\Delta p_i)} \tag{4.14}$$

Secondly, it is possible to increase the velocity of the air surrounding the injector. The 'air-containment' technique exploits the high pressure ratio that occurs across the throttle plate at light loads. Air from upstream of the throttle plate by-passes the inlet manifold and enters an annular cavity around the injector tip. At part-load conditions, the pressure drop across the throttle plate produces a pressure ratio, which causes a sonic velocity flow to leave the annulus surrounding

the injector tip. Figure 4.52 shows that this can lead to a droplet size distribution that is almost as good as the single-point injector or carburettor.

Phase Doppler Anemometry (PDA, see chapter 9, section 9.2.3) enables simultaneous measurements to be made of the droplet size and velocities. Figure 4.53 shows some of these measurements, along with air velocity measurements from tests on a motored engine. The measurements are on a diametral plane, 10 mm below the cylinder head, 75°ca after the start of the induction stroke. The hatched area represents the piston position, and for the droplet size distributions the Sauter mean diameter (six times the total volume of the droplets divided by their area, $\frac{4}{3}\pi R^3 \div 4\pi R^2$) is greater than the arithmetic mean diameter. Such measurements enable the distribution and vaporisation of the fuel to be studied during the compression stroke.

Meyer and Heywood (1997) have studied the fuel behaviour in SI engines during warm-up (using Phase Doppler Anemometry), for injection when the inlet valve is both open and closed. With injection when the inlet valve was closed (closed valve injection – CVI), the Sauter mean diameter was about 35 μm, compared to about 20 μm when injecting with the inlet valve open (open valve injection – OVI). In-cylinder liquid fuel films have been observed by Witze and Green (1997) using flood-illuminated laser-induced fluorescence, and this showed that CVI led to more extensive fuel films than OVI. Witze and Green (1997) also used direct video imaging to observe pool fires (combustion of liquid fuel that is vaporising from the piston), and noted that these pool fires are a likely source of particulate emissions.

4.7 | Electronic control of engines

There are two approaches to electronic control of engines or engine management. The first is to use a memory for storing the optimum values of variables, such as ignition timing and mixture strength, for a set of discrete engine-operating conditions. The second approach is to use an adaptive or self-tuning control system to continuously optimise the engine at each operating point. It is also possible to combine the two approaches.

The disadvantage of a memory system is that it cannot allow for different engines of the same type that have different optimum operating conditions because of manufacturing tolerances. In addition, a memory system cannot allow for changes due to wear of the build-up of combustion deposits. The disadvantage of an adaptive control system is its complexity. Instead of defining the operating conditions, it is necessary to measure the performance of the engine. Furthermore, it is very difficult to provide an optimum control algorithm, because of the interdependence of many engine parameters.

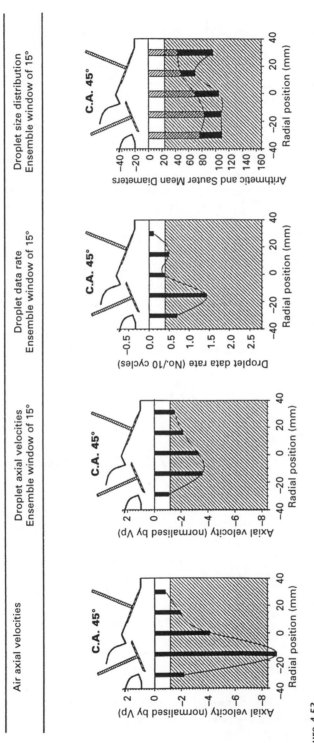

Figure 4.53

Air axial velocity measurements, and fuel droplet axial velocity and size measurements in a motored spark ignition engine (Arcoumanis, 1997).

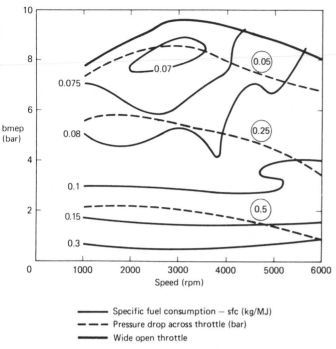

Specific fuel consumption – sfc (kg/MJ)
- - - Pressure drop across throttle (bar)
Wide open throttle

Figure 4.54
Engine map for fuel economy and manifold pressure (reprinted with permission, © 1982 Society of Automotive Engineers, Inc.).

The advantages of an electronic engine management system are the greater control it has on variables like ignition timing and mixture strength. The gains are manifest as reductions in both fuel consumption and emissions.

With electronic ignition and fuel injection, combining the electronic control is a logical step since the additional computing power is very cheap. In vehicular applications the natural extension will be control of transmission ratios in order to optimise the overall fuel economy.

With memory systems the engine-operating conditions are derived from engine maps. Figure 4.54 is an example of a typical engine map, derived from experimental results. The map shows contours for fuel economy and manifold pressure. Additional contours could be added for emissions, ignition timing and mixture strength, but these have been omitted to avoid confusion.

When an engine is tested the power output, emissions, manifold depression, optimum ignition timing and air/fuel mixture will all be recorded for each throttle setting and speed. The results are plotted against engine speed and bmep, since bmep is a measure of engine output, independent of its size. In a microprocessor-controlled system the optimum operating conditions will be stored in ROM (Read Only Memory) for each operating point. Since it is difficult to measure the output of an engine, except on a test bed, the operating point is identified by other parameters, such as engine speed and manifold pressure.

In an engine management system, some of the parameters that can be measured have been shown in figure 4.54. This information is then used by the engine management system to control the ignition timing, the exhaust gas recirculation valve and the fuel injection equipment. Since the engine will have been calibrated to operate with a particular schedule of air/fuel ratio, it is very important to know the air flow rate. This leads to two types of electronic fuel injection control, those which measure the air flow rate and those that deduce the air flow – the so-called speed–density systems. The speed–density systems measure the manifold pressure and air temperature, and then from the engine speed/manifold pressure relationship (stored in memory), the engine management system can deduce the air flow rate. This approach is less direct (and less accurate) than measuring the air flow into the engine. Nor can the speed–density system allow for exhaust gas recirculation. Two common flow measuring techniques are: (a) the use of a pivoted vane connected to a variable resistor, which is deflected by the air flow; and (b) the use of a hot wire anemometer. Unfortunately, the hot wire anemometer is insensitive to the flow direction, and as wide open throttle is approached, the flow becomes more strongly pulsating and there can be a backflow of air out of the engine (especially with a tuned induction system).

If a hot wire anemometer is used it is necessary to deduce the flow direction (by means of a pressure drop, or by looking for when the flow rate falls to zero), or alternatively to rely on the engine management system to correct the hot wire anemometer results for when backflow is present. Another advantage of measuring the air flow is that the speed–density system can still be used for either cross-checking or deducing the level of exhaust gas recirculation.

Other measurements that give an indication of the engine air flow (and operating point) are the combination of throttle angle and engine speed. The throttle angle is usually measured so that changes in demand can be detected immediately; for example, to inject extra fuel during throttle opening, or to stop injection when the engine is decelerating. The air and coolant temperature sensors also identify the cold-starting conditions, and identify the operating point during engine warm-up. The extra fuel required for cold-starting can be quantified, and perhaps be injected through an auxiliary injector.

The type of memory-based systems described so far can make no allowances for changes or differences in engine performance, nor can they allow for fuel changes. Suppose a fuel with a different density and octane rating was used, then the air/fuel ratio would be changed, and combustion knock might be encountered at certain full throttle conditions. Thus it is common practice to employ knock detection, and measurement of the exhaust gas oxygen level to deduce the air/fuel ratio.

If a knock detector is fitted, this can retard the ignition at the onset of knock, thereby preventing damage to the engine.

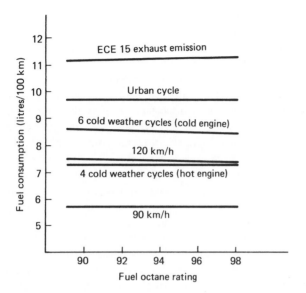

Figure 4.55
The effect on fuel consumption of reduced fuel quality on an engine designed for 98 octane fuel, but fitted with a knock sensor controlled electronic ignition (adapted from Meyer *et al.*, 1984).

Combustion knock causes the engine structure to vibrate, and an accelerometer can be used as a knock sensor. Forlani and Ferranti (1987) report that the signal is typically filtered with a pass band of 6–10 kHz, and examined in a window from tdc to 70° after tdc. Since the signal is examined for a particular time window, then the knocking cylinder can be identified and the ignition timing can be retarded selectively. The knock sensor provides a safety margin, which would otherwise be obtained by having a lower compression ratio or permanently retarded ignition. A very striking example of this is provided by Meyer *et al.* (1984), in which an engine fitted with a knock sensor is run with fuels of a lower octane rating than it had

been designed for. The results are shown in figure 4.55, and it can be seen that there is no significant change in the fuel consumption, for a range of driving conditions. Indeed, only a slight deterioration was found in the full-load fuel economy.

An exhaust gas oxygen sensor is shown in figure 4.56. When a three-way catalyst is to be used, then it is essential to use a feedback system incorporating such a sensor, to maintain an air/fuel ratio that is within about 1 per cent of stoichiometric (as discussed in section 4.3). The oxygen or lambda sensor has been described by Wiedenmann *et al.* (1984). One electrode is exposed to air, and the other electrode is exposed to the exhaust gas. The difference in the partial pressures of oxygen leads to a flow of electrons related to the difference in partial pressures. Since the platinum electrode also acts as a catalyst for the exhaust gases, then for rich or stoichiometric air/fuel ratios there is a very high output from the lambda sensor, since the partial pressure of the oxygen will be many orders of magnitude lower than for air. Since the sensor is used in a way that decides whether the mixture is rich or weak, a control system is needed that makes the air/fuel ratio perturbate around stoichiometric. As the lambda sensor will only work when it has reached a temperature of about 300°C, this feedback control system can only be used after the engine has started to warm-up (20–30 s).

This type of sensor can also be used for lean-burn engine control. However, for this application it is necessary to evaluate the partial pressure of the oxygen (as opposed to deciding whether or not there is oxygen present). As the voltage output is a function of both the oxygen level and the sensor temperature, then this led to the development of the electrically heated oxygen sensor described by Wiedenmann *et al.* (1984). A further refinement is the Universal Exhaust Gas Oxygen sensor (UEGO – sometimes prefixed with H as they are heated), which can also determine the mixture strength of rich mixtures. They also have to be able to withstand operating temperatures of about 1000°C.

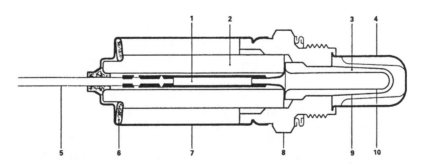

Lambda sensor
1 Contact element. 2 Protective ceramic element. 3 Sensor ceramic. 4 Protective tube (exhaust end). 5 Electrical connection. 6 Disc spring. 7 Protective sleeve (atmosphere end). 8 Housing (–). 9 Electrode (–). 10 Electrode (+).

Figure 4.56
Cross-section drawing of an exhaust oxygen sensor, reproduced by permission of Robert Bosch Ltd.

The UEGO sensor is shown in figure 4.57, and it is in effect a pair of lambda sensors, constructed from three layers of zirconia, all of which are heated, but only the top two layers have platinum electrodes and electrical connections. The UEGO sensor operates by seeking to maintain a constant low oxygen partial pressure in the cavity between the sensing cell and the pumping cell. The whole assembly is heated to maintain a constant temperature, since the voltage–oxygen partial pressure response is temperature dependent. The sensing cell measures the oxygen partial pressure, and the current imposed on the pumping cell seeks to maintain a constant oxygen partial pressure in the measurement cavity between the sensing cell and the pumping cell. When there is a weak mixture, the concentration gradient between the exhaust gas and the measurement cavity will cause oxygen to diffuse through the porous gas intake, and the electrical current in the pumping cell (that is removing oxygen) will be proportional to the oxygen concentration in the exhaust gas. With rich mixtures, the partial products of combustion (CO, H_2 and hydrocarbons) that have diffused into the measurement cavity will be oxidised, thus causing more oxygen to diffuse through the porous gas intake. The richer the mixture then the greater the flow rate of reactants diffusing into the measuring cavity, and the greater the current flow in the pumping cell that is adding oxygen (to maintain the constant oxygen level), so the current will be proportional to the richness of the exhaust gas. (Note that, as the oxygen ions have a negative charge, they flow in the opposite direction to the current.)

A closed-loop control system needs to be used when a carbon canister is used for the control of evaporative emissions. The major source of evaporative emissions from a passenger car is a consequence of the fuel tank being subject to diurnal temperature variations. The fuel tank has to be vented to atmosphere, to avoid a pressure build-up, as the fuel tank warms up. By venting the fuel tank through a canister containing active charcoal, the fuel vapour is absorbed. When the fuel tank cools down, air is drawn in through the carbon canister, and this removes some of the hydrocarbons from the active charcoal. However, to ensure adequate purging of the active charcoal, then it is necessary to draw additional air through the active charcoal. This is achieved by drawing some of the air flowing into the engine through the carbon canister. Clearly, the air/fuel ratio of this stream will be unpredictable, but a closed-loop engine management system will enable the engine to always adopt the intended air/fuel ratio.

It is also expected that the EC will adopt measures to control vapour emissions during fuel tank filling; these vapour emissions represent about a quarter of the evaporative emissions.

In an adaptive control system the operating point would be found by optimising the fuel economy or emissions. With the multitude of control loops the hierarchy has to be carefully defined, otherwise one loop might be working directly against another. The computational requirements are not a problem since they can be readily met by current microprocessors. Only this approach offers fully flexible systems.

An example of an adaptive engine control system is provided by Wakeman *et al.* (1987) and Holmes *et al.* (1988). Perturbations are applied to the ignition timing, and the slope of the ignition timing/torque curve is inferred from the response of the engine speed. This way the control system can find the MBT ignition timing for a particular operating condition, and then store this as a correction value from the ignition timing map. A similar system can be used for controlling EGR to minimise NO_x emissions. Holmes *et al.* also describe an approach to knock control in which the timing retard from the ignition map is stored as a function of operating point. The system is arranged so that when there is a change in fuel quality the ignition timing will converge to the new optimum timing schedule. The fuel economy benefits of an

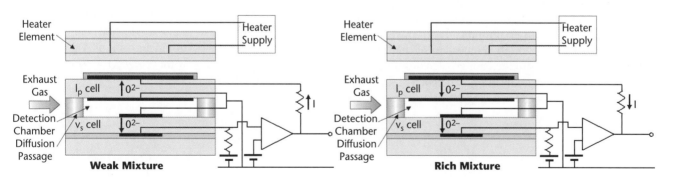

Figure 4.57
The Universal Exhaust Gas Oxygen (UEGO) sensor structure and operation, adapted from Zhao and Ladommatos (2001).

adaptive ignition system are shown by Wakeman *et al.* Four nominally identical vehicles were tested on the ECE 15 urban driving cycle, and it was found that the adapted ignition timing map was different for each vehicle (and different from the manufacturer's calibration). Fuel savings of up to 9 per cent were achieved.

An important method of improving part load fuel economy is cylinder disablement. In its simplest form this consists of not supplying a cylinder with fuel – a technique used on gas engines at the turn of the previous century. When applied to fuel-injected multi-cylinder engines a group of cylinders can be disabled, such as a bank of three cylinders in a V6. Alternatively, a varying disablement can be used; for example, every third injection pulse might be omitted in a four-cylinder engine. Slightly greater gains would occur if the appropriate inlet valves were not opened since this would save the pumping losses. This approach could also be used with single-point injection systems or carburettors.

In vehicular applications a very significant amount of use occurs in short journeys in which the engine does not reach its optimum temperature – the average British journey is about 10 miles. To improve the fuel economy under these conditions the throttle can be controlled by a stepper motor which responds fast enough to prevent the engine stalling at slower than normal idling speeds.

The questions about electronic engine management concern not just economics but also reliability and whether or not an engine is fail safe.

4.8 | Conclusions

Spark ignition engines can operate only within a fairly narrow range of mixture strengths, typically within a gravimetric air/fuel ratio range of 10–18:1 (stoichiometric 14.8:1). The mixture strength for maximum power is about 10 per cent rich of stoichiometric, while the mixture strength for maximum economy is about 10 per cent weak of stoichiometric. This can be explained qualitatively by saying that the rich mixture ensures optimum utilisation of the oxygen (but too rich a mixture will lead to unburnt fuel, which will lower the combustion temperatures and pressures). Conversely the lean mixture ensures optimum combustion of the fuel (but too weak a mixture leads to increasingly significant mechanical losses and ultimately to misfiring). The power output and economy at constant speed for a range of throttle settings are conveniently shown by the so-called 'fish-hook' curves (figure 4.2). From these it is self-evident that at part throttle it is always more economical to run an engine with a weak mixture as opposed to a slightly more closed throttle with a richer mixture.

The main combustion chamber requirements are compactness, sufficient turbulence, minimised quench areas, and short flame travel from the spark plug to the exhaust valve. Needless to say there is no unique solution, a fact demonstrated by the number of different combustion chamber designs. However, the high-turbulence, high compression ratio, lean-burn combustion chamber should be treated as a separate class. This type of combustion chamber is exemplified by the May Fireball combustion chamber. The high turbulence enables lean mixtures to be burnt rapidly, and the compact combustion chamber (to minimise heat transfer) is located around the exhaust valve. All these attributes contribute to knock-free operation, despite the high compression ratios; the compression ratio can be raised from the usual 9:1 to 15:1, so giving a 15 per cent improvement in fuel economy. However, such combustion chambers place a greater demand on manufacturing tolerances, mixture preparation and distribution, and ignition timing.

Mixture preparation is either by carburettor or by fuel injection. There are two main types of carburettor – fixed or variable venturi (or jet) – and both types can be used for single or multi-carburettor installations. Fuel injection can provide a much closer control on mixture preparation and injection is invariably at low pressure into the induction passage. When a single-point injection system is used the problems associated with maldistribution in the inlet manifold still occur, but are ameliorated by the better atomisation of the fuel. It is very difficult (maybe even impossible) to design an inlet manifold that gives both good volumetric efficiency and uniform mixture distribution.

The increasingly demanding emissions legislation requires ever more sophisticated engine management systems, and catalysts that are effective as soon as possible after an engine is started. In order to achieve better fuel economy, engines that can operate in a lean-burn mode have been developed. These are either port injected with a homogeneous charge and a method of controlling the combustion, or direct injection. In either case catalysts with an ability to reduce NO_x in an oxidising environment can provide scope for improving fuel economy without compromising the emissions requirements.

4.9 | Example

A variable jet carburettor is designed for a pressure drop (Δp) of 0.02 bar and an air/fuel ratio of 15:1. If the throttle is suddenly opened and the pressure drop quadruples, calculate the percentage increases in air flow and fuel, and the new air/fuel ratio. Neglect surface tension and the difference in height between the jet and the liquid level in the float chamber; atmospheric pressure is 1 bar, and $\gamma = 1.4$.

Solution:
Immediately after the throttle opens there will be no change in either orifice, as the piston and needle will not have moved.
Rewriting equations (4.12) and (4.13) gives

$$\dot{m}_f = k_f \sqrt{(\Delta p)}$$

$$\dot{m}_a = k_a \sqrt{(\Delta p)} \left[(r)^{1/\gamma} \sqrt{\left\{ \frac{\gamma}{\gamma - 1} \frac{1 - r^{(\gamma-1)/\gamma}}{\gamma - 1} \right\}} \right]$$

where $r = 1 - \Delta p/p$; k_f, k_a are constants.
If Δp quadruples, the increase in $\dot{m}_f$ is

$$\frac{\sqrt{(4\Delta p)} - \sqrt{(\Delta p)}}{\sqrt{(\Delta p)}} \times 100 = 100 \text{ per cent}$$

Using suffix 1 to denote conditions before the step change and suffix 2 to denote conditions after

$$r_1 = 1 - \frac{0.02}{1} = 0.98; \quad r_2 = 1 - \frac{0.08}{1} = 0.92$$

$$\dot{m}_{a1} = k_a \sqrt{(0.02)} 0.098^{(1/1.4)} \sqrt{\left(\frac{1.4}{1.4 - 1} \frac{1 - 0.98^{(1.4-1)/1.4}}{1 - 0.98} \right)}$$

$$= k_a \sqrt{(0.02)} 0.98^{0.714} \sqrt{\left(3.5 \frac{1 - 0.98^{0.286}}{1 - 0.98} \right)} = k_a \times 0.140$$

$$\dot{m}_{a2} = k_a \sqrt{(0.08)} 0.92^{0.714} \sqrt{\left(3.5 \frac{1 - 0.92^{0.286}}{1 - 0.92} \right)} = k_a \times 0.271$$

$$\% \text{ increase in } \dot{m}_a = \frac{0.271 - 0.140}{0.140} 100 = 93.6 \text{ per cent}$$

$$\text{New air/fuel ratio} = \frac{15 \times 1.936}{2.00} : 1 = 14.04 : 1$$

4.10 | Questions

4.1 Why does the optimum ignition timing change with engine-operating conditions? What are the advantages of electronic ignition with an electronic control system?

4.2 Explain the principal differences between fixed jet and variable jet carburettors. Why does the mixture strength become richer with increasing flow rate in a simple carburettor?

4.3 What are the air/fuel requirements for a spark ignition engine at different operating conditions? How are these needs met by a fixed jet carburettor?

4.4 List the advantages and disadvantages of electronic fuel injection.

4.5 Contrast high-turbulence, high compression ratio combustion chambers with those designed for lower compression ratios.

4.6 Two spark ignition petrol engines having the same swept volume and compression ratio are running at the same speed with wide open throttles. One engine operates on the two-stroke cycle and the other on the four-stroke cycle. State with reasons:

(i) which has the greater power output.
(ii) which has the higher efficiency.

4.7 The Rover M16 spark ignition engine has a swept volume of 2.0 litres, and operates on the four-stroke cycle. When installed in the Rover 800, the operating point for a vehicle speed of 120 km/h corresponds to 3669 rpm and a torque of 71.85 N m, for which the specific fuel consumption is 298 g/kWh.

Calculate the bmep at this operating point, the arbitrary overall efficiency and the fuel consumption (litres/100 km). If the gravimetric air/fuel ratio is 20:1, calculate the volumetric efficiency of the engine, and comment on the value.

The calorific value of the fuel is 43 MJ/kg, and its density is 795 kg/m^3. Ambient conditions are 27°C and 1.05 bar.

Explain how both lean-burn engines and engines fitted with three-way catalyst systems obtain low exhaust emissions. What are the advantages and disadvantages of lean-burn operation?

4.8 A spark ignition engine is to be fuelled by methanol (CH_3OH). Write down the equation for stoichiometric combustion with air, and calculate the stoichiometric gravimetric air/fuel ratio.

Compare the charge cooling effects of fuel evaporation in the following two cases:

(i) a stoichiometric air/methanol mixture in which 25 per cent of the methanol evaporates,
(ii) 60 per cent of a 14.5:1 gravimetric air/petrol mixture evaporates.

The enthalpy of evaporation for methanol is 1170 kJ/kg and that of petrol is 310 kJ/kg. State clearly any assumptions made.

If in a lean-burn engine the gravimetric air/fuel ratio is 8:1, rewrite the combustion equation, and determine the volumetric composition of the products that should be found from a dry gas analysis. What is the equivalence ratio for the air/fuel ratio? Why might methanol be used as a fuel for spark ignition engines?

4.9 The Rover K16 engine is a four-stroke spark ignition engine, with a swept volume of 1.397 litres and a compression ratio of 9.5. The maximum torque occurs at a speed of 4000 rpm, at which point the power output is 52 kW; the maximum power of the engine is 70 kW at 6250 rpm. Suppose the minimum brake specific fuel consumption is 261.7 g/kWh, using 95 RON lead-free fuel with a calorific value of 43 MJ/kg.

Calculate the corresponding Otto cycle efficiency, the maximum brake efficiency and the maximum brake mean effective pressure. Give reasons why the brake efficiency is less than the Otto cycle efficiency. Show the Otto cycle on the p–V state diagram, and contrast this with an engine indicator diagram – identify the principal features.

Explain why, when the load is reduced, the part-load efficiency of a diesel engine falls less rapidly than the part-load efficiency of a spark ignition engine.

4.10 Since the primary winding of an ignition coil has a finite resistance, energy is dissipated while the magnetic field is being established after switching on the coil. If the primary winding has an inductance of 5.5 mH, and a resistance of 1.9 ohms, and the supply voltage is 11.6V, show that;

(i) the energy supplied after 2 ms would be 39 mJ, increasing to 246 mJ after 6 ms,
(ii) the theoretical efficiency of energy storage is about 65 per cent for a 2 ms coil-on-time, falling to 32 per cent for a 6 ms coil-on-time.

What are the other sources of loss in an ignition coil?

Define MBT ignition timing, and describe how and why it varies with engine speed and load. Under what circumstances are ignition timings other than MBT used?

4.11 Redraw figure 4.9 using a scale of lambda (the relative air/fuel ratio, the reciprocal of equivalence ratio), and comment under what circumstances this is a more appropriate form for presenting the data.

CHAPTER 5

Direct injection spark ignition engines

5.1 | Introduction

The drive to produce engines of high specific power for aircraft applications in the 1930s led to the application of diesel direct injection technology to gasoline engines, most notably by Ricardo during the development of their two-stroke sleeve valve engine, which was the basis of the V12 Rolls-Royce 'Crecy' Gasoline Direct Injection (GDI) engine (Nahum *et al.*, 1994). The Crecy achieved a much higher output than other types of engine at the time, but was not implemented as a successor to the Merlin engine because of the advent of the gas turbine. This, along with improvements in carburation technology, postponed further application of direct injection technology.

In 1954, Mercedes-Benz introduced a direct injection system in their 300SL (Scherenberg, 1955) [cited by Iwamoto *et al.*, 1997], as a means of circumventing some of the inherent performance disadvantages of carburettor systems. This system was soon replaced by inlet Port Fuel Injection (PFI) systems, which were more robust and provided better mixing without excessive charge motion. Port Fuel Injection then gradually became the norm in automotive gasoline engines, initially using a single injector in the inlet manifold, and later moving to multipoint port injection (MPI) with fuel sprays onto the back of each inlet valve. Mitsubishi were the first to introduce a GDI engine in a modern car using reverse tumble and a spherically bowled piston, with fuel injected from the side of the cylinder (Kume *et al.*, 1996).

Gasoline Direct Injection (GDI) engines have the potential to improve the specific output of gasoline engines, yet with fuel economy that is said to be comparable to diesel engines. GDI engines operate at stoichiometric near full load, with early injection (during induction) so as to obtain a nominally homogeneous mixture. This gives a higher volumetric efficiency (by about 5 per cent) than a port-injected engine,

since any evaporative cooling is reducing the temperature of the air only, not the inlet port or other engine components as well. Furthermore, the greater cooling of the air means that at the end of compression the gas temperature will be about 30 K lower, and a higher compression ratio can be used (1 or 2 ratios higher) without the onset of combustion knock, so the engine becomes more efficient.

In contrast, at part load and low speed, direct injection engines have the potential to operate with injection during the compression stroke. This enables the mixture to be stratified, so that a flammable mixture is formed in the region of the spark plug, yet the overall air/fuel ratio is weak (and the three-way catalyst operates in an oxidation mode). The increased air flow (relative to stoichiometric operation) means a reduced throttling loss, so that the fuel consumption is lower. The leaner mixture also leads to a higher ratio of heat capacities so that the cycle efficiency is higher (see figure 2.10). However, in order to keep the engine-out NO_x emissions low, it is necessary to be very careful in the way that the mixture is stratified, and perhaps also to use some form of NO_x 'trap' (section 4.3.3).

GDI engines are also well suited to start/stop systems that stop the engine so as to avoid idling. This is particularly advantageous in the New European Drive Cycle for which the vehicle is stationary for significant periods, and idling can account for 10 per cent of the fuel consumption in an urban drive cycle. A Port Fuel Injection (PFI) system will require the equivalent of several cycles of operation to re-establish the fuel films in the inlet port, while in contrast a GDI engine can restart almost instantaneously. Systems have also been devised that record the crank position when the engine stops, so that starting can be by injecting fuel and igniting it (Ueda *et al.*, 2001). Fuel is injected and ignited in a cylinder that is on a compression stroke, causing the engine to reverse so that fuel can then be injected into a cylinder that was on a power stroke (but will contain air as the fuel will have been cut off before the engine stopped). The mixture in the 'expansion stroke' can then be ignited, thereby causing the engine to start operating conventionally. Starting can be achieved in about 3–400 ms, which is faster than a conventional starter motor, and comparable with the time it takes a driver to engage a gear.

Additional fuel-economy benefits can be achieved by using a downsized supercharged or turbocharged engine. The bmep is essentially proportional to inlet manifold pressure, so if the bmep is higher, then the pumping loss will be lower, and the fuel economy is improved. Also, the higher bmep will mean a higher mechanical efficiency as the fmep is mostly dependent on engine speed. Figure 5.1 illustrates how a boosted 1.6-litre engine can have the same maximum output (130 kW) as a naturally aspirated 2.5-litre engine, with the road load operating points moved to a more efficient part of the engine map. It should also be noted that the boosted engine has a higher maximum torque occurring at a lower speed, and this will improve the torque back-up and driveability of the vehicle. With supercharged or turbocharged engines, then with GDI it is also possible to delay fuel injection until after exhaust valve closure so that a scavenging flow of air can be used to reduce exhaust-residuals level, provide cooling, and reduce the tendency for knock to occur: knock is discussed later in section 5.6 in the context of highly boosted engines.

The terminology for Gasoline Direct Injection (GDI) engines needs some clarification, as the term Direct Injection Spark Ignition (DISI) engines is used interchangeably with GDI. There are also different combustion systems, and the two basic types (wall guided and spray guided) are shown in figure 5.2 along with a Port Fuel Injection (PFI) system for comparison. Wall-guided systems use a lower injection pressure (about 50 bar) and a single hole nozzle that is generally mounted at the side of the combustion chamber. Spray-Guided Direct Injection (SGDI) systems use a higher pressure (150 bar or higher) and injectors such as those with multi-hole nozzles that can produce finely dispersed fuel droplets for rapid evaporation. Spray-guided injectors are mounted at the side of the cylinder head or centrally. The two GDI concepts shown in figure 5.2 have different ways of achieving the correct mixture at the spark plug at the time of ignition. The wall-guided concept relies on interaction with the piston crown, while in a spray-guided GDI engine the mixture preparation is determined mainly by the characteristics of the injector, rather than the shape of the combustion chamber.

Some combustion systems are described as air-guided injection, and these systems seek to avoid wall-wetting and use the air flow to control the dispersion of the fuel spray. However, the air motion is very important in both wall-guided and spray-guided systems, so only the two basic GDI combustion systems will be discussed here. There are also air-assist injectors, such as those developed for two-stroke engines by the Orbital Corporation. These injectors use a high-velocity concentric flow of air to atomise the fuel and improve mixture preparation (chapter 8, section 8.6). An obvious disadvantage of the wall-guided systems is that a liquid film is formed, and this will be slow to evaporate and become a source of particulate matter emissions.

As wall-guided combustion systems were introduced first they will be described first (section 5.3). Spray-guided combustion systems (section 5.4) have also been used in highly turbocharged engines, and these are discussed in

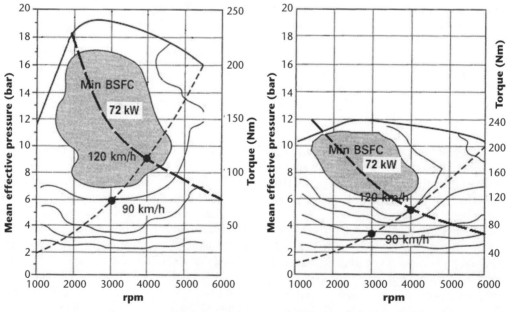

Figure 5.1 Downsizing, in which a boosted 1.6-litre engine can have the same maximum power output (130 kW) as a naturally aspirated 2.5-litre engine, but with lower throttling losses and a higher mechanical efficiency. Adapted from Zhao (2010).

Port Fuel injection   **Direct injection**

Wall Guided
Side Mounted injector

Spray Guided
Center Mounted Injector

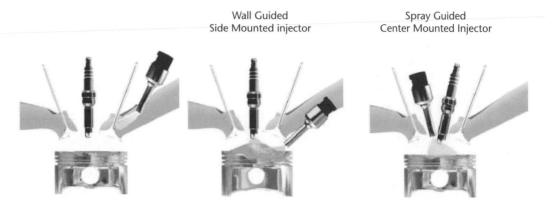

Figure 5.2 A comparison of Port Fuel Injection (PFI) with Direct Injection (DI), showing: a side-mounted injector with a wall-guided combustion system, and a centrally mounted spray-guided injector. Adapted from Wirth *et al.* (2004).

Swirl
Injector

Multihole
Injector

Outward Opening
Injector

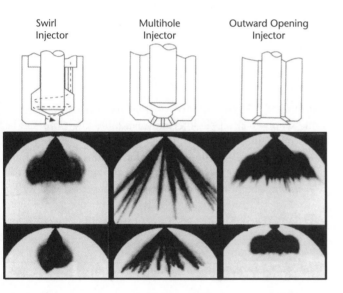

Figure 5.3 Swirl injectors, multi-hole injectors and outwardly opening injectors (adapted from Zhao, 2010) and fuel spray patterns at 1 bar (upper images) and 6 bar (lower images) with reduced spray penetration. Adapted from Achleitner *et al.* (2007).

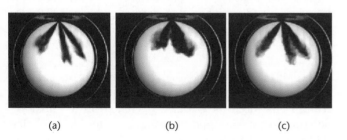

(a)   (b)   (c)

Figure 5.4 Spray development from a multi-hole injector at 0.5 bar gas pressure with 150 bar fuel pressure: (a) standard gasoline at 20°C and (b) at 120°C; (c) iso-octane at 120°C. Images courtesy of Dr Pavlos Aleiferis.

section 5.6. A very comprehensive review of all types of GDI engines has been provided by Zhao (2010). However, before any discussion of the combustion systems, it is important to understand the characteristics of the different fuel injection systems.

5.2 | Gasoline fuel injection systems

There are essentially three types of fuel injector, and these are shown in figure 5.3: swirl injectors, multi-hole injectors and outwardly opening injectors. Figure 5.3 shows that as the back-pressure is increased, so the spray penetration is reduced. In a swirl injector, the small annular gap between the needle and its seat leads to the fuel flowing in a sheet that forms a 'hollow' conical spray (as if the fuel was flowing on the surface of a cone). The swirl stabilises the flow in the sheet and leads to good dispersion. However, the angle of the cone depends on the injection pressure and density of the air in the cylinder. As the air density in the cylinder increases, the angle of the cone decreases and the spray becomes more like a single jet. In contrast, multi-hole injectors control the mixture distribution by means of the orientation of the holes. They have typically six holes and operate at a pressure of up to 150–200 bar (compared to 50 bar for a swirl injector). The size of the holes is limited by the tendency for deposits to form ('coking'), so the injector tip has to be kept below about 130°C; swirl injectors are less susceptible to combustion deposits. The spray tends to penetrate further than with a swirl injector, especially if the engine is cold or the fuel does not contain volatile components. The outwardly opening injector has a well-defined cone angle that is unaffected by the gas density, and is less affected by combustion deposits, so it is well suited to spray-guided combustion systems.

Figure 5.4 shows a comparison between gasoline being injected into conditions that are representative of a warm or cold engine, and iso-octane being injected into a warm engine. The sprays disperse much more readily with gasoline in a warm engine, and this promotes a more homogeneous mixture and avoids wall-wetting – the spray hitting either the piston or cylinder walls. Wall-wetting leads to much slower evaporation, and indeed the fuel can still be evaporating after the main combustion event, so that there is a diffusion flame in the burned gas. Not surprisingly, this is a source of unburned hydrocarbon and particulate matter emissions. Figure 5.4 also shows how iso-octane can be unrealistic of gasoline; adding 5 per cent pentane to the iso-octane would make the fuel disperse more realistically.

The nature of the spray varies with injection and fuel parameters. In the Rayleigh regime, at very low velocities, a laminar liquid flow leaves the nozzle and creates an undisturbed jet. Rayleigh discovered that after a certain jet length, as inertial forces and surface tension interact to cause axi-symmetric surface oscillations, droplets with a constant diameter of 1.98 times the jet diameter start to break away from the jet. With higher jet velocities there are wind-induced regimes with a jet that has longitudinal surface oscillations caused by shear forces from the surrounding gas. The length of the undisturbed liquid jet reduces as the velocity increases, and the droplets are of the order of the jet diameter. As the jet velocity increases the droplets become smaller than the jet diameter as the jet surface oscillations become less structured. The atomisation regime is when there is no intact jet at the nozzle tip, and droplets that are much smaller than the nozzle diameter start to separate from the surface of a dense or intact liquid core, to create a spray of conical shape, as already seen in figure 5.4.

Three dimensionless groups which incorporate the key fluid properties can be used to describe the atomisation process: the liquid Weber number $We_L = \rho_L v^2 d/\sigma$,

the Reynolds number $Re = \rho_L vd/\mu_L$, and

the Ohnesorge number $Oh = We_L^{\frac{1}{2}} Re$

Calculation of the Weber and Ohnesorge numbers for a multi-component fuel is difficult, so Aleiferis et al. (2010) considered a range of components that are found in gasoline. As the temperature rises the viscosity (μ) and surface tension (σ) fall significantly, so Aleiferis et al. (2010) considered a range of temperatures from 25°C with 20 K increments. Figure 5.5 shows that for a typical multi-hole gasoline injector (with 0.3 mm diameter holes and an injection pressure of 150 bar), the injection process stays within the atomisation regime.

The fuel sprays have a liquid core surrounded by a cloud of droplets that form a plume, and the number-density of the droplets makes it very difficult to measure the extent of the liquid core. Correlations for the length of the liquid core will not give reliable predictions if cavitation occurs in the injector

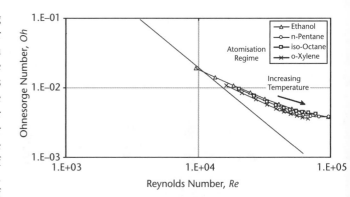

Figure 5.5 Ohnesorge plot adapted from Aleiferis et al. (2010).

nozzle holes, and this is discussed in more detail in chapter 6, section 6.5.2.

For the droplet size, Aleiferis et al. (2010) suggest a modification to the correlation developed by Hiroyasu et al. (1989) for diesel sprays:

$$\frac{SMD}{d} = 0.19 Re^{0.25} We_L^{-0.32} \left(\frac{\mu_L}{\mu_G}\right)^{0.37} \left(\frac{\rho_L}{\rho_G}\right)^{-0.47} \tag{5.1}$$

Gasoline injectors are mostly solenoid actuated, and this gives some scope for multiple injections. Multiple injections can be used to reduce spray penetration (so as to reduce the likelihood of wall-wetting) or provide some controlled form of stratification. Piezoelectric actuation provides a faster more repeatable operation, and this gives increased scope for controlling the mixture preparation by the use of multiple injections. Furthermore, with an outward opening injector, the lift can be controlled to determine the liquid sheet thickness and thence the droplet size in the spray.

5.3 | Wall-guided gasoline direct injection (GDI) engines

Zhao et al. (1999) have written a review of what are now termed wall-guided GDI engines. Mitsubishi were the first to introduce a GDI engine in a modern car, (Ando, 1997), and figure 5.6 shows some details of the air and fuel-handling systems; the spherically bowled piston is particularly important.

The Mitsubishi engine is able to operate in its stratified mode with the air/fuel ratio in the range of 30 to 40, thus reducing the need for throttled operation. In the homogeneous charge mode it mostly operates at stoichiometric, but (like the Honda VTEC engine, chapter 7, section 7.4.2) it can also operate lean at certain load conditions with air/fuel ratios in the range 20–25. With weak mixtures, the air/fuel ratio has to be lean enough for the engine-out NO_x emissions to need

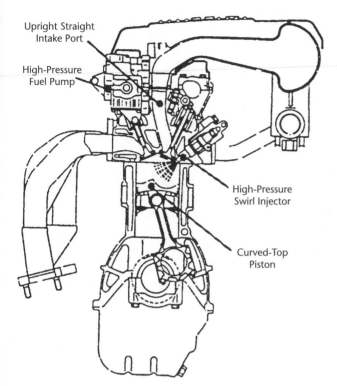

Figure 5.6 Mitsubishi GDI (Gasoline Direct Injection) engine, adapted from Ando (1997).

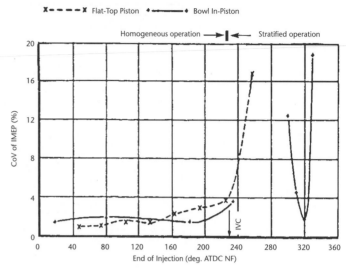

Figure 5.7 The effect of injection timing on the cycle-by-cycle variations in combustion for homogeneous and stratified charge operation (only the bowl-in-piston design would be used for stratified charge mode). Adapted from Jackson *et al.* (1997).

no catalytic reduction. Satisfactory operation of the engine is dependent on very careful matching of the in-cylinder air flow to the fuel injection. Reverse tumble (clockwise in figure 5.6, the opposite direction to a conventional homogeneous charge engine) has to be carefully matched to the fuel injection. The fuel injector is close to the inlet valves (to avoid the exhaust valves and their high temperatures), and the reverse tumble moves the fuel spray towards the spark plug, after impingement on the piston cavity. The injector in the Mitsubishi engine operates at pressures of up to 50 bar with a swirl-generating geometry that helps to reduce the droplet size, thereby facilitating evaporation.

For stratified charge operation, the fuel is injected during the start of the compression process, when the cylinder pressure is in the range 3–10 bar. The higher air density makes the spray less divergent than with homogeneous operation, which has injection when the gas pressure is about 1 bar. The greater spray divergence with early injection helps to form a homogeneous charge.

In addition to having properly controlled air and fuel motion, GDI performance is very sensitive to the timing of injection for stratified charge operation. Jackson *et al.* (1997) have found that cycle-by-cycle variations in combustion are very sensitive to the injection timing. The Ricardo combustion system is similar to the Mitsubishi system, and uses an injection pressure of 50–100 bar for stratified charge operation. Figure 5.7 shows that a bowl-in-piston design was

needed for stratified charge operation, and that the end of the injection timing window was only about 20°ca, if the cycle-by-cycle variations in combustion were to be kept below an upper limit for acceptable driveability (a 10 per cent Coefficient of Variation for the imep). Furthermore, the injection timing window narrows as the load is reduced, and it is the end of injection which is essentially independent of load. The reason for the sensitivity to injection timing has been explained by Sadler *et al.* (1998).

Figure 5.8 shows calculations of the fuel and piston displacements in a GDI Engine. Consider the fuel injected with a start of injection (SOI) of 310°ca after top dead centre on the non-firing revolution. The fuel strikes the piston (A), and flows to the rim of the piston (B), and is then swept towards the spark plug by the tumbling flow to arrive at (C). In figure 5.8, the horizontal bars represent the time between the start of injection and some mixture arriving at the spark plug. If there is too long a time, then the mixture will be over-diluted, while if there is too short a time, the mixture will be too rich. Results from in-cylinder sampling showed that the best combustion stability coincided with the richest mixture occurring in the region of the spark plug.

Exhaust gas recirculation (EGR) can be usefully applied to direct injection engines, since with lean operation there is a high level of oxygen and a low level of carbon dioxide in the exhaust gas. Jackson *et al.* (1997) show that for a fixed bmep of 1.5 bar at 1500 rpm, then applying 40 per cent EGR can lower: the fuel consumption by 3 per cent, the NO$_x$ emissions by 81 per cent and the unburned hydrocarbons by 35 per cent. Even with this level of EGR the cycle-by-cycle variations in combustion are negligible. They also point out that at some low load conditions it may be advantageous to throttle the

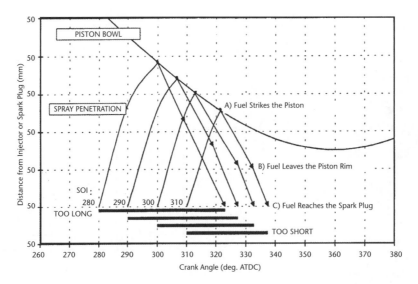

Figure 5.8 Fuel spray transport calculations for a gasoline direct injection spark ignition engine, showing the fuel and piston trajectories, adapted from Sadler *et al.* (1998).

engine slightly, since this will have a negligible effect on the fuel consumption, but reduce the unburned hydrocarbon emissions and cycle-by-cycle variations in combustion.

Figure 5.9 shows combustion images from a GDI engine when it was fuelled with unleaded gasoline (ULG), at stoichiometric with early-injection at wide open throttle and 1000 rpm; ignition was at 15°btdc. Start of injection (SOI) was at 680°abdc (start of compression, or 220°btdc compression), which was found to be the optimum setting from a volumetric efficiency study (Wyszynski *et al.*, 2002).

The high-speed cine images show: pre-mixed flame (blue) propagating across the combustion chamber, and droplets of fuel (yellow 'spots') evaporating and burning with a diffusion flame. The mixture near to a fuel droplet will be burning at a very rich AFR whereas that further away will be a much weaker mixture. It is this burning with rich AFRs that leads to glowing soot (the yellow 'flame'). There is also evidence of a pool fire, in which a pool of liquid fuel on the piston can be seen evaporating and burning. Glowing soot from the pool fire continues to combust long into the expansion stroke, after the apparent end of combustion as indicated by the mfb curve, and almost to exhaust-valve opening (100°atdc).

The key disadvantages of the wall guided systems are:

1 The limited speed range for stratified mixture preparation because the mixture preparation is determined by piston position, and this does not always allow the optimum combustion phasing;

2 and that a liquid film is formed, and this can become a source of particulate matter emissions if it is slow to evaporate.

3 The controlled air motion which leads to a reduction in volumetric efficiency, and this reduces the specific output and fuel economy.

These are disadvantages that can be avoided by using a Spray-Guided Direct Injection (SGDI) system.

5.4 | Spray-guided direct injection (SGDI) engines

As explained in section 5.1, the mixture preparation in SGDI engines is determined mainly by the characteristics of the injector, rather than the shape of the combustion chamber, and the use of multi-hole or outwardly opening injectors gives greater control of the spray pattern. Many SGDI engines operate only in a stoichiometric homogeneous mode (as a conventional three-way catalyst can be used), and the advantages over the wall-guided systems are: (a) there is a higher volumetric efficiency as there is no need to develop high-velocity in-cylinder flows, and (b) the better mixture preparation (smaller droplets from the higher injection pressures) leads to a more homogeneous mixture so there are lower emissions of particulate matter and unburned hydrocarbons. Also, SGDI engines can operate in stratified charge mode over a wider operating envelope: the maximum speed is increased from about 3000 rpm to 4000 rpm, and the maximum bmep from about 3.5 bar to 5 bar. Figure 5.10 shows how the wider operating envelope for a 3-litre engine readily encompasses the operating regime for the European Drive Cycle (NEDC) with its maximum speed of 120 km/h and a power requirement of about 30 kW.

The BMW engine uses an outwardly opening centrally-positioned piezoelectric injector that can provide a stratified mixture at the spark plug and reduce wall-wetting significantly (Schwarz *et al.*, 2006). Compared to a conventional

5.5 | Emissions

Table 5.1 *An operating regime for the Jaguar AJ133 engine during catalyst light-off*

Engine speed	1350 rpm	Manifold pressure	0.77 bar
Start of first injection	610°abdc	End of second injection	180–200°abdc
Ignition timing	188°abdc	Lambda	1.05
Inlet valve opening	574°abdc	Inlet valve closing	62°abdc
Exhaust valve opening	340°abdc	Exhaust valve closing	590°abdc

In the Jaguar AJ133 V8 SGDI engine, the NO_x versus HC trade-off at 1500 rpm and 2.62 bar bmep is slightly poorer for the SGDI engine compared with its PFI predecessor, but it is still below 5 g/kWh for both NO_x and HC. However, during vehicle operation the faster cam-timing response leads to a lower emissions level. Furthermore, injection control and valve timing control in the SGDI engine lead to faster catalyst light-off, as there is a threefold increase in the exhaust heat flow for a given level of engine stability.

Catalyst light-off is accelerated by operating the engine with very retarded ignition timing, so that although the engine is essentially idling (no net power output), the inlet manifold pressure can be very high (0.5 bar or higher). The strategy is described in detail by Chen *et al.* (2009), with additional detailed measurements reported by Twiney *et al.* (2010a and 2010b). The engine operating regime is specified in table 5.1, with figure 5.13 showing an example of the pressure traces from an extreme operating point that has been selected to include misfiring cycles.

The majority of the fuel is injected during the induction stroke so as to provide an overall homogeneous background mixture, and the second injection provides a locally rich mixture in the region of the spark plug at the time of ignition. The overall mixture is weak ($\lambda = 1.05$) so as to reduce CO and HC emissions. With the exhaust valve opening at 340°abdc, figure 5.13 clearly shows that combustion is continuing, but by 380°abdc the pressure in the exhaust stroke is essentially 1 bar. A single misfiring cycle would lead to unacceptable levels of HC emissions, but the HC emissions from the slow-burning cycles are essentially no different from those of the 'normal' cycles.

Figure 5.14 shows the second injection (fuel pressure of 150 bar), with four fuel plumes pointing downwards and two plumes closer to horizontal that pass either side of the spark plug; the fuel arrives at the spark plug about 1°ca after the end of injection. If injection is too late (relative to the spark), then there is no enrichment; while if the injection is too early, then the mixture at the spark plug can be too rich. The fuel plumes also induce an air flow, and this leads to stretching of the arc and an increase in the arc voltage.

Direct injection can lead to a less homogeneous mixture, which in turn will reduce combustion stability and increase

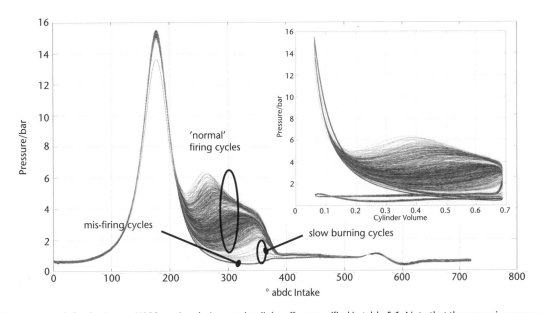

Figure 5.13 Pressure records for the Jaguar AJ133 engine during catalyst light-off, as specified in table 5.1. Note that the expansion pressure is lower than the compression pressure until the pressure has fallen to about 6 bar. The end of the second injection was 194°abdc, so as to generate misfiring cycles.

Figure 5.14 The second injection in the Jaguar AJ133 engine catalyst light-off mode, with the fuel plumes illuminated from the right; note that the central plane of the fuel jets is below the spark-plug gap. Combustion can be seen as the 'bright spots' immediately adjacent to the earth electrode of the spark plug.

The cylinder pressure limit was set at 120 bar for the average P_{max} and 140 bar for maximum P_{max}, which led to a maximum compression ratio of 9.7:1, although the combustion stability was better with a compression ratio of 9.2:1. The turbine inlet temperature was limited to 1025°C, but the exhaust lambda could be maintained at approximately stoichiometric except at the peak power point of 6000 rpm where enrichment was necessary to limit the turbine inlet temperature.

Figure 5.16 shows the high- and low-pressure compressor maps for the operating lines at 1500 rpm and bmep in the range 20–26 bar, but note that the overall pressure ratio does not relate to either compressor map. As the engine speed and air flow increase, the low-pressure compressor pressure ratio increases and the operating line moves into a more efficient part of the compressor map, while the operating line moves into a less efficient part of the high pressure compressor map.

particulate matter emissions. A solution adopted by Toyota has been to use both Port Fuel Injection and Direct Injection, leading to an engine with dual variable intake and exhaust valve timing, and a power output 66 kW/litre at 6400 rpm (Ikoma *et al.*, 2006). A homogeneous mixture is produced without any devices to generate intense in-cylinder air-motion at lower engine speeds. This gives an improved volumetric efficiency compared to engines with charge motion devices, resulting in improved full-load performance throughout the engine speed range. Simultaneous injection by the two injectors was effective in reducing HC emissions during cold start. At part load, using 40 per cent DI and 60 per cent PFI injection gave a 7 per cent improvement in fuel consumption, an 8-degree faster burn and a reduction in torque fluctuations.

5.6 | Downsizing

A good example of downsizing is the Mahle three cylinder 1.2-litre concept engine, which has an output that is comparable to a naturally aspirated 2.4-litre engine (Lumsden *et al.*, 2009). The philosophy behind this engine and details of its mechanical design are covered in great detail by Hancock *et al.* (2008). Figure 5.15 shows the twin turbocharger configuration and target torque curve of the Mahle demonstration engine with a target bmep of 30 bar.

The combustion system uses centrally mounted piezo fuel injectors, high-energy ignition coils, an integrated engine-mounted inter-cooler between the compressors, variable valve timing for both the inlet and exhaust valves, exhaust valves with sodium cooling, and carefully designed head cooling passages with high heat transfer coefficients on the exhaust side.

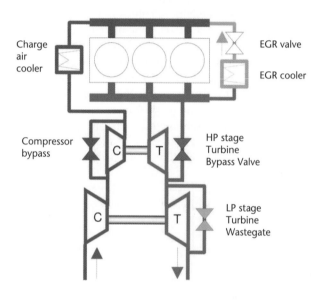

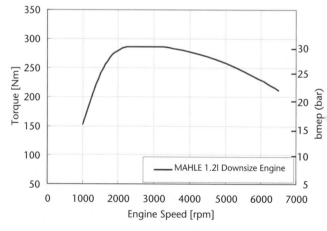

Figure 5.15 Twin turbocharger configuration and target torque curve of the Mahle demonstration engine, adapted from Lumsden *et al.* (2009).

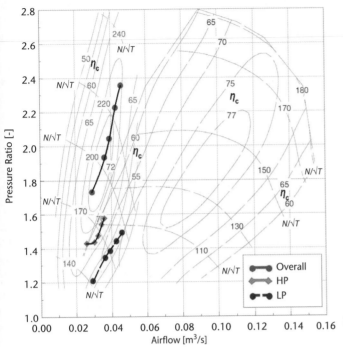

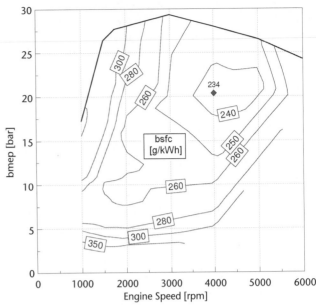

Figure 5.17 Brake specific fuel consumption map (g/kWh) for the Mahle demonstration engine, adapted from Lumsden et al. (2009).

Figure 5.16 High- and low-pressure compressor maps (showing the isentropic efficiency [per cent] and speed parameter) for the Mahle demonstration engine, with the operating lines at 1500 rpm and bmep in the range 20–26 bar. The overall pressure ratio does not relate to either compressor map. Adapted from Lumsden et al. (2009).

At 3000 rpm the low-pressure compressor speed is about 120 000 rpm ($N/\sqrt{T}$ = 115). When the turbochargers were tested in a single turbocharger configuration, the small ('high-pressure') turbocharger could provide sufficient boost at speeds up to 1800 rpm, while the larger ('low-pressure') turbocharger could provide sufficient boost at speeds above 3000 rpm. By using the sequential configuration of figure 5.14, the torque can be maintained between 1800 and 3000 rpm.

The brake specific fuel consumption map in figure 5.17 shows a minimum of 234 g/kWh at a speed of 4000 rpm and a bmep of 20 bar. In a naturally aspirated engine the minimum bsfc might be expected to occur just below the speed for maximum torque (say 2000 rpm) as the increasing torque increases the mechanical efficiency, but by (say) 3000 rpm the frictional losses are starting to rise rapidly. However, in the case of the turbocharged GDI engine there is a scavenging flow of air through the combustion chamber at close to full throttle, yet the overall mixture has to be stoichiometric for the three-way catalyst to operate correctly. This means that the trapped in-cylinder mixture will be rich of stoichiometric, leading to a lower combustion efficiency.

At part load and low speeds the engine can be optimised for fuel economy rather than volumetric efficiency. The ignition timing should always be either MBT or a fixed amount of retard for torque reserve – a retarded ignition timing leads to a hotter exhaust and greater turbine work, so that the transient response can be improved by both the immediate effect of advancing the ignition timing and the reduced turbo-lag. At part load and low speeds the engine can be optimised by control of the valve timing and this is illustrated in figure 5.18.

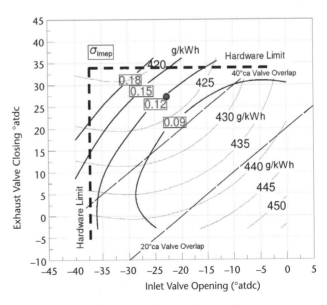

Figure 5.18 Effect of valve phasing on the bsfc (g/kWh) and combustion stability (standard deviation, σ, of the imep) at 2000 rpm, and 2 bar bmep, with a compression ratio of 9.2:1 and MBT spark timing. Adapted from Lumsden et al. (2009).

As the valve overlap increases, the residuals level increases since at low loads the inlet manifold pressure is lower than the exhaust pressure. The higher level of residuals increases the inlet manifold pressure, so that the throttling loss is reduced. However, increasing the residuals level reduces the burn rate, so the combustion stability decreases, and the limit here of 0.12 bar for the standard deviation of the imep corresponds to a CoV of 6 per cent.

An alternative to turbocharging is supercharging (chapter 10, section 10.2), and Nissan have produced an engine that is also 1.2-litre with three cylinders. The engine has a compression ratio of 13.0:1, and it too has continuously variable valve-timing controls on both the inlet and exhaust camshafts. The engine has a specific output of 60 kW/litre and a maximum bmep of 14.9 bar at 4400 rpm. The supercharger has a by-pass valve and it is also driven via an electromagnetic clutch, so that it is disengaged at low speeds and low load. The corresponding naturally aspirated engine has a maximum bmep of 11.5 bar, and has a higher fuel consumption leading to 115 gCO_2/km on the European Drive Cycle compared with 95 gCO_2/km for the supercharged engine (Isaji et al., 2011). With high boost systems a supercharger can be used in series with a turbocharger to replace the low-pressure turbocharger in compound systems (see chapter 10, section 10.6.2), and this will improve the transient response (Fraser et al., 2009).

Fraser et al. (2009) review other strategies for high boost engines, concluding that both two-stage turbocharging and an electric supercharger in series with a turbocharger have considerable potential. They also consider the influence of bore size (as the spaces needed for the injector and spark plug are essentially constant), and conclude that applications below a bore size of 70 mm are viable.

5.6.1 Superknock

With such high boost levels the likelihood of knock increases, and when knock does occur it will be particularly damaging in a highly boosted engine because the charge density is so high. Knock is most likely to occur at the low-speed and high-load operating points. Retarding the ignition timing from MBT reduces the likelihood of knock, but will lead to higher exhaust gas temperatures. The scavenging air flow removes hot residuals and helps to cool the combustion chamber components, and the use of cooled EGR (~15 per cent) also suppresses knock. Turner and Pearson (2010) provide a very useful review of turbo-charged GDI engines, and one of their conclusions is that gasolines that contain alcohol will perform particularly well. Their higher enthalpy of vaporisation (a factor of about 4.5 for a stoichiometric mixture of ethanol, and about 8 for methanol; see chapter 3, table 3.10) means that their anti-knock performance will be higher than that suggested by their octane ratings.

Highly boosted GDI engines sometimes encounter a very intense form of knock that is referred to as 'superknock'. With 'normal' knocking combustion, advancing the ignition timing leads to a gradual increase in the fraction of cycles that knock and the severity of the knock. With 'superknock' there are occasional but very severe knocking cycles, and combustion can be initiated before the spark. Kalghatgi and Bradley (2012) argue that this is related to pre-ignition. Historically, pre-ignition has been associated with surface hot-spots and deposits, but with modern engines (and fuel additives that inhibit the formation of deposits) it is more likely that ignition is initiated away from surfaces. Ignition might be centred around oil droplets or particulates, as oil will have a very low octane rating. Auto-ignition of the fuel mixture can be ruled out by analysis of the Livengood–Wu integral (chapter 3, section 3.5.2). Surface (catalytic) chemistry is likely to be important, and the metallic additives in the lubricant could be very significant. The action of the piston rings could cause oil droplets to mix with the bulk gas, and the question is – if they auto-ignite will they lead to flame propagation? Whether or not a flame will propagate can be related to its size, and Kalghatgi and Bradley (2012) show that the critical size is inversely proportional to the laminar burning velocity (as fast combustion means that the heat release will be faster than the rate at which heat diffuses away from the reaction zone). They also show that the Pre-ignition Rating for different hydrocarbon fuels (using data from Ricardo and Hempson, 1968) is inversely related to the laminar burning velocity, but that there is no correlation with either the RON or MON. As the laminar burning velocity reduces with increasing pressure, auto-ignition of an oil droplet (for example) is more likely to lead to pre-ignition as the pressure is increased. Once pre-ignition occurs, then the earlier combustion will mean increased combustion pressures (and thence temperatures in the unburned gas) so that auto-ignition (and thence knock) will occur earlier and thereby consume a larger amount of mixture.

Kalghatgi and Bradley (2012) explain how the nature of the subsequent auto-ignition that can follow pre-ignition depends on the auto-ignitive propagation velocities which can be estimated from assumed reactivity gradients in the fuel/air mixture at hot-spots and computed values of the ignition delay times. The nature of the combustion is then:

1 *Deflagaration* – 'normal' combustion with the pressure wave attenuating if the temperature falls quickly away from the hot-spot.

2 *Thermal Explosion* – which gives a high pressure rise rate if there is no inhomogeneity (zero gradient).

3 *Developing Detonation (DD)* – amplification of the pressure wave to give an extremely high local pressure pulse at moderate temperature gradients.

Developing Detonation (DD) is a mutual reinforcement when the auto-ignition front moves into the unburned mixture at approximately the local speed of sound, and the front of the pressure wave generated by the rate of heat release couples with the auto-ignition reaction front. The severity of the associated pressure pulse is shown by Kalghatgi and Bradley (2012) to be dependent on the ratios of:

1 the local speed of sound to the localised auto-ignitive velocity, and

2 the residence time of the acoustic wave in the hot-spot to the short excitation time in which most of the chemical energy is released.

5.7 | Particulate matter emissions

Figure 5.9 has shown the sources of particulate matter associated with pool fires resulting from wall-wetting in a wall-guided direct injection combustion system. Even if wall-wetting is avoided in Spray-Guided Direct Injection engines, the emissions or particulate matter are higher than those from PFI engines because the mixture is less homogeneous (as there is less time for evaporation and mixing).

Emissions legislation in Europe applies to particulate matter (PM) emissions from Gasoline Direct Injection Engines, and a comparison of GDI and other engined vehicles has been made by Braisher *et al.* (2010). This shows that SGDI-engined vehicles would readily comply with the EU5 mass limits (see chapter 3, table 3.21), but not necessarily the number limit that is current for diesel engines. Braisher *et al.* (2010) show that a Diesel Particulate Filter (DPF, chapter 6, section 6.6) equipped engine has PM number (Pn) emissions that are about an order of magnitude lower than those of a PFI gasoline engine, and both are below the EU5 Pn limit for diesel engines. A diesel engine without a DPF has number emissions about two orders of magnitude above the EU5 Pn limit. SGDI engines are up to an order of magnitude above the EU5 Pn limit, and wall-guided GDI engine Pn emissions are about an order of magnitude higher than those of SGDI engines. These GDI engines and vehicles have not yet been optimised for reducing the Pn emissions, but this will need to be done, preferably without recourse to filters or traps that add cost and fuel economy penalties (see chapter 6, section 6.6.3 for a discussion in the context of diesel engines). Analysis of the engine-out PM from a diesel engine shows that most of the mass is soot (carbon) with a small amount of condensed hydrocarbons (say 80 per cent 'elemental carbon' and 20 per cent 'organic carbon'). This is in complete contrast to GDI engines where analysis of the engine-out PM shows that most of the mass is condensed hydrocarbons with a small amount of soot (say 40 per cent 'elemental carbon' and 60 per cent 'organic carbon').

The mechanisms for the formation of PM are discussed in chapter 3, section 3.8.5, while chapter 14, section 14.4.5 discusses some of the measurement techniques.

5.7.1 Influence of GDI engine parameters on PM emissions

Figure 5.19 shows the PM emissions upstream and down-stream of a three-way catalyst (TWC) for a SGDI engine. With both the hot and cold coolant, it can be seen how the nucleation-mode dominates the particle number emissions, but these are reduced substantially by the catalyst, especially for the hot engine. The accumulation-mode dominates the mass emissions and there is also (as expected) a higher geometric mean diameter for the mass emissions than for the number emissions.

The PM emissions increase by an order of magnitude for both PFI and SGDI engines when the mixture is 10 per cent rich ($\lambda = 0.9$) compared with a stoichiometric mixture. The inherently higher levels of emissions with GDI engines are due to the lower mixture homogeneity and wall-wetting. If the mixture is not homogeneous, the rich pockets (which might remain from larger droplets that have been slower to evaporate) will contribute disproportionately to the PM emissions. When there is wall-wetting (as will occur with the first generation wall-guided injection systems), the liquid films will be slow to evaporate and rich regions will form that lead to diffusion-controlled combustion. A typical injection timing for an SGDI engine (operating at stoichiometric) is 240°ca btdc, which is 60°ca after the start of the induction stroke. If injection is later, then wall-wetting is even less likely, yet the PM emissions increase, so this can only be attributed to the reduced homogeneity. The comparison between hot (90°C) and cold (20°C) coolant in figure 5.19 also shows how inhomogeneity that is associated with a cold engine increases the Particulate Matter number (Pn) emissions by a factor of about 3. The increase is most marked for the accumulation-mode particles, and these are reduced less by the three-way catalyst and contribute most to the mass, so the Particulate Matter mass (Pm) emissions increase by a factor of about 10. The mixture homogeneity can also be improved if hot residuals are trapped in the cylinder by controlling the valve timing; this leads to a slight reduction in PM, though too high a level of residuals decreases the combustion stability.

As either the load or speed is increased then PM emissions increase. With an increase in engine speed there is less time for a homogeneous mixture to form. As load is increased, the residuals level will reduce, and the higher in-cylinder pressure will reduce the rate of droplet vaporisation, and there can be adverse effects on spray penetration. Post-TWC measurements show that the Pn emissions are approximately proportional to load, while the Pm emissions increase less markedly.

Figure 5.20 shows how retarding the ignition timing reduces the PM emissions. As with unburned hydrocarbons, the reduction in PM emissions is because (with less work and heat transfer) the hot gases later in the expansion stroke lead to more post-flame oxidation.

Figure 5.21 shows results from an SGDI vehicle on the New European Drive Cycle (NEDC), which are compared by Braisher *et al.* (2010) with other engine types.

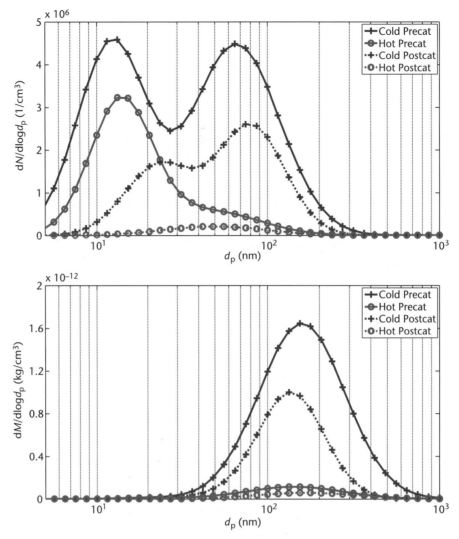

Figure 5.19 The effect of a three-way catalyst (TWC) and coolant temperature (90°C or 20°C) on the Particulate Matter number (Pn) and Particulate Matter mass (Pm) emissions from a SGDI engine.

The top subplot in figure 5.21 shows the particle-size distribution, giving particle-number concentration in the range 5–1000 nm (ordinate) for the duration of the NEDC (abscissa), as recorded by the Cambustion DMS500; the greyscale map represents the number concentration. The vehicle-speed trace is also plotted in this and the other subsequent subplots. The second subplot contains cumulative particle number (Pn), against time, as recorded by the Solid Particle Counting System (SPCS), which is compliant with the European measurement protocol and Cambustion DMS500 data that has been integrated over the accumulation-mode size range. As the SPCS is designed to remove volatile material through a combination of sample conditioning and its nominal 23 nm cut-off, the SPCS Pn is thought to be solid only, so this is why only the accumulation-mode fit is used for the DMS Pn value.

The third subplot contains cumulative particulate mass (Pm), recorded over the drive cycle by both the Photo-Acoustic Soot Sensor (PASS) and the DMS500. The DMS500 Pm calculation uses an empirically calibrated model to estimate mass which is applied to the accumulation-mode fit. The final subplot contains a trace of the cumulative mass for the three regulated gaseous emissions (HC, CO and NO_x); note that the HC results have been scaled up by a factor of 10.

If allowance is made for delay times in the sampling system, it can be seen in the top plot of figure 5.21 that there are bursts of PM following the accelerations, and these lead to the stepwise increase in both the cumulative Pn and Pm. The lowest plot shows the cumulative CO emissions, and these too tend to have a stepwise increase following the accelerations. The implication from these results is that there are rich-mixture excursions during the acceleration transients, and it is

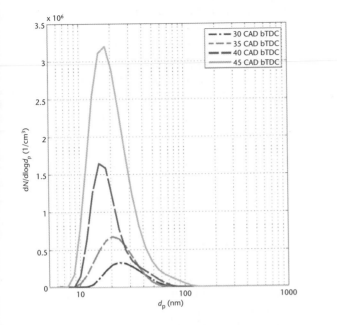

Figure 5.20 The effect of the ignition timing on the Particulate Matter number (Pn) emissions; homogeneous SGDI with stoichiometric toluene at an inlet manifold absolute pressure of 0.5 bar (MBT ignition is 35°bTDC).

these that lead to the increase in PM emissions. (Recall that, compared to stoichiometric mixtures, a 10 per cent rich mixture can give an order of magnitude increase in the PM emissions.)

5.7.2 Influence of fuel composition on PM emissions

Aikawa *et al.* (2010) report a PM index for PFI engines that is a function of the weight fraction (W_t) of components in the fuel, using the vapour pressure (VP, evaluated at 443 K) and the Double Bond Equivalent (DBE) of the components. The Double Bond Equivalent is the number of hydrogen molecules needed to saturate a hydrocarbon; for example toluene (C_7H_8) has a DBE of 4, as heptane (C_7H_{16}) is the corresponding saturated hydrocarbon.

$$\text{PM index} = \left[\frac{\text{DBE} + 1}{\text{VP}}\right] W_t \qquad (5.2)$$

The tests used a standard gasoline to which different components were added, and the vehicle was tested on the New European Drive Cycle (NEDC). Although this index was derived for PFI engines, similar trends can be expected with GDI engines.

Tests on an SGDI engine have been conducted at part load with a blend of iso-octane and n-octane that was designed to have the same volatility as toluene. At stoichiometric, toluene had about three times the Pn emissions of the octanes. With a 10 per cent rich mixture ($\lambda = 0.9$) there was negligible change

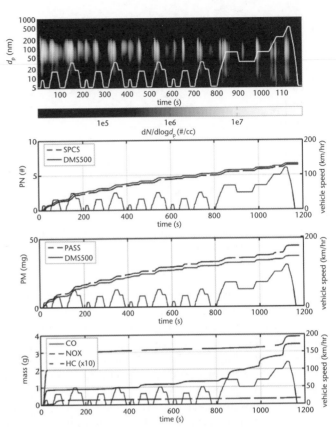

Figure 5.21 Results from an SGDI vehicle on the New European Drive Cycle (NEDC) reported by Braisher *et al.* (2010). SPCS – Solid Particle Counting System which is compliant with the European measurement protocol; PASS – Photo-Acoustic Soot Sensor.

in the octane Pn emissions, but the toluene Pn emissions increased by two orders of magnitude; the response to the toluene content was highly non-linear as 35 per cent toluene increased the Pn by about an order of magnitude compared with the octanes (35 per cent is the maximum level of aromatics allowed in European gasoline).

Chen *et al.* (2011) report on ethanol/gasoline blends that have been studied in the same SGDI engine. Tests were conducted under cold and warm conditions (20°C, 80°C). The spray images and combustion analysis show that ethanol addition results in protracted injection and an increased combustion duration. The mixture inhomogeneity increased as the ethanol content was increased, and this led to slower combustion and increased levels of cycle-by-cycle variations in combustion. Figure 5.22 shows the distillation curves for an unleaded gasoline (ULG) and its 10 per cent ethanol blend (E10).

Figure 5.22 shows how adding ethanol to gasoline increases its volatility, for example, at 60°C only 20 per cent of the ULG has evaporated while 32 per cent of the E10 has evaporated. As ethanol is polar and the hydrocarbons are non-polar they do not want to mix, and additional evidence of their

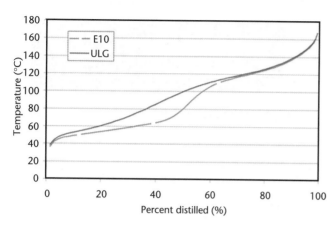

Figure 5.22 Distillation curves for an unleaded gasoline (ULG) and its 10 per cent ethanol blend (E10).

Table 5.2 *Particulate number and particulate mass emissions for a rich mixture ($\lambda = 0.9$, ignition 35° btdc); the base fuel is 65 per cent iso-octane/35 per cent toluene*

	Total number (particles/cm³)	Total mass (kg/cm³)
Base fuel	2.18×10^8	4.02×10^{-13}
E10 fuel	3.45×10^6	4.03×10^{-14}

emissions during rich-mixture operation will lead to an overall reduction in particulate emissions, even if the stoichiometric emissions are increased.

5.8 | Conclusions

It is difficult to apply GDI to small engines, and these are also the applications where manufacturers will be wanting to use low-cost technology. There are also greater incentives with emissions reduction and fuel economy for using GDI in larger engines. With larger displacement engines it can be more difficult to comply with emissions legislation that is based on distance travelled.

There are considerable challenges in making stratified charge engines operate with sufficiently low engine-out emissions, and there is also the challenge of switching between the stratified and homogeneous charge operating modes. The operating envelope for unthrottled stratified operation is still comparatively limited, and this means that many SGDI engines only operate in a homogeneous mode.

aspiration to be separate is the slight increase in volume when the two liquids are mixed and the cooling that occurs with mixing. The increased volatility does not lead to improved evaporation because of the increase in enthalpy of evaporation as ethanol is added to gasoline (see chapter 3, table 3.10). For stoichiometric mixtures, as the ethanol content was increased, the mixture homogeneity reduced and the particulate emissions increased. However, table 5.2 shows that for rich mixtures the particulate number emissions were at least an order of magnitude lower with E10 than for the base fuel.

As any rich mixture operation in a drive cycle is disproportionately responsible for particulate emissions, anything (such as ethanol addition) that reduces particulate

CHAPTER 6

Compression ignition engines

6.1 | Introduction

Satisfactory operation of compression ignition engines depends on proper control of the air motion and fuel injection. The ideal combustion system should have a high output (bmep), high efficiency, rapid combustion, a clean exhaust and be silent. To some extent these are conflicting requirements; for instance, engine output is directly limited by smoke levels. There are two main classes of combustion chamber: those with direct injection (DI) into the main chamber (figure 6.1) and those with indirect injection (IDI) into some form of divided chamber (figure 6.6). The fuel injection system cannot be designed in isolation since satisfactory combustion depends on adequate mixing of the fuel and air. Direct injection engines have inherently less air motion than indirect injection engines and, to compensate, high injection pressures (up to 2000 bar and higher) are used with multiple-hole nozzles. Even so, the speed range is more restricted than for indirect injection engines. Injection requirements for indirect injection engines are less demanding; single-hole injectors with pressures of about 300 bar can be used. However, indirect injection engines have a lower efficiency and have now been superseded by direct injection engines, since improvements with injection equipment and a better understanding of combustion have overcome the limitations of direct injection engines. Turbocharged direct injection engines are capable of achieving a specific output of 70 kW/litre with a bmep of over 25 bar (Körfer *et al.*, 2007).

There are two types of injector pump for multi-cylinder engines, either in-line or rotary. The rotary pumps are cheaper, but the originally limited injection pressure made them more suited to indirect injection engines.

Two alternatives to traditional injection pumps are unit injectors and common rail injection systems, both of which are discussed further in section 6.5.5. The unit injector has the pumping element integral to the injector, so that higher pressures can be used, which give better mixture preparation. Common rail injection systems can operate at even higher pressures, and have the advantage that the pressure can be controlled, so that there is control of the rate of injection as well as its timing.

The minimum useful cylinder volume for a compression ignition engine is about 400 cm^3, otherwise the surface-to-volume ratio becomes disadvantageous for the normal compression ratios. The combustion process is also slower than in spark ignition engines and the combined effect is that maximum speeds of compression ignition engines are much less than those of spark ignition engines. Since speed cannot be raised, the output of compression ignition engines is most effectively increased by turbocharging. The additional benefits of turbocharging are improvements in fuel economy, and a reduction in the weight per unit output.

The gains with turbocharging can be enhanced by inter-cooling (cooling the air leaving the compressor; more correctly after-cooling), which further increases the density of the air, and reduces the thermal loading on the engine. Further improvements in output can be achieved with four valves per cylinder in direct injection engines. The main motivation for using four valves per cylinder in direct injection engines is the ability to install the injector vertically at the centre of the combustion chamber. This leads to better mixture preparation and lower emissions.

The compression ratio of turbocharged engines has to be reduced, in order to restrict the peak cylinder pressure; the compression ratio is typically in the range 12–24:1. The actual value is usually determined by the cold starting requirements, and the compression ratio is often higher than optimum for either economy or power.

There are many different combustion chambers designed for different sizes of engine and different speeds, though inevitably there are many similarities. Very often it will be the application that governs the type of engine adopted. For automotive applications a good power-to-weight ratio is needed and some sacrifice to economy is accepted by using a high-speed engine. For marine or large industrial applications size and weight will matter less, and a large slow-running engine can be used with excellent fuel economy.

All combustion chambers should be designed to minimise heat transfer. This does not of itself significantly improve the engine performance, but it will reduce ignition delay. Also, in a turbocharged engine a higher exhaust temperature will enable more work to be extracted by the exhaust turbine. The so-called adiabatic engine, which minimises (not eliminates) heat losses, uses ceramic materials, and will have higher exhaust temperatures.

Another improvement in efficiency is claimed for the injection of water/fuel emulsions; for example, see Katsoulakos (1983). By using an emulsion containing up to 10 per cent water, improvements in economy of 5–8 per cent are reported. Improvements are not universal, and it has been suggested that they occur only in engines in which the air/fuel mixing has not been optimised. For a given quantity of fuel, a fuel emulsion will have greater momentum and this could lead to better air/fuel mixing. An additional mechanism is that when the small drops of water in the fuel droplets evaporate, they do so explosively and break up the fuel droplet. However, the preparation of a fuel emulsion is expensive and it can lead to problems in the fuel injection equipment. If a fuel emulsion made with untreated water is stored, bacterial growth occurs. Fuel emulsions should reduce NO_x emissions since the evaporation and subsequent dissociation of water reduce the peak temperature.

As in spark ignition engines, NO_x emissions can be reduced by exhaust gas recirculation since this lowers the mean cylinder temperature. Alternatively NO_x emissions can be reduced by retarded injection. However this has an adverse effect on output, economy and emissions of unburnt hydrocarbons and smoke.

Section 6.6 discusses the emissions from diesel engines and after-treatment systems, and summarises the relevant emissions legislation. The ways of predicting the ignition delay, and the combustion rate are discussed in chapter 11, section 11.2, where modelling techniques are discussed.

Before discussing direct and indirect diesel engines, it is instructive to compare their part-load efficiency with that of a spark ignition engine; this had been done in figure 6.1. The difference in efficiency of direct and indirect injection diesel engines is discussed later in section 6.3. When comparing the sfc of compression ignition and spark ignition engines, it must be remembered that on a specific basis the energy content of diesel fuel is about 4.5 per cent lower. Figure 6.1 shows that the diesel engines have a higher maximum efficiency than spark ignition engine for three reasons:

(a) The compression ratio is higher.

(b) During the initial part of compression, only air is present.

(c) The air/fuel mixture is always weak of stoichiometric.

Furthermore, the diesel engine is, in general, designed to operate at lower speeds, and consequently the frictional losses are smaller.

In a diesel engine the air/fuel ratio is always weak of stoichiometric, in order to achieve complete combustion. This is a consequence of the very limited time in which the mixture can be prepared. The fuel is injected into the combustion chamber towards the end of the compression stroke, and around each droplet the vapour will mix with air, to form a flammable mixture. Thus the power can be regulated by varying the quantity of fuel injected, with no need to throttle the air supply.

As shown in figure 6.1, the part-load specific fuel consumption of a diesel engine rises less rapidly than for a spark ignition engine. A fundamental difference between conventional spark

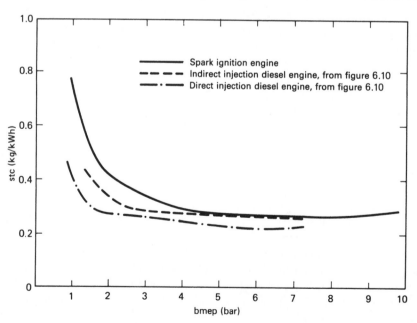

Figure 6.1
Comparison of the part-load efficiency of spark ignition and diesel engines at 2000 rpm (Stone, 1989b).

ignition and diesel engines is the manner in which the load is regulated. A conventional spark ignition engine always requires an air/fuel mixture that is close to stoichiometric. Consequently, power regulation is obtained by reducing the air flow as well as the fuel flow. However, throttling of the air increases the pumping work that is dissipated during the gas exchange processes.

Also, since the output of a diesel engine is regulated by reducing the amount of fuel injected, the air/fuel ratio weakens and the efficiency will improve. Finally, as the load is reduced, the combustion duration decreases, and the cycle efficiency improves. To summarise, the fall in part-load efficiency is moderated by;

(a) The absence of throttling.

(b) The weaker air/fuel mixtures.

(c) The shorter duration combustion.

Low heat loss (so-called adiabatic) diesel engines are discussed later in chapter 10, section 10.3, as the benefits are most significant for turbocharged engines.

6.2 Direct injection (DI) systems

Some typical direct injection combustion chambers given by Howarth (1966) are shown in figure 6.2. Despite the variety of shapes, all the combustion chambers are claimed to give equally good performance in terms of fuel economy, power and emissions, when properly developed. This suggests that the shape is less critical than careful design of the air motion and fuel injection. The most important air motion in direct injection diesel engines is swirl, the ordered rotation of air about the cylinder axis. Swirl can be induced by shrouded or masked inlet valves and by design of the inlet passage – see chapter 3, figure 3.2.

$$\text{swirl (ratio)} = \frac{\text{swirl speed (rpm)}}{\text{engine speed (rpm)}} \tag{6.1}$$

The swirl speed will vary during the induction and compression strokes and an averaged value is used.

The methods by which the swirl is measured, how the swirl ratio is defined, and how the averaging is undertaken all vary widely. Heywood (1988) summarises some of the approaches, but a more comprehensive discussion is provided by Stone and Ladommatos (1992b). For steady-flow tests, the angular velocity of the swirl will vary across the bore, but an averaged value is given directly by a paddle wheel (or vane) anemometer. Alternatively, a flow straightener can be used, and the moment or momentum of the flow corresponds to the torque reaction on the flow straightener.

If there is a uniform axial velocity in the cylinder, then

$$M = \frac{1}{8}\dot{m}\omega B^2 \tag{6.2}$$

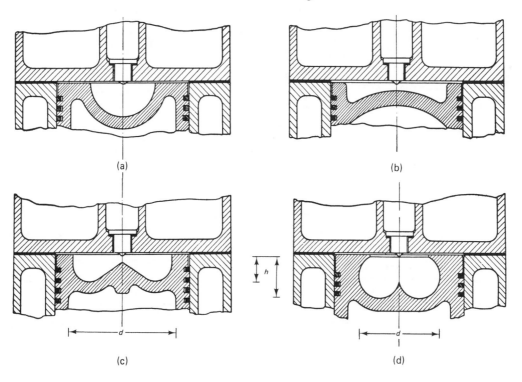

Figure 6.2
Different types of direct injection combustion chamber: (a) hemispherical combustion chamber; (b) shallow bowl combustion chamber; (c) shallow toroidal bowl combustion chamber ($d/h \approx 4$); (d) deep toroidal bowl combustion chamber ($d/h \approx 2$) (reproduced with permission from Ricardo and Hempson, 1968).

where M = moment of momentum flux (N m)
$\quad\quad m$ = mass flow rate (kg/s)
$\quad\quad \omega$ = swirl angular velocity (rad/s)
$\quad\quad B$ = cylinder bore (m).

If hot wire anemometry or laser doppler anemometry are used to define the flow field in the cylinder, and a single value is required to quantify the swirl, then

$$M = \int_0^{B/2} \int_0^{2\pi} \rho v_t(r,\theta) v_a(r,\theta) r^2 \, dr \, d\theta \tag{6.3}$$

where ρ = air density
$\quad\quad v_t$ = tangential velocity
$\quad\quad v_a$ = axial velocity
$\quad\quad r$ = radial coordinate
$\quad\quad \theta$ = angular coordinate.

Because of the different ways in which swirl can be measured and defined, published swirl data should be treated on a comparative basis, unless the measurement method has been fully defined. The means by which swirl is generated in a helical inlet port is illustrated by figure 6.3.

Ricardo and Hempson (1968) report extensive results from an engine in which the swirl could be varied. The results for a constant fuel flow rate are shown in figure 6.4. For these conditions the optimum swirl ratio for both optimum economy and power output is about 10.5. However, the maximum pressure and rate of pressure rise were both high, which caused noisy and rough running. When the fuel flow rate was reduced the optimum swirl ratio was reduced, but this incurred a slight penalty in performance and fuel economy. Tests at varying speeds showed that the optimum relationship of fuel injection rate to swirl was nearly constant for a wide range of engine speeds. Constant speed tests showed that the interdependence of swirl and fuel injection rate became less critical at part load. Increasing swirl inevitably increases the convective heat transfer coefficient from the gas to the cylinder walls. This is shown by a reduction in exhaust temperature and an increase in heat transfer to the coolant.

Care is needed in the method of swirl generation in order to avoid too great a reduction in volumetric efficiency, since this would lead to a corresponding reduction in power output. The kinetic energy associated with swirl has to come from the

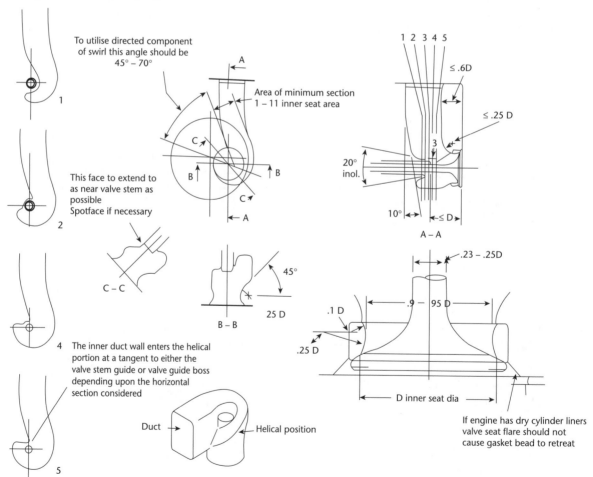

Figure 6.3
Geometric details of a helical inlet port (Beard, 1984). [Reproduced by kind permission of Butterworth–Heinemann Ltd, Oxford.]

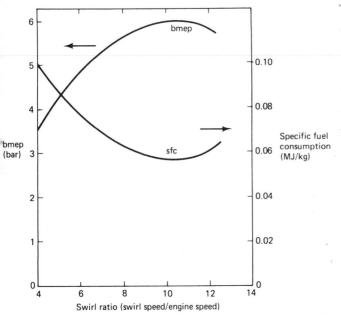

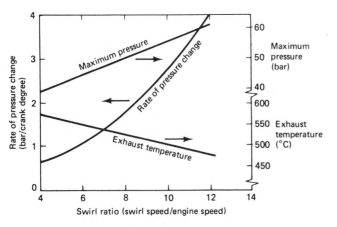

Figure 6.4
Variation of engine performance with swirl ratio at constant fuelling rate (adapted from Ricardo and Hempson, 1968).

pressure drop across the inlet port and valve, so clearly there has to be a trade-off between the swirl ratio and the volumetric efficiency. These volumetric efficiency/swirl ratio trade-offs are illustrated by Pischinger (1998), who compares the performance of two-valves-per-cylinder arrangements with various four-valve-per-cylinder arrangements. These show that the four-valve arrangements lead to lower swirl levels, unless a significant decrease in volumetric efficiency can be tolerated.

An additional advantage of four-valve-per-cylinder arrangements is that a throttle plate can be incorporated into one of the inlet ports. Closing the throttle plate deactivates the inlet port, thereby increasing the swirl level (this has already been described in chapter 4, section 4.2.3 in the context of spark ignition engines). This is useful at low engine speeds, where the higher swirl improves the air/fuel mixing, thereby

increasing the low speed torque, despite a reduction in the volumetric efficiency.

In two-stroke engines there are conflicting requirements between swirl and the scavenging process. The incoming air will form a vortex close to the cylinder wall, so trapping exhaust products in the centre.

The combustion chamber is invariably in the piston, so that a flat cylinder head can be used. Very often the combustion chamber will have a raised central portion in the piston, on the grounds that the air motion is minimal in this region. In engines with a bore of less than about 150 mm, it is usual for the fuel to impinge on the piston during injection. This breaks up the jet into droplets, and a fuel film forms on the piston. Although the fuel film helps to control the combustion rate, it also means that the mixing between the air and the fuel is not as good, so the smoke or particulate emissions are higher.

The role of squish, the inward air motion as the piston reaches the end of the compression stroke, remains unclear. Combustion photography suggests that the turbulence generated by squish does not influence the initial stages of combustion. However, it seems likely that turbulence increases the speed of the final stages of combustion. In coaxial combustion chambers, such as those shown in figure 6.2, squish will increase the swirl rate, by the conservation of angular momentum in the charge.

The compression ratio of direct injection engines is usually between 12:1 and 16:1. The compact combustion chamber in the piston reduces heat losses from the air, and reliable starting can be achieved with these comparatively low compression ratios.

The stroke-to-bore ratio is likely to be greater than unity (under-square engine) for several reasons. The longer stroke will lead to a more compact combustion chamber. The effect of tolerances will be less critical on a longer stroke engine, since the part of the clearance volume between the piston and cylinder head will be a smaller percentage of the total clearance volume. Finally, the engine speed is most likely to be limited by the acceptable piston speed. Maximum mean piston speeds are about 12 m/s, and this applies to a range of engines, from small automotive units to large marine units. Typical results for a direct injection engine are shown in figure 6.5 in terms of the piston speed. The specific power output is dependent on piston area since

$$\text{power} = \bar{p}_\text{b}LAN' \tag{2.38}$$

where $N' = $ number of firing strokes per second. For a four-stroke engine

$$N' = \frac{n\bar{v}_\text{p}}{4L}$$

where $\bar{v}_\text{p}$ is the mean piston velocity, and n is the number of cylinders. Thus

$$\text{brake power} = \bar{p}_\text{b}\bar{v}_\text{p}n\frac{A}{4} \tag{6.4}$$

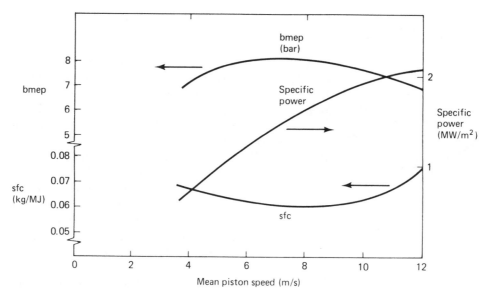

Figure 6.5
Typical performance for direct injection engine (reproduced from Howarth, 1966, courtesy of M. H. Howarth, Atlantic Research Associates).

But the torque output depends on the swept volume, V_s

$$\text{torque} = \frac{\text{power}}{\text{angular velocity}} = \frac{\bar{p}_b \cdot V}{4\pi} \qquad (6.5)$$

The fuel injector is usually close to the centre-line of the combustion chamber and can be either normal or angled to the cylinder head. To provide good mixing of the fuel and air during the injection period, all the air should move past a jet of fuel. By using multi-hole injectors the amount of swirl can be reduced. However, multi-hole injectors require smaller holes and higher injection pressures for the same flow rate and jet penetration.

Traditionally, low swirl or quiescent combustion chambers (such as those shown in figures 6.2b and 6.2c) tend to have been used in the larger diesel engines (say about 200 mm bore). With a high enough injection pressure (1500 bar) a better emissions/fuel consumption trade-off can be obtained, and this caused a trend towards quiescent combustion systems for small direct injection diesel engines (Frankl *et al.*, 1989).

Until the mid 1980s the maximum speed of small direct injection engines was limited to about 3000 rpm by combustion speed. High-speed direct injection engines have been developed by meticulous attention during the development of the air/fuel mixing; such an engine is discussed in chapter 15.

When combustion speed is not the limiting factor, there is none the less a limitation on the maximum mean piston speed $(\bar{v}_p)$, which limits the engine speed (N):

$$N = \frac{\bar{v}_p \times 60}{2 \times L} \qquad (6.6)$$

The maximum mean piston speed is a result of mechanical considerations and the flow through the inlet valve. As the bore, stroke, valve diameter and valve lift are all geometrically linked, then as the mean piston speed increases, so does the

air velocity past the inlet valve. This is discussed further in chapter 7, section 7.3, but 12 m/s is a typical maximum mean piston speed. For an engine speed of 3000 rpm this corresponds to a stroke of 0.12 m, or about a swept volume of 1 litre per cylinder. Traditionally the combustion speed has limited direct injection engines to 3000 rpm, so for engines of under one litre per cylinder swept volume, higher speeds could only be obtained by using the faster indirect injection combustion system. With careful development of the combustion system, direct injection engines can now operate at speeds of up to 5000 rpm.

6.3 Indirect injection (IDI) systems

Indirect injection systems have a divided combustion chamber, with some form of pre-chamber in which the fuel is injected, and a main chamber with the piston and valves. The purpose of a divided combustion chamber is to speed up the combustion process, in order to increase the engine output by increasing engine speed. There are two principal classes of this combustion system; pre-combustion chamber and swirl chamber. Pre-combustion chambers rely on turbulence to increase combustion speed and swirl chambers (which strictly are also pre-combustion chambers) rely on an ordered air motion to raise combustion speed. Howarth (1966) illustrates a range of combustion chambers of both types, see figure 6.6.

Pre-combustion chambers were not widely used, but a notable exception was Mercedes-Benz. The development of a pre-combustion chamber to meet emissions legislation is described by Fortnagel (1990), who discusses the influence of the pre-chamber geometry and injection parameters on emissions and fuel consumption.

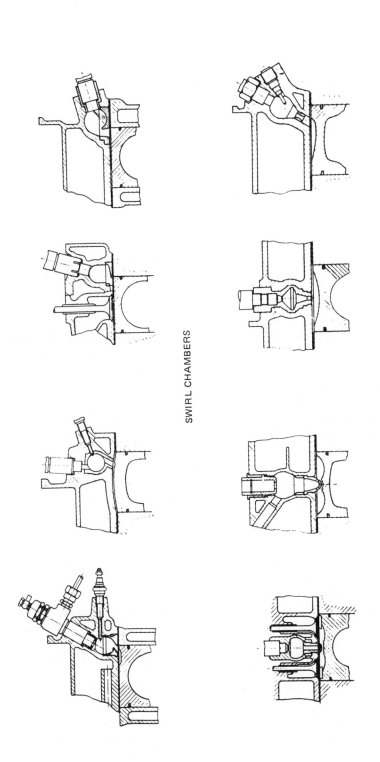

SWIRL CHAMBERS

PRE-COMBUSTION CHAMBERS

Figure 6.6
Different types of pre-combustion and swirl combustion chambers (reproduced from Howarth, 1966, courtesy of M. H. Howarth, Atlantic Research Associates).

Both types of pre-combustion chamber use heat-resistant inserts with a low thermal conductivity. The insert is quickly heated up by the combustion process, and then helps to reduce ignition delay. These combustion chambers are much less demanding on the fuel injection equipment. The fuel is injected and impinges on the combustion chamber insert, the jet breaks up and the fuel evaporates. During initial combustion the burning air/fuel mixture is ejected into the main chamber, so generating a lot of turbulence. This ensures rapid combustion in the main chamber without having to provide an ordered air motion during the induction stroke. Since these systems are very effective at mixing air and fuel, a large fraction of the air can be utilised, so giving a high bmep with low emissions of smoke. Unfortunately there are drawbacks. During compression the high gas velocities into the pre-chamber cause high heat transfer coefficients that reduce the air temperature. This means that compression ratios in the range 18–24:1 have to be used to ensure reliable ignition when starting. These compression ratios are higher than optimum for either power output or fuel economy, because of the fall-off in mechanical efficiency. The increased heat transfer also manifests itself as a reduction in efficiency.

Neither type of divided combustion chamber is likely to be applied to two-stroke engines since starting problems would be very acute, and the turbulence generated is likely to interfere with the scavenging process. In a turbocharged two-stroke engine the starting problem would be even worse, but the scavenging problem would be eased.

The fuel injection requirements for both types of divided combustion chamber are less severe, and lower fuel injection pressures are satisfactory. A single orifice in the nozzle is sufficient, but the spray direction should be into the air for good starting and onto the chamber walls for good running. Starting aids like heater plugs are discussed in the next section.

The disadvantages with divided combustion chambers increase in significance as the cylinder size increases. With large cylinders, less advantage can be taken of rapid combustion, and divided combustion chambers were only used in the range of 400–800 cm³ swept volume per cylinder. By far the most successful combustion chamber for this size range was the Ricardo Comet combustion chamber. The Comet combustion chamber dates back to the 1930s and the Mk V version is shown in figure 6.7. The volume of the pre-chamber is about half the total clearance volume. The two depressions in the piston induce two vortices of opposing rotation in the gas ejected from the pre-chamber. The insert or 'hot plug' has to be made from a heat-resistant material, since temperatures in the throat can rise to 700°C. Heat transfer from the hot plug to the cylinder head is reduced by minimising the contact area. The temperature of the plug should be sufficient to maintain combustion, otherwise products of partial combustion such as aldehydes would lead to odour in the exhaust.

The direction of fuel injection is critical, see figure 6.7; the first fuel to ignite is furthest from the injector nozzle. This fuel has been in the air longest, and it is also in the hottest air – that which comes into the swirl chamber last. Combustion progresses and the temperature rises; ignition spreads back to within a short distance from the injector. Since the injection is directed downstream of the air swirl, the combustion products are swept away from, and ahead of, the injection path. If the direction of injection is more upstream, the relative velocity between the fuel and the air is greater, so increasing the heat transfer. Consequently the delay period is reduced and cold starting is improved. Unfortunately, the combustion products are returned to the combustion zone; this decreases the efficiency and limits the power output.

The high heat transfer coefficients in the swirl chamber can cause problems with injectors; if the temperature rises above 140°C carbonisation of the fuel can occur. The injector temperature can be limited by increasing the cooling or, preferably, by using a heat shield to reduce the heat flow to the injector. Typical performance figures are shown in figure 6.8, which includes part-load results at constant speed. As with all compression ignition engines, the power output is limited by the fuelling rate that causes just visible (jv) exhaust smoke. The reduction in economy at part-load operation is much less than for a spark ignition engine with its output controlled by throttling.

The Comet combustion system was well suited to engines with twin overhead valves per cylinder. If a four-valve arrangement is chosen then a pre-chamber is perhaps more appropriate.

A final type of divided combustion chamber is the air cell, of which the MAN air cell is an example, see figure 6.9. Fuel is injected into the main chamber and ignites. As combustion proceeds, fuel and air will be forced into the secondary chamber or air cell, so producing turbulence. As the expansion stroke continues the air, fuel and combustion products will flow out of the air cell, so generating further turbulence. In comparison with swirl chambers, starting will be easier since the spray is directed into the main chamber. As the combustion-generated turbulence and swirl will be less, the speed range and performance will be more restricted than in swirl chambers; the air cell is not in common use.

Divided combustion chambers have reduced ignition delay, greater air utilisation and faster combustion; this permits small engines to run at higher speeds with larger outputs. Alternatively, for a given ignition delay lower quality fuels can be used. As engine size increases the limit on piston speed reduces engine speed, and the ignition delay becomes less significant. Thus, in large engines, direct injection can be used with low-quality fuels. The disadvantage of divided combustion chambers is a 5–15 per cent penalty in fuel economy, and the more complicated combustion chamber design.

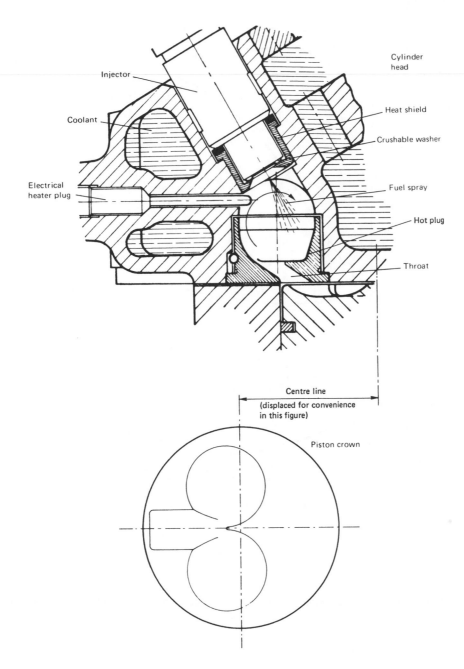

Figure 6.7
Ricardo Comet Mk V combustion chamber (reproduced with permission from Ricardo and Hempson, 1968).

A direct comparison between DI and IDI engines has been reported by Hahn (1986), and the differences in specific fuel consumption are shown by figure 6.10.

As energy costs rise the greater economy of direct injection engines has led to the development of small direct injection engines to run at high speeds. The fuel economy of high compression ratio, lean-burn spark ignition engines is comparable with that of indirect injection compression ignition engines. These factors have both led to the demise of indirect injection diesel engines.

6.4 Cold starting of compression ignition engines

Starting compression ignition engines from cold is a serious problem. For this reason a compression ratio is often used that is higher than desirable for either optimum economy or power output. None the less, starting can still be a problem, as a result of any of the following: poor-quality fuel, low temperatures, poorly seated valves, leakage past the piston

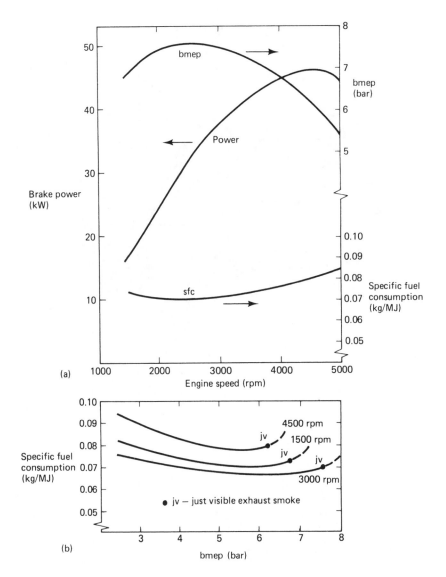

Figure 6.8
Typical performance for a 2-litre engine with Comet combustion chamber: (a) full load; (b) part load at different speeds (adapted from Howarth, 1966).

rings or low starting speed. One way of avoiding the compromise in compression ratio would be to use variable compression ratio pistons. In these pistons the distance between the top of the piston (the crown) and the piston pin (little end) can be varied hydraulically. So far these pistons have not been widely used. Ignition in a compression ignition engine relies on both a high temperature and a high pressure. The fuel has to evaporate and reach its self-ignition temperature with sufficient margin to reduce ignition delay to an acceptable level. The high pressure brings air and fuel molecules into more intimate contact, so improving the heat transfer. With too long an ignition delay the expansion stroke will have started before ignition is established.

While at low turning speeds there is greater time for ignition to be established, the peak pressures and temperatures are much reduced. The peak pressure is reduced by greater leakage past the valves and piston rings. The peak temperatures are reduced by the greater time for heat transfer. The situation is worst in divided chamber engines, since the air passes through a cold throat at a high velocity before the fuel is injected. The high velocity causes a small pressure drop, but a large temperature drop owing to the high heat transfer coefficient in the throat. The starting speed is not simply the mean engine speed; of greatest significance is the speed as a piston reaches the end of its compression stroke. For this reason it is usual to fit a large flywheel; starting

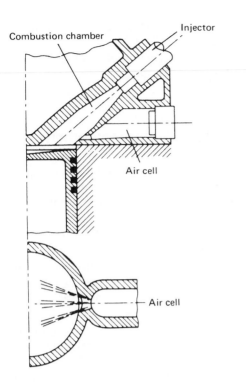

Figure 6.9
MAN air cell combustion chamber.

nozzles in the injector. Excess fuel injection is beneficial for several reasons: its bulk raises the compression ratio, any unburnt fuel helps to seal the piston rings and valves, and extra fuel increases the probability of combustion starting. It is essential to have an interlock to prevent the excess fuel injection being used during normal operation; although this would increase the power output the smoke would be unacceptable. Retarded injection timing means that fuel is injected when the temperature and pressure are higher. In systems where the fuel spray impinges on the combustion chamber surface it is particularly beneficial to have an auxiliary nozzle in the injector, in order to direct a spray of fuel into the air.

An alternative starting aid is to introduce with the air a volatile liquid which self-ignites readily. Ether (diethyl ether) is very effective since it also burns over a very wide mixture range; self-ignition occurs with compression ratios as low as 3.8:1.

The final type of starting aids is heaters. Air can be heated electrically or by a burner prior to induction. A more usual arrangement is to use heater plugs, especially in divided combustion chambers – for example, see figure 6.7. Heater plugs were originally exposed loops of thick resistance wire, but now sheathed heater plugs are used with finer multi-turned wire insulated by a refractory material. Exposed heater plugs had a low resistance and were often connected in series to a battery. However, a single failure was obviously inconvenient. The sheathed heater plugs can be connected to either 12 or 24 volt supplies through earth and are more robust. With electric starting the heater plugs are switched on prior to cranking the engine. Otherwise the power used in starting would prevent the heater plugs

systems cannot usually turn engines at the optimum starting speed. Sometimes a decompression system is fitted, which opens the valves until the engine is being turned at a particular starting speed.

Various aids for starting can be fitted to the fuel injection system: excess fuel injection, late injection timing and extra

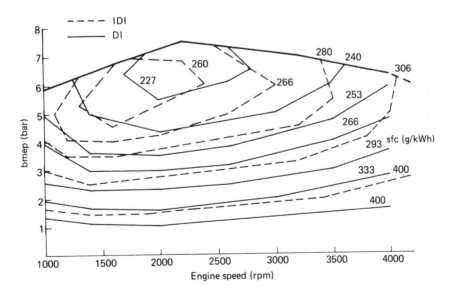

Figure 6.10
Comparison of specific fuel consumption (g/kWh) maps for four-cylinder 2.5-litre naturally aspirated DI and IDI engines (adapted from Hahn, 1986).

reaching the designed temperature. Ignition occurs when fuel is sprayed onto the surface of the heater; heater plugs do not act by raising the bulk air temperature. With electrical starting systems, low temperatures also reduce the battery performance, so adding further to any starting difficulties. A final possibility would be the use of the high-voltage surface discharge plugs used in gas turbine combustion chambers. Despite a lower energy requirement their use has not been adopted.

In multi-cylinder engines, production tolerances will give rise to variations in compression ratio between the cylinders. However, if one cylinder starts to fire, that is usually sufficient to raise the engine speed sufficiently for the remaining cylinders to fire.

The differences in the starting performance of DI and IDI diesel engines have been reported in detail by Biddulph and Lyn (1966). They found that the increase in compression temperature once IDI diesel engines had started was much greater than the increase in compression temperature with DI diesel engines. Biddulph and Lyn concluded that this was a consequence of the higher levels of heat transfer that occur in IDI diesel engines, and the greater time available for heat transfer at cranking speeds. They also pointed out that for self-ignition to occur in diesel engines, then a combination of sufficient time and temperature is required. With IDI engines that have started firing on one cylinder, the reduction in time available for ignition is more than compensated for by the rise in compression temperature. Thus once one cylinder fires, all the cylinders should fire at the idling speed. Biddulph and Lyn found that this was not the case for DI diesel engines. The smaller increase in compression temperature did not necessarily compensate for the reduced time available for self-ignition, and it was possible under marginal starting conditions for an engine to reach idling speeds with some cylinders misfiring, and for these cylinders not to fire until the engine coolant had warmed up. Such behaviour is obviously undesirable, and is characterised by the emission of unburned fuel as white smoke.

In turbocharged engines, the compression ratio is often reduced to limit peak pressures. This obviously has a detrimental effect on the starting performance.

6.5 | Fuel injection equipment

For many decades the most common arrangement for fuel injection in diesel engines was to use a plunger-in-barrel pump, connected by a thick-walled fuel line to a fuel injector. This is the type of system illustrated by figure 6.11 and discussed in sections 6.5.3 and 6.5.4. More recently, higher fuel injection pressures have led to an increasing use of unit injectors and common rail fuel injection systems, which are discussed in section 6.5.5. Unit injectors are ones in which the pumping element and injector are packaged together, with the pumping element operated from a camshaft in the cylinder head. This eliminates the high-pressure fuel line, and its associated pressure propagation delays and elasticity. Common rail fuel injection systems have a high-pressure fuel pump which produces a controlled and steady pressure, and the injector has to control the start and end of injection. Charlton (1998) provides a comprehensive overview of all types of electronically controlled fuel injection equipment.

After the overview in section 6.5.1 there is a discussion of the fuel injection nozzles (section 6.5.2), since these are common to all types of fuel injection system.

6.5.1 Injection system overview

A traditional pump–line–injector (PLI) system is shown in figure 6.11. Before discussing this, it is important to realise that the low-pressure part of the system is common to all diesel engines.

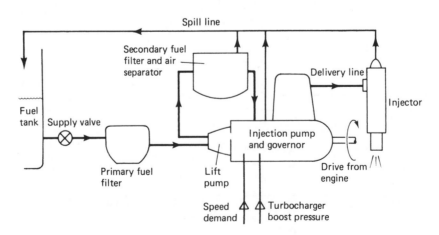

Figure 6.11
Fuel injection system for a compression ignition engine.

In general the fuel tank is below the injector pump level, and the lift pump provides a constant-pressure fuel supply (at about 0.75 bar) to the injector pump. The secondary fuel filter contains the pressure-regulating valve, and the fuel bleed also removes any air from the fuel. If air is drawn into the injection pump it cannot provide the correctly metered amount of fuel. It is essential to remove any water or other impurities from the fuel because of the fine clearances in the injection pump and injector. The injection pump contains a governor to control the engine speed. Without a governor the idle speed would vary and the engine could over-speed when the load on the engine is reduced.

The injection pump is directly coupled to the engine (half engine speed for a four-stroke engine) and the pump controls the quantity and timing of the fuel injection (figure 6.12). The quantity of fuel injected will depend on the engine load (figure 6.13). The maximum quantity of fuel that can be injected before the exhaust becomes smoky will vary with speed, and in the case of a turbocharged engine it will also vary with the boost pressure. The injection timing should vary with engine speed, and also load under some circumstances. As the engine speed increases, injection timing should be advanced to allow for the nearly constant ignition delay. As load increases, the ignition delay would reduce for a fixed injection timing, or for the same ignition delay the fuel can be injected earlier. This ensures that injection and combustion are complete earlier in the expansion stroke. In engines that have the injection advance limited by the maximum permissible cylinder pressure, the injection timing can be advanced as the load reduces. Injection timing should be accurate to $1°$ crank angle.

In multi-cylinder engines equal amounts of fuel should be injected to all cylinders. At maximum load the variation between cylinders should be no more than 3 per cent, otherwise the output of the engine will be limited by the first cylinder to produce a smoky exhaust. Under idling conditions the inter-cylinder variation can be larger (up to 15 per cent), but the quantities of fuel injected can be as low as 1 mm^3 per cycle. Ricardo and Hempson (1968) and Taylor (1985b) provide an introduction to fuel injection equipment, but a much more comprehensive treatment is given by Judge (1967), including detailed descriptions of different manufacturers' equipment. More recent developments include reducing the size of components, but not changing the principles, see Baranescu and Challen (1999) or Lai and Dingle (2005). The matching of fuel injection systems to engines is still largely empirical.

6.5.2 Fuel injectors

The most important part of the fuel injector is the nozzle; various types of injector nozzle are shown in figure 6.14. All these nozzles have a needle that closes under a spring load when they are not spraying. Open nozzles are used much less

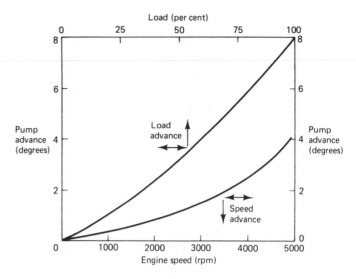

Figure 6.12
Typical fuel injection advance with speed and load.

than closed nozzles since, although they are less prone to blockage, they dribble. When an injector dribbles, combustion deposits build up on the injector, and the engine exhaust is likely to become smoky. In closed nozzles the needle-opening and needle-closing pressures are determined by the spring load and the projected area of the needle, see figure 6.15. The pressure to open the needle is greater than that required to maintain it open, since in the closed position the projected area of the needle is reduced by the seat contact area. The

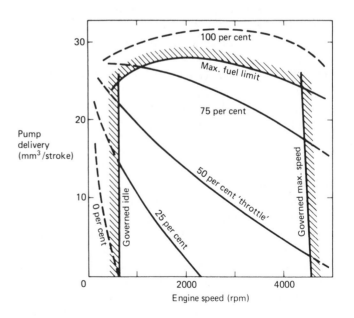

Figure 6.13
Fuel delivery map.

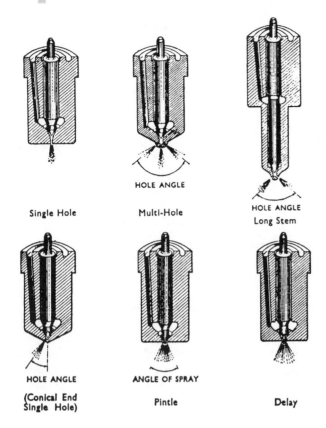

Figure 6.14
Various types of injector nozzle (courtesy of Lucas Diesel Systems).

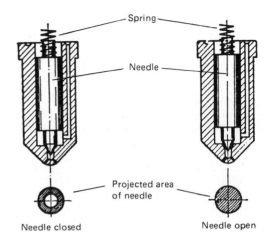

Figure 6.15
Differential action of the injector needle (from Judge, 1967).

differential pressures are controlled by the relative needle diameter and seat diameter. A high needle-closing pressure is desirable since it maintains a high seat pressure, so giving a better seal. This is also desirable, since it keeps the nozzle holes free from blockages caused by decomposition of leaked fuel. In automotive applications the nozzles are typically about 20 mm in diameter, 45 mm long, with 4 mm diameter needles.

The pintle nozzle (figure 6.16), has a needle or nozzle valve with a pin projecting through the nozzle hole. The shape of the pin controls the spray pattern and the fuel-delivery characteristics. If the pin is stepped, a small quantity of fuel is injected initially and the greater part later. Like all single-hole nozzles the pintle nozzle is less prone to blockage than a multi-hole nozzle. The Pintaux nozzle (PINTle with AUXiliary hole) injector (figure 6.17) was developed by Ricardo and CAV for improved cold starting in indirect injection engines. The spray from the auxiliary hole is directed away from the combustion chamber walls. At the very low speeds when the engine is being started, the delivery rate from the injector pump is low. The pressure rise will lift the needle from its seat, but the delivery rate is low enough to be dissipated through the auxiliary hole without increasing the pressure sufficiently to open the main hole. Once the engine starts, the increased

fuel flow rate will cause the needle to lift further, and an increasing amount of fuel flows through the main hole as the engine speed increases.

A problem with the type of injector nozzles shown in figures 6.14 and 6.15 is the volume of fuel downstream of the needle seat, in the so-called nozzle sac volume and in the nozzle holes. After the end of injection, this fuel is heated by the ensuing combustion, and some of this fuel will enter the combustion chamber late in the combustion process. This leads to combustion deposits forming on the nozzle, and partial combustion of the fuel with consequences on emissions that are discussed in section 6.6.2. Figure 6.18 shows a valve covered orifice (VCO) injector. The needle is seated over the injector holes and this reduces the amount of fuel that can leave the injector after the end of injection.

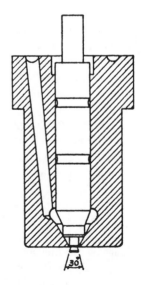

Figure 6.16
Enlarged view of a pintle nozzle (from Judge, 1967).

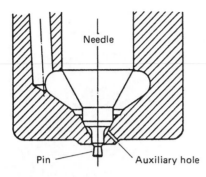

Figure 6.17
The Ricardo–CAV Pintaux nozzle (from Judge, 1967).

As explained in section 6.1, there is a move towards four valves per cylinder even in the smallest diesel engines. This gives a more central injector position for better control of combustion and reduced emissions. A further consequence of four valves per cylinder is an improvement in the volumetric efficiency but a reduced level of swirl. The lower level of swirl means that injector nozzles have to be used with a larger number of holes, in order to disperse the fuel as widely as possible in the air. If high injection pressures are to be maintained (for good fuel dispersion), then the nozzle holes have to be as small as possible. Nozzles with holes of 0.1 mm diameter are being used that have non-cylindrical profiles.

In all nozzles the fuel flow helps to cool the nozzle. Leakage past the needle is minimised by the very accurate fit of the needle in the nozzle. A complete injector is shown in figure 6.19, for use with traditional fuel injection pumps. The preload on the needle or nozzle valve from the compression spring is controlled by the compression screw. The cap nut locks the compression screw, and provides a connection to the spill line, in order to return any fuel that has leaked past the needle.

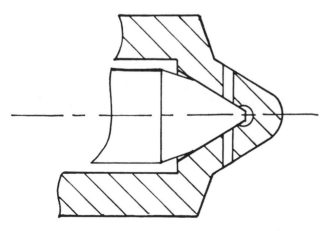

Figure 6.18
A valve covered orifice (VCO) injector.

Pilot injection, in which a small amount of fuel is injected during the ignition delay period, has already been identified in chapter 3, section 3.6 as a means of reducing diesel knock, by limiting the amount of mixture that is burnt rapidly. Pilot injection is most simply achieved by using a two-spring nozzle holder, as shown in figure 6.20, adapted from Warga (1994). The weak spring, acting through the central pressure pin, allows the nozzle to open a small amount (H_1, usually less than 0.1 mm) at low pressures, during the early part of the injection pump plunger stroke. During pilot injection, the nozzle needle is held against the stop sleeve which is held against the strong spring. As the pressure rises further, the nozzle needle and strong spring lift further ($H_1 + H_2$), for the main part of the injection. The pre-loads from the two springs are controlled by shims; this is an alternative to the screw arrangement seen in figure 6.19.

The spray pattern from the injector is very important, and high-speed combustion photography is very informative. The combustion can either be in an engine or in a special combustion rig.

An example of a computer model for diesel fuel sprays is the phenomenological model reported by Partridge and Greeves (1998). The model initially treated the fuel spray as a turbulent gas jet, but this was developed further by including models for fuel evaporation and two-phase flow. The spray is thus divided into gas and liquid regions. The liquid jet is assumed to have a uniform flow and, as fuel evaporates, its cross-sectional area is reduced. Momentum is transferred from the liquid to the gas phase by the evaporating fuel, and the relative shear between the liquid and gaseous phases. The momentum exchange causes the gas phase to accelerate, and the liquid flow to decelerate.

Partridge and Greeves (1998) have validated the model from spray measurements in an engine with optical access, and then used the model to explain how engine operating conditions and injection parameters (such as the nozzle hole size) affect combustion.

Increasing the injection pressure increases the spray penetration but above a certain pressure the spray becomes finely atomised and has insufficient momentum to penetrate as far. The aspect ratio of the nozzle holes (ratio of length to diameter) also affects the spray characteristics.

The mass flow rate ($\dot{m}_f$) of fuel through the injector can be modelled by

$$\dot{m}_f = nC_dA_n\sqrt{(2\rho_f\Delta p)} \tag{6.7}$$

where C_d = discharge coefficient
   A_n = nozzle flow area per hole (n holes)
   ρ_f = fuel density
   Δp = the pressure drop across the nozzle
   n = number of nozzle holes.

The nozzle hole diameters (d_n) are typically in the range of 0.15–1 mm with aspect ratios (length/diameter) from 2 to 8.

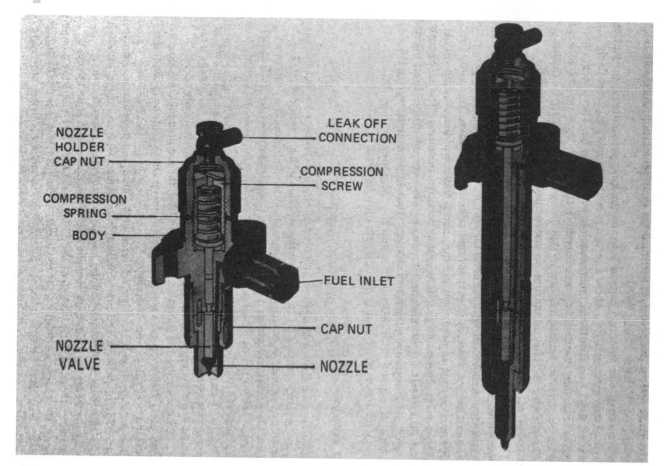

Figure 6.19
Complete fuel injectors (courtesy of Lucas Diesel Systems).

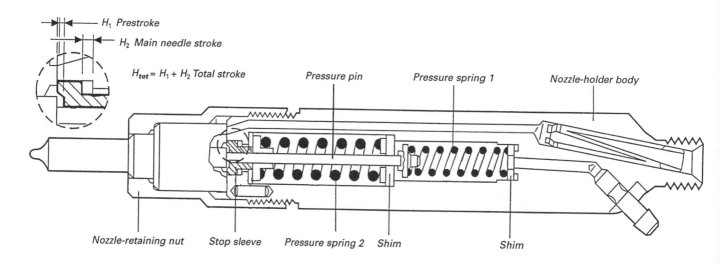

Figure 6.20
A two-spring nozzle holder for pilot injection, adapted from Warga (1994).

Increasing the aspect ratio produces a jet that diverges less and penetrates further. The differential pressure, which is invariably greater than 300 bar (and can be as high as 2000 bar), produces a high-velocity jet (typically faster than 250 m/s) which becomes turbulent. The jet spreads out as a result of entraining air, and breaks up into droplets; as the jet diverges the spray velocity decreases. Two important spray parameters are the droplet size distribution and the spray penetration distance (S). Heywood (1988) recommends a correlation by Dent:

$$S = 3.07 \left(\frac{p}{\rho_g}\right)^{1/4} (t d_n)^{1/2} \left(\frac{294}{T_g}\right)^{1/4} \tag{6.8}$$

where T_g = gas temperature (K)
ρ_g = gas density (kg/m^3)
t = time after the start of injection.

Later work by Arcoumanis et al. (1997) divided the spray penetration into two stages: that before the time the jet breaks up ($t_{breakup}$) in which the behaviour is that of a liquid jet, and the subsequent spray penetration.

$t < t_{breakup}$

$$S = C_d \left[\sqrt{(2\Delta p/\rho_f)}\right] t \tag{6.9}$$

$t > t_{breakup}$

$$S = \sqrt{\left[\alpha(2\Delta p/\rho_g)^{1/4}\right] \sqrt{(C_d d_n t)}} \tag{6.10}$$

With the breakup time given by:

$$t_{breakup} = \alpha \rho_f d_n / \left[C_d \sqrt{(2\Delta p \rho_g)}\right] \tag{6.11}$$

where ρ_g = gas density, the empirical constant (α) is evaluated as 15.7. Arcoumanis et al. (1997) noted that swirl reduced the tip penetration, and that a discharge coefficient of 0.39 gave the best fit to the experimental data.

During fuel injection it is possible for cavitation to occur in the nozzle holes, and this is a dominant factor in atomising the fuel spray. Cavitation occurs when the vapour pressure of the fuel is higher than the local static pressure, and it leads to clouds of vapour bubbles being formed. The nozzle tip temperature will be quite high (giving a high vapour pressure for the fuel), and in the nozzle hole there will be a region of recirculation in the *vena contracta*, as illustrated in figure 6.21. Although the injection process is at a high pressure, this pressure difference is used to produce a high velocity, so the pressure in the recirculation region is going to be very close to the in-cylinder pressure. Furthermore, the fuel in the recirculation region is resident for long enough to be heated up by the nozzle. Large-scale model tests by Soteriou et al. (1998), have established that there are several cavitation regimes:

1 At the onset of cavitation there is a bubbly annular flow that surrounds the liquid core.

2 With more severe cavitation, it is possible for cavitation to occur across the whole of the nozzle hole entrance, but after the *vena contracta* the cavitation becomes concentrated in the boundary layers, but with some faint cavitation in the centre of the flow.

Cavitation can be important for promoting good mixing of the fuel jet with the air.

Figure 6.21 has not assumed an axi-symmetric flow into the nozzle hole, because of the asymmetric nature of the geometry at the entry to the nozzle hole. Furthermore, there will be different flows through each of the nozzle holes, since there is normally only one fuel feed hole in the injector body.

The fuel jet breaks-up as a result of surface waves, and it can be argued that the initial average drop diameter (D_d) is proportional to the wavelength of the most unstable surface waves:

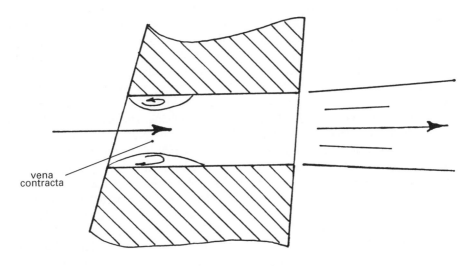

Figure 6.21
The flow inside a nozzle hole, showing the region of recirculation where cavitation is initiated.

$$D_d = C \frac{2\pi\sigma}{\rho_g v_j^2} \qquad (6.12)$$

where σ = liquid-fuel surface tension (N/m)
v_j = jet velocity (m/s)
C is a constant in the range 3–4.5.

In practice this is a simplification, since the droplets can break up further, and also coalesce. A convenient way of characterising a fuel spray is the Sauter mean diameter (SMD). This is the droplet size of a hypothetical spray of uniform size, such that the volume and overall volume to surface area ratio is the same:

$$\text{SMD} = \Sigma(nD^3)/\Sigma(nD^2) \qquad (6.13)$$

where n is the number of droplets in the size group of mean diameter D.

Heywood (1988) presents a useful discussion on droplet size distributions and correlations to predict droplet sizes.

An attempt to base the design process for DI engines on a more rational basis was introduced by Timoney (1985). In this approach, the fuel spray trajectory during injection, within the swirling air flow, is estimated from the laws of motion and semi-empirical correlations describing the air drag forces on the spray. From the spray trajectory and the air velocity, the relative velocity between the fuel spray and the swirling air is calculated in the tangential direction. This calculation is performed at the moment that the tip of the spray impinges on the walls of the piston bowl. Timoney provided some evidence from engine tests which showed that the magnitude of the tangential relative velocity (called the crosswind velocity) correlates well with the specific fuel consumption and smoke emissions.

To summarise, increasing the density of the fuel increases penetration, but increasing the density of the air reduces jet penetration. Also, denser fuels are more viscous and this causes the jet to diverge and atomise less. The jet penetration is also increased with increasing engine speed. The injection period with in-line or rotary injection pumps occupies an approximately constant fraction of the cycle for a given load, so as engine speed increases the jet velocity (and thus penetration) also increases.

The choice of injector type and the number and size of the holes are critical for good performance under all operating conditions. Indirect injection engines and small direct injection engines have a single injector. The largest direct injection engines can have several injectors arranged around the circumference of the cylinder.

6.5.3 Traditional injection pumps

Originally the fuel was injected by a blast of very high-pressure air, but this has long been superseded by 'solid' or airless injection of high-pressure fuel. The pumping element is invariably a piston/cylinder combination; the differences arise in the fuel metering. A possibility, mentioned already, is to have a unit injector – a combined pump and injector. The pump is driven directly from the camshaft so that it is more difficult to vary the timing.

The fuel can be metered at a high pressure, as in the common rail system, or at a low pressure, as in the jerk pump system. In the common rail system, a high-pressure fuel supply to the injector is controlled by a mechanically or electrically operated valve. As the speed is increased at constant load, the required injection time increases and the injection period occupies a greater fraction of the cycle.

The jerk pump system is widely used, and there are two principal types: in-line pumps and rotary or distributor pumps. With in-line (or 'camshaft') pumps there is a separate pumping and metering element for each cylinder.

In-line pumps

A typical in-line pumping element is shown in figure 6.22, from the Lucas Minimec pump. At the bottom of the plunger stroke (a), fuel enters the pumping element through an inlet port in the barrel. As the plunger moves up (b) its leading edge blocks the fuel inlet, and pumping can commence. Further movement of the plunger pressurises the fuel, and the delivery valve opens and fuel flows towards the injector. The stroke or lift of the plunger is constant and is determined by the lift of the cam. The quantity of fuel delivered is controlled by the part of the stroke that is used for pumping. By rotating the plunger, the position at which the spill groove uncovers the spill port can be changed (c) and this varies the pumping stroke. The spill groove is connected to an axial hole in the plunger, and fuel flows back through the spill port to the fuel gallery. The rotation of the plunger is controlled by a lever, which is connected to a control rod. The control rod is actuated by the governor and throttle. An alternative arrangement is to have a rack instead of the control rod, and this engages with gears on each plunger; see figure 6.23.

The delivery valve is shown more clearly in figure 6.22. In addition to acting as a non-return valve it also partially depressurises the delivery pipe to the injector. This enables the injector needle to snap onto its seat, thus preventing the injector dribbling. The delivery valve has a cylindrical section that acts as a piston in the barrel before the valve seats on its conical face; this depressurises the fuel delivery line when the pumping element stops pumping. The effect of the delivery valve is shown in figure 6.24.

Owing to the high pumping pressures the contact stresses on the cam are very high, and a roller type cam follower is used. The fuel pumping is arranged to coincide with the early part of the piston travel while it is accelerating. The spring decelerates the piston at the end of the upstroke, and accelerates the piston at the beginning of the downstroke.

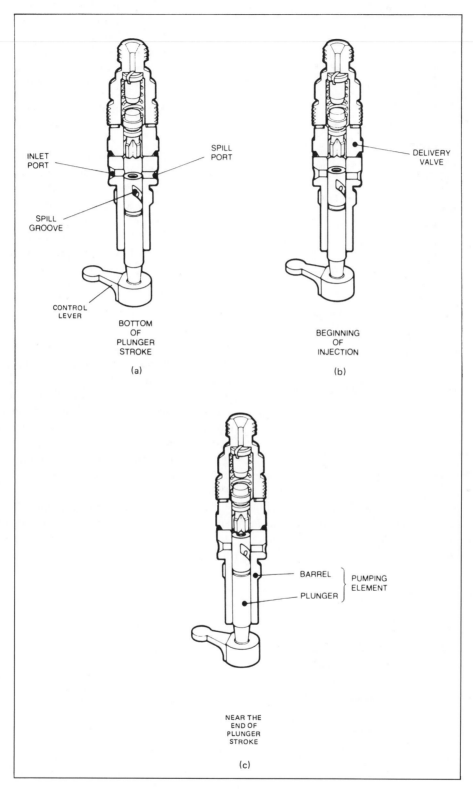

Figure 6.22
Fuel-pumping element from an in-line pump (courtesy of Lucas Diesel Systems).

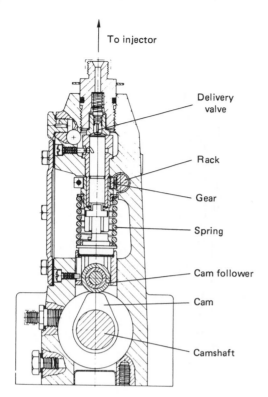

Figure 6.23
Cross-section of an in-line fuel pump (courtesy of Lucas Diesel Systems).

The cam profile is carefully designed to avoid the cam follower bouncing.

At high speed the injection time reduces, and the injection pressures will be greater. If accurate fuel metering is to be achieved under all conditions, leakage from the pump element has to be minimal:

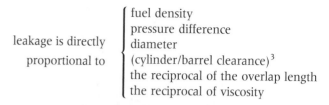

leakage is directly proportional to
$$\begin{cases} \text{fuel density} \\ \text{pressure difference} \\ \text{diameter} \\ \text{(cylinder/barrel clearance)}^3 \\ \text{the reciprocal of the overlap length} \\ \text{the reciprocal of viscosity} \end{cases}$$

The importance of a small clearance is self-evident, and to this end the barrel and piston are lapped; the clearance is about 1 μm.

A diagram of a complete in-line pump is shown in figure 6.25. The governor and auto-advance coupling both rely on flyweights restrained by springs. The boost control unit limits the fuel supply when the turbocharger is not at its designed pressure ratio. The fuel feed pump is a diaphragm pump operated off the camshaft. In order to equalise the fuel delivery from each pumping element, the position of the control forks can be adjusted on the control rod. The control forks engage on the levers that control the rotation of the plunger.

Rotary pumps

Rotary or distributor pumps have a single pumping element and a single fuel-metering element. The delivery to the appropriate injector is controlled by a rotor. Such units are more compact and cheaper than an in-line pump with several pumping and metering elements. Calibration problems are avoided, and there are fewer moving parts. However, rotary pumps cannot achieve the same injection pressures as in-line pumps, and were only developed for DI engines in the mid 1980s.

The fuel system for a rotary pump is shown in figure 6.26, and figure 6.27 shows the details of the high-pressure pump and rotor for a six-cylinder engine. The transfer pump is a sliding vane pump situated at the end of the rotor. The pump

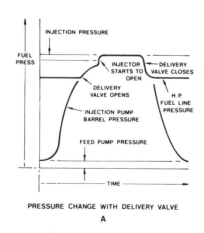

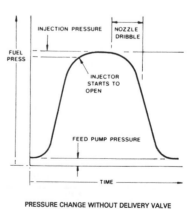

Figure 6.24
Effect of the injector delivery valve (courtesy of Lucas Diesel Systems).

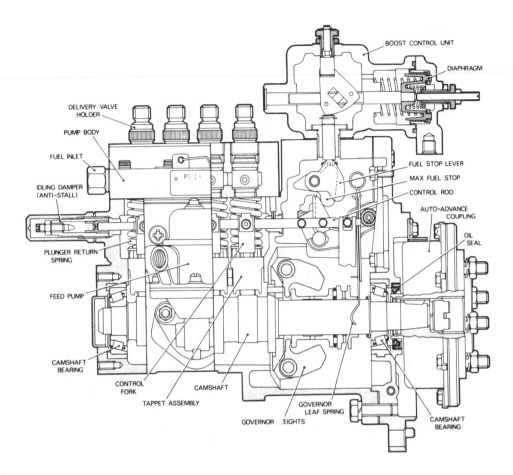

Figure 6.25
Sectional view of the CAV Minimec fuel pump (courtesy of Lucas Diesel Systems).

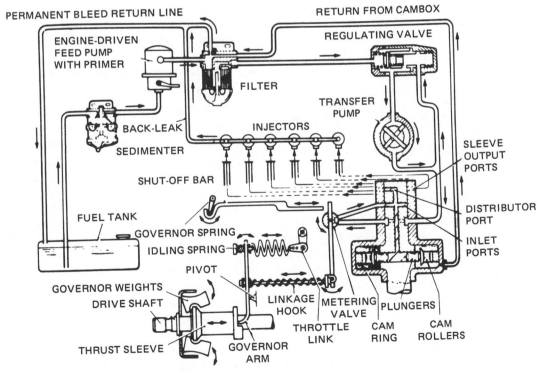

Figure 6.26
Rotary pump fuel system (with acknowledgement to Newton *et al.*, 1983).

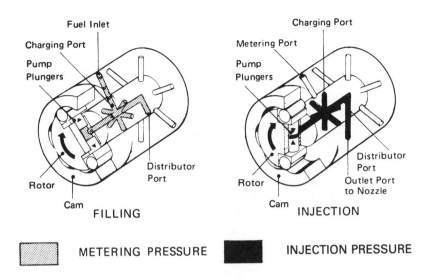

Figure 6.27
Rotor and high-pressure pump from a rotary fuel pump (courtesy of Lucas Diesel Systems).

output is proportional to the rotor speed, and the transfer pressure is maintained constant by the regulating valve. The metering value is regulated by the governor, and controls the quantity of fuel that flows to the rotor through the metering port, at the metering pressure. Referring to figure 6.27, the pump plungers that produce the injection pressures rotate in barrels in the rotor. The motion of the plungers comes from a stationary cam with six internal lobes. The phasing is such that the plungers move out when a charging port coincides with the fuel inlet; as the rotor moves round, the fuel is isolated from the inlet. Further rotation causes the distributor port to coincide with an outlet, and the fuel is compressed by the inward movement of the pump plungers.

Governing can be either mechanical or hydraulic, injection timing can be retarded for starting, excess fuel can be provided for cold start, and the turbocharger boost pressure can be used to regulate the maximum fuel delivery. Injection timing is changed by rotating the cam relative to the rotor. With a vane-type pump the output pressure will rise with increasing speed, and this can be used to control the injection advance. A diagram of a complete rotary pump is shown in figure 6.28.

Electronic control of rotary pumps

The technology for electronic control of injector pumps is well established – see Ives and Trenne (1981). Open-loop control systems can be used to improve the approximations used for pump advance (figure 6.13), but the best results are obtained with closed-loop control. An electronic control system described by Glikin (1985) is shown in figure 6.29. The signals from the control unit operate through hydraulic servos in otherwise conventional injector pumps. The optimum injection timing and fuel quantity are controlled by the microprocessor in response to several inputs: driver demand,

engine speed, turbocharger boost pressure, air inlet temperature and engine coolant temperature. A further refinement used by Toyota is to have an optical ignition timing sensor for providing feedback; this is described by Yamaguchi (1986). The ignition sensor enables changes in ignition delay to be detected, and the control unit can then re-optimise the injection timing accordingly. This system thus allows for changes in the quality of the fuel and the condition of the engine. An electronically controlled rotary pump that is capable of 1500 bar injection pressures has been described by Felton (1996). This has control of the injection timing, fuel cut-off and a fast-acting solenoid-controlled spill valve (for control of the quantity of fuel injected), which provides a rapid pressure collapse at the end of injection.

6.5.4 Interconnection of traditional pumps and injectors

The installation of the injector pump and its interconnection with the injectors are critical for satisfactory performance. In a four-stroke engine the injector pump has to be driven at half engine speed. The pump drive has to be stiff so that the pump rotates at a constant speed, despite variations in torque during rotation. If the pump has a compliant drive there will be injection timing errors, and these will be exacerbated if there is also torsional oscillation. The most common drive is either a gear or roller chain system. Reinforced toothed belts can be used for smaller engines, but even so they have to be wider than gear or roller chain drives. The drives usually include some adjustment for the static injection timing.

The behaviour of the complete injection system is influenced by the compressibility effects of the fuel, and the length and size of the interconnecting pipes.

The compressibility of the fuel is such that an increase in pressure of 180 bar causes a 1 per cent volume reduction.

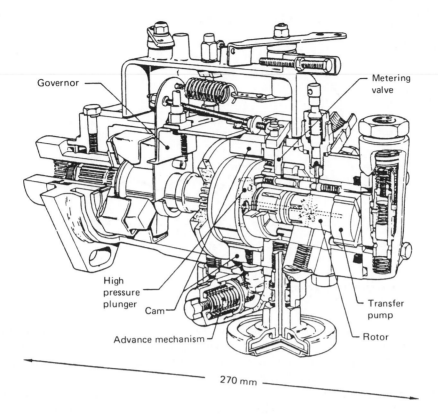

Figure 6.28
Typical rotary or distributor type fuel pump (from Ives and Trenne, 1981).

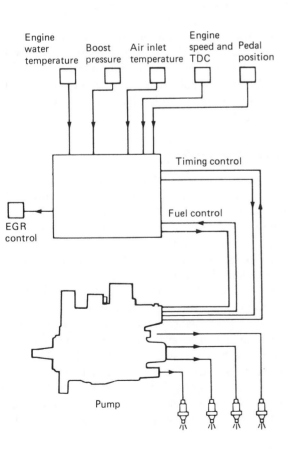

Incidentally, a 10 K rise in temperature causes about a 1 per cent increase in volume; since the fuel is metered volumetrically this will lead to a reduction in power output. Since the fuel pipes are thick walled, the change in volume is small compared with the effects due to the compressibility of fuel. Pressure (compression or rarefaction) waves are set up between the pump and the injector, and these travel at the speed of sound. The pressure waves cause pressure variations, which can influence the period of injection, the injection pressure and even cause secondary injection.

After the pump delivery valve has opened there will be a delay of about 1 ms per metre of pipe length before the fuel injection begins. At 4000 rpm, 1 ms corresponds to 24°ca, so to maintain the same fuel injection lag for all cylinders the fuel pipe lengths should be identical. The compression waves may be wholly or partially reflected back at the nozzle as a compression wave if the nozzle is closed, or as a rarefaction wave if the nozzle is open. During a typical injection period of a few milliseconds, waves will travel between the pump and injector several times; viscosity damps these pressure waves. The fuel-line pressure can rise to several times the injection

Figure 6.29
Schematic arrangement of electronic fuel injection control (reprinted from Glikin, 1985 by permission of the Council of the Institution of Mechanical Engineers).

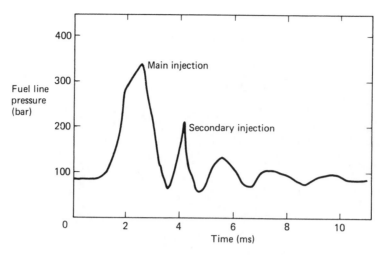

Figure 6.30
Pressure waves in the fuel line from the pump to the injector.

pressure during ejection, and the injection period can be extended by 50 per cent. The volumes of oil in the injector and at the pump have a considerable effect on the pressure waves. Fuel injection systems are prone to several faults, including secondary injection, and after-dribble.

Pressure wave effects can lead to secondary injection – fuel injected after the main injection has finished. Secondary injection can lead to poor fuel consumption, a smoky exhaust and carbon formation on the injector nozzle. Figure 6.30 shows a fuel-line pressure diagram in which there is a pressure wave after the main injection period that is sufficient to open the injector. Secondary injection can be avoided by increasing the fuel-line length, or changing the volumes of fuel at the pump or injector.

After-dribble is a similar phenomenon; in this case the pressure wave occurs as the injector should be closing. The injector does not fully close, and some fuel will enter the nozzle at too low a pressure to form a proper spray.

Problems with interconnecting pipework are of course eliminated with unit injectors, since the pump and injector are combined. Unit injectors can thus be found on large engines, since particularly long fuel pipes can be eliminated, and the bulk of the unit injector can be accommodated more easily. Techniques exist to predict the full performance of fuel injection systems in order to obtain the desired injection characteristics – see Knight (1960–61).

6.5.5 Common rail and electronic unit injector systems

Common rail (CR) and electronic unit injector (EUI) systems eliminate the problems associated with the interconnecting pipes in traditional pump–line–injector (PLI) systems. They also have scope for pilot injection (so as to control the amount of fuel injected during the ignition delay period, thereby

controlling combustion noise), and more desirable injection pressure characteristics. Figure 6.31 shows how injection pressure varies significantly with engine speed for pump–line–injector (PLI) systems, and that for low speeds only low injection pressures are possible. The low injection pressures limit the quantity of fuel that can be injected because of poor air utilisation, thereby limiting the low speed torque of the engine. With common rail injection systems there is independent control of the injection pressure within a wide operating speed range. The higher torque at low speeds means a hotter exhaust so that the turbine in a turbocharger produces more work and the compressor boost pressure will be higher. The higher boost pressure in turn means a higher torque, in effect there is positive feedback, and the low speed torque is increased and turbo-lag (see chapter 10, section 10.6.1) is reduced.

Electronic unit injectors (EUI)

Figure 6.32 shows the Lucas Diesel Systems electronic unit injector (EUI), which contains both the high-pressure fuel pumping element and the injector nozzle. The device is placed in the cylinder head, and is driven from a camshaft. The quantity and timing of injection are both controlled electronically through a Colenoid actuator. The Colenoid is a solenoid of patented construction that can respond very quickly (injection periods are of order 1 ms), to control very high injection pressures (up to 1600 bar or so).

The design and performance of the Colenoid controlled unit injector is presented in detail by Frankl *et al.* (1989). The Colenoid controls the spill valve, and is operated at 90 V, so as to reduce the wire gauge of the windings yet still have low resistive losses. The pumping plunger displaces diesel fuel, and as soon as the spill valve is closed, the fuel pressure builds up. Injection is ended by the spill valve opening again; the

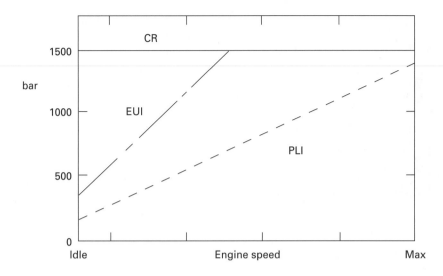

Figure 6.31
Typical maximum injection pressures for common rail (CR), electronic unit injector (EUI) and traditional pump–line–injector (PLI) systems with varying engine speed.

injection pressures can be as high as 1500 bar. The speed of response is such that pilot injection can be employed, to limit the amount of fuel injected during the ignition delay period. Frankl *et al.* point out that timing consistencies of about 5 microseconds are required; this corresponds to the time in which about 0.5 mm³ of fuel is injected. They also note that all units are checked to ensure a ±4 per cent tolerance in

fuelling level across the operating range, and that a major source of variation is attributable to the flow tolerance of the nozzle tip.

EUIs are widely used in heavy-duty truck applications, since they enable the best trade-offs to be obtained between low NO_x emissions, and low fuel consumption and particulate emissions, whether or not EGR is being used for NO_x reduction. An important part of this strategy is pilot injection (Tullis and Greeves, 1996).

An alternative approach to EUI is the Caterpillar Hydraulic Electronic Unit Injector (HEUI, also supplied to other manufacturers). Figure 6.33 shows the HEUI, which uses a hydraulic pressure intensifier system with a 7:1 pressure ratio to generate the injection pressures. The hydraulic pressure is generated by pumping engine lubricant to a controllable high pressure, so like common rail injection systems there is control of the injection pressure. The HEUI-B uses a two-stage valve to control the oil pressure, and this is able to control the rate at which the fuel pressure rises, thereby controlling the rate of injection, since a lower injection rate can help control NO_x emissions.

Common rail fuel injection systems

Common rail fuel injection systems decouple the pressure generation from the injection process, and have become popular because of the possibilities offered by electronic control. These developments have occurred mostly since the early 1990s, and are usefully summarised by Hawley *et al.* (1998). The design and control functionality of a common rail fuel injection system have been described in some detail by Guerrassi and Dupraz (1998).

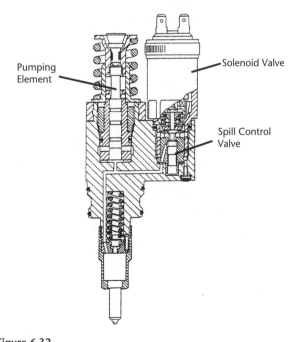

Pumping Element

Solenoid Valve

Spill Control Valve

Figure 6.32
Unit injector cross-section.

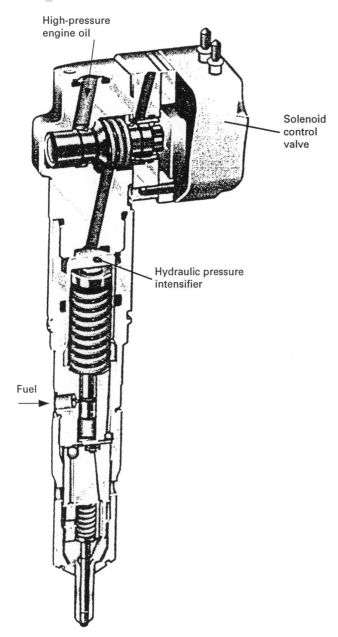

High-pressure
engine oil

Solenoid
control
valve

Hydraulic pressure
intensifier

Fuel

Figure 6.33
The Caterpillar Hydraulic Electronic Unit Injector (HEUI), showing the internal pressure intensifier (adapted from Walker, 1997).

Figure 6.34 shows the key elements of a common rail injection system:

(a) a high-pressure (controllable) pump,

(b) the fuel rail with a pressure sensor,

(c) electronically controlled injectors, and

(d) an engine management system (EMS).

The high-pressure pump (figure 6.35) can generate pressures in the range of 150–1600 bar, and this is achieved by control

of the inlet flow and by means of a high-pressure regulator. The inlet flow is controlled, since this reduces the pumping work compared to always pumping to the highest pressure and expanding the flow in a high-pressure regulator. The LDCR pump (figure 6.34) has two pairs of pumping plungers phased at 90°, so that the pressure fluctuations and peak driving torque are reduced. The driving torque is much smoother, and lower than for a conventional rotary pump. The inlet metering valve and high-pressure (HP) regulator are controlled by solenoids.

The electrohydraulic injector (figure 6.36a) is fitted within a 17 mm diameter body, and the key to its operation is the nozzle needle hydraulic control system (figure 6.36b). When the control valve solenoid is energised by the engine management system (EMS), the control valve lifts from its seat and the fuel pressure in the control chamber falls rapidly. Because the fuel pressure acting at the nozzle seat is equal to the rail pressure, the nozzle needle opens and injection starts. When the solenoid is de-energised the solenoid spring reseats the control valve. The pressure in the needle control chamber increases and, when it is larger than the pressure at the nozzle needle seat, the nozzle needle reseats. The selection of the orifice diameters is critical for optimal control.

Mock and Lubitz (2008) provide a detailed review of piezoelectric injection systems, with a very thorough review of the properties of the piezoelectric materials and their manufacture. Piezo-injectors for gasoline (up to 200 bar) can be direct acting, but diesel injection requires pressures of 2000 bar, so initially a servo-hydraulic system has been used – the piezoelectric actuator replaces the solenoid in figure 6.36. Piezoelectric materials expand when subject to an electric field; a typical working value is 2 kV/mm. Direct acting piezo-injectors are being developed with fuel rail pressure targets of 2500 bar. Since operational voltages need to be a maximum of about 200 V, it is necessary to have a multi-layer stack (with each element being about 0.1 mm thick). In order to obtain a displacement of 40 μm, then the stack will be about 30 mm long. A force of about 1 kN is needed, and this leads to a cross-sectional area of 50 mm^2.

Lee and Min (2005) compared a conventional solenoid-driven injector with a piezo-driven injector. The piezo-driven injector had a higher injection flow rate, a faster needle response time and the ability to control the injection rate. This allows smaller quantities of fuel to be injected during pilot injection, and a shorter time between injections. This leads to quieter combustion and better fuel economy.

The engine management system can divide the injection process into four phases:

(a) two pilot injections,

(b) the main injection, and

(c) the post-injection (for supplying a controlled quantity of hydrocarbons as a reducing agent for NO_x catalysts).

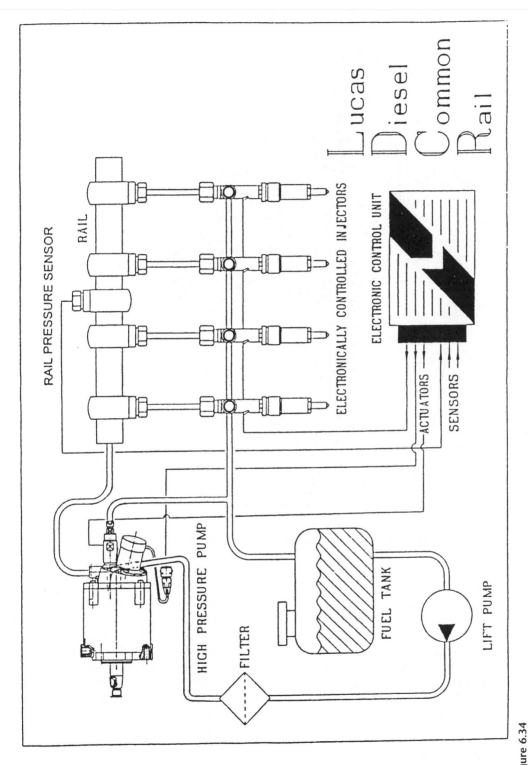

Figure 6.34
The key elements of a common rail injection system (courtesy of Lucas Diesel Systems).

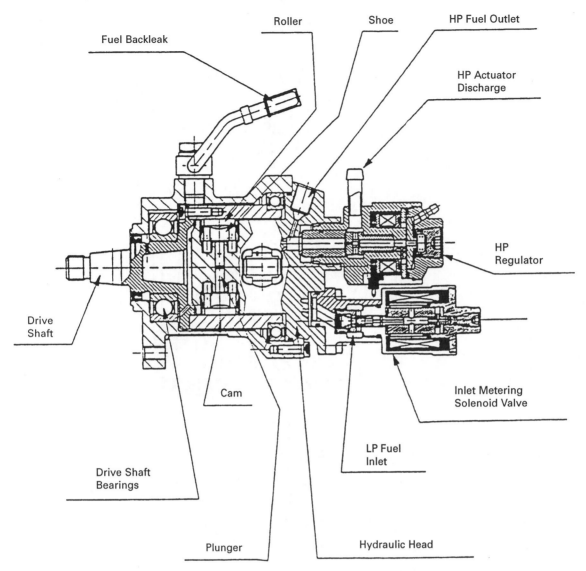

Figure 6.35
The Lucas Diesel Common Rail (LDCR) high-pressure pump (courtesy of Lucas Diesel Systems).

In addition to the usual inputs, the engine management system uses the fuel rail pressure as an input. It is also possible by monitoring the engine block accelerations to detect abnormal combustion, as would occur if dirt caused an injector needle not to seat properly (Guerrassi and Dupraz, 1998). Engine speed fluctuations within a cycle can also be used to correct the injection duration of individual injectors, so as to obtain as smooth a torque output as possible.

6.6   Diesel engine emissions

6.6.1   Emissions legislation

The worldwide variations in emissions legislation, and the introduction of stricter controls as technology improves, mean that it is difficult to give comprehensive and up-to-date information on emissions legislation. However, to provide a context in which to judge engine performance, the Western European legislation will be summarised here.

For light-duty vehicles, the same test cycle (figure 3.29) is used as for gasoline-powered vehicles, and table 6.1 summarises the requirements. Previously legislation had distinguished between emissions targets for type approval, and less onerous production conformity values. The emissions legislation now refers to all vehicles for 80 000 km, so in-house development targets have to be more stringent.

As heavy-duty vehicles tend to be manufactured on a bespoke basis, with the customer specifying the power-train units, then it is impractical to test all possible vehicle

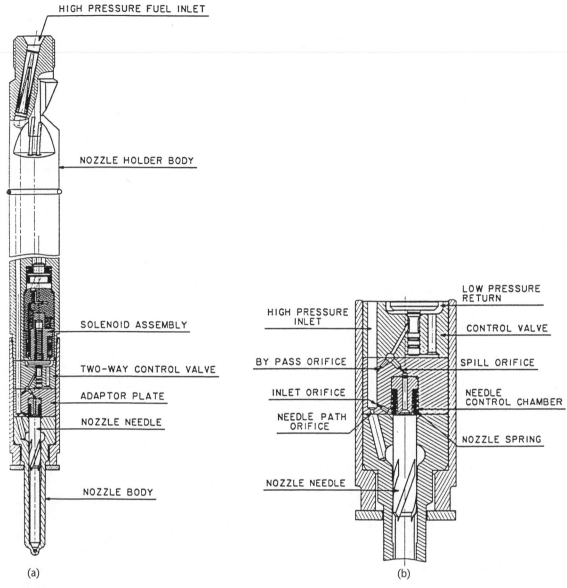

Figure 6.36

(a) The Lucas Diesel Common Rail (LDCR) electrohydraulic injector, and (b) nozzle needle hydraulic control system (courtesy of Lucas Diesel Systems).

combinations. Instead, the engines are submitted to dynamometer based tests, governed by ECE-R 49/01. This test is a series of 13 steady-state operating conditions (known as the 13-mode cycle) in which each operating point is held for 6 minutes. However, only the final minute at each operating point is used for evaluating the emissions. Table 6.2 summarises the ECE-R 49/01 13-mode cycle.

Rated speed means the speed at which the maximum power is produced, and intermediate speed means the speed at which maximum torque occurs, if this is 60–75 per cent of the rated speed, otherwise the intermediate speed is set to be 60 per cent of the rated speed. The most significant

shortcoming of this test is the lack of any transient, during which emissions might be expected to increase. In the USA a transient test is used that requires a computer-controlled transient dynamometer.

6.6.2 Sources and control of engine out emissions

Carbon monoxide emissions need not be considered here, as a correctly regulated diesel engine always has negligible emissions of carbon monoxide, since the air/fuel ratio is always lean.

Table 6.1 *European exhaust emissions standards for light-duty diesel-engined vehicles (g/km)*

Level	Date	CO	THC	NO$_x$	HC + NO$_x$	Pm	Pn
Euro 1	1992 (DI)[+]	2.75	–	–	1.36	0.19	–
Euro 1	1992 (IDI)[+]	2.75	–	–	0.97	0.14	–
Euro 2	1996 (DI)[+]	1.00	–	–	0.90	0.10	–
Euro 2	1996 (IDI)[+]	1.00	–	–	0.70	0.08	–
Euro 3	2000	0.64	–	0.50	0.56	0.05	–
Euro 4	2005	0.50	–	0.23	0.30	0.025	–
Euro 5	2009	0.50	–	0.18	0.230	0.005	–
Euro 6	2014	0.50	–	0.08	0.170	0.005	6×10^{11}

THC – Total hydrocarbons; NMHC – Non-methane hydrocarbons
Pm – Particulate (matter) mass; Pn – Particulate (matter) number
[+]Excludes the first 40 s after starting and the highway part of the current drive cycle

Table 6.2 *The ECE-R 49/01 13-mode cycle and emissions limits*

Mode no.	Engine speed	Per cent load	Mode weighting factor
1	idle	–	0.25/3
2	intermediate	10	0.08
3	intermediate	25	0.08
4	intermediate	50	0.08
5	intermediate	75	0.08
6	intermediate	100	0.25
7	idle	–	0.25/3
8	rated	100	0.10
9	rated	75	0.02
10	rated	50	0.02
11	rated	25	0.02
12	rated	10	0.02
13	idle	–	0.25/3

Emissions	Euro 2	Euro 3 (2000)	Euro 4 (2005)	Euro 5 (2009)
Carbon monoxide (g/kWh)	4.0	2.5	1.0	1.0
HC (g/kWh)	1.1	0.7	0.5	0.5
NO$_x$ (g/kWh)	7.0	5.0	3.5	2.0
Particulates (g/kWh)	0.15	0.10	0.02	0.02

The noise from diesel engines is usually mostly attributable to the combustion noise, which originates from the high rates of pressure rise during the initial rapid combustion. A typical relationship between combustion noise and the peak rate of combustion is shown in figure 6.37.

However, the more rapid the combustion, then the more efficient it is. There is thus a trade-off between the minimum combustion noise that can be achieved and the fuel economy. Data presented by Russell (1998) suggest an inverse relationship between combustion noise and fuel consumption, so that it is not possible to better:

$$(\text{isfc(g/kWh)} - 118)(\text{noise(dB(A))} - 44) = 3800 \qquad (6.14)$$

in the range $\qquad 201 < \text{isfc(g/kWh)} < 243$
or $\qquad\qquad 90 < \text{noise (dB(A))} < 65$

Russell (1998) shows how control of the injection rate and profiling influences combustion noise, and the formation of emissions such as NO$_x$ and particulate matter. This work demonstrates that pilot injection is important for controlling:

1 noise (by avoiding too much fuel being burnt at the end of the ignition delay period), and

2 hydrocarbon emissions (by avoiding too much fuel being over-diluted at the fringe of the spray during the ignition delay period).

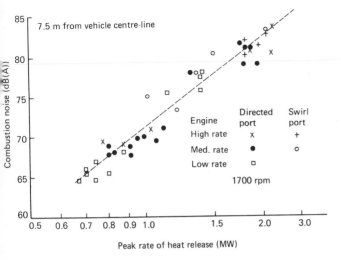

Figure 6.37
Relationship between combustion noise and peak rate of heat release (reprinted from Glikin, 1985 by permission of the Council of the Institution of Mechanical Engineers).

To understand the source of the rapid combustion, and how it can be controlled, it is necessary to consider the fuel injection and combustion processes.

As discussed earlier, the fuel is injected into the combustion chamber towards the end of the compression stroke. The fuel evaporates from each droplet of fuel, and mixes with air to form a flammable mixture. However, combustion does not occur immediately, and during the delay period (between the start of injection and the start of combustion) a flammable mixture is being continuously formed. Consequently, when ignition occurs it does so at many sites, and there is very rapid combustion of the mixture formed during the delay period. This rapid combustion produces the characteristic diesel knock. Evidently, the way to reduce the combustion noise is either to reduce the quantity of mixture prepared during the delay period, and this can be achieved by reducing the initial rate of injection (or by using pilot injection); or more commonly to reduce the duration of the delay period. The delay period is reduced by having:

(a) A higher temperature, which increases both the rate of heat transfer and the chemical reaction rates.

(b) A higher pressure, which increases the rate of heat transfer.

(c) A fuel that spontaneously ignites more readily (a higher cetane number).

A higher cetane fuel is unlikely to be feasible, unless more use is made of GTL fuels (chapter 3, section 3.7.6).

Higher temperatures occur in turbocharged engines and low heat-loss engines, but as will be seen later, this also leads to higher NO$_x$ emissions. An alternative approach is to retard the injection timing, so that injection occurs closer to the end of the compression stroke. This leads to an increase in the fuel consumption and other trade-offs that are discussed later in figure 6.38, for a fixed high load.

The trade-off between specific fuel consumption and injection timing for different injection rates is shown in figure 6.38. Ideally, combustion should occur instantaneously at top dead centre. In practice, combustion commences before top dead centre and continues afterwards. Advancing the start of injection (and thus combustion) increases the compression work, but the ensuing higher pressures and temperatures at top dead centre also increase the expansion work. However, if the injection timing is advanced too much, the increase in compression work will be greater than the increase in expansion work. Clearly, faster injection leads to more rapid combustion, and this results in less advanced injection timings. There is a rate of injection above which no further gains in fuel consumption occur, and the higher the swirl, the lower this injection rate.

The particulate matter from diesel engines originates from the fuel-rich side of the reaction zone in the diffusion-controlled combustion phase. After the rapid combustion at the end of the delay period, the subsequent combustion of the fuel is controlled by the rates of diffusion of air into the fuel vapour and vice versa, and the diffusion of the combustion products away from the reaction zone. Carbon particles are formed by the thermal decomposition (cracking) of the large hydrocarbon molecules, and the soot particles form by agglomeration. The soot particles can be oxidised when they enter the lean side of the reaction zone, and further oxidation occurs during the expansion stroke, after the end of the diffusion combustion phase.

Particulate matter generation is increased by high temperatures in the fuel-rich zone during diffusion combustion, and by reductions in the overall air/fuel ratio. The particulate matter emissions can be reduced by shortening the diffusion combustion phase, since this gives less time for soot formation and more time for soot oxidation. The diffusion phase can be shortened by increased swirl, more rapid injection, and a finer fuel spray. Advancing the injection timing also reduces the particulate matter emissions. The earlier injection leads to higher temperatures during the expansion stroke, and more time in which oxidation of the soot particles can occur. Unfortunately advancing the injection timing leads to an increase in noise. However, if the injection rate is increased and the timing is retarded, there can be an overall reduction in both noise and particulate matter. One such combination of points is shown as A and B in figure 6.38, and it can also be seen that the minimum specific fuel consumption has been reduced slightly, and that there is a significant reduction in nitrogen oxide emissions.

The formation of particulate matter is most strongly dependent on the engine load. As the load increases, more fuel is injected, and this increases the formation of particulate matter for three reasons:

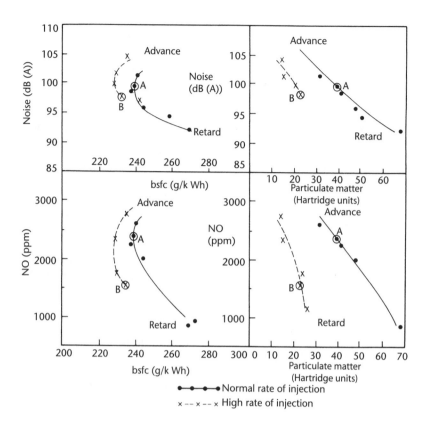

Figure 6.38

Trade-off curves between noise, particulate matter (Hartridge units), NO$_x$ and specific fuel consumption for different rates of injection and injection timing (reprinted from Glikin, 1985 by permission of the Council of the Institution of Mechanical Engineers).

(a) The duration of diffusion combustion increases.

(b) The combustion temperatures increase.

(c) Less oxidation of the soot occurs during the expansion stroke since there is less time after the end of diffusion combustion, and there is also less oxygen.

On naturally aspirated engines, it is invariably the formation of particulate matter that limits the engine output.

As explained in chapter 3, section 3.8, the formation of NO$_x$ is strongly dependent on temperature, the local concentration of oxygen and the duration of combustion. Thus in diesel engines, NO$_x$ is formed during the diffusion combustion phase, on the weak side of the reaction zone. Reducing the diffusion-controlled combustion duration by increasing the rate of injection leads to the reduction in NO$_x$ shown in figure 6.38. Retarding the injection timing also reduces the NO$_x$ emissions, since the later injection leads to lower temperatures, and the strong temperature dependence of the NO$_x$ formation ensures a reduction in NO$_x$, despite the increase in combustion duration associated with the retarded injection. The trade-off between NO$_x$ and particulate matter formation is also shown in figure 6.38.

At part load, particulate matter formation is negligible but NO$_x$ is still produced; figure 6.39a shows that NO$_x$ emissions

per unit output in fact increase. In general, the emissions of NO$_x$ are greater from an IDI engine than a DI engine, since the IDI engine has a higher compression ratio, and thus higher combustion temperatures. As with spark ignition engines, exhaust gas recirculation (EGR) at part load is an effective means of reducing NO$_x$ emissions. The inert gases limit the combustion temperatures, thereby reducing NO$_x$ formation. Pischinger and Cartellieri (1972) present results for a DI engine at about half load in which the NO$_x$ emissions were halved by about 25 per cent EGR, but at the expense of a 5 per cent increase in the fuel consumption.

Exhaust gas recirculation and NO$_x$ emissions

In a diesel engine, the EGR is displacing oxygen so only a limited amount of EGR can be used before there is insufficient oxygen for thorough combustion, with consequential rises in the emissions of carbon monoxide, particulates and unburnt hydrocarbons. Once these emissions rise, then since they are products of partial combustion, it is inevitable that the fuel consumption will also rise. The highest levels of EGR are used at low speeds and loads, and the EGR level has to be decreased as either load or speed is increased. The trade-offs between particulate emissions and NO$_x$ are shown in figure 6.40a,

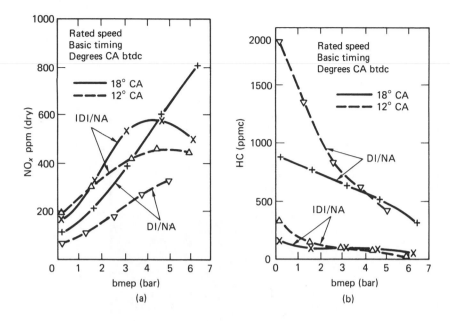

Figure 6.39
A comparison of emissions from naturally aspirated direct (DI) and indirect (IDI) injection diesel engines: (a) nitrogen oxide emissions; (b) hydrocarbon emissions. From Pischinger and Cartellieri (1972). [Reprinted with permission © 1972 Society of Automotive Engineers, Inc.]

which also shows that for a higher level of swirl (produced by closing one inlet port of a four-valve per cylinder combustion system), the higher swirl level gives a more advantageous trade-off between NO_x and particulates. Figure 6.40b shows that EGR is only used at the low-load and low-speed operating points, and that the higher swirl level is also beneficial at low speeds and higher loads. High injection pressure can also be used to obtain a better NO_x particulates trade-off, but Herzog (1998) argues that this leads to noisier combustion than if EGR is used.

Suzuki (1997) points out that EGR in diesel engines has only about half the effect that occurs with stoichiometrically operated spark ignition engines. In the spark ignition engine the stoichiometry is unaffected, and there is a significant increase in the heat capacity of the mixture. In contrast, EGR richens the mixture in a diesel engine, and has less effect on the heat capacity. A higher heat capacity mixture of course means lower combustion temperatures, which in turn means lower NO_x emissions. The mechanism for NO_x reduction by EGR is not obvious, but has been elucidated by some experiments in which the dilution, chemical and thermal effects have been isolated for the carbon dioxide present in EGR (Ladommatos *et al.*, 1998). Similar arguments would apply to the water vapour, since it has a similar heat capacity to carbon dioxide, and it too can dissociate:

1 The 'dilution' effect was examined by replacing oxygen with a nitrogen/argon mixture with the same heat capacity as the oxygen.

2 The 'chemical' effects were examined by replacing nitrogen with a carbon dioxide/argon mixture with the same heat

capacity as nitrogen, so that the carbon dioxide could lower the combustion temperature through dissociation.

3 The 'thermal' effects were examined by replacing nitrogen with a nitrogen/helium mixture with the same heat capacity as carbon dioxide, so that there would be the same cooling effect.

After making due allowance for changes in the ignition delay, Ladommatos *et al.* (1998) concluded that the main effect of the carbon dioxide was caused through reducing the oxygen availability, and that a small effect was attributable to the change in heat capacity. The dissociation of carbon dioxide was found to have a small contribution to the NO_x reduction, but it did contribute towards reducing particulate emissions.

A disadvantage of EGR is its tendency to increase engine wear rates, but low sulphur level fuels reduce the differences between wear rates with and without EGR.

Cooled EGR is also used, since if it displaces the same amount of air, it will represent a larger fraction of the charge and will increase the NO_x reductions as a result of the 'thermal' and 'chemical' effects. In practice, this means that a lower level of EGR can be used for a specified level of NO_x emissions, with a consequentially reduced increase in the particulate and other emissions. This also means that in a turbocharged engine at higher loads, the necessary levels of EGR can be achieved without recourse to devices such as inlet throttles. There are many different schemes for implementing EGR in a turbocharged engine, and several of these are reviewed by Suzuki (1997). Cooling EGR has its disadvantages, most notably a tendency to increase the ignition delay period and thereby increase the combustion noise.

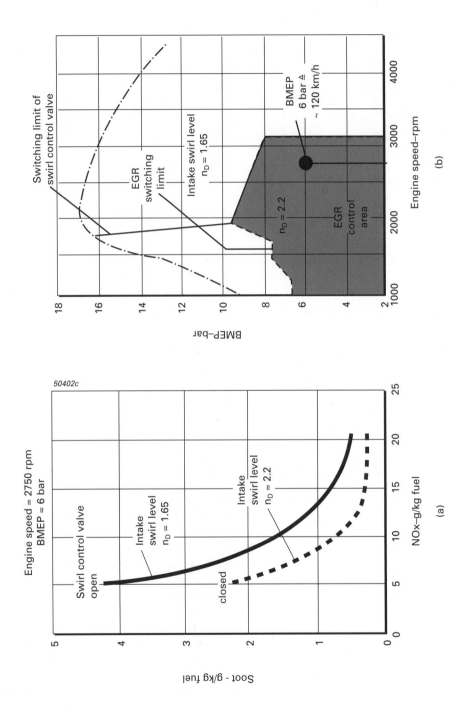

Figure 6.40

(a) The trade-offs between particulate emissions and NO_x for two levels of swirl, and (b) operating regimes where EGR and high swirl can be used (Herzog, 1998).
© IMechE/Professional Engineering Publishing Limited.

When EGR is used it is necessary to have some form of feedback system. The level of EGR achieved for a given EGR valve position will depend on the condition of many components, not least deposits in the EGR system. Feedback can be provided by measuring the oxygen level in the inlet manifold, or by measuring the air mass flow rate, and the inlet manifold temperature and absolute pressure. Diesel engine management systems are discussed further in chapter 15, section 15.4.

The trade-off between particulate matter (PM) and NO_x emissions, and the ever more demanding emissions limits, mean that an alternative technique to EGR is needed to reduce NO_x emissions. Chapter 4, section 4.3.3 has already discussed the options for NO_x reducing catalysts in lean combustion, and these technologies are directly applicable to diesel engines. Both Lean NO_x Trap (LNT) catalysts and Selective Catalytic Reduction (SCR) are used with diesel engines, and their relative merits are discussed in detail by Johnson (2008).

Lean NO_x Trap (LNT) catalysts are probably better suited to smaller engines and when a smaller reduction in NO_x is needed; they can achieve a 75–80 per cent conversion efficiency at full load with a 1–2 per cent fuel consumption penalty. The fuel consumption is increased because rich-mixture excursions are needed to regenerate the trap, and from time-to-time higher temperature excursions are needed to remove sulphate deposits (Johnson, 2008).

Selective Catalytic Reduction (SCR) can be used to reduce emissions of oxides of nitrogen from the exhaust of diesel engines. It is 'selective', because the oxidation has to be from the nitrogen oxides rather than from the much more plentiful oxygen. SCR uses an aqueous urea solution (32.5 per cent by mass, freezing point $-11°C$) that decomposes to produce ammonia (NH_3) and isocyanic acid (CHNO), followed by reaction with water to produce more ammonia and carbon dioxide. The second reaction requires a catalyst and a temperature of about 250°C. The overall reaction is

$$C(NH_2)_2O + H_2O \rightarrow NH_3 + CHNO + H_2O \rightarrow 2NH_3 + CO_2$$

This requires 77 per cent urea by mass, but 32.5 per cent has been chosen as this forms a eutectic mixture, so it gives the lowest freezing temperature and no phase separation. The ammonia reacts with the nitrogen oxide emissions to produce nitrogen and water within the catalytic converter (Frank et al., 2003):

$$4NH_3 + 4NO + O_2 \rightarrow 4N_2 + 6H_2O \quad (1)$$
$$2NH_3 + NO + NO_2 \rightarrow 2N_2 + 3H_2O \quad (2)$$
$$8NH_3 + 6NO_2 \rightarrow 7N_2 + 12H_2O \quad (3)$$

At low temperatures (200–300°C) reaction (2) dominates, so it is necessary to ensure that an oxidising catalyst converts the necessary NO to NO_2. When a vanadia catalyst is used, a temperature of about 250°C is needed for a 90 per cent conversion efficiency, but zeolite-based catalysts can achieve a higher efficiency at a slightly lower temperature (225°C) and be less sensitive to the $NO:NO_2$ composition (Johnson, 2008).

The urea solution is added into the SCR system at about 2–5 per cent of the rate of the diesel consumption. The exact quantity will have been established as part of the engine calibration process. If too much urea is added, then ammonia will 'leak' into the exhaust. A sensor can detect ammonia and modify the control strategy, or an additional oxidation catalyst can be added to oxidise the ammonia. Other options are to use solid urea (but this requires heating in the presence of water vapour to 180–200°C) or to store the ammonia in magnesium chloride ($MgCl_2$) cartridges (Johnson, 2008).

AdBlue and BlueTec are both aqueous urea systems that use AUS32 (Aqueous Urea Solution 32.5 per cent), which is known as *diesel exhaust fluid* in North America.

Hydrocarbon emissions

Unburnt hydrocarbons (HC) in a properly regulated diesel engine come from two sources. Firstly, around the perimeter of the reaction zone there will be a mixture that is too lean to burn, and the longer the delay period, the greater the amount of HC emissions from this source. However, there is a delay period below which no further reductions in HC emissions are obtained. Under these conditions the HC emissions mostly originate from a second source, the fuel retained in the nozzle sac (the space between the nozzle seat and the spray holes) and the spray holes. Fuel from these sources can enter the combustion chamber late in the cycle, thereby producing HC emissions.

Figure 6.39b shows that HC emissions are worse for DI engines than IDI engines, especially at light load, when there is significant ignition delay in DI engines. Advancing the injection timing reduces HC emissions, but this would lead to increased NO_x (figure 6.39b) and noise. The HC emissions increase at part load, since the delay period increases, and the quantity of mixture at the perimeter of the reaction zone that is too lean to burn increases. Finally, if a diesel engine is over-fuelled, there would be significant hydrocarbon emissions as well as particulate matter.

Particulate matter

The chemistry of particulate matter has already been discussed in chapter 3, section 3.8.5. For the legislative purposes, particulate matter is any substance, apart from water, that can be collected by filtering diluted exhaust at a temperature of 325 K. Since the particulates include soot and condensed hydrocarbons, any measure that reduces the HC emissions should also reduce the level of particulates.

Concern has also been expressed that the exhaust from diesel engines might contain carcinogens. Monaghan (1990) argues that the diesel engine is probably safe in this respect, but the exhaust does contain mutagenic substances. A technique for identifying mutagens is presented by Seizinger *et al.* (1985); these are usually polynuclear aromatic hydrocarbons (PNAH), also known as polycylic aromatic compounds (PAC).

There has been much debate about whether PAC come from unburnt fuel, unburnt lubricating oil or whether PAC can be formed during combustion. Work by Trier *et al.* (1990) has used radio-labelled (using carbon-14) hydrocarbons to trace the history of various organic fractions in the fuel after combustion. Trier *et al.* conclude that PAC can be formed during combustion.

Emissions from turbocharged engines and low heat loss engines are, in general, lower than from naturally aspirated engines; the exception is the increase in nitrogen oxide emissions. The higher combustion temperatures (and also pressure in the case of turbocharged engines) lead to a shorter ignition delay period, and a consequential reduction in the combustion noise and hydrocarbon emissions. The higher temperatures during the expansion stroke encourage the oxidation reactions, and the emissions of smoke and particulates both decrease. Nitrogen oxide (NO_x) emissions increase as a direct consequence of the higher combustion temperatures. However, when a turbocharged engine is fitted with an inter-cooler, the temperatures are all reduced, so that NO_x emissions also fall.

6.6.3   Catalytic after-treatment of diesel emissions

Catalysts have already been discussed in detail in chapter 4, section 4.3.2, but some additional discussion is needed for diesel engine applications. Diesel engines always operate weak of stoichiometric, so for NO_x reduction the systems already described in section 4.3.3 (Lean burn NO_x reducing catalysts) have to be applied.

Hawker (1995) points out that for diesel engines, a conventional platinum-based catalyst gives useful reductions in the gaseous unburnt hydrocarbons (and indeed any carbon monoxide as well), but has little effect on the particulate matter. However, before catalyst systems can be considered, the levels of sulphur in the diesel have to be 0.05 per cent by mass or less. This is because an oxidation catalyst would lead to the formation of sulphur trioxide and thence sulphuric acid. This is turn would lead to sulphate deposits that would block the catalyst. An additional advantage of using a catalyst is that it should lead to a reduction in the odour of diesel exhaust. Particulates can be oxidised by a catalyst incorporated into the exhaust manifold, in the manner described by Enga *et al.* (1982). However, for a catalyst to perform satisfactorily it has to be operating above its light-off temperature. Since diesel engines have comparatively cool exhausts, catalysts do not necessarily attain their light-off temperature.

Particulate traps are usually filters that require temperatures of about 550–600°C for soot oxidation. This led to the

development of electrically heated regenerative particulate traps, examples of which are described by Arai and Miyashita (1990) and Garrett (1990). The regeneration process does not occur with the exhaust flowing through the trap. Either the exhaust flow is diverted, or the regeneration occurs when the engine is inoperative. Air is drawn into the trap, and electrical heating is used to obtain a temperature high enough for oxidation of the trapped particulate matter. Pischinger (1998) describes how additives in the diesel fuel can be used to lower the ignition temperature, so that electrical ignition is only needed under very cold ambient conditions, or when the driving pattern is exclusively short distance journeys. Particulate traps have trapping efficiencies of 80 per cent and higher, but it is important to make sure that the back-pressure in the exhaust is not too high. An alternative to a filter is the use of a cyclone. In order to make the particulates large enough to be separated by the centripetal acceleration in a cyclone (figure 6.41), the particles have to be given an electrical charge so that they agglomerate before entering the cyclone (Polach and Leonard, 1994).

An oxidation catalyst and soot filter can be combined in a single enclosure, as shown in figure 6.42 (Walker, 1998). Hawker (1995) details the design of such a system. The platinum catalyst is loaded at 1.8 g/litre onto a conventional substrate with 62 cells/cm², and this oxidises not only the carbon monoxide and unburnt hydrocarbons, but also the NO_x to nitrogen dioxide (NO_2). The nitrogen dioxide (rather than the oxygen) is responsible for oxidising the particulates in the soot filter. The soot filter is an alumina matrix with 15.5 cells/cm², but with adjacent channels blocked at alternate ends. As the exhaust gas enters a channel, it then has to flow through the wall to an adjacent channel – hence the name of

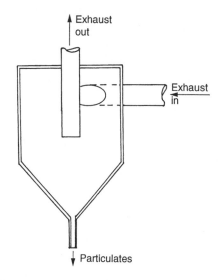

Figure 6.41

A cyclone for the collection of larger diesel particulates (to trap particles of about 10 microns and above, the swirl speed has to be about 1000 rpm).

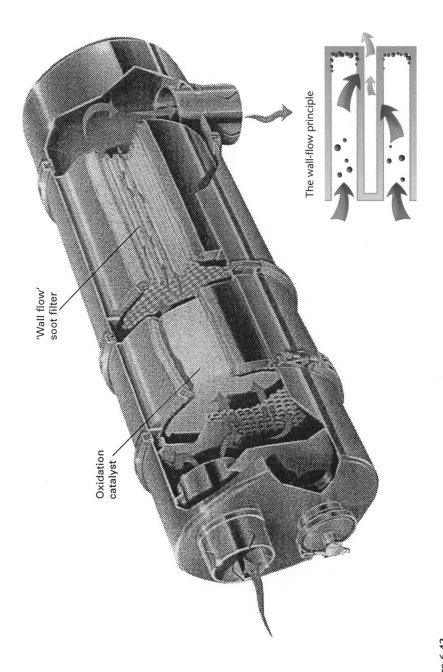

The wall-flow principle

'Wall flow' soot filter

Oxidation catalyst

Figure 6.42
An oxidation catalyst and soot filter assembly for use in diesel engines (adapted from Walker, 1998).

'wall flow' filter. With the presence of a platinum catalyst, the processes of soot trapping and destruction are continuous above temperatures of 275°C, and the system is known as a continuously regenerating trap (CRT). The system introduces a back-pressure of about 50 mbar, and the duty cycle of the vehicle has to be such as to ensure that a temperature of 275°C is regularly exceeded. Such an assembly can also be incorporated into a silencer (muffler), so that existing vehicles can be retrofitted (Walker, 1998). In 2005 the oxidation catalyst and diesel particulate filter were integrated into a single unit, and this leads to less energy for regeneration, since it can be mounted closer to the engine (Twigg and Phillips, 2009).

However, many light-duty diesel engines will not be operating with a hot enough exhaust for sufficient time. In urban driving the exhaust may only reach 150–200°C, so some form of active regeneration of diesel particulate filters (DPFs) is needed. Fuel is either injected during the expansion stroke or directly into the exhaust, in order to provide a high-temperature exhaust flow with high levels of unburned hydrocarbons and carbon monoxide. Another alternative is to throttle the inlet air flow to raise the exhaust temperature. Late injection requires use of common rail fuel injection systems, while injection into the exhaust requires only a low-pressure injector (but it usually needs a heater to assist with fuel vaporisation). Immediately before the diesel particulate filter (DPF) is an oxidation catalyst, and during regeneration this will complete the oxidation of the fuel that has been injected for regeneration, and raise the exhaust gas temperature. Post-combustion injection of fuel is increased to the full amount only after temperature sensors have detected light-off in the catalyst immediately before the DPF (Solvat et al., 2000). Without a catalyst, the filter regeneration requires a temperature of about 600°C, so this leads to the use of catalysts – either as a fuel additive or as an additive to the fuel to lower the regeneration temperature to 450–500°C (Solvat et al., 2000). The additive has to be stored in a separate tank and then added in controlled quantities each time the fuel tank is refilled. The additives are either iron or cerium based (for example, ceria – cerium oxide) and are trapped in the DPF as an intimate mixture with the soot. The soot will oxidise to carbon dioxide, but the additives will remain as an ash that will increase the exhaust back-pressure. The alternative is to catalyse the DPF (mainly platinum), although this is less effective as the catalyst is on the internal surfaces of the DPF, but it will lead to a lower back-pressure.

As particulate matter accumulates in the DPF the resistance to flow increases, and the pressure drop is measured so as to determine when the filter should be regenerated. The pressure drop (about 0.05 bar after regeneration, and 0.15 bar prior to regeneration) increases the pumping work in the engine, so fuel economy falls, and this has to be traded against the fuel that is used during regeneration. Regeneration takes about ten minutes, and oxidation within the DPF raises the

Table 6.3 *Diesel particulate filter (DPF) filtration efficiency versus soot loading for a 40 g capacity DPF (Thornton and Blohm, 2002)*

Carbon loading (g)	Filtration efficiency (per cent)
0	85
1.6	94
3.2	99
6.5	99
10	99

temperature to 900–1000°C; regeneration must not be too late, otherwise the DPF would be overheated. Table 6.3 shows how the DPF trapping efficiency improves as material accumulates and a 'cake' forms – a layer of particulate matter on top of the filter. Regeneration is required about once every 1000 km (600 miles). The pore sizes in a DPF are larger than the particulate matter, and particle-trapping is by a form of diffusion related to Brownian motion, and is explained fully by Hinds (1999). Brownian motion leads to random movement of particles, but in a region of high concentration falling to a low concentration there would (on average) be more particles moving away from the high-concentration region towards the low-concentration region than the other way around. This leads to particles diffusing from regions of high concentration to low concentration, and deposition onto surfaces.

The performance of diesel particulate filters is remarkable, as they can reduce particulate matter emissions by about two orders of magnitude. A comparison of particulate matter emissions has been made from a range of gasoline-engined vehicles (port fuel injection and direct injection) and diesel-engined vehicles (with and without a diesel particulate filter) by Braisher *et al.* (2010). The vehicles underwent the European drive cycle and measurements were taken of mass, number and size distributions.

Figure 6.43 shows a comparison between cycle average particle size distributions for diesel-engined vehicles with and without a diesel particulate filter (DPF) over the European drive cycle. A bi-log-normal fit has been applied to the data to separate the nucleation and accumulation modes. Only the number distribution for the vehicle without a DPF shows evidence of the nucleation mode, and since these particles are very small they have negligible mass. For both vehicles the number distribution maxima was for particles just below 100 nm and this leads to the maxima in the mass distributions occurring just above 100 nm, but note that there are about three orders of magnitude difference between the vehicles with and without the DPF.

With the DPF the emissions only occurred during acceleration transients within the first 200 s of the 1200 s test. In contrast, the vehicle without a DPF emitted particulate matter throughout the test, except during fuel cut-off when the

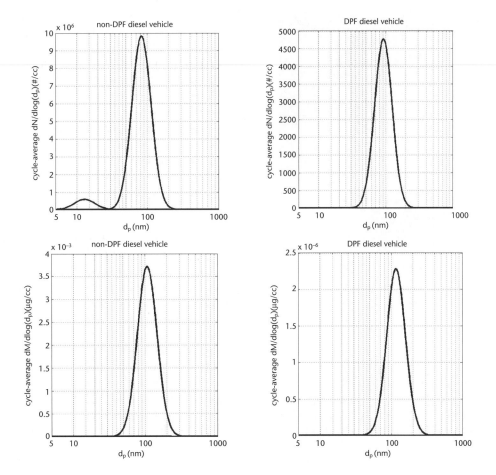

Figure 6.43
A comparison between cycle average particle size and mass distributions.

vehicle was decelerating. The emissions were essentially dependent on engine power output. Table 6.4 shows an end-of-cycle comparison of particulate number (Pn) and particulate mass (Pm) for 1.6-litre diesel-engined vehicles with and without a diesel particulate filter (DPF) over the European drive cycle. Note that there are about 2–3 orders of magnitude difference between the vehicles with and without the DPF.

6.7 | Homogeneous charge compression ignition (HCCI)

Homogeneous charge compression ignition (HCCI) is like a diesel engine, in that air and fuel are compressed to the point of auto-ignition but, unlike a diesel, the mixture is nominally homogeneous. It will be seen shortly that the mixture is not homogeneous, and that if it was then the combustion would

Table 6.4 *An end-of-cycle comparison of particulate number (Pn) and particulate mass (Pm) for 1.6-litre diesel-engined vehicles with and without a diesel particulate filter (DPF) over the European drive cycle.*

Vehicle	Particulate number (Pn) (particles/km)	Particulate mass (Pm) (mg/km)
Without DPF	4.87×10^{13}	21.3
With DPF	1.79×10^{10}	12.0×10^{-3}

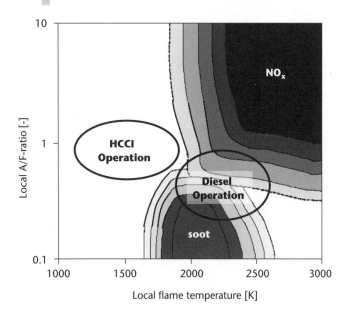

Figure 6.44
Comparison of the operating regime of an HCCI engine with a diesel engine, adapted from Otte *et al.* (2007).

be much too rapid; this leads to the alternative name of Controlled Auto-Ignition (CAI). HCCI can be applied to gasoline engines, and a comprehensive review of the technologies associated with both fuels are provided by Zhao (2007).

The very smallest diesel engines for model aircraft and similar applications employ HCCI and so did the very earliest diesel-type engine. Model aircraft engines rely on vaporising the fuel (usually diethylether based) with the air during induction, with subsequent ignition by combustion. As explained in chapter 1 (section 1.2, page 6), the Akroyd Stuart engine injected fuel into a large uncooled pre-chamber (heated externally to start the engine). During compression the temperature of the inducted air would rise (but the compression ratio was only about 3), and some of the air would enter the pre-chamber. However, the pre-chamber would also contain burned gases from the previous cycle, and these exhaust residuals would: firstly, help to heat the charge and secondly, ensure that after ignition the ensuing combustion occurred at a controlled rate. The Akroyd Stuart engine operated on paraffin (kerosene) and this would be more volatile than diesel fuel but more susceptible to auto-ignition than would gasoline. The motivation for HCCI with gasoline engines is to eliminate throttling losses, while for diesel engines the greater mixing of fuel and air leads to lower particulate emissions.

Figure 6.44 shows a comparison between the operating regime of an HCCI engine with that of a diesel engine. The diesel engine is operating with combustion at quite a high temperature (high enough to produce NO_x) and an air/fuel ratio across the full flammable range, with the result that both

NO_x and soot are generated. In contrast, the HCCI engine will be operating close to stoichiometric (so that soot is not generated) and with a significantly lower combustion temperature (so that NO_x is not produced).

As with a diesel engine, HCCI combustion begins once the mixture temperature has been high enough for sufficient time, so that there is no well-defined start of combustion that can be controlled directly. It is also necessary to have sufficient exhaust residuals to limit both the peak combustion temperatures and the rate of reaction (to limit the rate of pressure rise, as if combustion is too fast it will be noisy). Engines can be designed so that the ignition conditions occur at a chosen timing. The point of ignition can be controlled either by: inducted charge temperature, the compression ratio, the air/fuel ratio, or the quantity of exhaust residuals (either retained in the cylinder or returned to the induction system as EGR).

The compression ratio can be controlled either by changing the clearance volume or the stroke of the engine. Mechanisms have been devised for both options, but the favoured approach is to vary the inlet valve timing – if inlet valve closure is either delayed or is very early, the effective compression ratio is reduced. Controlling the incoming charge temperature is difficult for production engines, but has been widely used on experimental engines.

Exhaust residuals are a very important means of controlling: the maximum rate of pressure rise (and thus combustion noise), the equivalence ratio in an unthrottled engine (since the residuals are displacing air) and the time of ignition. The level of residuals can be controlled either internally or externally. External control is simplest but less accurate, so internal control by means of the valve timing is more versatile. A high level of exhaust residuals can be obtained by closing the exhaust valve significantly earlier or much later. In both cases the inlet valve opening is likely to be after the exhaust valve closure, and this is termed negative valve overlap (that is, no overlap). This is because conventional valve timing (chapter 7, section 7.4) has the inlet valve opening prior to exhaust valve closure. With early exhaust valve closure the retained gases will be compressed as the piston moves towards top dead centre, and then expanded on the start of the induction stroke. The inlet valve opening will be phased so that it opens as the pressure in the cylinder approaches the inlet manifold pressure. Too early, and there will be reverse flow into the induction system; too late, and the pumping losses will increase if the cylinder pressure falls below the inlet manifold pressure.

The need for high levels of exhaust residuals limits the load that can be achieved to about 25 per cent of the maximum bmep. The bmep level can be increased by a two-stage combustion and injection process. After HCCI-type combustion has initiated, a second fuel injection occurs, to give diffusion-controlled combustion (as with a conventional diesel engine) – Aoyagi, 2007. To achieve higher loads the engine operates as a conventional direct injection diesel

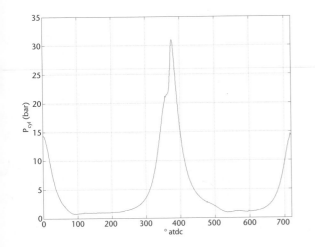

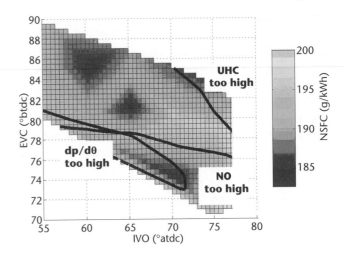

Figure 6.45

Trapping of residuals by early exhaust valve closure with recompression at the end of the exhaust stroke (0°ca) for HCCI operation with gasoline (Price *et al.*, 2007).

Figure 6.46

The HCCI operating envelope controlled by Exhaust Valve Closure (EVC) and Inlet Valve Opening (IVO) at an imep of about 4 bar with boundaries defined by acceptable NO, UHC and rate of pressure rise. Included is the net indicated specific fuel consumption (NSFC). Adapted from Price *et al.* (2007).

engine. Figure 6.44 has already illustrated the trade-off between particulate emissions and NO$_x$, but hydrocarbon emissions also have to be considered. Retarding the injection timing reduces the hydrocarbon emissions and, on a brake specific basis (gHC/kWh), increasing the exhaust residuals reduces the hydrocarbon emissions. Aoyagi (2007) points out that if the emissions are compared on a ppm basis then misleading results can be obtained, since with fewer exhaust residuals the air flow will be higher and there is a larger volume flow rate in the exhaust.

HCCI has also been applied to gasoline engines, and indeed the term was first used in this context (Thring, 1989), with the aim of eliminating throttling at part-load operation. HCCI required high exhaust gas recirculation (EGR) rates (13–33 per cent) and high intake temperatures. The HCCI system achieved fuel economy results comparable with a DI diesel engine (isfc in the range 180–200 g/kWh). More recent work with HCCI gasoline engines has used variable valve timing to control the residuals level and charge temperature.

Figure 6.45 shows how trapping residuals by early exhaust valve closure leads to a recompression at the end of the exhaust stroke (0°ca). That compression work is recovered (minus some heat and mass loss) by approximately symmetrically late opening of the inlet valves after top dead centre.

Figure 6.46 shows the HCCI operation regime at an imep of about 4.3 bar that can be achieved by valve timing control. Consider an inlet valve opening (IVO) of 65°atdc. When exhaust valve closure (EVC) is 77°btdc there is a low quantity of exhaust residuals trapped, and this leads to both a too high rate of pressure rise and a too high level of NO$_x$ emissions. When exhaust valve closure (EVC) is advanced to 86°btdc the level of exhaust residuals has been increased, but the unburned hydrocarbon emissions are becoming too high. A

high level of unburned hydrocarbon emissions will lead directly to a loss of efficiency, so the minimum indicated specific fuel consumption (ISFC) occurs with an exhaust valve closure (EVC) of about 82°btdc.

Control of HCCI engines is a major challenge. For example, if variable valve-timing systems are used they will have to be very fast acting. There is also the issue of transition to conventional operation. In the case of a gasoline engine this requires a step change, but in the case of a diesel engine it is more of a progression, since the split between early injection and conventional injection timing can be varied. In gasoline engines there is also the issue of fuel quality, since fuels with the same octane rating can have quite different performance in HCCI engines, and this is reviewed in detail by Kalghatgi (2007). The ignition process in HCCI engines depends on low-temperature reactions (about 1000 K, which is low compared with conventional combustion temperatures of about 2500 K). There is also a phenomenon known as negative temperature coefficient (NTC) behaviour. Reaction rates can decrease with increasing temperature, and an example of this is for n-heptane at a pressure of 55 bar in the temperature range of 800–1000 K.

Kalghatgi (2009) argues that the optimal fuel for HCCI engines could have a lower octane number and lower volatility (higher full boiling point) compared to gasoline, but be more volatile and have a lower cetane rating than diesel. The higher volatility than diesel would promote vaporisation and reduce particulate emissions, while the lower octane rating would facilitate self-ignition. Such a fuel could have lower refining costs, and might not be so different from paraffin, as used in the original Akroyd Stuart engine.

6.8 | Conclusions

In any type of compression ignition engine it is essential to have properly matched fuel injection and air motion. These requirements are eased in the case of indirect injection engines, since the pre-chamber or swirl chamber produces good mixing of the fuel and air. Since the speed range and the air utilisation are both greater in the indirect injection engine, the output is greater than in direct injection engines. However, with divided combustion chambers there is inevitably a pressure drop and a greater heat transfer, and consequently the efficiency of indirect injection engines is less than the efficiency of direct injection engines. Thus the development of small high-speed direct injection engines is very significant.

The compression ratio of compression ignition engines is often dictated by the starting requirements, and it is likely to be higher than optimum for either maximum fuel economy or power output. This is especially true for indirect injection engines where the compression ratio is likely to be in the range 18–24:1. Even so, additional starting aids are often used with compression ignition engines, notably, excess fuel injection, heaters and special fuels.

Apart from unit injectors there are two main types of injector pump: in-line pumps and rotary or distributor pumps. Rotary pumps are cheaper, but the injection pressures are lower than those of in-line pumps. Thus rotary fuel pumps are better suited to the less-demanding requirements of indirect injection engines. The fuel injectors and nozzles are also critical components, and like the injection pumps they are usually made by specialist manufacturers.

The correct matching of the fuel injection to the air flow is all important. The wide variety of combustion chambers for direct injection engines shows that the actual design is less important than ensuring good mixing of the fuel and air. Since the output of any compression ignition engine is lower than that of a similar-sized spark ignition engine, turbocharging is a very important means of raising the engine output. Furthermore, the engine efficiency is also improved; this and other aspects of turbocharging are discussed in chapter 10.

In conclusion, injection pressures are rising, so common rail injection systems and electronic unit injectors will be increasingly used when low emissions and a good fuel economy are important. In smaller diesel engine sizes the high speed direct injection (HSDI) engine is replacing the indirect injection (IDI) engine, because of its higher fuel economy. Many of the strategies for low emissions, good fuel economy and high specific output can be found in engines such as the General Motors Ecotec (Daniels 1996). This 2-litre turbocharged inter-cooled HSDI engine has a maximum power of 75 kW at 4000 pm, and a maximum bmep of 12.6 bar at 1800 rpm. The Ecotec has a four valve per cylinder combustion system with a progressively controlled swirl control valve (to increase the low-speed swirl), and injection pressures up to 1500 bar. The injectors are of the two-spring type to provide pilot injection, and the nozzle has five 0.18 mm holes. Up to 50 per cent EGR is used at light-load conditions to reduce NO_x emissions, and the EGR is routed internally through the cylinder head, which cools the EGR by typically 30 K.

6.9 | Example

Using the data in figure 6.5 estimate the specification for a four-stroke, 240 kW, naturally aspirated, direct injection engine, with a maximum torque of 1200 N m. Plot graphs of torque, power and fuel consumption against engine speed.

Solution:
First calculate the total required piston area (A_t) assuming a maximum of 2.0 MW/m^2:

$$A_t = \frac{240 \times 10^3}{2.0 \times 10^6} = 0.12 \text{ m}^2$$

Assuming a maximum bmep ($\bar{p}_b$) of 8×10^5 N/m^2

$$\text{Power} = T\omega = \bar{p}_b L A_t N^*$$

where $\omega = 2N^*.2\pi$ rad/s, and

$$\text{stroke, } L = \frac{T\omega}{\bar{p}_b A_t N^*} = \frac{1200 \times 4\pi N^*}{8 \times 10^5 \times 0.12 \times N^*} = \underline{0.157 \text{ m}}$$

The number of pistons should be such that the piston diameter is slightly smaller than the stroke. With an initial guess of 8 cylinders:

$$A = \frac{A_t}{8} = \frac{0.12}{8} = 0.015 \text{ m}^2$$

$$\text{bore} = \sqrt{\left(\frac{4 \times 0.015}{\pi}\right)} = \underline{0.138 \text{ m}}$$

a value that would be quite satisfactory for a stroke of 0.157 m:

swept volume $= A_t L = 0.12 \times 0.157 = 18.84$ litres

For a maximum mean piston speed (v_p) of 12 m/s, the corresponding engine speed is

$$\frac{v_p \times 60}{2L} = \frac{12 \times 60}{2 \times 0.157} = 2293 \text{ rpm}$$

The results from figure 6.5 can now be replotted as shown in figure 6.47. The final engine specification is in broad agreement with the Rolls-Royce CV8 engine. It should be noted that the power output is controlled by the total piston area. Increasing the stroke increases the torque, but will reduce the maximum engine speed, thus giving no gain in power.

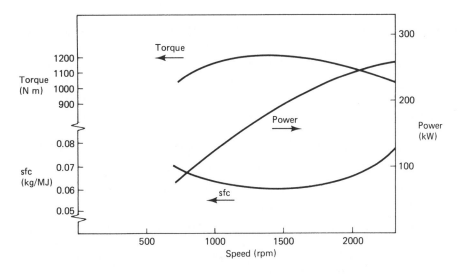

Figure 6.47
Performance curves for engine in the example; 18.8-litre naturally aspirated direct injection engine.

6.10 Questions

6.1 Contrast the advantages and disadvantages of indirect and direct injection compression ignition engines.

6.2 Discuss the problems in starting compression ignition engines, and describe the different starting aids.

6.3 Comment on the differences between in-line and rotary fuel injection pumps.

6.4 Describe the different ways of producing controlled air motion in compression ignition engines.

6.5 An engine manufacturer has decided to change one of its engines from a spark ignition type to a compression ignition type. If the swept volume is unchanged, what effect will the change have on:

(i)   maximum torque?
(ii) maximum power?
(iii) the speed at which maximum power occurs?
(iv) economy of operation?

6.6 Show that the efficiency of the Air Standard Diesel Cycle is:

$$\eta_{Diesel} = 1 - \frac{1}{r_v^{\gamma-1}} \cdot \left[\frac{\alpha^{\gamma} - 1}{\gamma(\alpha - 1)} \right]$$

where r_v = volumetric compression ratio
        α = volumetric expansion ratio during constant-pressure heat addition.

(i)   By considering the combustion process to be equivalent to the heat input of the Diesel cycle, derive an expression that relates the load ratio (α) to the gravimetric air/fuel ratio (F). Assume that the calorific value of the fuel is 44 MJ/kg, the compression ratio is 15:1, the air inlet temperature is 13°C, and for the temperature range involved, $c_p = 1.25$ kJ/kgK and $\gamma = 1.33$. State clearly any assumptions.
(ii) Calculate the Diesel cycle efficiencies that correspond to gravimetric air/fuel ratios of 60 : 1 and 20 : 1, and comment on the significance of the result.

6.7 Some generalised full load design data for naturally aspirated four-stroke DI diesel engines are tabulated below:

Mean piston speed, $\bar{v}_p$ (m/s)	4	6	8	10	12
Brake specific fuel consumption, bsfc (kg/MJ)	0.068	0.063	0.061	0.064	0.076
Brake mean effective pressure, $\bar{p}_b$ (bar)	7.24	8.00	8.08	7.70	6.93
Volumetric efficiency, η_{vol} (per cent)	89.8	88.3	82.1	77.7	76.2

Ambient conditions: $p = 105$ kN/m^2, $T = 17°$C.

Complete the table by calculating the brake specific power, bsp (MW/m^2), the air/fuel ratio, and the brake specific air consumption, bsac (kg/MJ). State briefly what is the significance of squish and swirl in DI diesel combustion. Why does the air/fuel ratio fall as the piston speed increases, what is the significance of the brake specific air consumption, and why is it a minimum at intermediate piston speeds? (It may be helpful to plot the data.)

6.8 A direct injection diesel engine with a swept volume of 2.5 litres has the full-load performance shown in figure 6.48. What is the maximum power output of the engine, and the corresponding air/fuel ratio?

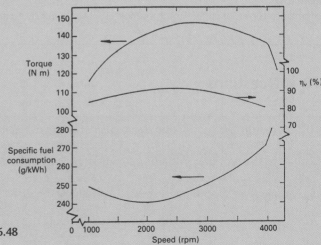

Figure 6.48

For both the maximum power and the maximum torque, calculate the brake mean effective pressure. What is the maximum brake efficiency?

The fuel has a calorific value of 44 MJ/kg, and the ambient conditions are a temperature of 17°C and a pressure of 1.02 bar.

Identify the differences between direct (DI) and indirect injection (IDI) diesel engines, and explain why the efficiency of the DI engine is higher. Why is IDI still used for the smallest high-speed diesel engines?

6.9 A proposal is being considered to convert a 2-litre automotive spark ignition engine to operation as a diesel engine. In spark ignition form the engine has a maximum bmep (with optimised air/fuel ratio and ignition timing) of 10.5 bar. A decision has to be made as to whether the diesel engine will use a direct (DI) or indirect (IDI) combustion system. List concisely the advantages and disadvantages of each type of combustion system, and state the bmep that might be expected in each case.

CHAPTER 7

Induction and exhaust processes

7.1 | Introduction

In reciprocating internal combustion engines the induction and exhaust processes are non-steady flow processes. For purposes such as some combustion modelling, the flows can be assumed to be steady. However, there are many cases for which the flow has to be treated as non-steady, and it is necessary to understand the properties of pulsed flows and how these can interact.

Pulsed flows are very important in the charging and emptying of the combustion chambers, and in the interactions that can occur in the inlet and exhaust manifolds. This is particularly the case for two-stroke engines where there are no separate exhaust and induction strokes. An understanding of pulsed flows is also needed if the optimum performance is to be obtained from a turbocharger; this application is discussed in chapter 10. In naturally aspirated engines it is also important to design the inlet and exhaust manifolds for pulsed flows, if optimum performance and efficiency are to be attained. However, inlet and exhaust manifold designs are often determined by considerations of cost, ease of manufacture, space and ease of assembly, as opposed to optimising the flow.

There is usually some form of silencing on both the inlet and exhaust passages. Again, careful design is needed if large pressure drops are to be avoided.

In four-stroke engines the induction and exhaust processes are controlled by valves. Two-stroke engines do not need separate valves in the combustion chamber, since the flow can be controlled by the piston moving past ports in the cylinder. The different types of valve (poppet, disc, rotary and sleeve) and the different actuating mechanisms are discussed in the next section.

The different types of valve gear, including a historical survey, are described by Smith (1967). The gas flow in the internal combustion engine is covered in some detail by Annand and Roe (1974).

7.2 | Valve gear

7.2.1 Valve types

The most commonly used valve is the mushroom-shaped poppet valve. It has the advantage of being cheap, with good flow properties, good seating, easy lubrication and good heat transfer to the cylinder head. Rotary and disc valves are still sometimes used, but are subject to heat transfer, lubrication and clearance problems.

The sleeve valve was once important, particularly for aero-engines prior to the development of the gas turbine. The sleeve valve consisted of a single sleeve or pair of sleeves between the piston and the cylinder, with inlet and exhaust ports. The sleeves were driven at half engine speed and underwent vertical and rotary oscillation. Most development was carried out on engines with a single sleeve valve. There were several advantages associated with sleeve valve engines, and these are pointed out by Ricardo and Hempson (1968), who also give a detailed account of the development work. Sleeve valves eliminated the hot-spot associated with a poppet valve. This was very important when only low octane fuels were available, since it permitted the use of higher compression ratios, so leading to higher outputs and greater efficiency. The drive to the sleeve could be at crankshaft level, and this led to a more compact engine when compared with an engine that used overhead poppet valves. The piston lubrication was also improved since there was always relative motion between the piston and the sleeve. In a conventional engine the piston is instantaneously stationary, and this prevents hydrodynamic lubrication at the ends of the piston stroke.

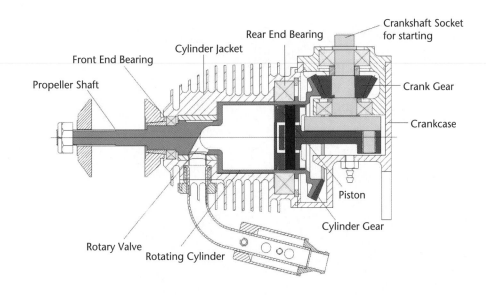

Figure 7.1
Cross-section of the RCV four-stroke engine, with acknowledgement to RCV Engines.

This problem is most severe at top dead centre where the pressures and temperatures are greatest.

The disadvantages of the sleeve valve were: the cost and difficulty of manufacture, lubrication and friction between the sleeve and cylinder, and heat transfer from the piston through the sleeve and oil film to the cylinder. Compression ignition engines were also developed with sleeve valves.

A more recent concept for small engines (up to 21 cm³ displacement) has been developed by RCV Engines (figure 7.1), in which a sleeve between the piston and cylinder barrel is driven at half engine speed. A port which is an extension at the top of the sleeve can align with either an inlet port, a spark plug or an exhaust port. The power take-off can be either from the crankshaft or the rotating sleeve. Since small engines can operate at very high speed (up to 12 000 rpm), the half engine speed is very useful; the high speed also leads to a high specific output (up to 75 kW/litre). Figure 7.2 shows a more recent design using a rotating valve in the cylinder head that is rotating about an axis parallel to the crankshaft (50 kW/litre); another option is rotation about the cylinder axis which has a specific output of 70 kW/litre (Lawes and Mason, 2011). When the valve in the cylinder head is rotating about an axis parallel to the crankshaft, then its drive can be by a toothed belt, and this leads to a low-cost design that is very tolerant of over-speeding compared with poppet valve designs. Emissions legislation for small engines (such as used in chainsaws) has led to the development of small four-stroke engines (30–40 cm³), and, like two-stroke engines, they can use an oil-in-fuel mix, so that lubrication is simplified and operation can be in any orientation.

7.2.2   Valve-operating systems

In engines with overhead poppet valves (ohv – overhead valves), the camshaft is either mounted in the cylinder block, or in the cylinder head (ohc – overhead camshaft). Figure 7.3

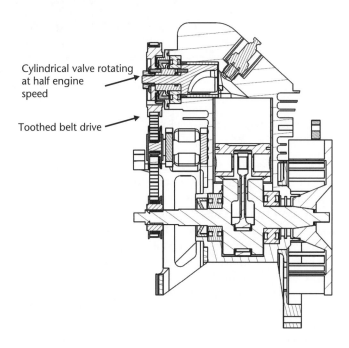

Figure 7.2
Cross-section of the HRCV four-stroke engine, with acknowledgement to RCV Engines.

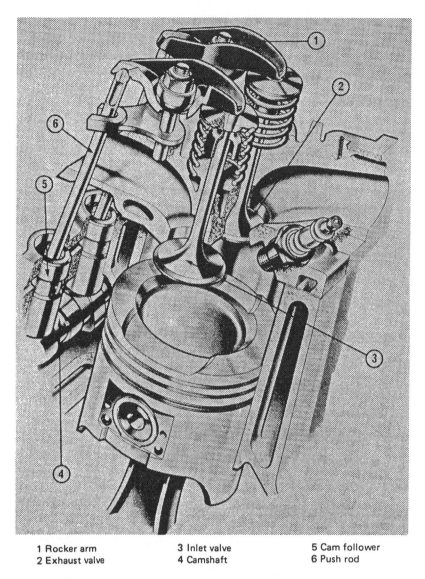

| 1 Rocker arm | 3 Inlet valve | 5 Cam follower |
| 2 Exhaust valve | 4 Camshaft | 6 Push rod |

Figure 7.3
Overhead valve engine (courtesy of Ford).

shows an overhead valve engine in which the valves are operated from the camshaft, via cam followers (tappets), push rods and rocker arms. This is a cheap solution since the drive to the camshaft is simple (either gear or chain), and the machining is in the cylinder block. In a 'V' engine this arrangement is particularly suitable, since a single camshaft can be mounted in the valley between the two cylinder banks.

In overhead camshaft (ohc) engines the camshaft can be mounted either directly over the valve stems, as in figure 7.4, or it can be offset. When the camshaft is offset the valves are operated by rockers and the valve clearances can be adjusted by altering the pivot height. The drive to the camshaft is usually by chain or toothed belt. Gear drives are also possible, but would be expensive, noisy and cumbersome with overhead camshafts. The advantage of a toothed belt drive is that it

can be mounted externally to the engine, and the rubber damps out torsional vibrations that might otherwise be troublesome.

Referring to figure 7.4, the cam operates on a follower or 'bucket'. The clearance between the follower and the valve end is adjusted by a shim. Although this adjustment is more difficult than in systems using rockers, it is much less prone to change. The spring retainer is connected to the valve spindle by a tapered split collet. The valve guide is a press-fit into the cylinder head, so that it can be replaced when worn. Valve seat inserts are used, especially in engines with aluminium alloy cylinder heads, to ensure minimal wear. Normally poppet valves rotate in order to even out any wear, and to maintain good seating. This rotation can be promoted if the centre of the cam is offset from the valve axis. Very often oil

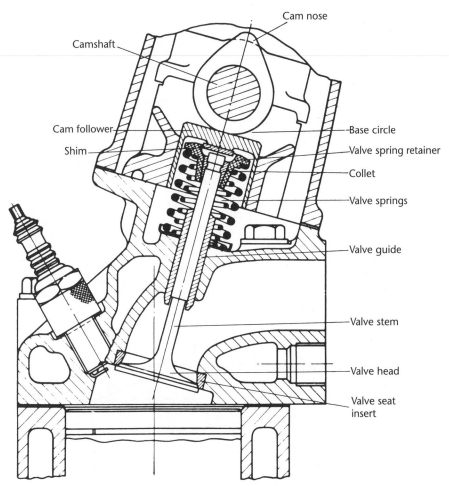

Figure 7.4
Overhead camshaft valve drive (reproduced with permission from Ricardo and Hempson, 1968).

seals are placed at the top of the valve guide to restrict the flow of oil into the cylinder. This is most significant with cast iron overhead camshafts which require a copious supply of lubricant.

Not all spark ignition engines have the inlet and exhaust valves in a single line. The notable exceptions are the high-performance engines with hemispherical or pent-roof combustion chambers. Valves in such engines can be operated by various push rod mechanisms, or by twin or double overhead camshafts (dohc). One camshaft operates the inlet valves, and the second camshaft operates the exhaust valves. The disadvantages of this system are the cost of a second camshaft, the more involved machining, and the difficulty of providing an extra drive. An ingenious solution to these problems is the British Leyland four-valve pent-roof head shown in figure 7.5. A single camshaft operates the inlet valves directly, and the exhaust valves indirectly, through a rocker. Since the same cam lobe is used for the inlet and exhaust valves, the

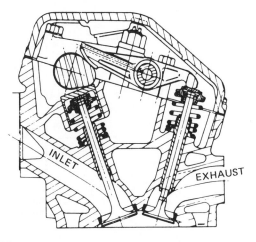

Figure 7.5
Four-valve per cylinder pent-roof combustion chamber (from Campbell, 1978).

valve phasing is dictated by the valve and rocker geometry. The use of four valves per combustion chamber is quite common in high-performance spark ignition engines, and widely used in the larger compression ignition engines. The advantages of four valves per combustion chamber are larger valve throat areas for gas flow, smaller valve forces and larger valve seat area. Smaller valve forces occur since a lighter valve with a less stiff spring can be used; this will also reduce the hammering effect on the valve seat. The larger valve seat area is important, since this is how heat is transferred (intermittently) from the valve head to the cylinder head.

7.2.3 Dynamic behaviour of valve gear

The theoretical valve motion is defined by the geometry of the cam and its follower. The actual valve motion is modified because of the finite mass and stiffness of the elements in the valve train. These aspects are dealt with after the theoretical valve motion.

The theoretical valve lift, velocity and acceleration are shown in figure 7.6; the lift is the integral of the velocity, and the velocity is the integral of the acceleration. Before the valve starts to move, the clearance has to be taken up. The clearance in the valve drive mechanism ensures that the valve can fully

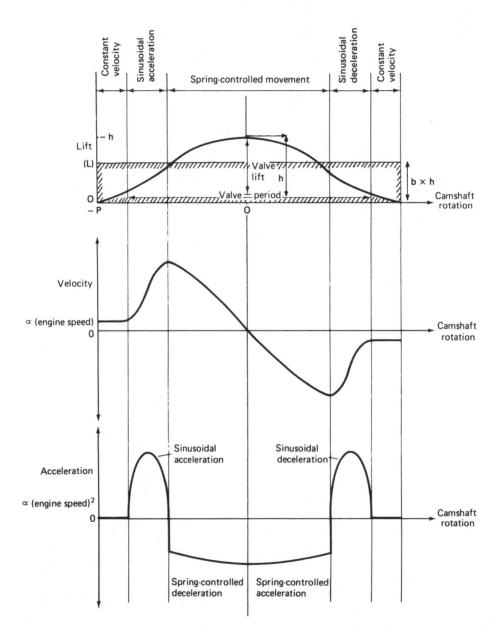

Figure 7.6
Theoretical valve motion.

seat under all operating conditions, with sufficient margin to allow for the bedding-in of the valve. To control the impact stresses as the clearance is taken up, the cam is designed to give an initially constant valve velocity. This portion of the cam should be large enough to allow for the different clearance during engine operation. The impact velocity is limited to typically 0.5 m/s at the rated engine speed.

The next stage is when the cam accelerates the valve. The cam could be designed to give a constant acceleration, but this would give rise to shock loadings, owing to the theoretically instantaneous change of acceleration. A better practice is to use a function that causes the acceleration to rise from zero to a maximum, and then to fall back to zero; both sinusoidal and polynomial functions are appropriate examples. As the valve approaches maximum lift the deceleration is controlled by the valve spring, and as the valve starts to close its acceleration is controlled by the valve spring. The final deceleration is controlled by the cam, and the same considerations apply as before. Finally, the profile of the cam should be such as to give a constant closing velocity, in order to limit the impact stresses.

Camshaft design is a complex area, but one which is critical to the satisfactory high speed performance of internal combustion engines. A review of some of the design considerations is given by Beard (1958), but this does not include the theory of a widely used type of cam – the polydyne cam (Dudley, 1948). The polydyne cam uses a polynomial function to define the valve lift as a function of cam angle, and selects coefficients that avoid harmonics which might excite valve spring oscillations:

$$L_v = f(\theta) = a + a_1\theta + a_2\theta^2 + a_3\theta^3 + \ldots + a_i\theta^i + \ldots \quad (7.1)$$

in which some values of a_i can be zero.

For a constant angular velocity of ω, differentiation gives:

velocity, $L_v' = f'(\theta) = \omega(a_1 + 2a_2\theta + 3a_3\theta^2 + \ldots + ia_i\theta^{i-1} + \ldots)$
$$(7.2)$$

acceleration, $L_v'' = f''(\theta) = \omega^2(2a_2 + 6a_3\theta + \ldots + i(i-1)a_i\theta^{i-2} + \ldots)$
$$(7.3)$$

jerk, $L_v''' = f'''(\theta) = \omega^3(6a_3 + \ldots + i(i-1)(i-2)a_i\theta^{i-3} + \ldots)$ $\quad (7.4)$

quirk, $L_v'''' = f''''(\theta) = \omega^4(24a_4 + \ldots + i(i-1)(i-2)(i-3)a_i\theta^{i-4} + \ldots)$
$$(7.5)$$

The dependence of the velocity on ω, the acceleration on ω^2, the jerk on ω^3 and the quirk on ω^4, explains why it is at high speeds that problems can occur with valve gear. It is normal practice to have the valve lift arranged symmetrically about the maximum lift, as shown in figure 7.6, and this is automatically satisfied if only even powers of θ are used in equation (7.1). This also ensures that the jerk term will be zero at the maximum valve lift (h).

There are various other boundary conditions to be considered.

When:

$$
\begin{array}{ll}
\theta = 0, & L_v = h \\
\theta = p, & L_v'' = 0 \\
\theta = p, & L_v''' = 0 \\
\theta = p, & L_v'''' = 0
\end{array}
$$

The valve lift 'area', A_θ, is a widely used concept to give an indication of the camshaft's ability to admit flow:

$$A_\theta = \int_{-p}^{p} L_v d\theta = 2bph \quad (7.6)$$

where b represents the effective mean height of the valve lift as a fraction of h

A_θ has units of *radians × metres*.

The valve lift 'area' can be specified by b (usually >0.5) and this, along with the boundary conditions, determine that the valve lift can be defined by equation (7.1) when four even-valued power terms are used. The selection of these power terms is a complex issue discussed in some detail by Dudley.

In practice, the valve lift characteristics will also be influenced by the stiffness of the valve spring, as this has to control the deceleration prior to the maximum lift, and the acceleration that occurs after maximum lift. Ideally the spring force should be uniformly greater than the required acceleration force at the maximum design speed. The acceleration is given by equation (7.3), and the mass needs to be referred to the valve axis. For a push-rod-operated valve system:

equivalent mass, $m_e =$

$$\frac{\text{tappet} + \text{push rod mass}}{(r_r/r_v)^2} + \frac{\text{polar inertia of rocker}}{r_v^2} + \text{valve mass}$$

$$+ \text{ upper spring retainer mass} + \frac{\text{spring mass}}{3}$$
$$(7.7)$$

where r_r = radius from rocker axis to cam line of action
r_v = radius from the rocker axis to the valve axis.

In practice, when either a rocker arm or a finger-follower system is used, the values of the radii r_r and r_v can change. Due account can be taken of this to convert the valve lift to cam lift, but it must be remembered that the equivalent mass (in equation 7.7) will also become a function of the valve lift.

Additional allowances in the spring load have to be made for possible overspeeding of the engine, and friction in the valve mechanism. The force (F) at the cam/tappet interface is given by equation (7.8):

$$F = \frac{r_r}{r_v}\left(m_e \times L_v + F_0 + kL_v + F_g\right) \quad (7.8)$$

where F_0 = valve spring pre-load
F_g = gas force on valve head (normally only significant for the exhaust valve)
k = valve spring stiffness.

Figure 7.7 shows the force at the cam/tappet interface for a range of speeds, along with the static force from the valve spring. At low speeds, the maximum force occurs at maximum valve lift, and this is because the valve spring force dominates. As the engine speed is increased, the acceleration terms dominate, and the largest force occurs just after the occurrence of the maximum acceleration.

A problem that can occur with valve springs is surge; these are intercoil vibrations. The natural frequency of the valve spring may be an order of magnitude higher than the camshaft frequency. However, as the motion of the cam is complex, there are high harmonics present and these can excite resonance of the valve spring. When this occurs the spring no longer obeys the simple force/displacement law, and the spring force will fall.

The natural frequency of a spring which has one end fixed is

$$\frac{1}{4}\sqrt{\left(\frac{k}{m}\right)} \text{ rad/s} \qquad (7.9)$$

which can be rewritten as

$$\frac{D}{d^3}k\sqrt{\left(\frac{2}{\pi^2 G}\right)} \qquad (7.10)$$

where D = coil mean diameter (m)
    d = wire diameter (m)
    G = bulk modulus (N/m²).

Examination of the standard equations for coil springs will show that if a higher natural frequency is required, then this will lead to higher spring stresses for a prescribed spring stiffness. Thus it may not be possible to avoid surge with a single valve spring. With two concentric springs, the frequencies at which surge occurs should differ, so that surge should be less troublesome. Surge can also be controlled by causing the coils to rub against an object to provide frictional damping, or by having a non-uniform coil pitch. If some coils in the spring close-up after lift commences, then the spring will no longer have a single natural frequency and this inhibits resonance.

The force at the cam/tappet interface (equation 7.8) also influences the contact stresses. For two elastic cylinders (1, 2) in contact, the Hertz stress (σ) is given by Roark and Young (1976) as

$$\sigma = \sqrt{\left(\frac{F}{\pi L} \times \frac{(1/r_1) + (1/r_2)}{[(1 - v_1^2)/E_1] + [(1 - v_2^2)/E_2]}\right)} \qquad (7.11)$$

where L = width of contact zone
    r = radius of curvature
    v = Poisson's ratio
    E = Young's modulus.

In the case of a flat follower, $1/r$ becomes zero. For surfaces with more than one curvature, appropriate formulae can be found in Roark and Young (1976). Equation (7.11) predicts line contact, and since the line is of finite width, this leads to

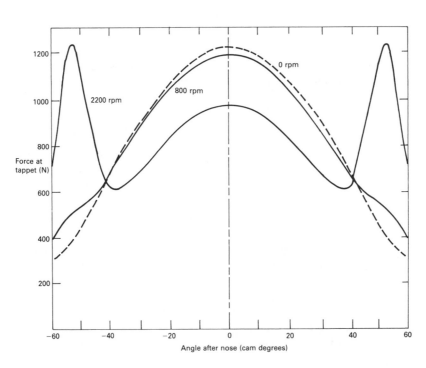

Figure 7.7
The force at the cam/tappet interface for a range of speeds.

discontinuities in the stress contours. These discontinuities have an adverse effect on the fatigue life of the contacting surfaces. The theoretical line contact is modified to a theoretical point contact, by having a tappet with a spherical (or barrelled) surface (but very high radius of curvature) or a cam with a second radius of curvature (again a very high value, lying in the plane of the camshaft axis). The theoretical point contact will spread to an elliptical contact zone, which then has no stress discontinuities.

The valve lift has now been specified as a function of cam angle, and all that remains is to decide on the base circle diameter. Increasing the base circle diameter increases the radius of curvature at the nose of the cam, and this reduces the contact stresses. Also, when a follower with a curved surface is used, increasing the base circle diameter will also increase the sliding velocity at the cam/tappet interface.

For the case of a flat follower moving orthogonally to the camshaft rotation, there is a remarkably simple equation for determining the radius of curvature (r) of the cam (Wilson et $al.$, 1983):

$$r = r_b + L_v + d^2 L_v / d\theta^2 \tag{7.12}$$

A flat-faced follower moving relative to a cam is shown in figure 7.8. This is an inversion of an engine cam-follower system, but it is kinematically equivalent and easier to analyse. The cam has a base circle radius of r_b and is assumed to rotate clockwise. For the inversion of the mechanism, as shown here, this corresponds to an anticlockwise rotation ω. The cartesian coordinates (x, y) are fixed in space, and are centred on the cam axis. The follower face has the equation of a straight line:

$$y = mx + b \tag{7.13}$$

and from inspection of figure 7.8, the gradient is given by:

$$m = \tan \theta \tag{7.14}$$

Point Q is the intercept of the follower line of action with its face, and its distance from the base circle represents the cam lift L_v. The coordinates of point Q are

$$x_Q = -(r_b + L_v) \sin \theta \tag{7.15}$$
$$y_Q = (r_b + L_v) \cos \theta \tag{7.16}$$

Substitution of equations (7.14)–(7.16) into equation (7.13) gives

$$(r_b + L_v) \cos \theta = [\tan \theta][-(r_b + L_v) \sin \theta] + b$$

or

$$(r_b + L_v)[\cos^2 \theta + \sin^2 \theta] = b \cos \theta$$

so

$$b = (r_b + L_v) / \cos \theta \tag{7.17}$$

Substitution of equations (7.17) and (7.14) into equation (7.13) gives

$$y = [x \sin \theta + (r_b + L_v)] / \cos \theta$$

or

$$y \cos \theta - x \sin \theta - r_b - L_v = 0 \tag{7.18}$$

Equation (7.18) represents the surface of the flat follower for any angle θ.

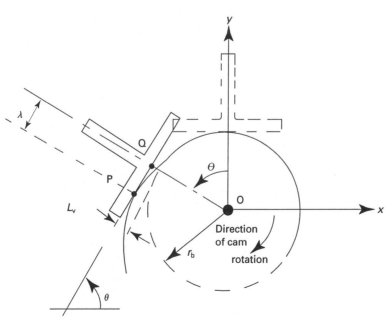

Figure 7.8
Geometry of a flat follower cam system.

Differentiation of equation (7.18) with respect to θ gives

$$-y\sin\theta - x\cos\theta - dL_v/d\theta = 0 \tag{7.19}$$

Use of the known lift/rotation characteristics enable $dL_v/d\theta$ to be determined, and this provides an equation for defining the point of contact, P. Point P is of course on the follower, so equation (7.18) also applies, and simultaneous solution with equation (7.19) will give the coordinates of the point of contact. Multiplication of equation (7.18) by $\sin\theta$, multiplication of equation (7.19) by $\cos\theta$, and adding gives

$$x_P(\sin^2\theta + \cos^2\theta) = x_P = -(r_b + L_v)\sin\theta - (dL_v/d\theta)\cos\theta \tag{7.20}$$

Multiplication of equation (7.18) by $\cos\theta$, multiplication of equation (7.19) by $-\sin\theta$, and adding gives

$$y_P(\cos^2\theta + \sin^2\theta) = y_P = (r_b + L_v)\cos\theta - (dL_v/d\theta)\sin\theta \tag{7.21}$$

The distance between the point of contact (P) and the centre-line of the follower (Q) is λ, and by Pythagoras:

$$\lambda = \sqrt{\left[(x_P - x_Q)^2 + (y_P - y_Q)^2\right]}$$

$$\lambda = \sqrt{\{[-(dL_v/d\theta)\cos\theta]^2 + [-(dL_v/d\theta)\sin\theta]^2\}} = dL_v/d\theta \tag{7.22}$$

The maximum value of λ can be found by looking for the maximum value of $dL_v/d\theta$, and this will define the minimum width of the follower. Differentiation of equation (7.22) will give the sliding velocity, and this is important for the hydrodynamic lubrication of the flat follower.

The parametric equation for the radius of curvature in cartesian coordinates is

$$r = \frac{[(dx/d\theta)^2 + (dy/d\theta)^2]^{3/2}}{(dx/d\theta)(d^2y/d\theta^2) - (dy/d\theta)(d^2x/d\theta^2)} \tag{7.23}$$

Differentiating equations (7.20) and (7.21) gives

$$dx_P/d\theta = -(r_b + L_v)\cos\theta - dL_v/d\theta\sin\theta + dL_v/d\theta\sin\theta - d^2L_v/d\theta^2\cos\theta$$
$$= -(r_b + L_v + d^2L_v/d\theta^2)\cos\theta \tag{7.24}$$
$$d^2x_P/d\theta^2 = (r_b + L_v + d^2L_v/d\theta^2)\sin\theta - (dL_v/d\theta + d^3L_v/d\theta^3)\cos\theta \tag{7.25}$$

and

$$dy_P/d\theta = -(r_b + L_v)\sin\theta + dL_v/d\theta\cos\theta + dL_v/d\theta\cos\theta - d^2L_v/d\theta^2\sin\theta$$
$$= -(r_b + L_v + d^2L_v/d\theta^2)\sin\theta \tag{7.26}$$
$$d^2y_P/d\theta^2 = -(r_b + L_v + d^2L_v/d\theta^2)\cos\theta - (dL_v/d\theta + d^3L_v/d\theta^3)\sin\theta \tag{7.27}$$

Substitution of equations (7.24)–(7.27) into equation (7.23) gives

$$r = \frac{[(r_b + L_v + d^2L_v/d\theta^2)^2(\cos^2\theta + \sin^2\theta)]^{3/2}}{(r_b + L_v + d^2L_v/d\theta^2)^2(\cos^2\theta + \sin^2\theta) + (r_b + L_v + d^2L_v/d\theta^2)(dL_v/d\theta + d^3L_v/d\theta^3)(\sin\theta\cos\theta - \cos\theta)}$$

or

$$r = r_b + L_v + d^2L_v/d\theta^2 \tag{7.12}$$

For satisfactory operation with a flat follower, the curvature (r) should always be positive. A negative value implies a concave surface (which a flat follower cannot operate on properly), and a zero value implies a cusp (that is, the cam comes to a point). Although it is self-evident that these equations have to be evaluated using radians, it is easy to forget this when programming, or inputting a tabulation of cam lift against degrees cam/crank angle.

As can be seen from figure 7.6, the theoretical valve-opening and valve-closing times will depend on the valve clearance. Consequently, the valve timing usually refers to the period between the start of the valve acceleration and the end of the valve deceleration. The valve lift refers to the lift in the same period, and is usually limited to about $0.25D_v$, to restrict the loads in the valve mechanism.

If the force required during the spring-controlled motion is greater than that provided by the spring, then the valve motion will not follow the cam, and the valve is said to 'jump'. The accelerations will increase in proportion to the square of the engine speed, and a theoretical speed can be calculated at which the valve will jump. The actual speed at which jumping occurs will be below this, because of the elasticity of other components and the finite mass of the spring.

The actual valve motion is modified by the elasticity of the components; a simple model is shown in figure 7.9. A comparison of theoretical and actual valve motion is shown in figure 7.10. Valve bounce can occur if the seating velocity is too great, or if the pre-load from the valve spring is too small.

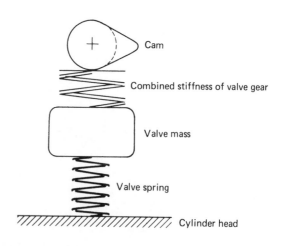

Figure 7.9
Simple valve gear model.

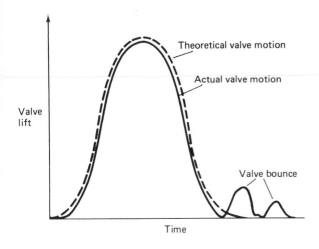

Figure 7.10
Comparison of theoretical and actual valve motion.

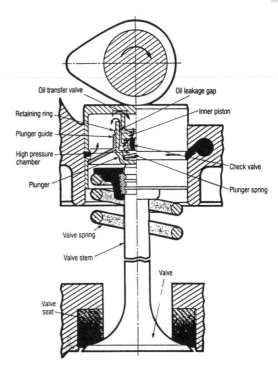

Figure 7.11
A hydraulic bucket tappet, with acknowledgement to Eureka.

This is likely to lead to failure, especially with the exhaust valve.

To minimise the dynamic effects, the valve should be made as light as possible, and the valve gear should be as stiff as possible. The camshaft should have a large diameter shaft with well-supported bearings, and the cams should be as wide as possible. Any intermediate components should be as light and stiff as possible.

More realistic models of valve systems can be used with a computer, which analyses the valve motion and then deduces the corrected cam profiles. The computer can then predict the response for a range of speeds.

The model should include the stiffness of the drive to the camshaft, the torsional stiffness of the camshaft, the stiffness of the camshaft mounting relative to the engine structure, the stiffness of any rocker and its mountings, as well as the stiffness of the valve train. The valve gear is likely to be modelled by a distributed mass (connected by stiff elements). Furthermore, wherever there is a stiffness there will be an associated damping factor. The damping is invariably difficult to evaluate, but small and highly significant in modifying resonances.

When valve spring surge is to be studied, it is necessary to subdivide the spring into a series of masses connected by massless springs and dampers. Valve spring manufacturers have developed such programs to aid engine designers (Hollingworth and Hodges, 1991).

Finally, it should be remembered that cams can only be manufactured to finite tolerances, and that oil films and deformation at points of contact and elsewhere will modify the valve motion. Beard (1958) considers both radial and lift manufacturing tolerances, and shows that both can be significant.

Sometimes hydraulic tappets (or cam followers) are used (figure 7.11). These are designed to ensure a minimum clearance in the valve train mechanism. They offer the advantages of automatic adjustment and, owing to the compressibility of the oil, they cushion the initial valve motion. This permits the use of more rapidly opening cam profiles. When the tappet is on the cam base circle, the engine lubricant flows past the check valve to fill the high-pressure chamber (figure 7.4). As soon as valve lift commences, the pressure in the high-pressure chamber will rise and close the check valve, so that the engine valve will then follow the motion of the cam follower. There will be some oil leakage during valve operation, and this ensures that when the valve closes there will be some clearance between the cam and the follower. This clearance allows for wear and differential thermal expansion, and is of course eliminated once the engine lubricant has filled the high-pressure chamber. The disadvantages are that they can stick (causing the valve to remain open), the valve motion is less well defined, and there are higher levels of friction unless a roller follower is also used, since the follower is always in contact with the cam. Hydraulic tappets (or lash adjusters) can be incorporated into the cam followers of overhead valve engines (item 5 in figure 7.3), or into the pivot post of overhead camshaft engines that use a finger-follower (see chapter 15, figure 15.10).

7.3 Flow characteristics of poppet valves

The shape of the valve head, port and seat are developed empirically or with CFD to produce the minimum pressure drop.

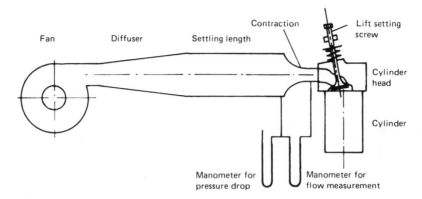

Figure 7.12
Air flow rig to determine the flow characteristics of an inlet valve (with acknowledgement to Annand and Roe, 1974).

Experiments are usually carried out on steady-flow air rigs similar to the one shown in figure 7.12. The flow from a fan is decelerated in a diffuser before entering a settling length. To help provide a uniform flow, the type of meshes used in wind tunnels may be useful. The contraction accelerates the flow, thus reducing the relative significance of any turbulence. It is essential for the contraction to match perfectly the inlet passage, otherwise turbulence and extraneous pressure drops will be introduced. The contraction also provides a means of metering the flow rate and of measuring the pressure drop across the valve. The lift setting screw enables the pressure drop to be measured for a range of valve lifts.

A similar arrangement can be used for measuring the flow characteristics of exhaust valves, the exhaust passage being connected to a suction system. Sometimes water rigs are used when the flow patterns are to be investigated. As in all experiments that measure pressure drops through orifices, slightly rounding any sharp corners can have a profound effect on the flow characteristics; care must be taken with all models.

The orifice area can be defined as a 'curtain' area, A_c:

$$A_c = \pi D_v L_v \tag{7.28}$$

where D_v = valve diameter
and L_v = axial valve lift.

This leads to a discharge coefficient (C_D); this is defined in terms of an effective area (A_e):

$$C_D = \frac{A_e}{A_c} \tag{7.29}$$

The effective area is a concept defined as the outlet area of an ideal frictionless nozzle which would pass the same flow, with the same pressure drop, with uniform constant-pressure flows upstream and downstream. These definitions are arbitrary, and consequently they are not universal.

For a given geometry, the discharge coefficient will vary with valve lift and flow rate. These quantities can be expressed conveniently as a non-dimensional valve lift (L_v/D_v), and Reynolds number (Re):

$$Re = \frac{\rho v x}{\mu} \tag{7.30}$$

where ρ = fluid density
v = flow velocity
x = a characteristic length
μ = fluid viscosity.

For the inlet valve it is common practice to assume incompressible flow since the pressure drop is small, in which case the ideal velocity would be

$$v_0 = \sqrt{(2\Delta p/\rho)} \tag{7.31}$$

where v_0 = frictionless velocity in the throat
Δp = pressure difference across the port and valve.

For compressible flow into the cylinder

$$v_0 = \left\{ \frac{2\gamma}{(\gamma-1)} \frac{p_0}{p_0} \left[\left(\frac{p_0}{p_c}\right)^{\frac{\gamma-1}{\gamma}} - 1 \right] \right\}^{1/2} \tag{7.32}$$

and

$$\frac{\dot{m}}{A} = \frac{p_0}{\sqrt{(RT_c)}} \left(\frac{p_0}{p_c}\right)^{\frac{1}{\gamma}} \left\{ \frac{2\gamma}{\gamma-1} \left[\left(\frac{p_0}{p_c}\right)^{\frac{\gamma-1}{\gamma}} - 1 \right] \right\}^{1/2} \tag{7.33}$$

where p_0 = upstream pressure
p_c = cylinder pressure
$\dot{m}/A$ = mass flow per unit area.

The discharge coefficient (equation 7.29) can now be defined in terms of the measured volume flow rate ($\dot{V}$) compared with the ideal volume flow rate ($\dot{V}_0$):

$$A_c C_D = A_e = \frac{\dot{V}}{\dot{V}_0/A_c} = \frac{\dot{V}}{v_0} \tag{7.34}$$

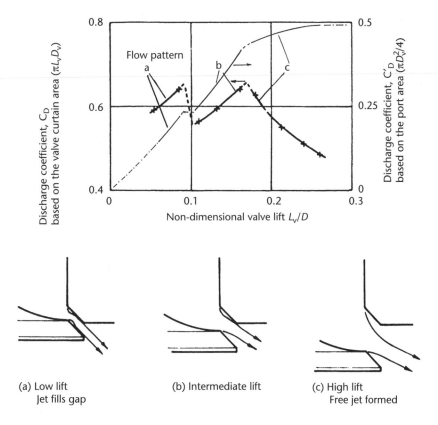

Figure 7.13
Flow characteristics of a sharp-edged inlet valve (with acknowledgement to Annand and Roe, 1974).

It can be less ambiguous to define the valve flow performance by the effective area as a function of valve lift. This avoids the problem of how the reference flow area has been defined. For example, does the curtain area correspond to the inner or outer valve seat diameter? Sometimes the reference area is the minimum cross-sectional area for the flow. The position of this can occur in three places according to the valve lift, and this geometry is discussed in detail by Heywood (1988). Ultimately the minimum flow area will be the annulus between the valve stem diameter (D_s) and the valve inner seat diameter (D_{vi}). This value of valve lift corresponds to

$$L'_v = \frac{D^2_{vi} - D^2_s}{4D_{vi}} \tag{7.35}$$

or a non-dimensional value lift of about 0.23.

The flow characteristics of a sharp-edged inlet valve are shown in figure 7.13, in terms of the discharge coefficient based on both the valve curtain area (C_D) and the port area (C'_D). These two discharge coefficients can be related by means of the effective flow area:

$$A_e = C_D \pi L_v D_v = C'_D \pi D^2_v / 4 \tag{7.36}$$

As predicted by equation (7.35), the effective flow area reaches a maximum when the non-dimensional valve lift is

about 0.23, since the smallest flow area is now the annular area between the valve seat and its stem.

At low lift the jet fills the gap and adheres to both the valve and the seat. At an intermediate lift the flow will break away from one of the surfaces, and at high lifts the jet breaks away from both surfaces to form a free jet. The transition points will depend on whether the valve is opening or closing. These points are discussed in detail by Annand and Roe (1974) along with the effects of sharp corners, radii and valve seat width. They conclude that a 30° seat angle with a minimum width seat and 10° angle at the upstream surface give the best results. In general, it is advantageous to round all corners on the valve and seat. For normal valve lifts the effect of Reynolds number on discharge coefficient is negligible.

When the discharge coefficient results from figure 7.13 are combined with some typical valve lift data, then it is possible to plot the effective flow area as a function of camshaft angle. This has been done in figure 7.14, and it should be noted that there is a broad maximum for the effective flow area – this is a consequence of the effective flow area being limited by the annular area (equation 7.35) and no longer being dependent on the valve 'lift. Also shown in figure 7.14 is a 'high performance' cam profile; this has an increased lift and valve open duration. However, the valve lift curve has been scaled so as to give the same maximum valve train acceleration.

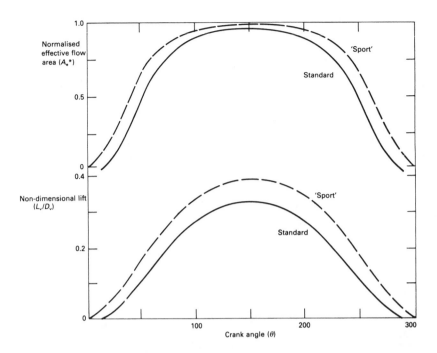

Figure 7.14
Comparison between a 'standard' cam profile and its 'sport' counterpart with an extended period and increased lift (but with the same maximum acceleration); also shown are the corresponding effective flow areas.

Increasing the valve lift has not significantly increased the maximum effective flow area, but a consequence of the longer duration valve event is an increase in the width of the maximum. In other words, it is the extended duration that will lead to an improvement in the flow performance of the valve.

Taylor (1985a) points out that if the pressure ratio across the inlet valve becomes too high, there will be a rapid fall in the volumetric efficiency. Taylor characterises the flow by an inlet Mach index, Z, which is the Mach number of a notional air velocity.

When the effective flow area in figure 7.14 is averaged, it can be divided by the reference area (based on the valve diameter D_v), to give a mean flow coefficient:

$$\bar{C}_d = \frac{\bar{A}_e}{\pi D_v^2/4} \tag{7.37}$$

The mean rate of change of the volume, is

$$\bar{v}_p \times \pi B^2/4 \tag{7.38}$$

and this leads to a notional mean velocity, which can be divided by the speed of sound (c), to give the Mach index, Z:

$$Z = \frac{\bar{v}_p \times \pi B^2/4}{c \times C_d \times \pi D_v^2/4} = \left(\frac{B}{D_v}\right)^2 \frac{\bar{v}_p}{c \times C_d} \tag{7.39}$$

Taylor found that, for a fixed valve timing, the volumetric efficiency was only a function of the Mach index, and to maintain an acceptable volumetric efficiency the Mach index should be less than 0.6; this is illustrated by figure 7.15. Taylor (1985a) also reports on the effects of the inlet valve timing on the volumetric efficiency. Most significant is the benefit of delaying the inlet valve closure for high values (> 0.6) of the Mach index. In practice, the volumetric efficiency will be modified by the effects of inlet system pressure pulsations, and this is discussed further in section 7.6. The selection of valve timing is discussed in section 7.4.

The effect of valve lift on the discharge coefficient is much smaller for the exhaust valve – see figure 7.16. The range of pressure ratios across the exhaust valve is much greater than that across the inlet valve, but the effect on the discharge coefficient is small. The design of the exhaust valve appears less critical than the inlet valve, and the seat angle should be between 30° and 45°.

When the exhaust valve opens (blowdown), there is a very high pressure ratio across the valve. Indeed the flow can become choked if

$$\frac{p_e}{p_c} \leq \left(\frac{2}{\gamma+1}\right)^{\frac{\gamma}{\gamma-1}} \tag{7.40}$$

where p_e = exhaust manifold pressure
p_c = cylinder pressure.

For exhaust gases this critical pressure ratio is about 3 (since γ depends on the temperature and composition of the gases), and when the flow is choked

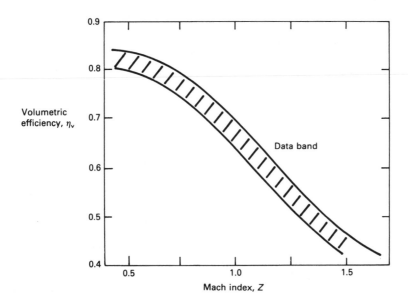

Figure 7.15
Volumetric efficiency as a function of the Mach index (adapted from Taylor, 1985a).

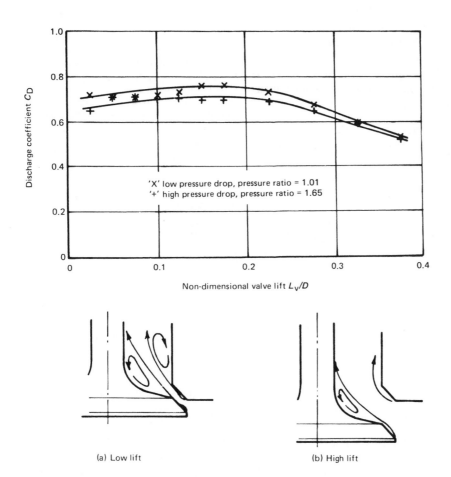

(a) Low lift   (b) High lift

Figure 7.16
Flow characteristics of a sharp-edged 45° seat angle exhaust valve (with acknowledgement to Annand and Roe, 1974).

$$\dot{m}/A = \frac{p_c}{\sqrt{(RT_c)}}\sqrt{\gamma\left(\frac{2}{\gamma+1}\right)^{\frac{\gamma+1}{2(\gamma-1)}}} \tag{7.41}$$

For the exhaust flow, p_c is substituted for p_0, and p_e is substituted for p_c in equations (7.32) and (7.33) to give

$$v_0 = \left\{\frac{2\gamma}{(\gamma-1)}\frac{p_c}{p_e}\left[\left(\frac{p_c}{p_e}\right)^{\frac{\gamma-1}{\gamma}}-1\right]\right\}^{1/2} \tag{7.32'}$$

and

$$\frac{\dot{m}}{A} = \frac{p_c}{\sqrt{(RT_c)}}\left(\frac{p_c}{p_e}\right)^{\frac{1}{\gamma}}\left\{\frac{2\gamma}{\gamma-1}\left[\left(\frac{p_c}{p_e}\right)^{\frac{\gamma-1}{\gamma}}-1\right]\right\}^{1/2} \tag{7.33'}$$

When these flow equations are being solved as part of a cylinder filling model, it is important to remember that reverse flows can occur through both inlet and exhaust valves. Therefore, it is necessary to check the pressure ratio across the valve and then use an equation for the appropriate flow direction.

In the absence of heat transfer, most of the flow through the valves can be treated as isenthalpic. The notable exception is the blowdown process immediately following the exhaust valve opening, when there is a rapid fall in the cylinder pressure. Figure 7.17 shows the separate processes that can be used to describe blowdown and the exhaust stroke.

Consider first the blowdown process (a → b) for which it will be assumed that the piston is stationary, in which a quantity of mass (δm_0) leaves the cylinder:

Decrease in stored energy in the cylinder	=	'Stored' energy (u_0) of δm_0 leaving the cylinder	+	Flow work ($p_0 v_0$) of δm_0 leaving the cylinder
$-dU$	=	(u_0	+	$p_0 v_0$) δm_0

$$-dU = h_0 \delta m_0 \tag{7.42}$$

If this is now 'integrated' for the whole mass leaving the cylinder during blowdown, with 1 referring to the state before blowdown, 2 referring to the state in the cylinder after blowdown, and 2″ referring to the state in the exhaust port after the end of blowdown, then

$$U_1 - U_2 = H_{2''} \tag{7.43}$$

Assuming perfect gas behaviour (for which $u = c_v T$, and $h = c_p T$), equation (7.43) becomes

$$m_1 c_v T_1 - m_2 c_v T_2 = (m_1 - m_2)c_p T_{2''}$$

or

$$T_{2''} = (m_1 T_1 - m_2 T_2)/\gamma(m_1 - m_2) \tag{7.44}$$

At the end of the exhaust stroke, before there is any mixing with the gases that left during blowdown, the situation would be as shown in figure 7.17c. Mass m_2 (designated as m_{2*}, since it now has a different state) has been displaced into the exhaust port, but at this stage without any mixing:

Decrease in stored energy in the cylinder	+	Work input	=	'Stored' energy (u_0) of δm_0 leaving the cylinder	+	Flow work of δm_0 leaving the cylinder
$U_2 - 0$	+	W	=	$\int(u_0$	+	$p_0 v_0)\delta m_0$

$$= H_{2*}$$

$$\tag{7.45}$$

The work input is $p_2 V_2$, thus

$$U_2 + p_2 V_2 = H_{2*}$$

As $m_2 = m_{2*}$, the $u_2 + p_2 v_2 = h_{2*}$, in other words

$$T_2 = T_{2*} \tag{7.46}$$

Thus the exhaust stroke is isothermal. Once the gases in the exhaust port are fully mixed (figure 7.17d), the final temperature becomes $T_{2'}$. From an energy balance:

$$m_{2'}c_p T_{2'} = m_{2''}c_p T_{2''} + m_2 c_p T_2$$

Noting that $m_{2'} = m_1$, and substituting for $T_{2''}$ from equation (7.44) gives

$$m_1 T_{2'} = (m_1 T_1 - m_2 T_2)/\gamma + m_2 T_2$$
$$T_{2'} = T_1/\gamma + (m_2/m_1)T_2(1 + 1/\gamma) \tag{7.47}$$

This result could also be obtained by considering the change from a → d directly:

Decrease in stored energy in the cylinder	+	Work input	=	'Stored' energy (u_0) of δm_0 leaving the cylinder	+	Flow work ($p_0 v_0$) of δm_0 leaving the cylinder
$-dU$	+	dW	=	(u_0	+	$p_0 v_0)\delta m_0$

$$\tag{7.48}$$

'Integrating' gives:

$$U_1 + p_2 V_2 = H_{2'} \tag{7.49}$$

or

$$m_1 c_v T_1 + m_2 R T_2 = m_1 c_p T_{2'}$$

thus

$$T_{2'} = T_1/\gamma + m_2 R T_2/m_1 c_p = T_1/\gamma_1 + (m_2/m_1)T_2(1 + 1/\gamma)$$
$$\tag{7.50 or 7.47}$$

since $R = c_p - c_v$, $\gamma = c_p/c_v$ and m_2 is computed from the volume V_2 and temperature T_2 (the isentropic expansion temperature).

Temperature T_2 is found by assuming an isentropic expansion from pressure p_1 to p_2. With this direct method, it has not been necessary to assume that the blowdown process is instantaneous. The first gas to leave during blowdown will be at a higher temperature than the gas that leaves

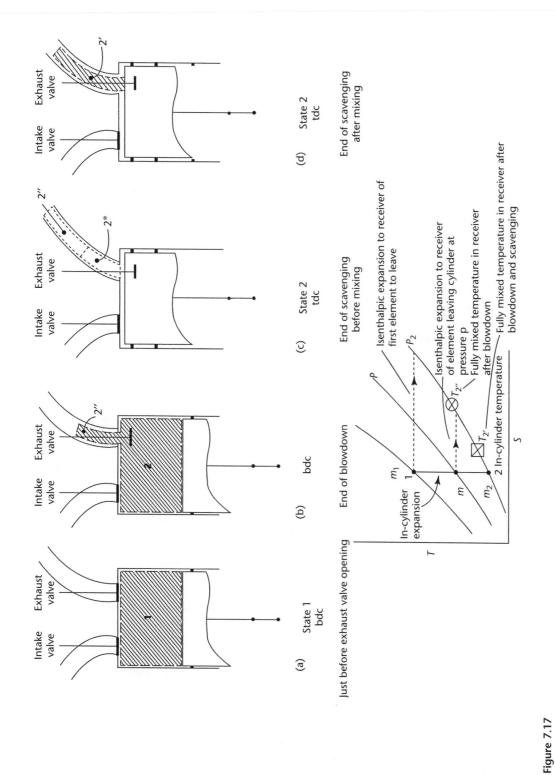

Figure 7.17
The exhaust blowdown and exhaust stroke processes for a four-stroke engine.

subsequently during blowdown; this is the subject of question 7.4.

In general, the inlet valve is of larger diameter than the exhaust valve, since a pressure drop during induction has a more detrimental effect on performance than a pressure drop during the exhaust stroke. For a flat, twin valve cylinder head, the maximum inlet valve diameter is typically 44–48 per cent and the maximum exhaust valve diameter is typically 40–44 per cent of the bore diameter. With pent-roof and hemispherical combustion chambers the valve sizes can be larger.

For a flat four-valve head, as might be used in a compression ignition engine, each inlet valve could be 39 per cent of the bore diameter, and each exhaust valve could be 35 per cent of the bore diameter. This gives about a 60 per cent increase in total valve circumference, or a 30 per cent increase in 'curtain' area (equation 7.1) for the same non-dimensional valve lift. The inlet and exhaust passages should converge slightly to avoid the risk of flow separation with its associated pressure drop. At the inlet side the division between the two valve ports should have a well-rounded nose; this will be insensitive to the angle of incidence of the flow. A knife-edge division wall would be very sensitive to flow breakaway on one side or the other. For the exhaust side the division wall can taper out to a sharp edge.

The port arrangements in two-stroke engines are discussed in chapter 8, section 8.5.3.

7.4 | Valve timing

7.4.1 Effects of valve timing

The valve timing dictated by the camshaft and follower system is modified by the dynamic effects that were discussed in section 7.2.3. Two timing diagrams are shown in figure 7.18. The first (figure 7.18a) is typical of a compression ignition engine or conventional spark ignition engine, while figure 7.18b is typical of a high-performance spark ignition engine. The greater valve overlap in the second case takes fuller advantage of the pulse effects that can be particularly beneficial at high engine speeds. Turbocharged diesel engines can also use large valve overlap periods. In compression ignition engines the valve overlap at top dead centre is often limited by the piston to cylinder-head clearance. Also the inlet valve has to close soon after bottom dead centre, otherwise the reduction in compression ratio may make cold starting too difficult. The exhaust valve opens about 40° before bottom dead centre (bbdc) in order to ensure that all the combustion products have sufficient time to escape. This entails a slight penalty in the power stroke, but 40°bbdc represents only about 12 per cent of the engine stroke. It should also be remembered that 5° after starting to open the valve may be 1 per cent of fully open, after 10°, 5 per cent of fully open, and not fully open until 120° after starting to open. If the exhaust valve is opened too late, although there is an increase in expansion work, this is more than offset by an increase in the pumping work – the cylinder pressure will be higher during the exhaust stroke as a greater mass has to be expelled.

In spark ignition engines with large valve overlap, the part throttle and idling operation suffers since the reduced induction manifold pressure causes back-flow of the exhaust. Furthermore, full load economy is poor since some unburnt mixture will pass straight through the engine when both valves are open at top dead centre. These problems are avoided in a turbocharged engine with in-cylinder fuel injection.

The level of exhaust residuals trapped in the cylinder has a significant effect on the cycle-by-cycle variations in combustion, and the emissions of NO_x. As with exhaust gas recirculation, high levels of exhaust residuals lead to lower emissions of NO_x and greater cycle-by-cycle variations in combustion.

The level of residuals increases with:

(a) decreasing absolute inlet manifold pressure
(b) reducing compression ratio
(c) increasing valve overlap
(d) decreasing speed
(e) increasing exhaust back-pressure.

If the level of exhaust residuals is increased at part load in a spark ignition engine, this means that the throttle will have to be more widely opened to admit a given mass of mixture. This in turn means that the cylinder pressure will be higher and that the pumping work will be reduced. So, variable valve timing systems (section 7.4.2) can be used to control the level of exhaust residuals (as an alternative to external exhaust gas recirculation – EGR) to reduce the pumping losses at part load.

An alternative method of retaining exhaust residuals is negative valve overlap, and this has already been mentioned in the context of homogeneous charge compression ignition engines in chapter 6, section 6.7. Negative valve overlap seems to be self-contradictory, but what it means is no overlap. The exhaust valve can be closed before tdc so that there is recompression of the burned gases, or the exhaust valve can be closed after tdc. In either case the normal aim would be to delay opening the inlet valve until the cylinder pressure has fallen to the inlet manifold pressure, so that as the inlet valve opens there is no reverse flow.

Comprehensive experimental results for the residual fraction have been reported by Toda *et al.* (1976). The exhaust residual (ER) levels can either be predicted by computer modelling or by extracting a gas sample during the compression process:

$$ER = \frac{\text{molar fraction of species } i \text{ during compression}}{\text{molar fraction of species } i \text{ in the exhaust}}$$

(7.51

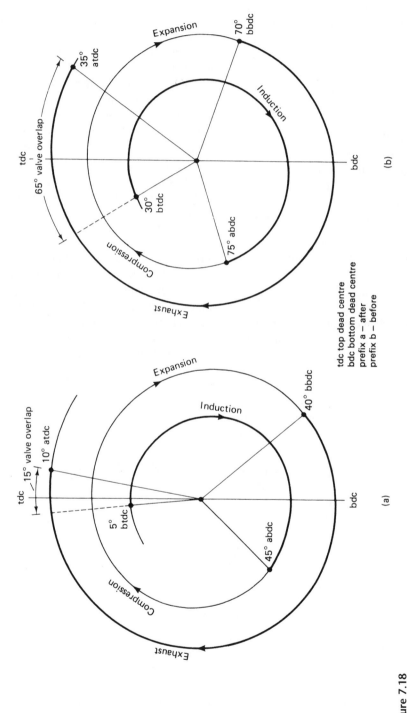

Figure 7.18
Valve-timing diagrams: (a) small valve overlap; (b) large valve overlap.

The molar fractions should be on a wet basis, and carbon dioxide is usually measured since it is the most plentiful species present. It should be remembered that this measurement will also include the exhaust residuals due to EGR. The methods for measuring the emissions and deducing the EGR level are discussed in chapter 14, section 14.4.7.

Fox *et al.* (1993) have produced a comprehensive model for predicting exhaust gas residual levels, in which the form of the equation used for correlating the data has been based on physical arguments:

$$ER = 76(OF/N)(p_i/p_e)^{-0.87}\sqrt{(|p_i - p_e|)} + 0.632(\phi/r_c)(p_i/p_e)^{-0.74}$$

$$(7.52)$$

where r_c is the compression ratio
ϕ is the equivalence ratio
p_e is the absolute exhaust pressure in bar
p_i is the absolute inlet pressure in bar
N is the speed in rpm and
OF is the overlap factor in °/m, as defined below.

The overlap factor (OF) can either be calculated, when the valve lift profiles are known, or estimated from equation (7.53):

$$OF = (D_i A_i D_e A_e)/V_s \qquad °/m \qquad (7.53)$$

where D is the valve inner seat diameter and A is defined in figure 7.19. In the absence of valve lift data, the overlap factor (OF) can be estimated from

$$OF = (155 + 11.3\Delta\theta + 1.45\Delta\theta^2)L_{v,max}D/B^3 \qquad °/m \qquad (7.54)$$

where $L_{v,max}$ is the mean of the inlet and exhaust maximum valve lift, and
D is the mean of the inlet and exhaust valve inner seat diameters (mm).

The trade-offs in selecting the valve timing for the 2.2-litre Chrysler engine (which is a case study in chapter 15, section 15.3) are discussed in detail by Asmus (1984). Asmus points out that early opening of the exhaust valve leads to a reduction in the effective expansion ratio and expansion work, but this is compensated for by reduced exhaust stroke pumping work. When an engine is being optimised for high speed operation this leads to the use of earlier exhaust valve opening; the usual timing range is 40–60° before bottom dead centre (bbdc). In the case of turbocharged engines, some of the expansion work that is lost by earlier opening of the exhaust valve is recovered by the turbine.

Exhaust valve closure is invariably after top dead centre (atdc), and the higher the boost pressure in turbocharged engines, or the higher the speed for which the engine performance is optimised, then the later the exhaust valve closure. The exhaust valve is usually closed in the range 10–60° atdc. The aim is to avoid any compression of the cylinder contents towards the end of the exhaust stroke. The exhaust valve closure time does not seem to affect the level of residuals trapped in the cylinder or the reverse flow into the inlet manifold. However, for engines with in-manifold

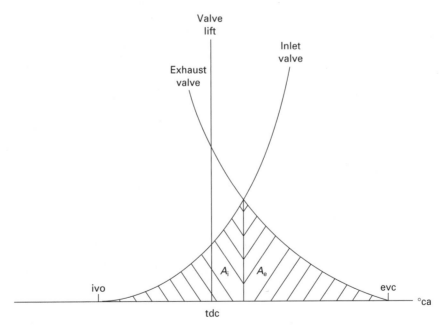

Figure 7.19
The definition of 'flow areas' during the valve overlap period for equation (7.52).

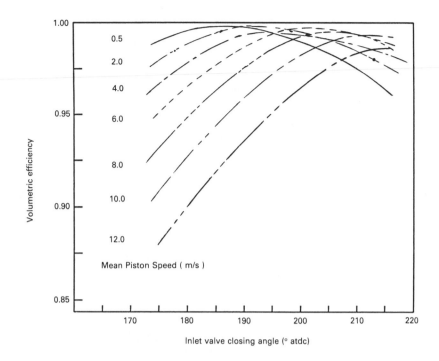

Figure 7.20
The effect of inlet valve closing angle (°atdc) on volumetric efficiency.

mixture preparation, a late exhaust valve closure can lead to fuel entering the exhaust manifold directly.

The inlet valve is opened before top dead centre (btdc), so that by the start of the induction stroke there is a large effective flow area. Engine performance is fairly insensitive to inlet valve opening in the range 10–25°btdc. For turbocharged engines at their rated operating point, then the inlet manifold pressure is greater than the cylinder pressure, which in turn is above the exhaust manifold pressure: under these circumstances even earlier inlet valve opening (earlier than 30°btdc) leads to good scavenging. However, at part load, for a turbocharged engine or a throttled engine, early inlet valve opening leads to high levels of exhaust residuals and back flow of exhaust into the inlet manifold. The results of this are most obvious with spark ignition engines, since the increased levels of exhaust residuals lead to increased cycle-by-cycle variations in combustion. The influence of valve overlap on the part-load performance of turbocharged diesel engines is illustrated in chapter 11, section 11.3.3.

Inlet valve closure is invariably after bottom dead centre (abdc) and typically around 40°abdc, since at bdc the cylinder pressure is still usually below the inlet manifold pressure. This is in part a consequence of the slider crank mechanism causing the maximum piston velocity to occur after 90°bbdc. Figure 7.20 illustrates the influence of inlet valve closure angle on the volumetric efficiency. A simple model has been used here, which ignores compressibility and dynamic effects. The mean piston speed has been used as a variable, since it

defines engine speed in a way that does not depend on the engine size. Figure 7.20 shows that at low speeds, a late inlet valve closure reduces the volumetric efficiency. In contrast, at high speeds an early inlet valve closure leads to a greater reduction in volumetric efficiency, and this limits the maximum power output.

7.4.2 Variable valve timing (VVT)

The discussion of valve timing in the previous section has shown that there are compromises in valve timing: high-speed versus low-speed performance and full-load versus part-load performance. Not surprisingly, there has been considerable effort devoted to developing variable valve timing (VVT) mechanisms, and some of this work has been reviewed by Stone and Kwan (1989) and by Ahmad and Theobald (1989). In addition to minimising valve timing compromises, variable valve timing can be used to reduce the throttling losses in spark ignition engines. If the inlet valve is closed before bdc, or significantly later than normal, then the trapped mass will be reduced without recourse to throttling. Figure 7.21 shows idealised p–V diagrams for throttling, early inlet valve closure and late (or delayed) inlet valve closure, and it can be seen that there is potential for reducing the pumping work or throttling loss. These modified inlet valve closure angles also reduce the effective compression ratio, and it might be thought that this will lead to a lower corresponding cycle efficiency. However, study of example 2.3 will show

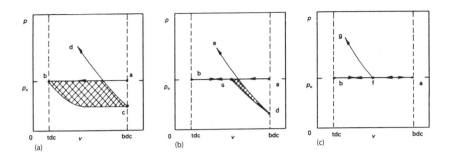

Figure 7.21
State diagrams for load control by (a) throttling, (b) early inlet valve closure and (c) late inlet valve closure. In reality all processes are irreversible; 'hatching' indicates dissipated work.

that the expansion ratio is a more important determinant of cycle efficiency, and this is unchanged by varying the inlet valve closure angle (Stone and Kwan, 1989). The lower compression ratio leads to a lower compression temperature, and this reduces the laminar burning velocity. Furthermore, Hara *et al.* (1985) also report lower levels of turbulence, so the combined effect is a reduced burn rate and a lower efficiency – this is probably why the theoretical predictions for an efficiency gain are higher than those achieved in practice. Most experimental studies of late and early inlet valve closure have used camshafts with special fixed timings, and the data from these sources have also been reviewed by Stone and Kwan (1989). The conclusion is, that in the bmep range of 3–4.5 bar, there is scope for a 5–10 per cent saving in fuel consumption.

Mechanical variable valve timing systems have been devised that enable the valve events to be phased (that is, both opening and closing events are moved equally), and the valve event duration to be modified (with usually the ability to control the relative amount that opening and closing are moved). Needless to say, it is the simplest devices that have been used in production engines. The first such use was by Alfa Romeo with the mechanism shown in figure 7.22 controlling the phasing of the inlet camshaft. An advantage of this phasing control is that at light-load and low speeds (very significant conditions in urban driving cycles) both delaying the inlet valve closure and delaying the inlet valve opening are beneficial. The delayed inlet valve closure reduces the throttling losses, and the reduced valve overlap at the light load also improves combustion. Ma (1988) quantified the effect of valve overlap, as enabling a 200 rpm reduction in idle speed with a 12 per cent reduction in fuel flow. He predicted that this would lead to a 6.1 per cent reduction in the ECE-15 urban cycle fuel consumption. The Alfa Romeo mechanism delayed the inlet valve events by 32°ca, with a 28 per cent reduction in the urban cycle fuel consumption (Anon, 1984b). However, it is likely that some of this reduction in fuel consumption is attributable to a change from a carburettor to fuel injection.

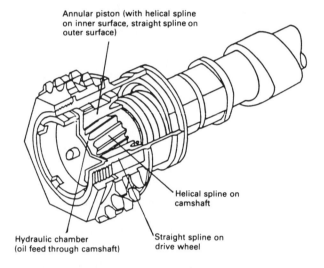

Figure 7.22
The Alfa Romeo variable phasing system (Anon, 1984b).

An example of a VVT mechanism in production that can control the duration and phasing of a camshaft is provided by the Rover VVC (Variable Valve Control) system, which is used on the inlet camshafts of some K16 engines (Whiteley, 1995). Figure 7.23 shows two VVC mechanisms; one is required for each cylinder. There is a drive to each end of the camshaft, and then the valves of the inner cylinders are operated by shafts that run through hollow camshafts for the outer cylinders. The mechanism is essentially a non-constant velocity drive system. Each cylinder has a drive ring in which two guide blocks can slide (one for the input and the other for the output). When this slot is offset from the camshaft axis, the output guide block causes the cam to rotate with a non-constant velocity. The slot offset is controlled by mounting the drive ring inside an eccentric sleeve: rotation of the sleeve (in response to an ECU input) controls the valve timing.

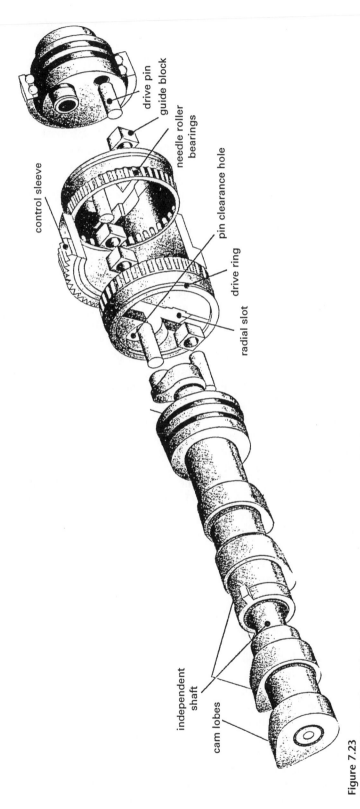

Figure 7.23
The Rover VVC variable valve timing system with control of the valve period and phasing.

Table 7.1 *Comparison of Rover K16 1.8-litre engines with and without VVC*

Engine	Maximum bmep (bar)	Speed (rpm) for maximum bmep	Maximum power (kW)	Speed (rpm) maximum power
Standard	11.6	3000	90	5500
VVC	12.2	4500	108	7000

The VVC mechanism properties depend on both the offset of the drive ring slot and the angular direction of this offset. The eccentric sleeve has only one input parameter (its rotation), and thus its effect on the drive ring is a 'programmed' series of magnitude and direction offsets. The VVC mechanism has been designed to control the valve duration from 220° to 295°ca, with a corresponding change in valve overlap from 21° to 58°ca. Thus with the maximum valve period, inlet valve closure has also been delayed by 38°ca. Table 7.1 compares the performance of the 1.8-litre VVC and conventional engines.

The VVC engine also had larger valves, with the inlet valves increased from 27.3 to 31.5 mm, and the exhaust valves increased from 24.0 to 27.3 mm. The K series engine family is discussed further in chapter 15, section 15.2.

Another approach to VVT is to switch between two different cam profiles, and this is how the Honda VTEC engine disables an inlet valve (chapter 4, section 4.2.3). Figure 7.24 shows the Honda VTEC engine valve gear, with a cross-sectioned diagram to show how the rocker arms for the two inlet valves can be locked together (Horie and Nishizawa, 1992). The engine has a central single overhead camshaft, with valve operation through rocker arms with roller followers. There are two separate profiles for the inlet cams, the secondary one having a very low lift (of about 0.5 mm, to prevent liquid fuel accumulating behind the valve). In normal operation, the hydraulically controlled piston locks the two inlet valve rocker arms together. For low-load operation, one inlet valve is controlled by the secondary cam: this can be much narrower since the lower lift requires much lower accelerations with correspondingly reduced contact forces. A similar result could be achieved for this application by having a butterfly valve (throttle plate) or swirl control valve in one of the inlet ports for each cylinder.

Another approach to cam profile switching has been developed for incorporation into hydraulic bucket tappets. Figure 7.25 shows how the primary cam profile acts directly on the valve, and the secondary cam lobes are only utilised when the hydraulically actuated locking pins connect the inner and outer tappet elements. (It should be noted that the designation of primary and secondary is the opposite from that used by Honda on the VTEC engine, and that in both cases the low lift cam profile has to operate within the envelope of the high lift cam profile.) Sandford *et al.* (1998) report on the use of these tappets for cylinder disablement, as a means of reducing part-load fuel consumption. Cylinder disablement can be achieved in a number of different ways. The simplest method is merely to turn the fuel and ignition off on the chosen cylinders, but unburnt hydrocarbons can pass from the inlet manifold to the exhaust manifold via the cylinder. Disablement of either the inlet valves or the exhaust valves prevents the direct passage of unburnt hydrocarbons into the exhaust, but the reduction in pumping work is no greater than that with fuel disablement. Pumping work in the disabled cylinder can be essentially eliminated if both the inlet and exhaust valves are disabled, and for a four-cylinder engine, disabling two-cylinders can reduce the fuel consumption by 15 per cent at a bmep of 2 bar. Sandford *et al.* (1998) point out that when a four-cylinder engine is switched to two-cylinder operation, the switching should occur in about 120°ca window, just prior to exhaust valve opening. This results in both cylinders subsequently compressing and expanding burnt gases. Valve reactivation has to be with the same phasing, so that the exhaust stroke can then occur prior to induction recommencing.

Not surprisingly the timing of valve events also has an impact on exhaust emissions. For example, control of valve overlap can be used to control the level of exhaust residuals, thereby regulating the emissions of NO_x. Stone and Kwan (1989) review the potential of variable valve timing for controlling emissions.

Variable valve lift and duration were first introduced by BMW with the Valvetronic system in 2001 with a consequential reduction in spark ignition engine fuel consumption of 12 per cent (Unger *et al.*, 2008). Figure 7.26 shows how the lift, duration and phasing of the inlet valve can vary; the valve lift can typically be varied in the range of 0.7–12.5 mm. Since the inlet valve can also be independently optimised for maximum power, this too can be improved.

Figure 7.27 shows how the reduction in pumping loss depends on the inlet valve lift strategy. With conventional throttling the inlet pressure in the cylinder falls rapidly towards the manifold pressure, and even with a late inlet valve closure strategy this might be 0.5 bar at a bmep of 2 bar. This leads to a large pumping loss compared to either of the low lift inlet valve strategies. By adopting a larger lift and earlier closure ('a' in figure 7.27), the cylinder pressure remains closer to atmospheric pressure and the pumping loss is smaller than if the lift alone was reduced ('b'). Fujita *et al.* (2008) also report that the low valve lift strategies improve

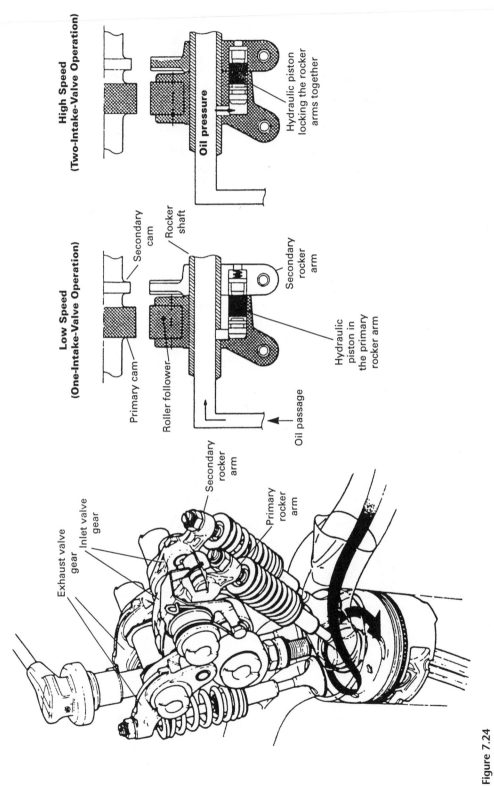

Figure 7.24
The Honda VTEC variable valve control system (adapted from Horie and Nishizawa, 1992).

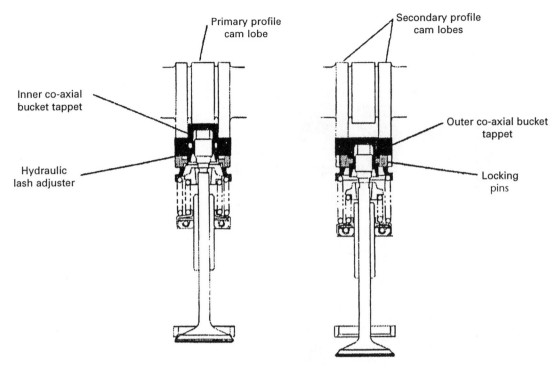

Figure 7.25
Trilobe tappet and cam for profile switching (courtesy of Lotus Engineering).

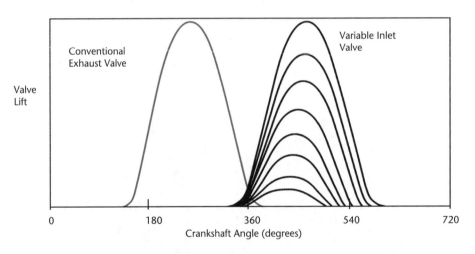

Figure 7.26
Lift, duration and phasing control of the inlet valve by the BMW Valvetronic system. Adapted from Unger *et al.* (2008).

mixture preparation during cold start operation of port fuel injection engines; this is because of the very high velocities past the valve seat when the valve lift is low. This improves the combustion stability so that ignition timing can be retarded more (to promote catalyst light-off), leading to the catalyst entry temperature being 300 K hotter after 15 s.

There are many systems for controlling the lift and duration

of the inlet valve event, and one approach is to use an intermediate rocker between the cam and the finger-follower (Unger *et al.*, 2008). The example in figure 7.28 is the Mechadyne VLD system.

The Mechadyne VLD system uses a spring-loaded rocker system to sum the lifts from two cam profiles (inner cam and outer cams) using rollers that are mounted on a rocker. The

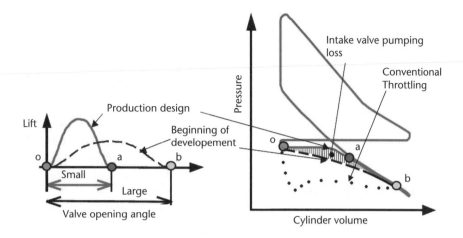

Figure 7.27
Schematic to show the reduction in pumping loss associated with different reduced inlet valve lift strategies. Adapted from Harada *et al.* (2008).

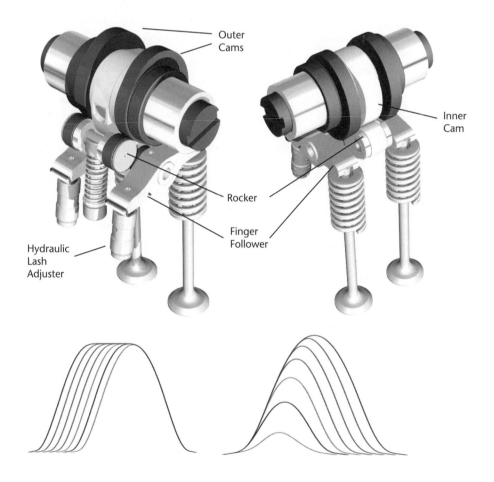

Figure 7.28
The Mechadyne Variable Lift and Duration (VLD) system that gives control of phase and lift. With acknowledgement to Mechadyne International.

rocker is supported on a shaft that is mounted on a pair of finger-followers. The finger-followers pivot on hydraulic lash adjusters that control the clearances in the system. By changing the phase between these two cam profiles, the valve lift and duration can be altered simultaneously. The two cam profiles are mounted on separate camshaft phasers, and this allows continuous control of their relative phase and the lift.

7.5 | Unsteady compressible fluid flow

Figure 7.29 shows that even for a single-cylinder engine operating at low speed (1000 rpm), there are significant pressure variations in the manifolds. These results are from the Rover K4 single-cylinder engine operating at 1000 rpm and half-load. The cylinder pressure has been fixed to an absolute datum by the barrel transducer, and this is explained in chapter 14, section 14.2.5. Inlet valve closure is at 52°ca, and from this time onwards the inlet pressure rises steadily towards atmospheric pressure. The inlet valve opens at 528°ca, and once the induction stroke has started, the inlet pressure falls rapidly to a minimum of about 0.5 bar. The exhaust valve closes at 552°ca, so there is 24°ca of valve overlap, during which time the inlet pressure is close to atmospheric pressure, and the exhaust and cylinder pressures are both below atmospheric pressure, because of wave-action effects in the exhaust. During most of the induction stroke there is about a 0.1 bar pressure drop across the inlet valve. The exhaust valve opens at 308°ca and the cylinder pressure at this point is about 2.2 bar, and this will lead to initially choked flow through the exhaust valve.

The pressure in the exhaust system rises to a maximum of 1.17 bar at 340°ca, and this pressure wave propagates down the exhaust system at the speed of sound. The exhaust pressure transducer is mounted 330 mm away from the exhaust valve, and for this operating point it takes the pressure wave 5°ca to travel from the exhaust valve to the transducer. When this pressure wave reaches an opening, it is reflected back as a rarefaction wave, and the minimum pressure is detected by the transducer at about 395°ca. This rarefaction wave also reduces the pressure in the cylinder. During the time that the exhaust valve is open, the cylinder pressure is always slightly higher than the exhaust system pressure, until the valve overlap period. During the valve overlap period the inlet pressure is above that of the cylinder and exhaust port, so this will tend to reduce the level of exhaust residuals in the cylinder. However, if this was a multi-cylinder engine with a conventional throttle at the inlet to the engine, then the inlet manifold pressure would fluctuate less and be much closer to 0.5 bar. In which case, the inlet pressure would be lower than the cylinder pressure during the valve overlap period, and exhaust gases would flow into the

inlet manifold, for subsequent return to the cylinder to become residuals.

Winterbone and Pearson (1999 and 2000) provide authoritative treatments on the modelling of unsteady flow in induction and exhaust systems, and these books have significantly greater coverage than Winterbone (1990a and 1990b) and are an extension of the work reported in Benson (1982). There are numerous commercial codes for modelling unsteady fluid flow, and well-known examples include GT-Power (Gamma Technologies), WAVE (Ricardo) and BOOST (AVL). These packages can link into CFD programs for more detailed modelling of in-cylinder processes, but include simpler in-cylinder modelling based on zero-dimensional or phenomenological modelling (see chapter 11); the geometric models that are needed can be incorporated from CAD systems. The main purpose is obviously the design of the intake and exhaust systems, and this includes turbocharger or supercharger modelling, and predictions of noise and sound quality (audio files can be generated). The predictions of combustion and emissions are based on simplified sub-models, but links can be made to chemical kinetics packages (such as CHEMKIN). These codes are also useful for control strategy development, real-time modelling of the engine for 'hardware in the loop' control systems and the modelling of complete vehicles. What follows here is an introduction to the physics of one-dimensional unsteady flow.

The derivation of the results for one-dimensional unsteady compressible fluid flow can be found in many books on compressible flow, for example Daneshyar (1976); the main results will be quoted and used in this section.

Unsteady flow is treated by considering small disturbances superimposed on a steady flow. For analytical simplicity the flow is treated as adiabatic and reversible, and thus isentropic. The justification for reversibility is that, although the flow may not in fact be frictionless, the disturbances or perturbations are small. Further, the fluid properties are assumed not to change across the perturbation.

By considering the conservation of mass, momentum and energy, the propagation speed for a perturbation or small pressure wave is found to be the speed of sound, c.

For a perfect gas

$$c = \sqrt{\left(\frac{\gamma p}{\rho}\right)} = \sqrt{\gamma R T} \tag{7.55}$$

In a simple wave of finite amplitude, allowance can be made for the change in properties caused by the change in pressure. In particular, an increase in pressure causes an increase in the speed of propagation. A simple wave can be treated by considering it as a series of infinitesimal waves, each of which is isentropic. If the passage of a wave past a point increases the pressure then it is a compression wave, while if it reduces the pressure then it is an expansion or rarefaction wave – see figure 7.30.

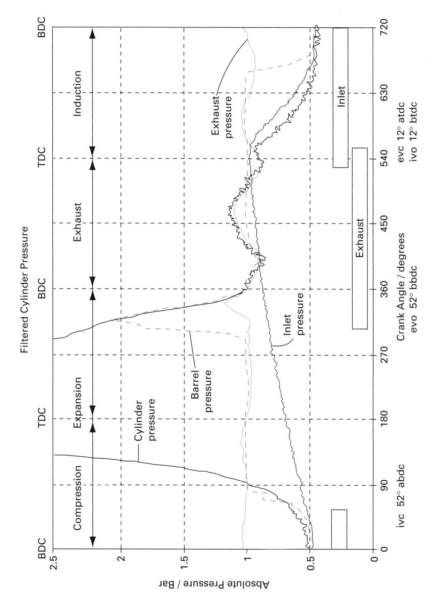

Figure 7.29

Absolute pressures in the Rover K4 single-cylinder research engine at a speed of 1000 rpm.

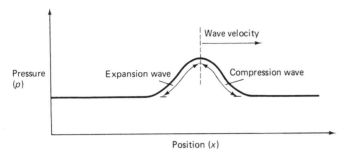

Figure 7.30

Simple pressure wave (adapted from Daneshyar, 1976, © 1968 Pergamon Press Ltd).

Recalling that an increase in pressure causes an increase in the propagation speed of a wave, then a compression wave will steepen, and an expansion wave will flatten. This is shown in figure 7.31 for a simple wave at four successive times. When any part of the compression wave becomes infinitely steep ($\partial p / \partial x = \infty$) then a small compression shock wave is formed. The shock wave continues to grow and the simple theory is no longer valid, since a shock wave is not isentropic. If the isentropic analysis was valid then the profile shown by the broken line would form, implying two values of a property (for example, pressure) at a given position and time.

The interactions between waves and boundaries can be determined from position diagrams and state diagrams. The state diagrams are not discussed here, but enable the thermodynamic properties, notably pressure, to be determined. The theory and use of state diagrams in conjunction with position diagrams is developed by books such as Daneshyar (1976).

Position diagrams are usually non-dimensional, using the duct length (L_D) to non-dimensionalise position, and the speed of sound (c) and duct length to non-dimensionalise time. The position diagram for two approaching compression waves is shown in figure 7.32; the slopes of the lines correspond to the local value of the speed of sound. The values of the pressure would be obtained from the corresponding state diagram.

For internal combustion engines it is important to know how waves behave at boundaries. There are open ends, junctions and closed ends. Examples of these are respectively an exhaust pipe entering an expansion box, a manifold, and a closed exhaust valve in another cylinder.

At a closed end, waves are reflected in a like sense; that is, a compression wave is reflected as a compression wave. This derives from the boundary condition of zero velocity.

At an open end there will be a complex three-dimensional motion. The momentum of the surrounding air from the three-dimensional motion causes pressure waves to be reflected in an unlike sense at an open (or constant-pressure) boundary. For example, a compression wave will be reflected as an expansion wave.

A thorough simulation of the flows in the inlet or exhaust system of a reciprocating engine would need a solution to the three-dimensional unsteady compressible fluid flow equations, taking due account of turbulence effects. Typically, the three-dimensional flow equation can only be solved for steady incompressible flow, and the solution of unsteady compressible flow is restricted to one-dimensional flow.

The pressure waves in the induction and exhaust systems of an engine are of such an amplitude that they cannot be treated as sound waves. Sound waves are of sufficiently low amplitude that the local conditions are not affected, and all the waves propagate at the speed of sound. With pressure waves, the local pressure level affects the propagation velocity of that part of the wave. If any compression or expansion process associated with the passage of the wave is isentropic, and the entropy is the same anywhere in the system, then the flow is homentropic. Thus, if the flow is frictionless, adiabatic and of fixed composition, then for a constant cross-sectional area pipe, the homentropic flow equations are as presented by Benson (1982) or Heywood (1988):

continuity

$$\frac{1}{\rho}\frac{\partial \rho}{\partial t} + \frac{u}{\rho}\frac{\partial \rho}{\partial x} + \frac{\partial u}{\partial x} = 0 \qquad (7.56)$$

momentum

$$\frac{1}{\rho}\frac{\partial p}{\partial x} + \frac{\partial u}{\partial t} + u\frac{\partial u}{\partial x} = 0 \qquad (7.57)$$

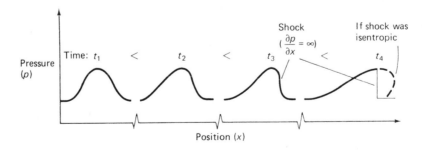

Figure 7.31

Spreading of an expansion wave, steepening of a compression wave (adapted from Daneshyar, 1976, © 1968 Pergamon Press Ltd).

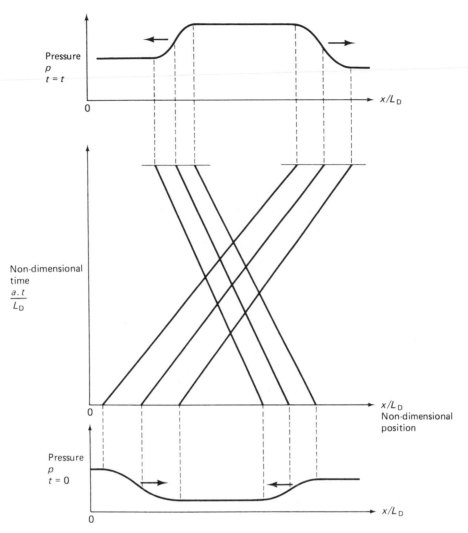

Figure 7.32
Interaction of two approaching compression waves (adapted from Daneshyar, 1976, © 1968 Pergamon Press Ltd).

energy and equation of state

$$c^2 = \left(\frac{\partial p}{\partial \rho}\right)_s \tag{7.58}$$

$$\frac{p}{p_{\text{ref}}} = \left(\frac{c}{c_{\text{ref}}}\right)^{2\gamma/(\gamma-1)} \tag{7.59}$$

where u = local flow velocity (m/s)
  x = position along pipe (m)
  ρ = local density (kg/m^3)
  p = local pressure (N/m^2).

These hyperbolic partial-differential equations can be solved by a number of techniques, and the first widely used approach was the method of characteristics.

In the method of characteristics, Riemann transformed the partial differential equations into total differential equations, which apply along so-called characteristic lines. The pressure waves propagate upstream and downstream relative to the flowing gas, at the local speed of sound. The local condition in the flow at any time and position are found from the so-called state diagram.

The method of characteristics was solved graphically by de Haller (1945), and this was first applied to an engine by Jenny (1950). The first numerical solution was by Benson *et al.* (1964), who used a rectangular grid in the x and t directions. A pipe is divided into sections joined at junctions that are represented by mesh points. Initial values have to be assigned to the mesh points at $t = 0$, and the values of the Riemann variables at each mesh point can be determined for subsequent time steps.

Finite difference methods have also been applied to the direct solution of the partial differential equations (7.56)–(7.59), and the one-step Lax–Wendroff technique is often used. Heywood (1988) summarises the methodology of the Lax–Wendroff solution, and points out that some form of damping (artificial viscosity) is required to prevent instabilities developing in the solution. However, the Lax–Wendroff method is significantly faster than the method of characteristics. A comparison of the Lax–Wendroff technique with the method of characteristics, by Poloni *et al.* (1987), has shown that both techniques give very similar results.

An alternative finite difference technique that has been applied by Iwamota and Deckker (1985) is the Random Choice Method: this does not require the addition of artificial viscosity terms to prevent instability in the solution. The method of characteristics can be extended to non-homentropic flow (to account for irreversibilities, such as heat transfer and fluid friction), by including a path line. This allows particle-related properties (namely temperature and entropy, but not pressure) to travel at the local fluid velocity.

Boucher and Kitsios (1986) describe how transmission line modelling techniques can be used to predict unsteady compressible flow. The right- and left-going waves in a transmission line undergo only a time delay in the line, and are otherwise unchanged in character. A fluid system is then modelled as a series of transmission lines and lumped elements (such as junctions, volumes or nozzles). The losses are most readily incorporated at the junctions, so as not to affect the transmission line solution. Boucher and Kitsios (1986) consider several applications, including the Helmholtz resonator, for which there was very close agreement in the prediction of the natural frequency.

However, it must be remembered that all solutions for unsteady flow depend on being able to define the initial and boundary conditions for the system. There are some simple boundaries, such as open pipes, closed pipes and pipe junctions; but many junctions are highly complex – an opening or closing valve, and the entry volute to a turbocharger.

In conclusion, the methods for solving unsteady one-dimensional flow may be complex, but the calculation methods are well established. It has also been shown by Poloni *et al.* (1987) that when the same boundary and initial conditions are used, then both the method of characteristics and the Lax–Wendroff technique give consistent answers. However, a more significant problem is the definition of the initial conditions and the boundary conditions. Some of the common assumptions and a discussion of their justification have been presented by Winterbone (1990a), who also identifies pipe boundaries (such as turbocharger entries and pipe junctions in manifolds) that require further study.

Owing to the complexity of solving these equations, the methods are probably better suited to analysis rather than design. In the next section on manifold design, acoustic methods will be described that enable resonating systems to be designed. However, these methods do not enable the volumetric efficiency to be quantified, so once a system has been identified, its performance can be evaluated by an unsteady-flow analysis.

7.6 | Manifold design

7.6.1 General principles

In designing the manifold system for a multi-cylinder engine, advantage should be taken of the pulsed nature of the flow. The system should avoid sending pulses from the separate cylinders into the same pipe at the same time, since this will lead to increased flow losses. However, it is sensible to have two or three cylinders that are out of phase ultimately connected to the same manifold. When there is a junction, a compression wave will also reflect an expansion wave back; this is shown in figure 7.33. If the expansion wave returns to the exhaust valve at the end of the exhaust valve opening, it will help to scavenge the combustion products; if the inlet valve is also open then it will help to draw in the next charge. Obviously the cancellation of compression and expansion waves must be avoided.

A typical exhaust system for taking advantage of the pulsations from a four-cylinder engine is shown in figure 7.34. The pipe length from each exhaust port to its first junction is the same, and the pipes will be curved to accommodate the specified lengths within the given distances. The length adopted will influence the engine speed at which

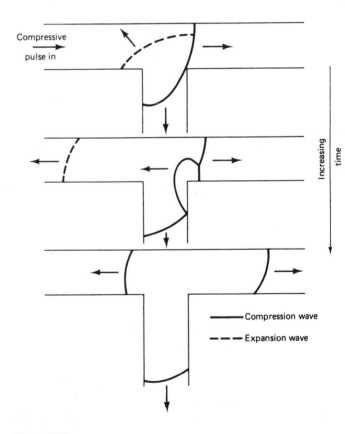

Figure 7.33
Pulsed flow at a junction (with acknowledgement to Annand and Roe, 1974).

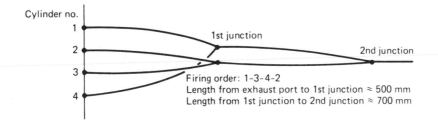

Figure 7.34
Exhaust system for a four-cylinder engine.

maximum benefit is obtained. The manifold is such that for the given firing order (cylinders 1–3–4–2), the pressure pulses will be out of phase.

Consider the engine operating with the exhaust valve just opening on cylinder 1. A compression wave will travel to the first junction; since the exhaust valve on cylinder 4 is closed an expansion wave will be reflected back to the open exhaust valve. The same process occurs 180° later in the junction connecting cylinders 2 and 3. At the second junction the flow is significantly steadier and ready for silencing.

Six-cylinder engines use a three-into-one connection at the first junction, with or without a second junction. Eight-cylinder engines can be treated as two groups of four cylinders. If a four-into-one system is used the benefits from pressure pulse interactions occur at much higher engine speeds. The choice of layout always depends on the firing order, and will be influenced by the layout of the engine – whether the cylinders are in-line or in 'V' formation. These points are discussed more fully by Annand and Roe (1974) and Smith (1968).

Induction systems are generally simpler than exhaust systems, especially for engines with fuel injection. The length of the induction pipe will influence the engine speed at which maximum benefit is obtained from the pulsating flow. The lengths shown in figure 7.35 are applicable to engines with fuel injection or a single carburettor per cylinder. In most cases it is impractical to accommodate the ideal length.

Inlet manifolds are usually designed for ease of production and assembly, even on turbocharged engines. When a single carburettor per cylinder is used, the flow pulsations will cause a rich mixture at full throttle as the carburettor will feed fuel for flow in either direction. In engines with a carburettor supplying more than one cylinder the flow at the carburettor will be steadier because of the interaction between compression and expansion waves. The remainder of this section will deal with manifolds for carburetted spark ignition engines.

In carburetted multi-cylinder engines the carburettor is usually connected by a short inlet manifold to the cylinder head. Although a longer inlet passage would have some advantages for a pulsed flow, these would be more than offset by the added delay in response to a change in throttle, caused by fuel lag or 'hold up'. For these reasons it is difficult to devise a central carburettor arrangement for engines with horizontally opposed cylinders.

Even with a four-cylinder in-line engine it is difficult to design a satisfactory single carburettor installation. The manifold shown in figure 7.36a will have a poor volumetric efficiency but can be arranged to give a uniform mixture distribution. In comparison, the manifold in figure 7.36b will have a good volumetric efficiency, but is unlikely to have a uniform mixture distribution. If a twin choke carburettor or two carburettors are fitted to this engine, two of the possible manifold arrangements are shown in figure 7.37. The first arrangement has uniform inlet passages and evenly pulsed flow for the common firing order of 1–3–4–2. The second system (figure 7.37b) is more widely used as it is simpler and equally effective. The pulsed flow will be uneven in each carburettor, so that there will be a tendency for maldistribution with each pair of cylinders. However, as the inlet passages are too short to benefit from any pulse tuning, the effect is not too serious, and is further mitigated by the balance pipe. The balance pipe usually contains an orifice, and the complete geometry has to be optimised by experiment. The same considerations apply to other multi-cylinder engine arrange-

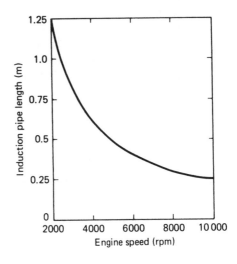

Figure 7.35
Induction pipe length for benefits from pulsating flow (based on Campbell, 1978).

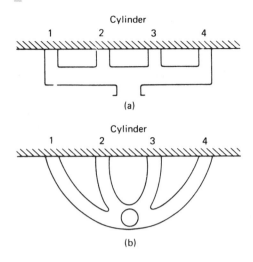

Figure 7.36
Inlet manifolds for single carburettor four-cylinder engine.

ments. Finally, it should be remembered that the throttle plate can have an adverse effect on the mixture distribution. For example, in figure 7.37a the throttle spindle axes should be parallel to the engine axis.

7.6.2 Acoustic modelling techniques

Spark ignition engines with multi-point or direct fuel injection provide scope for applying induction tuning. Induction tuning is also being applied to diesel engines, even those with turbochargers, since it is possible to design the induction system to improve the low-speed volumetric efficiency (the operating point where turbochargers are least effective).

It has already been explained (section 7.5) how wave modelling techniques can be used to analyse the induction system performance. But it is appropriate to explain here how acoustic models can be used to design induction systems. It is well established that induction tuning or 'ramming' can lead to an improvement in volumetric efficiency. The increase in the trapped mass of air allows more fuel to be burnt, and if the air/fuel ratio is maintained constant, then there is an increase

in power output. Typically an increase in output of 10–20 per cent can be obtained, with an accompanying reduction in the specific fuel consumption of 1–2 per cent. This reduction in fuel consumption occurs since the frictional losses do not increase in direct proportion to the work output. The gains are obviously dependent on valve timing, the interaction with the exhaust process, the effectiveness of the original induction system, and the speed at which the gains in volumetric efficiency are being sought.

The two ways of considering a tuned induction system are as an organ pipe or as a Helmholtz resonator. A Helmholtz resonator is shown in figure 7.38 and the resonant frequency (f_H) is given by

$$f_H = \frac{c}{2\pi}\sqrt{\left(\frac{A}{l \times V}\right)} \text{ (Hz)} \tag{7.60}$$

where c = speed of sound (m/s)
   A = pipe cross-sectional area (m^2)
   l = pipe length (m)
   V = resonator volume (m^3).

The way that these parameters are assigned to an engine and its induction system will be discussed later. However, even at this stage it is important to emphasise that when a system is tuned, the frequency of the oscillation is not necessarily the same as the frequency of the induction stroke ($N/120$, for a four-stroke engine, with N the engine speed (rpm)).

Alternatively the induction system can be modelled as an organ pipe, for which the fundamental resonant frequency (f_p) of a pipe closed at one end is given by

$$f_p = \frac{c}{4L} \text{ (Hz)} \tag{7.61}$$

and

$$L = l + 0.3d \text{ (m)}$$

where L = effective pipe length (m)
   l = pipe length (m)
   d = pipe diameter (m).

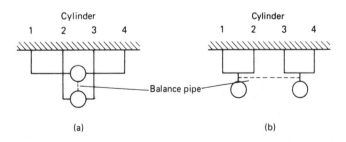

Figure 7.37
Twin carburettor arrangements for a four-cylinder engine.

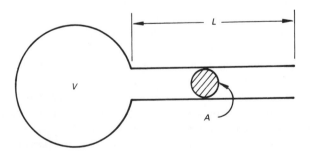

Figure 7.38
A Helmholtz resonator.

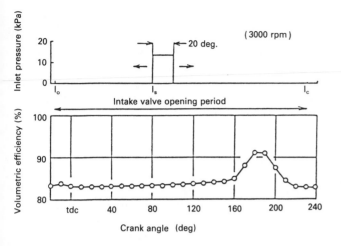

Figure 7.39
The effect of the timing of a hypothetical pressure pulse on the computed volumetric efficiency (Ohata and Ishida, 1982). [Reprinted with permission © 1982 Society of Automotive Engineers, Inc.]

Regardless of the model advocated, it is generally agreed that the tuned induction system benefits the volumetric efficiency, when a positive pressure pulse arrives at the inlet valve at or just before inlet valve closure. This has been demonstrated by Ohata and Ishida (1982) who modelled the effect of the inlet port pressure on the volumetric efficiency. First, Ohata and Ishida validated their cylinder filling models by using the recorded inlet port pressure as an input to their model. This gave very close agreement with the observed volumetric efficiency over the whole speed range of their four-cylinder four-stroke engine. Ohata and Ishida (1982) then considered hypothetical pressure pulse forms in the inlet system. This is

illustrated here by figure 7.39, and it can be seen that when the pressure pulse occurs around top dead centre, there is a negligible effect on the volumetric efficiency. However, when the pressure pulse arrives around bottom dead centre (ivc at about 40° abdc), then there is a significant gain in volumetric efficiency. There are many examples of tuned induction system reported in the literature, and there are almost as many different theories that have been contrived to explain the results. This literature (for organ pipe and Helmholtz systems) has been reviewed by Stone (1989c), so only the main conclusions will be summarised here.

When a pipe is connected to a single cylinder, the organ pipe theory can be used to define the resonant frequency of the pipe. Peaks in the engine volumetric efficiency are usually found when the pipe is resonating at close to 3, 4 or 5 times the engine cycle frequency. To maintain a resonating system when damping (viscosity) is present requires a forcing function, and this is provided by the part of the induction stroke in which a depression pulse is generated in the inlet port. The induction process period distorts the period of pressure oscillations in the inlet pipe. However, since the induction process period is ill-defined (the inlet valve being opened and closed at a finite rate), this is why resonance can occur at several engine speeds, none of which is an exact multiple of the pipe resonating frequency. Unfortunately, between the engine speeds at which resonance occurs there will be a minimum in the volumetric efficiency.

For multi-cylinder engines, Helmholtz resonator systems offer more versatile and compact induction tuning systems. In particular external resonating volumes can be used to introduce more than one degree of freedom. Figure 7.40 shows a tuned induction system with two degrees of freedom.

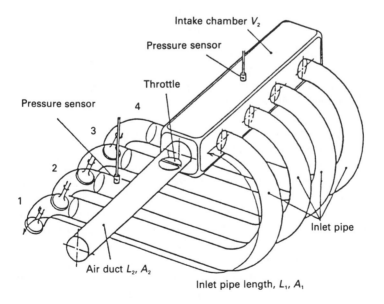

Figure 7.40
Physical arrangement of a tuned induction system with inlet pipes and a resonating volume (Ohata and Ishida, 1982). [Reprinted with permission © 1982 Society of Automotive Engineers, Inc.]

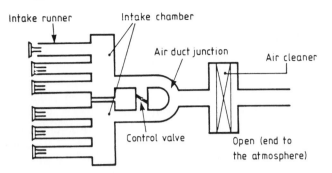

Figure 7.41
Schematic diagram of variable geometry intake manifold by Toyota (Winterbone, 1990b).

Equation (7.33) defines the resonating system for the geometry given in figure 7.40, where

$$ab L_1^2 V_1^2 \omega^4 - (ab + a + 1)(L_1/A_1) V_1 \omega^2 c^2 + c^4 = 0 \qquad (7.62)$$

for which there are two positive solutions that give the resonant frequencies:

$$f = \frac{c}{2\pi} \sqrt{ \left\{ \frac{(ab+a+1) \pm \sqrt{[(ab+a+1)^2 - 4ab]}}{2ab(L/A)_1 V_1} \right\} } \quad \text{Hz}$$

$$(7.63)$$

With reference to figure 7.40:

$$a = (L/A)_2/(L/A)_1 = \frac{L_2 A_1}{L_1 A_2}$$

$$b = V_2/V_1, \qquad V_1 = \frac{V_s}{2} \times \frac{r_v + 1}{r_v - 1}$$

$$c = \text{speed of sound}$$

where volume V_1 corresponds to the mean cylinder volume.

The use of resonating volumes also lends itself to engines with variable-geometry induction systems. Primary pipes of different lengths can be selected, resonating side limbs can be connected to the primary pipes, and resonating volumes can be connected together. Many practical examples are considered by Winterbone (1990b). A system that has been quite widely applied to six-cylinder engines is shown in figure 7.41. When the control valve is closed, the system has a lower resonant frequency than when the control valve is open.

A development of this system has been used on the Rover KV6 engine, for which the full-load operating modes and torque curve are shown in figure 7.42. Below about 2750 rpm both the connect valve and balance valve are closed, so that the engine behaves like two three-cylinder engines with long secondary pipes. Between 2750 and 3500 rpm the connect valve is opened, and above 3500 rpm the balance valve is also opened, to give the maximum torque at 4000 rpm, with a bmep of 12.1 bar. This compares with a maximum bmep of 12.2 bar obtained at 4500 rpm in the K16 engine with VVC. At full load the system operates to maximise the volumetric efficiency, while at part load the two valves are closed. This reduces the volumetric efficiency, so that wider open throttle settings are needed, thereby reducing the pumping losses and reducing the fuel consumption. The K series engine family is discussed further in chapter 15, section 15.2.

7.7 | Silencing

The human ear has a logarithmic response to the magnitude of the fluctuating pressures that are sensed as sound. The ear also has a frequency response, and is most sensitive to frequencies of about 1 kHz. The most effective approach to silencing is the reduction of the peaks, especially those in the most sensitive frequency range of the ear.

The noise from the engine inlet comes from the pulsed nature of the flow, and is modified by the resonating cavities in the cylinder and inlet manifold. A high-frequency hiss is also generated by the vortices being shed from the throttle plate. The inlet noise is attenuated by the air filter and its housing. In addition to its obvious role, the air filter also acts as a flame trap if the engine backfires.

Exhaust silencers comprise a range of types, as illustrated by Annand and Roe (1974) – see figure 7.43. In general, an exhaust system should be designed for as low a flow resistance as possible, in which case the constriction type silencer is a poor choice. Silencers work either by absorption, or by modifying the pressure waves in such a way as to lead to cancellation and a reduction in sound. Absorption silencers work by dissipating the sound energy in a porous medium. Silencers and their connecting pipes should be free of any resonances. Turbochargers tend to absorb the flow pulsations from the engine exhaust, but substitute a high-frequency noise generated by the rotating blades.

7.8 | Conclusions

To obtain the optimum performance from any internal combustion engine, great care is needed in the design of the induction and exhaust systems. Once the type and disposition of the valve gear have been decided, the valve timing has to be selected. The ideal valve behaviour is obviously modified by dynamic effects, owing to the finite mass and elasticity of the valve train components. The actual valve behaviour can be predicted by computer models. The valve timing will be determined by the application. The two extremes can be generalised as normal spark ignition engines or naturally aspirated compression ignition engines, and high output spark ignition engines or turbocharged compression ignition engines. The latter have the greater valve-opening periods. The chosen valve timing is also influenced by the design of the induction and exhaust passages.

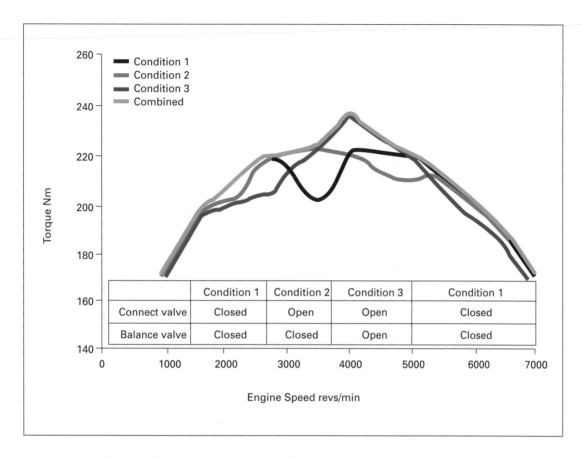

	Condition 1	Condition 2	Condition 3	Condition 1
Connect valve	Closed	Open	Open	Closed
Balance valve	Closed	Closed	Open	Closed

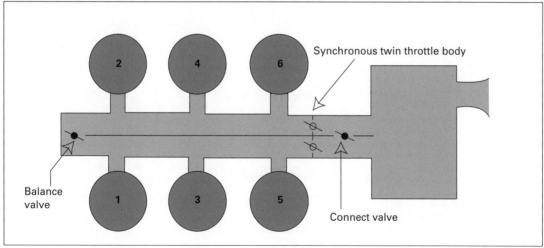

Figure 7.42
The full-load torque curves and operating modes of the Rover KV6 variable geometry induction system (adapted from Whiteley, 1996).

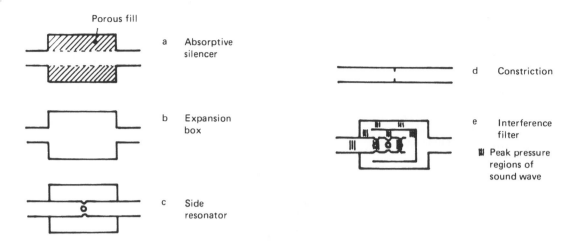

Figure 7.43
The basic silencer elements (with acknowledgement to Annand and Roe, 1974).

Successful design of the induction and exhaust processes depends on a full understanding of the pulsed effects in compressible flows. The first solution methods involved a graphical approach, but these have now been superseded by computer models. Computer models can also take full account of the flow variations during the opening and closing of valves, as well as interactions between the induction and exhaust systems. This approach is obviously very important in the context of turbochargers, the subject of chapter 10.

7.9 | Questions

7.1 Two possible overhead valve combustion chambers are being considered, the first has two valves and the second design has four valves per cylinder. The diameter of the inlet valve is 23 mm for the first design and 18.5 mm for the second design. If the second design is adopted, show that the total valve perimeter is increased by 60.8 per cent. If the valve lift is restricted to the same fraction of valve diameter, calculate the increase in flow area. What are the additional benefits in using four valves per cylinder?

7.2 Describe the differences in valve timing on a naturally aspirated diesel engine, a turbocharged diesel engine, and a high-performance petrol engine.

7.3 Devise an induction and exhaust system for an in-line, six-cylinder, four-stroke engine with a firing order of 1–5–3–6–2–4, using: (i) twin carburettors, (ii) triple carburettors.

7.4 During the exhaust blowdown process, sketch on a temperature–entropy diagram:

(a) the expansion of the first element of gas to leave the cylinder;
(b) the in-cylinder expansion;
(c) the expansion of an element of gas to leave the cylinder at pressure p.

Write down, in differential form, an equation for the energy associated with the expansion of an element of gas to leave the cylinder at pressure p. Then by integrating this equation confirm that the temperature of the fully mixed gas at the end of blowdown is given by

$$T_{2''} = (m_1 T_1 - m_2 T_2)/\gamma(m_1 - m_2)$$

where states 1 and 2 refer to before and after blowdown

7.5 A four-cylinder four-stroke engine is operating at a bmep of 2 bar. The pressure at inlet valve closure is 0.37 bar, and the fmep is 0.8 bar. Assuming that the engine gross indicated efficiency is unchanged, verify that disablement of the valves on two cylinders would reduce the fuel consumption by about 13 per cent compared with about an 8 per cent fuel saving if the fuel was not supplied to two cylinders. Assume that the gross imep is proportional to the pressure at inlet valve closure, and that the gross indicated efficiency is constant. State any additional assumptions that you make. By means of p–V diagram sketches, explain why disabling only the inlet or exhaust valves would have a fuel economy benefit comparable to fuel disablement to two cylinders.

CHAPTER 8

Two-stroke engines

8.1 | Introduction

The absence of the separate induction and exhaust strokes in the two-stroke engine is the fundamental difference from four-stroke engines. In two-stroke engines, the gas exchange or scavenging process can have the induction and exhaust processes occurring simultaneously. Consequently, the gas exchange processes in two-stroke engines are much more complex than in four-stroke engines, and the gas exchange process is probably the most important factor controlling the efficiency and performance of two-stroke engines.

Further complications with two-stroke engines are the way in which the gas flow performance is defined, and the terminology. The gas flow performance parameters are not always defined the same way, but the system described in section 8.2 is widely used. The terminology adopted here is shown in figure 8.1. The inlet port is sometimes called the transfer port, but this usage has been avoided here, since not all two-stroke engines use under-piston scavenging. The crankcase port can be called the inlet port, but the crankcase port is a more accurate description of its location and function.

Some background material on two-stroke engines can be found in Heywood (1988) and Taylor (1985a). The gas exchange processes are treated very comprehensively by Sher (1990). Hooper *et al.* (2011) provide a review of spark ignition and diesel engines, using crankcase, external and stepped piston scavenging that are intended for automotive, marine and defence applications. There is particular emphasis on the use of direct injection technology so as to lower both the fuel consumption and the unburned hydrocarbon emissions. Mention must also be made of the books written by Blair (1990 and 1996) that provide a very comprehensive treatment of the design, performance and modelling of two-stroke engines. The dynamic or wave effects in the inlet system and in particular the exhaust system have a profound effect on the gas exchange processes, but they are not discussed here, as their action is no different from their action in four-stroke engines.

Two-stroke engines can be either spark ignition or compression ignition, and currently the largest and smallest internal combustion engines utilise two-stroke operation. The smallest engines are those used in models, for which compression ignition (perhaps assisted by a glow plug for starting) is more common than spark ignition. The output of these engines is frequently much less than 100 W. Small two-stroke engines with outputs in the range of approximately 1–100 kW are usually naturally aspirated with spark ignition. The typical applications are where low mass, small bulk and high-power output are the prime considerations – for example motor cycles, outboard motors, chainsaws, small generators etc. These engines require the incoming charge to be pumped in, and this is achieved most simply by using the underside of the piston and the crankcase as the pump, a method adopted by Joseph Day in 1891. This arrangement is

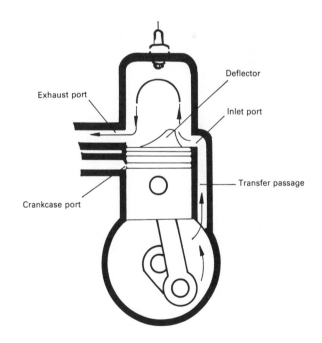

Figure 8.1
The elements of a two-stroke engine with under-piston or crankcase scavenging (adapted from Taylor, 1985a).

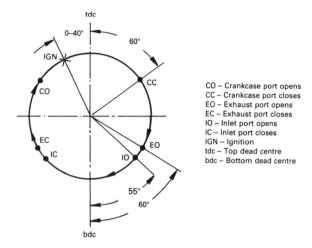

Figure 8.2
The timing diagram for a two-stroke engine.

still commonly used, and is illustrated by figure 8.1. Figure 8.2 shows the corresponding timing diagram for this under-piston or crankcase-scavenged engine. Figure 8.3 shows the calculated pressures for a motored (that is, non-fired) two-stroke engine. If the engine was fired, then the main difference would be a higher cylinder pressure at exhaust port opening. This would then lead to larger amplitude pressure oscillations in the exhaust system. Figure 8.4 is a representation of the charging process, and some of the terms involved.

With reference to figures 8.1, 8.2 and 8.3, the following processes occur in a two-stroke engine, at times determined by the piston covering and uncovering ports in the cylinder wall:

1 At about 60° before bottom dead centre (84°atdc in figure 8.3), the piston uncovers the exhaust port and exhaust blowdown occurs, such that the cylinder pressure approaches the ambient pressure (EO). This is the end of the power stroke.

2 Some 5–10° later the inlet port is opened (IO), 120°atdc in figure 8.3, and the charge compressed by the underside of the piston in the crankcase is able to flow into the cylinder. With this loop scavenge arrangement (figure 8.1), the incoming charge displaces and mixes with the exhaust gas residuals; some incoming charge will flow directly into the exhaust system.

3 The scavenge process ends with both the crankcase and cylinder pressure close to the ambient pressure level once the inlet port is closed (about 55° after bottom dead centre). Towards the end of the scavenge process, there can be a backflow of charge and exhaust gas residuals into the crankcase. The upward movement of the piston now reduces the pressure in the crankcase.

4 At about 60° after bottom dead centre (100°abdc in figure 8.3) the exhaust port is closed, and the charge is compressed by the upward motion of the piston.

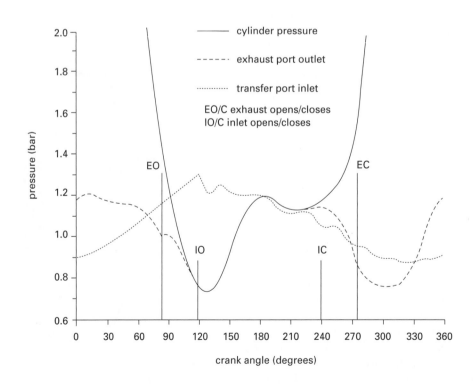

Figure 8.3
Calculated pressure histories for a motored under-piston scavenged two-stroke engine with a reed inlet valve (with acknowledgement to Tsai, 1994).

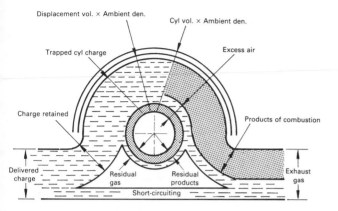

Figure 8.4
The composition of the flows in and out of a two-stroke engine and the cylinder. [Reprinted with permission from Sher, copyright 1990, Pergamon Press PLC).]

5 At about 60° before top dead centre the pressure in the crankcase is significantly below the ambient pressure, and the crankcase port opens to allow the incoming charge to flow into the crankcase.

6 Ignition occurs typically within the range of 10–40° before top dead centre. Work is done by the engine on the gas until top dead centre, at which point the power stroke starts, and continues until the exhaust port opens.

7 The crankcase port is closed at about 60° after top dead centre, but prior to this there will be some outflow of gas from the crankcase, as the crankcase pressure will have risen above the ambient pressure level around top dead centre.

The crankcase port system is frequently replaced by a reed valve connected directly to the crankcase. This simplifies the timing diagram (figure 8.2) as the crankcase port is eliminated. Instead the reed valve will open shortly after the inlet port is closed, and the reed valve will close when the piston is near top dead centre. Furthermore, the backflow in the crankcase port as the piston moves down the cylinder is eliminated. A reed valve was used by the engine modelled in figure 8.3, and this engine also had a power valve – a means of varying the exhaust port opening (and closing) from 84° to 101° away from top dead centre. Another alternative is to use a disc valve on the crankshaft; this is not quite the same as using a crankcase port, since the disc valve timing is not necessarily symmetric about top dead centre.

Two undesirable features of the two-stroke engine are the mixing of the incoming charge with the exhaust residuals, and the passage of the charge directly into the exhaust system. The fuel loss due to these shortcomings is largely eliminated in diesel engines (not considered here are model diesel engines), as they are invariably supercharged and there is in-cylinder fuel injection.

In order to discuss the gas exchange performance of two-stroke engines, it is necessary to define several gas flow parameters.

8.2 | Two-stroke gas flow performance parameters

It must be remembered that many engines are supercharged, and that for two-stroke engines it is not possible to make direct measurements for all the flow performance parameters. There are also different ways of defining the performance parameters, for example, cylinder volume can be defined as the swept volume, or as the trapped volume – the volume at the beginning of the compression process when the transfer port is closed.

The following definitions follow the SAE recommended practice (SAE Handbook, updated annually), and the term 'air' should be interpreted as 'mixture' for engines with external mixture preparation:

$$\text{delivery ratio (or scavenge ratio)} \quad \lambda_d = \frac{\text{mass of delivered air}}{\text{swept volume} \times \text{ambient density}} = \frac{M_i}{M_o} \tag{8.1}$$

(Ambient can refer to atmosphere conditions or, for the case of a supercharged engine inlet conditions, the delivery ratio can be calculated from direct measurements.)

$$\text{scavenging efficiency, } \eta_{sc} = \frac{\text{mass of delivered air retained}}{\text{mass of trapped cylinder charge}} = \frac{M_a}{M_a + M_b} \tag{8.2}$$

$$\text{trapping efficiency, } \eta_{tr} = \frac{\text{mass of delivered air retained}}{\text{mass of delivered air}} = \frac{M_a}{M_i} \tag{8.3}$$

The trapping efficiency is a measure of how much air flows directly into the exhaust system (short-circuiting) and how much mixing there is between the exhaust residuals and the air:

$$\text{charging efficiency, } \eta_{ch} = \frac{\text{mass of delivered air retained}}{\text{swept volume} \times \text{ambient density}} = \frac{M_a}{M_o} \tag{8.4}$$

Charging efficiency, trapping efficiency and delivery ratio are clearly related (equations 8.1, 8.3 and 8.4):

$$\eta_{ch} = \lambda_d \eta_{tr} \tag{8.5}$$

$$\text{relative charge (or volumetric efficiency)} \quad \lambda_c = \frac{\text{mass of trapped cylinder charge}}{\text{swept volume} \times \text{ambient density}} = \frac{M_a + M_b}{M_o} \tag{8.6}$$

The relative charge is an indication of the degree of super-charging, and it is the ratio of the scavenging efficiency and the charging efficiency. From equations (8.2), (8.4) and (8.6):

$$\lambda_d = \eta_{ch}/\eta_{sc} \tag{8.7}$$

If a sampling system could be used to measure the composition of the trapped charge, its pressure and temperature, and the trapped charge was homogeneous, then the trapped mass of the air and the residuals could be calculated, and all the performance parameters could be evaluated. As this is not practical, a variety of experimental techniques have been developed to evaluate the gas exchange performance of two-stroke engines, and these are discussed in section 8.5.

For two-stroke spark ignition engines with under-piston or crankcase scavenging, then typical scavenge performance results are

scavenge efficiency	$0.7 < \eta_{sc} < 0.9$
trapping efficiency	$0.6 < \eta_{tr} < 0.8$
charging efficiency	$0.5 < \eta_{ch} < 0.7$
delivery ratio	$0.6 < \lambda_d < 0.95$

8.3 | Scavenging systems

Scavenging is the simultaneous emptying of the burned gases and the filling with the fresh air or air/fuel mixture (operation 2 in the description of the two-stroke system in section 8.1). Ideally, the fresh charge would solely displace the burned gases, but in practice there is some mixing. There are many different scavenge arrangements, and some of the more common systems are shown in figure 8.5.

With the cross-scavenge arrangement (figure 8.5a) the charge could flow directly into the exhaust system, but this tendency is reduced by the deflector on the piston. The troublesome deflector on the piston was avoided by the loop scavenging system (figure 8.5b) with the exhaust ports above the inlet ports. The incoming air or mixture is directed towards the unported cylinder wall, where it is deflected upwards by the cylinder wall and piston. A modified form of loop scavenging was devised by Schnurle in 1920 (figure 8.5c), in which two pairs of inlet ports are located symmetrically around the exhaust ports. With this system, the flow forms a 'U'-shaped loop.

The final two types of scavenging system both employ a uniflow system, either by means of exhaust valve(s) (figure 8.5d), or with an opposed piston arrangement (figure 8.5e). Both these systems are particularly suited to diesel engines, since the inlet ports can be arranged to generate swirl. Swirl is very important in promoting effective diesel engine combustion, yet if swirl was used with either cross or loop scavenging, then there would be significant mixing of the inlet flow with the burned gases.

Figure 8.6 shows different scavenge pumping arrangements for two-stroke engines. The simplest arrangement for pump-

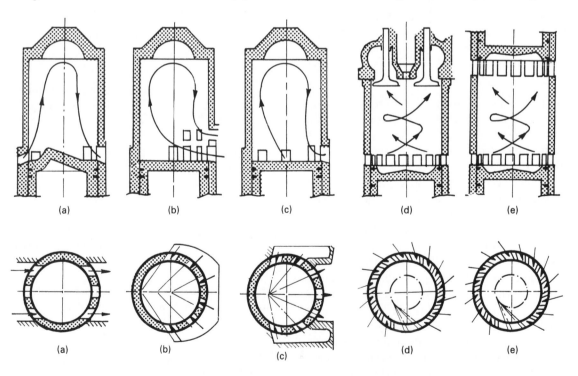

Figure 8.5
The different scavenging arrangements and the associated port geometry for two-stroke engines: (a) cross-scavenging; (b) loop scavenging; (c) Schnurle loop scavenging; (d) uniflow scavenging with poppet exhaust valves; (e) uniflow scavenging with opposed pistons (adapted from Heywood, 1988).

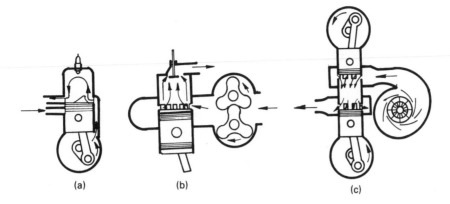

Figure 8.6
Scavenging pump arrangements: (a) under-piston or crankcase; (b) positive displacement (Roots blower); (c) radial compressor or fan (Taylor, 1985a).

ing the inlet gases into the cylinder is the under-piston pumping that has already been described. The swept volume of this pumping system is the same as that of the engine cylinder, yet better scavenging would be achieved with a larger displacement pump. The term 'blower' is used to describe here a low pressure ratio pump or fan that maintains close to atmospheric conditions in the cylinder. Supercharging refers to a higher pressure ratio system that generates in-cylinder pressures significantly above the atmospheric pressure. The supercharger can be a positive displacement device (such as a screw compressor) and/or a non-positive displacement device such as a radial or centrifugal compressor. These pumps can be driven from the crankshaft or by an exhaust gas turbine – that is turbocharging. Such systems are usually employed on compression ignition engines for non-automotive applications; they will not be discussed any further here.

8.4 | Scavenge modelling

The two simplest models for scavenging provide an upper and lower bound for the scavenging performance of an engine. They are the:

Perfect displacement model   – which assumes no mixing of the incoming gases with the burned gases.

Complete mixing   – which assumes the incoming gases mix entirely and instantaneously with the burned gases.

The performance of both models is shown in figure 8.7. In addition, this section will include a brief discussion of more complex scavenging models.

8.4.1   Perfect displacement scavenging model

The perfect displacement model assumes:

(i)   no mixing of the incoming gases with the burned gas

(ii)   the process occurs at a constant cylinder volume and pressure

(iii)   there is no heat transfer from either the burned gas or the cylinder walls

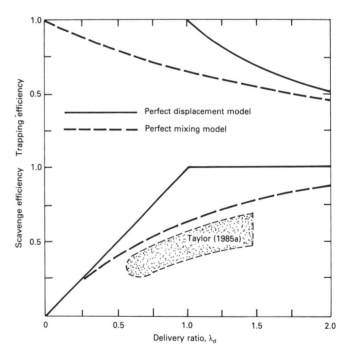

Figure 8.7
The trapping efficiency and scavenging efficiency as a function of the delivery ratio, for the perfect displacement and perfect isothermal mixing scavenging models. Scavenge efficiency data have been added from Taylor (1985a) for compression ignition engines.

(iv) there are no differences in density.

There are two cases to consider. Firstly, with the delivery ratio less than unity (equation 8.1), so that not all the exhaust gases are displaced. All the air admitted will be trapped, and the trapped mass will include some burned gases. In algebraic terms:

$$\left.\begin{array}{lll} \text{delivery ratio,} & \lambda_d \leq 1 & \text{(equation 8.1)} \\ \text{then trapping efficiency,} & \eta_{tr} = 1 & \text{(equation 8.3)} \\ \text{and scavenging efficiency,} & \eta_{sc} = \lambda_d & \text{(equation 8.2)} \end{array}\right\}$$

$$(8.8)$$

Secondly, with the delivery ratio greater than unity, then all the burned gases will be displaced, so that the scavenging efficiency is unity. However, some of the admitted air will also be exhausted, so the trapping efficiency will be less than unity. By inspection, for

$$\left.\begin{array}{lll} \text{delivery ratio,} & \lambda_d > 1 & \text{(equation 8.1)} \\ \text{then scavenging efficiency,} & \eta_{sc} = 1 & \text{(equation 8.2)} \\ \text{and trapping efficiency,} & \eta_{tr} = \lambda_d^{-1} & \text{(equation 8.3)} \end{array}\right\}$$

$$(8.9)$$

8.4.2 Perfect mixing scavenging model

The perfect mixing model assumes:

(i) instantaneous homogeneous mixing occurs within the cylinder

(ii) the process occurs at a constant cylinder pressure and volume

(iii) the system is isothermal, with no heat transfer from the cylinder walls

(iv) the incoming air (or mixture) and the burned gases have equal and constant ambient densities.

Thus, in the time interval dt at time t, a mass element dm_e enters the cylinder, and an equal amount of mixture dm_1 leaves the cylinder. The gas leaving the cylinder comprises a mixture of previously admitted air and burned gases, of instantaneous composition x:

$$x = \frac{\text{instantaneous mass or volume of air in cylinder}}{\text{total mass or volume in cylinder}} = \frac{m_a}{M}$$

$$(8.10)$$

Differentiating equation (8.10) gives the changes in the mass of trapped air:

$$M dx = dm_a \qquad (8.11)$$

Also, the change in the mass of trapped air is the difference between the inflow and the outflow:

$$dm_a = dm_e - x dm_1 \qquad (8.12)$$

Combining equations (8.11) and (8.12), and recalling that $dm_e = dm_1$:

$$M dx = dm_e(1 - x)$$

or

$$\frac{dx}{(1 - x)} = \frac{dm_e}{M} \qquad (8.13)$$

The total mass in the cylinder M is a constant and equation (8.13) can be integrated, noting the limits:

(i) At the start of scavenge there is wholly burned gas, so $x = 0$ and $M_e = 0$.

(ii) At the end of scavenge $x = \eta_{sc}$ and $M_e = \lambda_d M$.

$$\int_0^{\eta_{sc}} \frac{dx}{1 - x} = \int_0^{\lambda_d M} \frac{dm_e}{M} \qquad (8.14)$$

Integrating gives

$$-\ln(1 - \eta_{sc}) = \lambda_d \qquad (8.15)$$

or

$$\eta_{sc} = 1 - \exp(-\lambda_d)$$

and from equations (8.5) and (8.7):

$$\eta_{tr} = 1/\lambda_d[1 - \exp(-\lambda_d)] \qquad (8.16)$$

The assumption of no density differences between the incoming gas and the cylinder contents at anytime, may appear to make this result of limited use, when there is obviously a large temperature difference between the inlet gases and the burned gas. However, Sher (1990) presents an analysis that recognises such a temperature difference, but obtains the same results for the charging efficiency, while the scavenging efficiency becomes

$$\eta_{sc} = \frac{T_s}{T_e} \eta_{ch} \qquad (8.17)$$

where T_s = temperature at the end of the scavenge period. Sher assumes no variation in the specific heat capacities at constant pressure and constant volume.

The prediction of the scavenging efficiency and the trapping efficiency are plotted in figure 8.7 as a function of delivery ratio, for both the perfect displacement and the perfect isothermal mixing models. If gas dynamic effects are ignored, then for an engine with under-piston scavenging the delivery ratio has to be less than unity. This implies the potential for a high trapping efficiency but a poor scavenging efficiency. For an engine with external mixture preparation, then the high trapping efficiency is important if fuel is not to flow directly into the exhaust system. However, the low scavenging efficiency implies high levels of exhaust gas residuals which necessitate the use of rich mixtures to ensure combustion.

With externally pumped scavenge systems, the delivery ratio can be greater than unity. This leads to high scavenge efficiencies but poorer trapping efficiencies. The fuel consumption and unburnt hydrocarbon emissions will only then be acceptable if in-cylinder fuel injection is used.

Unfortunately, there are two further phenomena that compromise yet further the scavenging and trapping efficiency. Firstly, there can be pockets of burned gas that are neither diluted nor displaced, thus lowering both the scavenging efficiency and the trapping efficiency. Secondly, there can be a flow directly from the inlet port to the exhaust port; this is known as short-circuiting. The short-circuit flow does not necessarily mix with or displace the burned gas, and again, both the scavenging efficiency and the trapping efficiency are reduced.

Incorporated into figure 8.7 is a region that envelops some experimental data obtained by Taylor (1985a) from some diesel engines in the 1950s. In all cases the scavenging efficiency is lower than that predicted by the mixing theory but as the engine speed was increased the discrepancy always reduced. However, Sher (1990) states that for modern engines the perfect mixing model underestimates the scavenging performance.

8.4.3 Complex scavenging models

Complex scavenging models divide the cylinder into two or more zones, and usually allow for a fraction of the flow to be short-circuited; this requires at least two empirical constants to be included in any such correlation. Typical assumptions are:

(i) the in-cylinder pressure and volume are invariant

(ii) there is no heat or mass transfer between the zones

(iii) in each zone the temperatures may be different but are uniform

(iv) no variation in gas properties.

Maekawa (1957) proposed an isothermal two-zone model in which the entering stream was divided into three parts: a flow that short-circuits to the exhaust port, a pure charge zone and a mixing zone.

Benson and Brandham (1969) also used a two-zone model, but the inlet flow was divided into only two flows: a flow that short-circuited to the exhaust port, and a flow that enters a mixing zone. The second zone was a burned gas zone that would be displaced through the exhaust port.

Benson (1977) extended his two zone model to include a third zone of fresh charge. He also divided the scavenging process into the three phases shown in figure 8.8:

(i) *Phase I* 95–155°atdc, exhaust is displaced and there is mixing at the interface of the incoming charge and the burned gas.

(ii) *Phase II* 155–200°atdc, mixing at the interface continues, but some of the incoming charge is now allowed to short-circuit through the exhaust port.

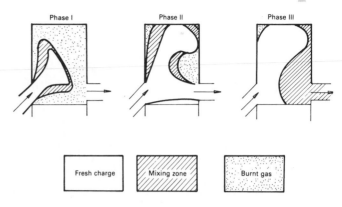

Figure 8.8
The Benson (1977) three-phase scavenging model.

(iii) *Phase III* 200–265°atdc, the short-circuit flow stops, and a homogeneous mixture of the fresh charge and the burned gas leaves the cylinder; as with the perfect mixing model, the composition is time dependent.

Sher (1990) has proposed a model that assumes the fresh charge content in the exhaust gas stream to be represented by a sigmoid ('S' type) curve. With this model, scavenging is represented by a combination of perfect displacement scavenging, followed by an isothermal mixing process. The selection of the shape factors for the sigmoid curve allows results to be produced that lie anywhere between the perfect displacement model and the perfect mixing model.

Numerical techniques have also been employed to model the scavenging process, and a short review is presented by Sher (1990). Typically, turbulence is modelled by the K–ε model; diffusion of mass heat and momentum all have to be incorporated, along with appropriate sub-models for jet mixing and flame propagation.

Figure 8.9 illustrates some CFD predictions for the flows in an under-piston scavenged two-stroke engine (Ng *et al.*, 1996). Experimental data were used to provide the boundary conditions at the ports, and the CFD study investigated the effect of the transfer port orientations on the scavenging process. The STAR-CD package was used with 23 040 cells, and the different transfer port configurations were modelled by specifying the jet velocities at inlet, with a uniform velocity profile. Piston movement was accounted for by a combination of mesh stretching/shrinking and cell layer addition/removal.

The transfer ports open at 120°atdc and, by 180°atdc in figure 8.9, the velocity vectors show that a loop vortex has been formed. When the transfer port closes (240°atdc) there is a small backflow from the exhaust port (which does not close until 270°atdc), and this enhances the vortex in the cylinder. The contours of the fresh charge concentration show how the overall purity of the cylinder contents increases during the scavenging process, and that significant stratification persists in the cylinder. These contours also show evidence of a vortex about the cylinder axis, since at 240°atdc, the region of

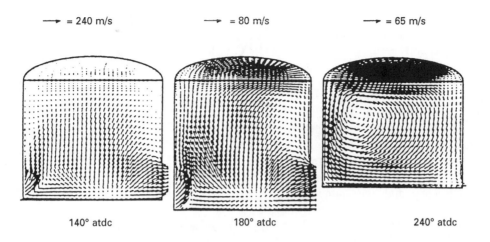

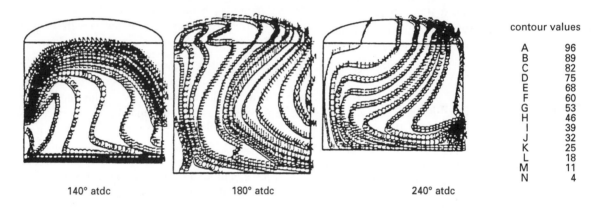

Figure 8.9
CFD predictions of scavenging in an under-piston scavenged two-stroke engine, showing the flow and fresh charge concentration (Ng *et al.*, 1996). © IMechE/Professional Engineering Publishing Limited.

highest purity has rotated to the exhaust port side. CFD models are thus able to predict the short-circuit loss of fresh charge, the scavenging efficiency and the trapping efficiency.

8.5 | Experimental techniques for evaluating scavenge and results for port flow coefficients

Experimental measurements have two vital roles. Firstly, to evaluate the performance of particular engines, and secondly, to enable empirical or numerical models of scavenging to be calibrated. Experiments can be conducted on either firing or non-firing engines. The techniques that can be applied to firing engines are more restricted, but they are discussed first here.

8.5.1 Firing engine tests

In section 8.2, in-cylinder sampling has already been suggested as a means of evaluating the trapped mass of the air and residuals. However, assumptions have to be made about how representative the sample of the gas is, and it is preferable to take a large sample which might affect the engine performance; it is also difficult to measure the pressure and temperature of the sample. A better technique (that can only be applied to engines with in-cylinder fuel injection) is the exhaust gas sampling system described by Taylor (1985a). The sampling valve shown in figure 8.10 is placed close to the exhaust port, and is connected to a receiver in which the pressure is controlled by the flows to a gas analyser and a bleed valve. The receiver pressure is set at a level such that the sampling valve opens during the exhaust blowdown period. This way the burned gas can be analysed, and the air/fuel ratio can be deduced. Since the quantity of fuel injected into

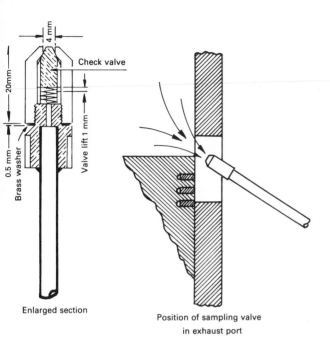

Figure 8.10
Exhaust sampling valve (Taylor, 1985a).

8.5.2 Non-firing engine tests

There are two types of non-firing test, those that use continuous steady flow and those that motor the engine or a model – these dynamic tests are usually limited to one cycle.

The simplest steady-flow tests are those with the piston fixed at bottom dead centre, and the cylinder head removed. Such tests are widely used to investigate the flow of air into the cylinder, the aim being to have a flow that interacts with the piston and cylinder to produce the maximum axial velocity at the opposite diameter to the exhaust port. The results can be plotted as velocity contours in the manner of figure 8.11. This technique was first proposed by Jante (1968) and its subsequent development has been considered by Sher (1990).

A range of flow measurement techniques has been applied to steady-flow tests with the cylinder head in place; these techniques include laser doppler anemometry, hot wire anemometry and the flow visualisation of scavenging by water and coloured salt solutions.

Sher (1990) also presents a comprehensive review of the flow visualisation and measurement techniques applied to dynamic tests of unfired engines – mostly for a single cycle. In the tests with liquids, only the parts of the cycle can be investigated in which the ports are open. None the less, such tests produce useful qualitative information. For example, short-circuiting flows and stagnant regions can be found. With gaseous systems, combinations of air, carbon dioxide, helium, ammonia and Freon-22 have been used to model the fresh charge and the burned gases. Some techniques give the instantaneous concentration of the species at discrete locations, although the techniques have only computed the mean concentration at the end of scavenge. Sher lists five criteria for similarity apart from geometric similarity, so it is probably impossible to obtain complete similarity between the model and the engine. Nor, of course, do these techniques provide any insight into the wave action effects that are so important in successful engine operation.

the engine is known, then it is possible to calculate the trapped mass of air. Since the mass of delivered air can be measured directly, then it is possible to calculate the delivery ratio, the scavenging efficiency, the trapping efficiency and the charging efficiency (equations 8.1–8.4).

Another technique for evaluating the scavenging process is the tracer gas method. In this method, a tracer gas is introduced continuously into the incoming air so as to give a fixed percentage of the tracer gas in the air. The tracer gas has to be selected so that it will be completely burnt at the combustion temperature, but not at the temperatures during compression or once the exhaust ports are open. Sher (1990) provides a comprehensive review of this technique, and the two most widely used gases are monomethylamine (CH_3NH_2) and carbon monoxide (CO). With monomethylamine, care is needed to ensure it is not absorbed by condensate in the exhaust. The carbon monoxide is easier to analyse (with an infra-red absorption technique) and it is also more stable in the exhaust; however carbon monoxide is also a product of combustion.

If: y is the mass fraction of tracer gas in the inlet air
     z is the mass fraction of unburned tracer gas in the exhaust and
     λ_d, the delivery ratio, is found by direct measurement (equation 8.1):

charging efficiency, $\eta_{ch} = \lambda_d \eta_{tr}$ (equation 8.5)

From the definition of trapping efficiency (equation 8.3):

$$\eta_{ch} = \lambda_d \frac{y - (M_i/M_e)z}{y} \qquad (8.18)$$

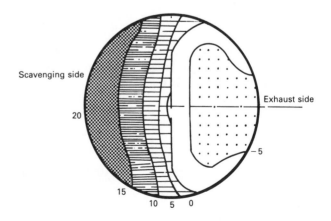

Figure 8.11
Velocity contours from a steady flow test at the open cylinder head (adapted from Sher, 1990).

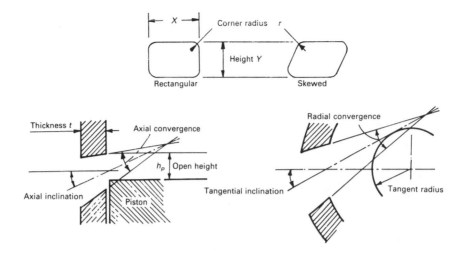

Figure 8.12
The geometric features of cylinder ports (Annand and Roe, 1974).

8.5.3 Port flow characteristics

A comprehensive summary of the flow phenomena through piston-controlled ports is presented by Annand and Roe (1974), who also provide a thorough description of the geometric features (figure 8.12). Beard (1984) states that up to 60–70 per cent of the bore periphery can be available as port width, but the remainder is required for carrying the axial loads and ensuring adequate lubrication for the piston and ring pack. Nor must individual ports be made too wide, otherwise the piston rings are likely to be damaged.

With reference to figure 8.12 the effective flow area when the piston completely uncovers the ports is:

$$A = XY - 0.86r^2 \qquad (8.19)$$

and when the port is partially covered by the piston (but $hp > r$), then

$$A = xhp - 0.43r^2 \qquad (8.20)$$

Figure 8.13 shows how the port geometry influences the inlet port flow – the flow coefficients are higher than those for poppet valves; some typical variations with the port open fraction are shown in figure 8.14.

The pressure ratio across the exhaust port varies considerably; it is high during the exhaust blowdown, but around unity during the scavenge period. Figure 8.15 shows the effect of the pressure ratio and piston position on the discharge coefficient for a single rectangular exhaust port. It should be noted that even when the port is fully open, the piston position influences the flow pattern and thus the discharge coefficient.

8.6 Engine performance and technology

Some typical full throttle performance curves for a two-stroke spark ignition engine are presented in figure 8.16. The maximum bmep is 6.4 bar at about 4000 rpm. The minimum fuel consumption of 400 g/kWh occurs at a lower speed (about 3000 rpm). For a naturally aspirated four-stroke spark ignition engine (in fact the Rover M16 multi-point injection engine) the minimum specific fuel consumption is less than 275 g/kWh – some 31 per cent lower than that for the two-stroke engine. The maximum bmep approaches 11 bar, so if torque or power per unit volume is considered, then the two-stroke engine is only about 17 per cent higher. The

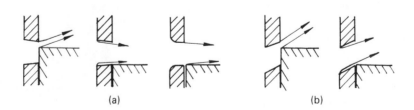

(a)　　　　　　　　　　　　　(b)

Figure 8.13
The influence of inlet port geometry on the flow patterns: (a) port axis perpendicular to wall; small opening and large opening with sharp and rounded entry; (b) port axis inclined (Annand and Roe, 1974).

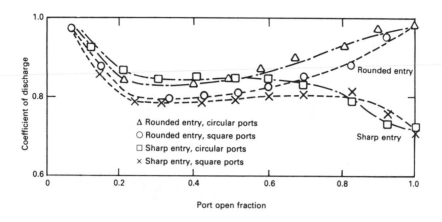

Figure 8.14
The discharge coefficient as a function of the port open fraction, for different inlet ports (Wallace, 1968). [Reprinted with permission © 1968 Society of Automotive Engineers, Inc.]

construction of a typical two-stroke engine is shown in figure 8.17.

As mentioned in section 8.1, one of the disadvantages of the externally carburated or fuel injected two-stroke engine is the passage of the air/fuel mixture into the exhaust system (by short-circuiting the mixing). This leads to the high fuel consumption and high emissions of unburned hydrocarbons. Furthermore, the high level of exhaust residuals in the trapped charge (low scavenging efficiency) dictates that a rich mixture is supplied, in order for combustion to be initiated

and propagated: this is detrimental to both the fuel consumption and emissions of carbon monoxide. However, the high level of exhaust residuals and the rich air/fuel ratio do lead to lower emissions of nitrogen oxides than those from untreated four-stroke engines. As would be expected, throttling reduces the delivery ratio and thus the charging efficiency is also reduced. Figure 8.18 shows how the engine output (bmep) is almost solely dependent on the charging efficiency. On the plot of trapping efficiency against delivery ratio, the charging efficiency contours are rectangular hyperbolae (equation 8.5). In figure 8.18, it can be seen that at low speeds the trapping efficiency is low, and this accounts for both the low maximum output and the poor fuel consumption at low speeds.

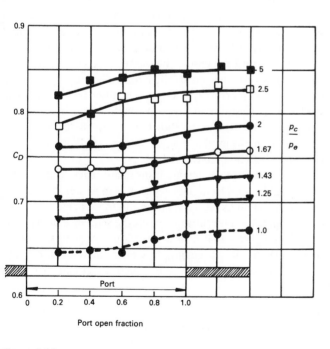

Figure 8.15
The effect of pressure ratio and port open fraction on the discharge coefficient for a simple rectangular exhaust port (Benson, 1960). [Reprinted with permission from *The Engineer*.]

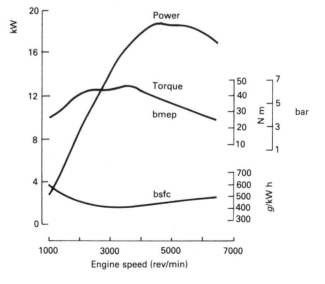

Figure 8.16
Performance characteristics of a three-cylinder, 450-cm³, two-stroke-cycle spark ignition engine. Bore = 58 mm, stroke = 56 mm (Uchiyama *et al.*, 1977).

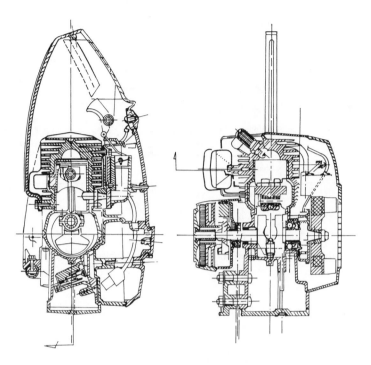

Figure 8.17
McCulloch chainsaw engine (Taylor, 1985b).

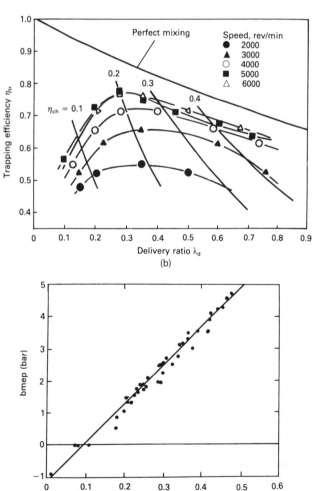

There are two designs of two-stroke engine that overcome some of the disadvantages associated with the conventional spark ignition two-stroke engine. Figure 8.19 shows the Orbital Combustion Process two-stroke engine that uses in-cylinder fuel injection. The novelty lies in the fuel injection system, which uses fuel at a pressure of 5.75 bar and air blast from a supply at 5 bar to help atomise the fuel to a spray that is mostly below 10 microns. The mixture preparation allows a weak mixture to be burnt, thus controlling nitrogen oxide emissions at source. The engine is fitted with a simple oxidation catalyst. Some fuel consumption and hydrocarbon emission comparisons are shown in figure 8.20. The maximum torque from this engine occurs at 4000 rpm, and corresponds to a bmep of 7.1 bar.

The idea of using a stepped piston dates back to the 1920s and as can be seen from figure 8.21, it is most sensibly applied to pairs of cylinders with the piston motion separated by 180° crank angle. With this arrangement, the sequence of events is no different from that of a conventional two-stroke engine with under-piston scavenging. However, the area ratio of the piston affects the delivery ratio. Such an engine has been

Figure 8.18
(a) The dependence of the brake mean effective pressure on the charging efficiency (that is, the trapped mass of fuel).
(b) The trapping and charging efficiency as function of the delivery ratio for a two-stroke spark ignition two-cylinder engine with a 347 cm³ swept volume (Tsuchiya and Hiramo, 1975). [Reprinted with permission © 1972 Society of Automotive Engineers, Inc.]

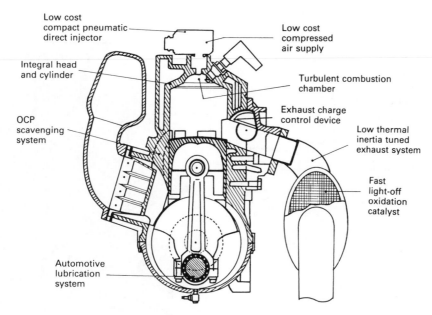

Figure 8.19

The Orbital Combustion Process two-stroke spark ignition engine which utilises in-cylinder fuel injection (courtesy of Orbital Engine Corporation Ltd).

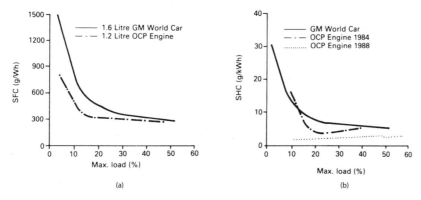

Figure 8.20

A comparison between a 1.6-litre four-stroke engine and the Orbital Combustion Process 1.2-litre two-stroke engine: (a) the specific fuel consumption as a function of load; (b) the specific hydrocarbon emissions as a function of load (courtesy of Orbital Engine Corporation Ltd).

developed to give an output of 6 bar bmep at a speed of 5500 rpm (Dunn, 1985). Another advantage of this design is that conventional four-stroke wet sump type lubrication can be adopted. A subsequent development has been to employ low-pressure fuel injection into the transfer port; this is shown in figure 8.22. By phasing the injection towards the end of scavenging, the trapped charge can be stratified and losses of fuel direct into the exhaust system are eliminated. Stepped piston engines have been developed for unmanned air vehicles which require a high specific output, a low mass and a low fuel consumption. Multi-cylinder engines now achieve a specific power of 61 kW/litre, and are comparable with the high power achievable with conventional two-stroke engines but with better fuel economy, and with size and weight factors that lie between two- and four-cycle engines

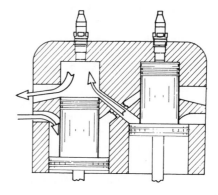

Figure 8.21

The stepped piston two-stroke engine applied to a two-cylinder engine (Dunn, 1985). [Reprinted with permission from *The Engineer*.]

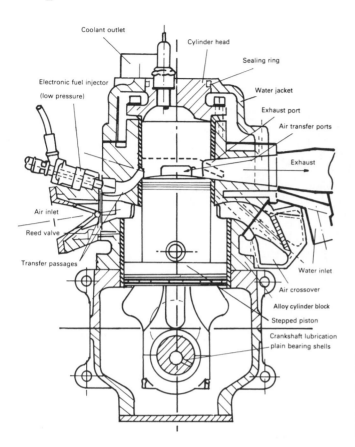

Figure 8.22
A prototype stepped piston two-stroke engine with wet-sump lubrication, and fuel injection into the transfer passage.

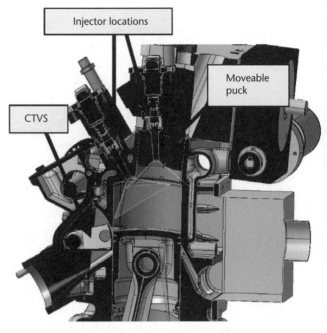

Figure 8.23
The Lotus Omnivore engine, with a movable puck to vary the compression ratio, and a Charge Trapping Valve System (CTVS) that oscillates at engine-speed, to allow asymmetric timing of the exhaust port opening and exhaust port closing, with acknowledgement to Lotus Engineering.

(Hooper, 2005). The wide open throttle (WOT) performance has an sfc of 304 g/kWh, with 286 g/kWh at cruise.

Valved two-stroke engines have also been developed so as to provide better control of scavenging. One of these, the Ricardo Flagship engine (Stokes *et al.*, 1992), became the starting point for the 2/4 SIGHT engine that can switch between two- and four-stroke operation (Osborne *et al.*, 2008). The engine operates in four-stroke mode for part load operation, and can achieve a bmep of 15 bar in two-stroke mode. The minimum bsfc was 235 g/kWh and the bsfc at 2000 rpm and 2 bar bmep was 350 g/kWh. Simulation results showed a fuel economy benefit of 27 per cent on the European Drive Cycle compared with a conventional engine of the same power. There are significant challenges with the valve gear, and the two-stroke operation is limited to speeds of about 4000 rpm.

Lotus have developed a very ingenious two-stroke engine, intended for spark ignition or HCCI operation (see section 5.7) with a wide range of fuels, known as Omnivore, which is comprehensively described by Blundell *et al.* (2010). The mechanical design and the flow and combustion philosophy are described in detail by Turner *et al.* (2010). Figure 8.23 is a

cross-section of the Omnivore engine, showing a movable puck that enables the compression ratio to be changed, and a CTVS – Charge Trapping Valve System. The CTVS oscillates at engine-speed, to allow asymmetric timing of the exhaust port opening and exhaust port closing. The engine has been operated with compression ratios in the range of 10 to 40:1 using 98 RON gasoline (injector in the puck), E85 (injector at the side) and diesel fuel with the same Orbital Flex DI fuel injector. The lowest isfc recorded on 98 RON gasoline was 218 g/kWh at 2.69 bar imep at 2000 rpm.

Automotive two-stroke compression ignition engines are uncommon outside North America but a Detroit Diesel is shown in figure 8.24. This V8 engine has a Roots-type scavenge pump operating in series with a turbocharger. The bore of 108 mm and stroke of 127 mm give a total displacement of 9.5 litres. At its rated speed of 2100 rpm the output of 276 kW corresponds to a bmep of 8.5 bar, with a specific fuel consumption of 231 g/kWh. The specific fuel consumption is perhaps 10 per cent worse than a four-stroke engine of comparable swept volume, but allowing for the two-stroke operation, then the specific output is about 40 per cent higher.

Automotive two-stroke diesel engines have also been developed in smaller sizes. AVL have produced a prototype engine (shown in figure 8.25) that has uniflow scavenging

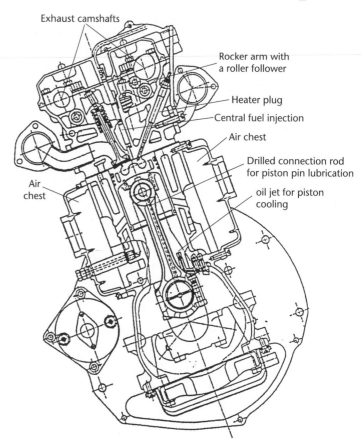

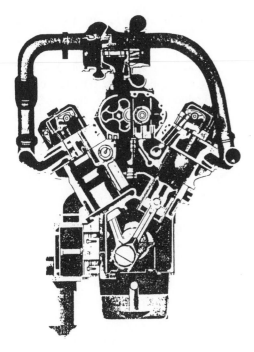

Figure 8.24

General Motors Detroit Diesel series 71 60° V two-stroke compression ignition engine, employing uniflow scavenging, with four exhaust valves per cylinder (Taylor, 1985b).

Two-Stroke-Diesel, Full Load, Development Steps

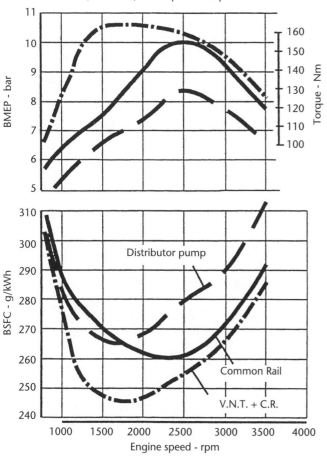

Figure 8.25

AVL automotive two-stroke diesel engine (adapted from Knoll, 1996).

(see figure 8.5d) (Knoll, 1996). This three-cylinder engine with a displacement of 0.98 litre has four exhaust valves per cylinder. Balance weights on the counter-rotating camshaft are used to balance the primary moments (the primary and secondary forces being inherently balanced by the 120° firing intervals, see chapter 12, section 12.2). As a result of engine simulation studies, the engine uses a combination of a Roots blower and a turbocharger. Figure 8.26 shows the performance of the engine in various builds. Initially the engine was run with a conventional distributor type fuel injection pump, but when this was changed to a common rail (CR) system the bmep was increased to 10 bar, and the full-load minimum brake specific fuel consumption was reduced by about 5 g/kWh. When a better turbocharger match was achieved by using a variable nozzle turbine (VNT) turbocharger, the bmep was increased yet further, with a particularly significant

Figure 8.26

Full-load performance of the AVL two-stroke diesel engine, showing the performance improvements associated with common rail (CR) fuel injection and a variable nozzle turbine (VNT) turbocharger.

increase in the low-speed bmep and the fuel economy. The minimum bsfc target of 215 g/kWh corresponds to a brake efficiency of just over 40 per cent, and this would require a variable drive ratio to the compressor. This engine was capable of meeting the contemporary emissions legislation, and had a very high specific output (45 kW) with the same level of torque fluctuations as a six-cylinder four-stroke engine.

Figure 8.27 is from the review paper by Hooper *et al.* (2011), and it shows a comparison between the specific output and fuel consumption for a selection of two-stroke diesel and spark ignition engines.

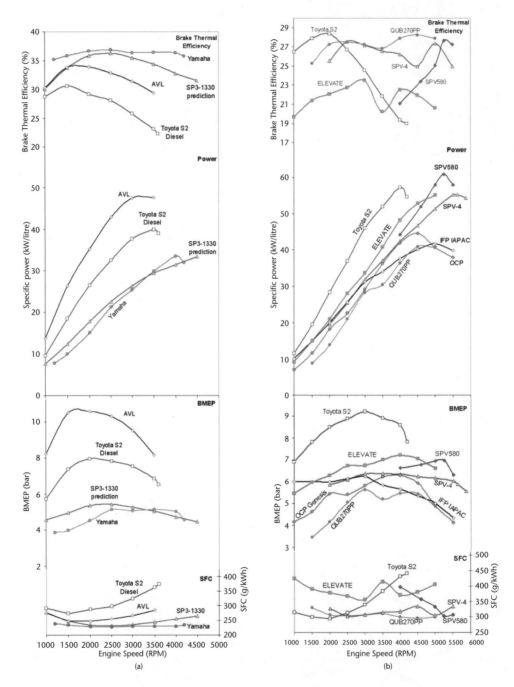

Figure 8.27

Comparison between the specific output and fuel consumption for a selection of two-stroke diesel and spark ignition engines. (a) Diesel: SP – Stepped Piston, Toyota and AVL – external blower for scavenging, Yamaha – crankcase scavenging. (b) Spark Ignition: SPV – Stepped Piston, S2 and ELEVATE – supercharged, QUB – loop scavenged. With acknowledgement to Peter Hooper.

8.7 | Concluding remarks

The two-stroke engine offers the advantages of mechanical simplicity, high specific output and compact size. In automotive applications, the potential for reducing the parts count in the engine and for reducing the volume and height of the engine compartment are sufficient to warrant serious investigations. Without doubt, the four-stroke engine has been easier to develop. The use of valves for controlling the essentially separate induction and exhaust processes has made it much easier to control the mixture motion and composition within the cylinder of four-stroke engines. However, the understanding of the phenomena that occur during scavenging is continually increasing. Much of this understanding has come from experimental work, but numerical modelling is also used (though this of course depends on experimental data from rig tests and firing engines for validation).

In its most familiar form, the two-stroke spark ignition engine has crankcase or under-piston scavenging with external mixture preparation. Such engines have a poor reputation for high fuel consumption and emissions levels. The under-piston scavenging limits the delivery ratio and this can lead to high levels of exhaust residuals in the cylinder, which then dictate the induction of a rich mixture to ensure

combustion. Furthermore, as the trapping efficiency is less than unity, unburnt fuel flows directly into the exhaust, with concomitant high emissions of unburnt hydrocarbons and a high fuel consumption. At part throttle the delivery ratio is reduced, and this decreases the scavenging efficiency and the charging efficiency. Indeed, it is not unknown for some two-stroke engines to fire only on alternate revolutions; this is known as 'four-stroking'. In other words, the scavenging is so poor that in alternate revolutions the mixture in the cylinder cannot be ignited.

In section 8.6 it was discussed how these shortcomings could be overcome. The use of a stepped piston allows the designer to increase the delivery ratio. When this is coupled with in-cylinder fuel injection, or fuel injection that is synchronised towards the end of induction, then the passage of fuel directly through the exhaust port can be controlled.

The automotive applications of two-stroke compression ignition engines are limited. However, in stationary and marine applications large two-stroke engines are widely available (see chapter 1, figure 1.5). The in-cylinder fuel injection ensures proper fuel utilisation, and the use of scavenge pumps and supercharging enables good scavenging and high outputs to be attained.

8.8 | Questions

8.1 The Ford two-stroke engine has a swept volume of 1.2 litres. At a speed of 5500 rpm the full-load torque is 108 N/m and the brake specific fuel consumption is 294 g/kWh. Calculate the brake mean effective pressure, and the brake efficiency (assume a calorific value of 43 MJ/kg). Give five reasons why the brake efficiency is lower than the value predicted by the Otto cycle analysis.

If the torque is reduced to 27 Nm (at the same speed), suppose that the brake specific fuel consumption rises to 325 g/kWh. Calculate the brake efficiency and explain why it is lower than at full load. What is the key attribute of the Ford Orbital two-stroke engine?

8.2 The perfect mixing scavenging model assumes:

(i)   instantaneous homogeneous mixing occurs within the cylinder,
(ii) the process occurs at a constant cylinder pressure and volume,
(iii) the system is isothermal, with no heat transfer from the cylinder walls, and
(iv) the incoming air (or mixture) and the burned gases have equal and constant ambient densities.

Show that the scavenging efficiency (η_{sc}) is given by

$$\eta_{sc} = 1 - \exp(-\lambda_d)$$

where λ_d is the delivery ratio. Sketch the variation in the scavenge efficiency with the delivery ratio, and for comparison include the results from the perfect displacement scavenging model.

CHAPTER 9

In-cylinder motion and turbulent combustion

9.1 | Introduction

The induction of air or an air/fuel mixture into the engine cylinder leads to a complex fluid motion. There can be an ordered air motion such as swirl, but always present is turbulence. The bulk air motion which approximates to a forced vortex about the cylinder axis is known as axial swirl, and it is normally associated with direct injection diesel engines. When the air motion rotates about an axis normal to the cylinder axis (this is usually parallel to the cylinder axis), the motion is known as barrel swirl or tumble. Such a motion can be found in spark ignition engines (normally those with two inlet valves per cylinder). Barrel swirl assists the rapid and complete combustion of highly diluted mixtures, for instance those found in lean-burn engines, or engines operating with stoichiometric mixtures but high levels of exhaust gas recirculation (EGR).

Turbulence can be pictured as a random motion in three dimensions with vortices of varying size superimposed on one another, and randomly distributed within the flow. Viscous shear stresses dissipate the vortices, and in doing so also generate smaller vortices, but overall the turbulent flow energy is dissipated by viscous effects. Thus energy is required to generate the turbulence, and if no further energy is supplied, then the turbulence will decay. In steady flows the turbulence can be characterised by simple time averages. The local flow velocity fluctuations (u) are superimposed on the mean flow ($\bar{U}$). For the x direction:

$$U(t) = \bar{U} + u(t) \tag{9.1}$$

For a steady flow, the mean flow is found by evaluating:

$$\bar{U} = \lim_{T \to \infty} \frac{1}{T} \int_{t_0}^{t_0+T} U(t)\mathrm{d}t \tag{9.2}$$

In other words the mean velocity ($\bar{U}$) is found by averaging the flow velocity ($U(t)$) over a sufficiently long time for the average to become stationary. As it will be the energy associated with the turbulence that will be of interest, the velocity fluctuations are defined by their root mean square (rms) value (u'). This is known as the turbulence or the turbulence intensity:

$$u' = \lim_{T \to \infty} \sqrt{\left(\frac{1}{T} \int_{t_0}^{t_0+T} [u(t)]^2 \mathrm{d}t \right)} \tag{9.3a}$$

or, as the time average $2u(t)$ $\bar{U}$ is zero:

$$u' = \lim_{T \to \infty} \sqrt{\left(\frac{1}{T} \int_{t_0}^{t_0+T} \left\{ [U(t)]^2 - U^2 \right\} \right) \mathrm{d}t} \tag{9.3b}$$

Unfortunately, in reciprocating engines the flow is not steady, but is varying periodically. Thus measurements have to be made over a large number of engine cycles, and measurements made at a particular crank angle (or crank angle window) are averaged. This approach is known as ensemble averaging or phase averaging. Before discussing turbulence in engines further (section 9.3), it is useful to know how the flows are measured, and this is the subject of section 9.2. Section 9.4 describes how turbulence measurements are used in models that predict turbulent combustion.

The way in which the flows can be predicted computationally is beyond the scope of this chapter. Fortunately Gosman (1986) has presented a very comprehensive review of flow processes in cylinders, and this encompasses turbulence definitions, flow measurement techniques and multi-dimensional computational fluid dynamics (CFD) techniques for flow prediction. Section 9.3 ends with a comparison between flow field measurements and CFD predictions.

9.2 | Flow measurement techniques

9.2.1 Background

The main flow measuring techniques for providing quantitative information are hot wire anemometry (HWA) and laser Doppler anemometry or velocimetry (LDA/V); the use of these techniques has been reviewed thoroughly by Witze (1980). Other techniques that are employed are smoke and tufts for gas motion studies, and water flow rigs with gas bubbles to identify the particle paths. In their basic forms, these techniques only provide qualitative information. However, these techniques are amenable to refinement. If the flow is illuminated by a sheet of light, then photographs can be taken as a multiple image or as a video. When multiple images (usually two) are taken, the technique is known as Particle Image Velocimetry (PIV), and the multiple images are usually obtained by pulsing the light source; this technique is described more fully in section 9.2.4, and is compared with other anemometry techniques in section 9.2.5.

Another technique used Freon as the working fluid so that a model engine could be run at reduced speed, yet retain dynamic similarity for the flow. The flow can be visualised by hollow phenolic spheres (microballoons) illuminated by sheets of light. The motion can then be tracked by video cameras linked to a computer-based system to compute the flow velocities. An example of this type of work is presented by Hartmann and Mallog (1988). However, when steady-flow rigs are used for visualising or measuring flows in an engine cylinder, consideration must always be given to how the results might relate to the unsteady flow in a reciprocating engine.

9.2.2 Hot wire anemometry

Hot wire anemometry (HWA) relies on measuring the change in resistance of a fine heated wire. The usual arrangement is a constant-temperature anemometer, in which a feedback circuit controls the current through the wire to maintain a constant resistance, and thus a constant temperature. The current flow then has to be related to fluid velocity. Unfortunately, the direction of the flow cannot be determined, as the wire will respond to normal flows, and to a lesser extent to an axial flow. Thus, to determine a three-

dimensional flow, three independent measurements are needed, and these are most conveniently analysed if the wire orientations are orthogonal, in which case for the arrangement shown in figure 9.1:

$$V_{\text{eft},1}^2 = V_x^2 + (K_2 V_y)^2 + (K_3 V_z)^2$$
$$V_{\text{eft},2}^2 = V_y^2 + (K_3 V_z)^2 + (K_2 V_x)^2 \qquad (9.4)$$
$$V_{\text{eft},3}^2 = (K_3 V_x)^2 + (K_2 V_z)^2 + V_y^2$$

where $V_{\text{eft},1}$, $V_{\text{eft},2}$ and $V_{\text{eft},3}$ are the effective velocities calculated from the anemometer output, and K_2 and K_3 are the pitch and yaw factors that account for the reduced sensitivity in the pitch and yaw directions.

In a time-varying flow, the three measurements have to be made simultaneously, and this implies a three wire probe. Each measurement corresponds to a mean over the length of the wire, and obviously the three measurements cannot refer to exactly the same point. As a hot wire anemometer cannot deduce the direction of the flow, the computed velocity vector can lie in any one of the eight octants. Furthermore, the presence of the wires and their supports will modify the flow structure. If the wires are made too short, then there will also be heat transfer to the wire supports, as well as to the fluid stream.

Ideally, the probes have to be calibrated for the range of pressures and temperatures that will be encountered. However, this is not convenient if measurements are to be made in a motored engine, since it implies making calibrations over a wide range of pressures and temperatures. Instead, a common approach is to make corrections to a baseline calibration, using analytical models that have been previously substantiated experimentally.

A widely used model is:

$$Nu = A + B\,Re^n \qquad (9.5)$$
$$\text{where} \qquad Nu = \frac{hd}{k} = \frac{Q}{\pi kl(T_w - T_g)}$$

A, B, n = constants
h = heat transfer coefficient (W/m²K)
d = wire diameter (m)
k = thermal conductivity (W/mK)
Q = heat flow (W)
l = wire length (m)
suffixes: w – wire, g – free stream gas.

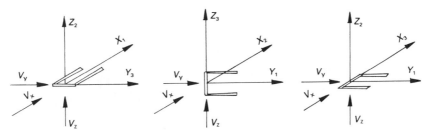

Figure 9.1
Probe positions for determining the three orthogonal velocity components in a fixed coordinate system (X, Y, Z), with probe coordinates: 1, 2, 3.

This model was first proposed by King (1914) and with $n = 0.5$, it is known as King's Law:

$$Nu = A + B\,Re^{0.5} \tag{9.6}$$

As with any correlation for heat transfer, there is the question of what temperature to evaluate the transport properties at. The options include the free stream gas temperature, the wire surface temperature, or some weighted average such as the film temperature (T_f):

$$T_f = (T_w + T_g)/2 \tag{9.7}$$

Davies and Fisher (1964) found that

$$Nu_w = 0.425 Re_g^{1/3} \tag{9.8}$$

Collis and Williams (1959) concluded that

$$Nu_f \left(\frac{T_f}{T_g}\right)^{-0.17} = 0.24 + 0.56 Re_f^{0.45} \quad \text{for } 0.02 < Re < 44$$
$$= 0.48 Re_f^{0.51} \qquad \text{for } 44 < Re < 140 \tag{9.9}$$

Witze (1980) recommended

$$Nu_g = A + B\,Re_g^{n} \tag{9.10}$$

Witze points out that analytical solutions for forced convection through a laminar boundary layer invariably yield $n = 0.5$, while the fit to experimental calibration gives

$$0.24 < n < 0.47$$

the exact value being a function of the gas velocity and wire temperature.

The implication from these results is that, whenever possible, the hot wire probes should be calibrated for the range of velocities being encountered, with the tests being conducted at different pressures and temperatures.

When analytical-based corrections are being made for pressure and temperature, the transport properties that have to be predicted are density, thermal conductivity and viscosity. Correlation for the temperature correction of dynamic viscosity (μ) and thermal conductivity (k) have been made by Collis and Williams (1959):

$$k \propto T^{0.8} \tag{9.11}$$
$$\mu \propto T^{0.76} \tag{9.12}$$

9.2.3 Laser Doppler anemometry

Laser Doppler anemometry (LDA) is illustrated by figure 9.2. A pair of parallel coherent laser beams are arranged to cross at the focal point of a lens. Interference fringes are created throughout the beam intersection region or measuring volume; the fringe lines (or really planes) are parallel to the lens axis and perpendicular to the plane of the incident laser beams. When particles carried by the fluid pass through the interference fringe, they scatter light. The scattered light is detected by a photomultiplier tube, and the discontinuous output from the scattered light is called the Doppler burst. The Doppler frequency equals the component of the velocity perpendicular to the fringes (that is, perpendicular to the lens axis) divided by the distance between the fringes. The spacing of the fringes (Δx) is related to the wavelength of the laser light (λ) and the half-angle of the two intersecting beams ($\phi = \theta/2$):

$$\Delta x = \frac{\lambda}{2 \sin \phi} \tag{9.13}$$

If the 514.4 nm wavelength light from an argon-ion laser is used with a half-angle of 7.37°, then the fringe spacing is 2 microns.

The measuring volume is an ellipsoid whose diameter and length depend on the half-angle and the laser beam diameter. Typically, the length (along the lens axis) is about 600 microns with a diameter of 80 microns; this will give a maximum of 40

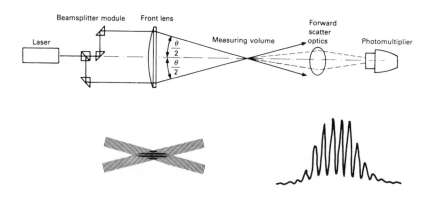

Figure 9.2

Optical arrangement for laser Doppler anemometry (LDA) showing the interference fringe pattern, collection of the light scattered by particles in the forward scatter mode, and a typical output from the photo-multiplier.

fringes. When a particle passes through the control volume it will only pass through some of the fringes. If the velocity component of the particle is 10 m/s, the frequency of the Doppler burst will be 5 MHz with a maximum possible duration of 8 μs.

The particle size should preferably be about a quarter of the light wavelength, and it is assumed that this is sufficiently small to move with the flow. The flow has to be seeded with particles, for example titanium dioxide (TiO_2) is used in non-reacting flows, and zirconium fluoride (ZrF_4) is used when combustion occurs. The particles arrive in the measuring volume at 0.1–10 ms intervals, and the output from the photodetector can vary in amplitude and dc offset, as shown in figure 9.2. The frequency of the Doppler burst corresponds to the magnitude of the velocity component, but there is no indication whether the velocity is positive or negative. However, if a frequency shifter is installed in the optical path of one of the beams, the fringe pattern will move with a constant velocity. If this velocity is greater than the negative component of any particle velocity, the flow direction can always be resolved.

The signal processing requirements are considerable. The signal is discontinuous, varying in amplitude and dc offset; it also has a very high frequency and a short duration. A commonly used system is a frequency tracking demodulator, in which a phased locked loop latches onto the Doppler frequency. This requires a nearly continuous Doppler signal, with a limited dynamic range; as neither of these conditions occurs with internal combustion engines, then other techniques are needed. Witze mentions other signal processing techniques (namely filtering and photon correlation) but concludes that period counting seems the most appropriate technique for internal combustion engines. Period counting entails the use of a digital clock to measure the transit time of a specified number of fringe transits in a single burst. Since the counter output is digital and randomly intermittent, further digital processing is essential. For example, it can be necessary to record 2000 or more velocity/crank angle data pairs.

The optical arrangement shown in figure 9.2 is a forward scatter system, which requires a second optical access and receiving lens for focusing the scattered light onto the photomultiplier (or photodector). A second system that only requires a single optical access employs backscatter (figure 9.3). While backscatter systems are easier to implement, the signal strength is significantly weaker. Other optical arrangements can be used, for example the scattered light can be collected through a window at right angles to the main beams. In order to obtain three-dimensional velocity measurements, multiple measurements (most convenient if orthogonal) need to be made, perhaps simultaneously.

An extension of LDA is Phase Doppler Anemometry (PDA), which uses the same principle as LDA, but makes use of three photo-detectors at different angles to the measurement volume (Bachalo *et al.*, 1991). Each detector receives the

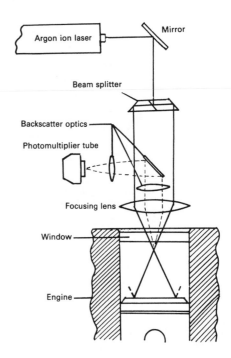

Figure 9.3
LDA system using backscatter, such that only a single optical access is required (from Witze, 1980). [Reprinted with permission © 1980 Society of Automotive Engineers, Inc.]

same modulated signal as in LDA, but with a time delay. The time delay is a phase shift which is proportional to the particle diameter. Thus simultaneous measurements can be made of the droplet size and velocities. Examples of such measurements have already been seen in chapter 4, figure 4.53. Similar measurements have been reported for the Honda VTEC engine using a steady-flow rig, with data presented for the axial and radial directions (Carabateas *et al.*, 1996).

9.2.4 Particle image velocimetry (PIV)

Particle imaging techniques for flow determination have been reviewed by Adrian (1991). With particle tracking and streak photographic techniques a sheet of light can be used to obtain flow information in a two-dimensional field. However, these techniques are limited to rather low densities of tracer particles, so as to avoid excessive computation time for the identification of individual particles, or the intersection or merging of streaks. This limits the number of flow field vectors that can be obtained from a single image. PIV does not rely on the identification of individual particles, so that a higher density of tracing particles can be used. With laser illumination, particle sizes of order 1 μm can be used (oil droplets are convenient, since they assist with in-cylinder lubrication), and these are well able to follow the flows.

With PIV, the flow-tracing particles are illuminated by a double-pulsed sheet of light, and the images are recorded by a camera normal to the light plane. The subsequent image is

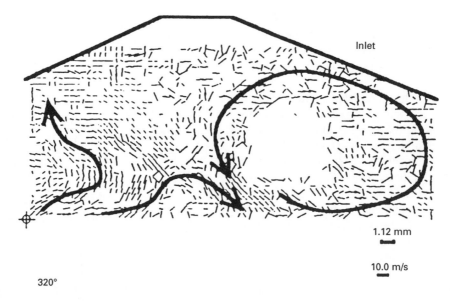

Inlet

1.12 mm

10.0 m/s

320°

Figure 9.4

Particle image velocimetry (PIV) vector map in a plane through the inlet and exhaust valve, in a motored engine at 1000 rpm, full throttle. The arrows have been added (not to any scale) to show the general flow direction (Reeves *et al.*, 1996).

© IMechE/Professional Engineering Publishing Limited.

analysed by a spatial autocorrelation of small regions of the image. The resulting autocorrelation plane consists of:

1   a large central spike, the radius of which determines the minimum particle displacement that can be measured, and

2   several non-central peaks, of which the centroid of the highest is located at the most likely mean particle displacement.

There will of course be two such peaks diametrically opposite each other, since there is not always a means of knowing which of the two images were recorded first, in other words, there is a 180° directional ambiguity. However, in many cases there is other information that can be used to eliminate this ambiguity.

Some results from PIV are shown in figure 9.4, which have been obtained from a motored pent-roof combustion chamber engine, very similar to that shown in chapter 4, figure 4.11 (Reeves *et al.*, 1996). In this case a sheet of laser light (of pulse duration 8 ns and separation 25 μs) was reflected from the mirror to illuminate a vertical plane in the cylinder. The whole of the cylinder liner was fused silica, and a specially developed optical correction system was used to remove the optical distortion introduced by the cylinder liner. Figure 9.4 shows the flow in a plane through the inlet and exhaust valves 40°ca before the end of compression. This is at the stage when the barrel vortex is starting to break up and generate turbulence.

9.2.5   Comparison of anemometry techniques

Witze (1980) has provided a comprehensive comparison of hot wire and laser Doppler anemometry. The advantages and disadvantages of each technique are almost entirely complementary.

Laser Doppler anemometry is non-intrusive, applicable at all temperatures and pressures, linear, direction sensitive and can be used when there is no mean flow. The disadvantages of LDA are its cost, the need for (multiple) optical access, the intermittent and poor signal-to-noise ratio of the signal and its high frequency, and the need to seed the flow. With LDA, it is also necessary to isolate the optical system from any engine vibration, and to ensure that the motion of the combustion chamber (due to vibration) is minimised – otherwise the engine motion will be superimposed on the measured flow velocity.

Laser Doppler anemometry is considered to be more accurate than hot wire anemometry. The inaccuracy of LDA is considered to be 5–10 per cent under favourable conditions and perhaps 10–20 per cent under adverse conditions. An overriding advantage of LDA can be its ability to make measurements in a firing engine. Unfortunately LDA equipment is an order of magnitude more expensive than HWA equipment.

Particle image velocimetry (PIV) provides full flow field measurements of the mean flows, and it is thus a complementary technique to LDA or HWA which provide point measurements of the mean flow and turbulence.

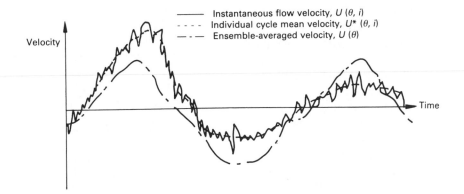

Figure 9.5
The separation of the instantaneous flow velocity signal into an individual cycle mean $U^*(\theta, i)$, turbulent velocity fluctuations, $U(\theta, i)$, and an ensemble-averaged velocity $U(\theta)$.

9.3 | Turbulence

9.3.1 Turbulence definitions

In the introductory section 9.1, equations (9.1)–(9.3) were used to define turbulence in steady flow. In a reciprocating engine there will be a periodic component in the flow that can be related to the crank angle. Thus instead of expressing the velocities as functions of time, they can be expressed as functions of crank angle (θ) and cycle number (i).

Equation (9.1) becomes

$$U(\theta, i) = \bar{U}(\theta) + u(\theta, i) \tag{9.14a}$$

Figure 9.5 shows that there can be cycle-to-cycle variations in the mean flow, thus equation (9.1) can alternatively be written in a way that recognises these variations separately from the turbulent fluctuations:

$$U(\theta, i) = \bar{U}(\theta, i) + u(\theta, i) \tag{9.14b}$$

or

$$U(\theta, i) = \bar{U}(\theta) + U^*(\theta, i) + u(\theta, i)$$

The $U^*(\theta, i)$ term is to account for the cycle-by-cycle variations in the mean flow. The $u(\theta, i)$ terms will be different in equations (9.14a) and (9.14b), and a dilemma in studies of turbulence is whether or not the low-frequency fluctuations $U^*(\theta, i)$ (that is, cycle-by-cycle variations in the mean flow) should be considered separately from the turbulent fluctuations – $u(\theta, i)$. Of course $U(\theta, i)$ is the only velocity that exists in the flow; the other velocity terms are introduced to enable the flow to be characterised.

At any chosen crank angle, the ensemble-averaged velocity can be found by averaging as many as possible velocity measurements corresponding to that crank angle:

$$U_{ea}(\theta) = \lim_{n \to \infty} \left[\frac{1}{n} \sum_{i=1}^{n} U(\theta, i) \right] \tag{9.15}$$

where n is the number of cycles.

The fluctuations from the mean $u(\theta, i)$ can be expressed in terms of their rms value:

$$u'_{ea}(\theta) = \lim_{n \to \infty} \left[\frac{1}{n} \sum_{i-1}^{n} U(\theta, i)^2 - U(\theta)^2 \right]^{1/2} \tag{9.16}$$

When the velocity measurements have been taken with an LDA system, it must be remembered that the signals are not continuous. It is thus necessary to: specify a crank angle window ($\pm \Delta\theta/2$) centred on a mean value of crank angle, $\bar{\theta}$, and have a system to assign the measurements to the appropriate windows – this can be called a phase resolver. The width of the crank angle window selected will influence the number of velocity measurements; in some cycles there will be no velocity measurements while in other cycles there will be several velocity measurements. Thus in equations (9.14)–(9.16)

θ is replaced by $\bar{\theta} \pm \Delta\theta/2$

and n becomes the number of velocity measurements made, not the number of cycles.

If the crank angle window is made too small, then the data rate (the rate at which velocity measurements are made) becomes too low, and a long time is required to obtain a statistically valid sample. The convergence of the averaging process within a window varies as the inverse of the square root of the number of points averaged. This implies that more measurements in total are needed as the window size is reduced. Typically, data from a few thousand engine cycles are recorded.

Conversely, if the crank angle window is too wide an error is introduced called crank angle broadening; this is due to the change in mean velocity during the crank angle window. The crank angle broadening error is discussed by Witze (1980), who suggests that a 10° window is satisfactory unless there is a dynamic event (such as combustion or fuel injection), in which case the crank angle window needs to be reduced (to say 2°).

Equations (9.15) and (9.16) can be used with equation (9.14a) to characterise the flow. However, this ignores the cycle-to-cycle variations in the mean flow that were identified by the additional term in equation (9.6). This variation is thought to be significant in the cycle-to-cycle variation in combustion observed in spark ignition engines. Gosman (1986) discusses the different techniques that can be used for separating the cycle-by-cycle variation in the mean flow (this is the $U^*(\theta, i)$ term in equation 9.14b). One approach is a hybrid time/ensemble-averaging scheme, in which the averaging time (T, in equations 9.2 and 9.3) is limited to a specified crank period (say 45°). However, this period is arbitrary, and during the period the mean velocity can vary significantly. An extension of this approach is to use a low pass filter to remove the cyclic variations in the mean velocity. It is said that this makes the signal time independent (stationary), so that simple time-averaging techniques can be used. Unfortunately, this makes it impossible to resolve individual cycles.

An approach devised by Rask (1981) fits smooth curves to the velocity data from each cycle, in order to define the variation in the mean velocity, $\bar{U}(\theta, i)$. The instantaneous turbulent velocity component, $u'(\theta, i)$, is then defined as the difference between the velocity signal and the smoothed curve. This enables the following to be determined:

(a) the turbulent velocity fluctuations $u'(\theta, i)$;

(b) the cycle-by-cycle variation in the mean velocity $\bar{U}^*(\theta, i)$;

(c) the ensemble-averaged mean velocity $\bar{U}(\theta)$.

Unfortunately, the choice of the curve fitting to the original signal can slightly influence the results obtained, in an analogous manner to filtering. Some results obtained by Rask (1981) are shown in figure 9.8, which is discussed in section 9.3.2.

Gosman (1986) concludes that there is uncertainty about how to characterise the air motion in engines. This point is also made by Heywood (1988). Now that the means of obtaining turbulence data have been described, it is necessary to discuss how the measurements are used, by defining some length and time scales that are used to characterise turbulence. The largest eddies are limited by the cylinder boundaries at one extreme, and at the other by molecular diffusion. The eddies that are responsible for producing most of the turbulence during induction arise from the jet-like flow that emerges from the inlet valve. The radial and axial velocity components in the jet are an order of magnitude greater than the piston velocity, and the initial width of the jet approximates to the valve lift. Figure 9.6 shows how the shear between the jet and the cylinder contents leads to eddies. These large eddies are unstable and break down into a cascade of smaller eddies. The length and time scales in turbulence are defined and discussed comprehensively by Gosman (1986) and Heywood (1988);

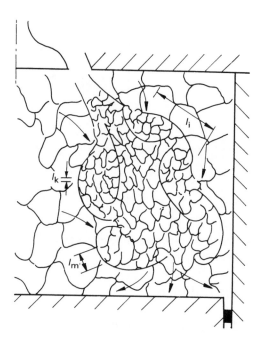

Figure 9.6

A representation of the induction process, to show the generation of eddies by shear action, and the characteristic length scales of the turbulence (from Gosman, 1986, by permission of Oxford University Press).

since the definitions are important they are also developed here.

The integral scale, l_i, is a measure of the largest scale structure. If velocity measurements are made a distance apart that is significantly less than the integral scale (l_i), then the velocity measurements will correlate with each other. The spatial correlation coefficient R_x is defined in terms of the velocity fluctuations at a location x_0 and the velocity fluctuations at a distance x away:

$$R_x = \frac{\overline{u(x_0)u(x_0 + x)}}{u'(x_0)u'(x_0 + x)} \tag{9.17}$$

or $\quad R_x = \dfrac{1}{n-1} \sum_{i=1}^{n} \dfrac{u(x_0)u(x_0 + x)}{u'(x_0)u'(x_0 + x)} \quad$ for n measurements

The variation of the correlation coefficient with distance is shown in figure 9.7. The area under the curve is used to define the integral length scale:

$$l_i = \int_0^\infty R_x \, dx \tag{9.18}$$

In other words, the area under the curve is equivalent to the rectangle bounded by the axes and the coordinates (l_i, 1).

A second scale for characterising turbulence is the Taylor microscale (l_m), and this has been interpreted as the spacing between the smallest eddies. The Taylor microscale (l_m) is defined by the intercept of a parabola with the x-axis, for a

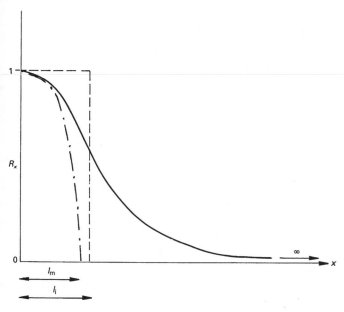

Figure 9.7
The variation of the correlation coefficient (R_x) with distance to illustrate the definitions of the integral length scale (l_i) and the Taylor microscale (l_m).

symmetrical parabola that matches the autocorrelation function in height and curvature at $x = 0$. This is also illustrated in figure 9.7. From the equation for a parabola, it can be shown that

$$l_m^2 = -2/(\partial^2 R_x/\partial x^2)_{x=0} \tag{9.19}$$

The third and smallest length scale is the Kolmogorov scale (l_k). Since the smallest eddies respond most quickly to changes in the local flow, it is assumed that turbulence at this level is isotropic (the same in all directions). It is at this level that viscosity dissipates the kinetic energy to heat. Dimensional analysis is used to relate the Kolmogorov scale to the energy dissipation rate per unit mass, ε:

$$l_k = \left(\frac{\nu^3}{\varepsilon}\right)^{1/4} \tag{9.20}$$

Owing to the difficulty in obtaining simultaneous flow measurements from two positions in an engine, an alternative and easier approach is to find characteristic time scales, and then see how these might be related to the length scales.

The temporal auto-correlation function (R_t) is defined in a similar way to the spatial correlation function, except that the velocity fluctuations are compared at the same point, but measurements are taken at time t apart. With n measurements:

$$R_t = \frac{1}{n-1}\sum_{i=1}^{n}\frac{u(t_0)u(t_0+t)}{u'(t_0)u'(t_0+t)} \tag{9.21}$$

This leads to a similar definition for the integral time scale, t_i:

$$t_i = \int_0^\infty R_t \mathrm{d}t$$

If the turbulence is relatively weak, and the flow is convected past the measuring point without significant distortion, then it has been found that the integral time scales and the integral length scales are related by the mean flow velocity ($\bar{U}$):

$$l_i = \bar{U}t_i \tag{9.22}$$

For flows without a mean motion, the integral time scale (t_i) is an indication of the eddy lifetime.

For turbulence that is homogeneous and isotropic the Taylor microscale is related in a similar manner:

$$l_m = \bar{U}t_m \tag{9.23}$$

The Kolmogorov time scale, t_k, is defined as

$$t_k = \left(\frac{\nu}{\varepsilon}\right)^{1/2} \tag{9.24}$$

and it characterises the momentum diffusion within the smallest flow structures.

Heywood (1988) explains how the turbulent kinetic energy per unit mass in the large eddies is proportional to u'^2, and that as these eddies lose a substantial fraction of their energy in one 'turnover' time l_i/u' then:

$$\varepsilon \approx u'^3/l_i \tag{9.25}$$

and from equation (9.20)

$$\frac{l_k}{l_i} \approx \left(\frac{u'l_i}{\nu}\right)^{-3/4} = Re_t^{-3/4} \tag{9.26}$$

where Re_t is the turbulent Reynolds number.

Heywood (1988) also states that

$$\frac{l_m}{l_i} = \left(\frac{15}{A}\right)^{1/2} Re_t^{-1/2} \tag{9.27}$$

where A is a constant of order unity.

9.3.2 In-cylinder turbulence

The origins of turbulence during the induction stroke were described in the introductory section 9.1, namely turbulence is generated by the shear between the jet flowing from the inlet valve and the cylinder contents. As this jet has a mean velocity that is an order of magnitude greater than the piston velocity, then the jet will impinge on the piston, as well as the cylinder walls. This leads to a toroidal vortex below the inlet valve, but offset towards the axis of the cylinder. The result of these flows is a high level of turbulence generated during induction, which then decays. However, further turbulence is generated by squish: the inward radial movement of air

Table 9.1 *Turbulence parameters in a motored CFR engine at 2000 rpm (Lancaster, 1976)*

	u' (m/s)	$\bar{U}$ (m/s)	l_i (mm)	l_m (mm)	l_k (mm)	t_i (ms)	t_m (ms)	t_k (ms)
Mid induction	5.0	20	4.0	1.0	0.02	0.4	0.07	0.04
Late compression	15	10	4.0	1.0	0.03	0.8	0.20	0.12

caused by the small clearances between the piston crown and cylinder head.

Some tabulated values of turbulence parameters from Lancaster (1976) are presented in table 9.1.

These results show that the integral, Taylor microscale and Kolmogorov lengths are almost independent of crank angle, and are all much smaller than the bore diameter or stroke (about 100 mm). The integral length scale corresponds to just under half the maximum valve lift. The time scales are in a similar ratio to one another as the length scales, and the longest time scale (the integral time scale) is significantly shorter than the duration of the induction stroke (15 ms).

Some flow measurements made at two locations by Rask (1981) are shown in figure 9.8. The engine was motored at 300 rpm, and this gave a mean piston speed of 0.76 m/s. As the engine was motored, the measurements during the expansion and exhaust strokes will not be representative of those in a fired engine. In figures 9.8d and 9.8e the differences between the smoothed ensemble and the ensemble-averaged individual-cycle variations in the mean flow represent the cycle-by-cycle variations in the mean flow. This variation can be seen to be small in comparison with the turbulence levels during induction, but the difference becomes significant during compression. Thus the cycle-by-cycle variations in the mean flow are also significant during compression. The differences in flow for locations *b* and *c* in figure 9.8 show that the flow is not homogeneous, nor is the flow isotropic (Rask, 1981). However, these results are for an engine in which there is a strongly directed flow from the inlet valve.

Bracco and his co-workers have conducted studies with a variety of port configurations and disc-shaped combustion chambers (Liou *et al.*, 1984; Hall and Bracco, 1986). Like Rask (1981), these studies have separated the velocity measurements into:

(i) a mean flow

(ii) a cycle-by-cycle variation in the mean flow and turbulence.

The separation was achieved through converting the velocity data into the frequency domain by a Fourier transformation, and identifying a cut-off frequency (which was a function of engine speed), and then applying inverse Fourier transformations to the separated signals. Above the cut-off frequency the velocity data represented turbulence, below the cut-off frequency, the velocity data represented the mean flow and the cycle-by-cycle variations in the mean flow.

Liou *et al.* (1984) report data from three different engine builds: a valved four-stroke engine (without swirl), a ported two-stroke engine with swirl, and a ported two-stroke engine without swirl. Since the combustion chambers were essentially disc-shaped, the turbulence is mostly generated by the induction process. If the cycle-by-cycle variations in the mean flow and the turbulence intensity were not separated, the turbulence intensity was overestimated by as much as 300 per cent in the valved engine (no swirl), 100 per cent in the ported engine (no swirl), and 10 per cent in the ported engine with swirl. This implies that the presence of swirl reduces the cycle-by-cycle variations in the mean flow, and that this should lead to less cyclic variation in spark ignition engine combustion.

Most important, so far as combustion is concerned, is the nature of the flow around top dead centre. Liou *et al.* (1984) show that with open disc-shaped combustion chambers (for measurements at different locations in the clearance volume around top dead centre, at the end of compression) the turbulence is relatively homogeneous (within ± 20 per cent). In the absence of any swirl, the turbulence was also essentially isotropic (within ± 20 per cent) near top dead centre. Data presented by Liou *et al.* (1984) in figure 9.9 show that for individual engines, the turbulence intensity at top dead centre increases linearly with engine speed:

$$u'_{tdc} \approx k\bar{v}_p \tag{9.28}$$

with the constant of proportionality in the range $0.25 < k < 0.5$ depending on the type of induction system. Liou *et al.* (1984) conclude that the maximum turbulence intensity that can be obtained at tdc in an open chamber without swirl is equal to half the mean piston speed. Turbulence intensity as high as 75 per cent of the mean piston speed has been reported, but this presumably includes cycle-by-cycle variations in the mean flow.

Heywood (1988) also notes that when swirl is present, the turbulence intensity at top dead centre is usually higher than when swirl is absent. The variation in individual-cycle mean velocity at the end of compression also scales with the piston speed, and this variation can be comparable in magnitude with the turbulence intensity. With swirling flows and bowl-in piston combustion chambers, the conservation of the moment of momentum means that the swirl speed will increase. The increase in angular velocity (swirl speed) causes shear that can cause an increase in the turbulence level.

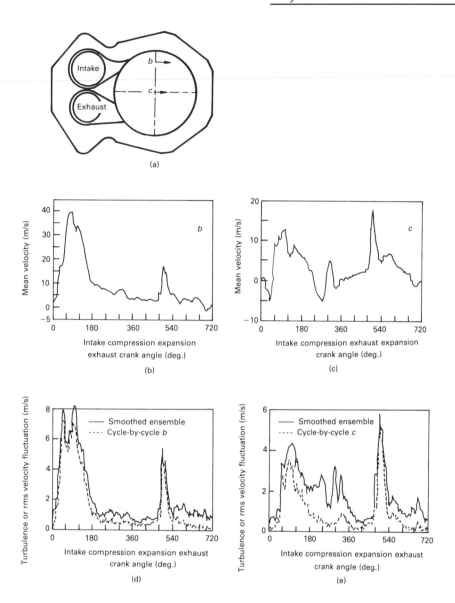

Figure 9.8

Turbulence measurements from a side valve engine motored at 300 rpm (mean piston speed of 0.76 m/s): (a) location and direction of the velocity measurements, note that the inlet port will generate swirl; (b) and (c) show the ensemble-averaged mean velocities at locations *b* and *c*; (d) and (e) show the turbulence intensity (u'), calculated by neglecting the cycle-by-cycle variations in the mean flow (—), and the turbulence intensity calculated when allowance is made for the cycle-by-cycle variations in the mean flow (---) (Rask, 1981, reproduced with permission of ASME).

The results discussed so far have been obtained from motored engines, in other words there has been no combustion. Hall and Bracco (1986) took LDA measurements in an engine with a disc-shaped combustion chamber, and a port inlet that generated swirl; the engine was both motored and fired. In the motoring tests they took readings right across the combustion chamber, and found that the tangential velocity at a location 0.5 mm from the wall (bore diameter of 82.6 mm) was within 90 per cent of the maximum tangential velocity. They also found that the turbulence intensity increased near the wall, and this was explained by shear at the wall generating turbulence. From the tests with combus-

tion, Hall and Bracco (1986) found that there was little increase in turbulence across the flames at all the spatial locations examined. The turbulence in the burned gas was homogeneous, except at the wall, and the turbulence was isotropic (within 20 per cent) to within 1.5 mm of the wall.

The role of turbulence in spark ignition engine combustion is discussed by Kyriakides and Glover (1988) who found a strong correlation between turbulence intensity and the 10–90 per cent burn time. They investigated various means of producing air motion, and concluded that tumble (or barrel swirl) was a more effective way for generating turbulence at top dead centre than axial swirl. In this work the cycle-by-

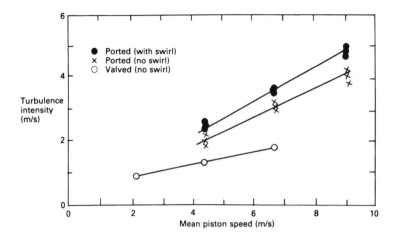

Figure 9.9
Relationship between the mean piston speed and the turbulence intensity at top dead centre for different induction systems (cycle-by-cycle variations in the mean flow are not included here in the turbulence intensity) (data adapted from Liou *et al.*, 1984).

cycle variations in mean flow were not separated from the turbulence intensity.

From a knowledge of turbulence intensity and length scales it is possible to construct a turbulent entrainment model of the combustion in a spark ignition engine; this is discussed in the next section.

Computational Fluid Dynamics (CFD) is now well able to predict both the mean flow and the turbulence intensity for in-cylinder flows. Figure 9.10 shows steady-flow comparisons between measurements and predictions of the flow in the axial direction, for both (a) the mean flow and (b) the turbulence intensity (Chen *et al.*, 1995). The inlet port, valve

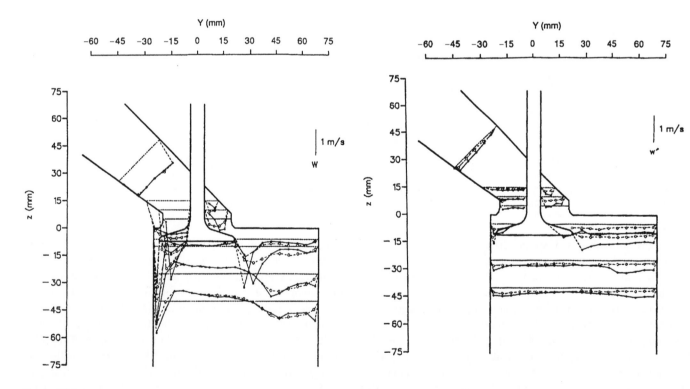

Figure 9.10
Steady flow comparisons between measurements (—o—) and predictions (---o---) of the flow in the axial direction, for both
(a) the mean flow, and (b) the turbulence intensity (Chen *et al.*, 1995). [Copyright John Wiley & Sons Limited. Reproduced with permission.]

and cylinder were manufactured from a transparent acrylic plastic, and the working fluid was a mixture of turpentine and tetraline at a carefully controlled temperature, such as to match the refractive index of the plastic and eliminate any optical distortion. The steady flow was simulated by the STAR-CD CFD program, with 77 260 cells and a k–ε turbulence model. The mean flow was studied by laser sheet illumination of small (about 30 μm) air bubbles, and LDA was used for mean flow and turbulence measurements. There is mostly good agreement, apart from the predictions of the turbulence intensity in the right-hand part of the cylinder which is under-predicted. Part of the explanation for this is that the intake jet was observed to 'flap' intensely (that is, its direction oscillates), and this would not be predicted by the CFD. The variations in the mean flow thus contribute to the measured turbulent fluctuations.

9.4 | Turbulent combustion modelling

9.4.1 A turbulent entrainment model of combustion

The combustion in a homogeneous charge spark ignition engine is commonly divided into three parts:

(a) An initial laminar burn, before the flame kernel is large enough to be influenced by turbulence; this can be considered as corresponding to the first few per cent mass fraction burned.

(b) Turbulent burning, with a comparatively wide flame front, and pockets of unburned mixture entrained behind the flame front.

(c) A final burn period ('termination period' or 'burn-up'), in which the mixture within the thermal boundary layer is burned at a slow rate, because of a reduced fluid motion and a lower temperature.

Many different worker have published phenomenological combustion models, including Tabaczynski (1976), Tabaczynski et al. (1977) and (1980), Keck (1982), Beretta et al. (1983), Borgnakke (1984), Tomita and Hamamoto (1988), Keck et al. (1987) and James (1990).

For the majority of cases, the combustion chamber will have a complex geometric shape, in which it will be necessary to define the enflamed volume, the flame front area, and the area wetted by the enflamed volume. Two different approaches to this problem are provided by Poulos and Heywood (1983) and Cuttler et al. (1987). Poulos and Heywood divide the combustion chamber surface into a series of triangular elements, and then check for interception by vectors, of random direction, radiating from the spark plug. Cuttler et al. fill the combustion with tetrahedra, and employ vector algebra.

The turbulent combustion models used by Keck and co-workers and Tabaczynski and co-workers both seem capable of giving good predictions of turbulent combustion. Since the information published by Tabaczynski and co-workers is more explicit (notably Tabaczynski et al., 1977 and 1980), their model will be described here.

The flame is assumed to spread by a turbulent entrainment process, with burning occurring in the entrained region at a rate controlled by turbulence parameters. The flame front is assumed to entrain the mixture (that is, spread into the unburned mixture) at a rate that is governed by the turbulence intensity and the local laminar burn speed. The rate at which the mass is entrained into the flame front is given by

$$\frac{dm_e}{dt} = \rho_u A_f(u' + S_L) \qquad (9.29)$$

where
m_e = mass entrained into the flame front
ρ_u = density of the unburned charge
A_f = flame front area (excluding area in contact with surfaces)
u' = turbulence intensity
S_L = laminar flame speed.

Natural light photography of the flame front, as for example in **Plate 1**, will show the extremes of the flame front in any of the planes that are normal to the line of sight, so such images only give an indication of the volume of mixture (and thence the mass) that has been entrained. This is also true of Schlieren photography, but if a laser sheet illumination is used which distinguishes between burnt and unburnt gases, the volume of burnt gas can be deduced. One technique is to use Mie scattering, in which the laser light is reflected from particles that are larger than the wavelength of the light. More light is reflected from the unburnt gas because its density (and the concentration of the particles) is higher. Typically sub-micron titanium dioxide particles are used, produced by the hydrolysis of titanium tetrachloride. Hicks et al. (1994) employed a system with four sheets of laser light, each at different frequencies, so that by means of filters and a CCD camera, four simultaneous images were obtained from different planes. Pockets of burnt gas can be ahead of the flame front (as seen in chapter 3, figure 3.3). The multiple sheet images show how these pockets of burnt gas are connected to burnt gas in other planes.

The turbulence intensity is assumed to be proportional to the mean piston speed. The rate at which the mixture is burned is assumed to be proportional to the mass of unburned mixture behind the flame front:

$$\frac{dm_b}{dt} = \frac{m_e - m_b}{\tau} \qquad (9.30)$$

where
m_b = mass burned behind the flame front
τ = characteristic burn time for an eddy of size l_m
l_m = the Taylor microscale, which characterises the eddy spacing.

Since the eddies are assumed to be burned up by laminar flame propagation:

$$\tau = \frac{l_m}{S_L} \qquad (9.31)$$

This model assumes isotropic homogeneous turbulence throughout the combustion chamber at the time of ignition. After ignition the integral scale (l_i) and the turbulence intensity (u') are assumed to be governed by the conservation of the moment of momentum for coherent eddies. Thus

$$l_i = (l_i)_o \left(\frac{\rho_{ui}}{\rho_u}\right)^{1/3} \qquad (9.32)$$

and

$$u' = u'_o \left(\frac{\rho_u}{\rho_{ui}}\right)^{1/3} \qquad (9.33)$$

with the suffix 'o' referring to the values at ignition.

For isotropic turbulence:

$$l_m = \sqrt{\frac{(15 \nu l_i)}{u'}} \qquad (9.34)$$

where ν = dynamic viscosity.

By using equations (9.32), (9.33) and (9.34), the eddy burn-up term can be calculated at each time step (equation 9.32). However, this pre-supposes a knowledge of the kinematic viscosity, and the burned and unburned mixture density.

The mass within the combustion chamber (m) can be assumed constant, and this means that the only unknown is the ratio of unburned to burned gas density (D):

$$m = \rho_o V_o = \rho_u V_u + \rho_b V_b, \qquad \text{let } D = \frac{\rho_u}{\rho_b}$$

$$\rho_u = \frac{\rho_o V_o}{V - V_b + \dfrac{V_b}{D}} \qquad (9.35)$$

where V = volume
   V_u = volume of unburned mixture
   V_b = volume of burned mixture
   V_o = volume at ignition
   ρ_o = unburned mixture density at ignition

$$\text{mfb} = \frac{m_b}{m} = \frac{\rho_b V_b}{\rho_o V_o} \qquad (9.36)$$

where   mfb = mass fraction burnt
   m_b = mass burnt.

To find V_b:

$$V_b = \frac{m_b}{m} \cdot \frac{\rho_o V_o D}{P_u} \qquad (9.37)$$

The density ratio has to be calculated from the combustion

thermodynamics, and the background to this has been introduced in chapter 3, section 3.3, and will be discussed further in chapter 11, section 11.2.2.

To calculate the density of the unburned mixture the compression process can be modelled on a stepwise basis, using a heat transfer correlation to predict the heat flow from the unburned mixture; this is discussed later in chapter 11, section 11.2.4. However, an alternative and simpler method is to use a polytropic process to describe the compression of the unburned mixture (with a user-supplied value of the polytropic index).

Thus

$$P = P_o \left(\frac{\rho_u}{\rho_o}\right)^k \qquad (9.38)$$

and the temperature of the unburned gas is

$$T_u = \frac{P \rho_o T_o}{P_o \rho_u} \qquad (9.39)$$

where T_o = temperature of (unburned) mixture at ignition.

The kinematic viscosity can be calculated from an empirical correlation proposed by Collis and Williams (1959) for the dynamic viscosity (equation 9.12):

$$\nu = \frac{\nu_o \rho_o}{\rho_u} \left(\frac{T_u}{T_o}\right)^{0.76} \qquad (9.40)$$

It now remains to calculate the laminar flame speed from the burning velocity.

9.4.2   Laminar burning velocities

The laminar burning velocity is a function of the equivalence ratio, pressure, initial temperature of the reactants and exhaust residuals. The burning velocity can be represented as a parabolic function of equivalence ratio for any substances, and is given by

$$S_U^* = B_m + B_\phi (\phi - \phi_m)^2 \qquad (9.41)$$

where S_u^* = burning velocity at datum conditions
          (298 K, 1 atm)
   B_m = maximum burning velocity, occurring at $\phi = \phi_m$
   B_ϕ = empirical constant.

Values of B_m, B_ϕ and ϕ_m are presented in table 9.2.

Figure 9.11 shows the laminar burning velocities evaluated from table 9.2 at 298 K and 1 atm for the equivalence ratio range 0.7–1.2, for methanol, propane, iso-octane and gasoline.

The data at higher temperatures and pressures can be fitted to a power law:

$$S_U = S_{U,o} \left(\frac{T_u}{T_o}\right)^\alpha \left(\frac{P}{P_o}\right)^\beta \qquad (9.42)$$

Table 9.2 *Laminar burning velocity parameters for the range of 300–700 K and 1–8 bar*

Fuel	ϕ_m	B_m (m/s)	B_ϕ (m/s)	Reference
Methanol	1.11	0.369	−1.41	Metghalchi and Keck (1982)
Propane	1.08	0.342	−1.39	Metghalchi and Keck (1982)
Iso-octanes	1.13	0.263	−0.85	Metghalchi and Keck (1982)
Gasoline	1.21	0.305	−0.55	Rhodes and Keck (1985)

where $\alpha = 2.18 - 0.8(\phi - 1)$
$\beta = -0.16 + 0.22\ (\phi - 1)$.

For gasoline, Rhodes and Keck (1985) propose:

$$\alpha_g = 2.4 - 0.271\phi^{3.15}$$
$$\beta_g = -0.357 + 0.14\phi^{2.77} \tag{9.43}$$

Rhodes and Keck (1985) also found that the proportional reduction in burning velocity caused by the presence of residuals, was essentially independent of equivalence ratio, pressure and temperature. They developed the following correlation:

$$S_U(x_b) = S_U(x_b = 0)(1 - 2.06x_b^{0.77}) \tag{9.44}$$

where x_b = mole fraction of burned gas diluent.

The laminar burning velocity is the speed at which the flame is advancing into the unburnt mixture. For engine combustion, the unburnt mixture immediately ahead of the flame front will be moving because of the 'expansion' of the burnt gases. The density of the burnt gases (ρ_b) is less than that of the unburnt gases (ρ_u), because of the higher temperature of the burnt gas. Thus

$$S_L = S_U(1 + \rho_b/\rho_u) \tag{9.45}$$

A model using these methods has been developed by Brown (1991) to investigate the combustion of different fuels under various operating conditions. He has used this model to show how movement of the flame nucleus around the combustion chamber can lead to cycle-by-cycle variation in combustion. Figure 9.12 shows predictions of how the variations in burn rate, imep, maximum cylinder pressure and the maximum rate of cylinder pressure rise are affected by ignition timing.

9.4.3 The effect of turbulence on flame behaviour

The effect of turbulence on combustion is not always beneficial, as too much turbulence can lead to the extinction of a flame. To understand this, it is first necessary to examine the effects of the Karlovitz number (*Ka*), and (to a lesser extent) the Lewis number (*Le*) on flames. The Lewis number affects both laminar and turbulent combustion, and it is the ratio of thermal diffusivity (α) to mass diffusivity (D):

$$Le = \alpha/D \tag{9.46}$$

When the Lewis number is less than unity, combustion tends to be intensified since the reactants diffusing into the flame have a greater effect than the heat diffusing away from the reaction zone. Thus the flame temperature and burning velocity would both tend to increase. When a plane flame is subject to disturbances, then whether or not the disturbances decay depends strongly on the Lewis number. For $Le < 1$ (favouring mass diffusion) the combustion becomes more intense at the crest and less intense at the troughs, thereby wrinkling the flame. In contrast, when $Le > 1$ the perturbations will tend to decay. For laminar flames with $Le < 1$ the wrinkling can grow to form a cellular surface, in which the flame front area is much greater than that of a smooth surface (Lewis and von Elbe, 1961).

The Karlovitz number (*Ka*) is the flame strain rate (u'/l_m) normalised by (δ_l/S_L):

$$Ka = (u'/l_m)(\delta_l/S_L) \tag{9.47}$$

where δ_l is the laminar flame thickness
u' is the turbulence intensity
l_m is the Taylor microscale, and
S_L is the laminar burning velocity.

The Karlovitz number can also be thought of as a ratio of chemical (δ_l/S_L) to eddy (l_m/u') lifetimes. The eddy lifetime is the inverse of the turbulent flame stretch rate.

Turbulent flames can be seen as an assembly of stretched laminar flamelets, so long as molecular diffusion and chemical reaction rates dominate over the turbulent fluctuations, see Griffiths and Barnard (1995). The turbulent entrainment (not burning) velocity (u_e) is a commonly used method of characterising turbulent combustion, and it relates the rate at which the flame advances into the unburnt mixture (there are obvious similarities with equation 9.29):

$$dm_e/dt = \rho_u A_f u_e \tag{9.48}$$

where m_e is the mass entrained
ρ_u is the density of the unburned charge
A_f is the (smooth) flame front area (excluding any area in contact with surfaces) and
u_e is the turbulent entrainment velocity.

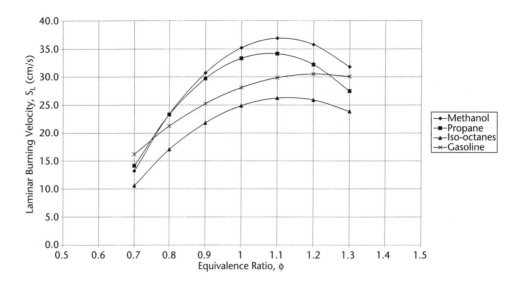

Figure 9.11
The laminar burning velocities evaluated from table 9.2 at 300 K and 1 bar for methanol, propane, iso-octane and gasoline.

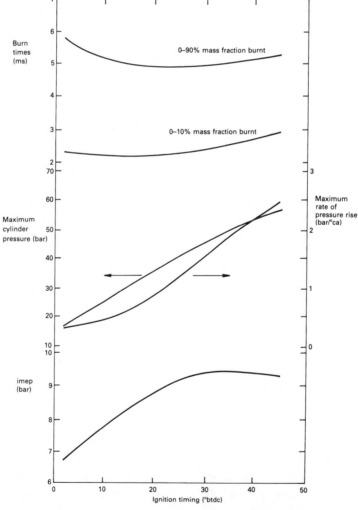

Figure 9.12
Predictions of how the burn rate, imep, maximum cylinder pressure, and the maximum rate of cylinder pressure rise are affected by ignition timing, for a Ricardo E6 engine operating at 1500 rpm with wide open throttle and an equivalence ratio of 1.0 using iso-octane.

Figure 9.13 is a Bradley diagram, which shows correlations between measurements of the turbulent entrainment velocity normalised by the laminar burning velocity (u_e/u_l) with a 'wrinkling' factor (the effective turbulence intensity normalised by the laminar burning velocity – u'_e/u_l). The correlations are based on the product of the Karlovitz number and the Lewis number ($Ka\,Le$). In turbulent combustion the Markstein number (Ma) is more relevant than the Lewis number, since it describes the response of a flame to stretching. However, $Ka\,Le$ has been used in preference to $Ka\,Ma$, because of difficulties in the evaluation of the Markstein number (Bradley et al., 1992). In figure 9.13 the effective turbulence intensity is that part of the turbulent spectrum that wrinkles a flame – the flame has to be larger than the turbulence scales for it to be wrinkled by the turbulence; the lower frequency components will just move the flame bodily.

For $Ka\,Le < 0.15$ there is a continuous flame sheet, and at higher values the flame sheet will begin to break up; above 6, the flame is likely to be completely quenched. In spark ignition engine combustion the locus of the flame on figure 9.13 will depend on the stage of combustion, the engine speed, load and mixture strength (Merdjani and Sheppard, 1993). Initially combustion will always be laminar, and in some cases (such as with very weak mixtures), the value of $Ka\,Le$ can be high enough to quench the flame. At a higher engine speed the weak mixture would become much closer to the extinction limit. Figure 9.13 also shows the loci for flames at three different equivalence ratios. The theoretical starting point for each flame is at (0, 1), since the combustion is purely laminar (so $u_t = u_l$), and the flame will be so small that it is not affected by the turbulence (that is the effective turbulence u'_e is zero). As a flame grows it is subject to more of the turbulence spectrum and it burns faster, so the locus moves up and to the right, approximately along a constant value of $Ka\,Le$. In the later stages of combustion the turbulence intensity is decreasing, and the higher pressures and

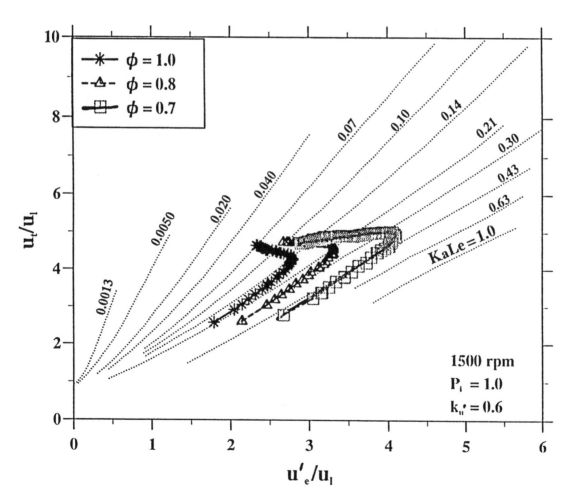

Figure 9.13
Correlations between measurements of the turbulent burning velocity normalised by the laminar burning velocity (u_t/u_l) with a 'wrinkling' factor (the effective turbulence intensity normalised by the laminar burning velocity, u'_e/u_l). The correlations are based on the product of the Karlovitz number and the Lewis number ($Ka\,Le$). Also shown are the operating loci for three different equivalence ratios, from an engine operating at 1500 rpm and full throttle (Sheppard, 1998).

temperatures in the unburnt gas increase the laminar burning velocity. The locus now moves to the left, with $Ka\,Le$ becoming smaller. The weakest mixture ($\phi = 0.7$) has the lowest laminar burning velocity, and its loci are thus closest to the quench limit. Figure 9.13 can be used to identify the conditions under which the flame might fail to propagate (partial burn).

9.5 | Conclusions

This chapter has firstly shown how the in-cylinder flow can be measured, and then that CFD is able to make realistic predictions of the steady and turbulent flow components.

Secondly, it has been shown how the turbulent burning velocity can be estimated from a knowledge of the turbulence and laminar burning velocity. Finally, flame stretch has been introduced as a means of deducing how a propagating flame will be affected by turbulence, and indeed whether or not it will be able to propagate.

The discussion of in-cylinder flow measurement techniques has shown that particle image velocimetry (PIV) is a good technique for the full field determination of the mean flows. For turbulence measurements it is necessary to use laser Doppler anemometry (LDA) or hot wire anemometry (HWA). Both are for point measurements, with LDA having about twice the accuracy of HWA, but discontinuous signals.

The parameters for characterising turbulence have been discussed, and this has led to a development of the eddy entrainment model for turbulent combustion. Correlations have been presented for estimating the laminar burning velocity of different air/fuel mixtures in terms of the unburnt gas temperature and pressure, the equivalence ratio and the residuals level.

9.6 | Questions

9.1 Verify from figure 9.7 that $l_{\mathrm{m}}^2 = -2/(\partial^2 R_x/\partial x^2)_{x=0}$.

9.2 Evaluate the laminar burning velocity for methanol, propane, iso-octane and gasoline at

(a) 300 K and 1 bar;
(b) 500 K and 1 bar; and
(c) 500 K and 5 bar,

for the equivalence ratio range 0.7–1.2, using the data from table 9.2.

CHAPTER 10

Turbocharging and supercharging

10.1 | Introduction

Turbocharging is a particular form of supercharging in which a compressor is driven by an exhaust gas turbine, so the presence of 'turbocharging' in the chapter is superfluous, but it does emphasise the most important form of supercharging. The concept of supercharging, supplying pressurised air to an engine, dates back to the beginning of the twentieth century. By pressurising the air at inlet to the engine the mass flow rate of air increases, and there can be a corresponding increase in the fuel flow rate. This leads to an increase in power output and usually an improvement in efficiency since mechanical losses in the engine are not solely dependent on the power output. Whether or not there is an improvement in efficiency ultimately depends on the efficiency and matching of the turbocharger or supercharger. Turbocharging does not necessarily have a significant effect on exhaust emissions.

Compressors can be divided into two classes: positive displacement and non-positive displacement types. Examples of positive displacement compressors include Roots, sliding vane, screw, reciprocating piston and Wankel types; some of these are shown in figure 10.1. The axial and radial flow compressors are dynamic or non-positive displacement compressors – see figure 10.2. Because of the nature of the internal flow in dynamic compressors, their rotational speed is an order of magnitude higher than that of internal combustion engines or positive displacement compressors. Consequently, positive displacement compressors are more readily driven from the engine crankshaft, an arrangement usually referred to as a 'supercharger'. Axial and radial compressors can most appropriately be driven by a turbine, thus forming a turbocharger. Again the turbine can be of an axial or radial flow type. The thermodynamic advantage of turbochargers over superchargers stems from their use of the exhaust gas energy during blow-down (chapter 2, figure 2.5).

Another type of supercharger is the Brown Boveri Comprex pressure wave supercharger shown in figure 10.3. The paddle-wheel type rotor is driven from the engine crankshaft, yet the air is compressed by the pressure waves from the exhaust. Some mixing of the inlet and exhaust gases will occur, but this is not significant.

The characteristics of turbochargers are fundamentally different from those of reciprocating internal combustion engines, and this leads to complex matching problems when they are combined. The inertia of the rotor also causes a delay in response to changes in load – turbolag. Superchargers have the added complication of a mechanical drive, and the compressor efficiencies are usually such that the overall economy is reduced at full load. However, the flow characteristics are better matched, and the transient response is good because of the direct drive. The Comprex supercharger absorbs minimal power from its drive, and has a good transient response; but it is expensive to make and requires a drive. The fuel economy is worse than a turbocharger, and its thermal loading is higher. The development theory and application of the Comprex supercharger are covered in the *Brown Boveri Review*, Vol. 7, No. 8, August 1987.

Comprex superchargers have not been widely used, and superchargers are used on spark ignition engines only where the main consideration is power output. Turbochargers have been used for a long time on larger compression ignition engines, and are now being used increasingly on automotive compression ignition and spark ignition engines.

Compound engines are also likely to gain in importance. A compound engine has a turbine geared to the engine crankshaft, either the same turbine that drives the compressor or a separate power turbine. The gearing is usually a differential epicyclic arrangement, and if matching is to be optimised over a range of speeds a variable ratio drive is also needed. Such combinations are discussed by Wallace

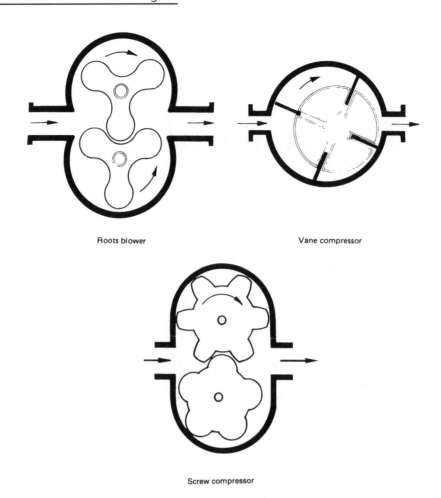

Roots blower　　　　　Vane compressor

Screw compressor

Figure 10.1
Types of positive displacement compressor (reproduced from Allard, 1982, courtesy of the publisher Patrick Stephens Ltd).

et al. (1983) and by Watson and Janota (1982). Compound engines offer improvements in efficiency of a few per cent compared with conventional turbocharged diesel engines.

Another development that is most relevant to turbocharged engines is the low heat loss (so called 'adiabatic') diesel engine.

In a naturally aspirated engine, the higher combustion chamber temperature will lead to a reduction in the volumetric efficiency, and this will offset some of the gains from the increased expansion work. The fall in volumetric efficiency is less significant in a turbocharged engine, not least since the higher exhaust temperature will lead to an increase

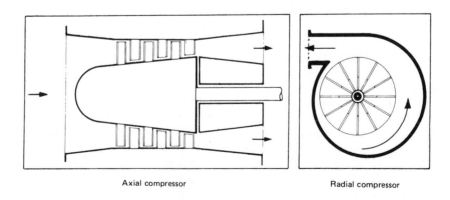

Axial compressor　　　　Radial compressor

Figure 10.2
Types of dynamic or non-positive displacement compressor (reproduced from Allard, 1982, courtesy of the publisher Patrick Stephens Ltd).

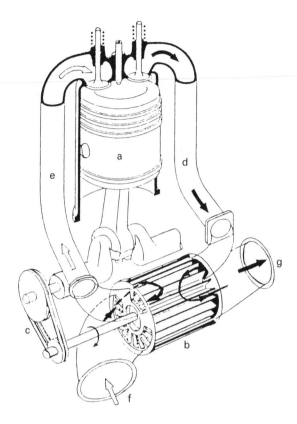

Figure 10.3
Brown Boveri Comprex pressure wave supercharger: (a) engine; (b) cell-wheel; (c) belt drive; (d) high-pressure exhaust; (e) high-pressure air; (f) low-pressure air; (g) low-pressure exhaust.

sors (figure 10.1). High-performance aircraft engines can use radial flow compressors (see figure 10.2), but their performance will be discussed later in the context of turbochargers in section 10.3.1. The screw compressor and vane compressor (figure 10.1) have some internal compression, but the Roots blower has no internal compression; all three are used for supercharging engines. This will lead to a lower efficiency at high pressure ratios, but this is less of a disadvantage than if a compressor with internal compression is compressing to too high a pressure; with supercharged spark ignition engines, the pressure ratios are mostly below 2. This discussion of efficiencies can be quantified once the performance of the different compressor types has been analysed.

Figure 10.4 shows how the Roots blower operation depends on a volume of gas trapped (V_t) between the rotor and the casing, being transported to a region of high pressure. Compression is by the reverse flow from the high-pressure region expanding irreversibly into the blower outlet. As this can be difficult to comprehend, pictures of an equivalent reciprocating compressor have been added to the state

in the work available from the turbine. This is discussed further at the end of section 10.4.

Commercial and marketing factors also influence the use of turbochargers. A turbocharged engine will fit in the existing vehicle range, and would not need the new manufacturing facilities associated with a larger engine.

Allard (1982) provides a practical guide to turbocharging and supercharging, and Watson and Janota (1982) give a rigorous treatment of turbocharging. Most of this chapter is devoted to turbocharging, and there is also a case study that considers a turbocharged engine at the end of chapter 11 on Computer modelling. Computer modelling is of great value in establishing the performance of turbocharged engines, since the complex performance characteristics of turbochargers make it difficult to predict their performance in conjunction with an engine.

10.2 | Supercharging

10.2.1 Thermodynamics of supercharging

Supercharging is mostly associated with spark ignition engines, and invariably uses positive displacement compres-

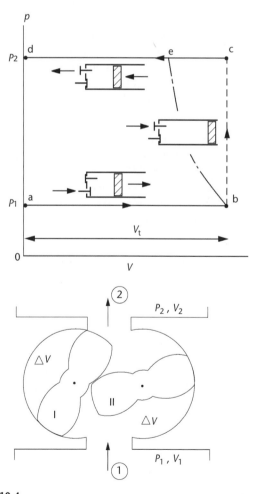

Figure 10.4
Operation of a Roots blower – a compressor with no internal compression.

diagram in figure 10.4, and its operating sequence is in parentheses after the description of the Roots blower, which operates as follows:

a → b   A volume V_t of gas at pressure p_1 enters the blower, and as the rotor tip passes the inlet port it is trapped. (The inlet valve is open and the piston travels down the cylinder.)

b → c   At the instant when the trapped volume communicates with the high-pressure port (at pressure p_2), the gas is compressed to pressure p_2 by the reverse flow from the high-pressure region expanding irreversibly into the volume V_t. The volume of the high-pressure reservoir is assumed to be sufficiently large so that the pressure p_2 remains constant. (At the end of the stroke the inlet valve is closed and the outlet valve is opened; the piston does not move.)

c → d   Further rotation of the rotor causes the volume V_t of high-pressure gas to be displaced into the high-pressure system, so long as there is a work input to the rotors. (The outlet valve remains open and the piston travels up the cylinder.)

d → a   The rotors are designed so that no pockets of gas are trapped between the rotors as they move from the high-pressure region to the low-pressure region. (At the end of the stroke the outlet valve is closed and the inlet valve is opened; the piston does not move.)

This description (and the subsequent analysis that follows) assumes that the downstream volume is sufficiently large for pressure p_2 to remain constant. Analysis for when this is not the case is discussed by Stone (1988), where there is also a discussion of the irreversibilities with over and under compression.

An examination of figure 10.4 shows that the work input required is

$$W = V_t(p_2 - p_1) = p_1 V_t(p_2/p_1 - 1) = mRT_1(p_2/p_1 - 1) \tag{10.1}$$

So the power requirement for a mass flow rate, $\dot{m}$, of a perfect gas is

$$\dot{W} = \dot{m}RT_1(p_2/p_1 - 1) \tag{10.2}$$

If the Roots blower is adiabatic, then the Steady Flow Energy Equation reduces to $w = h_2 - h_1$, and by assuming ideal gas behaviour ($h = c_p T$), equation (10.1) can be used to calculate the temperature rise:

$$W = V_t(p_2 - p_1) = p_1 V_t(p_2/p_1 - 1) = mRT_1(p_2/p_1 - 1)$$
$$= mc_p(T_2 - T_1)$$

or $T_2 = T_1(1 + [p_2/p_1 - 1][R/c_p]$ $\tag{10.3}$

To analyse compressors with internal compression, use the analysis for a steady flow process:

$$w = -\int_1^2 vdp \tag{10.4}$$

If the compression process in a compressor with internal compression is polytropic ($pv^n = k$), equation (10.4) can be rewritten and evaluated:

$$w = -\int_1^2 vdp = -\int_1^2 (k/p)^{1/n}dp = -(k)^{1/n}\int_1^2 p^{-1/n}dp$$
$$= -(1 - 1/n)^{-1}[k^{1/n} \times p^{1-1/n}]_1^2 = -(1 - 1/n)^{-1}[p^{1/n}v \times p^{1-}]$$
$$w = -\frac{n}{n-1}[pv]_1^2 = \frac{n}{n-1}(p_1v_1 - p_2v_2) = -\frac{nR}{n-1}(T_1 - T_2)$$
$$\text{for } n \neq 1 \tag{10.5}$$

For a polytropic process, $T_2 = T_1(p_2/p_1)^{(n-1)/n}$, so equation (10.5) becomes

$$w = -\frac{nR}{n-1}(T_1 - T_2) = \frac{nRT_1}{n-1}((p_2/p_1)^{(n-1)/n} - 1) \quad \text{for } n \neq 1 \tag{10.6}$$

Comparison of equation (10.6) with equation (10.2), and knowing that $(n - 1)/n < 1$, means that the polytropic compression process will always require less work than the compression process with a Roots blower. The polytropic compression process is shown as a broken line (be) in figure 10.4, so it is immediately apparent that the work input to the Roots blower (area abcd) is greater than into a compressor with internal compression (area abed), and as the pressure ratio increases so the difference increases more quickly.

At a particular engine speed the Roots blower will induct a specific flow rate of air. If this is more than the engine requires (for example, at part load), some of the outlet flow can be returned to the inlet and the correct delivery pressure will always be achieved. With a compressor that has a fixed internal pressure ratio, then too high a pressure at delivery will be followed by unresisted expansion and release of unwanted air, with no reduction in compression work. This loss can be reduced by throttling the inlet to the compressor, and this is all best illustrated by an example.

Consider an engine (with 1 bar ambient pressure) that is operating at part load, and only requires air to be delivered at a pressure of 1.5 bar, and the options are to use a Roots blower or a compressor with an internal pressure ratio of 2:1.

Table 10.1 shows that when the engine is at part load, throttling the compressor reduces the work input, but the Roots blower requires less power even though it is inherently less efficient than a compressor with internal compression – this is because it is only compressing over a 1.5:1 pressure ratio, so there are neither throttling losses nor unresisted expansion losses. If the Roots blower is throttled, the mass flow rate is reduced, but the pressure ratio is higher, so overall there is an increase in the compression work.

Table 10.1 *Comparison of the power (kW) required for a Roots blower (equation 10.2) and a vane compressor with an internal pressure ratio of 2:1 (equation 10.6). The polytropic compression index for air (k) is assumed to be 1.3, ambient conditions are 1 bar and 300 K, R = 287 J kg^{-1} K^{-1} and c$_p$ = 1010 J kg^{-1} K^{-1}. Consider two cases: engine inlet pressure of 2 bar (with 200 g/s) and 1.5 bar (with 150 g/s delivery to the engine, but the intake to the compressor or blower remains at 200 g/s unless throttled)*

Pressure ratio	Roots blower		Compressor -- internal pressure ratio of 2 : 1	
	Throttled at intake	Not throttled	Throttled at intake	Not throttled
2:1	Not applicable	17.22	Not applicable	12.96
1.5	(0.75 bar) 12.92	8.61	(0.75 bar) 9.72	12.96

It is also important that high-pressure gas is not returned to the inlet, and figure 10.5 shows the consequences of a 'pocket' of gas being trapped in a clearance volume and then expanding back into the inlet of the compressor.

Inspection of figure 10.5 shows that the work input will be reduced (it is now $\Delta V'(p_2 - p_1)$), but there is an even greater reduction in the freshly trapped charge (it is now V_t'). The expansion d' → a' can be treated as polytropic (so as to define the end state), so pV^k = constant, and if the clearance volume is known, the reduction in the intake flow can be readily calculated if there is a value for the polytropic exponent. The relevant process has already been discussed in chapter 7 in the context of exhaust blowdown at the end of section 7.3; see equation (7.50). The first gas to leave will expand isothermally at temperature $T_{d'}$, but the gas remaining in the pocket will expand isentropically until it too leaves isothermally.

The expanded gas will be at a lower temperature than ambient, but when it is fully mixed, the increase in volume of the expanded gas is exactly matched by the reduction in volume of the ambient air, as the air can be treated as a perfect gas. The effect of a clearance volume is best illustrated by an example:

If the clearance volume is $y\,\Delta V$, the work input is reduced by a factor $(1 - y)$.

If the pressure ratio is π, the reduction in the inducted volume is $y\pi^{1/k}$, and the inducted volume is reduced by a factor $(1 - y\pi^{1/k})$.

The efficiency is then reduced by a factor $(1 - y\pi^{1/k})/(1 - y)$

$$(10.7)$$

As would be expected from the form of equation (10.7), table 10.2 shows that as either the pressure ratio or clearance volume increases, the efficiency reduction factor increases in an ever-increasing manner.

The supercharger power requirements are not entirely parasitic, since the greater-than-ambient pressure during induction will lead to crankshaft work being produced during induction. If the air enters the engine at a temperature of 350 K, the density at 1.5 bar will be 1.49 kg/m^3, and at 2.0 bar 1.99 kg/m^3, so in both cases discussed in table 10.1 the flow rate is about 100 litre/s, and the power recovered will be potentially 5 kW at 1.5 bar (compressor work of 8.67 kW) and 10 kW at 2.0 bar (compressor work of 17.22 kW). With a turbocharged engine, the turbine might be 'undersized' so as to give a good transient response, but this will lead to a higher back-pressure in the exhaust and increased pumping work. McAllister and Buckley (2009) suggest that on a drive cycle there will only be a penalty of about 1 per cent in fuel consumption when using a supercharger compared to a turbocharger for a modestly boosted engine (up to a 1.5:1 pressure ratio), but there will be a better transient response.

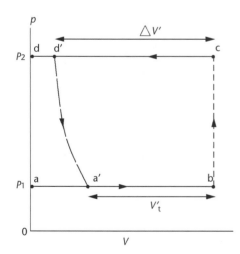

Figure 10.5
A 'pocket' of gas trapped in a clearance volume is expanding back into the inlet of the compressor.

Table 10.2 *The efficiency reduction factor caused by a clearance volume (equation 10.7) if k = 1.25*

Clearance volume (per cent)	Pressure ratio	
	1.5	2.0
5	0.980	0.961
10	0.957	0.918

10.2.2 Design and performance of superchargers

The design of Roots-type superchargers is discussed in great detail by Swartzlander (2006 and 2010). The Eaton supercharger uses helical multilobe rotors (twist of 160°) with an abradable graphite coating. The rotors in a Roots blower have an involute profile (as do gears), but the two rotors require external gearing if there is not to be touching of the rotors and sliding. The gears have a large number of teeth (as many as 100) so that gear whine is mostly above the audible range. There is an obvious trade-off when decreasing the rotor clearances between reducing leakage but increasing friction. The graphite coating abrades during the first few turns of the rotor and subsequently forms a low-friction surface, with operation at temperatures up to 150°C.

Increasing the rotor twist from 60° to 160° led to a better match of flow velocity to rotor velocity and avoided excessive flow velocities. However, increasing the rotor twist also decreases the sealing area for a given length rotor, but increasing the number of lobes or reducing the centre distance allows a greater amount of twist (Swartzlander, 2006). The increase in the rotor twist also improves other areas of the supercharger, as the inlet expansion occurred over a longer time period. This reduces the need to have a large dwell time (when the volume is sealed) during which the flow stagnates in the transfer volume. Swartzlander explains that the dwell and the seal events should be as short as possible. In the fifth-generation design, expansion occurs over 180° of rotation and a dwell of 40° was needed to ensure proper filling at high speeds, while in the sixth-generation (TVS design), expansion

occurred over 250° with only 10° of dwell (Swartzlander, 2006). The high-pressure side incorporates a 'blowhole' that introduces the high-pressure air to the low-pressure trapped volume so that its momentum adds to that of the internal flow. The high-pressure air is introduced immediately after the inlet is closed and at the inlet side. Swartzlander (2010) shows the results from CFD modelling, and further explains how: higher helix angles improve the high-speed performance; that lowering the L/d ratio improves the sealing properties (which in turn enhances the low-speed efficiency region); while reducing the number of lobes improves the sealing properties of the supercharger (thus enhancing the low-speed efficiency). The axial inlet and outlet help maintain the flow field of the transferring air which increases the pressure ratio and higher speed capabilities.

Figure 10.6 shows the performance map of an Eaton Roots blower. As expected, the isentropic efficiency falls as the pressure ratio increases (the isentropic efficiency is based on the adiabatic work case of equation 10.6, for which the polytropic index, n, is equal to the ratio of heat capacities, γ). At a pressure ratio of 2:1 the theoretical isentropic efficiency is 77 per cent (compared to a maximum of about 72 per cent in figure 10.6) and at a pressure ratio of 1.5:1 the theoretical isentropic efficiency is 86 per cent (compared to a maximum of about 75 per cent in figure 10.6). At a given pressure ratio there is a trade-off between the effects of leakage and flow losses. At a given pressure ratio the leakage flow is, to a first-order approximation, independent of the rotational speed and flow rate, so as the flow rate increases, the effect of leakage is less significant. In contrast, as the flow rate increases, the

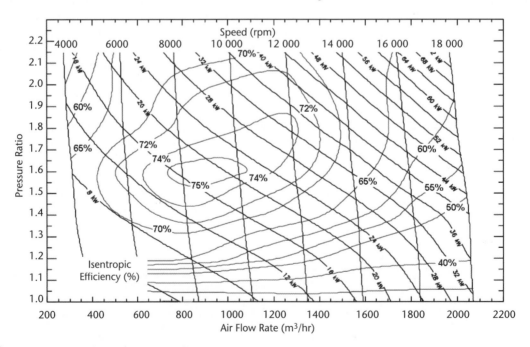

Figure 10.6
Performance of an Eaton sixth-generation R1900 TVS Roots compressor (swept volume of 1.9 litres/rev), the size used on the Jaguar AJ133 V8 SGDI engine (section 5.4). Adapted from Swartzlander (2006).

effects of fluid friction become more significant. So, for a given pressure ratio, the trade-off between these two effects leads to maximum efficiency. The entry flow rate only reduces slightly as the pressure ratio is increased, and this is a consequence of an increased leakage flow.

10.2.3 Supercharging case study

Figure 10.7 shows the Audi V6 3.0-litre TFSI engine which is a good example of a supercharged engine; it has been described in detail by Fitzen *et al.* (2008), including many mechanical design details. It is a direct injection gasoline engine with a 10.5:1 compression ratio, and a side-mounted solenoid-operated fuel injector; the six holes in the nozzle can flow up to 23 m^3/s.

Figure 10.8 shows how the Eaton Generation 6, R1320 Roots-type supercharger is accommodated within the valley of the V engine. The inlet manifold volume is reduced to below 2 litres, and this ensures a good transient response. The supercharger is shown in more detail within figure 10.8.

The throttle, supercharger, by-pass valve and inter-coolers (one for each bank of the V6) are all housed within a single module. The supercharger is driven at 2.5 times the engine speed by its own six-ribbed Poly-V belt, and is decoupled from the crankshaft torsional oscillations by means of the rubber

layer in the crankshaft torsional vibration damper. The supercharger also has a torsion spring element incorporated into the supercharger drive flange. It is carried between two hubs which also limit the dynamic movement with and against the direction of rotation of the supercharger. The design challenge is to allow the spring element to be soft enough to decouple torsional vibrations efficiently, yet at the same time to be able to prevent hammering of the end stops during dynamic operation. The pulsed nature of the Roots compressor delivery can also lead to noise, but this can be minimised by careful design of the inlet and outlet ports. The module casing has a damping panel and is ribbed to increase its stiffness so as to minimise radiated noise. A broadband damper has been integrated between the air filter and the main throttle, so that the pressure pulses emanating from the plenum volume are damped close to where they are generated. Specific resonances in the intake duct have been eliminated by means of an appropriately tuned Helmholtz resonator.

Figure 10.9 shows the air flow inside the supercharger module at part load with the by-pass valve open; some of the air volume delivered is fed back to the inlet side via the open by-pass valve. Using by-pass valves and a permanent drive to the supercharger incurs a fuel-consumption penalty of about 1 per cent compared to a clutched-drive. When the inlet

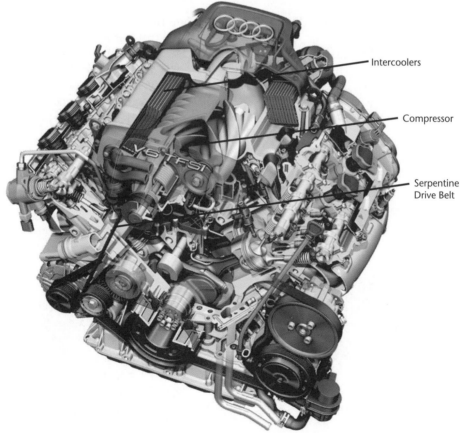

Intercoolers

Compressor

Serpentine
Drive Belt

Figure 10.7
The Audi V6 3.0-litre TFSI supercharged engine – Fitzen *et al.* (2008), © VDI_AUDI.

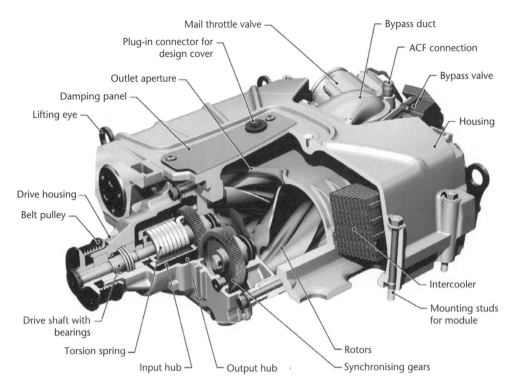

Mail throttle valve
Plug-in connector for design cover
Outlet aperture
Damping panel
Lifting eye
Drive housing
Belt pulley
Drive shaft with bearings
Torsion spring
Input hub
Output hub

Bypass duct
ACF connection
Bypass valve
Housing
Intercooler
Mounting studs for module
Rotors
Synchronising gears

Figure 10.8
The Eaton Generation 6, R1320 Roots-type supercharger used in the Audi V6 3.0-litre TFSI supercharged engine – Fitzen *et al.* (2008) , © VDI_AUDI.

manifold pressure is below atmospheric pressure, the by-pass valve is fully open and the air flow into the engine is controlled by the main throttle. When the inlet manifold pressure is above atmospheric pressure, the main throttle valve is fully open and the air flow into the engine is controlled by the by-pass valve.

Figure 10.10 shows the fuel consumption map of the Audi

V6 3.0-litre TFSI supercharged engine. If this is compared with the Jaguar SGDI Engine (chapter 5, figure 5.11), there are remarkable similarities below 10 bar, with both engines having a very similar minimum bsfc island of 240 g/kWh. At 120 km/h (75 mph) in the A6 saloon, the engine speed is 2160 rpm, the bmep is 5.1 bar and the bsfc is about 275 g/kWh. This implies a fuel consumption of about 9.1 litres/100 km.

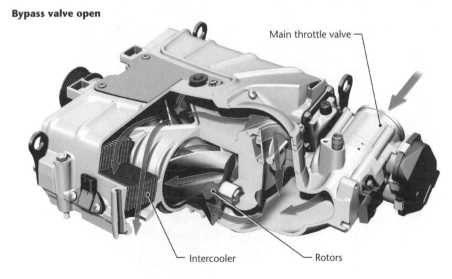

Bypass valve open

Main throttle valve

Intercooler
Rotors

Figure 10.9
Air flow inside the supercharger module with the bypass valve open – Fitzen *et al.* (2008), © VDI_AUDI.

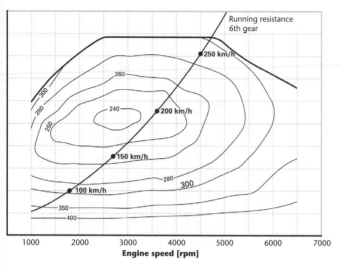

Figure 10.10
Fuel consumption map (g/kWh) of the Audi V6 3.0-litre TFSI supercharged engine, and the road load curve of the Audi A6 – Fitzen *et al.* (2008) , © VDI_AUDI.

10.3 | Radial flow and axial flow machines

10.3.1 Introduction and fluid mechanics

The turbomachinery theory applied to turbochargers is the same as for gas turbines, and is covered in books on gas turbines such as Harman (1981), and in books on turbocharging, see Watson and Janota (1982). As well as providing the

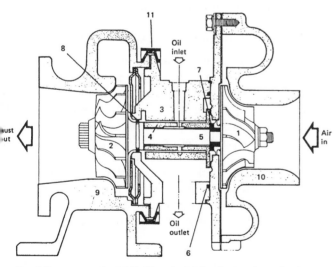

Key: 1 *Compressor wheel.* 2 *Turbine wheel.* 3 *Bearing housing.* 4 *Bearing.* 5 *Shaft.* 6 *Seal ('O' ring).* 7 *Mechanical face seal.* 8 *Piston ring seal.* 9 *Turbine housing.* 10 *Compressor housing.* 11 'V' band clamp.

Figure 10.11
Automotive turbocharger with radial compressor and radial turbine (reproduced from Allard, 1982, courtesy of the publisher Patrick Stephens Ltd).

theory, gas turbines also provided the materials technology for the high temperatures and stresses in turbochargers. Provided that the turbocharger is efficient enough to raise the engine inlet pressure above the exhaust pressure of the engine, the intake and exhaust processes will both benefit. This is particularly significant for engines with in-cylinder fuel injection (since unburnt fuel will not pass straight through the engine), and for two-stroke engines (since there are no separate induction and exhaust strokes).

The efficiency of turbines and compressors depends on their type (axial or radial flow) and size. Efficiency increases with size, since the losses associated with the clearances around the blades become less significant in large machines. These effects are less severe in radial flow machines, so although they are inherently less efficient than axial machines their relative efficiency is better in the smaller sizes.

Compressors are particularly difficult to design since there is always a risk of back-flow, and a tendency for the flow to separate from the blades in the divergent passages. Dynamic compressors work by accelerating the flow in a rotor, giving a rise in total or dynamic pressure, and then decelerating the flow in a diffuser to produce a static pressure rise. Radial compressors are more tolerant of different flow conditions, and they can also achieve pressure ratios of 4.5:1 in a single stage; an axial compressor would require several rotor/stator stages for the same pressure ratio.

A typical automotive turbocharger is shown in figure 10.11, with a radial flow compressor and turbine. For the large turbochargers used in marine applications, the turbine is large enough to be designed more efficiently as an axial flow turbine – see figure 10.12.

Immediately obvious in figure 10.11 is the absence of any stator blades; this is in contrast to the larger turbocharger shown in figure 10.12, or indeed almost any gas turbine. The absence of stator blades simplifies small turbochargers, and means that the turbocharger will operate satisfactorily over a wider flow range, albeit with a lower peak efficiency. This is illustrated by the direct comparison between vaned and vaneless diffusers by Fisher (1986). When stator blades are present, the flow angle from the rotor to the stator (or vice versa) has to match the blade angle, and this means that for every volumetric flow rate, there will only be one rotor speed at which there is proper matching. This is illustrated in figure 10.13, for a flow leaving an axial flow compressor rotor and entering stator blades that are intended to diffuse (decelerate) the flow to produce a pressure rise.

Radial flow compressors work well without stator blades because of the conservation of the moment of momentum in the diffuser and the increase of the flow area as the radius increases; this can be explained with reference to figure 10.14. In figure 10.14, the flow has a velocity relative to the blade of C_{rel}, and this can be resolved into two components: the radial component (C_r) and a tangential component (C_t). The tangential component has to be converted to an absolute

a)

b)

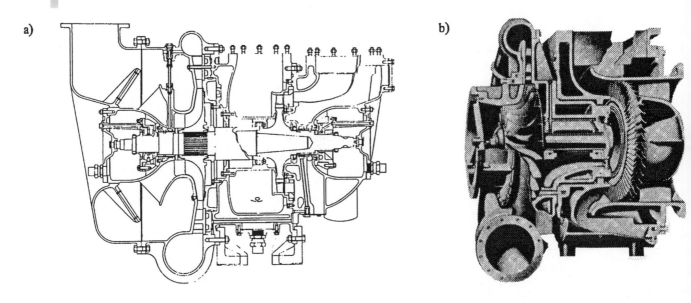

Figure 10.12
Marine turbochargers with radial compressors and axial turbines: (a) Napier; (b) Elliot (with acknowledgement to Watson and Janota, 1982).

velocity, by subtracting the blade tip velocity (ωr), to give the absolute tangential (or whirl) velocity (C_θ). The flow in the stator, immediately after leaving the rotor, thus has absolute velocity components comprising the radial component (C_r) and the tangential velocity (C_θ). These absolute velocities can also be defined as the velocity C with a flow angle α, and this is tangential to the particle path, which is shown as the chain dotted line. However, the action of the diffuser is best separately explained in terms of the radial and tangential velocity components. For convenience, the flow will be considered as frictionless. Firstly, even with a constant depth (*b*) diffuser, as the radius is increased there is a larger circumferential area, so that the radial velocity component will be reduced. (The mass flow is the product of the density, circumferential area and the *radial* velocity component.) Secondly, the tangential velocity component will reduce because of the conservation of the moment of momentum, which states that the product of the radius and the tangential velocity will be a constant. Thus it can be seen how the flow is diffused (decelerated) which will lead to a rise in pressure. Finally, this rise in pressure will increase the density, and this too contributes to the lower radial velocity component.

Radial flow turbochargers can be fitted with a number of variable-geometry devices, and the simplest is the variable scroll area turbine shown in figure 10.15a. For low flow rates the small nozzle area gives higher gas velocities, to maintain the turbocharger rotor speed and give a higher compressor boost pressure. Figure 10.15b shows a more complex arrangement in which the rotating nozzle vanes (the stator) are in a position that reduces the effective flow area for improved low flow rate performance. With higher flow rates the vanes are opened, so that the flow becomes more radial. The impact of rotating nozzle vanes on the turbocharger and engine performance has been reported by Yokota *et al.* (1986). The variable-geometry turbocharger, and other developments, led to an improved low speed torque and a specific fuel consumption of 197 g/kWh or less around 1200 rpm in the bmep range 7–16 bar. Figure 10.16 shows the Holset moving sidewall or sliding nozzle variable-geometry turbine. In the position shown there is a small flow area for the exhaust gases, but for higher flow rates the slide nozzle ring moves to the right, and exposes more of the nozzle guide vanes, thereby increasing the flow area. The performance of this turbocharger is discussed further in section 10.6.2.

10.3.2 Thermodynamics of turbochargers

The operation of a compressor or turbine is most sensibly shown on a temperature/entropy (T–s) plot. This contrasts with the Otto and Diesel cycles which are conventionally drawn on pressure/volume diagrams. The ideal compressor is both adiabatic and reversible and is thus isentropic – a process represented by a vertical line on the T–s plot (figure 10.17). The suffix s denotes an isentropic process. Real processes are of course irreversible, and are associated with an increase in entropy; this is shown with broken lines on figure 10.17. Expressions for work (per unit mass flow) can be found using the simplified version of the steady-flow energy equation

$$h_{in} + q = h_{out} + w$$

and since the processes are treated as adiabatic

$$w = h_{in} - h_{out}$$

No assumptions about irreversibility have been made in applying the steady-flow energy equation; thus

$$\text{turbine work, } w_t = h_3 - h_4 \tag{10.8}$$

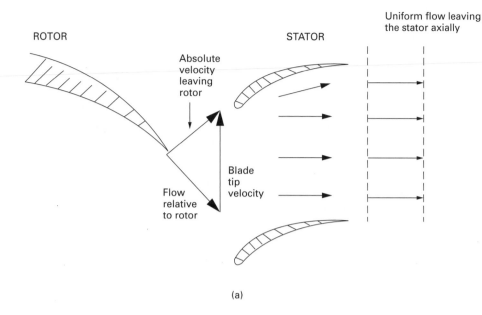

(a)

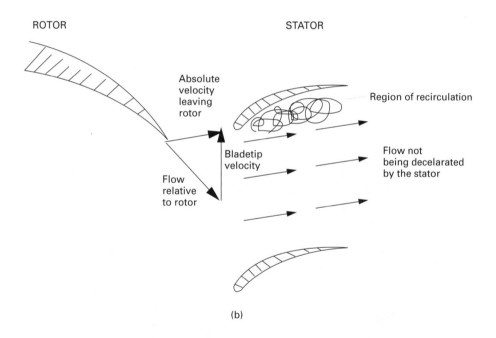

(b)

Figure 10.13
Flow leaving an axial compressor rotor and entering stator blades: (a) a well-matched flow; (b) a poorly matched flow (due to a lower rotor velocity for the same volume flow rate), in which separation from the stator blade leads to no deceleration of the flow.

and defining compressor work as a negative quantity

$$w_c = h_2 - h_1$$

For real gases, enthalpy is a strong function of temperature and a weak function of pressure. For semi-perfect gases, enthalpy is solely a function of temperature, and this is sufficiently accurate for most purposes. Thus

$$w_c = c_p(T_2 - T_1) \text{ kJ/kg} \tag{10.9}$$

and

$$w_t = c_p(T_3 - T_4) \text{ kJ/kg} \tag{10.10}$$

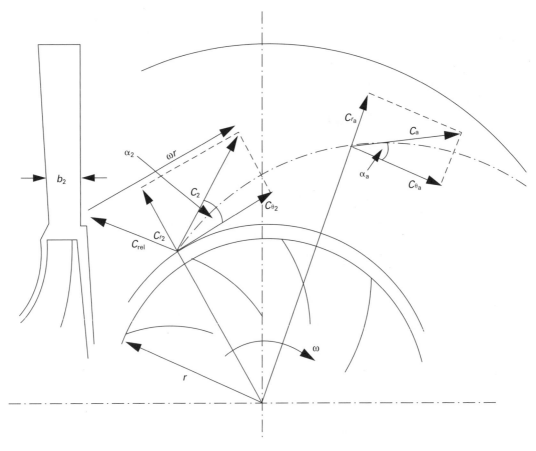

Figure 10.14
Velocity triangles for the flows in a vaneless radial flow compressor diffuser (adapted from Watson and Janota, 1982).

where c_p is an appropriate mean value of the specific heat capacity. The mean specific heat capacity can be evaluated from the information in Appendix A, table A.1, and such an exercise has been conducted in producing table 14.1. Consequently the T–s plot gives a direct indication of the relative compressor and turbine works.

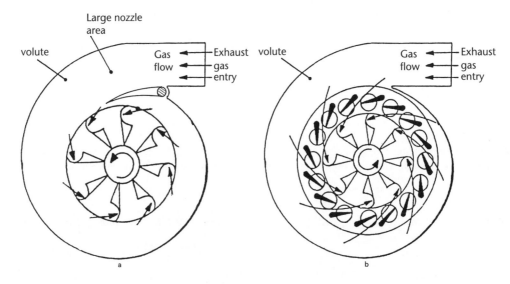

Figure 10.15
Variable-geometry radial flow turbines (adapted from Heisler, 1995).

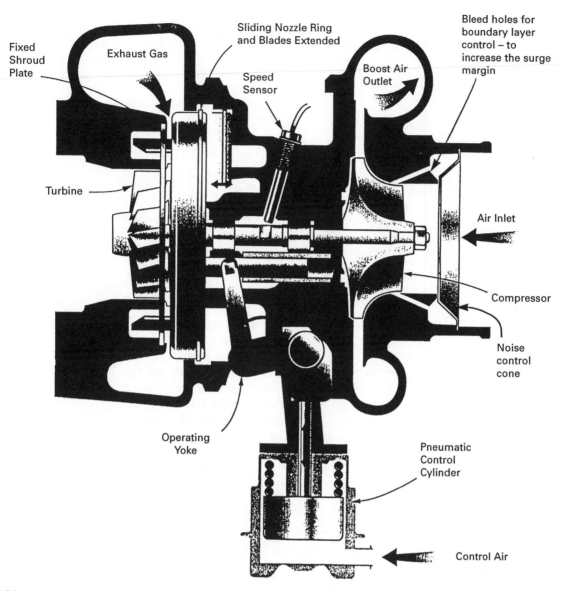

Figure 10.16
A moving sidewall variable-geometry radial flow turbine (courtesy of Cummins Turbo Technologies Ltd).

This leads to isentropic efficiencies that compare the actual work with the ideal work:

$$\text{compressor isentropic efficiency, } \eta_c = \frac{h_{2s} - h_1}{h_2 - h_1} = \frac{T_{2s} - T_1}{T_2 - T_1}$$

$$\text{and turbine isentropic efficiency, } \eta_t = \frac{h_3 - h_4}{h_3 - h_{4s}} = \frac{T_3 - T_4}{T_3 - T_{4s}}$$

It may appear unrealistic to treat an uninsulated turbine that is incandescent as being adiabatic. However, the heat transferred will still be small compared to the energy flow through the turbine. Strictly speaking, the kinetic energy terms should be included in the steady-flow energy equation. Since the kinetic energy can be incorporated into the enthalpy term, the preceding arguments still apply by using stagnation or total enthalpy with the corresponding stagnation or total temperature.

The shape of the isobars (lines of constant pressure) can be found quite readily. From the Second Law of Thermodynamics

$$T\mathrm{d}s = \mathrm{d}h - v\mathrm{d}p$$

Thus $\left(\dfrac{\partial h}{\partial s}\right)_p = T$

or $\left(\dfrac{\partial T}{\partial s}\right)_p \propto T$ that is, on the $T-s$ plot isobars have a positive slope proportional to the absolute temperature

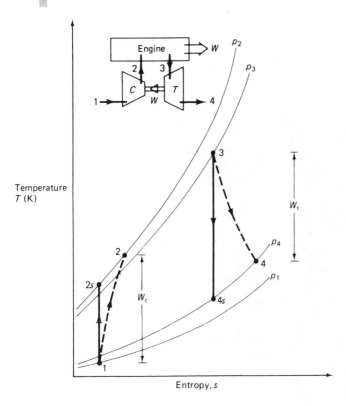

Figure 10.17
Temperature/entropy diagram for a turbocharger.

and $\left(\dfrac{\partial h}{\partial p}\right)_s = v$ that is, the vertical separation between isobars is proportional to the specific volume, and specific volume increases with temperature

Consequently the isobars diverge in the manner shown in figure 10.17.

In a turbocharger the compressor is driven solely by the turbine, and a mechanical efficiency can be defined as

$$\eta_m = \frac{w_c}{w_t} = \frac{m_{12}c_{p12}(T_2 - T_1)}{m_{34}c_{p34}(T_3 - T_4)} \tag{10.11}$$

As in gas turbines, the pressure ratios across the compressor and turbine are very important. From the pressure ratio the isentropic temperature ratio can be found:

$$\frac{T_{2s}}{T_1} = \left(\frac{p_2}{p_1}\right)^{(\gamma-1)/\gamma} \quad \text{and} \quad \frac{T_3}{T_{4s}} = \left(\frac{p_3}{p_4}\right)^{(\gamma-1)/\gamma} \tag{10.12}$$

The actual temperatures, T_2 and T_4, can then be found from the respective isentropic efficiencies.

In constant-pressure turbocharging it is desirable for the inlet pressure to be greater than the exhaust pressure ($p_2/p_3 > 1$), in order to produce good scavenging. This imposes limitations on the overall turbocharger efficiency ($\eta_m \cdot \eta_T \cdot \eta_c$) for different engine exhaust temperatures (T_3). This is shown in figure 10.18. The analysis for these results originates from the above expressions, and is given by Watson and Janota (1982).

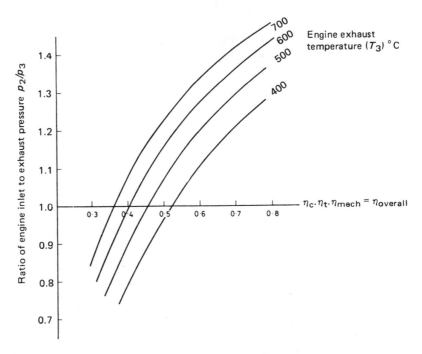

Figure 10.18
Effect of overall turbocharger on the pressure ratio between engine inlet and exhaust manifold pressures, for a 2:1 compressor pressure ratio ($p_1/p_2 = 2$) with different engine exhaust temperatures (with acknowledgement to Watson and Janota, 1982).

Example 10.1 also illustrates the work balance in a turbocharger.

The flow characteristics of an axial and radial compressor are compared in figure 10.19. The isentropic efficiencies would be typical of optimum-sized machines, with the axial compressor being much larger than the radial compressor. Since the turbocharger compressor is very small the actual efficiencies will be lower, especially in the case of an axial machine. The surge line marks the region of unstable operation, with flow reversal etc. The position of the surge line will also be influenced by the installation on the engine. Figure 10.19 shows the wider operating regime of a radial flow compressor. The isentropic efficiency of a turbocharger radial compressor is typically in the range 65–76 per cent.

The design of turbines is much less sensitive and the isentropic efficiency varies less in the operating range, and rises to over 90 per cent for aircraft gas turbines. The isentropic efficiency of turbocharger turbines is typically 70–85 per cent for radial flow and 80–90 per cent for axial flow machines. These are optimistic 'total to total' efficiencies that assume recovery of the kinetic energy in the turbine exhaust.

A detailed discussion of the internal flow, design and performance of turbochargers can be found in Watson and Janota (1982).

The flow from an engine is unsteady, owing to the pulses associated with the exhaust from each cylinder, yet turbines are most efficient with a steady flow. If the exhaust flow is smoothed by using a plenum chamber, then some of the energy associated with the pulses is lost. The usual practice is to design a turbine for pulsed flow and to accept the lower turbine efficiency. However, if the compressor pressure ratio is above 3:1 the pressure drop across the turbine becomes excessive for a single stage. Since a multi-stage turbine for pulsed flow is difficult to design at high pressure ratios, a steady constant-pressure turbocharging system should be adopted. The effect of flow pulsations on turbine performance is discussed in chapter 11, section 11.3.2.

In pulse turbocharging systems the area of the exhaust pipes should be close to the curtain area of the valves at full valve lift. Some of the gain in using small exhaust pipes comes from avoiding the expansion loss at the beginning of blowdown. In addition, the kinetic energy of the gas is preserved until the turbine entry. To reduce frictional losses the pipes should be as short as possible.

For four-stroke engines no more than three cylinders should feed the same turbine inlet. Otherwise there will be interactions between cylinders exhausting at the same time. For a four-cylinder or six-cylinder engine a turbine with two inlets should be used. The exhaust connections should be such as to evenly space the exhaust pulses, and the exhaust pipes should be free of restrictions or sharp corners. Turbines with four separate entries are available, but for large engines it can be more convenient to use two separate turbochargers. For a 12-cylinder engine two turbochargers, each with a twin entry turbine, could each be connected to a group of six cylinders. This would make installation easier, and the frictional losses would be reduced by the shorter pipe lengths. For large marine diesel engines, there can be one turbocharger per pair of cylinders. While there are thermodynamic advantages in lagging the turbines and pipework, the ensuing reduction in engine room temperature may be a more important consideration.

The pressure pulses will be reflected back as compression waves and expansion waves. The exact combination of reflected waves will depend on the pipe junctions and turbine entry. The pipe lengths should be such that there are no undesirable interactions in the chosen speed range. For example, the pressure wave from an opening exhaust valve will be partially reflected as a compression wave by the small turbine entry. If the pipe length is very short the reflected wave will increase the pressure advantageously during the initial blowdown period. A slightly longer pipe, and the delayed reflected wave will increase the pumping during the exhaust stroke – this increases the turbine output at the expense of increased piston work in the engine. An even longer pipe would cause the reflected wave to return to the exhaust valve during the period of valve overlap – this would impair the performance of a four-stroke engine and could ruin the performance of a two-stroke engine. If the pressure wave returns after the exhaust valve has closed, then it has no effect. Evidently great care is needed on engines with long exhaust pipes and large valve overlaps.

An alternative to multi-entry turbines is the use of pulse converters. An early pulse converter system is shown in figure 10.20; the idea was to use the jet from the nozzle to produce a low-pressure area around each exhaust port. The principal disadvantages are:

1 insufficient length between the ports for efficient diffusion

2 high frictional losses

3 each nozzle has to be larger than the last, resulting in high manufacturing cost.

A more realistic approach is to use pulse converters to connect groups of cylinders that would otherwise be separate. For example, four cylinders could be connected to a single turbocharger entry, figure 10.21. The steadier flow can also lead to an improvement in turbine performance. The design of the pulse converter is a compromise between pressure loss and unwanted pulse propagation. Reducing the throat area increases the pressure loss, but reduces the pulse propagation from one group of cylinders to another. The optimum design will depend on the turbine, the exhaust pipe length, the valve timing, the number of cylinders, the engine speed etc.

Constant-pressure turbocharging (that is, when all exhaust ports enter a chamber at approximately constant pressure) is best for systems with a high pressure ratio. The dissipation of the pulse energy is offset by the improved turbine efficiency.

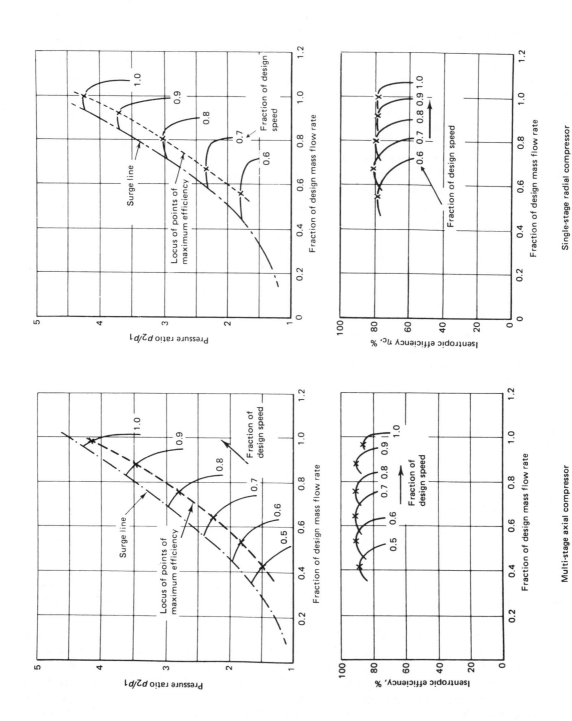

Figure 10.19
Flow characteristics of axial and radial compressors (reproduced with permission from Cohen *et al.*, 1972).

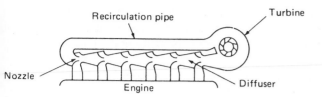

Figure 10.20
Early pulse converter system (with acknowledgement to Watson and Janota, 1982).

Furthermore, during blowdown the throttling loss at the exhaust valve will be reduced. However, the part load performance of a constant-pressure system is poor because of the increased piston pumping work, and the positive pressure in the exhaust system can interfere with scavenging.

10.4 | Turbocharging the compression ignition engine

The purpose of turbocharging is to increase the engine output by increasing the density of the air drawn into the engine. The pressure rise across the compressor increases the density, but the temperature rise reduces the density. The lower the isentropic efficiency of the compressor, the greater the temperature rise for a given pressure ratio.

Substituting for T_{2s} from equation (10.12) into equation (10.10) and rearranging gives

$$T_2 = T_1 \left[1 + \frac{(p_2/p_1)^{(\gamma-1)/\gamma} - 1}{\eta_c} \right] \qquad (10.13)$$

This result is for an ideal gas, and the density ratio can be found by applying the Gas Law, $\rho = p/RT$. Thus

$$\frac{\rho_2}{\rho_1} = \frac{p_2}{p_1} \left[1 + \frac{(p_2/p_1)^{(\gamma-1)/\gamma} - 1}{\eta_c} \right]^{-1} \qquad (10.14)$$

The effect of compressor efficiency on charge density is shown in figure 10.22; the effect of full cooling (equivalent to isothermal compression) has also been shown. It can be seen that the temperature rise in the compressor substantially decreases the density ratio, especially at high pressure ratios. Secondly, the gains in the density ratio on cooling the compressor delivery can be substantial. Finally, by ensuring that the compressor operates in an efficient part of the regime, not only is the work input minimised but the temperature rise is also minimised. Higher engine inlet temperatures raise the temperature throughout the cycle, and while this reduces ignition delay it increases the thermal loading on the engine.

The advantages of charge cooling lead to the use of inter-coolers. The effectiveness of the inter-cooler can be defined as

$$\varepsilon = \frac{\text{actual heat transfer}}{\text{maximum possible heat transfer}}$$

For the cooling medium it is obviously advantageous to use a medium (typically air or water) at ambient temperature (T_1), as opposed to the engine cooling water.

If T_3 is the temperature at exit from the inter-cooler, and the gases are perfect, then

$$\varepsilon = \frac{T_2 - T_3}{T_2 - T_1} \qquad (10.15)$$

or

$$T_3 = T_2(1 - \varepsilon) + \varepsilon T_1$$

In practice it is never possible to obtain heat transfer in a heat exchanger without some pressure drop. For many cases the two are linked linearly by Reynolds' analogy – that is, the heat transfer will be proportional to the pressure drop. In the following simple analysis the pressure drop will be ignored.

Substituting for T_2 from equation (10.13), equation (10.15) becomes

$$T_3 = T_1 \left\{ \left[1 + \frac{(p_2/p_1)^{(\gamma-1)/\gamma} - 1}{\eta_c} \right](1 - \varepsilon) + \varepsilon \right\}$$

$$= T_1 \left[1 + (1 - \varepsilon) \frac{(p_2/p_1)^{(\gamma-1)/\gamma} - 1}{\eta_c} \right] \qquad (10.16)$$

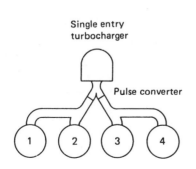

Single entry turbocharger

Pulse converter

1 2 3 4

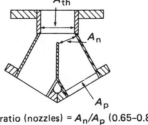

A_{th}

A_n

A_p

Area ratio (nozzles) = A_n/A_p (0.65–0.85)
Area ratio (throat) = $A_{th}/2A_p$ (0.5–1.0)

Figure 10.21
Exhaust manifold arrangement (four-cylinder engine) and pulse converter details (with acknowledgement to Watson and Janota, 1982).

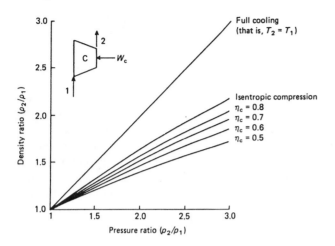

Figure 10.22

Effect of compressor efficiency on air density in the inlet manifold (with acknowledgement to Watson and Janota, 1982).

Neglecting the pressure drop in the inter-cooler, equation (10.14) becomes

$$\frac{\rho_3}{\rho_1} = \frac{p_2}{p_1}\left[1 + (1-\varepsilon)\frac{(p_2/p_1)^{(\gamma-1)/\gamma} - 1}{\eta_c}\right]^{-1} \qquad (10.17)$$

The effect of charge cooling on the density ratio is shown in figure 10.23 for a typical isentropic compressor efficiency of 70 per cent, and an ambient temperature of 20°C.

Despite the advantages of inter-cooling it is not universally used. The added cost and complexity are not justified for medium output engines, and the provision of a cooling source is troublesome. Gas-to-gas heat exchangers are bulky and in automotive applications would have to be placed upstream of the radiator. An additional heat exchanger could be used with an intermediate circulating liquid, but with yet more cost and

complexity. In both cases energy could be needed to pump the flows. Finally, the added volume of the inter-cooler will influence the transient performance of the engine.

The effect of inter-cooling on engine performance is complex, but two cases will be considered: the same fuelling rate and the same thermal loading. Inter-cooling increases the air flow rate and weakens the air/fuel ratio for a fixed fuelling rate. The temperatures will be reduced throughout the cycle, including the exhaust stage. The turbine output will then be reduced, unless it is rematched, but the compressor pressure ratio will not be significantly reduced. The reduced heat transfer and changes in combustion cause an increase in bmep and a reduction in specific fuel consumption. Watson and Janota (1982) estimate both changes as 6 per cent for a pressure ratio of 2.25 and inter-cooler effectiveness of 0.7. The gains are greatest at low flow rates where the inter-cooler is most effective.

If the fuelling rate is increased to give the same thermal loading Watson and Janota (1982) estimate a gain in output of 22 per cent. The specific fuel consumption will also be improved since the mechanical losses will not have increased so rapidly as the output.

Low heat loss engines also offer scope for improving the performance of diesel engines. Firstly, and most widely quoted, is the improvement in expansion work and the higher exhaust temperature. This leads to further gains when an engine is turbocharged. Secondly, the reduced cooling requirements allow a smaller capacity cooling system. The associated reduction in power consumed by the cooling system is, of course, most significant at part load.

Reducing the heat transfer from the combustion chamber also leads to a reduced ignition delay and hence reduced diesel knock. However, the high combustion temperatures will lead to an increase in NO_x emissions.

Reductions in heat transfer from the combustion chamber can be obtained by redesign with existing materials, but the

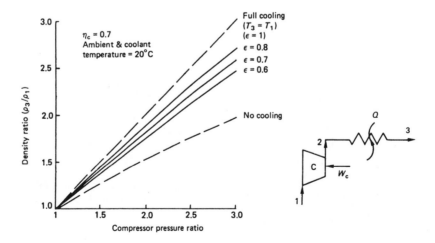

Figure 10.23

Effect of charge cooling on inlet air density (with acknowledgement to Watson and Janota, 1982).

greatest potential here is offered by ceramics. Ceramics can be used as an insulating layer on metallic components, or more radically as a material for the complete component. These heat transfer aspects are discussed further in chapter 13, section 13.2.3.

Differences between the thermal coefficients of expansion of metals and ceramics mean that great care is needed in the choice of the ceramic and the metal substrate, if the insulating layer is not to separate from its substrate. None the less, ceramic insulation has been used successfully in diesel engines – for example, the work reported by Walzer *et al.* (1985). In this turbocharged engine, 80 per cent of the combustion chamber surface was covered to an average depth of 3 mm by aluminium titanate or zirconium dioxide insulation. These measures led to a 13 per cent reduction in heat flow to the coolant, and a 5 per cent improvement in the urban cycle fuel economy (this was without re-optimisation of the cooling system).

A significant example of using ceramics in a diesel engine has been provided by Isuzu, with a 2.9-litre V6, 24-valve engine (Anon, 1990). This engine is shown in figure 10.24, and it can be seen that there is no cooling system.

A study of a low heat loss engine by Hay *et al.* (1986) suggested that a 30 per cent reduction in heat transfer would lead to a 3.6 per cent reduction in the specific fuel consumption at the rated speed and load. Furthermore, the 2 kW reduction in the cooling fan power would lead to an additional 2.7 per cent reduction in the fuel consumption, at the rated speed and load, and correspondingly greater percentage improvements at part load. Work by Wade *et al.* (1984) in a low heat loss engine suggests that the greatest gains in efficiency are at light loads and high speeds (figure 10.25).

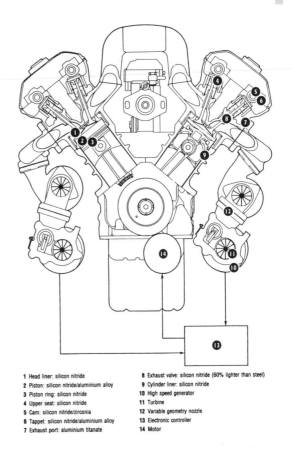

1 Head liner: silicon nitride
2 Piston: silicon nitride/aluminium alloy
3 Piston ring: silicon nitride
4 Upper seat: silicon nitride
5 Cam: silicon nitride/zirconia
6 Tappet: silicon nitride/aluminium alloy
7 Exhaust port: aluminium titanate
8 Exhaust valve: silicon nitride (60% lighter than steel)
9 Cylinder liner: silicon nitride
10 High speed generator
11 Turbine
12 Variable geometry nozzle
13 Electronic controller
14 Motor

Figure 10.24
Isuzu 2.9-litre V6 diesel engine with many ceramic components (Anon, 1990).

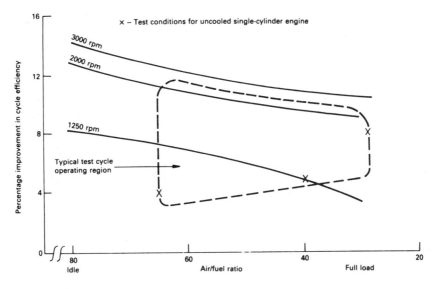

Figure 10.25
Calculated improvement in cycle efficiency for a fully insulated DI diesel engine relative to a baseline water-cooled DI diesel engine as a function of air/fuel ratio and engine speed. Bore 80 mm, stroke 88 mm, compression ratio 21:1 (adapted from Wade *et al.*, 1984).

Alternatively, the higher exhaust temperature will allow a smaller turbine back-pressure for the same work output. Work by Hoag *et al.* (1985) indicates that a 30 per cent reduction in heat transfer would lead to a 3.4 per cent fall in volumetric efficiency, a 70 K rise in the exhaust gas temperature, and reductions in fuel consumption of 0.8 per cent for a turbocharged engine, and 2 per cent for a turbocompound engine.

So far no mention has been made of matching the turbocharger to the engine. Reciprocating engines operate over a wide speed range, and the flow range is further extended in engines with throttle control. In contrast, turbomachinery performance is very dependent on matching the gas flow angles to the blade angles. Consequently, a given flow rate is correct only for a specific rotor speed, and away from the 'design point' the losses increase with increasing incidence angle. Thus turbomachines are not well suited to operating over a wide flow range. However, they do have high design point efficiencies, and are small because of the high speed flows.

The first stage in matching is to estimate the air flow rate. The compressor delivery pressure will be determined by the desired bmep. The air density at entry to the engine (ρ_2) can then be calculated from equation (10.17), and this leads to the air mass flow rate:

$$\dot{m}_a \approx \rho_2 \cdot N^* \cdot V_s \cdot \eta_{vol}$$

where $\dot{m}_a$ = air mass flow rate (kg/s)
        N^* = number of cycles per second (s^{-1})
        V_s = swept volume (m^3)
        η_{vol} = volumetric efficiency.

This will enable a preliminary choice of turbocharger to be made in terms of the 'frame size'. Within a given 'frame size', a range of compressor and turbine rotors and stators can be fitted. The compressor will be chosen in the context of the speed and load range of the engine, so that the engine will operate in an efficient flow regime of the compressor, yet still have a sufficient margin from surge. Once the compressor has been matched the turbine can be chosen. The turbine is adjusted by altering its nozzle ring, or volute if it is a radial flow machine. The turbine output is controlled by the effective flow area, hence also controlling the compressor boost pressure. Although calculations are possible, final development is invariably conducted on a test bed in the same manner as for naturally aspirated engines.

The flow characteristics (figure 10.19) can be conveniently combined by plotting contours of efficiency. The engine operating lines can then be superimposed (figure 10.26). The x-axis would be dimensionless mass flow rate if multiplied by $R/A\sqrt{c_p}$, but since these are constants for a given machine they are omitted. If the engine is run at constant speed, but increasing load, then the mass flow rate will increase almost proportionally with the increasing charge density or pressure

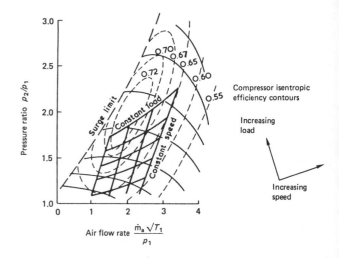

Figure 10.26
Superimposition of engine running lines on compression characteristics – constant engine load and speed lines (with acknowledgement to Watson and Janota, 1982).

ratio. This is shown by the nearly vertical straight line in figure 10.26.

If an engine is run at constant load but increasing speed, the volumetric flow rate of air will also increase. Since the effective flow area of the turbine remains almost constant, the turbine inlet pressure rises, so increasing the turbine work. The increased turbine work increases the compressor pressure ratio. This is shown by the gently rising lines in figure 10.26.

There must be sufficient margin between surge and the nearest operating point of the engine to allow for two factors. Firstly, the pulsating nature of the flow is likely to induce surge, and secondly the engine operating conditions may change from the datum. For example, a blocked air filter or high altitude would reduce the air flow rate, so moving the operating points closer to surge.

The turbine is tolerant of much wider flow variations than the compressor, and it is unrealistic to plot mean values for a turbine operating on a pulsed flow. Even for automotive applications with the widest flow variations, it is usually sufficient to check only the compressor operation.

The matching of two-stroke engines is simpler, since the flow is controlled by ports. These behave as orifices in series, and they have a unique pressure drop/flow characteristic. This gives an almost unique engine operating line, regardless of speed or load. However, the performance will be influenced by any scavenge pump.

In automotive applications the wide flow range is made yet more demanding by the requirement for maximum torque (or bmep) at low speeds. High torque at low speed reduces the number of gearbox ratios that are needed for starting and hill climbing. However, if the turbocharger is matched to give high torque at low speeds, then at high speeds the pressure ratio will be too great, and the turbocharger may also over-speed.

This problem is particularly severe on passenger car engines and an exhaust by-pass valve (waste gate) is often used (see figure 10.30b). The by-pass valve is spring regulated and, at high flow rates when the pressure rises, it allows some exhaust to by-pass the turbine, thus limiting the compressor pressure ratio.

Turbocharging is particularly popular for automotive applications since it enables smaller, lighter and more compact power units to be used. This is essential in cars if the performance of a compression ignition engine is to approach that of a spark ignition engine. In trucks the advantages are even greater. With a lighter engine in a vehicle that has a gross weight limit, the payload can be increased. Also, when the vehicle is empty the weight is reduced and the vehicle fuel consumption is improved. The specific fuel consumption of a turbocharged compression ignition engine is better than that for a naturally aspirated engine, but additional gains can be made by retuning the engine. If the maximum torque occurs at an even lower engine speed, the mechanical losses in the engine will be reduced and the specific fuel consumption will be further improved. However, the gearing will then have to be changed to ensure that the minimum specific fuel consumption occurs at the normal operating point. Ford (1982) claim that turbocharging can reduce the weight of truck engines by 30 per cent, and improve the specific fuel consumption by from 4 to 16 per cent. Figure 10.27 shows a comparison of naturally aspirated and turbocharged truck engines of equivalent power outputs.

In passenger cars a turbocharged compression ignition engine can offer a performance approaching that of a comparably sized spark ignition engine; its torque will be greater but its maximum speed lower. Compression ignition engines can give a better fuel consumption than spark ignition engined vehicles, but this will depend on the driving pattern (Radermacher, 1982) and whether the comparison uses a volumetric or gravimetric basis (see chapter 3, section 3.7).

The emissions from a diesel engine are affected by both the turbocharger and the inter-cooler. At a given stoichiometry, the NO_x emissions will increase with a turbocharged engine, because the higher temperatures at the start of compression are magnified by the subsequent compression and combustion processes. Consequently inter-cooling will reduce the inlet temperature to the engine and the NO_x emissions. With turbocharging, the higher temperatures towards the end of compression will reduce the ignition delay time. For a single injection system, the shorter delay time will reduce combustion noise and unburned hydrocarbon emissions, as there is less time for over-dilution at the fringe of the spray. Inter-cooling will have the opposite effect, but if split injection is used then only a small amount of fuel can be injected prior to combustion starting, and this too will reduce combustion noise and hydrocarbon emissions. The carbon monoxide emissions are governed by the stoichiometry and the partial combustion of hydrocarbons towards the end of the expan-

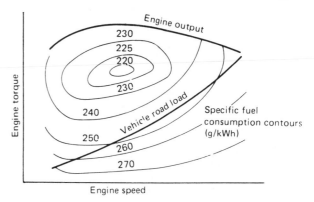

Naturally aspirated 8-cylinder Diesel engine

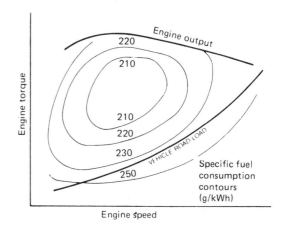

Turbocharged 6-cylinder Diesel engine

Figure 10.27
Comparison of comparably powerful naturally aspirated and turbocharged engines (Ford, 1982).

sion stroke, so a reduction in hydrocarbon emissions will lead to a reduction in carbon monoxide emissions.

10.5 | Turbocharging the spark ignition engine

Turbocharging the spark ignition engine is more difficult than turbocharging the compression ignition engine. The material from the previous section applies, but in addition spark ignition engines require a wider air flow range (owing to a wider speed range and throttling), a faster response, and more careful control to avoid either pre-ignition or self-ignition (knock). The fuel economy of a spark ignition engine is not necessarily improved by turbocharging. To avoid both knock and self-ignition it is common practice to lower the compression ratio, thus lowering the cycle efficiency. This may or may not be offset by the frictional losses representing a smaller fraction of the engine output.

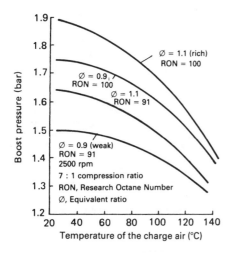

Figure 10.28
Influence of charge temperature on charge pressure (knock-limited) with different air/fuel ratios and fuel qualities (with acknowledgement to Watson and Janota, 1982).

The turbocharger raises the temperature and pressure at inlet to the spark ignition engine, and consequently pressures and temperatures are raised throughout the ensuing processes. The effect of inlet pressure and temperature on the knock-limited operation of an engine running at constant speed, with a constant compression ratio, is shown in figure 10.28. Higher octane fuels and rich mixtures both permit operation with higher boost pressures and temperatures. Retarding the ignition timing will reduce the peak pressures and temperatures to provide further control on knock. Unfortunately there will be a trade-off in power and economy and the exhaust temperature will be higher; this can cause problems with increased heat transfer in the engine and turbocharger. Reducing the compression ratio is the commonest way of inhibiting knock and retarding the ignition is used to ensure knock-free operation under all conditions.

Inter-cooling may appear attractive, but in practice it is not necessarily used. Compared with a compression ignition engine, the lower pressure ratios cause a lower charge temperature, which would then necessitate a larger inter-cooler for a given temperature drop. Furthermore, the volume of the inter-cooler impairs the transient response, and this is more significant in spark ignition engines with their low inertia and rapid response. Finally, a very significant temperature drop occurs through fuel evaporation, a process that cannot occur in compression ignition engines.

Water injection has been used in pressure-charged military spark ignition engines as a means of cooling the charge, so as to increase the charge density and inhibit the onset of knock. Saab have demonstrated the use of water injection in their 2.3-litre turbocharged spark ignition engine. Water injection (at up to 0.5 litre/min) is used at full throttle conditions above 3000 rpm,

permitting an increase in power from 120 kW at 4200 rpm to 175 kW at 5600 rpm for stoichiometric operation.

The fuel/air mixture can be prepared by either carburation or fuel injection, either before or after the turbocharger. Fuel injection systems are simplest since they deduce air mass flow rate and will be designed to be insensitive to pressure variations. In engines with carburettors it may appear more attractive to keep the carburettor and inlet manifold from the naturally aspirated engine. However, the carburettor then has to deal with a flow of varying pressure. The carburettor can be rematched by changing the jets, and the float chamber can be pressurised. Unfortunately, it is difficult to obtain the required mixture over the full range of pressures and flow rates. In general it is better to place the carburettor before the compressor for a variety of reasons. The main complication is that the compressor rotor seal needs improvement to prevent dilution of the fuel/air mixture at part load and idling conditions. The most effective solution is to replace the piston ring type seals with a carbon ring lightly loaded against a thrust face. A disadvantage of placing the carburettor before the compressor is that the volume of air and fuel between the carburettor and engine is increased. This can cause fuel hold-up when the throttle is opened, and a rich mixture on over-run when the throttle is closed, as discussed in chapter 4, section 4.6.1.

The advantage of placing the carburettor or single-point fuel injection before the compressor are:

(i)   the carburettor operates at ambient pressure

(ii) there is reduced charge temperature

(iii) compressor operation is further from the surge limit

(iv) there is a more homogeneous mixture at entry to the cylinders.

If the carburettor operates at ambient pressure then the fuel pump can be standard and the carburettor can be re-jetted or changed to allow for the increased volumetric flow rate.

The charge temperature will be lower if the carburettor is placed before the compressor. Assuming constant specific heat capacities, and a constant enthalpy of evaporation for the fuel, then the temperature drop across the carburettor (ΔT_{carb}) will be the same regardless of the carburettor position. The temperature rise across the compressor is given by equation (10.13):

$$T_2 = T_1 \left[1 + \frac{(p_2/p_1)^{(\gamma-1)/\gamma} - 1}{\eta_c} \right]$$

The term in square brackets is greater than unity, so that ΔT_{carb} will be magnified if the carburettor is placed before the compressor. In addition, the ratio of the specific heat capacities (γ) will be reduced by the presence of the fuel, so causing a further lowering of the charge temperature. This is illustrated by example 10.2, which also shows that the compressor work will be slightly reduced. The reduced charge temperature is very important since it allows a wider knock-free operation – see figure 10.28.

In spark ignition engines the compressor operates over a wider range of flows, and ensuring that the operation is always away from the surge line can be a greater problem than in compression ignition engines. If the carburettor, and thus the throttle, is placed before the compressor the surge margin is increased at part throttle. Consider a given compressor pressure ratio and mass flow rate and refer back to figure 10.24. The throttle does not change the temperature at inlet to the compressor (T_1), but it reduces the pressure (p_1) and will thus move the operating point to the right of the operating point when the throttle is placed after the compressor and p_1 is not reduced.

By the time a fuel/air mixture passes through the compressor it will be more homogeneous than at entry to the compressor. Furthermore, the flow from the compressor would not be immediately suitable for flow through a carburettor.

Performance figures vary, but typically a maximum boost pressure of 1.5 bar would raise the maximum torque by 30 per cent and maximum power by up to 60 per cent. Figure 10.29 shows the comparative specific fuel consumption of a turbocharged and naturally aspirated spark ignition engine. The turbocharged engine has improved fuel consumption at low outputs, but an inferior consumption at higher outputs. The effect on vehicle consumption would depend on the particular driving pattern, but also see chapter 5, section 5.2.

10.6 | Practical considerations and systems

10.6.1 Transient response

The transient response of turbocharged engines is discussed in detail by Watson and Janota (1982). The problems are most severe with spark ignition engines because of their wide speed range and low inertia; the problems are also significant with the more highly turbocharged compression ignition engines. The poor performance under changing speed or load conditions derives from the nature of the energy transfer between the engine and the turbocharger. When the engine accelerates or the load increases, only part of the energy available at the turbine appears as compressor work, the balance is used in accelerating the turbocharger rotor. Additional lags are provided by the volumes in the inlet and exhaust systems between the engine and turbocharger; these volumes should be minimised for a good transient response. Furthermore, the inlet volume should be minimised in spark ignition engines to limit the effect of fuel hold-up on the fuel-wetted surfaces. Turbocharger lag cannot be eliminated without some additional energy input, but the effect can be minimised. One approach is to under-size the turbocharger, since the rotor inertia increases with $(length)^5$, while the flow

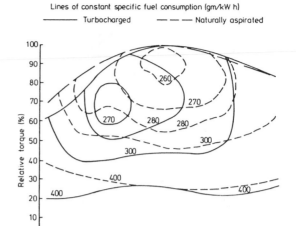

Figure 10.29
Comparative specific fuel consumption of a turbocharged and naturally aspirated engine scaled for the same maximum torque (with acknowledgement to Watson and Janota, 1982).

area increases with $(length)^2$. Then to prevent undue back-pressure in the exhaust, an exhaust by-pass valve can be fitted. An alternative approach is to replace a single turbocharger by two smaller units.

The same matching procedure is used for spark ignition engines and compression ignition engines. However, the wider speed and flow range (due to throttling) of the spark ignition engine necessitate greater compromises in the matching of turbomachinery to a reciprocating engine. If the turbocharger is matched for the maximum flow then the performance at low flows will be very poor, and the large turbocharger size will give a poor transient response. When a smaller turbocharger is fitted, the efficiency at low flow rates will be greater and the boost pressure will be higher throughout the range; the lower inertia will also reduce turbocharger lag. However, at higher flow rates the boost pressure would become excessive unless modified; two approaches are shown in figure 10.30.

The compressor pressure can be directly controlled by a relief valve, to keep the boost pressure below the knock-limited value. The flow from the relief valve does not represent a complete loss of work since the turbine work derives from energy that would otherwise be dissipated during the exhaust blowdown. The blow-off flow can be used to cool the turbine and exhaust systems. If the carburettor is placed before the compressor, the blow-off flow has to be returned to the compressor inlet, which results in yet higher charge temperatures.

The exhaust waste-gate system (figure 10.30b) is more attractive since it also permits a smaller turbine to be used, because it no longer has to be sized for the maximum flow. Turbocharger lag is reduced by the low inertia, and the control

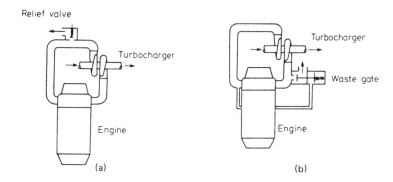

Figure 10.30

(a) Compressor pressure-relief valve control system. (b) Boost pressure-sensitive waste control system (with acknowledgement to Watson and Janota, 1982).

system ensures that the waste gate closes during acceleration. The main difficulty is in designing a cheap reliable system that will operate at the high temperatures. Variable-area turbines, compressor restrictors and turbine outlet restrictors can also be used to control the boost pressure; restrictors of any form are an unsatisfactory solution.

The transient response of an engine can be modelled by an extension to the type of simulation described in chapter 11 (Charlton *et al.*, 1991). This requires a knowledge of the engine, load and turbocharger inertias, and the governor and fuel injection pump (or engine management system) dynamic response. At the end of each cycle a torque balance needs to be made on each rotating component, so that any torque surplus (or deficit) can be used to accelerate (or decelerate) the relevant shaft. A change in a parameter, such as the boost pressure, can then be used to determine the new fuelling level and injection timing. Transient simulations are a useful method of identifying the strategies that might lead to better load acceptance on turbocharged engines.

10.6.2 Variable-geometry turbochargers, superchargers and two-stage turbocharging

Variable-geometry turbochargers

Variable-geometry turbines have already been discussed in section 10.3.1, and illustrated by figures 10.15 and 10.16. Figure 10.31 illustrates the benefits to the low speed torque of using the Holset moving sidewall variable-geometry turbine

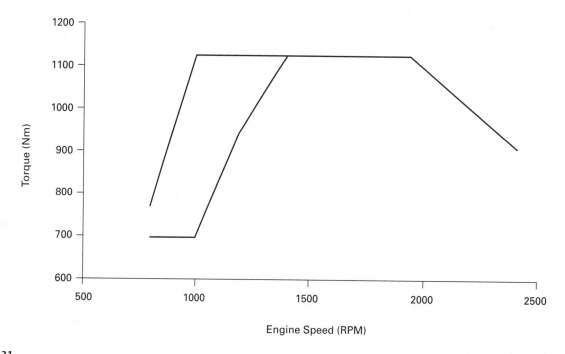

Figure 10.31

The increase in low engine speed torque, through the use of the Holset moving sidewall variable-geometry radial flow turbine (courtesy of Cummins Turbo Technologies Ltd).

Table 10.3 *Performance comparison of the waste-gate and variable-geometry turbocharger Fiat JTD direct injection diesel engines*

Model	1.9 JTD	2.4 JTD
Swept volume (cm^3)	1910	2387
Number of cylinders	4	5
Turbocharger system	Waste-gate	Variable-geometry
Maximum torque (N m)	255	304
bmep (bar)	16.8	16.0
at (rpm)	2000–2500	2000–2500
Maximum power (kW)	77	100
bmep (bar)	12.4	12.6
at (rpm)	3900	4000

of figure 10.16 in a truck engine application. The maximum torque engine speed range is extended by 40 per cent, and there is a 43 per cent improvement in the torque at 1000 rpm.

The nozzle position controls the flow area, and thus the turbine back-pressure, and the work output. This in turn determines the compressor boost pressure and the engine torque that is possible with an appropriate fuelling level. It is thus possible to control the turbocharger speed and boost pressure without using a waste gate. The nozzle position is varied by a pneumatic actuator, controlled by the engine management system, in response to:

1 engine speed,

2 throttle demand,

3 inlet manifold temperature and pressure,

4 exhaust manifold pressure (optional),

5 ambient pressure,

6 turbocharger speed, and

7 fuelling.

The variable-geometry turbine area can also be used to increase the turbine back-pressure, thereby increasing the engine braking.

Variable-geometry turbochargers are also used on smaller automotive diesel engines, and in the case of the Fiat JTD engines, the variable-geometry turbocharger is used to maintain a high bmep at high engine speed (Piccone and Rinolfi, 1998). The JTD engines are direct injection, with a bore of 82.0 mm, a stroke of 90.4 mm, an air-to-air inter-cooler, two valves per cylinder, an electronically controlled common rail fuel injection system, EGR, and a compression ratio of 18.5:1. Table 10.3 compares the performance of two JTD engines.

Table 10.3 shows how the variable-geometry turbocharger has led to a higher specific output by a combination of a higher bmep occurring at a higher speed.

Supercharging

One way of eliminating turbo-lag is to use a supercharger (a mechanically driven compressor), and this is the approach adopted by Jaguar for a spark ignition engine (Joyce, 1994). The Jaguar engine has a swept volume of 4 litres, so in normal use it will be operating under comparatively low load and speed conditions, the very conditions from which turbo-lag would be most significant. Obviously a supercharger will lead to worse fuel economy than a turbocharger, since there is no recovery of the exhaust gas expansion work. However, because of the limit to boost pressure imposed by fuel quality and knock, the pressure ratio will be comparatively low, and the fuel economy penalty will not be unacceptable. In any case, people who want a supercharged 4-litre engine will probably be more concerned with performance than economy. Jaguar adopted a Roots compressor with a pressure ratio of 1.5 and an air/coolant/air inter-cooler. Apart from the effect of compression irreversibilities increases, the other main loss is leakage past the rotors, which depends solely on the pressure ratio and seal clearances, and not on the flow rate through the compressor. Leakage losses are most significant at low speeds since the flow rates are low and the running clearances are greatest (due to the comparatively low temperature of the compressor); Jaguar found that inlet system deposits on the Roots blower rotors led to an in-service reduction in the leakage loss.

At part-load operation the supercharger is unnecessary, so one option is to use a clutch to control its use. However, to avoid a loss of refinement through engaging and disengaging the supercharger, Jaguar adopted a permanent drive with a step-up speed ratio of 2.5 (the drive has to meet a requirement of 34 kW). (If a continuously variable ratio drive is available, then throttling losses can be reduced at part-load operation by running the supercharger more slowly, and using it as an expander.) So as to avoid unnecessary compression in the supercharger, a by-pass valve opens at part load (in response to manifold pressure); the main throttle is upstream of the supercharger.

An inter-cooler is used since otherwise the performance gains from supercharging could be reduced by about 50 per cent. Without an inter-cooler, the supercharger outlet temperature would be limited to about 80°C (compared to 120°C with the inter-cooler), and after the inter-cooler the

Table 10.4 *The naturally aspirated and supercharged Jaguar AJ6 engines*

	Naturally aspirated	Supercharged
Compression ratio	9.5	8.5
Maximum torque (N m)	370 at 3500 rpm	510 at 3000 rpm
bmep (bar)	11.6	16.0
Maximum power (kW)	166 at 5000 rpm	240 at 5000 rpm
bmep (bar)	10.0	14.4

Table 10.5 *Comparison of the brake specific fuel consumptions (kg/kWh) for naturally aspirated and supercharged Jaguar engines at 2000 rpm, for a fixed bmep and a fixed torque*

Engine type	2 bar bmep	64 N m torque
6.0 litre V12	0.414	0.54
4.0 litre supercharged	0.412	0.41
4.0 litre naturally aspirated	0.390	0.39

temperature might be about 50°C. The inter-cooler thus increases the output of the engine in two ways. Firstly, the higher pressure ratio and the cooler air temperature both increase the air density, and secondly, the lower temperature allows a higher boost pressure or compression ratio for knock-free operation with a given quality fuel.

Table 10.4 compares the naturally aspirated and super-charged Jaguar AJ6 engines. Both have an aluminium cylinder head with 4 valves per cylinder, a bore of 91 mm and a stroke of 102 mm.

The supercharged engine uses the same camshafts (242°ca valve open period) as the naturally aspirated engine, but with zero overlap at tdc, so as to avoid short-circuiting loss of the mixture at high-load conditions, and to give good idle stability.

The supercharged engine has a higher power and torque output than the naturally aspirated 6-litre V12 engine, and table 10.5 compares the brake specific fuel consumptions of the engine at both the same bmep and the same torque (corresponding to a bmep of 2 bar for the 4-litre engine and 1.33 bar for the 6-litre engine).

The supercharged engine shows a slight loss of fuel economy compared to the naturally aspirated engine, but a substantial fuel economy benefit compared to the V12 engine, especially when the comparison is on the basis of a specified torque. The greater output of the supercharged engine allows a higher gearing compared to the naturally aspirated engine, which partially compensates for the lower fuel economy of the engine.

Two-stage supercharging

Figure 10.32 illustrates an arrangement for two-stage turbo-charging, in which the smaller high-pressure turbocharger has a turbine by-pass which opens as the inlet manifold boost pressure rises (Pfluger, 1997). At low mass flows the by-pass is

closed, so that best use is made of the available expansion in the small turbine. As the gas flows increase with load and speed, progressively more expansion work is extracted in the larger turbine that drives the low-pressure compressor.

Figure 10.33 shows how two-stage turbocharging leads to an improvement in the low speed torque, an increase in the maximum power output, and a reduction in the brake specific

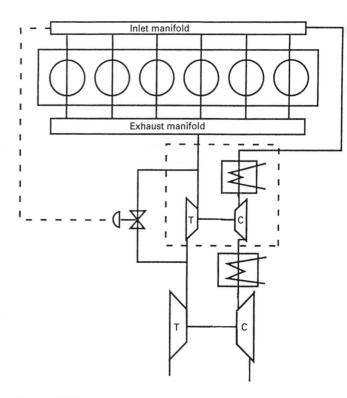

Figure 10.32
Arrangement for two-stage turbocharging, in which the smaller high-pressure turbocharger has a turbine by-pass which opens as the inlet manifold boost pressure rises (adapted from Pfluger, 1997).

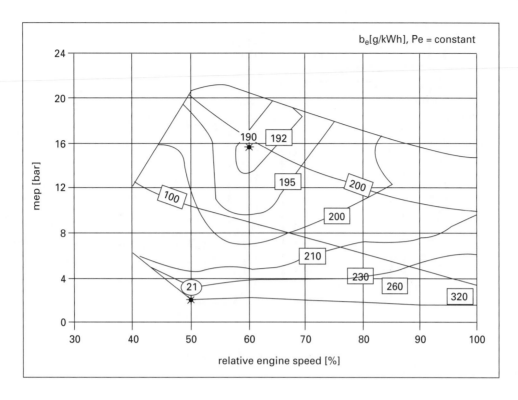

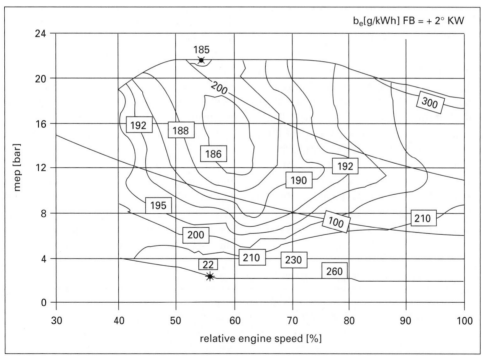

Figure 10.33
Torque and fuel economy comparisons between single-stage and two-stage turbocharging (adapted from Pfluger, 1997).

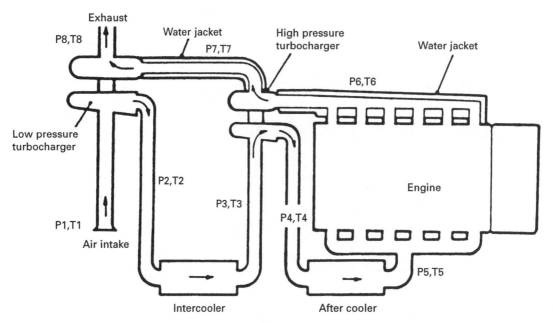

Figure 10.34
Arrangement of the turbochargers and inter-coolers for the Sabre Marathon diesel engine at 2400 rpm (Kaye, 1989).
© IMechE/ Professional Engineering Publishing Limited.

fuel consumption. Two-stage turbocharging also leads to a better transient response.

Two-stage turbocharging is also used for very high specific output applications, and a very well-documented example is the marine racing engine based on the Ford Dover diesel engine (Kaye, 1989). The Dover engine is also discussed in figures 1.10, 12.9 and 12.18. This engine has a maximum bmep of 33 bar and a specific output of 66 kW/litre. Figure 10.34 shows the arrangement of the turbochargers and inter-coolers; in marine applications inter-cooling is assisted by the typical sea water temperature of 12°C. Figure 10.35 shows the operating points on the compressor operating maps, for the maximum bmep of 33.0 bar and for 27.4 bar. Table 10.6 summarises the full-load operating conditions, and some analysis of these and other data forms the basis of question 10.16.

The maximum cylinder pressure has increased from 125 to 162 bar, and this has been limited by reducing the compression ratio from about 15:1 to 10.5:1, and by operating with relatively retarded injection timing. The dynamic start of injection is about 5°btdc with a duration of 36°ca, so combustion only commences after the start of the expansion stroke. The reduced compression ratio led to cold starting and warm-up difficulties; these were overcome by the use of ether (not just for starting) and other techniques detailed by Kaye (1989).

The basic engine structure required remarkably little modification for the higher loadings. The crankshaft main bearing webs were thickened, and used with steel bearing caps. A steel camshaft was used without a keyway to locate the drive gear, owing to the higher loads imposed on this from the fuel injection pump drive gear.

Table 10.6 *Operating conditions for the Sabre Marathon 6.8-litre diesel engine (Kaye, 1989)*

Power	448 kW at 2400 rpm	compression ratio	10.5 : 1
Torque	1780 N m	bmep	33 bar
bsfc	268 g/kWh	air/fuel ratio	21.9
Air flow rate	590 litres/s	Bosch Smoke Number	2.0

Turbocharger performance	Low pressure	High pressure
Temperature after compressor	168°C	217°C
Temperature after cooler	64°C	49°C
Compressor pressure ratio	2.14	2.50
Overall pressure ratio (after cooler losses)	5.1	
Turbine entry temperature	716	790°C
Exhaust temperature	601°C	

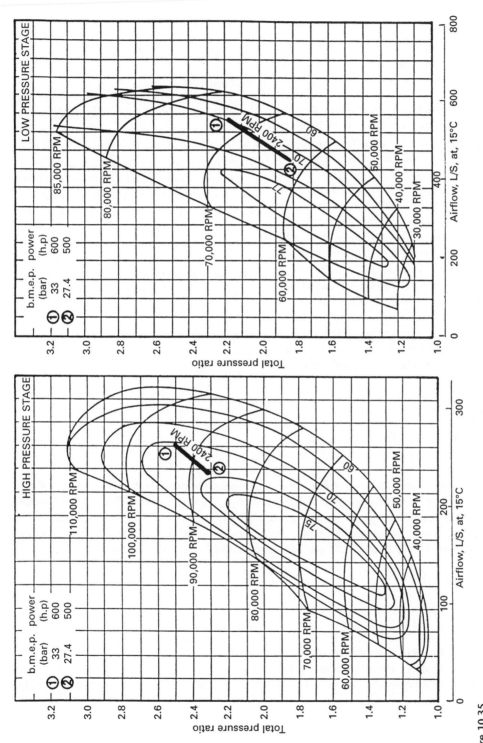

Figure 10.35
Operating points on the compressor operating maps for the Sabre Marathon diesel engine at 2400 rpm (adapted from Kaye, 1989).

10.7 | Conclusions

Turbocharging is a very important means of increasing the output of internal combustion engines. Significant increases in output are obtained, yet the turbocharger system leads to only small increases in the engine weight and volume.

The fuel economy of compression ignition engines is usually improved by turbocharging, since the mechanical losses do not increase in direct proportion to the gains in power output. The same is not necessarily true of spark ignition engines, since turbocharging invariably necessitates a reduction in compression ratio to avoid knock (self-ignition of the fuel/air mixture). The reduction in compression ratio reduces the indicated efficiency and this usually negates any improvement in the mechanical efficiency.

The relatively low flow rate in turbochargers leads to the use of radial flow compressors and turbines. In general, axial flow machines are more efficient, but only for high flow rates. Only in the largest turbochargers (such as those for marine applications) are axial flow turbines used. Turbochargers are unlike positive displacement machines, since they rely on dynamic flow effects; this implies high velocity flows, and consequently the rotational speeds are an order of magnitude greater than reciprocating machines. The characteristics of reciprocating machines are fundamentally different from those of turbochargers, and thus great care is needed in the matching of turbochargers to internal combustion engines.

The main considerations in turbocharging matching are:

(i) to ensure that the turbocharger is operating in an efficient regime;

(ii) to ensure that the compressor is operating away from the surge line (surge is a flow reversal that occurs when the pressure ratio increases and the flow rate decreases);

(iii) to ensure a good transient response.

Turbochargers inevitably suffer from turbo-lag; when either the engine load or speed increases, only part of the energy available from the turbine is available as compressor work – the balance is needed to accelerate the turbocharger rotor. The finite volumes in the inlet and exhaust system also lead to additional delays that impair the transient response.

As well as offering thermodynamic advantages, turbochargers also offer commercial advantages. In trucks, the reduced weight of a turbocharged engine gives an increase in the vehicle payload. Manufacturers can add turbocharged versions of an engine to their range more readily than producing a new engine series. Furthermore, turbocharged engines can invariably be fitted into the same vehicle range – an important marketing consideration.

10.8 | Examples

EXAMPLE 10.1

A diesel engine is fitted with a turbocharger, which comprises a radial compressor driven by a radial exhaust gas turbine. The air is drawn into the compressor at a pressure of 0.95 bar and at a temperature of 15°C, and is delivered to the engine at a pressure of 2.0 bar. The engine is operating on a gravimetric air/fuel ratio of 18:1, and the exhaust leaves the engine at a temperature of 600°C and at a pressure of 1.8 bar; the turbine exhausts at 1.05 bar. The isentropic efficiencies of the compressor and turbine are 70 per cent and 80 per cent, respectively. Using the values

$$c_{P_{air}} = 1.01 \text{ kJ/kg K}, \gamma_{air} = 1.4$$

and

$$c_{P_{ex}} = 1.15 \text{ kJ/kg K}, \gamma_{ex} = 1.33$$

calculate (i) the temperature of the air leaving the compressor

(ii) the temperature of the gases leaving the turbine

(iii) the mechanical power loss in the turbocharger expressed as a percentage of the power generated in the turbine.

Solution:
Referring to figure 10.36 (a new version of figure 10.17), the real and ideal temperatures can be evaluated along with the work expressions.

(i) If the compression were isentropic, $T_{2s} = T_1 \left(\dfrac{p_2}{p_1} \right)^{(\gamma-1)/\gamma}$

$$T_{2s} = 288\left(\frac{2.0}{0.95}\right)^{(1.4-1)/1.4} = 356 \text{ K, or } 83°C$$

From the definition of compressor isentropic efficiency,

$$\eta_c = \frac{T_{2s} - T_1}{T_2 - T_1}$$

$$T_2 = \frac{T_{2s} - T_1}{\eta_c} + T_1 = \frac{83 - 15}{0.7} + 15 = 133°C$$

(ii) If the turbine were isentropic, $T_{4s} = T_3\left(\dfrac{p_4}{p_3}\right)^{(\gamma-1)/\gamma}$

$$T_{4s} = 873\left(\frac{1.05}{1.8}\right)^{(1.33-1)/1.33} = 762.9 \text{ K, or } 490°C$$

From the definition of turbine isentropic efficiency,

$$\eta_t = \frac{T_3 - T_4}{T_3 - T_{4s}}$$

$$T_4 = T_3 - \eta_t(T_3 - T_{4s}) = 600 - 0.8(600 - 490) = 512°C$$

(iii) Compressor power $\dot{W}_c = \dot{m}_{air}\, c_{p_{air}}(T_2 - T_1)$

$$= \dot{m}_{air}\, 1.01(113 - 15) \text{ kW}$$

$$= \dot{m}_{air}\, 98.98 \text{ kW}$$

from the air/fuel ratio

$$\dot{m}_{ex} = \dot{m}_{air}\left(1 + \frac{1}{18}\right)$$

and turbine power

$$\dot{W}_t = \dot{m}_{ex}\, c_{p_{ex}}(T_3 - T_4)$$

$$= \dot{m}_{air}\, 1.056 \times 1.15(600 - 512)$$

$$= \dot{m}_{air}\, 106.82 \text{ kW}$$

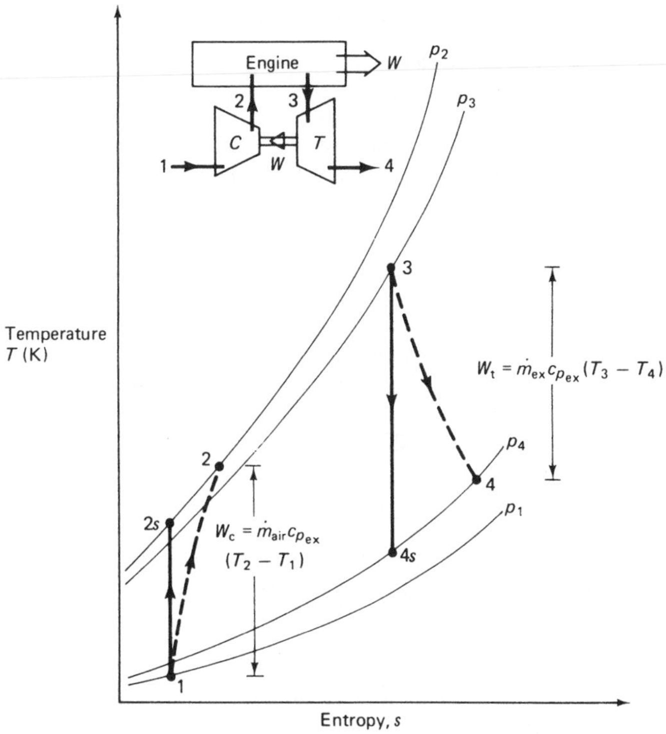

Thus, the mechanical power loss as a percentage of the power generated in the turbine is

$$\frac{106.82 \times 98.98}{106.82} \times 100 = 7.34 \text{ per cent}$$

This result is in broad agreement with figure 10.18, which is for a slightly different pressure ratio and constant gas flow rates and properties.

Figure 10.36
Temperature/entropy diagram for a turbocharger.

EXAMPLE 10.2

Compare the cooling effect of fuel evaporation on charge temperature in a turbocharged spark ignition engine for the following two cases:

(a) the carburettor placed before the compressor

(b) the carburettor placed after the compressor.

The specific heat capacity of the air and the latent heat of evaporation of the fuel are both constant. For the air/fuel ratio of 12.5:1, the evaporation of the fuel causes a 25 K drop in mixture temperature. The compressor efficiency is 70 per cent for the pressure ratio of 1.5, and the ambient air is at 15°C. Assume the following property values:

for air $c_p = 1.01$ kJ/kg K, $\gamma = 1.4$
for air/fuel mixture $c_p = 1.05$ kJ/kg K, $\gamma = 1.34$

Finally, compare the compressor work in both cases.

Solution:
Both arrangements are shown in figure 10.37.

(a) $T_1 = 15°C = 288$ K
 $T_2 = T_1 - 25 = 263$ K

If the compressor were isentropic, $T_{3s} = 263 \, (1.5)^{(1.34-1)/1.34} = 291.5$ K
  From the definition of compressor isentropic efficiency

$$T_3 = \frac{T_{3s} - T_2}{\eta_c} + T_2 = \frac{291.5 - 263}{0.7} + 263 = 303.7 \text{ K}$$

(b) $T_4 = 288$ K

From isentropic compression $T_{5s} = 288 \, (1.5)^{(1.4-1)/1.4} = 323.4$ K
  From the definition of compressor isentropic efficiency

$$T_5 = \frac{T_{5s} - T_4}{\eta_c} + T_4 = \frac{323.4 - 288}{0.7} + 288 = 338.5 \text{ K}$$

$$T_6 = T_5 - 25 = 338.5 - 25 = 313.5 \text{ K}$$

Since $T_6 > T_3$, it is advantageous to place the carburettor before the compressor.
  Comparing the compressor power for the two cases:

$$(W_c)_a = \dot{m}_{mix} \, c_{P_{mix}} (T_3 - T_2)$$
$$= 1.08 \, \dot{m}_{air} \, 1.05(303.7 - 263)$$
$$= 46.15 \, \dot{m}_{air} \text{ kW}$$

$$(W_c)_b = \dot{m}_{air} \, c_{P_{air}} (T_5 - T_4)$$
$$= \dot{m}_{air} \, 1.01(338.5 - 288)$$
$$= 51.01 \, \dot{m}_{air} \text{ kW}$$

Thus placing the carburettor before the compressor offers a further advantage in reduced compressor work.
  The assumptions in this example are somewhat idealised. When the carburettor is placed before the compressor, the fuel will not be completely evaporated before entering the compressor, and evaporation will continue during the compression process. However, less fuel is likely to enter the cylinders in droplet form if the carburettor is placed before the compressor rather than after.

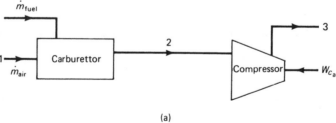

(a)

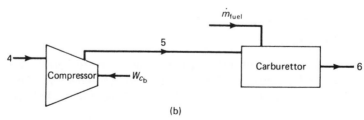

(b)

Figure 10.37
Possible arrangement for the carburettor and compressor in a spark ignition engine: (a) carburettor placed before the compressor; (b) carburettor placed after the compressor.

10.9 | Questions

10.1 A spark ignition engine is fitted with a turbocharger that comprises a radial flow compressor driven by a radial flow exhaust gas turbine. The gravimetric air/fuel ratio is 12:1, with the fuel being injected between the compressor and the engine. The air is drawn into the compressor at a pressure of 1 bar and at a temperature of 15°C. The compressor delivery pressure is 1.4 bar. The exhaust gases from the engine enter the turbine at a pressure of 1.3 bar and a temperature of 710°C; the gases leave the turbine at a pressure of 1.1 bar. The isentropic efficiencies of the compressor and turbine are 75 per cent and 85 per cent, respectively.

Treating the exhaust gases as a perfect gas with the same properties as air, calculate:

(i) the temperature of the gases leaving the compressor and turbine
(ii) the mechanical efficiency of the turbocharger.

10.2 Why is it more difficult to turbocharge spark ignition engines than compression ignition engines? Under what circumstances might a supercharger be more appropriate?

10.3 Why do compression ignition engines have greater potential than spark ignition engines for improvements in power output and fuel economy as a result of turbocharging? When is it most appropriate to specify an inter-cooler? [Consider equation (10.17) to illustrate the answer.]

10.4 Derive an expression that relates compressor delivery pressure (p_2) to turbine inlet pressure (p_3) for a turbocharger with a mechanical efficiency η_{mech}, and compressor and turbine isentropic efficiencies η_c and η_t, respectively. The compressor inlet conditions are p_1, T_1, the turbine inlet temperature is T_3 and the outlet pressure is p_4. The air/fuel ratio (AFR) and the differences between the properties of air (suffix a) and exhaust (suffix e) must all be considered. Assume $p_4 = p_1$.

10.5 Why do turbochargers most commonly use radial flow compressors and turbines with non-constant pressure supply to the turbine?

10.6 Why does turbocharging a compression ignition engine normally lead to an improvement in fuel economy, while turbocharging a spark ignition engine usually leads to decreased fuel economy?

10.7 Show that the density ratio across a compressor and inter-cooler is given by

$$\frac{\rho_3}{\rho_1} = \frac{P_2}{P_1} \left[1 + (1 - \varepsilon)\frac{(P_2/P_1)^{\frac{\gamma-1}{\gamma}} - 1}{\eta_c} \right]^{-1}$$

where 1   refers to compressor entry
       2   refers to compressor delivery
       3   refers to inter-cooler exit
       η_c = compressor isentropic efficiency
       ε = inter-cooler effectiveness = $(T_2 - T_3)/(T_2 - T_1)$.

Neglect the pressure drop in the inter-cooler, and state any assumptions that you make.

Plot a graph of the density ratio against effectiveness for pressure ratios 2 and 3, for ambient conditions of 1 bar, 300 K, if the compressor isentropic efficiency is 70 per cent.

What are the advantages and disadvantages in using an inter-cooler? Explain under what circumstances it should be used.

10.8 A turbocharged diesel engine has an exhaust gas flow rate of 0.15 kg/s. The turbine entry conditions are 500°C at 1.5 bar, and the exit conditions are 450°C at 1.1 bar.

(a) Calculate the turbine isentropic efficiency and power output.

The engine design is changed to reduce the heat transfer from the combustion chamber, and for the same operating conditions the exhaust temperature becomes 550°C. The pressure ratio remains the same, and assume the same turbine isentropic efficiency.

(b) Calculate the increase in power output from the turbine.

How will the performance of the engine be changed by reducing the heat transfer, in terms of economy, power output and emissions?

Assume ratio of specific heat capacities = 1.3, and $c_p = 1.15$ kJ/kg K.

10.9 A compressor with the performance characteristics shown in figure 10.38 is operating with a mass flow rate (m) of 49.5 g/s at an isentropic efficiency of 60 per cent. The compressor is fitted to a turbocharged and inter-cooled diesel engine. Assume

p (the compressor inlet pressure) is 95 kN/m^2
T (the inlet temperature) is 291 K
γ (the ratio of gas specific heat capacities) is 1.4
$c_p = 1.01$ kJ/kg K.

Calculate the pressure ratio, the compressor speed, the compressor delivery temperature and the compressor power.

If the inter-cooler is removed and the air/fuel ratio is kept constant, how would the compressor operating point be affected? Neglecting any change in the compressor efficiency and pressure ratio, estimate the maximum increase in output that the inter-cooler could lead to. State clearly any assumptions that you make.

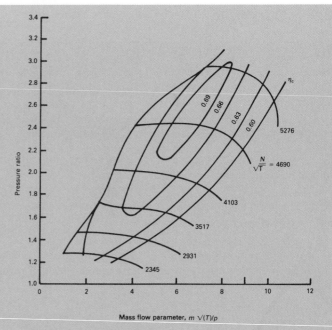

Figure 10.38
Compressor map.

10.10 A turbocharged six-cylinder four-stroke diesel engine has a swept volume of 39 litres. The inlet manifold conditions are 2.0 bar and 53°C. The volumetric efficiency of the engine is 95 per cent, and it is operating at a load of 16.1 bar bmep, at 1200 rpm with an air/fuel ratio of 21.4. The power delivered to the compressor is 100 kW, with entry conditions of 25°C and 0.95 bar. The fuel has a calorific value of 42 MJ/kg.

Stating any assumptions, calculate:

(a) the power output of the engine
(b) the brake efficiency of the engine
(c) the compressor isentropic efficiency
(d) the effectiveness of the inter-cooler.

Estimate the effect of removing the inter-cooler on the power output and emissions of the engine, and the operating point of the turbocharger.

10.11 A turbocharged diesel engine has a compressor operating point that is marked by a cross on figure 10.39. If the compressor entry conditions are a pressure of 1 bar and a temperature of 20°C, determine the volume flow rate out of the compressor and the power absorbed by the compressor. Assume the following properties for air: $\gamma = 1.4$; $c_p = 1.01$ kJ/kg K.

The fuelling rate to the engine is increased and when the air mass flow rate into the engine is increased by 50 per cent the volume flow rate out of the compressor increases by 22 per cent. Stating any assumptions, establish approximately the new operating point for the compressor, its rotor speed and the power that it is absorbing.

List briefly the advantages and disadvantages of turbocharging a diesel engine, and how the disadvantages can be ameliorated.

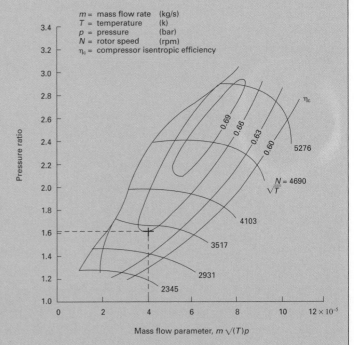

Figure 10.39

10.12 The Rolls-Royce Crecy engine (for which some development was undertaken in the University of Oxford, Department of Engineering Science during the 1939–45 war by Ricardo) was intended for aircraft use. The Crecy was a supercharged two-stroke engine with a swept volume of 26 litres. The predicted performance at an altitude of 4500 m ($p = 0.578$ bar, $T = 259$ K) and a speed of 792 km/h was

    supercharger pressure rise   1.05 bar
    supercharger isentropic efficiency   0.70
    engine speed   3000 rpm
    trapped gravimetric air/fuel ratio   15:1
    brake specific fuel consumption   225 g/kWh
    brake power output   1740 kW
    air flow rate   5.4 kg/s
    exhaust temperature   750°C
    exhaust duct outlet area   0.042 m²
    mean molar mass of the exhaust products   28.5 kg/kmol

The engine had in-cylinder fuel injection, and the valve timing was such that a large flow of air (the scavenge flow) would not be trapped in the cylinder. You may neglect the pressure rise in the inlet system to the supercharger due to the forward motion of the aircraft. Stating any other assumptions that you make:

(1) Calculate the scavenge flow as a percentage of the total flow into the supercharger.
(2) If the frictional losses in the engine are equivalent to a frictional mean effective pressure (fmep) of 1 bar, calculate the indicated mean effective pressure (imep).
(3) Determine the volumetric efficiency based on the inlet manifold conditions and the trapped mass in the cylinder.
(4) Calculate the thrust from the exhaust, and compare this 'jet power' to the brake power of the engine.

List four of the advantages/disadvantages of in-cylinder petrol injection.

10.13 A Sulzer RTA two-stroke diesel engine has a bore of 0.84 m and a stroke of 2.09 m; the bmep is 15.53 bar. If the engine operates at a speed of 70 rpm, calculate the power output per cylinder. If the brake specific fuel consumption is 167 g/kWh, calculate the fuel mass flow rate and the brake efficiency (assuming the fuel to have a calorific value of 42 MJ/kg).
    Such engines are turbocharged and inter-cooled. Assume the following data:

    turbine isentropic efficiency   0.90
    turbocharger mechanical efficiency   0.98
    specific heat capacity of air   1.01 kJ/kg K
    specific heat capacity of the exhaust products   1.20 kJ/kg K
    ratio of the heat capacities of air (γ_a)   1.4
    ratio of the heat capacities of the exhaust products (γ_{ex})   1.3
    compressor entry pressure   1.0 bar
    compressor entry temperature   300 K

Stating clearly any assumptions, determine the relationship between the compressor pressure ratio and the turbine entry temperature, such that the compressor delivery pressure is always greater than the turbine entry pressure. Plot the results for pressure ratios in the range 2–3 (using a scale length of 100 mm) for compressor isentropic efficiencies of 0.65 and 0.75 – use a scale of 1 K/mm for the temperature axis.
    Why, especially in a two-stroke engine, is it desirable for the compressor delivery pressure to be greater than the turbine entry pressure?

10.14 A turbocharged 2-litre direct injection diesel engine operates on a four-stroke cycle. At 2900 rpm and a bmep of 9.7 bar (full load), it is operating with a 22:1 gravimetric air/fuel ratio and a brake specific fuel consumption of 230 g/kWh. The turbocharger is fitted with a waste gate to regulate the pressure ratio to 2.0, and the compressor map is shown on figure 10.40, for which the pressure units are kN/m², the mass flow is in g/s, and the temperature units are K. The turbine entry temperature is 850 K, its pressure ratio is also 2.0, and the turbine isentropic efficiency is 0.75.

The compressor entry conditions are 1 bar and 298 K, and you should assume the following thermodynamic properties:

	Air	Exhaust
Specific heat capacity at constant pressure, c_p	1.01	1.12 kJ/kg K
Ratio of heat capacities, γ	1.4	1.33

(a) Stating any assumptions that you make, calculate the brake power output of the engine, its volumetric efficiency (based on inlet manifold conditions) and the fraction of the exhaust gas that passes through the turbine.

(b) Explain, by means of an annotated sketch, how the waste gate operates to control the boost pressure.

(c) Suggest ways of increasing the bmep of this engine in order of increasing complexity, with an indication of the likely increase in output (and how this would be calculated). Comment on how other aspects of the engine performance would be affected.

10.15 In a turbocharged engine, it is desirable for the compressor boost pressure to be greater than the back-pressure from the turbine. Assuming steady flow, and the following locations:

entry to the compressor	1
exit from the compressor	2
entry to the turbine	3
exit from the turbine	4

devise an expression for the pressure ratio p_2/p_3 in terms of

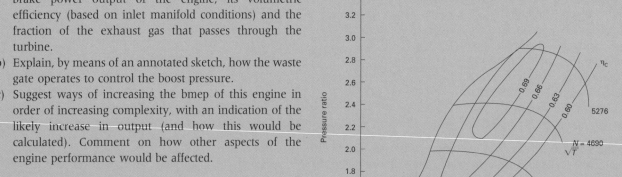

Figure 10.40

(a) the temperature ratio, T_3/T_1
(b) the pressure ratio, p_2/p_1
(c) the compressor isentropic efficiency, η_c
(d) the turbine isentropic efficiency, η_t
(e) the turbocharger mechanical efficiency, η_m and
(f) the air/fuel ratio, AFR.

Assume that the air and exhaust products behave as perfect gases with the same property values, but allow for any differences in the mass flows through the compressor and turbine. State clearly any additional assumptions that you make.

Comment briefly on all of the assumptions. What might the effect of a compact exhaust manifold be when using steady-flow turbocharger performance data for predicting the performance of a turbocharged diesel engine?

10.16 The Sabre Marathon engine is a two-stage turbocharger conversion of the Ford Dover diesel engine, with inter-coolers after each compressor. Figure 10.34 shows the arrangement of the turbochargers and inter-coolers, figure 10.35 shows the compressor maps. The engine has an output of 448 kW at 2400 rpm, from a swept volume of 6.8 litre. At this operating point the bsfc is 268 g/kWh, the air/fuel ratio is 21.9:1, and the operating points of the turbochargers are as follows:

Turbocharger performance	Low pressure	High pressure
Temperature before compressor	25	64°C
Temperature after compressor	168	217°C
Temperature after cooler	64	49°C
Compressor pressure ratio	2.14	2.50
Turbine entry temperature	716	790°C
Exhaust temperature		601°C

(a) Calculate the power used by each compressor and its isentropic efficiency.
(b) Compare the compressor operating points with those shown on the compressor maps in figure 10.35 (note that the air flow rates need to be pseudo non-dimensionalised as $(m\sqrt{T})/p$); assume that the compressor maps have a datum pressure of 1 bar.
(c) Determine the effectiveness of the inter-coolers, assuming a cooling medium is available at 15°C.
(d) If the overall pressure ratio is 5.1 (after allowing for pressure drops in the inter-coolers), what is the volumetric efficiency of the engine?
(e) Assuming no mechanical losses in the turbochargers, calculate the apparent value for the heat capacity of the exhaust gases in each turbocharger, and comment on these values.

Data: Calorific value of diesel fuel = 42 MJ/kg.
For air, take c_p = 1.01 kJ/kg K and c_v = 0.721 kJ/kg K.

10.17 The Ford Essex V6 spark ignition engine was the subject of a turbocharging study, using an Airesearch T-04B turbocharger; the carburettor was placed downstream of the compressor. The compression ratio was reduced from 9.1:1 to 7.6:1, so as to avoid combustion 'knock', and it may be assumed that the reduction in the indicated efficiency is 1.5 times that of the reduction in the corresponding Otto cycle. At 3000 rpm, the bmep was increased from 9.6 bar (when naturally aspirated) to 11.1 bar when turbocharged with a boost pressure ratio of 1.4. The fmep is (independent of load) 1.2 bar, and the air flow rate into the engine is 105 g/s.

Assuming that the air/fuel ratio, the pumping work and the volumetric efficiency are unchanged, calculate the density of the air leaving the compressor and the compressor isentropic efficiency (assuming an ambient temperature of 25°C and pressure of 1 bar). Determine and comment on the change in the brake specific fuel consumption, and any other assumptions that have been made.

10.18 A 4-litre swept volume spark ignition engine is supercharged using a Roots blower to give a pressure ratio of 1.5. At maximum power, the engine output is 240 kW at 5000 rpm.

Show that the type efficiency of the Roots blower (the work required from a perfect Roots blower divided by the work required in an isentropic compressor) is

$$\frac{\gamma}{\gamma-1} \frac{(p_2/p_1)^{\frac{\gamma-1}{\gamma}} - 1}{p_2/p_1 - 1}$$

Because of losses (including internal leakage), the volume flow rate into the compressor is reduced, but the work is still area abcd in figure 10.41. If the air mass flow rate is 243 g/s at ambient conditions of 1 bar and 25°C, calculate the power requirement of a Roots compressor with a pressure ratio of 1.5, for mechanical and volumetric efficiencies both of 80 per cent. What would the exit air temperature be from the Roots blower?

After inter-cooling the air temperature is reduced to 50°C. Neglecting any pressure drop in the inter-cooler, calculate the volumetric efficiency of the engine.

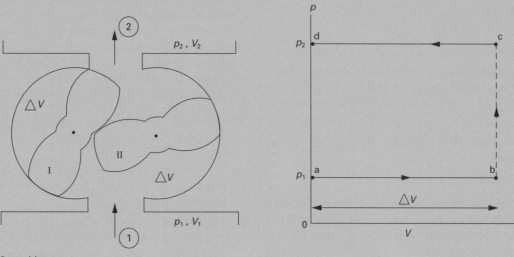

Figure 10.41
Operation of a Roots blower.

CHAPTER 11

Engine modelling

11.1 | Introduction

The aim with modelling internal combustion engines is twofold:

1 To predict engine performance without having to conduct tests.

2 To deduce the performance of parameters that can be difficult to measure in tests, for example, the trapped mass of air in a two-stroke engine or a turbocharged engine.

It is obviously an advantage if engine performance can be predicted without going to the trouble of first building an engine, then instrumenting it, testing it and finally analysing the results. Modelling should lead to a saving of both time and money. Unfortunately, the processes that occur in an internal combustion engine are so complex that most of the processes cannot be modelled from first principles. Consider a turbocharged diesel engine, for which it might be necessary to model:

1 the compressor (and inter-cooler if fitted)

2 unsteady-flow effects in the induction system

3 flow through the inlet valve(s)

4 air motion within the cylinder

5 dynamics of the injection system

6 fuel jet interaction with the trapped air to form a spray

7 combustion (including the effects of the ignition delay and turbulent combustion, and possibly including the modelling of the gaseous and particulate emissions)

8 noise generated mechanically and by combustion

9 heat flow within the combustion chamber and to the cooling media

10 turbine performance.

Unsteady fluid flow modelling has already been discussed in chapter 7, section 7.5, and is not discussed further here. Consider the air motion within the cylinder. It is now possible to use CFD techniques to model the gross flow details, such as the distribution of swirl and axial velocities. Predicting the

turbulence levels is more difficult and requires experimental data for validation. However, great improvements have been made with CFD and the number of nodes and cells is now such that thermal boundary layers can be modelled (with a resolution of about 1 μm). Previously it was necessary to rely on correlations such as the 'Law of the Wall', but these were not applicable to closed-cylinder processes since they assumed that pressure was constant. The challenges in modelling turbulent combustion are immense, because of the range of timescales and length scales – their multi-scale nature. Turbulent flows have a broad range of length and timescales, and these are further extended when combustion is included.

Veynante and Vervisch (2002) provide a very comprehensive review of the different approaches to turbulent combustion modelling. With hydrocarbon combustion, there are short-lived species, such as CH_2, that have lifetimes of the order of 10^{-10} s; while other reactions can be much slower – a notable example is the extended Zeldovich mechanism for the formation of NO, which can be of the order of milliseconds (Echekki, 2009). Length scales range from 10^{-9} to 10^{-1} m. It is also necessary to include modelling of soot formation and destruction (from the formation of soot precursors to primary particles), not least because of their role in radiative heat transfer from flames. Finally, molecular processes are coupled with turbulent transport and there can be considerable overlap between their corresponding scales (Echekki, 2009).

Echekki (2009) explains that it is necessary to make a distinction between different types of coupling among physical processes, and defines four types of coupling:

(i) *Type A.* Strong physical coupling and scale coupling; for example, diffusion and reaction within a laminar premixed or diffusion flame.

(ii) *Type B.* Weak physical coupling and scale coupling; for example, the extended Zeldovich mechanism for NO formation in flames.

(iii) *Type C.* Strong physical coupling and no scale coupling; for example, thermo-acoustic coupling in a combustion chamber, or soot formation and soot radiation in turbulent flames.

(iv) *Type D*. Weak physical coupling and no scale coupling.

Echekki (2009) reviews these different cases and the approaches that can be used in their solution.

Information on the air flow is vital, since it interacts with the fuel during the spray formation and its subsequent combustion. The turbulence affects the heat transfer, the fuel-burning rate and thus the noise originating from combustion.

In view of these complications, it is not surprising that engine models rely heavily on experimental data and empirical correlations. For example, the turbocharger performance and the valve flow characteristics can be determined from steady-flow tests, and look-up tables can be built into the engine model, along with appropriate interpolation routines. However, it must be remembered that the flow through the engine is unsteady, and this can have a significant effect. An example of this is that the turbine operating point is non-stationary under steady-state conditions (Dale *et al.*, 1988).

Empirical correlations are used for predicting processes such as the heat transfer, the ignition delay and the burn rate. Engine simulations that follow the approach described here are known as *zero dimensional, phenomenological* or *filling and emptying* models. Cycle calculation models for spark ignition and compression ignition engines are listed in FORTRAN by Benson and Whitehouse (1979) and by Ferguson (1986). As better understanding is gained of the sub-processes (such as ignition delay), then it is possible to replace the corresponding empirical model. For example, simulations that predict turbulent combustion within a defined combustion chamber are known as *multi-dimensional* models.

A treatment of the thermodynamics and fluid mechanics of internal combustion engines, and how this knowledge is applied to engine modelling, is presented very comprehensively by Benson (1982) and Horlock and Winterbone (1986).

Even if a multi-dimensional model was available that predicted performance by solving the underlying physical phenomenon, its use may not be that great for two reasons:

1 the computational requirements might be prohibitive

2 most engine design is derivative or evolutionary.

Even when a manufacturer introduces a new engine, it is likely to have a similar combustion system to an existing engine, for which there is already experimental data, and a calibrated computer model. However, most changes to an engine can be described as development. Perhaps a different turbocharger is being considered to give a higher boost pressure and output. An existing engine model can then be adapted, to investigate how the compression ratio, valve timing, fuel injection and other variables might be selected to give the best performance (trade-off between fuel economy, emissions and output) within the constraints of peak cylinder pressure (typically 150 bar) and the maximum exhaust temperature (about 950°C).

Although only thermodynamic modelling is being considered here, the mechanical design of the engine also benefits from modelling. The mechanical model requires inputs of the pressure loading and thermal loading from the thermodynamics model. The mechanical model can then predict bearing loads from the engine dynamics. A finite element model is usually used for the engine structure, so that the heat flow, the thermal and mechanical strains, the vibration modes, assembly strains and the noise transmission can be predicted. If the motion of the valve train is to be modelled, then in addition to the information on the engine structure, the dynamic model of the valve train would need to include the stiffness of the components and the load/speed-dependent stiffness of the oil films.

When the crankshaft is being modelled it is necessary to consider the interactions with the engine block. The engine block and crankshaft both deform under load, and this affects the bearing oil film thickness. The attitude of the piston in the bore will also be affected, and both these effects modify the damping of the crankshaft torsional oscillations. The piston and bore have to be modelled, using inputs of the temperature distributions and pressure loading, so that the piston ring/liner clearances can be predicted. This then enables the ring pack to be analysed, so that the oil film thicknesses and their contribution to friction can be found. This overview of the mechanical modelling should be sufficient to show that it is at least as complex as the thermodynamic modelling.

In subsequent sections there will be a description of how the laws of energy and mass conservation are applied to an engine, how individual processes are applied to an engine, and how individual processes can be modelled. This will be followed by an example in which SPICE (the Simulation Program for Internal Combustion Engines, written by Dr S. J. Charlton, formerly of the University of Bath) is used to investigate the influence of valve timing on engine performance.

11.2 | Zero-dimensional modelling

11.2.1 Thermodynamics

As analytic functions cannot be used to describe engine processes, it is necessary to solve the governing equations on a step-wise basis; often with increments of 1° crank angle. The application of the First Law of Thermodynamics in modelling internal combustion engine processes is described in complementary ways by Watson and Janota (1982), Wallace (1986a and 1986b) and Heywood (1988).

The First Law of Thermodynamics can be written in differential form for an open thermodynamic system, and if changes in potential energy are neglected, then

$$\frac{dU}{dt} = \frac{dQ}{dt} - \frac{dW}{dt} + \sum_i \frac{dH_{oi}}{dt} \qquad (11.1)$$

where subscript i refers to control volume entries. Or

$$\frac{d(mu)}{dt} = \sum_s \frac{dQ_s}{dt} - p\frac{dV}{dt} + \sum_i h_{oi}\frac{dm_i}{dt} \qquad (11.2)$$

where subscript s refers to surfaces at which heat transfer occurs and h_{oi} is the specific stagnation enthalpy of flows entering or leaving the system.

Consider the left-hand side term in equation (11.2). During combustion, the composition of the species present will change, but, of course, the mass is conserved. However, by expressing the internal energy of the reactants and products on an absolute basis (as in figure 3.7), the chemical energy transferred to thermal energy during combustion does not need to be included as a separate term. (*Note*: in the tables prepared by Rogers and Mayhew (1980a), the internal energy of the reactants and products is not given on an absolute basis; a datum is used of zero enthalpy at 25°C, and the enthalpy of reaction (ΔH_0) is presented for various reactions also at 25°C, so it is possible to compute the internal energy of the reactants and products on an absolute basis; see chapter 3, section 3.3 and examples 3.4a and 3.5a.)

Differentiating the left-hand side of equation (11.2) and dividing the combustion chamber into a series of zones gives

$$\frac{d(mu)}{dt} = \sum_j \left(m\frac{du}{dt} + u\frac{dm}{dt} \right) \qquad (11.3)$$

with subscript j referring to different zones within the combustion chamber.

Within each zone, if dissociation is neglected, it can be assumed that the internal energy is only a function of the temperature and the equivalence ratio (ϕ). Thus $u = u(T, \phi)$ and equation (11.3) becomes

$$\frac{d(mu)}{dt} = \sum_j \left(m\frac{\partial u}{\partial T}\frac{dT}{dt} + m\frac{\partial u}{\partial \phi}\frac{d\phi}{dt} + u\frac{dm}{dt} \right) \qquad (11.4)$$

For diesel engines, the combustion process is frequently modelled as a single zone, and for spark ignition engines a two-zone model is used in which a flame front divides the unburned and burned zones. Thus, for a two-zone model there will be mass transfer between the zones, but for a single zone the dm/dt term will be zero.

For many purposes in diesel engine simulation, the assumption of no dissociation with a single-zone model is acceptable. The assumptions reduce the computation time significantly without a serious loss of accuracy. As the combustion should always be weak of stoichiometric, this leads to temperatures at which dissociation does not have much effect on the thermodynamic performance of the engine. However, if emissions are to be considered, then a multi-zone model is required that incorporates the spatial variations in air/fuel ratio. The properties of each zone (temperature and composition) then have to be computed, as part of the modelling process for emissions.

It is less satisfactory to neglect dissociation in spark ignition engine combustion, as the mixtures are normally close to stoichiometric, and the combustion temperatures make dissociation significant. Methods for computing dissociation are presented by Benson and Whitehouse (1979), Ferguson (1986) and Baruah (1986). The prediction of emissions is also dependent on a combustion model that includes dissociation. The thermal boundary layer has to be modelled, as it quenches the flame front to generate unburnt hydrocarbon emissions. It may also be necessary to use a multi-zone model for the combustion, even if the charge is homogeneous, as the temperature of the unburnt gas will vary spatially for several reasons.

Firstly, the gas that is burned first will end up being hotter than the gas that is burned subsequently. If constant-volume combustion is considered, then the gas to be burned first contributes only a small pressure rise, while the gas burned later will contribute a greater pressure rise. Thus more work is done on the first burned gas by subsequent combustion than is done by the first burned gas on the unburned gas. This leads to temperature gradients in the burned gas that can influence emissions, especially NO_x for which the formation is strongly temperature dependent. Secondly, the combustion chamber surface temperature will vary, and this will result in different levels of heat transfer from the different regions of the combustion chamber.

However, for the current purpose a single-zone combustion model without dissociation is to be developed, which can be used for simulating a diesel engine. Substituting equation (11.4) into equation (11.2), assuming the gas behaves as a perfect gas ($pV = mRT$), gives

$$m\frac{\partial u}{\partial t}\frac{dT}{dt} + m\frac{\partial u}{\partial \phi}\frac{d\phi}{dt} + u\frac{dm}{dt} = -\frac{mRT}{V}\frac{dV}{dt} + \sum_s \frac{dQ_s}{dt} + \sum_i h_{oi}\frac{dm}{dt}$$
$$(11.5)$$

This equation has to be solved iteratively, and to do so, it is necessary to know:

1 The gas properties, to calculate the internal energy as a function of temperature and equivalence ratio, $u = u(T, \phi)$. Thence

$$m\frac{\partial u}{\partial T}\frac{dT}{dt} \quad \text{and} \quad m\frac{\partial u}{\partial \phi}\frac{d\phi}{dt}$$

can be found. Implicit in this, is a knowledge of how the mass fraction burned (mfb) will vary as a function of time, as the internal energies of the burned and unburned mixtures will be different functions of temperature and equivalence ratio.

2 The mass transfer between zones (dm/dt), which will be zero here, as a single-zone combustion model is being considered.

3 The rate of change of volume, so that the displacement work can be computed:

$$-\frac{mRT}{V} = \frac{dV}{dt} \qquad \text{(see equation 11.6b)}$$

4 The heat flows

$$\sum_s \frac{dQ_s}{dt} \qquad \text{(see section 11.2.4)}$$

5 The stagnation enthalpy of any flows in or out of the control volume, for the evaluation of

$$\sum_i h_{o,i} \frac{dm_i}{dt}$$

Equation (11.5) is a differential equation that has to be solved to give temperature as a function of time or crank angle, so equation (11.5) has to be applied to each control volume. Once the temperature has been calculated, so long as the instantaneous volume and mass in the control volume are known, the pressure can be calculated from the equation of state ($pV = mRT$).

If the control volume is of fixed volume (for example, a manifold), then the dV/dt term is zero. For the engine cylinder (see figure 12.6):

$$V = V_c + A\left[r(1 - \cos\theta) + \{l - \sqrt{(l^2 - r^2\sin^2\theta)}\}\right] \qquad (11.6a)$$

and by differentiation:

$$\frac{dV}{dt} = A\left[r\sin\theta\frac{d\theta}{dt} + (l^2 - r^2\sin^2\theta)^{-1/2}r^2\sin\theta\cos\theta\frac{d\theta}{dt}\right] \qquad (11.6b)$$

where V_c = clearance volume at tdc
  A = piston area
  θ = crank angle measured from tdc
  l = connecting-rod length
  r = crank throw (stroke/2).

11.2.2 Gas properties

The gas properties are required as a function of temperature and composition.

For individual species, the internal energy can be expressed as a function of temperature by means of a polynomial

expansion with either a molar or specific basis:

$$u(T) = u_0 + u_1T + u_2T^2 + u_3T^3 + \ldots \qquad (11.7)$$

This approach is developed fully in Appendix A, section A.2.1. However, for diesel engines, a simpler and acceptable alternative (that is widely used) is to consider the internal energy of the reactants and products on an absolute basis, as a function of air/fuel ratio and temperature. This is satisfactory, as most fuels have similar hydrogen/carbon ratios, and in any case the major constituent of both the reactants and products is nitrogen. Such data are presented by Gilchrist (1947) with allowance for dissociation, for a fuel containing 84 per cent carbon and 16 per cent hydrogen by mass. Gilchrist gives separate tabulations of internal energy for the reactants and products, as a function of the mixture strength and absolute temperature.

A similar approach was adopted by Krieger and Borman (1986), who provided polynomial coefficients from a curve-fit to combustion product calculations for weak mixtures ($\phi \leq 1$) of C_nH_{2n} with air (85.6 per cent carbon and 14.4 per cent hydrogen by mass):

$$u = K_1(T) - K_2(T)\phi \text{ kJ/(kg of original air)} \qquad (11.8)$$

where $K_1 = 0.692T + 39.17 \times 10^{-6}T^2 + 52.9 \times 10^{-9}T^3$
$\qquad - 228.62 \times 10^{-13}T^4 + 277.58 \times 10^{-17}T^5$

and $K_2 = 3049.33 - 5.7 \times 10^{-2}T - 9.5 \times 10^{-5}T^2 + 21.53 \times 10^{-9}T^3$
$\qquad - 200.26 \times 10^{-14}T^4$

with the gas constant given by

$$R = 0.287 + 0.020\phi \text{ kJ(kg of original air)}^{-1}\text{K}^{-1} \qquad (11.9)$$

Krieger and Borman also suggest a modification to account for dissociation if the temperature is above 1450 K:

$$u_{corr} = u + 2.32584\exp(A + B + C) \text{ kJ/kg of original air}$$

where $\quad A = 10.41066 + 7.85125\phi - 3.71257\phi^3$
$\qquad B = (-15.001 - 15.838\phi + 9.613\phi^3) \times 10^3/T$
$$C = \left[0.154226\phi^3 - 0.38656\phi - 0.10329\right.$$
$$\left. + \frac{118.27\phi - 14.763}{T}\right] \times \ln(p \times 14.503)$$
$$(11.10)$$

and

$$R_{corr} = R + 0.004186\exp\left\{\left[11.98 - \frac{25442}{T} - 0.4354\ln(p \times 14.503)\right]\phi\right.$$
$$\left. + 0.2977\ln(\phi)\right\} \text{ kJ (kg original air)}^{-1}\text{ K}^{-1} \qquad (11.11)$$

For a fuel of composition C_nH_{2n} the stoichiometric gravimetric fuel/air ratio is 0.0676. Thus if the internal energy of the products is wanted on a basis of per unit mass of products, it is necessary to divide equations (11.8)–(11.11) by $(1 + 0.0676\phi)$.

During combustion there will be both burned and unburned mixture present, and the next section (11.2.3) discusses how the mass fraction burned as a function of time (mfb(t)) is modelled. Prior to this, it is sensible to discuss how the internal energy of the burned and unburned mixtures are computed. Referring back to equation (11.5), it will be recalled that the internal energy of the control-volume contents is needed as a function of temperature, so that equation (11.5) can be solved in terms of temperature.

The simplest model is to have a single zone, in which the burned and unburned mixture is considered to be mixed with a uniform temperature. This was assumed in the derivation of equation (11.5). Using equation (11.8) for predicting the internal energy of the burned (suffix b) and unburned mixture (suffix u):

$$u = [(1 - \text{mfb})K_1(T)]_u + \text{mfb}[K_1(T) - K_2(T)\phi]_b$$
$$u = K_1(T) - \text{mfb}K_2(T)\phi \quad \text{kJ/kg of original air} \quad (11.12)$$

The gas constant is used to compute the cylinder pressure (using the equation of state $pV = mRT$), from equation (11.9):

$$R = 0.287 + 0.020\phi \, \text{mfb} \quad \text{kJ(kg of original air)}^{-1} \, \text{K}^{-1}$$
$$(11.13)$$

It should be noted that in equation (11.12), the internal energy of the unburned fuel has not been considered, neither has the effect of fuel vaporisation been considered on the energy balance or the molecular balance.

11.2.3 Burn rate

The burn rate in a spark ignition engine can be modelled by a Wiebe function (chapter 3, section 3.9.2) which requires the selection of two shape parameters, and specification of the ignition timing and the combustion duration. Combustion in compression ignition engines is more complex, as it is influenced by:

1 the delay period – the time between the start of injection and the start of combustion

2 the amount of flammable mixture prepared during the delay period that burns rapidly once ignited

3 the combustion of the remainder of the fuel.

Ignition delay

Ignition delay data are usually modelled by an equation that has origins in the Arrhenius equation for reaction rate:

reaction rate $\propto \exp(-E_a/RT)$

Ignition delay (t_{id}) is correlated by an equation of the form

$$t_{id} = Ap^{-n} \exp(E_a/R_o T) \quad (11.14)$$

where E_a is an apparent activation energy
A, n are constants, but functions of fuel type and the air/fuel mixing/motion.

Ignition delay data can be obtained from constant-volume combustion bombs, rapid compression machines and steady flowrigs. However, Heywood (1988) shows that different workers find widely differing values for the parameters in equation (11.14):

$$0.757 < n < 2$$
$$0.4 \times 10^{-9} < A < 0.44 \quad (11.15)$$
$$4650 < E_a/R_o < 20926$$

Not surprisingly, ignition delay values predicted for engine operation also vary widely. Belardini *et al.* (1983) reviewed many correlations for the ignition delay in direct injection engines, and made comparisons with ignition delay measurements detected by an optical start of combustion sensor. This leads to the question of how to measure the ignition delay period in an engine.

The first question is how to define the start of injection. The injector can be fitted with a needle-lift transducer, and two popular approaches are to use the change in coupling in a coil that is part of a resonating circuit, or to use a Hall effect transducer. A typical needle-lift signal is shown in figure 11.1. It should be noted that there can be noise associated with the signal (most readily seen during the period the needle is closed). Furthermore, the dc level of the signal can vary, especially with the resonating circuit system. Figure 11.1 shows that initially the injector needle starts to lift very slowly, and such movement can be difficult to distinguish from noise. This leads to two approaches that are used to define the start of injection:

1 as when the injector needle has lifted to (say) 10 per cent of its maximum lift (point A and time t_a in figure 11.1)

2 or, as when the slope at point A in figure 11.1 is extrapolated back to zero lift, to give an intercept at time t_o.

This second method is more difficult to implement, as the gradient has to be evaluated, and this can introduce errors whether undertaken digitally or with analogue electronics. Consequently, the start of injection is usually defined as when the injector needle has lifted a specified distance from its seat.

The start of combustion is more difficult to define. If combustion analysis equipment is available, it can be used to compute an apparent heat release rate that is shown in figure 11.2. (It is termed an apparent heat release rate, as no allowance is made for heat transfer. The same assumptions about gas properties have been used as described in section 11.2.2.) The heat release rate becomes negative, owing to heat

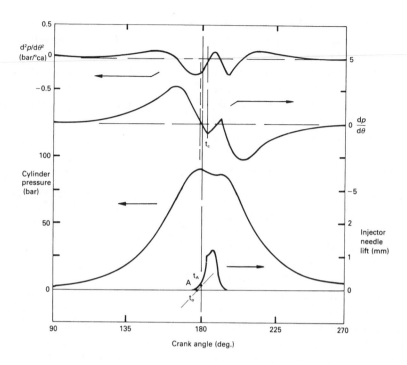

Figure 11.1
The injector needle lift, cylinder pressure, first and second derivative of cylinder pressure from a 2.5-litre diesel engine operating at a speed of 2000 rpm and a bmep of 3 bar.

being transferred from the gas to the combustion chamber. The start of combustion can be defined as when the heat release rate becomes zero (that is, when the cumulative heat release is at a minimum), and this corresponds to 2°atdc in figure 11.2. Alternatively, the time when the cumulative heat release has returned to zero can be used to define the start of combustion; this is discussed further in chapter 14, section 14.5.3.

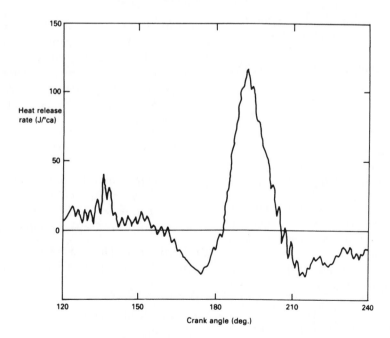

Figure 11.2
The apparent heat release rate for the pressure data shown in figure 11.1.

A simpler alternative is to differentiate the pressure with respect to time to obtain the result already shown in figure 11.1; the second derivative of pressure has also been plotted here. The start of combustion can be defined as the minimum that occurs in the first derivative after the start of injection, time t_c in figure 11.1. The start of combustion in figure 11.1 is just after $2°$atdc, and is marginally later than the start of combustion determined from the heat release rate. If the start of combustion is being evaluated by computer, it is necessary to check that a minimum has indeed been found by checking that the second derivative is zero, and the third derivative of pressure is positive.

The pressure will not have been recorded on a continuous basis, but at discrete values of time (that is, crank angle). Thus the derivatives are not continuous functions, and the minimum in the first derivative has to be identified by a change in sign of the second derivative in going from negative to positive. Unless there is a correction, by means of interpolation, this implies that t_i will be detected slightly after the start of combustion.

Alternatively, the start of combustion can be sensed by an optical transducer that detects the radiation from the flame front. It has generally been found that the start of combustion detected by the optical sensor is after that detected from the pressure trace. This will no doubt be influenced by the frequency response of the optical sensors, which should be biased towards the infra-red part of the visible light spectrum. As with the trace from the needle-lift transducer, there is the question of how to define the start of combustion. This problem can be by-passed if a high gain amplification is used, and the signal is allowed to saturate. In effect, the optical sensor is then being used to generate a digital signal.

The only remaining problem is that the ignition delay period is subject to cycle-by-cycle variations. It is thus necessary to record data from a sufficiently large number of consecutive cycles, such that the statistical analysis is not influenced by the sample size.

A widely used correlation for ignition delay was deduced by Watson (1979):

$$t_{id} = \frac{3.52 \exp(2100/T)}{p^{1.022}} \tag{11.16}$$

where t_{id} = ignition delay (ms)
  p = mean pressure during ignition delay (bar)
  T = mean temperature during ignition delay (K).

This correlation has the disadvantage that mean values of temperature and pressure are required. This will require an iterative solution to be used, so as to evaluate the mean properties during the delay period.

Heywood (1988) recommends a correlation for ignition delay developed by Hardenberg and Hase (1979):

$$t_{id} = (0.36 + 0.22v_p) \exp\left[E_a\left(\frac{1}{R_0 T} - \frac{1}{17190}\right)\left(\frac{21.2}{p - 12.4}\right)^{0.63}\right] \tag{11.17}$$

where t_{id} = ignition delay (degrees crank angle)
  v_p = mean piston velocity (m/s)
  T = temperature at tdc (K)
  p = pressure at tdc (bar)
and E_a = 618 840/$(CN + 25)$
  where CN = fuel cetane number.

This correlation has several advantages. Firstly it recognises that there is an engine speed dependence, and a dependence on the self-ignition quality of the diesel fuel. Furthermore, the temperature and pressure are determined at a single condition. If this correlation is not being applied within a computer simulation of an engine, then the temperature and pressure can be estimated by assuming the compression can be modelled by a polytrophic process:

$$T = T_i r_i^{k-1}, \quad p = p_i r_i^k \tag{11.18}$$

where suffix i refers to conditions at inlet valve closure (ivc), r_i is the volumetric compression ratio from ivc to the start of injection, and k is the polytropic index (typically 1.3).

However, the dependence on the cetane number should be treated with caution, as there is not necessarily a relation between ignition delay and cetane number for a given engine at a particular operating point. The problem arises from the way in which the cetane number is evaluated, by ignition delay measurements made in a CFR engine, with an indirect injection (IDI) combustion system. Typical values of ignition delay are within the range of 0.5–2.0 ms, though under light-load low-speed operation longer ignition delays occur, especially if the engine is cold.

Diesel combustion

As explained in section 11.2.3, it is common practice to divide the combustion into two parts: the *pre-mixed* burning phase and the *diffusion* burning phase. The *pre-mixed* phase is a consequence of the mixture prepared during the ignition delay period burning rapidly; the *diffusion* burning phase accounts for the remainder of combustion. Watson *et al.* (1981) proposed:

$$\text{mfb}(t) = \beta f_1(t) + (1 - \beta)f_2(t) \tag{11.19}$$

where mfb(t) = mass fraction burned
  β = fraction of fuel burned in the *pre-mixed* phase
  $f_1(t)$ = *pre-mixed* burning function
  $f_2(t)$ = (*diffusion*) burning function.

The fraction of fuel burned in the *pre-mixed* phase (β) will be a function of the ignition delay period:

$$\beta = 1 - a\phi^b/(t_{id})^c \tag{11.20}$$

  ϕ = equivalence ratio
  t_{id} = ignition delay (ms)

and for turbocharged DI engines:

$$0.8 < a < 0.95$$
$$0.25 < b < 0.45$$
$$0.25 < c < 0.5$$

with

$$f_1(t) = 1 - (1 - t^{K_1})^{K_2}$$
$$f_2(t) = 1 - \exp(-K_3 t^{K_4})$$

in which $K_1 = 2.00 + 1.25 \times 10^{-8}(t_{id} \times N)^{2.4}$
$K_2 = 5000$
$K_3 = 14.2/\phi^{0.644}$
$K_4 = 0.79 K_3^{0.25}$

and N is the engine speed (rpm).

A more fundamental approach has been proposed by Whitehouse and Way (1968), and although it was derived for a single-zone combustion model, it can be extended to a multi-zone combustion model. The Whitehouse–Way combustion model comprises two parts. The first part is based on the Arrhenius equation, and it predicts the fuel burn rate (FBR). The second part predicts the fuel preparation rate (FPR), based on the diffusion of oxygen into the fuel jet. The fuel burn rate is modelled by:

$$FBR = \frac{K}{N} \times \frac{p'_{O_2}}{\sqrt{T}} \exp(-E_a/T) \int (FPR - FBR) d\theta \quad (11.21)$$

and the fuel preparation rate is given by

$$FPR = K' m_i^{(1-x)} m_u^x (p'_{O_2})^z \quad (11.22)$$

where p'_{O_2} is the partial pressure of oxygen
m_i is the mass of fuel injected
m_u is the mass of fuel yet to be injected.

Benson and Whitehouse (1979) also discuss the derivation of this model, and quote values of the parameters for specific two- and four-stroke engines. They lie within the following ranges:

$$0.01 < x < 1$$
$$z = 0.4$$
$$0.008 < K < 0.020 \ (bar^{-2})$$
$$1.2 \times 10^{10} < K' < 65 \times 10^{10} \ (K^{1/2}/bar \ s)$$
$$E_a = 1.5 \times 10^4 \ (K)$$

Winterbone (1986) presents cylinder pressure versus crank angle plots to show the effect of systematic variation of these parameters. The mass fraction burned (mfb(t)) is found by integration of equation (11.21).

If emissions are to be predicted, then a more sophisticated combustion model is needed. This can be provided by CFD, but it is also possible to use a phenomenological model, for example Bazari et al. (1996). This model includes the effects of fuel atomisation, air mixing, ignition, heat release and emissions formation. Fuel injection is treated as a distribution of fuel segments within the spray angle. Combustion zones are formed (of varying air/fuel ratio) at the start of atomisation, with air entrainment calculated from a jet model that applies the conservation of momentum, taking due account of swirl and wall impingement of the fuel. The combustion is modelled by a simple reaction model that accounts for the fuel vapour and oxygen concentrations. Nitric oxide emissions are modelled by the extended Zeldovich scheme, and likewise the soot and carbon monoxide formation are also assumed to be kinetically controlled (in other words, equilibrium values are not assumed).

11.2.4 Engine gas-side heat transfer

Engine heat transfer is discussed later (chapter 13) in the context of heat transfer to the engine cooling media (usually aqueous ethylene glycol and lubricating oil). Typically 20–35 per cent of the fuel energy passes to the engine coolant. With rich mixtures in spark ignition engines, as little as 15 per cent of the fuel energy passes to the coolant, but with spark ignition engines at low loads a much higher proportion (over 40 per cent) of the fuel energy passes to the coolant. Of the heat flow to the coolant, about half comes from in-cylinder heat transfer, and most of the balance flows from the exhaust port. The heat flow from the exhaust port depends on its geometry, and the extent of its passage through the coolant. Exhaust ports are now sometimes insulated, and this reduces the heat flow to the coolant. The higher exhaust temperature promotes further oxidation of the combustion products and increases the energy that an exhaust turbine can convert to work.

In-cylinder heat transfer

The in-cylinder heat transfer coefficient will vary with position and time. A knowledge of these spatial variations is needed if thermal stresses are to be calculated. At the opposite extreme, a simple correlation giving a position- and time-averaged heat transfer coefficient will be satisfactory for predicting the heat transfer to the coolant. For engine modelling, the variation in heat transfer with time is needed: the variation with position is not necessary if the emissions are not being considered.

The heat transfer from the combustion products occurs by convection and radiation. In spark ignition engines radiation may account for up to 20 per cent of the heat transfer, but it is usually subsumed into a convective heat transfer correlation (Annand, 1986). However, for compression ignition engines, the radiation from soot particles during combustion can be significant, and some correlations allow for the radiation in a separate form from the contribution by convection.

Until computational fluid dynamic (CFD) techniques permit a full prediction of the in-cylinder gas motion, it is necessary to predict the in-cylinder heat transfer coefficient, by using correlations that have been derived from experimental measurements. Fortunately, the predictions of the engine output and efficiency are not very sensitive to the predictions of heat transfer. Typically, a 10 per cent error in the prediction of in-cylinder heat transfer leads to a 1 per cent error in the engine performance prediction.

One of the earliest correlations for in-cylinder heat transfer was developed by Eichelberg (1939):

$$Q_s/A_s = 2.43 v_p^{\frac{1}{3}} (pT)^{\frac{1}{2}} \{T - T_s\} \quad \text{W/m}^2 \tag{11.23}$$

where v_p = mean piston speed (m/s)
p = instantaneous cylinder pressure (bar)
T = instantaneous bulk gas temperature (K)
T_s = mean surface temperature (K)
A_s = instantaneous surface area (m^2)
Q_s = instantaneous heat flow rate (W)

thus the heat transfer coefficient (h) is given by

$$h = \frac{Q_s}{A_s(T - T_s)} \tag{11.24}$$

This correlation has the advantage of simplicity, as it only requires the user to specify a surface temperature (a value of around 350 K is often assumed).

However, the Eichelberg correlation is not dimensionally consistent, and it has been argued that its generality is therefore suspect. Furthermore, great care is needed in using exactly the units specified, otherwise the constant has to have a different value. This has led to correlations of the form

$$Nu = a\,Re^b \tag{11.25}$$

where Nu = Nusselt number = hx/k
Re = Reynolds number = $\rho ux/\mu$
k = gas thermal conductivity (W/m K)
ρ = gas density (kg/m^3)
μ = gas dynamic viscosity (kg/m s)
x = characteristic dimension (m)
u = characteristic velocity (m/s).

As the details of the fluid motion are not known, then arbitrary but convenient definitions are used, with the characteristic length (x) assumed to be the bore diameter (B), and the characteristic velocity (u) assumed to be the mean piston speed ($\bar{v}_p$).

When a separate radiation term is included, an equation of the following form results. This was first proposed by Annand (1963) and then refined further by Annand and Ma (1971):

$$\frac{Q_s}{A_s} = c\frac{k}{B}Re^b(T - T_s) + d(T^4 - T_s^4) \tag{11.26}$$

Watson and Janota (1982) suggested that for a compression ignition engine:

$$b = 0.7$$
$$0.25 < c < 0.8$$
$$d = 0.576\sigma$$
$$\sigma = \text{Stefan–Boltzmann constant.}$$

During the intake compression and exhaust processes the radiation term should be zero ($d = 0$) and for a spark ignition engine $d = 0.075\sigma$. Watson and Janota comment that the range of values for c suggests that some important underlying variables are being ignored.

Woschni (1967) pointed out that equation (11.25) could be expanded:

$$h = (ak/x)(\rho ux/\mu)^b$$
$$= (ak)\frac{p^b}{RT}(k/\mu)^b u^b x^{b-1}$$

Woschni also assumed that

$$k \propto T^{0.75} \quad \text{and} \quad \mu \propto T^{0.62}$$

thus

$$h \propto p^b u^b x^{b-1} T^{(0.75-1.62b)} \tag{11.27}$$

Woschni assumed and subsequently justified the value of b as 0.8, and took the characteristic length (x) to be the piston bore (B). He argued that the characteristic velocity would comprise two parts:

(i) a contribution due to piston motion

(ii) a contribution due to combustion.

Woschni expressed the contribution due to combustion as a function of the pressure rise due to combustion, that is

$$p - p_m$$

where p_m is the pressure (motoring pressure) that would occur without combustion.

Equation (11.27) becomes:

$$h = 129.8 p^{0.8} u^{0.8} B^{-0.2} T^{-0.55} \text{ W/m}^2\,\text{K} \tag{11.28}$$

where p = instantaneous cylinder pressure (bar)
B = bore diameter (m)
T = instantaneous gas temperature (K)

and $u = C_1\bar{v}_p + C_2\dfrac{V_s T_r}{p_r V_r}(p - p_m)$

with V_s = swept volume
V_r, T_r, p_r evaluated at any reference condition, such as inlet valve closure.

Watson and Janota (1982) suggest that the motoring pressure is evaluated by assuming the compression and expansion to be modelled by a polytropic process:

$$p_m = p_r \left(\frac{V_r}{V} \right)^k \tag{11.29}$$

a typical value of k being around 1.3.

Values suggested by Woschni for C_1 and C_2 are

For gas exchange $\qquad\qquad C_1 = 6.18 \quad C_2 = 0$
For compression $\qquad\qquad\; C_1 = 2.28 \quad C_2 = 0$
For combustion and expansion $\; C_1 = 2.28 \quad C_2 = 3.24 \times 10^{-3}$
For IDI engines $\qquad\qquad\qquad\qquad\qquad C_2 = 6.22 \times 10^{-3}$

Sihling and Woschni (1979) suggested modified values for the coefficients C_1 for higher speed direct injection engines with swirl:

For gas exchange $\quad C_1 = 6.18 + 0.417\dfrac{B\omega_p/2}{\bar{v}_p}$

elsewhere $\qquad\qquad C_1 = 2.28 + 0.308\dfrac{B\omega_p/2}{\bar{v}_p}$

where ω_p = paddle wheel angular velocity (rad/s) in steady-flow swirl tests.

Hohenberg (1979) proposed a simplified form of equation (11.28):

$$h = \frac{129.8p^{0.8}}{V_s^{0.06}T^{0.4}}(\bar{v}_p + 1.4) \tag{11.30}$$

for which the piston area A_{pis} is given by

$$A_{pis} = A(\text{piston crown}) + A^{0.3}(\text{piston top land})$$

Not surprisingly, the different correlations for heat transfer coefficient can lead to widely varying predictions of the instantaneous and the mean heat transfer coefficient; this is illustrated by figure 11.3. However, the variation is not too important, since (as pointed out in the introduction to this section) errors in predicting the heat transfer coefficient only have a small effect on the engine performance prediction.

Heat transfer during the gas exchange processes

The heat transfer from the combustion chamber walls to the incoming charge is small compared with the total heat transfer. However, as with heat transfer from the inlet valve and inlet port, it can have a significant effect on the volumetric efficiency. The significance of the heat flow to the exhaust valve and port (up to half the heat flow to the coolant) has already been discussed in the introduction to this section.

The Woschni correlation (equation 11.28) includes terms for the heat flow within the cylinder. There appear to be few correlations and data for predicting the heat transfer in other parts of the gas exchange process, but Annand (1986) provides a comprehensive review.

A summary of Annand's recommendations follows:

1 Heat flow to the incoming flow from an inlet valve:

$$Nu = a\,Re^b \tag{11.31}$$

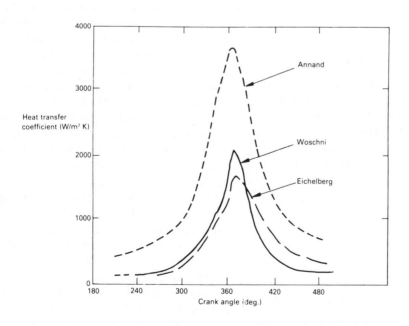

Figure 11.3
Comparison between the heat transfer correlations predicted by different correlations (adapted from Woschni, 1967).

where $Nu = \dfrac{\dot{Q}}{0.25\pi d_v^2 \Delta T}$

and $Re = \dfrac{4\dot{m}}{\pi d_v \mu}$

The value of a is high at low value lifts, but decreases to 0.9 in the range $0.1 < L_v/d_v < 0.4$. The Reynolds number index is in the range of 0.6–0.9, and this might be a function of the port geometry.

2 Heat flow to the exhaust valve from an outflowing charge. Annand suggests using equation (11.31) but with

$$a = 0.4d_v/L_v \qquad (11.32)$$

3 Heat flow to the exhaust port. Annand recommends

$$Nu = a\,Re^{0.8} \qquad (11.33)$$

with the value of a depending on the valve seat angle ($a = 0.258$ for a 45° seat angle), but with no apparent dependence on the port curvature.

11.2.5 Induction and exhaust processes

Flow through the valves

The flow processes through the inlet and exhaust valves have already been discussed in chapter 7, section 7.3. It will be recalled that the flow coefficients are a function of:

flow direction
valve lift
pressure ratio
valve seat angle
port geometry
valve face angle
whether or not the flow is treated as compressible

For inlet valves the pressure ratio is usually close to unity, but for exhaust valves the pressure ratio can vary widely, and during exhaust blowdown the flow can be supersonic. Care also needs to be taken in deciding how the reference area for the discharge coefficient has been defined. For this reason, it is less ambiguous to talk about effective flow areas, even though this does not make comparisons between different engines very easy. Values for flow coefficients have been presented in chapter 7, section 7.3.

Scavenging

In the case of naturally aspirated four-stroke engines scavenging is not usually significant. However, in the cases of:

(i) two-stroke engines, naturally aspirated or turbocharged, and

(ii) four-stroke engines, turbocharged with large valve overlap periods

then scavenging is significant. The different ways of modelling scavenging have been reviewed in chapter 8, section 8.4, and comprehensive treatments of scavenging in the context of engine modelling are provided by Horlock and Winterbone (1986) and Wallace (1986a and 1986b). Watson and Janota (1982) suggest that using a perfect mixing model is the best compromise. This gives a pessimistic estimate of the scavenging performance, but it seems that the energy available at the turbine is largely independent of the model used. The displacement scavenging model predicts a smaller exhaust flow with a higher temperature, while the perfect mixing model predicts a larger mass flow but with a lower temperature.

Scavenging will also be influenced by the flow pulsations that occur in the inlet and exhaust systems. These pulsations have the greatest effects on the scavenging of two-stroke engines, but they also affect the gas exchange processes in four-stroke engines. Their treatment has been discussed in chapter 8, section 8.5.

Turbochargers

Turbochargers are most easily modelled by using turbine and compressor maps, which show the speed and isentropic efficiency as functions of the pressure ratio and mass flow.

The compressor maps are usually produced by conducting steady-flow experiments, while the turbine maps are often produced by modelling techniques. The turbine and compressor data are stored in arrays, typically as a series of discrete turbocharger speeds for which the pressure ratio, mass flow parameter and isentropic efficiency are tabulated.

An immediate question is: how representative is the assumption of steady flow in the turbocharger? The conditions in the compressor are steady, certainly in comparison with the pulsations that occur in the exhaust system of an engine using pulse turbocharging. The usual compromise is to calculate the instantaneous pressure difference across the turbine at each crank angle increment, and to compute the turbine performance. In other words, the flow is treated as quasi-steady, such that the turbine performs under nonsteady flow conditions as it would if the instantaneous flow was a steady flow. This leads to turbocharger speed fluctuations (thus it is necessary to know the moment of inertia of the turbocharger) and a locus of operating points for a single engine operating condition. There are thus flow variations in the compressor within an engine cycle. Watson and Janota (1982) suggest that these errors will not exceed 5 per cent.

Another source of errors occurs with multi-entry turbochargers for which there will be a different pattern of flow pulsations at each entry – this can lead to so-called partial admission losses. Dale et al. (1988) describe a test facility for measuring the instantaneous performance in a turbine subject to pulsating flow; they could also investigate the effect of partial admission losses. Dale *et al.* (1988) found that the operating locus with pulsating flow gave a smaller mean mass flow for a given pressure ratio (figure 11.4), and that the efficiency was also lower (figure 11.5). They found that for a twin-entry turbine with flow pulsations arriving out-of-phase at each inlet, the instantaneous operating locus was closer to the operating line for partial admission steady flow than the operating line for steady full admission (figure 11.6).

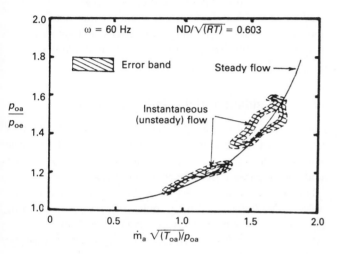

Figure 11.4
Locus of instantaneous flow under pulsed inlet conditions, superimposed on steady-flow data (for two mean inlet pressures, single-entry turbine) (Dale *et al.*, 1988, reproduced by permission of the Council of the Institution of Mechanical Engineers).

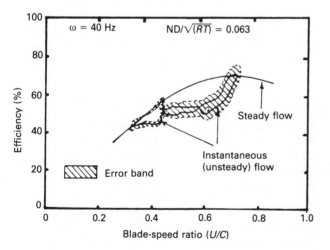

Figure 11.5
Locus of instantaneous efficiency under pulsed inlet conditions, superimposed on steady-flow data (for two mean inlet pressures, single-entry turbine) (Dale *et al.*, 1988, reproduced by permission of the Council of the Institution of Mechanical Engineers).

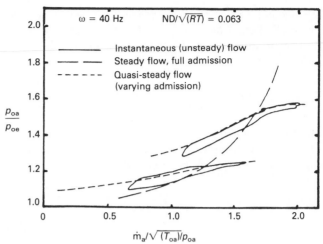

Figure 11.6
Instantaneous flow locus, at one entry, with out-of-phase flows at each entry of a twin-entry turbine (for two mean inlet pressures) (Dale *et al.*, 1988, reproduced by permission of the Council of the Institution of Mechanical Engineers).

Energy balance

With a knowledge of the flow processes that occur during the gas exchange process, it is possible to solve the energy equation. Since there is no combustion, equation (11.5) becomes

$$m\frac{\partial u}{\partial t}\frac{dT}{dt} + u\frac{dm}{dt} = -\frac{mRT}{V}\frac{dV}{dt} + \sum_s \frac{dQ_s}{dt} + \sum_i h_{oi}\frac{dm_i}{dt}$$

(11.34)

Equation (11.34) has to be applied to the cylinder and each of the manifolds; for the manifolds the dV/dt term is zero. It also has to be remembered that reverse flows can occur, in particular from the cylinder to the inlet manifold, and less significantly from the exhaust manifold into the cylinder. Also needed is an equation for mass conservation, and this is solved simultaneously with equation (11.34), so that the mass, $m(i)$, and temperature, $T(i)$, can be found on a step-by-step basis. The modelling of the gas exchange process is discussed comprehensively by Wallace (1986a and 1986b), and it also forms a significant part of the discussion of modelling by Watson and Janota (1982).

11.2.6 Engine friction

So far no account has been taken of engine frictional losses; the results from modelling will be in terms of indicated performance parameters. To convert the indicated performance to brake performance, it is necessary to predict the frictional losses. The difficulties in determining the frictional losses have been explained elsewhere, it is thus necessary to use the following predictions with caution.

Heywood (1988) suggests the use of the following correlation from Barnes-Moss (1975), for the frictional mean effective pressure (fmep) for automotive four-stroke spark ignition engines:

$$\text{fmep} = 0.97 + 0.15\left(\frac{N}{1000}\right) + 0.05\left(\frac{N}{1000}\right)^2 \qquad (11.35)$$

with fmep (bar) and N (rpm).

Millington and Hartles (1968) developed a similar correlation from motoring tests on diesel engines:

$$\text{fmep} = (r_\text{v} - 4)/14.5 + 0.475\left(\frac{N}{1000}\right) + 3.95 \times 10^{-3}\bar{v}_\text{p}^2 \qquad (11.36)$$

However, these correlations were for motored engines in which the pumping work and frictional losses will both be underestimated. This means that the earlier correlation proposed by Chen and Flynn (1965), which accounts for the pressure loading, has merit:

$$\text{fmep} = 0.137 + \frac{p_\text{max}}{200} + 0.162\bar{v}_\text{p} \qquad (11.37)$$

with fmep (bar), p_max (bar) and $\bar{v}_\text{p}$ (m/s).

Winterbone (1986) discusses a correlation that has the same form as equation (11.37):

$$\text{fmep} = 0.061 + \frac{p_\text{max}}{60} + 0.294\frac{N}{1000} \qquad (11.38)$$

Winterbone also discusses the way in which the frictional losses increase during transients.

11.3 Application of modelling to a turbocharged medium-speed diesel engine

11.3.1 Introduction

With naturally aspirated engines, it is quite easy to envisage the trade-offs associated with varying parameters such as ignition timing or valve timing. For example, earlier opening of the exhaust valve will reduce the expansion work produced by the engine, but it will also lead to a reduction in the pumping work. With a turbocharged engine the interactions are more complex. Again considering the effect of exhaust valve opening, earlier opening will increase the work done by the turbine and thus the pressure ratio across the compressor. The higher inlet pressure increases the work done on the piston during induction, and for many operating points the pumping work will be a positive quantity. (That is, the work done by the engine during the exhaust stroke is less than the work done by the incoming charge on the engine.) But what will be the effect on a turbocharger operating point be? The

turbocharger will be operating faster, but will the compressor be closer to surge, will the turbine inlet temperature become too high, will the response to transients be improved?

So, for turbocharged engines it is difficult to predict what happens when parameters are changed, unless a modelling technique is used. Even when predictions have been made once, it can be dangerous to generalise, as the efficiencies of the compressor and turbine are highly dependent on their operating points, and the turbocharger efficiencies will in turn affect the engine operating point.

To illustrate the use of a parametric investigation, a medium-speed turbocharged diesel engine has been simulated using SPICE (Simulation Program for Internal Combustion Engines). SPICE is a filling-and-emptying type of model, and its use and background theory are documented by Charlton (1986).

11.3.2 Building and validating the model

The engine to be modelled is described in table 11.1 and an essential part of the modelling process is the validation of the model results against test data. The purpose is to use the model for a parametric investigation of valve timing.

The engine has a compact water-cooled exhaust manifold, and is turbocharged and inter-cooled. A pulse turbocharging approach is employed, with cylinders 1, 2, 3 and 4, 5, 6 exhausting into separate manifolds. In this particular case the ignition delay was a user input, but it was found subsequently that the Watson correlation (equation 11.16) gave an accurate prediction. The combustion was modelled by the Watson correlation (equations 11.19 and 11.20), and the empirical constants were varied with engine load. Heat transfer within the cylinder was predicted by the Woschni (1967) correlation, and the heat transfer within the water-cooled exhaust manifold was predicted using a heat transfer coefficient to match experimental data. The Chen and Flynn (1965) model was used for predicting the friction levels, as it has a dependence on cylinder pressure; but under some low-load

Table 11.1 *Medium-speed diesel engine specification*

Configuration	6 cylinder in-line
Bore	200 mm
Stroke	215 mm
Swept volume	40 litres
Compression ratio	12.93
Firing order	153624
Rated output	840 kW at 1500 rpm
Compressor pressure ratio	2.6
Standard valve timing:	
inlet opens	60°btdc
inlet closes	45°abdc
exhaust opens	65°bbdc
exhaust closes	60°atdc
Weight	4363 kg

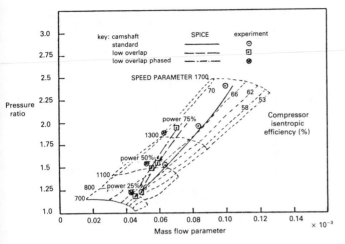

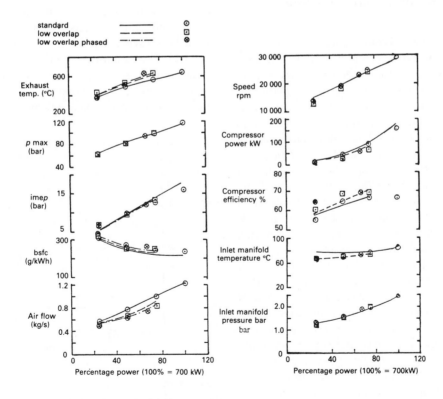

Figure 11.7

A comparison between the experimental and simulated operating points on the compressor map.

compressor data were from steady-flow tests, while the turbine data were one-dimensional steady-flow predictions using the Ainley and Mathieson method, as modified by Dunham and Came. The inter-cooler was modelled from manufacturer's data. To predict the flow through the valves, the effective flow areas were tabulated as a function of crank angle.

Model validation is an important yet difficult part of any simulation. There is never as much experimental data as might be desired, and some measurements are notoriously difficult or even impossible to make (for example, measuring the trapped mass in the cylinder). However, by conducting systematic variations of parameters in the sub-model correlations (for heat transfer, combustion etc.) it is possible to obtain a good match with experimental data.

There is also inevitably some concern over ignoring wave action effects, and in using steady-flow turbomachinery performance measurements or predictions. Watson and Janota (1982) suggest that wave action effects can be neglected if the pressure wave takes less than 15–20° crank angle to travel the length of the manifold and back. In the case of this medium-speed engine the exhaust manifold is compact, and a typical wave travel time is about 40° crank angle. It was none the less thought to be acceptable to ignore the wave action effect, as this is still a small duration compared with the valve period (305° crank angle).

conditions friction was predicted by the Millington and Hartles (1968) correlation.

In SPICE, the turbocharger is modelled from compressor and turbocharger maps that are constructed from tabulated data. The compressor map is shown in figure 11.7, and various operating points have been identified on it. The

Figure 11.8

Comparison between experimental and simulated operating points with three different camshaft timings for the engine operating at 1500 rpm for a range of loads.

Another difficulty is the modelling of the turbocharger. Firstly, the compressor performance computed from engine test data was less efficient than that predicted by the map. This was accounted for in part by the map data not including the performance of the compressor air filter and inlet. The remaining differences were attributed to the pressure drop in the inter-cooler. The compressor map was thus modified by reducing all efficiencies by a 90 per cent scaling factor to agree with the performance when measured on the engine. Secondly, the turbine is subject to a pulsating flow, which both reduces the turbine efficiency, and leads to a non-stationary operating locus. Again the turbine map was modified so as to give a closer agreement with the turbine performance data obtained from the engine.

Figure 11.8 shows a comparison of the model predictions with the test data. The matching of the bmep and bsfc could have been improved by modifying the frictional losses. This has not been done as the purpose of this model was to predict the effects of valve timing variations. Figure 11.8 also shows a comparison between experimental and predicted engine and turbocharger performance for a range of loads. The model was matched for the 120° overlap camshaft data, and then not changed when used to predict the performance with the other camshaft timings.

11.3.3 The effect of valve overlap on engine operation

In general, highly turbocharged compression ignition engines have lower compression ratios in order to limit the peak pressures and temperatures. The lower compression ratios increase the clearances at top dead centre, thus permitting greater valve overlaps. In the case of large engines with quiescent combustion systems, cut-outs in the piston to provide a valve clearance have a less serious effect on the combustion chamber performance. Thus large turbocharged engines designed for specific operating conditions can have a valve overlap of 150° or so. At a full-load operating condition, the boost pressure from the compressor will be greater than the back-pressure from the turbine. Consequently the large valve overlap allows a positive flow of air through the engine; this ensures excellent scavenging and cooling of the components with high thermal loadings (that is exhaust valves, turbine, combustion chamber). However, such a turbocharged engine running at reduced speeds or at part load is likely to have a boost pressure lower than the turbine back-pressure, in which case a reduced valve overlap is probably desirable, to reduce the back-flow of exhaust into the induction system. Too much reverse flow can lead to combustion deposits fouling the inlet port and thereby throttling the air flow. The purpose of this type of investigation is to decide whether or not a system that could control the valve timing might improve the engine performance.

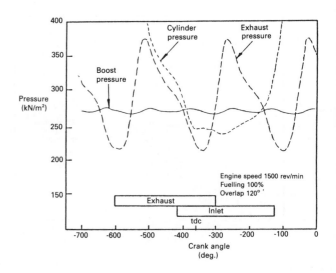

Figure 11.9
The cylinder, inlet manifold and exhaust manifold pressures during the gas exchange process at full load (750 kW) at 1500 rpm for the standard camshaft.

Figure 11.9 shows the cylinder, exhaust and inlet pressures, along with the valve events for the standard camshaft (ivo 60° btdc) at full load. It can be seen that initially the cylinder pressure is greater than the inlet manifold pressure, but as the inlet valve opens slowly there is only a small amount of reverse flow for about 25° crank angle (some 0.4 per cent of the ultimate trapped mass). Figure 11.10 is equivalent to figure 11.9, except that the fuelling has been reduced to 25 per cent. The compressor boost pressure level has fallen significantly (from 2.7 to 1.4 bar) and, although the cylinder pressure is lower, it does not equal the inlet manifold pressure until about tdc. An immediate consequence is that there is

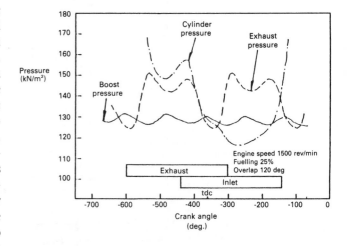

Figure 11.10
The cylinder, inlet manifold and exhaust manifold pressures during the gas exchange process at 1500 rpm for the standard camshaft at the 25 per cent load level.

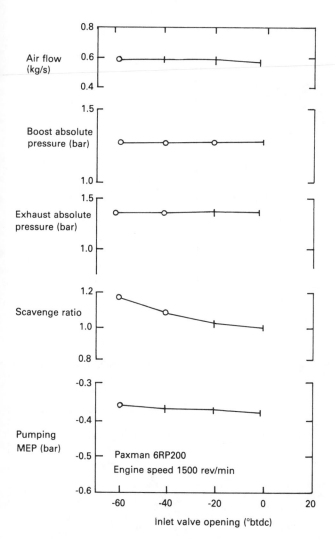

Figure 11.11
The effect on overall engine performance of delay in opening the inlet valve at the 25 per cent load level, 1500 rpm.

reverse flow of exhaust gases into the inlet manifold for about 60° crank angle, and this amounts to about 3.3 per cent of the ultimate trapped mass. It is this flow of exhaust gases into the inlet system that can cause inlet port fouling problems.

The effect of delaying the inlet valve opening is shown in figure 11.11 for the 25 per cent fuelling case. There are negligible changes in the mean boost pressure and the mean exhaust pressure, but there is a slight increase in the pumping work. Of greater importance than the mean pressure levels in the manifolds is the instantaneous pressure difference across the inlet valve; this is shown in figure 11.12 for a range of inlet valve opening angles.

Figure 11.12 shows that the back-flow into the inlet system would only be eliminated if the inlet valve opening was delayed to around top dead centre. The reduction in the inlet valve period also reduces the inflow into the cylinder, and the overall result is the slight reduction in volumetric efficiency

shown in figure 11.11. If the inlet valve opening is delayed to 20° btdc, then the reverse flow is reduced to 0.33 per cent of the ultimate trapped mass at 25 per cent fuelling (compared with 3.3 per cent with the standard inlet valve opening of 60° btdc).

Figure 11.13 summarises the effects of the different inlet valve timings at the 25 per cent fuelling level. The substantial reduction in the reverse flow at inlet valve opening can be seen when the inlet valve opening is delayed by 40° crank angle. When the inlet valve closure is also delayed by 40° crank angle, there is an even larger reverse flow, but as this is essentially air, it should not lead to any inlet port fouling. This suggests that if the inlet valve opening is to be delayed, then it is probably preferable (and certainly easier) to vary the phasing of the inlet valve events. The results from the simulation are discussed further, along with a discussion of how the variable valve timing might be achieved, by Charlton *et al.* (1990).

11.4 Conclusions

This discussion of engine modelling commenced with a review of the requirements of a comprehensive engine model. Only the requirements of a thermodynamic model have been treated here, although the introduction did summarise some of the requirements for a model to predict the mechanical performance of an engine.

Only the simplest type of thermodynamic model has been described here, that is, a model which makes use of a series of empirical sub-models to describe individual processes (for example, the burn rate), and treats the gas as a single zone of uniform composition and temperature. However, when the energy equation and combustion models were being described, an explanation was given of how the concepts are extended to multi-zone combustion models. In order to generate even a simple model of engine behaviour, it is necessary to: be able to determine the gas properties (section 11.2.2); predict the ignition delay and combustion rate (section 11.2.3); model the heat transfer (section 11.2.4); predict the flow through the engine and any associated components – notably a turbocharger and inter-cooler (section 11.2.5); and estimate the friction levels if brake performance parameters are required (section 11.2.6).

The application of this type of model has been illustrated by investigating the effect of valve overlap, on the part-load performance of a highly rated medium-speed diesel engine. This has illustrated how variables that might be difficult to change in an experiment can be investigated, and also how parameters that are difficult to measure (such as the reverse flow into the inlet manifold) can be predicted. If emissions predictions are needed, then a more sophisticated model is needed for combustion, and that developed by Bazari *et al.* (1996) has been described in section 11.2.3.

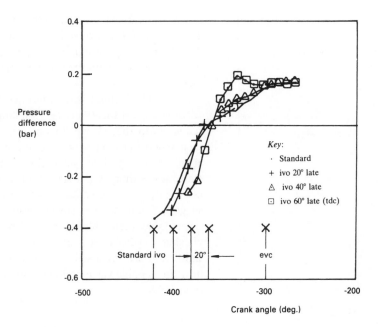

Figure 11.12
The effect on the instantaneous pressure difference across the inlet valve of delaying the inlet valve opening at the 25 per cent load level, 1500 rpm.

Unfortunately, there are still limitations for even the most complex models. The difficulties in modelling turbulence were discussed in the introduction. Further limitations in the fluid dynamics modelling are encountered in the induction system and for the in-cylinder flows. The unsteady flow in the induction system has to be modelled one-dimensionally, and the in-cylinder flow can only be modelled in a steady flow. These limitations are a consequence of finite computing power.

The most complex models also have limitations in the way in which combustion is modelled. Until the turbulence is fully understood and its effects on combustion quantified, then even multi-zone combustion models will require an empirical model for combustion. This is especially so for engines with in-cylinder fuel injection, where it is necessary to model the interaction of the fuel jet with the air. Once the fluid flow and temperature fields can be predicted before, during and after combustion, then it should be possible to predict heat transfer on a more fundamental basis.

Clearly there is scope for much further work in the area of thermodynamic and fluid mechanic engine modelling.

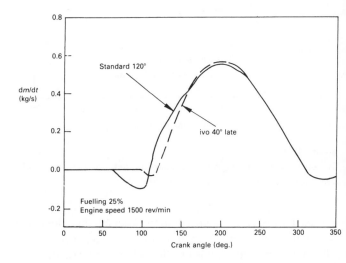

Figure 11.13
The air flow rate through the inlet valve, with a fuelling level for 25 per cent load and a speed of 1500 rpm for the standard camshaft timing and the inlet valve opening delayed by 40°ca.

CHAPTER 12

Mechanical design considerations

12.1 | Introduction

Once the type and size of engine have been determined, the number and disposition of the cylinders have to be decided. Very often the decision will be influenced by marketing and packaging considerations, as well as whether or not the engine needs to be manufactured with existing machinery.

The engine block and cylinder head are invariably cast, the main exception being the fabrications used for some large marine diesel engines. The material is usually cast iron or an aluminium alloy. Cast iron is widely used since it is cheap and easy to cast; once the quenched outer surfaces have been penetrated it is also easy to machine. Aluminium alloys are more expensive but lighter, and are thus likely to gain in importance as designers seek to reduce vehicle mass.

Pistons are invariably made from an aluminium alloy, but in higher-output compression ignition engines the piston crown needs to be protected by either a cast iron or steel top. The piston rings are often cast iron, sometimes with a chromium-plated finish. The valves are made from one or more alloy steels to ensure adequate life under their extreme operating conditions.

Engine bearings are invariably of the journal type with a forced lubrication system. To economise on the expensive bearing alloys, thin-wall or shell-type bearings are used; these have a thin layer of bearing metal on a strip steel backing. These bearings can easily be produced in two halves, making assembly and replacement of all the crankshaft bearings (main and big-end) very simple. For the more lightly loaded bearings the need for separate bearing materials can be eliminated by careful design. The use of roller or ball bearings in crankshafts is limited because of the ensuing need for a built-up crankshaft; the only common applications are in some motor-cycle and two-stroke engines.

The role of the lubricant is not just confined to lubrication. The oil also acts as a coolant (especially in some air-cooled engines), as well as neutralising the effects of any corrosive combustion products.

Only an outline of the main mechanical design considerations can be given here. Further information can be found in the SAE publications, and books such as those detailed in Baker (1979), Newton *et al.* (1983), Taylor (1985b) and Hoag (2006).

12.2 | The disposition and number of the cylinders

The main constraints influencing the number and disposition of the cylinders are:

1 the number of cylinders needed to produce a steady output

2 the minimum swept volume for efficient combustion (say 400 cm^3)

3 the number and disposition of cylinders for satisfactory balancing

4 the number of cylinders needed for an acceptable variation in the torque output.

For a four-stroke engine with five or more cylinders, there can always be a cylinder generating torque. Figure 12.1 shows the variation in the instantaneous torque associated with different numbers of cylinders. A six-cylinder four-stroke engine is of course equivalent to a three-cylinder two-stroke engine.

The most common engine types are: the straight or in-line, the 'V' (with various included angles) and the horizontally opposed – see figure 12.2.

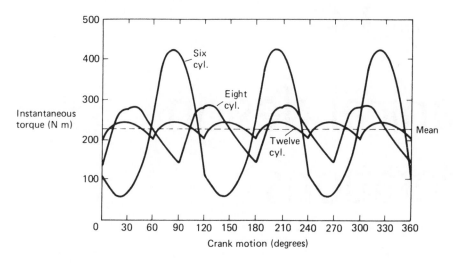

Figure 12.1

Torque characteristics of six-cylinder, eight-cylinder and twelve-cylinder four-stroke engines (from Campbell, 1978).

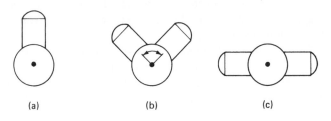

Figure 12.2

Common engine arrangements: (a) in-line; (b) 'V'; (c) horizontally opposed ('boxer').

'V' engines form a very compact power unit; a more compact arrangement is the 'H' configuration (in effect two horizontally opposed engines with the crankshafts geared together), but this is an expensive and complicated arrangement that has had limited use. Whatever the arrangement, it is unusual to have more than six or eight cylinders in a row because torsional vibrations in the crankshaft then become much more troublesome. In multi-cylinder engine configurations other than the in-line format, it is advantageous if a single crankpin can be used for a connecting-rod to each bank of cylinders. This makes the crankshaft simpler, reduces the number of main bearings, and facilitates a short crankshaft that will be less prone to torsional vibrations. None the less, the final decision on the engine configuration will also be influenced by marketing, packaging and manufacturing constraints.

In deciding on an engine layout there are two interrelated aspects: the engine balance and the firing interval between cylinders. The following discussions will relate to four-stroke engines, since these only have a single firing stroke in each cylinder once every two revolutions. An increase in the number of cylinders leads to smaller firing intervals and smoother running, but above six cylinders the improvements are less noticeable. Normally the crankshaft is arranged to give equal firing intervals, but this is not always the case. Sometimes a compromise is made for better balance or simplicity of construction; for example, consider a twin cylinder horizontally opposed four-stroke engine with a single throw crankshaft – the engine is reasonably balanced but the firing intervals are 180°, 540°, 180°, 540° etc.

The subject of engine balance is treated very thoroughly by Taylor (1985b) with the results tabulated for the more common engine arrangements in table 12.1.

When calculating the engine balance, the simplest approach is to treat the connecting-rod as two masses concentrated at the centre of the big-end and the centre of the little-end – see figure 12.3. For equivalence

$$m_1 = m_1 + m_2$$
$$m_1 r_1 = m_2 r_2 \tag{12.1}$$

The mass m_2 can be considered as part of the mass of the piston assembly (piston, rings, gudgeon-pin etc.) and be denoted by m_r, the reciprocating mass. The crankshaft is assumed to be in static and dynamic balance (figure 12.4). For static balance $Mr = Ba$, where B is the balance mass. For dynamic balance the inertia force from the centripetal acceleration should act in the same plane; this is of importance for crankshafts since they are relatively long and flexible. As a simple example, consider a planar crankshaft for an in-line four-cylinder engine, as shown diagrammatically in figure 12.5. By taking moments and resolving at any point on the shaft, it can be seen that there is no resultant moment or force from the individual centipetal forces $mr\omega^2$.

The treatment of the reciprocating mass is more involved. If the connecting-rod were infinitely long the reciprocating mass would follow simple harmonic motion, producing a primary

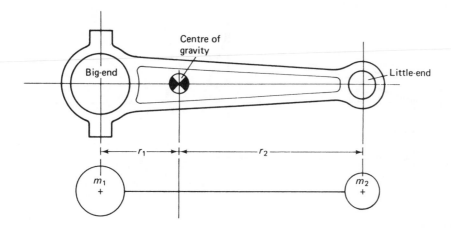

Figure 12.3
Connecting-rod and its equivalent.

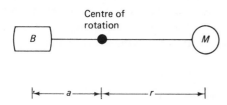

Figure 12.4
Balancing arrangements.

out-of-balance force. However, the finite length of the connecting-rod introduces higher harmonic forces.

Figure 12.6 shows the geometry of the crank slider mechanism, when there is no offset between the little-end (or gudgeon-pin or piston-pin) axis and the cylinder axis. The little-end position is given by:

$$x = r\cos\theta + l\cos\phi \qquad (12.2)$$

Inspection of figure 12.6 indicates that

$$r\sin\theta = l\sin\phi$$

and recalling that $\cos\phi = \sqrt{(1 - \sin^2\phi)}$, then

$$x = r(\cos\theta + l/r\sqrt{\{1 - (r/l)^2\sin^2\theta\}}) \qquad (12.3)$$

The binomial theorem can be used to expand the square root term:

$$x = r\{\cos\theta + l/r[1 - \tfrac{1}{2}(r/l)^2\sin^2\theta - \tfrac{1}{8}(r/l)^4\sin^4\theta + \ldots]\} \qquad (12.4)$$

The powers of $\sin\theta$ can be expressed as equivalent multiple angles:

$$\sin^2\theta = \tfrac{1}{2} - \tfrac{1}{2}\cos 2\theta$$
$$\sin^4\theta = \tfrac{3}{8} - \tfrac{1}{2}\cos 2\theta + \tfrac{1}{8}\cos 4\theta \qquad (12.5)$$

Substituting the results from equation (12.5) into equation (12.4) gives

$$x = r\{\cos\theta + l/r[1 - \tfrac{1}{2}(r/l)^2(\tfrac{1}{2} - \tfrac{1}{2}\cos 2\theta) - \tfrac{1}{8}(r/l)^4(\tfrac{3}{8} - \tfrac{1}{2}\cos 2\theta + \tfrac{1}{8}\cos 4\theta) + \ldots]\} \qquad (12.6)$$

The geometry of engines is such that $(r/l)^2$ is invariably less than 0.1, in which case it is acceptable to neglect the $(r/l)^4$ terms, as inspection of equation (12.6) shows that these terms will be at least an order of magnitude smaller than the $(r/l)^2$ terms.

The approximate position of the little-end is thus:

$$x \approx r\{\cos\theta + l/r[1 - \tfrac{1}{2}(r/l)^2(\tfrac{1}{2} - \tfrac{1}{2}\cos 2\theta)]\} \qquad (12.7)$$

Equation (12.7) can be differentiated once to give the piston velocity, and a second time to give the piston acceleration (in

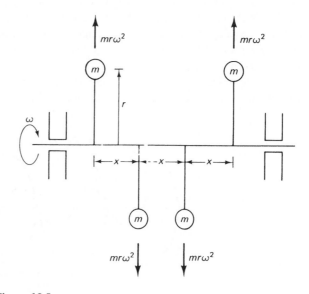

Figure 12.5
Diagrammatic representation of a four-cylinder in-line engine crankshaft.

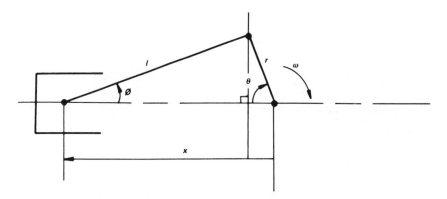

Figure 12.6
Geometry of the crank slider mechanism.

both cases the line of action is the cylinder axis):

$$\dot{x} \approx -r\omega(\sin\theta + \tfrac{1}{2}r/l\sin 2\theta) \tag{12.8}$$

$$\ddot{x} \approx -r\omega^2(\cos\theta + r/l\cos 2\theta) \tag{12.9}$$

This leads to an axial force

$$F_r \approx m_r\omega^2 r\left(\cos\theta + \frac{r}{l}\cos 2\theta\right) \tag{12.10}$$

where F_r = axial force due to the reciprocating mass
m_r = equivalent reciprocating mass
ω = angular velocity, $d\theta/dt$
r = crankshaft throw
l = connecting-rod length
$\cos\theta$ = primary term
$\cos 2\theta$ = secondary term.

In other words there is a primary force varying in amplitude with crankshaft rotation and a secondary force varying at twice the crankshaft speed; these forces act along the cylinder axis. Referring to figure 12.5 for a four-cylinder in-line engine it can be seen that the primary forces will have no resultant force or moment.

By referring to figure 12.7, it can be seen that the primary forces for this four-cylinder engine are 180° out of phase and thus cancel. However, the secondary forces will be in phase, and this causes a resultant secondary force on the bearings. Since the resultant secondary forces have the same magnitude and direction there is no secondary moment, but a resultant force of

$$4m_r\omega^2\frac{r^2}{l}\cos 2\theta \tag{12.11}$$

In a four-stroke four-cylinder engine with a typical firing order of 1–3–4–2, if the piston in cylinder 1 is at top dead centre, then so is piston 4, while pistons 2 and 3 would be at bottom dead centre. If moments are taken about the centre of the crankshaft, the primary moments cancel between cylinders 2 and 3, and 1 and 4. It can also be seen from figure 12.7 that the secondary moments will cancel.

For multi-cylinder engines in general, the phase relationship between cylinders will be more complex than the four-cylinder in-line engine. For cylinder n in a multi-cylinder engine:

$$F_{r,n} \approx m_{r,n}\omega^2 r[\cos(\theta + a_n) + r/l\cos(2\theta + 2a_n)] \tag{12.12}$$

where a_n is the phase separation between cylinder n and the reference cylinder.

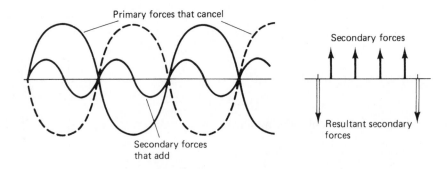

Figure 12.7
Secondary forces for the crankshaft shown in figure 12.5.

It is then necessary to evaluate all the primary and secondary forces and moments, for all cylinders relative to the reference cylinder, to find the resultant forces and moments. A comprehensive discussion on the balancing and firing orders of multi-cylinder in-line and 'V' engines can be found in Taylor (1985b), and table 12.1 summarises the balancing results for in-line, horizontally opposed and V cylinder configurations.

In multi-cylinder engines the cylinders and their disposition are arranged to eliminate as many of the primary and secondary forces and moments as possible. Complete elimination is possible for: in-line six-cylinder or eight-cylinder engines, horizontally opposed eight-cylinder or twelve-cylinder engines, and twelve-cylinder or sixteen-cylinder 'V' engines. Primary forces and moments can be balanced by masses running on two contra-rotating countershafts at the engine speed, while secondary out-of-balance forces and moments can be balanced by two contra-rotating countershafts running at twice the engine speed – see figure 12.8. Such systems are rarely used because of the extra cost and mechanical losses involved, but examples can be found on engines with inherently poor balance such as in-line three-cylinder engines or four-cylinder 'V' engines.

In vehicular applications the transmission of vibrations from the engine to the vehicle structure is minimised by the careful choice and placing of flexible mounts.

Table 12.1 *Summary of engine balance for engines of different configuration, with cylinder spacing a, crank throw r, and connecting-rod length L*

Number of cylinders	Arrangement of crank-pins[1]	Firing interval (°ca)	Primary		Secondary	
			Forces $m_r\omega^2 r$	Moments $m_r\omega^2 ra$	Forces $m_r\omega^2 r^2/L$	Moments $m_r\omega^2 r^2 a/L$
In-line cylinder engines						
1	1	720	$V = \cos\theta$	$M = 0$	$V = \cos 2\theta$	$M = 0$
2	1–2	180, 540	$V = 0$	$M = \cos\theta$	$V = 2\cos 2\theta$	$M = 0$
3	1–2–3	240	$V = 0$	$M = \sqrt{3}\sin\theta$	$V = 0$	$M = \sqrt{3}\sin 2\theta$
4	1, 4–2, 3	180	$V = 0$	$M = 0$	$V = 4\cos 2\theta$	$M = 0$
5	1–5–2–3–4	144	$V = 0$	$M = 0.449\cos\theta$	$V = 0$	$M = 4.98\cos 2\theta$
6	1, 6–2, 5–3, 4	120	$V = 0$	$M = 0$	$V = 0$	$M = 0$
8	1, 8–4, 5–2, 7–3, 6	90	$V = 0$	$M = 0$	$V = 0$	$M = 0$
Opposed cylinder engines						
2	1, 2	180, 540	$V = 2\cos\theta$	$M = 0$	$V = 0$	$M = 0$
2	1–2	360	$V = 0$	$M = \cos\theta$	$V = 0$	$M = \cos 2\theta$
4	1, 4–2, 3	180	$V = 0$	$M = 0$	$V = 0$	$M = 2\cos 2\theta$
4	1, 2–3, 4	180	$V = 0$	$M = 2\cos\theta$	$V = 0$	$M = 0$
6	1–4–5–2–3–6	120	$V = 0$	$M = 0$	$V = 0$	$M = 0$
8	1, 2, 7, 8–3, 4, 5, 6	180	$V = 0$	$M = 0$	$V = 0$	$M = 0$
8	1, 2–3, 4–7, 8–5, 6	90	$V = 0$	$M = 6\cos\theta - 2\sin\theta$	$V = 0$	$M = 0$
V cylinder engines[2]						
2	1, 2 2α block angle	2α, 720 – 2α	$V = 2\cos\theta\cos^2\alpha$ $H = 2\sin\theta\sin^2\alpha$	$M = 0$	$V = 2\cos 2\theta\cos^2\alpha\cos 2\alpha$ $H = 2\sin 2\theta\sin^2\alpha\sin 2\alpha$	$M = 0$
4	1–3–2–4; V60°, 4 throws	180	$V = 0$	$M = \cos\theta$	$V = 2\sqrt{3}\cos 2\theta$	$M = 3\cos 2\theta$
6	1–4–5–2–3–6 V60°, 6 throws	120	$V = 0$	$M = \frac{3}{2}\cos\theta$	$V = 0$	$M = \frac{3}{2}\cos 2\theta$
8	1, 2, 7, 8–3, 4, 5, 6 V90°	90	$V = 0$ $H = 0$	$M = 0$	$V = 0$ $H = 4\sqrt{2}\cos 2\theta$	$M = 0$
8	1, 2–5, 6–7, 8–3, 4 V90°	90	$V = 0$ $H = 0$	$M = 0$	$V = 0$ $H = 0$	$M = 0$
12	1, 2, 11, 12–5, 6, 7, 8–3, 4, 9, 10; V60°	60	$V = 0$ $H = 0$	$M = 0$	$V = 0$ $H = 0$	$M = 0$

Notes: [1]The crank-pins are distributed equally around a circle; 1, 4–2, 3 means that crank-pins 1 and 2 are at 12 o'clock when pins 3 and 4 are at 6 o'clock. Consecutively numbered crank-pins can of course share a journal.
[2]Odd-numbered pins will correspond to one bank of cylinders, and even-numbers to the other bank.

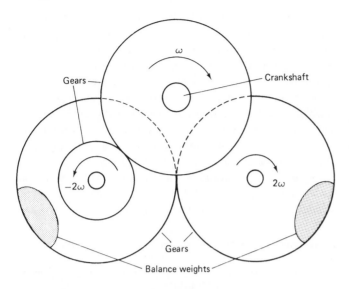

Figure 12.8
Countershafts for balancing secondary forces.

Two additional forms of engine that were very popular as aircraft engines were the radial engine and the rotary engine. The radial engine had stationary cylinders radiating from the crankshaft. For a four-stroke cycle an odd number of cylinders is used in order to give equal firing intervals. If there are more than three rows of cylinders there are no unbalanced primary or secondary forces or moments. In the rotary engine the cylinders were again radiating from the crankshaft, but the cylinders rotated about a stationary crankshaft.

12.3 | Cylinder block and head materials

Originally the cylinder head and block were often an integral iron casting. By eliminating the cylinder head gasket, problems with distortion, thermal conduction between the block and head, and gasket failure were avoided. However, manufacture and maintenance were more difficult. The most widely used materials are currently cast iron and aluminium alloys. Typical properties are shown in table 12.2.

There are several advantages associated with using an aluminium alloy for the cylinder head. Aluminium alloys have the advantage of lightness in weight and ease of production to close tolerances by casting – very important

considerations for the combustion chambers. The high thermal conductivity also allows higher compression ratios to be used, because of the reduced problems associated with hot-spots. The main disadvantages are the greater material costs, the greater susceptibility to damage (chemical and mechanical), and the need for valve seat inserts and valve guides (see chapter 7, figure 7.4). Furthermore, the mechanical properties of aluminium alloys are poorer than those of cast iron. The greater coefficient of thermal expansion and the lower Young's Modulus make the alloy cylinder head more susceptible to distortion. None the less, aluminium alloy is increasingly being used for cylinder heads.

When aluminium alloy is used for the cylinder block, cast iron cylinder liners are invariably used because of their excellent wear characteristics. The principal advantage of aluminium alloy is its low weight, the disadvantages being the greater cost and lower stiffness (Young's Modulus). The reduced stiffness makes aluminium alloy cylinder blocks more susceptible to torsional flexing and vibration (and thus noisier). Furthermore, it is essential for the main bearing housings to remain in accurate alignment if excessive wear and friction are to be eliminated. These problems are overcome by careful design, with ribs and flanges increasing the stiffness.

In order to facilitate design, much use is now made of finite element methods. These enable the design to be optimised by carrying out stress analysis and vibration analysis with different arrangements and thicknesses of ribs and flanges. In addition, the finite element method can be applied to heat transfer problems, and the thermal stresses can be evaluated to complete the model.

An interesting example of an aluminium cylinder block is the Chevrolet Vega 2.3-litre engine. The open-deck design reduces the torsional stiffness of the block but enables the block to be diecast, thus greatly easing manufacture. The aluminium alloy contains 16–18 per cent silicon, 4–5 per cent copper and 0.45–0.65 per cent magnesium. Cast iron cylinder liners are not used; instead the cylinder bore is treated to form a wear-resistant and oil-retaining surface by electrochemical etching to expose the hard silicon particles. To provide a compatible bearing surface the piston skirts are electroplated successively with zinc, copper, iron and tin. The zinc bonds well to the piston alloy, and the copper protects the zinc; the iron provides the bearing material, while the tin protects the iron and facilitates the running-in.

Table 12.2 *Properties of cast iron and aluminium alloy*

	Cast iron	Aluminium alloy
Density (kg/m^3)	7270	2700
Thermal conductivity (W/m K)	52	150
Thermal expansion coefficient (10^{-6}/K)	12	23
Young's Modulus (kN/mm^2)	115	70

In larger (non-automotive) engines, steel liners are often used because of their greater strength compared to that of cast iron. To provide an inert, oil-retaining, wear-resistant surface a carefully etched chromium-plated finish is often used.

The majority of engines use a water-based coolant, and the coolant passages are formed by sand cores during casting. Water is a very effective heat transfer medium, not least because, if there are areas of high heat flux, nucleate boiling can occur locally, thereby removing large quantities of energy without an excessive temperature rise. Because of the many different materials in the cooling system it is always advisable to use a corrosion inhibitor, such as those that are added to ethylene glycol in antifreeze mixtures. Coolants and cooling systems are discussed extensively in chapter 13.

A typical cooling-water arrangement is shown in figure 12.9. This is for the Ford direct injection compression ignition engine that is discussed in chapter 1, section 1.3, and shown in figure 1.10. At the front of the engine is a water pump which enhances the natural convection flow in the cooling system. The pump is driven by a V-belt from the crankshaft pulley. The outflow from the pump is divided into two flows which enter at opposite ends of the cylinder block. On four-cylinder engines the flow would not normally be divided in this way. The flow to the far end of the engine first passes through the oil cooler. Once the water enters the cylinder block it passes around the cylinders, and rises up into the cylinder head. While the engine is reaching its working temperature the flow passes straight to the pump inlet since the thermostats are closed (see also chapter 13, section 13.3.1).

When the engine approaches its working temperature first one thermostat opens and then, at a slightly higher temperature (say 5 K), the second thermostat opens. Once the second thermostat has opened about 50 per cent of the flow passes through the radiator. The use of two thermostats prevents a sudden surge of cold water from the radiator, and also provides a safety margin should one thermostat fail. The optimum water flow pattern can be obtained from experiments with clear plastic models, but greater use is now being made of CFD.

Air-cooling is also used for reasons of lightness, compactness and simplicity. However, the higher operating temperatures require a more expensive construction, and there is more noise from the combustion, pistons and fan.

12.4 | The piston and rings

The design and production of pistons and rings is a complicated job, which is invariably carried out by specialist manufacturers; a piston assembly is shown in figure 12.10.

Pistons are mostly made from aluminium alloy, a typical composition being 10–12 per cent silicon to give a relatively low coefficient of thermal expansion of 19.5×10^{-6} K^{-1}. The

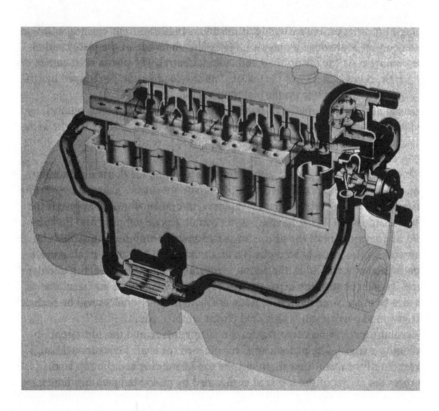

Figure 12.9
Cooling-water system; see also figures 1.10 and 12.18 (courtesy of Ford).

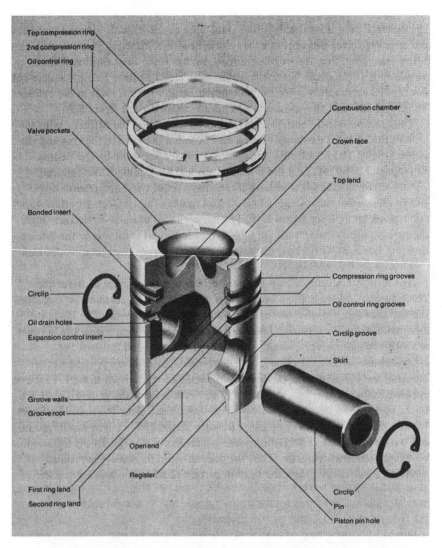

Figure 12.10
Piston assembly (with acknowledgement to GKN Engine Parts Division).

low density reduces the reciprocating mass, and the good thermal conductivity avoids hot-spots. The temperature of the piston at the upper ring groove should be limited to about 200°C, to avoid decomposition of the lubricating oil and softening of the alloy.

The high-temperature strength of the piston can be improved by increasing the copper content from 1 per cent to between 3 and 5 per cent. At a temperature of 200°C (typical of the piston pin boss) the strength is unchanged, but at diesel piston crown operating temperatures of 350°C there is an increase in strength of about 25 per cent (Mullins, 1995). Such alloys allow the position of the top ring groove to be raised in spark ignition engines, thereby reducing the crevice volume and unburned hydrocarbon emissions. In high output spark ignition engines (and invariably in diesel engines for improved durability), there can be a top ring insert as shown in figure 12.10. An alternative is to use a squeeze cast piston

with local ceramic fibre reinforcement. A similar reinforcement can be used to improve the strength and durability of the re-entrant bowl used in some diesel combustion chamber bowls (Mullins, 1995). The alumina fibre has to be formed into a shape (a preform) that can be placed in the die before the casting process.

Squeeze casting is a patented process in which a pressure in the region of 1000 bar is applied to the molten metal in a die by a hydraulically actuated die plug. The pressure is maintained during solidification, so the casting remains in contact with the die to give a faster cooling rate, and the pressure also reduces porosity. The combined effect is a significant improvement in the fatigue strength.

In high-output engines, additional piston cooling is provided by an oil spray to the underside of the piston; otherwise cooling is via the piston rings and cylinder barrel. The piston skirt carries the inertial side loading from the piston, and this loading can be reduced by offsetting the gudgeon pin (piston pin) from the piston diameter.

The gudgeon-pin is usually of hollow, case hardened steel, either retained by circlips or by an accurate diametral fit. The centre hole reduces the weight without significantly reducing the stiffness. The piston is reinforced by bosses in the region of the gudgeon-pin.

One of the key problems in piston design is allowing for thermal expansion and distortion. The thermal coefficient of expansion for the piston is greater than that of the bore, so that sufficient clearances have to be allowed to prevent the piston seizing when it is at its maximum possible service temperature. Furthermore, the asymmetry of the piston leads to non-uniform temperature distributions and asymmetrical expansions. To ensure minimal but uniform clearances under operating conditions, the piston is accurately machined to a non-circular shape. To help control the expansion, carefully machined slots and steel inserts can also be used. None the less, it is inevitable that the clearances will be such that piston slap can occur with a cold engine.

Combustion chambers are often in the piston crown, and the additional machining is trivial. The piston can also influence the engine emissions through the extent of the quench areas around the top piston ring and the top land. However the extent of the top land is governed by piston-temperature limitations.

More complex pistons include those with heat-resistant crowns, articulated skirts, and raised pads on the skirt to reduce the frictional losses.

Very high output diesel engines sometimes use cast steel or cast iron pistons. Such engines can also use an integral annular cooling gallery that surrounds the piston bowl. The cooling gallery is fed with oil through drillings in the connecting-rod and gudgeon-pin. Cast iron or steel pistons have a greater tolerance of higher temperatures than aluminium alloy pistons, and they also have a lower coefficient of thermal expansion. With their low expansion, iron and steel pistons can be designed to have lower clearances with the piston bore. This leads to reduced exhaust emissions and piston slap; piston slap can be a significant source of noise in engines during warm-up.

Rings

The three main roles of the piston rings are:

1 to seal the combustion chamber
2 to transfer heat from the piston to the cylinder walls
3 to control the flow of oil.

The material used is traditionally a fine grain alloy cast iron, with the excellent heat and wear resistance inherent in its graphitic structure. Piston rings are usually cast in the open condition and profile-finished, so that when they are closed their periphery is a true circle. Since the piston rings tend to rotate a simple square-cut slot is quite satisfactory, with no tendency to wear a vertical ridge in the cylinder. Numerous different ring cross-sections have been used – see figure 12.11.

The cross-sectional depth is dictated by the required radial stiffness, with the proviso that there is adequate bearing area between the sides of the ring and the piston groove. The piston ring thickness is governed primarily by the desired radial pressure; by reducing the thickness the inertial loading is also reduced.

Narrower piston rings are now being used, since they reduce the frictional losses and have better conformity to out-of-round bores. The narrow rings are fabricated from a nitrided chromium steel sheet as thin as 1.0 mm, and this compares with a typical 1.5 mm cast iron piston ring. The requirement to reduce lubricant consumption means that the oil film thickness on the bore is reduced, and a more wear-resistant coating is needed for the piston rings. AE Goetze have developed a process that results in microscopic ceramic particles being embedded during the chromium plating process. This is said to have more than double the wear resistance of a conventional chromium plating (Scott, 1996).

Conventional practice is to have three piston rings: two compression rings and an oil control ring. A typical oil control ring is of slotted construction with two narrow lands – see figure 12.12. The narrow lands produce a relatively high pressure on the cylinder walls, and this removes oil that is surplus to the lubrication requirements. Otherwise the pumping action of the upper rings would lead to a high oil consumption.

The Hepolite SE ring shown in figure 12.12 is of a three-piece construction. There are two side rails and a box section with a cantilever torsion bar; this loads the side rails in a way that can accommodate uneven bore wear. The lands are chromium-plated to provide a hard wearing surface. For long engine life between reboring and piston replacement, the material selection and finish of the cylinder bore are also very important. Chromium-plated bores provide a very long life, but the process is expensive and entails a long running-in time. Cast iron cylinder bores are also very satisfactory, not least because the graphite particles act as a good solid lubricant. The bore finish is critical, since too smooth a finish would fail to hold any oil. Typically, a coarse silicon carbide hone is plunged in and out to produce two sets of opposite hand spiral markings. This is followed by a fine hone that removes all but the deepest scratch marks. The residual scratches hold the oil, while the smooth surface provides the bearing surface.

12.5 | The connecting-rod, crankshaft, camshaft and valves

Connecting-rods are invariably steel stampings, with an 'H' cross-section centre section to provide high bending strength.

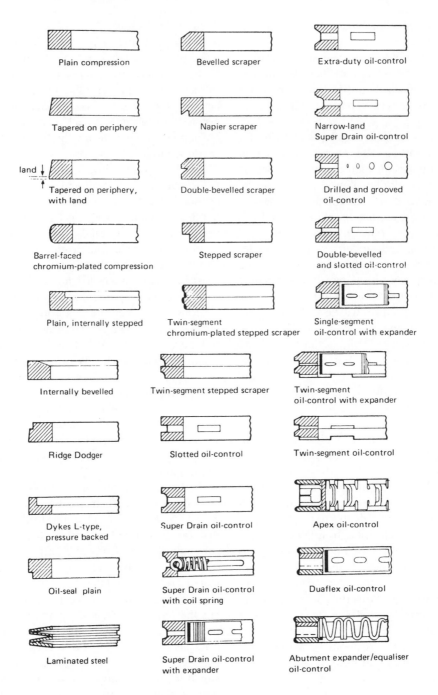

Figure 12.11
Different types of piston ring (with acknowledgement to Baker, 1979; and the source of data – AE Group).

Titanium, aluminium alloys and cast irons have all been used for particular applications, with the manufacture being by forging and machining. The big-end bearing is invariably split for ease of assembly onto the crank pin. Sometimes the split is on a diagonal to allow the largest possible bearing diameter. The big-end cap bolts are very highly loaded and careful design, manufacture and assembly are necessary to minimise the risk of fatigue failure. The little-end bearing can be a force-fit bronze bush.

Connecting-rods should be checked for the correct length, the correct weight distribution, straightness and freedom from twist.

Crankshafts for many automotive applications are now made from SG (spheroidal graphite) or nodular cast iron as

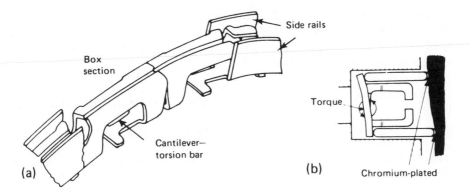

Figure 12.12
Oil control ring construction (with acknowledgement to Newton *et al.*, 1983).

opposed to forged steel. The cast iron is cheaper to manufacture and has excellent wear properties; the lower stiffness makes the shaft more flexible but the superior internal damping properties reduce the dangers from torsional vibrations. In normal cast iron the graphite is in flakes which are liable to be the sources of cracks and thus reduce the material's strength. In SG cast iron, the copper, chromium and silicon alloying elements make the graphite particles occur as spheres or nodules; these are less likely to introduce cracks than are flakes of graphite with their smaller radii of curvature.

A five-bearing crankshaft for a four-cylinder engine is shown in figure 12.13. The drilled oil passages allow oil to flow from the main bearings to the big-end bearings. The journals (bearing surfaces) are usually hardened and it is common practice to fillet-roll the radii to the webs. This process puts a compressive stress in the surface which inhibits the growth of cracks, thereby improving the fatigue life of the crankshaft. The number of main bearings is reduced in some instances. If the crankshaft in figure 12.13 had larger journal diameters or a smaller throw, it might be sufficiently stiff in a small engine to need only three main bearings. Whatever the bearing arrangement, as the number of main bearings is increased it becomes increasingly important for the journals and main bearings to be accurately in-line.

To reduce torsional vibration a damper can be mounted at the front end of the crankshaft. A typical vibration damper is shown in figure 12.14, where a V-belt drive has also been incorporated. An annulus is bonded by rubber to the hub, and the inertia of the annulus and the properties of the rubber insert are chosen for the particular application. The torsional energy is dissipated as heat by the hysteresis losses in the rubber. The annulus also acts as a 'vibration absorber' by changing the crankshaft vibration characteristics.

Camshafts are typically made from hardened steel, hardened alloy cast iron, nodular cast iron or chilled cast iron. Chill casting is when suitably shaped iron 'chills' are inserted into the mould to cause rapid cooling of certain parts. The rapid cooling prevents some of the iron carbide dissociating, and thus forms a very hard surface. A variety of surface hardening techniques are used, including induction hardening, flame hardening, nitriding, Tufftriding and carburising. The material of the cam follower has to be carefully selected since the components are very highly loaded, and the risk of surface pick-up or cold welding must be minimised.

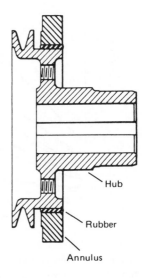

Figure 12.14
Torsional vibration damper (with acknowledgement to Newton *et al.*, 1983).

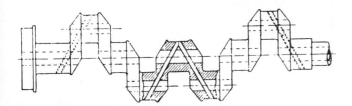

Figure 12.13
Five-bearing crankshaft for a four-cylinder engine (with acknowledgement to Newton *et al.*, 1983).

The inlet valves and in particular the exhaust valves have to operate under arduous conditions with temperatures rising above 500°C and 800°C, respectively. To economise on the exhaust valve materials a composite construction can be used; a Nimonic head with a stellite facing may be friction-welded to a cheaper stem. This also allows a material with a low coefficient of thermal expansion to be used for the stem. The valve guide not only guides the valve, but also helps to conduct heat from the valve to the cylinder head. In cast iron cylinder heads the guide is often an integral part of the cylinder head, but with aluminium alloys a ferrous insert is used.

In spark ignition engines with high outputs (say over 60 kW/litre), then sodium-cooled exhaust valves are used. These valves have a hollow exhaust valve stem, and as the sodium melts (98°C) the liquid is shaken inside the chamber. This provides a very high heat transfer coefficient between the valve head and the valve stem.

Valve seat inserts have to be used in aluminium cylinder heads, while with cast iron cylinder heads the seats can be induction-hardened. In spark ignition engines running on leaded fuel, the lead compounds lubricate the valve seat, so obviating the need for surface hardening.

The factors which affect valve gear friction and wear have been reviewed by Narasimhan and Larson (1985), along with a comprehensive overview of the materials that are used in

the valve gear. Measurements, and the associated measuring techniques for the strain and temperature distribution are discussed by Worthen and Rauen (1986).

12.6 | Lubrication and bearings

12.6.1 Lubrication

The frictional energy losses inside an engine arise from the shearing of oil films between the working surfaces. By motoring an engine (the engine is driven without firing) and sequentially dismantling the components it is possible to estimate each frictional component. The results are not truly representative of a firing engine since the cylinder pressures and temperatures are much reduced. None the less, the results in figure 12.15 derived from Whitehouse and Metcalfe (1956) indicate the trends. A useful rule of thumb is that two-thirds of engine friction occurs in the piston and rings, and two-thirds of this is friction at the piston rings. The frictional losses increase markedly with speed, and at all speeds the frictional losses become increasingly significant at part load operation. Currently much work is being conducted with highly instrumented engines and special test rigs in order to evaluate frictional losses over a range of operating conditions. For example, valve train losses can be derived from strain

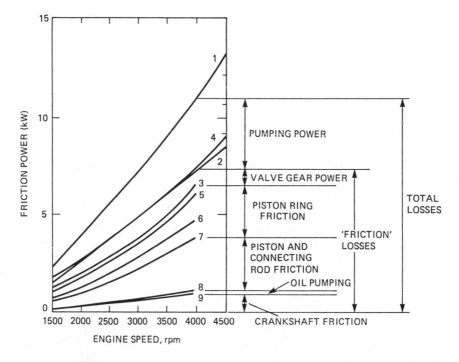

Figure 12.15
Analysis of engine power loss for a 1.5-litre engine with oil viscosity of SAE 30 and jacket water temperature of 80°C. Curve 1, complete engine; curve 2, complete engine with push rods removed; curve 3, cylinder head raised with push rods removed; curve 4, as for curve 3 but with push rods in operation; curve 5, as for curve 3 but with top piston rings also removed; curve 6, as for curve 5 but with second piston rings also removed; curve 7, as for curve 6 but with oil control ring also removed; curve 8, engine as for curve 3 but with all pistons and connecting-rods removed; curve 9, crankshaft only (adapted from Blackmore and Thomas, 1977).

gauges measuring the torque in the camshaft drive, and piston friction can be deduced from the axial force on a cylinder liner.

The scope for improving engine efficiency by reducing the oil viscosity is limited, since low lubricant viscosity can lead to lubrication problems, high oil consumption and engine wear. The SAE oil viscosity classifications are widely used: there are four categories (5W, 10W, 15W, 20W) defined by viscosity measurements at −18°C (0°F), and a further four categories (20, 30, 40, 50) defined by measurements at 99°C (210°F). Multigrade oils have been developed to satisfy both requirements by adding polymeric additives (viscosity index improvers) that thicken the oil at high temperatures; the designations are thus SAE 10W/30, SAE 20W/40 etc. Multigrade oils give better cold start fuel economy because the viscosity of an SAE 20W/40 oil will be less than that of an SAE 40 oil at ambient conditions.

A disadvantage of the SAE classification is that viscosity measurements are not made under conditions of high shear. When multigrade oils are subject to high shear rates the thickening effects of the additives are temporarily reduced. Thus, any fuel economy results quoted for different oils need a more detailed specification of the oil; a fuller discussion with results can be found in Blackmore and Thomas (1977).

There are three lubrication regimes that are important for engine components; these are shown on a Stribeck diagram in figure 12.16. Hydrodynamic lubrication is when the load-carrying surfaces of the bearing are separated by a film of lubricant of sufficient thickness to prevent metal-to-metal contact. The flow of oil and its pressure between the bearing surfaces are governed by their motion and the laws of fluid mechanics. The oil film pressure is produced by the moving surface drawing oil into a wedge-shaped zone, at a velocity high enough to create a film pressure that is sufficient to separate the surfaces. In the case of a journal and bearing the wedge shape is provided by the journal running with a slight eccentricity in the bearing. Hydrodynamic lubrication does not require a supply of lubricant under pressure to separate the surfaces (unlike hydrostatic lubrication), but it does require an adequate supply of oil. It is very convenient to use a pressurised oil supply, but since the film pressures are very much greater the oil has to be introduced in such a way as not to disturb the film pressure.

As the bearing pressure is increased and either the viscosity or the sliding velocity is reduced, then the separation between the bearing surfaces reduces until there is contact between the asperities of the two surfaces – point A on figure 12.16. As the bearing separation reduces, the solid-to-solid contact increases and the coefficient of friction rises rapidly, so leading ultimately to the boundary lubrication mode shown in figure 12.17. The transition to boundary lubrication is controlled by the surface finish of the bearing surfaces, and the chemical composition of the lubricant becomes more important than its viscosity. The real area of contact is governed by the geometry of the asperities and the strength of the contacting surfaces. In choosing bearing materials that have boundary lubrication, it is essential to choose combinations of material that will not cold-weld or 'pick-up' when the solid-to-solid contact occurs.

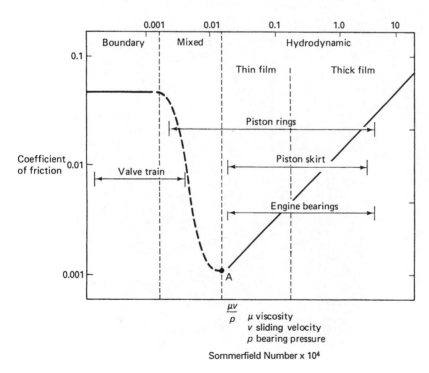

Figure 12.16
Engine lubrication regimes on the Stribeck diagram.

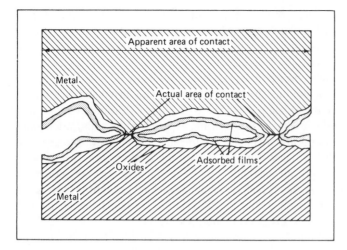

Figure 12.17
Contact between surfaces with boundary lubrication.

The lubricant also convects heat from the bearing surfaces, and there will be additives to neutralise the effect of acidic combustion products.

Part of an engine lubrication system is shown in figure 12.18; this is the same engine as is shown in figures 1.10 and 12.9. Oil is drawn from the sump through a mesh filter and into a multi-lobe positive displacement pump. The pump is driven by spur gears from the crankshaft. The pump delivery passes through the oil cooler, and into a filter element. Oil filters should permit the full pump flow, and incorporate a pressure-relief system, should the filter become blocked. Filters can remove particles down to 5 μm and smaller.

The main flow from the filter goes to the main oil gallery, and thence to the seven main bearings. Oil passes to the main bearings through drillings in the crankshaft; these have been omitted from this diagram for clarity. The oil from the main bearings passes to the camshaft bearings, and is also sprayed into the cylinders to assist piston cooling. A flow is also taken from the centre camshaft bearing to lubricate the valve gear in the cylinder head.

An alternative lubricant circuit (from a 5-litre V12 engine) is shown in figure 12.19 (Mundy, 1972). The crescent type oil pump is mounted directly on the crankshaft, in the space created at the front of the crankcase by the offset between the two banks of cylinders. With any fixed displacement pump driven directly by the engine, the pump delivery is too great at high engine speeds if the pump is of adequate size to provide sufficient lubricant at low speeds. This characteristic is turned to advantage, by using the flow from the relief valve to divert through the oil cooler. Thus at high engine speeds, when oil cooling is most needed, the flow will be highest through the oil cooler. The oil return from the cooler is returned to the pump inlet, thus minimising the oil flow through the pick-up pipe, and minimising the oil temperature in the pump. The oil is cooled by the coolant that comes direct from the radiator,

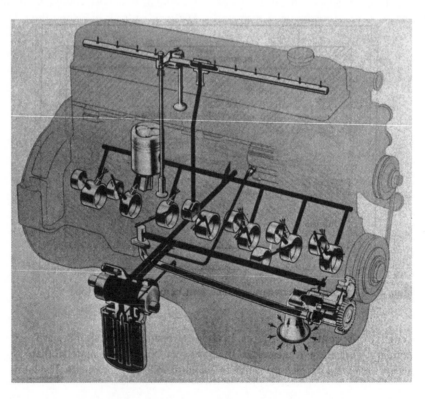

Figure 12.18
Engine lubrication system; see also figures 1.10 and 12.9 (courtesy of Ford).

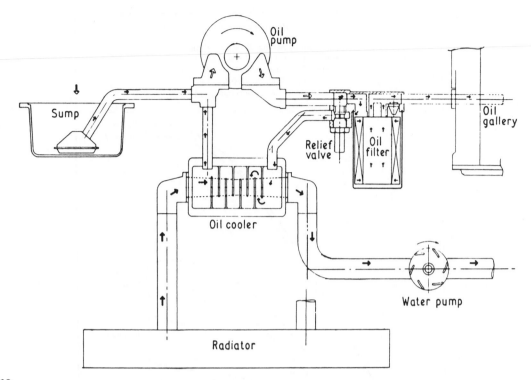

Figure 12.19
Diagrammatic layout of the oil cooler circuit from a 5-litre V12 spark ignition engine (from Mundy, 1972).

before it enters the water pump. Figure 12.20 shows the oil flow as a function of engine speed. The pump delivery increases almost linearly with engine speed (thus showing almost no dependence on pressure), and the flow is largely unaffected by the oil viscosity. The oil flow to the engine also increases with engine speed, but with high oil temperatures the reduced viscosity increases the oil demand. The difference between these two flows is the relief flow that passes through the oil cooler.

12.6.2 Bearing materials

There are two conflicting sets of requirements for a good bearing material:

1 the material should have a satisfactory compressive and fatigue strength

2 the material should be soft, with a low modulus of elasticity and a low melting point.

Soft materials allow foreign particles to be absorbed without damaging the journal. A low modulus of elasticity enables the bearing to conform readily to the journal. The low melting point reduces the risk of seizure that could occur in the boundary regime – all bearings at some stage during start-up will operate in the boundary regime. These conflicting requirements can be met by the steel-backed bearings discussed below.

Initially Babbit metal, a tin–antimony–copper alloy, was widely used as a bearing material in engines. The original composition of Babbit metal or white metal was 83 per cent tin, 11 per cent antimony and 6 per cent copper. The hard copper–antimony particles were suspended in a soft copper–tin matrix, to provide good wear resistance plus conformability and the ability to embed foreign particles. The disadvantages were the expense of the tin and the poor high-temperature performance, consequently lead was substituted for tin. The white metal bearings were originally cast in their housings, or made as thick shells sometimes with a thick bronze or steel backing. In all cases it was necessary to fit the bearings to the engine and then to hand-scrape the bearing surfaces. This technique made the manufacture and repair of engines a difficult and skilled task.

These problems were overcome by the development of thin-wall or shell-type bearings in the 1930s. Thin-wall bearings are made by casting a thin layer of white metal, typically 0.4 mm thick, onto a steel strip backing about 1.5 mm thick. The manufacture is precise enough to allow the strip to be formed into bearings which are then placed in accurately machined housings. These bearings kept all the good properties of white metal, but gained in strength and fatigue life from the steel backing. To provide bearings for higher loads a lead–bronze alloy was used, but this required hardening of the journals – an expensive process. To overcome this difficulty a three-layer bearing was developed. A thin layer of white metal was cast on top of the lead–bronze lining. To prevent diffusion of this tin into the lead–bronze layer, a plated-nickel barrier was necessary. The expense of

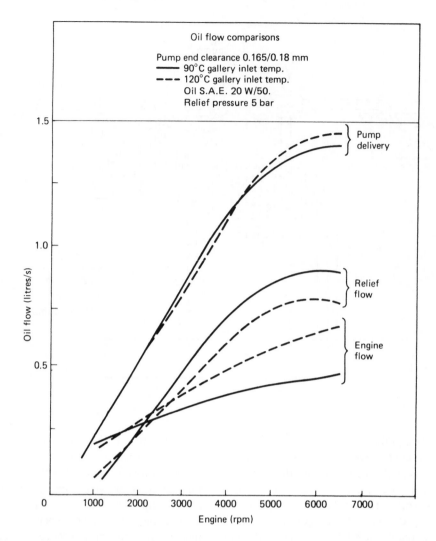

Figure 12.20
Oil pump flow characteristics with oil temperature changes for a 5-litre V12 spark ignition engine (from Mundy, 1972).

three-layer bearings led to the development of single-layer aluminium–tin bearings with up to 20 per cent tin. More recently an 11 per cent silicon–aluminium alloy has been developed for heavily loaded bearings.

The manufacture of such bearings is a specialist task carried out by firms such as Glacier Vandervell. There are essentially three different types of bearing materials, and in decreasing order of strength they are copper-based, aluminium-based and white metals (tin or lead based). In their simplest form, thin wall bearings are of bimetal construction – a steel backing with a bearing material on top. Trimetal bearings are when there is an overlay material above the bearing material.

Copper-based bearings are mostly lead bronzes, in which the lead is dispersed in a bronze matrix, which is either cast or sintered onto the steel backing. Copper-based materials are relatively hard, and are overplated with a soft phase material (such as lead–indium) for crankshaft applications. The soft phase material provides conformability to the journal, and

embeddability, so that wear debris from elsewhere can be absorbed into the bearing without harm. Copper-based bearings are used in high-performance racing engines, and heavy-duty diesel engines.

Aluminium bimetal bearings are widely used for medium-duty engine applications, and consist of tin or tin and silicon in an aluminium matrix. Tin gives the bearing good soft phase properties (for conformability and embeddability). These bearings are economical and have good corrosion resistance; they are manufactured by roll bonding onto a steel backing with an aluminium foil interlay.

White metals are either lead- or tin-based, and have been traditionally used because of their excellent soft phase properties. They have a comparatively low strength and are thus rarely used now; they are made by casting onto the steel backing.

In trimetal bearings the overlay is very thin (14–33 μm) and was originally applied by electroplating. The overlay leads to improvements in corrosion resistance, reduced friction wear and seizure resistance (especially in the early engine life). Thicker overlays provide greater conformability and embeddability, while thinner overlays promote a higher load-carrying capacity. The overlay is applied only after the bimetal bearing has been finish bored.

During electroplating it is possible to incorporate inert particles which have been in suspension in the plating bath. Alumina particles of 1 μm or smaller can be incorporated in the metal matrix this way (up to 2.0 per cent by mass), and this leads to reduced wear rates. Adding 1 per cent alumina can halve the wear rate. An alternative to electroplating is to sputter the overlay onto the bearing. In the case of aluminium–tin overlays, this leads to a greater mechanical strength (Eastham *et al.*, 1995).

The desire to reduce engine frictional losses had led to smaller area bearings being used, and this in turn places a greater loading on the bearings. This makes the selection of the bearing materials yet more critical, along with the design of the bearing clearances, the bearing housing and journals. The bearing housings must not distort under load, and the journals must remain in alignment – in the case of a crankshaft this means it should have a high bending stiffness.

Table 12.3 lists some typical bearing materials.

12.7 Advanced design concepts

The Steyr–Daimler–Puch M1 engine will be used here to illustrate various advanced concepts (Freudenschuss, 1988). This engine is a turbocharged, high-speed, direct injection diesel engine, with many innovative features; these include:

(a) monoblock construction

(b) acoustic mounting and enclosure

(c) unit injectors.

Table 12.3 *A selection of typical thin-wall bearing materials (courtesy of Glacier Vandervell)*

Designation and nominal composition		Comments
Lead bronzes		
Vandervell VP1		
Lead	17%	Used in high loading applications
Tin	5%	such as little-end bearings, and
Copper to	100%	the highest loaded big-ends
Vandervell VP2 (SAE 49)		
Lead	23%	Used in connecting-rod and
Tin	1.5%	main bearings
Copper to	100%	
Vandervell VP10		
Lead	10%	Very widely used in high loading
Tin	10%	applications such as little-end
Copper to	100%	bearings
Aluminium-based		
Glacier AS1241		
Tin	12%	Used in big-end and main
Silicon	4%	bearings especially with
Copper	1%	cast iron journals
Aluminium to	100%	
Glacier AS15 (SAE 783)		
Tin	20%	Used in big-end and main
Copper	1%	bearings with a medium loading,
Aluminium to	100%	and camshaft bearings
White metals		
Glacier GM130 (SAE 12)		
Antimony	7.5%	Camshaft bearings, and low-loaded
Copper	3%	main bearings
Tin to	100%	
Glacier GM155 (SAE 13)		
Antimony	10%	Camshaft bearings, and low-loaded
Tin	6%	main bearings
Lead to	100%	

Table 12.4 *Specification of the Steyr–Daimler–Puch M1 turbo-charged and aftercooled diesel engine*

Arrangement	6-cylinder in-line	
Swept volume	3.2	litres
Bore	85	mm
Stroke	94	mm
Maximum output	136; 4300	kW; rpm
Maximum torque	390; 2800	N m; rpm
Minimum bsfc	218	g/kWh

The engine is available as a four- or six-cylinder unit, with a range of ratings for automotive and marine applications. A cross-section of the engine is shown in figure 12.21, and the specification of the marinised engine (the version with the highest rating) is given in table 12.4.

The details of the monoblock construction and the installation of the unit injector can be seen more clearly in figure 12.22. Monoblock construction means that the cylinder head is integral with the cylinder liner, and it is a type of construction that has been used occasionally for many years, Monoblock construction has several advantages:

(a) The absence of cylinder head bolts gives greater freedom in optimising the positions of the inlet and exhaust ports, and the fuel injector.

(b) Some machining operations are eliminated, and there can be no cylinder head gasket problems.

(c) The absence of cylinder head clamping forces (which cannot be uniform circumferentially) means that the cylinder bore is distorted less, and expansion due to thermal and pressure effects is also uniform.

The reduction of bore distortion is important, since this can be a major cause of piston blow-by and oil consumption. Figure 12.22 also shows drilled coolant passages around the combustion chamber. These ensure that the coolant flow can be directed into regions with high thermal loadings, namely around the injector and the valve bridge area.

The unit injector is inclined at about 15° to the cylinder axis, and the fuel feed and return are by drilled holes in the cylinder head. This eliminates much pipework, and simplifies installation of the injectors. The fuelling is regulated by a 'rack' that is controlled by a computer-based engine management system. A single overhead camshaft operates the valve gear through finger-followers, and the same camshaft operates the pump within the unit injector by a rocker arm with a roller follower. The roller follower is used to minimise frictional losses on what is the most heavily loaded cam. The injector uses a two-spring system that gives two stages to the injection (also known as pilot injection); this helps to reduce the ignition delay and the combustion noise (chapter 6, section 6.6.2).

The acoustic mounting and enclosure can be seen in figure 12.21. The main structure of the engine (the cylinder block

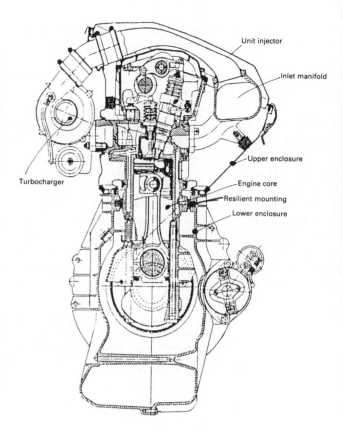

Figure 12.21

Cross-section of the Steyr–Daimler–Puch M1 monoblock DI diesel engine, showing the acoustic enclosures (adapted from Freudenschuss, 1988).

and crankshaft bearings) is made from cast iron, and this is encased by an acoustic enclosure in two parts. The lower part of the enclosure is diecast from aluminium alloy in two parts, and it acts as the sump, and the mounting for auxiliary equipment and any gearbox. This lower enclosure is mounted through elastic support rings to provide acoustic isolation. The upper housing is made from sheet metal, and it is attached to the lower enclosure. All apertures in the housing are carefully sealed, to minimise the transmission of noise. The maximum noise level (at 1 m distance) is 94 dB(A) for the six-cylinder engine at 4300 rpm. On average, the reduction in noise attributable to the acoustic enclosure is 8 dB(A).

A prototype Ford engine has been developed that uses a similar monoblock construction, but with three fibre-reinforced plastic panels to encase the engine, and a dough moulding compound for the camshaft cover. The side panels are also used as the outer surface of the coolant jacket. The engine is reported as producing 30 per cent less noise than an equivalent all-metal unit. Fibre-reinforced plastics have also been used for experimental connecting-rods. These offer potential for significant mass reductions. However, their manufacture and design are both involved, since the orientation of the fibre reinforcement is critical.

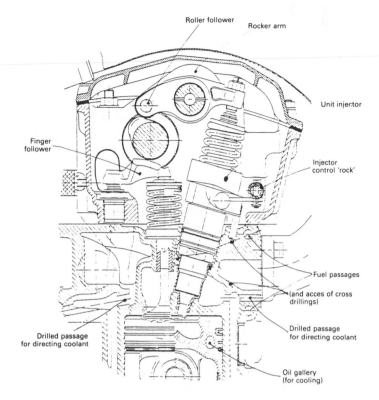

Figure 12.22
Cross-section of the unit injector, its fuel supply and cooling details of the Steyr–Daimler–Puch M1 monoblock DI diesel engine (adapted from Freudenschuss, 1988).

A low heat loss engine was discussed in chapter 10, section 10.4 (and shown in figure 10.24). This Isuzu engine makes significant use of ceramics. The silicon nitride piston crown, bonded to an alloy piston, reduces heat transfer and the reciprocating mass. The silicon nitride tappet face has a lower mass and better wear properties than conventional metal cam followers. However, most significant is the use of ceramics in the turbocharger. The lower density of the rotor leads to a better transient response, and the high temperature strength permits gas temperatures as high as 900°C.

12.8 | Conclusions

By careful design and choice of materials, modern internal combustion engines are manufactured cheaply with a long reliable life. Many ancillary components such as carburettors, radiators, ignition systems, fuel injection equipment etc. are made by specialist manufacturers. Many engine components such as valves, pistons, bearings etc. are also manufactured by specialists. Consequently, engineers also seek advice from specialist manufacturers during the design stage.

12.9 | Questions

12.1 The Rover K4 single-cylinder research engine has a reciprocating mass of 1.53 kg in optical mode, the stroke is 89 mm, and the connecting-rod length/crank throw ratio (L/R) is 3.6. What product of mass (m) and eccentricity (e) should be used for the primary and secondary balance masses, and how should they be deployed?

12.2 A three-cylinder two-stroke engine uses uniflow scavenging through overhead valves. The reciprocating mass for each cylinder is 0.3 kg, the stroke is 70 mm, the connecting-rod length/crank throw ratio (L/R) is 3.5, and the inter-cylinder bore spacing is C. Show that, with a 120° firing interval, the primary and secondary forces are balanced. If the balance masses for eliminating the primary moments are a distance D from the bore of the centre cylinder, what product of mass (m) and eccentricity (e) should be used, and what will the unbalanced secondary moments be? How should the balance masses be configured?

CHAPTER 13

Heat transfer in internal combustion engines

There are two aspects to heat transfer within internal combustion engines. Firstly there is heat transfer from within the combustion chamber to its boundaries (discussed in chapter 11, section 11.2.4), and secondly there is heat transfer from the combustion chamber to its cooling media – this aspect is discussed here.

First of all, the cooling requirements are considered on a global basis, as a function of engine type, load and speed, and cooling system type (that is, liquid- or oil-cooled). Also included here is a discussion of the heat flow in low heat loss diesel engines (so-called 'adiabatic' engines) and the use of ceramic components or coatings as insulators. Such engines have already been discussed in chapter 10, section 10.4 and chapter 12, section 12.7.

Secondly, a conventional liquid coolant system is described, and this leads to a review of alternative coolants and cooling systems that might cause a lower part-load fuel consumption. There is also a discussion of engine performance during warm-up.

13.1 | In-cylinder heat transfer

There are two approaches to measuring the in-cylinder heat transfer. The first to be discussed here is when a pair of temperatures are measured, sufficiently far from the combustion gases for the values to be considered as steady. The steady heat flow equation can then be used to deduce the heat flow. The second approach is to measure the instantaneous variation in the surface temperature, and a second reference temperature, and then to solve the unsteady heat flow equations. This is obviously more difficult (experimentally and analytically), but it does reveal the instantaneous heat flow as well as the mean heat flow.

Thermocouples are usually used for temperature measurements because they are robust and comparatively cheap; their disadvantage is their small signal, of about 40 μV/K. In the steady-state approach either a pair of thermocouples is installed a known distance apart, or a thermocouple is traversed in a hole normal to the combustion chamber (French and Atkins, 1973). As well as determining the heat flow, such measurements enable the mean surface temperature to be found at the gas and coolant faces. Often it is necessary to section an engine after tests, in order to determine the thermocouple positions with sufficient accuracy. The analysis assumes one-dimensional heat flow, and there is little alternative to this. The exception is when a finite element model of the complete engine is being developed, and the model is being tuned to individual temperature measurements. Temperature differences will be of the order of 10 K, so extreme care is needed with the instrumentation.

The results from such steady-state studies are reviewed in detail by Taylor (1985a), who adopts a dimensionally rigorous approach, and considers the independent influence of parameters such as ignition timing, mixture strength, engine size and speed. A less rigorous approach was adopted by French and Atkins (1973) who concluded that

$$\phi \, [\text{kW/m}^2] =$$
$$(\text{heat flux factor}) \times (5.13 \times \text{fuel flow [kg/s]/piston area [m}^2])^{0\cdot}$$

$$(13.1)$$

The heat flux factor is clearly not dimensionless, and varies with position and engine type, being about 100 for the cylinder liner, and in the range of 300–700 for the cylinder head.

The second approach requires a very fast-acting surface-mounted thermometer, and Gatowski et al. (1989) reviewed a number of technologies (figure 13.1):

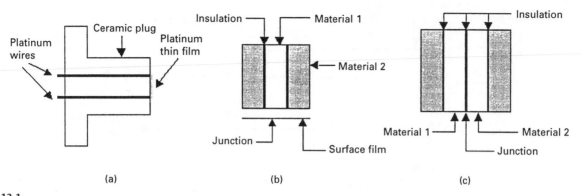

Figure 13.1
Three types of fast response surface thermocouple: (a) a platinum resistance thermometer; (b) a Bendersky-type thermocouple probe; (c) an 'erodable' thermocouple.

(a) A platinum resistance thermometer, with a thin platinum film on a ceramic plug.

(b) A Bendersky-type thermocouple probe, in which the thermocouple junction is formed by vacuum deposition of a thermocouple material onto an insulated wire of a second thermocouple material.

(c) An 'erodable' thermocouple, in which the thermocouple material is insulated from the parent metal, and the thermocouple junction is formed by abrading the exposed ends of the thermocouple, so that the thermocouple material is smeared (as a result of plastic flow) to form a junction.

Gatowski *et al.* (1989) concluded from shock tube experiments that the erodable thermocouple had the fastest response, which was much shorter than time constants associated with engine heat flux measurements. Although the platinum resistance thermometer will have a large output, it is likely that the ceramic substrate will interfere with the heat flow and invalidate the assumption of one-dimensional heat flow. The Bendersky-type probes are less durable than the erodable thermocouples, since if the junction fails on the erodable thermocouple it can be reinstated by abrading the surface.

A second temperature measurement is needed (referred to here as the reference temperature), and ideally this should be far enough from the surface temperature fluctuations to be a steady temperature, since this enables the temperature difference between the two probes to be amplified. Even if the second junction could be made close enough to the first junction to assume that the steady heat flow equation could be used, the temperature difference would be very small, and errors associated with measuring the separation would also be significant. It is thus necessary to solve the unsteady heat flow equation

$$\partial T / \partial t = \alpha \partial^2 T / \partial x^2 \tag{13.2}$$

where $\alpha = k/\rho c$, the thermal diffusivity.

Ferguson (1986) and Heywood (1988) both explain how a Fourier series can be fitted to the surface temperature variation, from which the surface heat flux can then be used. If a Fast Fourier Transform (FFT) is to be used, then there must be 2^n samples/cycle. Otherwise a slower Discrete Fourier Transform (DFT) has to be used. An alternative approach, which is perhaps conceptually simpler, is to use a finite difference method, and these methods are treated very comprehensively by Smith (1985). The forward difference approximation for the first derivative is

$$\partial T / \partial t \approx \left[T(x, t)_{t+\Delta t} - T(x, t)_t \right] / \Delta t \tag{13.3}$$

and the central difference approximation for the second derivative is

$$\partial^2 T / \partial x^2 \approx \left[T(x, t)_{x+\Delta x} - 2T(x, t)_x + T(x, t)_{x-\Delta x} \right] / \Delta x^2 \tag{13.4}$$

Substitution of equations (13.3) and (13.4) into (13.2) gives

$$\left[T(x, t)_{t+\Delta t} - T(x, t)_t \right] / \Delta t \approx \alpha \left[T(x, t)_{x+\Delta x} - 2T(x, t)_x + T(x, t)_{x-\Delta x} \right] / \Delta x^2 \tag{13.5}$$

Rearranging gives

$$T(x, t)_{t+\Delta t} \approx T(x, t)_t + r \left[T(x, t)_{x+\Delta x} - 2T(x, t)_x + T(x, t)_{x-\Delta x} \right] \tag{13.6}$$

where $r = \alpha \Delta t / \Delta x^2$.

This is an explicit formulation, since the value at the new time step $(t + \Delta t)$ depends solely on the known values at the current time step. There is a step size limit and $r \leq 0.5$ for a stable solution. With $r = 0.5$, equation (13.6) reduces to

$$T(x, t)_{t+\Delta t} \approx \left[T(x, t)_{x+\Delta x} + T(x, t)_{x-\Delta x} \right] / 2 \tag{13.7}$$

The time step is fixed by the sampling rate, and the thermal diffusivity is also fixed, so there will be a pre-determined number of grid points for the thermocouple separation.

The temperature distribution in the solid for a cycle of engine data has to be found by iteration. To start with, a linear

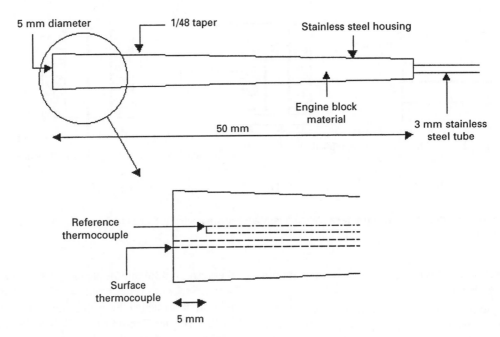

Figure 13.2
A NANMAC Corporation erodable surface thermocouple and reference thermocouple in a tapered housing, for installation in a combustion chamber.

temperature distribution can be assumed, and then the first iteration is completed by applying equation (13.7) to every grid point for each sample within the cycle. The iteration has to continue until, from one iteration to the next, there is an acceptably small variation in the temperature distribution.

A more detailed review of the transient behaviour of an eroding type of surface thermocouple has been reported by Wang *et al.* (2006). A two-dimensional finite element model was used to analyse the unsteady heat conduction behaviour of the thermocouple, and the impulse response of the thermocouple was found by a laser impulse excitation experiment. Comparisons between the experimental impulse measurements with the one-dimensional and two-dimensional models concluded that the two-dimensional model was more accurate. However, if only a one-dimensional model is available then the thermal properties of the thermocouple should be used, and not those of the surrounding material.

Figure 13.2 shows the construction of a type K thermocouple assembly, which relies on its taper for installation in a combustion chamber. The surface temperature variations are quite small, as can be seen from the results in figure 13.3. Figure 13.4 shows the ensemble-averaged heat flux corresponding to figure 13.3. However, each cycle was analysed for its instantaneous heat flux; it would not be correct to calculate the heat flux from the ensemble-averaged temperature variation. Both figures 13.3 and 13.4 show interference from the ignition system, so this should be ignored. As would be expected, the earlier ignition timing leads to an earlier and higher heat flux.

13.2 | Engine cooling

13.2.1 Background

There are three reasons for cooling engines: firstly to promote a high volumetric efficiency, secondly to ensure proper combustion, and thirdly to ensure mechanical operation and reliability. The cooler the surfaces of the combustion chamber, the higher the mass of air (and fuel) that can be trapped in the cylinder. In general, the higher the volumetric efficiency, the higher the output of the engine.

In the case of spark ignition engines, cooling of the combustion chamber inhibits the spontaneous ignition of the air/fuel mixture. Since spark ignition engines have an essentially homogeneous mixture of fuel and air, spontaneous ignition can affect a significant quantity of mixture, and the subsequent rapid pressure rise, or so-called detonation, generates the characteristic 'knocking' sound. This process destroys the thermal boundary layer, and can lead to overheating of components and ensuing damage.

There are three ways in which overheating can affect the mechanical performance of an engine. Firstly, overheating can lead to a loss of strength. For example, aluminium alloys soften at temperatures over about 200°C, and the piston ring grooves can be deformed by a creep mechanism. Furthermore, if the spontaneous ignition is sufficiently severe, the piston can be eroded in the top-land region. This is usually the hottest region of the piston, and it also coincides with the end-gas region, the region where spontaneous ignition most

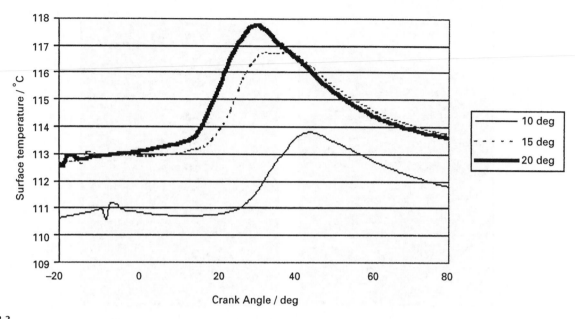

Figure 13.3
Ensemble (that is, data from a particular crank angle)-averaged temperature variations from 150 cycles for three different ignition timings in the Rover K4 single-cylinder engine; engine operating at 1000 rpm, with 1.04 litres/s of air, and a relative air/fuel ratio of 0.9 with iso-octane as the fuel.

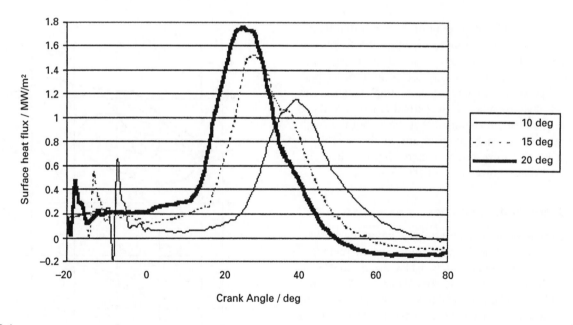

Figure 13.4
Ensemble-averaged heat flux variations from 150 cycles for three different ignition timings in the Rover K4 single-cylinder engine; engine operating conditions as in figure 13.3.

frequently occurs. Figure 13.5 shows that such 'detonation' damage can also occur with diesel engines, if the rapid combustion following the ignition delay period is too severe. Secondly, the top piston ring groove temperature must also be limited to about 200°C if the lubrication is to remain satisfactory. Above this temperature lubricants can degrade, leading to both a loss of lubrication, and packing of the piston ring groove with products from the decomposed oil. Finally, failure can result through thermal strain. Data for material failure are usually expressed in terms of stress, either for a single load application, or alternatively as fatigue data where the number of load applications also has to be specified. Such

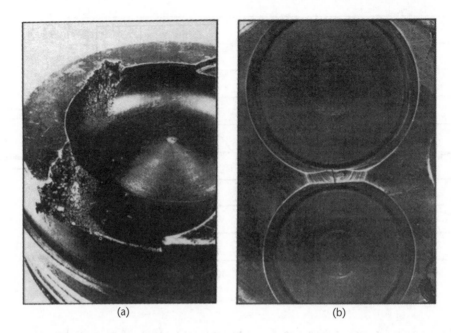

(a)   (b)

Figure 13.5
Examples of thermal loading: (a) detonation damage on the edge of a diesel engine piston; (b) crack across the valve bridge (French, 1984). [Reproduced by kind permission of Butterworth–Heinemann Ltd, Oxford.]

data can also be considered in terms of the strain that would cause failure: strain can be caused by either mechanical or thermal loading. The thermal strain is directly proportional to the temperature gradient. Failure is not likely from a single occurrence, but as a consequence of thermal fatigue. The regions most likely to suffer from thermal fatigue are those within the combustion chamber that have both a high temperature, and a high temperature gradient. This is exemplified by the valve bridge region, the area between the inlet and exhaust valve seats (figure 13.5).

13.2.2 Spark ignition engines

The heat rejected to the coolant in spark ignition engines is a function of the speed, load, ignition timing and air/fuel ratio. The large number of dependent variables means that even when comprehensive energy balance data are published, not all the variables are likely to have been investigated systematically, nor will the test conditions have necessarily been fully defined.

Gruden and Kuper (1987) present a series of contour plots for the different energy flows (fuel in, brake power, coolant, oil, exhaust) as functions of bmep and engine speed for a 2.5-litre spark ignition engine. They also present contour plots of the brake, mechanical and indicated efficiency. Figure 13.6 shows the brake and mechanical efficiency as a function of the engine load and speed. The engine has been tuned for stoichiometric operation at part load, while at full throttle the mixture has been enriched to give the maximum power. The

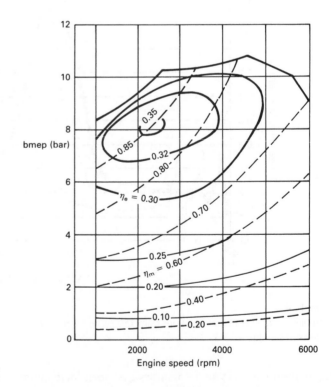

Figure 13.6
The brake efficiency and mechanical efficiency contours of the four-cylinder 2.5-litre spark ignition engine, whose thermal performance is illustrated by figures 13.7 and 13.8 (adapted from Gruden and Kuper, 1987).

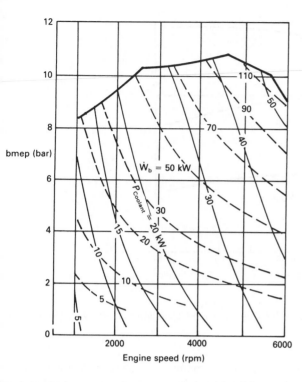

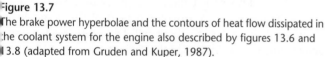

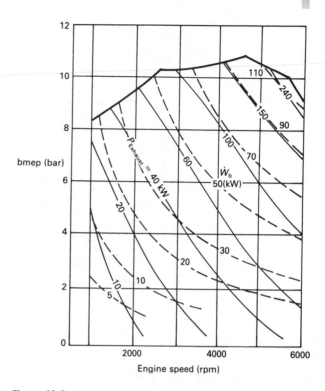

Figure 13.7

The brake power hyperbolae and the contours of heat flow dissipated in the coolant system for the engine also described by figures 13.6 and 13.8 (adapted from Gruden and Kuper, 1987).

Figure 13.8

The brake power hyperbolae and the contours of thermal energy in the exhaust, for the engine already described by figures 13.6 and 13.7 (adapted from Gruden and Kuper, 1987).

mechanical efficiency is directly affected by the load (with zero mechanical efficiency by definition at no load). Figure 13.6 also shows that the mechanical efficiency at full load falls from about 90 per cent at 1000 rpm to 70 per cent at 6000 rpm; at 6000 rpm the frictional losses represent about 34 kW. Friction dissipates useful work as heat, some of which appears in the coolant and some in the oil. The heat loss recorded to the oil is almost solely a function of speed, with about 5 kW dissipated at 3000 rpm, and 15 kW dissipated at 6000 rpm.

Figure 13.7 shows the contours of the energy flow to the coolant as a function of the load and speed. For convenience, the brake power output hyperbolae (calculated from bmep and speed) have also been added. At a bmep of about 1 bar, the energy flow to the coolant is about twice the brake power output, while at a load of 3 bar bmep, the energy flow to the coolant is comparable to the brake power output. In the load range of 8–10 bar bmep, the energy flow to the coolant is about half the brake power output. However, of greater importance to the vehicle cooling system are the absolute values of the heat rejection, and figure 13.7 shows that heat rejected to the coolant is a stronger function of speed than load.

Figure 13.8 is the counterpart to figure 13.7, as it shows contours of the exhaust gas energy superimposed on the brake power output contours. Almost without exception, the energy flow in the exhaust is greater than the brake power output.

However, the exhaust energy comprises in part the chemical energy of the partial combustion products: at full load the chemical energy is comparable to the exhaust thermal energy. The use that can be made of the exhaust energy will be discussed later in the context of engine warm-up.

The effect of the air/fuel ratio or equivalence ratio on the energy flow to the coolant is illustrated by some data obtained from a gas engine (Brown, 1987). The gas engine has an entirely homogeneous air/fuel mixture, so that it is possible to run with a very wide range of equivalence ratios. The equivalence ratio of 0.75 corresponds to an air/fuel ratio of about 20:1 for a gasoline fuelled engine. There are several ways of expressing the energy flow to the coolant as a function of the equivalence ratio, and these are illustrated by figure 13.9. Firstly, in terms of the total fuel energy supplied, the heat flow to the coolant is nearly constant at 28 per cent of the supplied energy, with a slight fall with rich mixtures to 25 per cent at an equivalence ratio of 1.2. Next, the energy flow to the coolant can be considered as a fraction of the brake power output, but as this remains close to unity (within experimental tolerances), it has not been plotted here. Finally, the absolute values of the energy flow to the coolant have been plotted. From the foregoing discussion, it can be seen that the energy flow to the coolant is a reflection of the way in which the engine brake output responds to the variation of the equivalence ratio. The explanation is that both the work

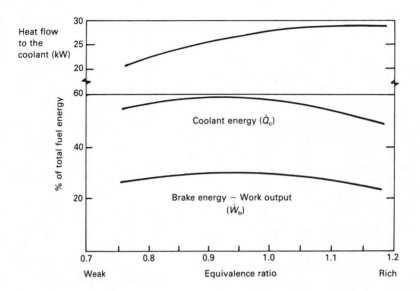

Figure 13.9

A comparison of the energy flow to the coolant and the brake output for a Waukesha VRG220 gas engine, operating at 1500 rpm full throttle with varying air/fuel ratio and MBT ignition timing (Brown, 1987).

output and the heat rejected are functions of the combustion temperature.

Advancing the ignition timing leads to an increased absolute value of heat rejected to the coolant, if the throttle, speed and air/fuel ratio are fixed. Earlier ignition causes higher temperatures in both the burned and unburned gas, and this leads to higher levels of heat rejection from the combustion chamber.

Raising the compression ratio also increases the in-cylinder gas temperatures, but this does not necessarily lead to an increase in the heat flow to the coolant. The higher compression ratio increases the work output from each charge that is ignited, and thus the exhaust temperature is lowered. Consequently, the heat rejected to the exhaust valve and exhaust port is reduced, and this can offset any increase

in heat flow from the combustion chamber. As the compression ratio continues to be increased, the gains in work output reduce and the surface-to-volume ratio of the combustion chamber deteriorates, so there will be a compression ratio above which the heat flow to the coolant increases. However, this is not necessarily a compression ratio that is likely to be encountered in practice, for the reasons discussed in chapter 4, section 4.2.1.

The material of the engine, and in particular the cylinder head, can affect the engine performance. Tests reported by Gruden and Kuper (1987) included comparisons of the energy balance with cast iron and aluminium alloy cylinder heads that were otherwise identical. Some results for a speed of 2000 rpm are presented in figure 13.10, showing the brake power ($\dot{W}_b$); the power flow to the coolant ($\dot{Q}_c$), oil ($\dot{Q}_{oil}$),

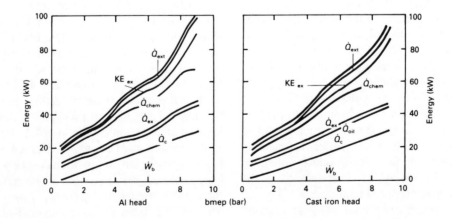

Figure 13.10

A comparison between aluminium alloy and cast iron cylinder heads, showing the effect of load on the energy balance of a 2-litre spark ignition engine at a speed of 2000 rpm (adapted from Gruden and Kuper, 1987).

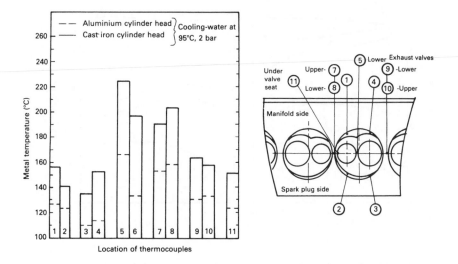

Figure 13.11
Illustrates the lower metal temperatures that occur in aluminium alloy cylinder heads for full-load operation at 2000 rpm (adapted from Finlay *et al.*, 1985).

exhaust ($\dot{Q}_{ex}$); the chemical energy flow in the exhaust ($\dot{Q}_{chem}$); and the kinetic or dynamic energy flow in the exhaust (KE_{ex}); the balance was assigned to extraneous heat transfer by radiation and convection ($\dot{Q}_{ext}$). Gruden and Kuper (1987) concluded that the energy flows were identical for a wide range of loads. However, at full load the aluminium cylinder is less susceptible to self-ignition, so that ignition timing is not knock limited. The explanation for this is given by Finlay *et al.* (1985), who also made direct comparisons between cast iron and aluminium alloy cylinder heads. Figure 13.11 shows that for identical full load operation at 2000 rpm, the aluminium alloy cylinder head had metal temperatures that were in the range 17–60 K lower than the cast iron cylinder head. Aluminium alloys have a thermal conductivity that is typically three times that of cast iron, and the lower surface temperatures reduce the likelihood of knock in the end-gas. However, the overall energy flow is not affected significantly, as the highest thermal resistance is in the gaseous thermal boundary layer of the combustion chamber.

Air cooling is most frequently encountered on engines of low output, where simplicity in the absence of an additional cooling medium is an advantage. Small two-stroke engines are frequently air cooled. Air-cooled engines tend to be noisier for several reasons:

(i) the combustion chamber walls can radiate sound

(ii) the engine structure tends to be less rigid, as liquid cooling passages result in a box type construction

(iii) the cooling fan is generating significant air motion.

The inefficiency of simple cooling fans also means that the cooling power requirements of air-cooled engines are greater than those of liquid-cooled engines. When spark ignition engines were the common form of aircraft propulsion, it was accepted that air-cooled engines had a slightly lower output and efficiency. However, the weight saving associated with an air-cooled engine meant that for journeys of up to five or six hours' duration, the overall 'engine plus fuel' weight of an air-cooled engine was lower than the weight of a liquid-cooled engine and its fuel (Judge, 1967). A direct comparison has been made by Gruden and Kuper (1987) between liquid- and air-cooled engines; they concluded that there were no significant differences. Figure 13.12 shows an energy balance for 2000 rpm, in which the following energy flows were evaluated: the brake power ($\dot{W}_b$); the power flow to the coolant ($\dot{Q}_c$), oil ($\dot{Q}_{oil}$), exhaust ($\dot{Q}_{ex}$); the chemical energy flow in the exhaust ($\dot{Q}_{chem}$); and the kinetic or dynamic energy flow in the exhaust (KE_{ex}); the balance was assigned to extraneous heat transfer by radiation and convection ($\dot{Q}_{ext}$). Both the air-cooled and the liquid-cooled engine have a maximum brake efficiency of 32 per cent in the range of 6–7 bar bmep. With the air-cooled engine there is a greater energy flow to the oil, and somewhat surprisingly the energy flow to the coolant is also greater. Figure 13.12 also shows how the chemical energy in the exhaust increases rapidly above a bmep of 7 bar. At this stage the engine is likely to be operating at wide-open throttle, and the output is increased by enriching the mixture. For the liquid-cooled engine operating close to full load (8–9 bar bmep) the fraction of the total energy input to the coolant falls, and on an absolute basis the heat flow to the coolant is almost constant. This would correspond to the coolant power contours on figure 13.7 becoming vertical at full load. Thus for a given engine output at full throttle, a larger capacity engine operating with a weak mixture would reject more heat to the coolant than a smaller capacity engine operating with a rich mixture.

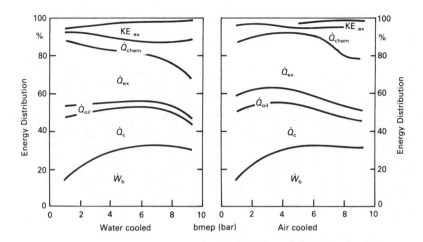

Figure 13.12
A comparison between air- and liquid-cooled engines, showing the effect of load on the energy balance of a 2-litre spark ignition engine at a speed of 2000 rpm (adapted from Gruden and Kuper, 1987).

In conclusion, it is important to emphasise that caution must be used in applying any of the energy balance data for spark ignition engines reported here to other engines, since the ignition timing and the compression ratio both influence the energy balance. It is unfortunate that comprehensive energy balance data are not readily available in the literature. However, so far as the engine designer is concerned it is the maximum thermal loading that is the main interest, and this occurs at full throttle and (usually) maximum speed.

13.2.3 Compression ignition engines

Energy balance data for compression ignition engines are equally sparse. The heat flow to the coolant is higher in indirect injection (IDI) engines than with direct injection (DI) engines, as there is a high heat transfer coefficient in the pre-chamber throat, and this leads to a higher heat loss from the cylinder contents. Direct injection diesel engines frequently have an oil cooler, which employs the engine coolant as the cooling medium; consequently the heat flow to the coolant usually includes the heat removed by the oil. Indirect injection engines are usually limited to smaller power outputs, for which there often is no need to employ an oil cooler.

The heat flow to the coolant is frequently expressed as a function of the fuel flow rate. Alcock *et al.* (1958) present results from a single-cylinder direct injection engine with a swept volume of 1.78 litres. They concluded that

$$Q_c \propto (m_f)^{0.64} \tag{13.8}$$

However, within the data used for deriving this correlation, there are clearly additional dependencies: the constant of proportionality increases with increasing speed and reducing boost pressure. The heat flow to the coolant was always within the range of 18–34 per cent of the fuel energy, with the

highest percentage heat rejection occurring at no load with atmospheric induction.

Watanabe *et al.* (1987) also present results from a turbocharged direct injection engine with a cylinder capacity of 1.945 litres; their data suggest a linear dependence on load with about 14 per cent of the fuel energy entering the coolant.

Taylor (1985a) presents cooling heat flow data from large two-stroke and four-stroke engines, which are shown here in figure 13.13. The heat flow to the coolant is shown as a function of the engine load, and is in the range 17–32 per cent of the fuel energy, and this is consistent with the Alcock *et al.* data.

French (1984) presents some data that compares the heat rejected to the coolant for both direct and indirect injection engines as a function of the piston speed: this is presumably full-load data and is shown in figure 13.14.

In recent years, there has been much interest in the application of ceramics to compression ignition engines. The simple argument is that ceramics have a much lower thermal conductivity than metals (one to two orders of magnitude) so that the energy flow to the coolant will be reduced, and the higher combustion temperatures will lead to more expansion work. However, the largest thermal resistance is in the gas-side thermal boundary layer adjacent to the combustion chamber, and this will not be affected much. Figure 13.15 shows a representation of the temperature distribution: the gas temperature (T_g) might be as high as 2500 K, the surface temperature on the gas side ($T_{w,g}$) might be 600 K, the surface temperature on the coolant side ($T_{w,c}$) might be 400 K with a coolant temperature (T_c) of 360 K. This system can be modelled as a series of thermal resistances to represent the thermal boundary layers on the gas side and the coolant side, and a thermal resistance to control the heat flow through the combustion chamber wall. If the heat flow is considered to be

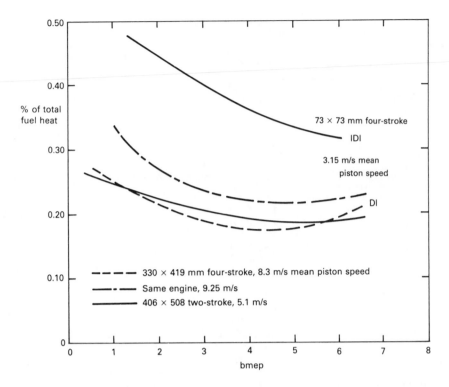

Figure 13.13
The heat flow to the coolant as a function of load for direct and indirect injection compression ignition engines (data from Taylor, 1985a and Brunel University).

steady, the thermal resistances are proportional to the temperature differences (compare voltage differences). If the heat flow (compare current) is to be changed then the largest effect will be obtained by changing the largest thermal resistance, that is the gas-side heat transfer coefficient. Thus an order of magnitude change to the thermal conductivity of the combustion chamber wall does not lead to an order of magnitude change in the heat flow. The thermal resistances used by Moore and Hoehne (1986) to simulate the effect of a

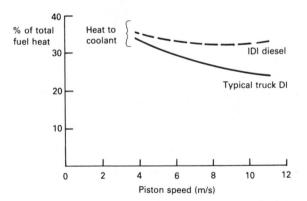

Figure 13.14
The heat flow to the coolant as a function of piston speed at full load for direct and indirect injection compression ignition engines (adapted from French, 1984).

1.25 mm polystabilised zirconia (PSZ) coating on a cylinder head are summarised in table 13.1.

It should be noted that the gas-side heat transfer coefficient is increased, so that some of the gain in cylinder head insulation is offset. However, because of the time-dependent variation of gas temperature within the combustion chamber, there are other factors controlling the heat flow which will be considered later.

Some predictions of surface and gas temperatures have been made by Assanis and Heywood (1986) for a partially insulated turbocompounded direct injection engine. Figure 13.16 shows that the insulation causes a significant increase in the piston surface temperature. This leads to heat transfer to the induction gas (thus lowering the volumetric efficiency) and heat transfer to the gas much further into the compres-

Table 13.1 *Thermal resistances (K/kW) for a conventional and insulated (1.25 mm of PSZ) cylinder head (Moore and Hoehne, 1986)*

Region	Uninsulated	Insulated
Coolant side	7.9	7.9
Cylinder head	13.6	12.1
Insulation	–	80.3
Gas side	67.3	54.4
TOTAL	88.5	155

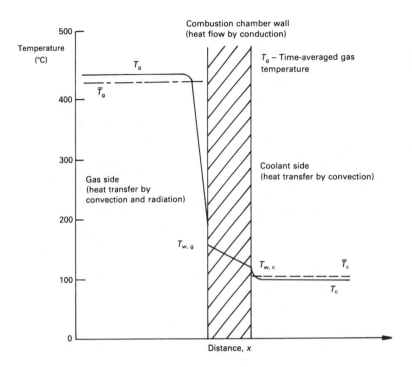

Figure 13.15
Schematic diagram of the temperature distribution, showing the large temperature drop in the gaseous thermal boundary layer.

sion stroke. This raises the compression and combustion temperatures, so that more compression work is required. Also, the raised in-cylinder temperatures moderate the effect of the thermal resistance that is reducing the heat flow.

However, what is of interest here, is the effect of insulating the combustion chamber on the heat flow to the coolant, and this is influenced by several factors. Firstly, the cylinder head face represents a small proportion of the bore area (especially with four valve-per-cylinder arrangements). Secondly, heat transfer from the exhaust valve and port is not affected.

When Moore and Hoehne (1986) applied a 1.25 mm PSZ coating to a cylinder head, there was no net change in the fuel energy rejected to the coolant. The reduction of 0.2 percentage points in the heat rejected to the cylinder head coolant (to 7.4 per cent of the fuel energy) was exactly balanced by the increase in the heat rejected to the oil (to 6.9 per cent of the fuel energy). When a similar coating was applied to the piston crown and top land, the results were as summarised in table 13.2.

In absolute terms there was no significant change in the heat flow to the coolant, as the engine brake efficiency fell. The piston insulation has a direct effect on the heat flow to the oil, as the undersides of the piston were cooled by an oil spray. These results should be sufficient to demonstrate that the use of ceramics does not produce a significant change in the heat flow to the coolant.

Some compression ignition engines are air cooled, and their use is most prevalent in off-highway applications. The

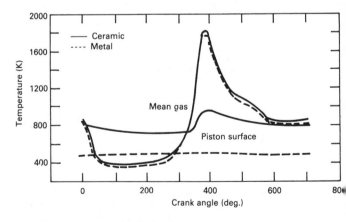

Figure 13.16
The effect of combustion chamber insulation on the piston surface temperature and the mean gas temperature (Assanis and Heywood, 1986). [Reprinted with permission © 1986 Society of Automotive Engineers, Inc.]

Table 13.2 *The effect of a 1.25 mm PSZ piston coating on fuel energy distribution at full load (Moore and Hoehne, 1986)*

	Percentage of fuel energy	
	Uninsulated pistons	Insulated pistons
Cylinder head	8.4	8.2
Oil	6.3	5.4

development of an engine family that comprises both air- and liquid-cooled options is presented by Stevens and Tawil (1989).

13.3 | Liquid coolant systems

13.3.1 Conventional coolant systems

The cooling circuit from a typical automotive application is shown in figure 13.17. Originally a coolant pump was not used, instead natural convection led to a thermosiphon effect. Initially with a cold engine, the thermostat is closed and the pump circulates the coolant within the primary circuit, which is completed by the internal passages within the engine. The interior heater matrix is part of the primary coolant circuit, along with the inlet manifold (when it is coolant heated), so that these items reach their proper working temperature as quickly as possible.

Some typical results recorded during engine warm-up are plotted in figure 13.18. These results were obtained from a Rover M16 spark ignition engine installed on a dynamometer, and operated at fixed throttle and speed. There was no interior heater matrix, but an oil cooler was installed in the primary coolant circuit. This in fact has a beneficial effect in causing the engine oil to warm up more rapidly. It should be noted that the exhaust temperature rises very rapidly, and this suggests that it would be a useful energy source which might be employed to accelerate the engine warm-up. The engine load or torque (bmep) rises quite quickly immediately after starting, though it took over 20 minutes for the engine to achieve its steady-state output under these conditions with a bmep of 2 bar.

This particular version of the engine had a single-point injection system with a coolant heated inlet manifold: the mixture temperature is seen to track the coolant temperature, but with an initial fall that is attributable to the evaporative cooling of the fuel. The engine management system was arranged to give a fixed fuel flow rate into the engine. Soon after starting there is a 10 per cent fall in the air flow rate, but this only partly accounts for the 24 per cent richening of the burnt air/fuel ratio. The remaining difference is probably due to changes in the liquid fuel film on the wall of the inlet manifold.

Once the engine has reached its operating temperature (usually around 90°C) the thermostat opens so that some coolant flows through the main circuit, and is cooled by the radiator. Figure 13.18 shows that once the thermostat opens, there are fluctuations in the temperature of the coolant in the primary circuit. This is due to cold coolant entering from the main circuit; these oscillations would be smaller if the thermostat opened more slowly, but there would then be a risk of overheating if the thermostat did not open quickly enough at a high engine load. Even when the thermostat is fully open, there will still be a significant flow through the primary circuit.

The only remaining element to discuss in figure 13.17 is the expansion tank, which is at the highest elevation. The vent

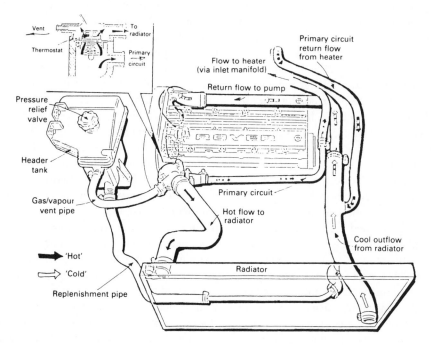

Figure 13.17
Schematic diagram of the Rover M16 cooling system.

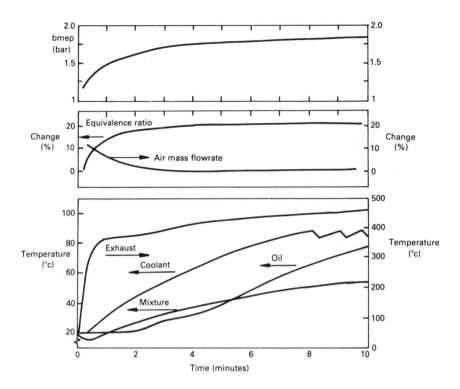

Figure 13.18
The performance of the Rover M16 SPi 2.0-litre spark ignition engine during warm-up at a speed of 1500 rpm and fixed throttle, which gives an eventual bmep of 2 bar.

pipe allows any gas to leave the cooling system and to be separated from the coolant. Such gas comes from the degassing of dissolved gases in the coolant, and under some circumstances there might be a leakage of combustion gases through the cylinder head gasket. (This can be identified by using an exhaust gas analyser.) The expansion tank has a pressure relief valve to limit the system pressure (usually about 1 bar gauge pressure), and there is also an outlet into the main circuit so that make-up liquid can enter the engine. Raising the system pressure by 1 bar raises the boiling point by about 20 K, and this is further increased by the presence of antifreeze (see table 13.4). The expansion tank should be large enough to accommodate expansion of the coolant, and have a large enough gas/vapour space so that the gas/vapour is not compressed to a high enough pressure to open the pressure relief valve. Otherwise, when the system cools down the pressure would fall below atmosphere pressure.

The coolant pump is usually located at a low part of the cooling system with a cooled inlet flow; this is to minimise the risk of cavitation. Frequently the pump is partly formed by the cylinder block, as this simplifies assembly and reduces the component costs. Traditionally, the coolant pump is of a very simple design with a correspondingly low efficiency. The rotor may be stamped from sheet metal, or be a simple casting. Fisher (1989) argues that the coolant pump can be made much more efficient, and that it can then be driven by an electric motor. This gives independent control of the coolant

flow, and the overall power requirement can be less, even allowing for the low efficiency of the motor and alternator.

The coolant flow pattern within a Ford Dover direct injection compression ignition engine has already been seen in chapter 12, figure 12.9.

The outflow from the pump is divided into two flows which enter at opposite ends of the cylinder block. On four-cylinder engines the flow would not normally be divided in this way. The flow to the far end of the engine first passes through the oil cooler. Once the water enters the cylinder block it passes around the cylinders, and rises up into the cylinder head. The flow rate to different parts of the cylinder head is frequently controlled by adjusting the size of the coolant holes in the cylinder head gasket. The coolant distribution is often arrived at with the help of clear plastic models. The flow can be visualised by means of particles or small gas bubbles, and photographic techniques can be used to record the flow pattern, for example continuous ciné or single frame but with illumination from controlled duration flash(es). Laser Doppler anemometry and hot wire anemometry techniques can be utilised for generating more quantitative information.

The traditional coolant passage in the cylinder head is a void, which has been defined by the walls of the combustion chambers, the inlet and exhaust ports, the top deck and other features. Some control of the flow is achieved by regulating the flow from around the cylinder liners, and in selecting the outflow from the cylinder head. This somewhat haphazard arrangement is satisfactory in many cases, because overheating of the combustion chamber is prevented by nucleate boiling. If boiling occurs, then the heat transfer coefficient can increase by an order of magnitude, and the metal temperatures are then linked to the saturation temperature of the coolant.

With compression ignition engines, it is quite common to have a drilled passage for coolant passing through the valve bridge area or some other means of directing the coolant. Drilled passages are preferable to relying on holes made by cores, as such cores have a very high aspect ratio, and are thus susceptible to damage or movement during the casting process. Drilled coolant passages have already been seen in chapter 12, figure 12.22 (see also section 13.3.3).

13.3.2 Cooling media performance

Water is a very effective cooling medium, with a high enthalpy of vaporisation, high specific heat capacity and a high thermal conductivity. Its saturation temperature of 99.6°C at 1 bar is also very convenient; less convenient are its freezing point of 0°C and its contribution towards corrosion – especially when there are different metals in the cooling system.

Removal of heat from metal into a liquid coolant can be by forced convection, subcooled boiling (where the bubbles detach when small and collapse into the bulk fluid that is below its saturation temperature) and saturated boiling (which has large bubbles and no condensation in the bulk fluid). With saturated boiling there is the risk of vapour blankets and film boiling – this can cause overheating which leads to thermal fatigue, fracture of the component, or some form of abnormal combustion.

Rowe (1980) provides a very comprehensive summary of the requirements of automobile engine coolants and their methods of evaluation, and he precedes his treatment with a comprehensive historical overview. The requirements of an engine coolant are summarised by Rowe in table 13.3.

Rowe (1980) concludes that an aqueous ethylene glycol solution provides a sound basis for an automotive engine

Table 13.3 *Characteristics and requirements of an engine coolant (Rowe, 1980)*

High specific heat and good thermal conductivity
Fluidity within the temperature range of use
Low freezing point
High boiling point
Non-corrosive to metals; minimum degradation of non-metals
Chemical stability over the temperature range and conditions of use
Non-foaming
Low flammability; high flash point
Reasonable compatibility with other coolants or oil
Low toxicity; no unpleasant odour
Reasonable cost; available in large quantities

coolant. However, to prevent corrosion and other shortcomings, it is necessary to have a range of additives in the ethylene glycol concentrate, and these are typically:

(a) inhibitors to prevent metal corrosion
(b) alkaline substances to provide a buffering action against acids
(c) an antifoam additive
(d) a dye for ready identification
(e) a small amount of water to dissolve certain additives.

Mercer (1980a) describes the background to the selection of corrosion inhibitors. Sodium benzoate and sodium nitrite were found to be suitable for engines of cast iron and steel construction. In parallel, a combination of triethanolammonium phosphate (TEP) and sodium mercaptobenzothiazole (NaMBT) was found to be suitable for engines of aluminium alloy construction with copper radiators. However, it was subsequently found that both corrosion inhibitor systems were acceptable for either type of engine construction.

Mercer (1980b) discusses laboratory tests for determining the heat transfer coefficients of different coolants under nucleate boiling conditions, with and without flow. (Nucleate boiling is where vapour bubbles are nucleated and grow from separate sites, and liquid remains in contact with part of the heated surface.) Mercer makes comparisons between as-cast and machined surfaces for cast irons, and also investigates ageing effects.

The condition of the metal/coolant interface can have a profound effect on the heat transfer coefficient (and thereby the metal temperatures). O'Callaghan (1974) reviews some of these issues, but a very comprehensive summary is presented by French (1970). French considered a wide range of different cast steels, cast irons, and coolants (including coolant contaminated by sea water) with different corrosion inhibitors. He conducted rig tests for flow and pool boiling, as well as presenting data from diesel engine tests. French (1970) also conducted some long-term rig tests of up to 2500 hours' duration, with a fixed heat flux of 520 kW/m² and a coolant temperature of 70°C. He recorded the change in wall superheat (the temperature difference between the metal surface and the saturation temperature), and found that initially the wall superheat was in the range 6–32 K. However, after prolonged boiling, the wall superheat rose by as much as a further 68 K. For one particular additive, the wall superheat rose from 30 K to 70 K after only 300 hours of testing, and was still increasing when the test was stopped. These results emphasise the care which is needed in selecting additives if they are not to lead to elevated metal temperatures. French (1970) also reported on a 'thermal barrier' that could be found with some cast irons. The thermal barrier leads to an increase in the wall superheat, and French found that it was a property of the surface sub-layer (of about 1 mm thickness), and this sub-layer had to be completely machined away to obtain a normal value of the wall superheat. It appears that this was not an effect of the surface finish.

As described in section 13.3.1, boiling is an important heat transfer mechanism in liquid-cooled internal combustion engines. When boiling occurs, its high heat transfer coefficient restricts the increase in the metal temperature. Boiling is a very complex phenomenon that defies a thorough analytical approach. However, for certain geometries there are correlations that have been found to give good predictions; such a geometry is a drilled hole that might occur in the valve bridge area (or in a precision cooled system that is discussed later in section 13.3.3).

Finlay *et al.* (1987a) found that the heat transfer coefficient for a drilled hole could be predicted, by using the Dittus–Boelter correlation for forced convection, and the Chen correlation for nucleate boiling. Heat transfer coefficients have been evaluated here using these correlations, and although the relevant geometry is not widely encountered in engines, the correlations will illustrate the increase in heat transfer coefficient that is associated with nucleate boiling.

If the heat transfer mechanism is assumed to be a combination of convection and nucleate boiling, then the heat flux is given by the following equation:

$$q = h_c(T_w - T_l) + h_b(T_w - T_s) \qquad (13.9)$$

where T_w is the surface temperature
  T_l is the liquid bulk temperature
  T_s is the saturation temperature of the fluid.

For turbulent convection the heat transfer coefficient (h_c) is predicted with the Dittus–Boelter correlation for the Nusselt number (Nu):

$$Nu = 0.023 Re_l^{0.8} Pr^{0.4} \qquad (13.10)$$

$$Nu = \frac{h_c d}{k_l} \quad Re_l = \frac{\rho_l v_l d}{\mu_l} \quad \text{and} \quad Pr = \frac{\mu_l c_{p,l}}{k_l}$$

where v_l = liquid flow velocity (m/s)
  d = hole diameter (m)
  k_l = liquid thermal conductivity (W/m K).

The boiling heat transfer coefficient (h_b) in equation (13.9) is predicted using the Chen (1966) correlation:

$$h_b = 0.00122 \left(\frac{k_l^{0.79} c_{p,l}^{0.45} \rho_l^{0.49}}{\sigma^{0.5} \mu_l^{0.29} h_{fg}^{0.29} \rho_v^{0.24}} \right) (T_w - T_s)^{0.24} \Delta p_s^{0.75} S$$

$$(13.11)$$

where σ = surface tension (N/m)
  h_{fg} = enthalpy of vaporisation (J/kg K)
  $\Delta p_s = (p_{sat} \text{ at } T_w) - p$
  S = suppression factor.

The suppression factor has been plotted as a function of Reynolds number by Finlay *et al.* (1987a), and this empirical result is shown in figure 13.19.

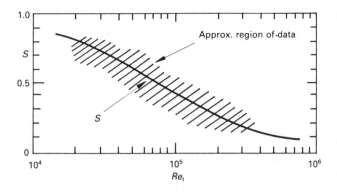

Figure 13.19

The suppression factor (S) used in the Chen correlation for the prediction of convective boiling heat transfer (Finlay *et al.*, 1987a).

Some results obtained from this model (equations 13.9, 13.10 and 13.11) for a 50 per cent by volume ethylene glycol/water mixture are plotted in figure 13.20 for two flow velocities. In the convective heat transfer regime, the heat transfer coefficient rises as the surface temperature rises. This is because the dominant effect is the fall in the mixture viscosity as the temperature rises. It should be noted that for the lower flow rate (0.25 m/s), the flow is laminar until the viscosity has been reduced sufficiently by the rise in temperature. For each flow rate, raising the system pressure

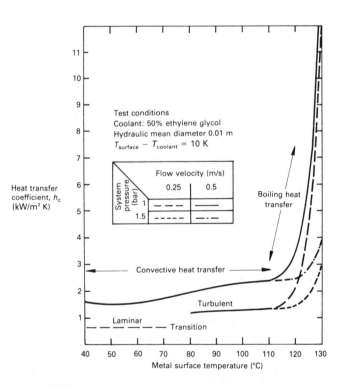

Figure 13.20

The heat transfer coefficient as a function of surface temperature, for different pressure and velocity coolant flows; note the effect on the transition to boiling heat transfer.

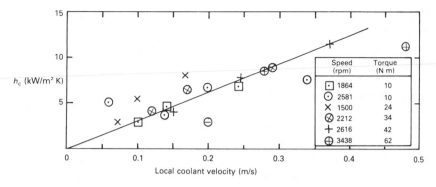

Figure 13.21
The heat transfer coefficient as a function of load coolant velocity around the combustion chamber and cylinder liner for a Ford Valencia 1.1-litre spark ignition engine with 50 per cent water/ethylene glycol coolant (derived from Sorrell, 1989).

by 0.5 bar delays the onset of nucleate boiling by about 10 K. The nucleate boiling heat transfer coefficient is so high that once boiling occurs, then there will only be a slight difference in surface temperature (the combustion chamber can be considered as a constant heat flux source). In other words, once boiling occurs the flow velocity is of secondary importance, although the higher flow velocity does delay slightly the onset of nucleate boiling.

It must not be assumed that this model (equations 13.9–13.11) can be applied to conventional cooling systems. Tests were conducted by Sorrell (1989) in which the flow velocity and the coolant side heat transfer coefficient were measured separately, in a Ford Valencia 1.1-litre spark ignition engine. Data for the heat transfer coefficients and the local flow velocity are plotted for five locations around the cylinder head and cylinder liner in figure 13.21. The data are from a range of operating points up to a torque of 62 N m at 3488 rpm, and the heat transfer coefficients are about an order of magnitude higher than those predicted by the Dittus–Boelter correlation.

Presumably the complex geometry of the flow passages defeats correlations for heat transfer coefficients, which in general assume some form of fully developed flow. Willumeit and Steinburg (1986) suggest that the heat transfer coefficient for the cylinder liner to the coolant can be predicted by correlation for tube bundles; a result not supported by the work of Sorrell (1989).

Aqueous mixtures of ethylene glycol are now well established as an engine coolant, with the percentage of ethylene glycol ranging from typically 25 to 60 per cent on a volumetric basis (that is, 26.9–61.9 per cent by weight, or 9.7–32.1 per cent on a molar basis). A 50 per cent by volume water/ethylene glycol mixture has an initial boiling point of 111°C at 1 bar. None the less, many alternative coolants have also been considered, but in general their heat transfer properties are inferior to those of water, or water/ethylene glycol mixtures. For example oil, which is frequently used for splash cooling the undersides of pistons, has been suggested for cooling the cylinder liner. Propylene glycol, Fluorinet (a fluid developed by the 3M company) and pure ethylene glycol have all been proposed as engine coolants. Their properties are compared with the properties of water and water/ethylene glycol in table 13.4.

Table 13.4 *Properties of liquid coolants*

Property	Water	Propylene glycol	Ethylene glycol/ water (50/50)	Ethylene glycol	Fluorinet FC-77
Boiling point, 1 bar (°C)	100	187	111	197	97
Freezing point (°C)	0	–14	–37	–9	–110
Enthalpy of vaporisation (MJ/kmol)	44.0	52.5	41.2	52.6	–
Specific heat capacity (kJ/kg K)	4.25	3.10	3.74	2.38	1.05
Thermal conductivity (w/m K)	0.69	0.15	0.47	0.33	0.06
Density 20°C (kg/m³)	998	1038	1057	1117	1780
Viscosity, 20°C (cS, 10^{-6} m²/s)	0.89	60	4.0	20	0.80

Pure propylene glycol has been advocated as an engine coolant by Light (1989). Mark and Jetten (1986) also discuss the use of propylene glycol as a base fluid for automotive coolants. They conclude that aqueous solutions of propylene glycol and aqueous solutions of ethylene glycol have similar heat transfer properties, but propylene glycol has environmental advantages, and a greater resistance to cavitation. Pure propylene glycol has a high boiling point (187°C), and thus is likely to make any boiling subcooled, in which case the vapour bubbles tend to collapse, thereby avoiding any risk of vapour pockets 'insulating' the components that are to be cooled. Table 13.4 also shows that the molar enthalpy of vaporisation is high for propylene glycol; this implies that a small volume of vapour is formed per unit heat input. However, the high boiling point of propylene glycol might also lead to component overheating.

In conclusion, water has exceptional properties as a heat transfer medium for both convective and boiling heat transfer; these properties are moderated when water/organic mixtures are used as a coolant. However, the heat transfer characteristics are still almost an order of magnitude better than those of pure organic liquids. This implies that water-based systems are likely to be needed for cooling the cylinder head. Organic coolants (such as oil or propylene glycol) might be suitable for cooling the cylinder liner.

13.3.3 Advanced cooling concepts

There are two main disadvantages with conventional cooling systems. Firstly the large volume of coolant in the primary circuit can lead to a slow engine warm-up, and a high short-journey fuel consumption. Secondly, the cooling system tends to overcool parts of the engine, especially the cylinder liners at light loads. To overcome these two shortcomings, many alternative cooling concepts have been proposed. These include:

(a) precision cooling

(b) dual circuit cooling

(c) controlled component temperature cooling

(d) evaporative cooling.

These systems all aim at making the engine temperature more uniform, and less sensitive to the operating point (in terms of the torque and speed). For example, if the cylinder liner temperature is raised, then the piston and ring pack friction are reduced. The significance of reductions in the mechanical losses increases as the load is reduced, and this leads to corresponding reductions in the fuel consumption. The tendency for the combustion to be quenched in the thermal boundary layer is reduced, as is the tendency for any hydrocarbons to be absorbed in the oil film; there is thus a reduction in the emissions of unburnt hydrocarbons (Went-

worth, 1971). Unfortunately, the rise in the in-cylinder temperature might also lead to an increase in the emissions of nitrogen oxides (the formation of which is strongly temperature dependent). The emissions of NO_x (nitrogen oxides) can be reduced by either retarding the ignition timing or employing exhaust gas recirculation (EGR). However, both these remedies impose a fuel consumption penalty that might or might not be offset by the reduction in fuel consumption due to the reduced frictional losses. It should be remembered that for diesel engines, the NO_x emissions are only likely to be problematic close to full load.

With spark ignition engines, another problem that can occur at full load is the reduction in the knock margin caused by the higher in-cylinder temperatures. Again, this can be remedied by EGR or retarding the ignition timing (usually retarding the ignition timing).

Precision cooling

The principle of precision cooling, originally developed for diesel engines, is to provide cooling only where it is needed, at a rate proportional to the local heat flux. Figure 13.22 shows a precision cooled cylinder head (Priede and Anderton, 1974). A precision cooled system has small local passages which are used to cool critical regions such as the injector nozzle, valve bridge, exhaust valve region and valve guides, with much space in the head filled with air. A similar result is achieved by the Ruston flame plate system (figure 13.23). The application of precision cooling to an aluminium alloy cylinder head is described by Ernest (1977).

With precision cooling, many regions are not directly cooled and their temperatures rise, but the system is designed to keep the temperature within safe limits. The corresponding benefit of this is a more even temperature distribution producing less thermal strain, and the possible elimination of hot-spots to allow higher compression ratios in spark ignition engines.

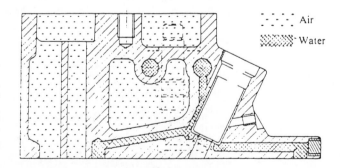

Figure 13.22

A precision cooled cylinder head (Priede and Anderton, 1974). [Reproduced by permission of the Council of the Institution of Mechanical Engineers.]

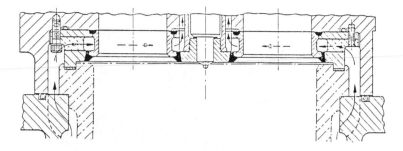

Figure 13.23
The Ruston flame plate cylinder head design (French, 1970). [Reproduced by permission of the Council of the Institution of Mechanical Engineers.]

The design of a precision cooling system is difficult because the heat flow patterns in the engine are complex and poorly defined, it is difficult to obtain the correct flow in the cooling passages, and it is difficult to predict the boiling heat transfer performance. In particular, a phenomenon known as dry-out can occur. If a vapour bubble grows to such a size so as to fill the passage, then it is possible for the thin liquid layer between the bubble and surface (the microlayer) to 'dry out'. If the surface is not re-wetted, then the vapour effectively insulates the surface, and the metal temperatures will rise in an uncontrolled manner. Also, the flow in the passages is driven by a pressure differential (it is not a fixed volume flow rate), and Finlay *et al.* (1987a) have identified an oscillatory flow caused by bubbles growing and collapsing. This leads to 'dry-out' occurring at comparatively low heat fluxes.

Dual circuit cooling

This process is characterised by separate head and block cooling circuits, with the head coolant temperature being lower than that of the block. The higher block temperature reduces the frictional losses associated with the piston and ring pack, and this leads to a reduced fuel consumption, especially at part load. The lower coolant temperature in the cylinder head tends to reduce the indicated efficiency, but as the risk of knock in spark ignition engines is also reduced, a higher compression ratio can be used. The overall result is an improved efficiency. The effect on emissions is more complex. If the combustion temperatures decrease (either through a lower compression ratio or overall cooler combustion chamber surfaces), then the NO_x emissions should decrease significantly. Consequently, the effect of dual circuit cooling on emissions is likely to be highly engine dependent.

Figure 13.24 summarises the results from tests reported by Kobayaski *et al.* (1984), in which the cylinder head coolant temperature and the compression ratio were varied. At full load (WOT), the ignition timing was limited by the onset of knock at 1200 rpm. If a cylinder head coolant temperature of 50°C could be maintained, then the compression ratio could be increased to from 9 to 12:1 yet still maintain the same output at 1200 rpm. The raised compression ratio gives about

a 5 per cent reduction in the fuel consumption at 20 N m torque. At higher engine speeds the ignition timing will not be knock limited, so the raised compression ratio will also improve the full throttle output and efficiency.

Work on dual circuit cooling by Finlay *et al.* (1987b) concentrated on raising the cylinder block coolant temperature as high as 150°C. This led to reductions in the fuel consumption and hydrocarbon emissions, but an increase in the NO_x emissions.

Controlled component temperature cooling (CCTC)

This process keeps a particular component temperature constant at all speeds and loads by varying the amount of cooling. The system was first described by Willumeit *et al.*

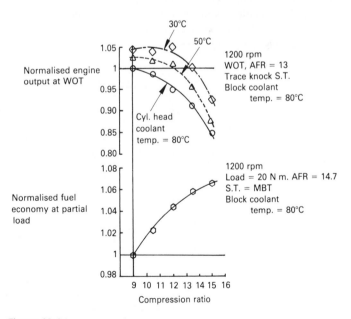

Figure 13.24
The effect of compression ratio on WOT output and part-load fuel economy for various cylinder head coolant temperatures (Kobayashi *et al.*, 1984). [Reprinted with permission © 1984 Society of Automotive Engineers, Inc.]

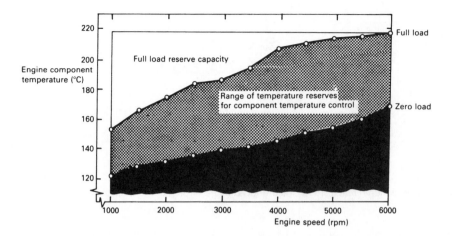

Figure 13.25

Component temperature as a function of engine speed for full load and zero load (Willumeit *et al.*, 1984). [Reprinted with permission © 1984 Society of Automotive Engineers, Inc.]

(1984), and the temperature used was that of the cylinder liner at the top ring reversal point at tdc. This philosophy could also be applied to a dual circuit cooling system.

Figure 13.25 shows the cylinder liner temperature at the upper ring reversal position for varying engine speeds at zero and full loads; the temperature only reaches a maximum at maximum power, and is overcooled at other points.

Figure 13.26 shows the reductions in fuel consumption that are obtained when the controlled component temperature concept is applied. Up to 20 per cent fuel savings can be made at low loads. Other benefits of high metal temperatures (as discussed with precision cooling above) are also obtained, such as reduced unburnt hydrocarbon emissions. It is also

suggested that with these high temperatures, it is possible to run the engine leaner, and to reduce emissions further. However, at full load with low speeds, raising the component temperatures towards their rated load and speed values will promote knock, thereby reducing the benefits of CCTC at full load.

A not unimportant problem with the system is its response to transients; when the throttle is opened suddenly the cooling system must be able to react fast enough to prevent metal temperatures rising – otherwise knock is likely to occur (especially at low speeds). However, the role of the engine management system could be extended to control the cooling system. Coolant temperatures would be

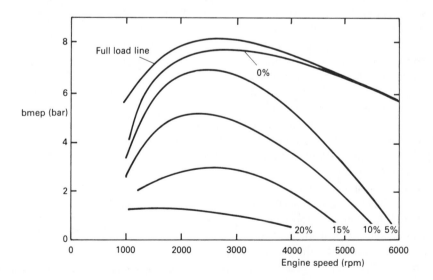

Figure 13.26

The percentage reduction in fuel consumption obtained by using the controlled component temperature cooling concept (Willumeit *et al.*, 1984). [Reprinted with permission © 1984 Society of Automotive Engineers, Inc.]

monitored, and used in conjunction with a map of the cooling requirements (analogous to mixture or ignition timing control). The control system could then respond to transients before any change might be detected in the metal temperature. The metal temperature measurement would then provide a check on the cooling system performance, and allowance could be made automatically for the fouling of the water/metal interface.

Evaporative cooling

The main heat transfer method to the coolant with evaporative cooling is through boiling, the liquid usually being at the saturation temperature. Perhaps the most basic system is the total loss system, where the coolant continually needs to be replenished as it boils away. Leshner (1983) describes some early versions of evaporative cooling systems and presents a comprehensive overview of the system requirements. Leshner quotes examples of evaporative cooling that have been used in automotive and stationary applications. However, evaporative cooling has also been used in high-performance aircraft engines by Rolls-Royce (McMahon, 1971).

Most evaporative cooling systems are closed, with a condenser and a method of returning the condensed coolant to the cooling jacket. One method is a gravity system, in which the vapour rises into the condenser and the liquid drops back into the engine. The problem with this is the space required, so most systems are likely to pump the condensate back into the engine, and look very similar to the standard convective cooling system with the condenser at the same level as the engine (figure 13.27).

Leshner (1983) points out that the high enthalpy of vaporisation reduces the pumping requirements by at least an order of magnitude. Furthermore, the high heat transfer coefficients associated with boiling lead to smaller tempera-

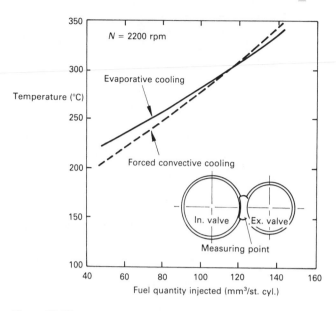

Figure 13.28
The effect of load on the valve bridge temperature in a four-cylinder 11.7-litre diesel engine, with evaporative and convective cooling systems (Watanabe *et al.*, 1987). [Reprinted with permission © 1987 Society of Automotive Engineers, Inc.]

ture differences between the coolant and the components that it cools. Data are also presented that show smaller variations in coolant temperature and metal temperatures for the evaporative cooling system.

Watanabe *et al.* (1987) applied an evaporative cooling system to a diesel engine and some results obtained are shown in figures 13.28 and 13.29. Firstly, with increasing fuel injection (analogous to the indicated power), the valve bridge temperature did not vary so greatly with the evaporative system, as compared with the standard forced convective system, and at higher powers the temperature was actually lower. Secondly, at a given speed and load, the lower part of the cylinder liner is at a higher temperature with evaporative cooling than with forced convection, though towards the top of the liner where there are higher temperatures the two systems have the same temperature.

The advantages of evaporative cooling are higher block temperatures and a more uniform temperature distribution in the cylinder head – the advantages of this have been discussed already. The maximum engine temperatures do not necessarily increase over the convective system, even though the coolant temperature is around 20°C higher. The higher temperatures in the condenser will allow its size to be reduced, perhaps by around 30 per cent, so allowing improvements in aerodynamics and reducing the fan power.

In conclusion, novel cooling system designs would seem to offer more potential than novel coolants, for controlling the engine temperature, reducing the fuel consumption and reducing the emissions.

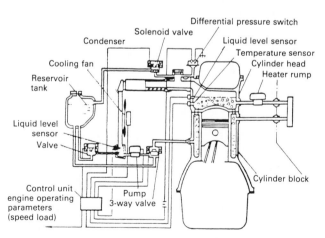

Figure 13.27
The elements of an evaporative cooling system (Leshner, 1983). [Reprinted with permission © 1983 Society of Automotive Engineers, Inc.]

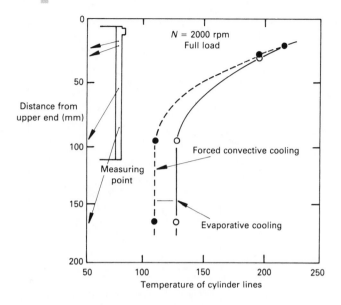

Figure 13.29
The temperature distribution in the cylinder liner at full load on a
four-cylinder 11.7-litre diesel engine (Watanabe *et al.*, 1987). [Reprinted
with permission © 1987 Society of Automotive Engineers, Inc.]

13.4 Conclusions

In engine design, many decisions have to be taken that could
affect the cooling system. Principally, is the engine to be made
from cast iron or aluminium alloy? Is the engine to be liquid or
air cooled? Data have been presented that show typical
variations in the energy balance for compression and spark
ignition engines as functions of the speed and load. The effect of
air/fuel ratio, the ignition timing and the compression ratio have
also been discussed in the context of spark ignition engines.

Liquid-cooling systems have been discussed in some detail,
as their use is widespread. This led to a discussion of boiling
and convective heat transfer performance, and how it is
affected by the choice of coolant. Finally, there has been a
discussion of novel cooling systems, which offer potential for
reducing the emissions and fuel consumption of spark ignition
and diesel engines.

CHAPTER 14

Experimental facilities

14.1 | Introduction

The testing of internal combustion engines is an important part of research, development and teaching of the subject. Engine test facilities vary widely. The facilities used for research can have very comprehensive instrumentation, with computer control of the test and computer data acquisition. On the other hand, a more traditional test cell with the engine controlled manually, and the data recorded by the operator, can be better for educational purposes. This second type of test cell is covered in some detail by Greene and Lucas (1969), and is dealt with first in this chapter. A more recent treatment of engine testing by Plint and Martyr (1995) also includes substantial coverage of the building and infrastructure requirements for engine testing.

Some of the instrumentation described in this chapter is unlikely to be encountered by students. The instrumentation most likely to be met by students is introduced first. Thus, the emissions analysis equipment and combustion analysis techniques are introduced at the end of the chapter. Within the sections dealing with each measurement, the simplest instrumentation is discussed first. For example, the section on fuel flow measurement starts with control volumes, and ends with flowmeters that give a continuous reading of the instantaneous flow rate.

The chapter ends with a case study of an advanced engine test system using computer-based engine control and data acquisition. A final class of test facility not separately discussed here is those used for acceptance tests. Most engines are tested immediately after manufacture to check power output and fuel consumption; the main requirement here is ease of installation.

Before dealing with any test facility it is important to remember that there are certain advantages in using single-cylinder engines for research and development purposes:

1 No inter-cylinder variation. The manufacturing and assembly tolerances in multi-cylinder engines cause performance differences between cylinders. This is attributable to differences in compression ratio, valve timing etc.

2 No mixture variation. With fuel injection systems it is difficult to calibrate pumps and injectors to give identical fuel distribution. In carburated engines it is difficult to design the inlet manifold to give the same air/fuel mixture to all cylinders for all operating conditions.

3 For a given cylinder size the fuel consumption will be less and a smaller capacity (cheaper) dynamometer can be used.

4 Lower fuel consumption. This is a consideration for very large engines, or when pure fuel components are being tested.

Single-cylinder engines are almost invariably used when optical access is required into the combustion chamber. With overhead valve engines optical access through the piston crown is used in studies of flow, mixture preparation and combustion. A typical arrangement is shown in figure 14.1, and this was the approach used for the combustion photographs in this book. Above the conventional piston crown is a slotted tubular extension that supports the window in the piston crown. The piston slots enable an elliptical mirror to be fixed at 45°, so that an image of the combustion chamber can be viewed through the opposite slot. The piston crown window is typically 75 per cent of the bore diameter. The piston rings in the combustion chamber are often based on graphite and fibre reinforced polymers, so as to eliminate the need for any lubrication. Such rings can run on cast iron cylinder liners or indeed an annular window. The lower part of the piston will have a conventional oil control ring, to avoid oil contamination of the mirror and piston window.

The windows are usually made of fused silica (often referred to as 'quartz') because of its high strength and low coefficient of thermal expansion, which leads to a good resistance to thermal shock. However, fused silica is very brittle, so great care is needed in the installation and support of the fused silica windows. Fully annealed copper gaskets are a widely used method in firing engines, while silicone sealant is sometimes used in non-firing applications. The choice of window material is governed by the radiation wavelengths that are of interest. Sapphire is used when ultra-violet

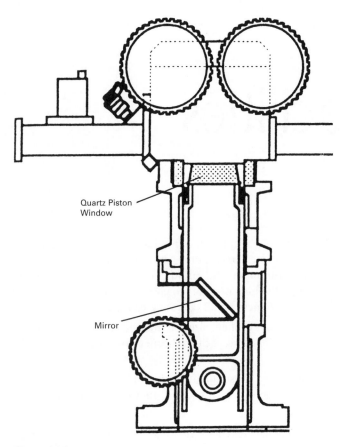

Figure 14.1
A single-cylinder engine with optical access through the piston crown, and an annulus below the cylinder head.

radiation is being studied, while germanium or silicon is used when infra-red radiation needs to be transmitted (Zhao *et al.*, 1994). In diesel engines the windows need frequent cleaning, so less durable materials can be used, such as polymethyl-methacrylate (known as Perspex and Plexiglass). There have been many photographic studies to elucidate the different processes in engines, and much of this work has been reviewed by Winterbone *et al.* (1997) with comprehensive illustrations.

14.2 | Quasi-steady engine instrumentation

Figure 14.2 shows the schematic arrangement of a simple engine test rig for a Ruston Oil Engine. This single-cylinder compression ignition engine has the advantage of simplicity and ruggedness. The engine has a bore of 143 mm and a stroke of 267 mm, giving a displacement of 4.29 litres. The engine is governed to operate at 450 rpm, but fine speed adjustment is still necessary. The slow operating speed means that a particularly simple engine indicator can be used to determine the indicated work output of the engine.

14.2.1 Dynamometers

The dynamometer is perhaps the most important item in the test cell, as it is used to measure the power output of the engine. The term 'brake horse power' (BHP) derives from the simplest form of engine dynamometer, the friction brake.

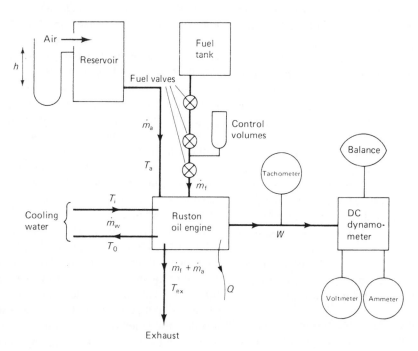

Figure 14.2
Schematic test arrangement for a Ruston Oil Engine.

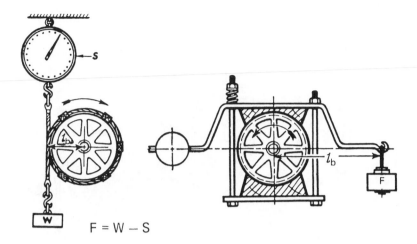

$$F = W - S$$

Figure 14.3
Friction brake dynamometers (courtesy of Froude Hofmann Ltd).

Typically the engine flywheel has a band of friction material around its circumference, and the torque reaction on the friction material corresponds to the torque output of the engine – see figure 14.3.

Another type of dynamometer is the electric dynamometer which acts as a generator to absorb the power from the engine. An advantage of this type of dynamometer is that it can also be used as a motor for starting the engine, and for motoring tests (when the engine is run at operating speeds without combustion) to determine the mechanical losses in the engine, or to investigate the thin-cylinder flows. The torque output or load absorbed by the dynamometer is controlled by the dynamometer field strength.

Electrical dynamometers were traditionally DC machines because of the difficulty of controlling the speed of AC machines. However, with modern electronic controllers, speed and torque control of both AC and DC machines is commonplace, and the controllers are often capable of four-quadrant control (rotation in either direction, with torque absorption or production) and regeneration (when firing, the electricity generated by the dynamometer is returned to the electrical distribution system). AC machines tend to be cheaper, to have a higher speed rating and to be more durable, but their controllers are more complex when regeneration is required. Unless such a dynamometer is to be heavily utilised, it might be more economic to have four-quadrant control without regeneration, and to dissipate the electrical energy in heater banks. With computer-controlled systems it is possible to compensate for the high inertia of electrical dynamometers, and this can be very important in transient testing.

When dynamometers are fitted with voltmeters and ammeters, these must not be used for calculating power unless the dynamometer efficiency (and power factor for AC dynamometers) is known for all operating conditions.

Another common type of dynamometer is the water brake (figure 14.4). A vaned rotor turns adjacent to a pair of vaned stators. The sluice gates separate the stators from the rotor, and these control the load absorbed by the dynamometer. Figure 14.5 shows typical absorption curves for a hydraulic dynamometer; only engine-operating points between the upper and lower solid lines can be attained.

Figure 14.5 also shows some dynamometer operating lines for constant sluice gate settings and some fixed throttle operating lines. The dynamometer operating lines show how the torque absorbed by the dynamometer increases with speed, and the throttle operating lines show how the torque varies at a fixed throttle setting if the speed is allowed to vary. By varying the throttle and sluice gate setting, any operating point should be attainable. For stable operation the dynamometer operating lines and throttle lines should intersect as close as possible to 90°. If both the operating lines are nearly parallel there will be poor stability. For example, if the operating lines are nearly parallel to the speed axis, a small change in torque will allow a large variation in speed. This is illustrated by figure 14.6 where the effects of perturbations from the throttle setting and dynamometer operating lines are considered. Such problems might occur, for example, with an engine operating at full throttle, when coupled to an electrical dynamometer that has a flat torque/speed characteristic.

Two other dynamometer types that might be encountered are the hydraulic dynamometer and the eddy current dynamometer. The hydraulic dynamometer consists of a high-pressure hydraulic pump connected to the engine. The hydraulic pump will need a control system that also controls the expansion of the high-pressure oil; the system is completed by an oil cooler. Since hydraulic dynamometers utilise standard components they can be made cheaply.

Eddy current dynamometers have a thin rotor that rotates within a magnetic field (the flux lines being parallel to the

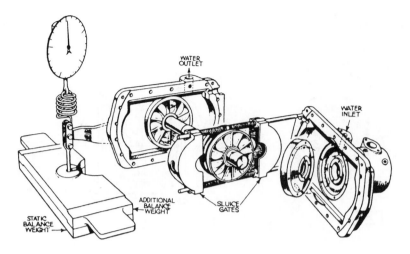

Figure 14.4
Hydraulic dynamometer (courtesy of Froude Hofmann Ltd).

dynamometer axis). The torque reaction is controlled by the strength of the magnetic field, and the system is suited to electronic control. The energy is dissipated by the eddy currents (short circuited) within the rotor, and the rotor is invariably water cooled.

Most dynamometers measure the load absorbed by the torque reaction on the dynamometer casing. The dynamometer is mounted in bearings co-axial with the shaft, so that the complete dynamometer is free to rotate, but usually within a restricted range (figure 14.7). The torque reaction (T) is equal to the product of the effective lever arm length (l_b) and the net force on the lever arm (F):

$$T = F . l_b \qquad (\text{N m}) \qquad (14.1)$$

The force F must be measured in the datum position for two reasons:

1 The effective lever arm length becomes $l_b \cos \phi$

2 Away from the datum position a torque will be exerted by the connections for the cable or hose.

Usually a dashpot is connected to the lever arm to damp any oscillations. All these problems are avoided if a load cell is connected to the lever arm, although calibration problems are introduced instead, and some form of low pass filter will be needed.

Dynamometers usually include a tachometer for measuring engine speed; the principles of operation are the same as those for car speedometers. Alternatively, revolution counters, either mechanical or electrical, can be used. These give accurate results, especially when the engine is operating steadily and the count is timed over a long period.

Where a tachometer is not fitted a stroboscope provides a convenient means of determining the engine speed, by illuminating a marker on the flywheel or crankshaft. However, care must be taken or otherwise a submultiple ($\frac{1}{2}$, $\frac{1}{3}$ etc.) of the speed will be found, as the marker will be illuminated once every second, third etc. revolution. This problem can be avoided by starting at the highest strobe frequency, and reducing the frequency until the first steady image of the marker is obtained.

After the measurements of torque and speed the fuel-consumption measurements are next in importance.

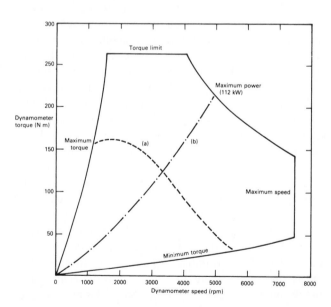

Figure 14.5
Absorption characteristics of a broad DPX2 dynamometer, showing the operating envelope with operating lines for: (a) constant engine throttle position, but varying the sluice gate position; (b) constant sluice gate position but varying the engine throttle position.

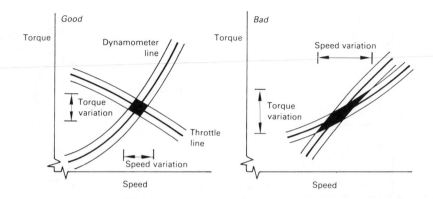

Figure 14.6
The influence of throttle and dynamometer operating line perturbations on the stability of the operating point for well- and ill-conditioned engine and dynamometer operating lines.

14.2.2 Fuel-consumption measurement

A common measurement system for fuel consumption is to time the consumption of a fixed volume. This has to be converted to a gravimetric consumption by using the density as determined from a separate test, or deduced from fuel temperature for a specific fuel. A typical arrangement is shown in figure 14.8. In normal operation valve 1 is open and fuel flows directly to the engine. The calibrated volumes are filled when valve 2 is open. If the vent pipe ends below the level of the fuel tank, care is needed in filling the control volumes. To measure the fuel flow rate, valve 1 is closed and any fuel to the engine is drawn from the calibrated volumes; valve 2 must be open. Usually there are a range of volumes to give the best compromise between accuracy and speed of taking the readings. A common problem is when a vapour-bubble travels back down the fuel line; this obviously invalidates a reading. In a fuel-injection system, if the spill flow is fed back to the tank this must be measured separately; such fuel is much more prone to vapour-bubbles. Although this method gives accurate results, the readings are not instantaneous.

Figure 14.9 shows a flowmeter that does give instantaneous readings, though the accuracy is less than that of the flowmeter previously described. The float chamber provides a constant head of fuel, and the pressure drop across the orifice is proportional to the square of the volumetric flow rate. The pressure drop is measured by the difference in head between the float chamber and the sight glass; alongside is a scale calibrated directly in volumetric flow rate. Again, problems can occur with fuel injection systems.

Both these flowmeters can be adapted for automatic flow measurement. The control volume system can be automated by solenoid valves, opto-electronic liquid level sensors and a timer. The orifice flowmeter could be automated by a differential pressure transducer, but there is of course a

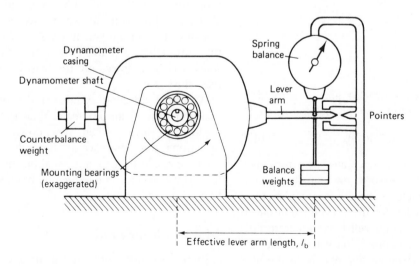

Figure 14.7
Dynamometer mounting system.

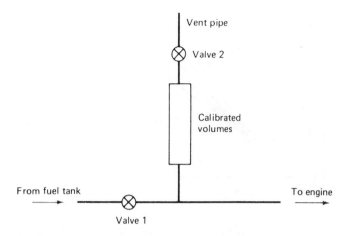

Figure 14.8
Fuel flow measurement.

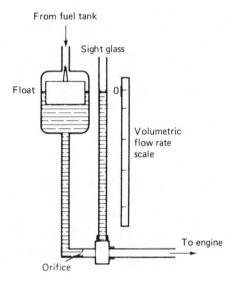

Figure 14.9
Orifice-type flowmeter.

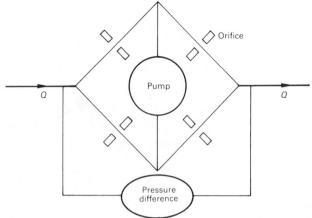

Figure 14.10
Flow system in the Flotron mass flowmeter.

non-linear relation between flow ($\dot{V}_f$) and pressure drop (Δp), and it is also necessary to know the fuel density (ρ_f):

$$\dot{V}_f \text{ is directly proportional to } \sqrt{(\Delta p/\rho_f)} \qquad (14.2)$$

Furthermore, the range of fuel flow rates (the turndown ratio) that might have to be measured on a single engine is about 50:1. If a single flowmeter is used, then equation (14.2) implies a 2500:1 range needed for the pressure transducer! Many fuel flowmeters for engines will have duplicate flow measuring systems, and the fuel flowmeter will automatically select the flow measuring elements appropriate to the flow rate. Turbine flowmeters or positive displacement flowmeters can be used, but their turndown ratio is usually only about

10:1, and an accuracy of 1 per cent across the full flow range can be difficult to achieve.

Ideally, a system is required that will give a direct reading of the fuel mass flow rate on a continuous basis. A quasi-continuous reading is given by systems that allow fuel to be drawn from a measuring volume. Readings of mass flow rate can be deduced from either measurements of the change in hydrostatic pressure at the base of a cylindrical volume, or the change in buoyancy on a float immersed in the measuring volume. Obviously, measurements cannot be taken while the measuring volume is being refilled.

There are two widely used systems for continuous measurement of the fuel mass flow rate. The theory of both systems is complex, but is presented in detail by Katys (1964). The first system employs a hydraulic equivalent of the Wheatstone bridge. Figure 14.10 shows the four orifices and the pump that provides a reference flow. The pressure drop is proportional to the mass flow rate, with a fast response to changes in flow, and no dependence on the fuel density or viscosity. The second system employs a vibrating 'U'-shaped tube through which the flow passes. Coriolis accelerations cause a twisting force, and the twist corresponds to the fuel mass flow rate.

With fuel injection systems, the return flow from the injection system can be greater than the fuel flow into the engine. So, even when two flowmeters are available, measuring the difference in flow would not lead to an accurate result. Instead, it is usual practice to cool the return fuel flow, and connect it to the fuel system downstream of the flow measuring system.

14.2.3 Air flow rate

A simple system to measure the air flow rate is obtained by connecting the air intake to a large rigid box with an orifice at its inlet. The box should be large enough to damp out the

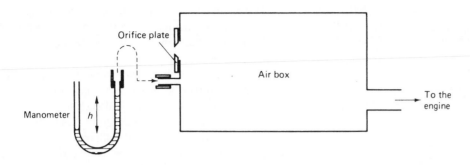

Figure 14.11
Air flow rate tank.

pulsations in flow and be free of resonances in the normal speed range of the engine. The pressure drop across the orifice can be measured by a water tube manometer, as shown in figure 14.11. For incompressible flow

$$\dot{m}_a = C_d A_o \sqrt{(2gh\rho_f\rho_a)} \tag{14.3}$$

where $\dot{m}_a$ = mass flow rate of air (kg/s)
C_d = discharge coefficient of the orifice
A_o = cross-sectional area of the orifice (m)
g = acceleration due to gravity (m/s²)
h = height difference between liquid levels in the manometer (m)
ρ_f = density of manometer fluid (kg/m³)
ρ_a = density of air ($= p/R_a T$) (kg/m³).

The accuracy depends on knowing the discharge coefficient for the orifice; this should be checked against a known standard.

In practice, it can be very difficult to make the air box work satisfactorily. Apart from making the air box act as a Helmholtz resonator (chapter 7, section 7.6.2), coupling a long tube to the inlet of an engine can have a significant effect on the engine performance. It should be remembered that even without a tuned induction system, there can be flow reversals occurring at the air inlet to the engine. A way of minimising any coupling effects and checking for their effect is shown in figure 14.12. If the flexible pipe is of large enough diameter, and the air entry protrudes far enough into the pipe, then the unsteady flow sensor will show no change to the air flow pattern when the flexible pipe is connected. The unsteady flow sensor is most likely to be a form of hot wire anemometer (see chapter 9, section 9.2.2).

An alternative approach is the viscous flowmeter. As no damping is required it is much more compact than an air box. To obtain viscous flow a matrix of passages is used in which the length is much greater than a typical diameter. Since the flow is viscous the pressure drop is proportional to the velocity or volumetric flow rate. The flowmeter has to be calibrated against a known standard.

However, in its usual form the viscous flowmeter is not well suited to pulsating flows. The instantaneous pressure drop will be a function of the unsteady flow, and the pressure-measuring system will not necessarily indicate a pressure drop that corresponds to the mean flow. With a pulsating flow, the acceleration and deceleration of the flow cause an additional pressure difference term. These and other effects have been considered by Stone and Wright (1994), who show how the instantaneous pressure difference can be analysed, so as to give the instantaneous volume flow rate and the mean flow rate.

Another flowmeter that gives an instantaneous measurement is the Lucas–Dawe air mass flowmeter. This was originally intended for use with engine management systems, however it is better suited to laboratory use. The principle is illustrated by figure 14.13. The central electrode is maintained at about 10 kV so that a corona discharge is formed. The exact voltage is varied so that the sum of the currents flowing to the two collector electrodes is constant. When air flows through the duct, the ion flow is deflected, and this causes an imbalance in the current flowing to the two collector electrodes. Cockshott *et al.* (1983) show that the difference in current flow is proportional to the air mass flow rate. This flowmeter has a response time of about 1 ms and is bi-directional, but the length of the measuring section leads to a slight averaging effect. The disadvantage of this meter is its sensitivity to air temperature and humidity.

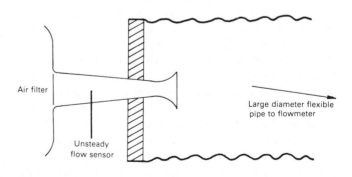

Figure 14.12
Connection of an air flowmeter to minimise any effects on engine performance.

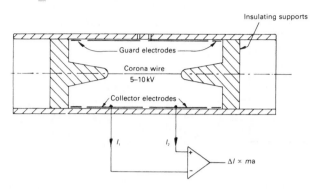

Figure 14.13
The Lucas–Dawe air mass flowmeter.

Finally, it is also possible to use a positive displacement flowmeter. These flowmeters often have a geometry like a Roots blower, and are capable of a 1 per cent or better accuracy over a turndown ratio of 50. Positive displacement meters cause a small pressure drop in the flow, but this can be eliminated if a servo-drive system is used to maintain the rotor speed at the correct value. The inertia of the rotors in positive displacement flowmeters leads to a poor transient response.

In general, gas flow rate is more difficult to determine than liquid flow rate. The calibration of liquid flowmeters can always be checked gravimetrically.

Under many circumstances, the most accurate means of evaluating the air flow rate is to measure the fuel flow rate, and to calculate the air/fuel ratio from an exhaust gas analysis. The relevant techniques have already been introduced by chapter 3, example 3.2, but the methods are discussed further in section 14.4.8.

14.2.4   Temperature and pressure

Mercury-in-glass thermometers and thermocouples both provide economical means of measuring temperature, with the potential of achieving an accuracy of about ±0.1 K. However, if this level of accuracy is required, it can be cheaper to use platinum resistance thermometers. A disadvantage of thermocouples is their low output, about 40 μV/K. This can lead to the use of thermistors for temperatures up to 400°C. Thermocouples can be used at temperatures above 1000°C, so there is no restriction on their use in engines. However, at the high temperatures in the exhaust system, care is needed to avoid errors due to radiation losses from the thermocouple (this is quantified in section 14.3). Care is also required when trying to measure the temperature of a pulsating flow. Caton and Heywood (1981) argue that the thermocouple temperature corresponds to a time-averaged temperature, and this does not correspond to the average energy of the flow. For the exhaust port of an engine, the mass flow rate varies widely, and the highest temperatures occur at a time when the mass flow rate is high. The mass-averaged temperature can be

100 K higher than the time-averaged temperature. This is why the exhaust temperature at entry to a turbocharger can be higher than any of the temperatures indicated in the exhaust ports.

If a thermocouple junction could be made sufficiently small, then it would be possible to measure the instantaneous temperature. However, even if such a thermocouple could be made, it would not be sufficiently robust. Instead, by using a pair of thermocouples with different junction diameters, it is possible to correct for the time constants of the thermocouples. Tagawa and Ohta (1997) have used thermocouples with junction diameters of 25 and 60 μm, and developed a method that allows the time constant fluctuations to be determined.

Bourdon pressure gauges and manometers provide a cheap and accurate means of measuring steady pressures. By selecting the manometer fluid and its angle of inclination, pressures in the range of 1–100 kN/m^2 can be measured with an accuracy of about 1 per cent. When the manometer fluid is not a single compound, care must be used in selecting the fluid, as otherwise vaporisation of the more volatile component will change the fluid density.

Most pressure transducers utilise a piezo-resistive element, a notable exception being the piezoelectric transducers used for measuring the in-cylinder pressure (section 14.2.5). A common arrangement is to have a diaphragm etched with strain gauges in a Wheatstone bridge configuration. For low pressures a silicon diaphragm is used, while for higher pressures stainless steel is more likely to be used (for example, in measuring the injector fuel line pressure).

14.2.5   In-cylinder pressure measurement

Engine indicators record the pressure/volume history of the engine cylinder contents. The simplest form of engine indicator is the Dobbie McInnes mechanical indicator shown in figure 14.14. A piece of paper is attached to a drum, and the rotation of the drum is linked to the piston displacement by a cord wrapped around the drum. The pressure in the cylinder is recorded by a linkage attached to a spring-loaded piston in a cylinder. The indicator cylinder is connected to the engine cylinder by a valve. Unless the indicator cylinder has a much smaller volume than the engine cylinder, the indicator will affect the engine performance.

When the paper is unwrapped from the drum the area of the diagram can be found, and this corresponds to the indicated work per cylinder per cycle. The area can be determined by 'counting squares', by cutting the diagram out and weighing it, or by using a planimeter. The planimeter is a mechanical device that computes the area of the diagram by tracing the perimeter. Practice is necessary in order to obtain reliable results, and accuracy can be improved by tracing round the diagram several times. To convert area to work a calibration constant is needed; alternatively imep can be found more directly. The diagram area is divided by its length

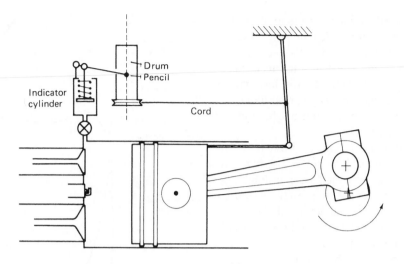

Figure 14.14
Mechanical indicator.

to give a mean height. When this height is multiplied by the indicator spring constant (bar/mm) the imep can be found directly:

$$\text{imep} = kh_d = k\frac{A_d}{l_d} \tag{14.4}$$

where A_d = diagram area
l_d = diagram length
h_d = mean height of diagram.

Because of the inertia effects in moving parts – friction, backlash and finite stiffness – mechanical indicators can be used only at speeds of up to about 600 rpm. Also, this simple type of mechanical indicator is not sensitive enough to record the 'pumping losses' during the induction and exhaust strokes.

Electronic systems are now very common for recording indicator diagrams. Care is needed in their interpretation since the pressure is plotted on a time instead of a volumetric basis. As with any electronic equipment, care is also needed in the calibration. The output from the pressure transducer is connected to the y-channel of an oscilloscope and the time base is triggered by an inductive or equivalent pick-up on the crankshaft – see figure 14.15. The inductive pick-up should also be connected to a second y-channel so that the position of tdc can be accurately recorded. Since tdc occurs during the period of maximum pressure change, a 1° error in position of tdc can cause a 5 per cent error in imep.

The location of tdc is not straightforward, not least because of the finite stiffness of the crank slider mechanism. The flexibility of the crankshaft is such that at full load there can be about 1° of twist at certain speeds. Ultrasonic techniques

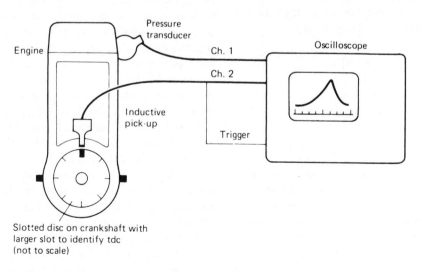

Figure 14.15
Electronic engine indicator system.

have been used to determine the dynamic position of tdc, but more usually the measurements are made when the engine is stationary. If the cylinder head is removed, a dial gauge can be used to measure the piston height. The dial gauge needs to act in-line with the gudgeon-pin, otherwise the readings will be influenced by the piston rocking. Two angular positions are found either side of tdc with the same piston height, and the angle is then bisected to give tdc.

The included angle needs to be as large as possible to give accurate results. Some engines have valves whose axes are parallel to the cylinder axis, in which case it might be acceptable to let the valve act as a 'follower' on the piston, to obviate the need to remove the cylinder head. With the appropriate piston close to tdc, the valve spring is removed, and the valve is allowed to hit the piston.

The position of tdc has to be recorded by the data acquisition system, and the options are:

(a) use a separate marker

(b) use a wider crank angle marker

(c) add extra markers around tdc.

One of the most popular arrangements is to have markers every 10°ca, with extra markers at ±5°tdc. When an oscilloscope is being used, the time base can be adjusted so that a division on the graticule corresponds to a convenient crank angle increment (5° or 10° etc.). If there is a shortage of channels, the crank markers can be used on the z-input to modulate the trace brightness. The output from the oscilloscope can be recorded photographically. If a transient recorder is connected before the oscilloscope, the output can alternatively be directed to an x–y plotter.

To convert the time base to a piston displacement base it is usual to assume constant angular velocity throughout each revolution. Assuming that the gudgeon-pin or 'little-end' is not offset (that is, assuming that the line of motion of the pivot in the piston intersects the axis of rotation of the crankshaft), the piston displacement (x) is given by

$$x = r(1 - \cos\theta) + \left(l - \sqrt{(l^2 - r^2\sin^2\theta)}\right) \qquad (14.5)$$

where θ = crank angle measured from tdc
$\quad r$ = crank radius (half piston stroke)
$\quad l$ = connecting-rod length.

When $l >> r$ the motion becomes simple harmonic.

A voltage can be generated that corresponds to equation (14.4), by either analogue or digital electronics (Hudson *et al.*, 1988). When this is correctly phased, it can be used as the x-input to an oscilloscope operating in its x–y mode. This is illustrated by figure 14.16, which shows three successive cycles superimposed to give an indication of the cycle-by-cycle variation in combustion. Figure 14.16 also shows an enlargement of the pumping loop, and this is useful when

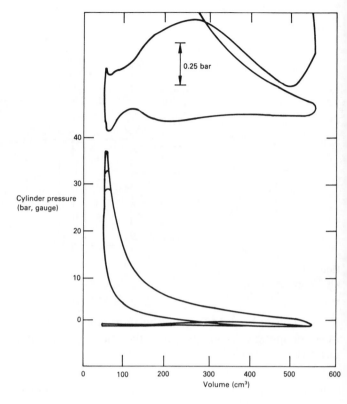

Figure 14.16

The pressure/volume or indicator diagram from a Rover M16 engine operating at 2000 rpm; with an enlargement of the pumping loop. bmep = 3.8 bar, imep = 4.6 bar (including the pumping work of 0.45 bar pmep).

tuned induction or exhaust systems and valve timing are being investigated.

Figure 14.17 is from the same operating point as figure 14.16, except that a time base has been used. The cylinder pressure trace and tdc markers are self-explanatory. When additional channels are available on the oscilloscope or data logging system, they can be used for a variety of purposes. With spark ignition engines it is useful to record the ignition timing. This can be done by wrapping a wire around the appropriate spark plug lead, or by monitoring the coil LT voltage. The voltage should be measured at the switched side of the coil (negative terminal for negative earth systems). The signals in figure 14.17 indicate when the coil is switched on and off, and the spark duration.

For diesel engines it is useful to measure the fuel line pressure and the injector needle lift. The needle lift can be measured by a proprietary Hall effect transducer or an FM (frequency modulated) system that records the change in inductance of a coil. Figure 14.18 shows the construction of a needle-lift transducer. The armature connected to the injector needle extends about half-way into the coil. The coil forms part of a tuned circuit that is resonating at close to 2 MHz. As

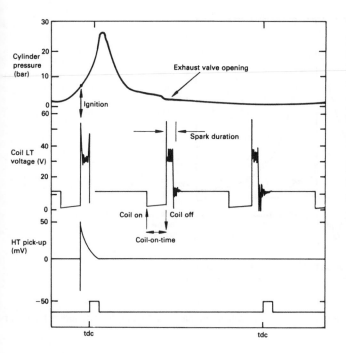

Figure 14.17
A pressure trace, along with tdc markers for the same operating point as figure 14.16. Also shown are the signals from an HT pick-up and the coil LT winding.

the armature position moves the inductance of the coil changes and the resonant frequency changes. This frequency modulation is converted to an analogue voltage that is proportional to needle lift for a uniformly wound coil. The construction of such a system is described by Osborne (1985). Figure 14.19 shows the output from a needle-lift transducer, along with the cylinder pressure (a), the fuel line pressure (c) and crank angle markers (d).

The in-cylinder pressure transducer requirements are very demanding because of the high temperature and pressures, and the need for a high-frequency response. The transducers

usually have a metal diaphragm which is displaced by the pressure. The displacement can be measured, inductively, by capacitance, by a strain gauge or by a piezoelectric crystal. Piezoelectric transducers are common, but have the disadvantage that they respond only to the rate of change in pressure; thus they have to be used in conjunction with a charge amplifier that integrates the signal.

The piezoelectric transducer produces an electrical charge that is proportional to pressure (typically between 2 and 50 pC/bar). When the signal is integrated it is necessary to define the pressure/voltage datum, and this can be done in a variety of ways:

(a) By assuming that the minimum pressure recorded in the cycle corresponds to a particular value, for example the inlet manifold pressure.

(b) By adjusting the datum value so that the compression process is described by a polytropic process (pv^k = constant).

(c) By using a cycle simulation program to define the pressure at a particular crank angle.

(d) By using a clipper adaptor – a pressure transducer mounting designed to record a datum pressure, and pressures either above or below the datum pressure (according to the clipper adaptor design).

(e) By installing a pressure transducer towards the bottom of the cylinder barrel, so that when the piston is close to bdc the transducer records the cylinder pressure.

Figure 7.29 has already illustrated the use of a cylinder-barrel-mounted pressure transducer. The barrel was fitted with a water-cooled piezo-resistive pressure transducer, so that when the piston was within about 65°ca of bdc the transducer recorded the cylinder pressure. However, a settling time of up to 30°ca is required for the transducer to respond to the pressure change, because of the limited clearance around the piston top land. Elsewhere the barrel pressure transducer senses atmospheric pressure (or in some installations the

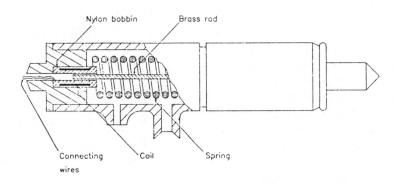

Figure 14.18
The installation of an injector needle-lift transducer.

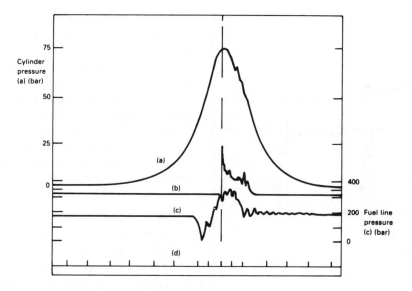

Figure 14.19
Cylinder pressure (a), injector needle lift (b), fuel line pressure (c) and crank marker (d) (10°ca with additional ±5°ca marks) from an indirect injection diesel engine operating at 1000 rpm, no load.

crankcase pressure). The barrel pressure transducer can fix the pressure datum for the piezoelectric pressure transducer in the cylinder head, by averaging the two signals for say 10°ca either side of bdc. This determines the offset that has to be applied to the piezoelectric transducer signal. A piezo-resistive transducer is likely to have a better signal-to-noise ratio than a piezoelectric transducer.

For calculations such as imep it is not necessary to know the absolute values of pressure, and for peak pressures (invariably above 30 bar) uncertainty of 0.3 bar or so in the datum will not be significant. However, for combustion analysis it is important to know the absolute pressure more accurately (section 14.5.2).

The electrical charge is prone to both leakage and accumulation, and this causes the voltage output from the charge amplifier to drift. By using a coupling with an appropriate time constant, the effect of the drift is eliminated. However, when the pressure transducer is calibrated by static pressures (often by means of a dead-weight pressure tester), a long time constant is needed, so that a steady pressure corresponds to a steady voltage. This places considerable demands on the input resistance of the charge amplifier, the internal resistance of the pressure transducer and the interconnecting cable. To prevent the electrical charge being dissipated, a resistance of greater than 10^{14} ohms is required. The construction of a piezoelectric pressure transducer is shown in figure 14.20.

The pressure transducer response should be independent of temperature, and the calibration should be free from drift. The pressure transducer should be mounted flush with the cylinder wall or as close to the engine cylinder as possible through a small communicating passage. This minimises the lag in the pressure signal and should avoid introducing any resonances in the connecting passage.

In general, the larger the transducer and the better the cooling then the less susceptible it will be to thermal effects. Also, with larger transducers there is a larger output. Finally, great care is needed to avoid using too short a time constant on the charge amplifier, otherwise the amplified signal will droop after the maximum pressure.

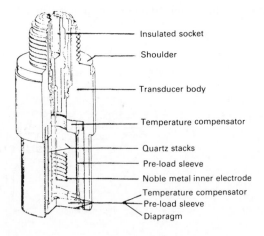

Figure 14.20
Sectional drawing of a Kistler 601A pressure transducer.

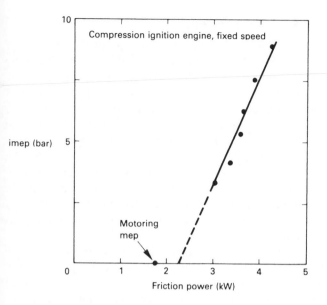

Figure 14.21
Dependence of friction power on indicated power at constant speed.

14.2.6   Techniques for estimating indicated power

Very often a pressure transducer cannot be readily fitted to an engine, so alternative means of deducing imep are useful. The difference between indicated power and brake power is the power absorbed by friction, and this is often assumed to be dependent only on engine speed. Unfortunately, the friction power also depends on the indicated power since the increased gas pressures cause increases in piston friction etc.; this is shown by figure 14.21. When extrapolated to zero imep, fmep is about 2.25 bar. This can be compared to a 1.75 bar motoring mep – the equivalent of the power output of the electric dynamometer turning the engine at the same speed.

If the friction power is assumed to be independent of the indicated power, the friction power can be deduced from the Morse test. This is applicable only to multi-cylinder engines (either spark or compression ignition), as each cylinder is disabled in turn. When a cylinder is disabled the load is reduced so that the engine returns to the test speed; the reduction in power corresponds to the indicated power of that cylinder. For a n-cylinder engine

$$\sum^{n} \text{indicated power} - \sum^{n} \text{friction power} = (\text{brake power})_n$$

$$(14.6)$$

With one-cylinder disabled

$$\sum^{n-1} \text{indicated power} - \sum^{n} \text{friction power} = (\text{brake power})_{n-1}$$

$$(14.7)$$

Subtracting:

indicated power of disabled cylinder = reduction in brake power   (14.8)

This underestimates the friction power since the disabled cylinder also has reduced friction power. However, the test does check that each cylinder has the same power output.

With spark ignition engines, the traditional approach to a Morse test was to disconnect the spark plug lead. The high voltages are dangerous, and there will also be unburnt mixture entering the exhaust system. However, with multi-point fuel injection engines it is easy to electrically isolate an injector. With diesel engines the injectors can be disabled, either mechanically (by disconnecting the fuel supply) or electrically as appropriate.

A method for estimating the friction power of compression ignition engines is Willans' line. Again it is assumed that at constant speed the friction power is independent of indicated power, but in addition it is assumed that the indicated efficiency is constant. This is a reasonable assumption away from maximum power. Figure 14.22 shows a plot of fuel consumption against power output. Willans' line is when the graph is extrapolated to zero fuel flow rate to determine friction power. Figure 14.21 is for the same engine as shown

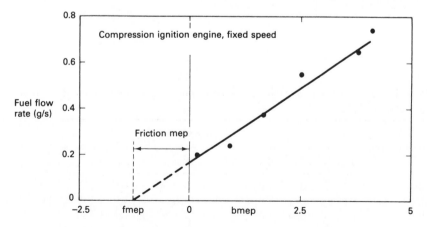

Figure 14.22
Willans' line for a diesel engine.

in figure 14.22, but Willans' line suggests a friction mep of only 1.25 bar.

14.2.7 Engine test conditions

Various standards authorities (BS, DIN, SAE) are involved with specifying the test conditions for engines, and how allowance can be made for variations in ambient conditions.

In the past a wide range of performance figures could be quoted for a given engine, depending on which standard was adopted and how many of the engine ancillary components were being driven (water pump, fan, alternator etc.). Obviously it is essential to quote the standard being used, and to adhere to it.

Corrections for datum conditions vary, and in general they are more involved for compression ignition engines, whether turbocharged or naturally aspirated. Corrections for spark ignition engines in the *SAE Handbook* (SAE, 1987) are as follows:

$$\left.\begin{array}{l} \text{ambient test conditions} \\ \text{should be in the range} \end{array}\right\} \quad \begin{array}{l} 95 < p < 101 \text{ kN/m}^2 \\ 15 < T < 43°C \end{array}$$

where p is ambient pressure and T is ambient temperature. The corrections are applied to indicated power, where $(\dot{W}_i)_o$ is observed value and $(\dot{W}_i)_c$ is the corrected value:

$$(\dot{W}_i)_c = (\dot{W}_i)_o \left(\frac{99}{p_d}\right) \sqrt{\left(\frac{T + 273}{298}\right)} \tag{14.9}$$

where $p_d = p - p'_{water}$, the partial pressure of dry air (kN/m^2). Values for brake power $(\dot{W}_b)$ are found from a knowledge of the friction power $(\dot{W}_f)$:

$$\dot{W}_i = \dot{W}_b + \dot{W}_f \tag{14.10}$$

This approach relies on knowing the friction power. If this has not been found, an alternative approach is

$$(\dot{W}_b)_c = (\dot{W}_b)_o \left[1.18\left(\frac{99}{p_d}\right)\sqrt{\left(\frac{T + 273}{298}\right)} - 0.18\right] \tag{14.11}$$

14.2.8 Energy balance

Experiments with engines very often involve an energy balance on the engine. Energy is supplied to the engine as the chemical energy of the fuel and leaves as energy in the cooling water, exhaust, brake work and extraneous heat transfer. Extraneous heat transfer is often termed 'heat loss', but this usage is misleading as heat is energy in transit and the First Law of Thermodynamics states that energy is conserved.

The heat transfer to the cooling water is found from the temperature rise in the coolant as it passes through the engine and the mass flow rate of coolant. The temperature rise is measured with mercury in glass thermometers or thermocouples. The mass flow rate of coolant is usually derived from the volumetric flow rate. Common flow-measuring devices include tanks, weirs and variable-area flowmeters such as the Rotameter. The Rotameter has a vertical-upwards flow through a diverging graduated tube; a float rises to an equilibrium position to indicate the flow rate. To conserve water the coolant is usually pumped in a loop with some form of heat exchanger. The heat exchanger should be regulated to control the maximum engine-operating temperature. The engine coolant flow can be adjusted to make the temperature rise sufficiently large to be measured accurately without making the flow rate too small to be measured accurately.

The energy leaving in the exhaust is more difficult to determine. If the gas temperature is measured, the mean specific heat capacity can be estimated in order to calculate the enthalpy in the exhaust.

The enthalpy of the exhaust can be calculated from the polynomial functions that define the enthalpy of the constituents in the exhaust (see Appendix A, section A.2.1). This requires a knowledge of the temperature and composition. Table 14.1 presents the mean specific heat capacity of exhaust products in terms of the temperature and the equivalence ratio. The fuel composition has been taken as C_nH_{2n}, and it has been assumed that no oxygen is present in rich mixtures, and no carbon monoxide is present in weak mixtures.

Table 14.1 *Mean specific heat capacity of exhaust products, for idealised combustion of C_nH_{2n} (datum of 25°C)*

Temperature K	Equivalence ratio							
	0	0.2	0.4	0.6	0.8	1.0	1.2	1.4
400	1.010	1.024	1.037	1.050	1.063	1.075	1.100	1.124
600	1.024	1.041	1.057	1.072	1.087	1.102	1.133	1.164
800	1.045	1.064	1.082	1.100	1.117	1.132	1.150	1.173
1000	1.067	1.087	1.107	1.126	1.145	1.162	1.179	1.194
1200	1.088	1.110	1.131	1.151	1.172	1.191	1.206	1.221

Sometimes a known flow rate of water is sprayed into the exhaust, and the temperature is measured after the water has evaporated. The exhaust can include partially burnt fuel, notably with spark ignition engines operating on rich mixtures; this can make a significant difference to an energy balance.

The significance of chemical energy has already been seen in chapter 13, figures 13.11 and 13.12. When emissions data are available, then the chemical energy associated with the partially burnt fuel can be evaluated. For most purposes the energy associated with unburned hydrocarbons can be neglected, so it is only necessary to consider the chemical energy associated with the carbon monoxide and hydrogen.

Rogers and Mayhew (1980a) provide molar enthalpies of reaction for carbon monoxide and hydrogen at 25°C:

$$(\Delta H_0)_{CO} = -283.0 \text{ MJ/kmol}$$
$$(\Delta H_0)_{H_2} = -241.8 \text{ MJ/kmol}$$

Consider the reaction of a generalised fuel with air, that produces 100 kmol of dry products:

$$x(CH_yO_z) + a(O_2 + 79/21\,N_2) \rightarrow bCO + cCO_2 + dH_2 + \dots$$

where $b \equiv$ percentage of CO, $c \equiv$ percentage of CO_2, $d \equiv$ percentage of H_2 etc.

For a fuel with known composition and calorific value (CV in units of MJ/kg), a carbon balance can be used to establish the mass of fuel to produce 100 kmol of dry products:

C balance	$x = b + c$	(14.12)
mass of fuel	$m = x(12 + y + 16z)$	(14.13)

Thus, the percentage of the fuel energy that remains as chemical energy in the exhaust is

$$\frac{283.0 \times b + 241.8 \times d}{CV \times (b + c) \times (12 + y + 16z)} \times 100 \text{ per cent} \quad (14.14)$$

The data in chapter 3, figure 3.19 suggest that for a typical hydrocarbon fuel with an equivalence ratio of 1.1, about 15 per cent of the fuel energy remains as chemical energy in the exhaust. If the fuel is a hydrocarbon but its composition is unknown, then the emissions data can be used to find the fuel composition; this is discussed further in section 14.4.8. The carbon balance has been used here, since it is the simplest and usually the most accurate; of course other atomic balances could be used to determine the mass of fuel needed to produce 100 kmol of dry exhaust products.

Heat transfer from the engine cannot be readily determined from temperature measurements of the engine. If the engine is totally enclosed, the temperature rise and mass flow rate of the cooling air can be used to determine the heat transfer.

Finally, brake power should be used in the energy balance, not indicated power. The power dissipated in overcoming friction degenerates to heat, and this is accounted for already.

14.3 | Experimental accuracy

Whenever an experimental reading is taken there is an error associated with this reading. Indeed, it can be argued that any reading is meaningless unless it is qualified by a statement of accuracy. There are three main sources of error:

1 the instrument is not measuring what is intended

2 the instrument calibration is inaccurate

3 the instrument output is incorrectly recorded by the observer.

To illustrate the different errors, consider a thermocouple measuring the exhaust gas temperature of an engine. Firstly, the temperature of the thermocouple may not be the temperature of the gas. Heat is transferred to the thermocouple by convection, and is transferred from the thermocouple by radiation, and to a lesser extent by conduction along the wires. If the gas stream has a temperature of 600°C and the pipe temperature is about 400°C then the thermocouple could give a reading that is 25 K low because of radiation losses (Rogers and Mayhew, 1980a).

Secondly, there will be calibration errors. Thermocouple outputs are typically 40 μV/K and very close to being linear; the actual outputs are tabulated as functions of temperature for different thermocouple combinations. The output is very small, so amplification is often needed; this can introduce errors of gain and offset that will vary with time. Thermocouples also require a reference or cold junction, and this adds scope for further error whether it is provided electronically, or with an additional thermocouple junction in a water/ice mixture. The output has to be indicated on some form of meter, either analogue or digital, with yet further scope for errors.

Thirdly, errors can arise through misreading the meter; this is less likely with digital meters than with analogue meters.

In serious experimental work the instrument has to be checked regularly against known standards to determine its accuracy. Where this is not possible, estimates have to be made of the accuracy; this is easier for analogue instruments than for digital instruments. In a well-designed analogue instrument (such as a spring balance, a mercury in glass thermometer etc.), the scale will be devised so that full benefit can be obtained from the instrument's accuracy. In other words, if the scale of a spring balance has 1 Newton divisions and these can be subdivided into quarters, then it is reasonable to do so, and to assume that the accuracy is also $\pm\frac{1}{4}$ Newton. Of course a good spring balance would have some indication of its accuracy engraved on the scale.

This approach obviously cannot be applied to digital instruments. It is very tempting, but wrong, to assume that an instrument with a four-digit display is accurate to four significant figures. The cost of providing an extra digit is much less than that of improving the accuracy of the instrument by an order of magnitude! For example, most thermocouples with a digital display will be accurate to only one degree.

There are many books, such as Adams (1975), that deal with instrumentation and the handling of results. One possible treatment of errors uses binomial approximations. If a quantity u is dependent on the quantities x, y and z such that

$$u = x^a y^b z^c \tag{14.15}$$

then for sufficiently small errors

$$\frac{\delta u}{u} \approx a\left(\frac{\delta x}{x}\right) + b\left(\frac{\delta y}{y}\right) + c\left(\frac{\delta z}{z}\right) \tag{14.16}$$

where δu denotes the error associated with u etc.

As an example consider equation (10.12):

$$T_{2s} = T_1 \left(\frac{p_2}{p_1}\right)^{(\gamma-1)/\gamma} \tag{14.17}$$

If the error in T_1 is $\pm\frac{1}{2}$ per cent and the error associated with the pressure ratio (p_2/p_1) is ± 5 per cent, then the error associated with the isentropic compressor temperature T_{2s} is

$$\pm\tfrac{1}{2} \pm \frac{1.4 - 1}{1.4} \times 5 \text{ per cent} \tag{14.18}$$

which is ± 1.9 per cent. This is a pessimistic estimate of the error, since it assumes a worst possible combination of errors that is statistically unlikely to occur.

Sometimes it is possible for the experimenter to minimise the effect of errors. Consider the heat flow to the engine coolant:

$$\dot{Q} = \dot{m}c_p(T_{out} - T_{in}) \tag{14.19}$$

Suppose the mass flow rate ($\dot{m}$) of coolant is 4.5 kg/s and the inlet and outlet temperatures are 73.2°C and 81.4°C respectively. If the errors associated with mass flow rate are ± 0.05 kg/s and the errors associated with the temperature are ± 0.2 K, then

$$Q = (4.5 \pm 0.05)\, c_p[(81.4 \pm 0.2) - (73.2 \pm 0.2)]$$
$$= (4.5 \pm 0.05)c_p\,(8.2 \pm 0.4)$$
$$= (4.5 \pm 1.11 \text{ per cent})c_p\,(8.2 \pm 4.88 \text{ per cent})$$
$$= 4.5 \times c_p \times 8.2 \pm 6 \text{ per cent} \tag{14.20}$$

It can be readily shown by calculus that the errors would be minimised if the percentage error in each term were equal. Denoting the optimised values of mass flow rate as $\dot{m}'$ and temperature difference as $\Delta T'$, then

$$\frac{0.05}{\dot{m}'} = \frac{0.4}{\Delta T'}, \quad \Delta T' = 8\dot{m}' \tag{14.21}$$

and for the same heat flow

$$\dot{m}' . \Delta T' = 4.5 \times 8.2$$

Combining

$$\dot{m}' = \sqrt{\left(\frac{4.5 \times 8.2}{8}\right)}$$
$$= 2.15 \text{ kg/s}$$

and $\quad \Delta T' = 17.18 \tag{14.22}$

The error is now reduced to

$$\pm\left(\frac{0.05}{2.15}\right) \pm \left(\frac{0.4}{17.18}\right) \tag{14.23}$$

or ± 4.65 per cent.

Since the effect of errors is cumulative, always identify the weakest link in the measurement chain, and see if it is possible to make an improvement.

14.4 | Measurement of exhaust emissions

Exhaust gas emissions need to be measured because of legislation, and also because of the insights the measurements provide into engine performance. The emissions governed by legislation are carbon monoxide (measured by infra-red absorption), nitrogen oxides (measured by chemiluminescence), unburnt hydrocarbons (measured by flame ionisation detection) and particulates. If carbon dioxide (measured by infra-red absorption) and oxygen (measured by a chemical cell, or more accurately by paramagnetism) are also analysed, then it is possible to calculate the air/fuel ratio. Each of these measurement techniques will be described in the following sections, and there also follows a discussion of how to compute the air/fuel ratio and other parameters such as exhaust gas recirculation (EGR).

With any emissions-measuring equipment it is important to test the calibration against (bottled) mixtures of accurately known composition. As a minimum the zero reading and full-scale reading need to be checked. An economical method for checking the linearity of a gas analyser is shown in figure 14.23. The second calibration mixture can of course be air or whatever is being used for the zero checking. The flowmeters have to be very accurate over a wide range of flows, so positive displacement domestic gas meters are a good choice. If a second analyser is available (preferably of a completely different type), the gas flowmeters can be dispensed with, at the expense of some compromise in the certainty of the results. If both analysers indicate a linear response then it is reasonable to assume that they are both indeed linear, since they are unlikely to be non-linear in the same way. This sort of approach is more likely to be acceptable in an academic than an industrial environment.

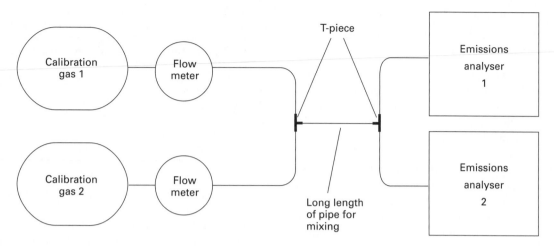

Figure 14.23
A flow system for checking the linearity of emissions analysers.

14.4.1 Infra-red absorption

Infra-red radiation is absorbed by a wide range of gas molecules, each of which has a characteristic absorption spectrum. The fraction of radiation transmitted (τ_λ) at a particular wavelength (λ) is given by Beer's Law:

$$\tau_\lambda = \exp(-\rho\alpha_\lambda L) \tag{14.24}$$

where ρ is the gas density, α_λ is the absorptivity and L is the path length.

Figure 14.24 shows the key components in a non-dispersive infra-red gas analyser.

Figure 14.25 shows the absorption spectra of carbon monoxide and carbon dioxide. This shows that in the region

of 4.4 microns, infra-red radiation is absorbed by both carbon dioxide and carbon monoxide. In other words, for the simple arrangement shown in figure 14.24, when carbon dioxide is present in the sample, then this will affect slightly the readings of carbon monoxide, and vice versa when carbon dioxide is being measured. This problem can be eliminated by using a filter cell between the infra-red source and the detector.

The windows in the analyser have to be transparent to infra-red, so are made from materials such as mica and quartz. Obviously readings would be invalidated by fouling of the windows in the sample cell. To minimise this risk, the sample should be filtered to remove particulates, and condensation (of water vapour or high molecular weight hydrocarbons) is avoided by either:

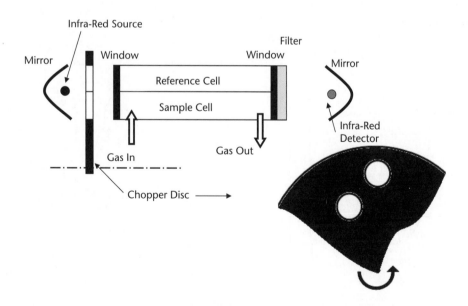

Figure 14.24
Diagram of a non-dispersive infra-red (NDIR) gas analyser.

(i) heating the sample lines and analyser, or

(ii) cooling and removing the condensate, then warming the sample to ambient temperature.

Infra-red from a source (such as a heated filament) is focused by the mirror and passes through both the sample cell and a reference cell with a known concentration of the gas to be measured. The radiation that passes through the cells is focused by a second mirror onto the infra-red detector, which is a semi-conductor such as lead selenide. The rotating chopper disc has apertures, so that the detector receives infra-red from either the reference cell (filled with a known concentration of the gas that is being measured) or the sample cell. When no direct infra-red is transmitted through the chopper disc then the background signal level can be found, so that a comparative measurement can be made between the reference cell and the sample cell. As the windows in the sample cell will get dirty, a calibration gas can be passed through the sample cell to enable compensation to be made for the dirt.

Non-dispersive infra-red absorption (NDIR) can be used for measuring the unburned hydrocarbons. However, this is not entirely satisfactory, as different hydrocarbon species have different absorption spectra. Ideally, when quantitative measurements of hydrocarbons are required a flame ionisation detection system should be used.

14.4.2 Flame ionisation detection (FID)

When hydrocarbons are burned, electrons and positive ions are formed. If the unburned hydrocarbons are burned in an electric field, the current flow corresponds very closely to the number of carbon atoms present.

Table 14.2 shows that the response is slightly dependent on the molecular structure, and furthermore the presence of some atoms in a molecule can suppress the ionisation current from the carbon atoms. Table 14.2 also shows that, for a

Table 14.2 *Typical responses of a flame ionisation detector to different molecular structures, normalised with respect to propane*

Molecular structure	Relative response
Alkanes	0.97–1.05
Aromatics (benzene rings)	0.97–1.12
Alkynes	0.99–1.03
Alkenes	1.07
Carbonyl radical (CO^-)	0
Oxygen in primary alcohol	0.23–0.68

particular species type, there is a dependence on the molecule size. A review of these dependencies has been compiled by Cheng *et al.* (1998), along with a comprehensive discussion of the theory of conventional and fast response flame ionisation detectors.

A flame ionisation detector is shown in figure 14.26. The sample is mixed with the fuel and burned in air. The fuel should not cause any ionisation, so a hydrogen or hydrogen/helium mixture is used. The air should be of high purity, again to reduce the risk of introducing hydrocarbons or other species. The fuel and sample flows have to be regulated, as the instrument response is directly related to the sample flow rate, and dependent on the fuel flow rate as this influences the burner temperature (and thence the sensitivity). The flows are regulated by maintaining fixed pressure differences across devices such as:

(a) capillary tubes

(b) porous sintered metallic elements

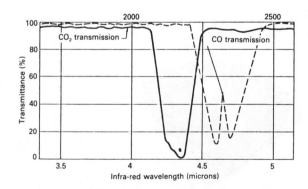

Figure 14.25
Transmittance of infra-red through gaseous carbon dioxide and carbon monoxide. [Adapted from Ferguson (1986) and reprinted by permission of John Wiley and Sons Inc.]

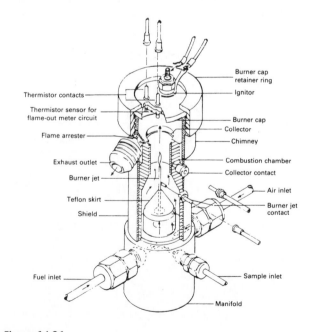

Figure 14.26
The burner and electrode system for a flame ionisation detector (Beckman).

(c) critical flow orifices – when the pressure ratio is above about 2, the flow through the orifice is supersonic, and dependent only on upstream conditions.

In figure 14.26, the burner jet and the annular collector form the electrodes, and a potential of about 100 volts is applied between them. The signals have to be amplified, and calibration is achieved by:

(a) zeroing the instrument with a sample containing no hydrocarbons (such as pure nitrogen),

(b) using calibration gases of known hydrocarbon concentration (such as 0.1 per cent C_3H_8 in N_2).

As with NDIR analysers it is necessary to minimise the risk of sample deposition occurring in the sample line; the usual arrangement is to have a heated sample line.

A particularly ingenious FID system has been developed by Collings and Willey (1987). This system has a very fast response (about 1 ms), since the flame ionisation detector has been miniaturised and can be sited remotely from the analyser, and adjacent to the sample source. The system is capable of making cycle-resolved measurements, and has been used for:

(a) In-cylinder sampling to look at cycle-by-cycle variations in unburned hydrocarbons, before and after combustion (Cheng et al., 1990).

(b) Sampling of the pre-combustion chamber gases in a lean-burn gas engine to establish the air/fuel ratio distribution (Tawfig et al., 1990).

(c) Measurement of the unburned hydrocarbons in the exhaust port of a gasoline engine (Finlay et al., 1990).

(d) Measurement of in-cylinder hydrocarbons to deduce the effects of mixture preparation and trapped residuals on the performance of a port-injected spark ignition engine (Brown and Ladommatos, 1991).

The flame ionisation detector cannot distinguish between different hydrocarbon species. However gas chromatography can be used for identifying particular hydrocarbon species in the exhaust gas. The gas sample and carrier gas are passed into the column – a long tube which contains a medium (liquid or solid) that tends to absorb the constituents in the sample. Since different molecules pass through the column at different rates, when the carrier gas leaves the column it contains the different molecules in discrete groups, and the different molecules can be identified by their residence time. The exit of the species from the column is detected by measuring the ionisation in a flame. The chromatograph is calibrated by injecting samples of known gases into the carrier gas.

With gas chromatography the output is a relative concentration against (elution) time. Consequently there is no explicit identification of the species. This can be resolved by calibrating the system to establish the elution times of particular species. A more sophisticated approach is to combine gas chromatography with mass spectrometry, since the mass spectrometer determines the mass/charge ratio of different components, from which it is possible to deduce what a particular species is. Lehrle et al. (1995) have applied these techniques to an engine exhaust and provide a good explanation of the underlying physics of the measuring system, and why it is advantageous to use a combination of gas chromatography and mass spectrometry.

14.4.3 Chemiluminescence

The chemiluminescence technique depends on the emission of light. Nitric oxide (NO) reacts with ozone (O_3) to produce nitrogen dioxide in an activated state (NO_2^*), which in due course can emit light as it reverts to its normal state:

$$NO + O_3 \rightarrow NO_2^* + O_2 \rightarrow NO_2 + O_2 + \text{photon}$$

The nitrogen dioxide can also be deactivated by a collision with another molecule. For example, the presence of 15 per cent carbon dioxide should alter the NO reading by less than 2 per cent. Ferguson (1986) shows that if

(a) the reactor is sufficiently large

(b) the ozone flow rate is steady and high compared with a steady sample flow

(c) the reactor is at a fixed temperature

then the light emitted is proportional to the concentration of nitric oxide in the sample stream.

Both nitric oxide and nitrogen dioxide can exist in the exhaust of an engine, and NO_x is used to denote the sum of the nitrogen oxides. Nitrogen dioxide can be measured by passing the sample through a heater that decomposes the nitrogen dioxide to nitric oxide. By switching the converter in and out of the sample line, the concentrations of NO and (NO + NO_2) can be found in the exhaust sample.

Figure 14.27 shows the arrangement of an NO_x analyser. The vacuum pump controls the pressure in the reaction chamber, and is responsible for drawing in the ozone and exhaust sample. The ozone is generated by an electrical discharge in oxygen at low pressure, and the flow of ozone is controlled by the oxygen supply pressure and the critical flow orifice (a short length of capillary tube). The sample can either by-pass or flow through the nitrogen dioxide converter. The sample flow rate is regulated by two critical flow orifices. The by-pass flow is drawn through by a sample pump. This arrangement ensures a high flow rate of sample gas, so as to minimise the instrument response to a change in NO_x concentration in the sample. The flow of sample into the reactor is controlled by the pressure differential across the critical flow orifice upstream of the NO_x converter; this pressure differential is controlled by a differential pressure regulator.

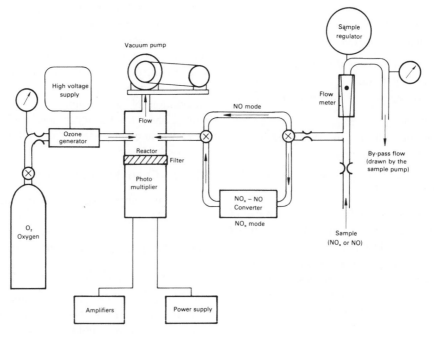

Figure 14.27
Key elements in a chemiluminescence NO_x analyser.

The light emission in the reactor is measured by a photomultiplier, and then amplified. In view of variations in gain that might occur with the photomultiplier, and in the other parameters that affect the light emission, it is essential to calibrate the NO_x analyser regularly. This is similar to calibrating a NDIR analyser, since calibration gases are used to set the zero and check the span.

As with an FID, a fast response chemiluminescence NO analyser (with a response of a few ms) can be constructed by miniaturising the reaction chamber and installing it close to the sampling point (Reavell *et al.*, 1997). When this is used for exhaust gas sampling the interpretation of the data is more straightforward than for unburnt hydrocarbons, since the NO composition varies less during the exhaust process. Ball *et al.* (1999) have shown that there is mostly only about a 5 per cent difference between a mass-weighted average of the signal and a simple time-based average.

14.4.4 Oxygen and air/fuel ratio analysers

Oxygen is obviously not an undesirable exhaust emission, but its measurement is very useful when evaluating the air/fuel ratio.

The lowest cost oxygen analysers are usually based around a galvanic cell. A typical galvanic cell comprises a PTFE membrane with a gold coating that acts as the cathode. Also immersed in the electrolyte (potassium chloride gel) is a silver or lead anode. A potential is applied across the electrodes. When the oxygen diffuses through the membrane it is reduced electrochemically, and a current flows that is proportional to the partial pressure of the oxygen in the sample. The galvanic cell also responds to other gases, and the gas of greatest significance with combustion is carbon dioxide. However, a 12 per cent carbon dioxide concentration would only give an output equivalent to that of 0.1 per cent oxygen.

Paramagnetic oxygen analysers are probably the most accurate. Paramagnetism occurs in oxygen because two of the electrons in the outer shell of the oxygen molecule are unpaired, and in consequence the molecule is attracted by a magnetic field. Other gas molecules can also be attracted (and conversely some are repelled) and most significant are NO_x. Nitric oxide has 43 per cent of the magnetic susceptibility of oxygen, and nitrogen dioxide has 28 per cent of the magnetic susceptibility. However, the NO_x concentration is usually an order of magnitude lower than the oxygen concentration so its effect can be ignored. (Furthermore, the effect on any atomic balance would be even smaller.) There are two main types of oxygen analyser using the paramagnetic principle: thermomagnetic analysers and magneto-dynamic analysers.

The principle of the thermomagnetic oxygen analyser is illustrated by figure 14.28. The heated filament in the cross-tube forms part of a Wheatstone bridge. Oxygen is attracted by the magnetic field, but when it is heated by the filament its paramagnetism is reduced, so the oxygen flows away from the magnetic field. The flow of gas cools the filament, thereby changing its resistance and leading to an imbalance on the Wheatstone bridge, which can be related to the oxygen level. Care must be taken with this type of analyser for several reasons:

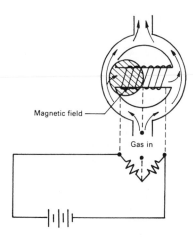

Figure 14.28
Basis of the thermomagnetic paramagnetic oxygen analyser.

(a) The heated filament is affected by changes in the transport properties of the gases in the sample.

(b) Hydrocarbons and other combustible gases can react on the filament causing changes in the filament temperature.

(c) The cross-tube must be mounted horizontally to avoid the occurrence of natural convection.

Magneto-dynamic analysers use a diamagnetic body (usually dumb-bell shaped) located in a strong non-uniform magnetic field; such an arrangement is shown in figure 14.29. The spheres are repelled by the magnetic field, and will reach equilibrium when the repulsion force is balanced by the torque from the fibre torsion suspension system. When the oxygen level in the cell changes, then the magnetic field will also change, and the dumb-bell will reach a new equilibrium position. Alternatively, a coil attached to the dumb-bell can be energised, and electro-magnetic feedback used to restore the dumb-bell to its datum position. The current required to maintain the dumb-bell in its datum position is directly related to the oxygen partial pressure of the sample. If the oxygen

analyser is operated at a constant pressure then the analyser can be calibrated directly in terms of the oxygen concentration. Unlike thermodynamic analysers, the magneto-dynamic analyser is not influenced by changes in the transport properties of the sample gas or the influence of hydrocarbons.

Exhaust gas oxygen sensors have already been described in chapter 4, section 4.7 (see figure 4.56), and these can be used as the basis of an oxygen analyser with a comparatively fast response. The two options are either to mount the sensor in the exhaust stream, or to take a sample of the exhaust to the analyser. If a sample of the exhaust is fully oxidised, it is also possible to deduce the air/fuel ratio.

In the Cussons Lamdascan, a sample of the exhaust is taken to an analyser. If a rich mixture is being burned, a controlled amount of oxygen is blended into the exhaust. This way, there is always oxygen present in the sample being analysed. The sample is passed through a heated catalyst to fully oxidise any partial products of combustion. For rich mixtures, one of several levels of dope air is added, so that there is a low level of oxygen in the diluted sample. If the hydrogen/carbon/oxygen ratio of the fuel is known, and also the quantity of any dope air, the equivalence ratio or air/fuel ratio can be deduced. Consider first the combustion of such a fuel as a weak mixture with an equivalence ratio ϕ:

$$CH_xO_y + \frac{(1 + x/4 - y/2)}{\phi}\left(O_2 + \frac{79}{21}N_2\right) \rightarrow$$
$$CO_2 + (x/2)H_2O + (1/\phi - 1)(1 + x/4 - y/2)O_2$$
$$+ 1/\phi(1 + x/4 - y/2)\frac{79}{21}N_2 \qquad (14.25)$$

Examination of the right-hand side of equation (14.25) shows that the oxygen level in the exhaust gases enables the equivalence ratio to be found. Since the system is maintained at a temperature for which the water is vaporised:

$$\%O_2 = \frac{21(1 - \phi)(1 + x/4 - y/2)}{21\phi + 21x\phi/2 + (100 - 21\phi)(1 + x/4 - y/2)} \times 100 \qquad (14.26)$$

for which ϕ is the only unknown variable.

When a rich mixture is burned, dope air is added, denoted here by $d(O_2 + 79/21\,N_2)$:

$$CH_xO_y + \frac{(1 + x/4 - y/2)}{\phi}(O_2 + 79/21\,N_2) \rightarrow$$

partial combustion products $+ d(O_2 + 79/21\,N_2)$
$$\rightarrow CO_2 + (x/2)H_2O + \{(1/\phi - 1)(1 + x/4 - y/2) + d\}O_2$$
$$+ \{(1 + x/4 - y/2) + d\}79/21\,N_2 \qquad (14.27)$$

As with equation (14.26), the equivalence ratio can be expressed in terms of the oxygen level in the exhaust and the fuel composition, but it is now also in terms of the dope air level (d):

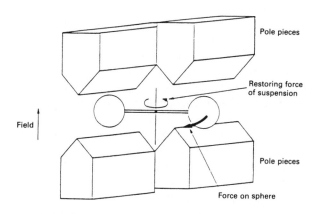

Figure 14.29
Basis of the magneto-dynamic paramagnetic oxygen analyser.

$$\%O_2 = \frac{(1/\phi - 1)(1 + x/4 - y/2) + d}{1 + x/2 + (1/\phi + 58/21)(1 + x/4 - y/2) + 100d/21} \times 100$$

$$(14.28)$$

from which ϕ can be evaluated if d is known. The flows of exhaust and dope air are controlled by flows through critical orifices, such that the ratio of molar flows is constant. The ratio of flows is evaluated by heating nitrogen to the same temperature as the exhaust stream, and measuring the level of the oxygen in the mixture of dope air and nitrogen:

$$N_2 + e(O_2 + 79/21\, N_2) \rightarrow eO_2 + (1 + 79e/21)N_2 \qquad (14.29)$$

The percentage of oxygen in this mixture can be measured, and is denoted here by f:

$$f\% = 100e/(1 + 100\, e/21) \qquad (14.30)$$

Equation (14.30) can be evaluated to give e, and this leads to a determination of d in equation (14.27):

$$d = e \times (\text{number of kmols of partial combustion products}) \qquad (14.31)$$

The partial combustion products will include nitrogen, water vapour, carbon dioxide, carbon monoxide, hydrogen and a small amount of oxygen. The oxygen level will be small, especially with regard to the other constituents, and it has to be assigned a value which is here taken to be zero. The partial products of combustion (from equation 14.27) are thus:

$$\rightarrow (a + b)CO_2 + (1 - a - b)CO + bH_2 + (x/2 - b)H_2O$$
$$+ 79/21\phi(1 + x/4 - y/2) \qquad (14.32)$$

Fortunately the temporary variables (a and b) in equation (14.32) cancel when the number of kmols are being evaluated. Equations (14.29), (14.30) and (14.31) enable d to be found in terms of known variables and ϕ. This result can then be substituted into equation (14.28) to give a solution for the equivalence ratio ϕ.

The oxygen sensor is calibrated by means of atmospheric air, and bottled nitrogen. The calibration is then checked by means of a nitrogen/oxygen calibration gas, with a low level of oxygen (as might be encountered in the exhaust). The performance of the catalyst and the dope air system can be checked by a calibration gas that represents the combustion products of a rich mixture, with for example 10 per cent carbon dioxide, 1 per cent carbon monoxide and 0.1 per cent propane in nitrogen. Such a reference gas is also useful in checking other analysers.

The Cussons Lamdascan is a unit that has a fast response (significantly less than a second) and is able to analyse combustion over a wide range of equivalence ratios (from about 0.3 to 3).

When the exhaust gas oxygen sensor is fitted in the exhaust system, there is a faster response (about 100 ms) but with a more limited range of equivalence ratios. The importance of operating the sensor at the correct temperature has already been discussed in chapter 4, section 4.7. Also described here was how the platinum acts as an electrode and as a catalyst to fully oxidise the exhaust products. It is self-evident how such an exhaust gas oxygen sensor operates with weak or stoichiometric air/fuel ratios. However, the sensor can also be contrived for use with mixtures slightly rich of stoichiometric (up to about an equivalence ratio of 1.4). As explained in chapter 4, near the end of section 4.7, by using a second zirconia sensor, oxygen can be either 'pumped' into or out of the exhaust gas sample. The detecting cell is kept at stoichiometric, and the amount of oxygen that has to be pumped can be measured, since it is proportional to the current flow in the pumping cell.

14.4.5 Exhaust smoke

A variety of systems has been developed for measuring the smoke level in diesel exhaust. Unfortunately, the systems respond to the particle size distributions in different ways, so that it is not possible to make comparisons between different measuring systems when either the operating condition or the engine is changed. Simple measurement systems either measure the obscuration of a light beam (for either the whole flow or part of the flow) or the fouling of a filter paper. An alternative approach described by Kittelson and Collings (1987) involves measuring the electrical charge associated with the smoke particles.

A very widely used system has been the Bosch Smokemeter, in which a controlled volume of exhaust is drawn through a filter paper. The change in the reflectance of the paper then corresponds to the smoke level. A value of zero is assigned to a clean filter paper, and a value of 10 is assigned to a piece of paper that reflects no light. The calibration of intermediate values can be checked by placing a perforated piece of non-reflecting paper over a filter paper. The output is in units of Bosch Smoke Units (BSU) which is also known as the Filter Smoke Number (FSN).

Exhaust particulates are defined as material that can be collected on a filter at specified conditions of flow and temperature. Under the E/ECE/324 Regulation No. 83, the filter face velocity should be set to a single value within the range of 20–80 cm/s, while the value is within 0–100 cm/s under the EPA Regulations which specify a temperature of $47°C \pm 5$ K (European legislation merely specifies an upper limit of 52°C). The filter face velocity has a slight influence on the filtration efficiency, and this is discussed in detail, along with details of the deposition mechanisms, by Hinds (1999). Filter measurements are now very challenging, since the EU PM limit corresponds to about 250 μg (allowing for dilution) on a filter that might weigh 100 mg. A balance is needed with a readability of 0.1 μg, and corrections even need to be made for buoyancy. The filters have to be weighed in a room with

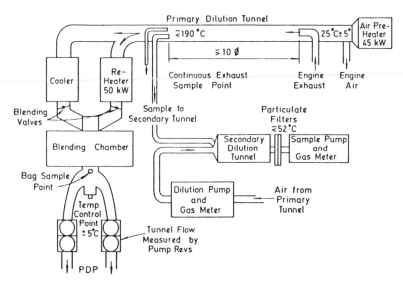

Figure 14.30
Schematic diagram of a Federal Test Procedure (FTP) emissions sampling system.

temperature and humidity control; the permitted variation in humidity (from 37 to 53 per cent) can lead to a change in mass of 7 μg – Andersson *et al.* (2007). Since it is impractical to pass the whole of the exhaust stream through a filter, a sample of the exhaust is drawn off and cooled by dilution with air. This leads to a complex system, such as that illustrated by figure 14.30, which enables evaluation of the fraction of the exhaust being filtered. By weighing the filter before and after use, the mass of the particulates is evaluated. The particulates consist of carbon particles with adsorbed and condensed hydrocarbons on their surface, sulphur-containing material (such as condensed sulphuric acid) and metallic ash that probably originates from the lubricant. In modern engines with low sulphur fuel and minimal oil consumption, the vast majority of the particulates comprises organic material and carbon.

A simpler way of estimating the mass of particulates has been proposed by Greeves and Wang (1981). They argued that as particulates comprise soot and unburned hydrocarbons, it should be possible to deduce the mass of particulates from a smoke reading and unburned hydrocarbon reading. Greeves and Wang tested direct and indirect injection engines, and proposed the following correlation:

$$\text{particulates (g/m}^3) = 1.024 \times \text{ smoke (g/m}^3)$$
$$+ 0.505 \times \text{HC (g/m}^3) \qquad (14.33)$$

Conversion of an FID reading of unburned hydrocarbons to g/m^3 is straightforward, but the smoke reading cannot be calculated from a Smokemeter reading. Greeves and Wang (1981) used the results of Fosberry and Gee (1961), whose correlation for the Bosch Smoke Units is presented in figure 14.31.

14.4.6 Exhaust particulate matter measurement

Particulate matter in an engine exhaust is an example of an aerosol, and the measurement of aerosols is much more complex than the measurement of material in the gaseous phase. Aerosol measurement is covered in great detail by Hinds (1999), who includes all the underlying physics, and an even more comprehensive treatment has been produced by Baron and Willeke (2001). Eastwood (2008) also discusses the measurement of particulate matter, along with a thorough discussion of the sources and nature of the particulate matter (which have been discussed earlier in this book in chapter 3, section 3.8.5). The scope of the discussion on sample collection, particle loss mechanisms and analysis techniques is intended to provide an overview, as it is limited to identifying the mechanisms and their likely significance; for a detailed discussion of the underlying science it is necessary to consult specialist texts such as Hinds (1999), Baron and Willeke (2001) or Eastwood (2008).

Sampling of exhaust particulate matter

The first requirement in any measurement is to obtain a representative sample, and for it to then be transferred without loss to a measurement device. These two issues are discussed first, before considering different aerosol/particulate matter measurement techniques.

For a sample to be representative, it must accurately reflect the characteristics of the bulk flow from which it was taken. As most probes are single-point samplers, the first requirement is that the bulk flow is homogeneous. In addition the sample should be extracted isokinetically – the flow velocity in the sampling system needs to match the bulk flow (Hinds, 1999). The requirement for isokinetic sampling is because the

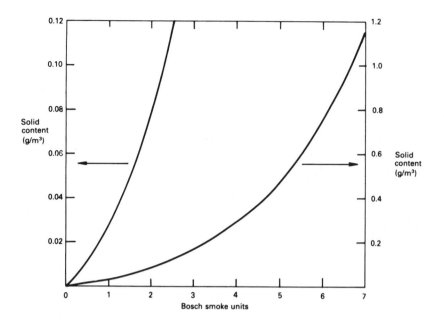

Figure 14.31
Relation between the Bosch Smoke Number and the solid content in the exhaust flow (Fosberry and Gee, 1961).

particles have mass, and if the flow has to move around an object, such as a sample probe, then some particles can have sufficient momentum to cause them to deviate from their original streamlines. Consider a probe facing into the flow that it is sampling:

(i) If the velocity at the sample probe entrance is less than the bulk free stream velocity (sub-isokinetic sampling), the streamlines diverge around the probe and particles with sufficient momentum will be unable to negotiate the flow path around the probe, and the probe will over-sample: sufficiently large particles have been entrained into the sample that otherwise should not have.

(ii) If the probe velocity is greater than the bulk flow, streamlines are required to converge and some particles with sufficient momentum are unable to follow the flow, and this causes under-sampling.

Eastwood (2008) notes that isokinetic sampling is often considered unnecessary for particles that are smaller than 1000 nm, but Eastwood reviews research that shows this to be optimistic, and also argues that these issues are likely to be most significant when sampling from an engine exhaust (as opposed to a dilution tunnel). A mass loss of order 10 per cent can occur with 500 nm particles, but the corollary of this is that the number loss will be negligible (see figure 3.25).

A directly related question is – what happens if the probe is not facing into the flow? Figure 14.32 shows results from experiments with a Spray-Guided Direct Injection (SGDI) gasoline engine which has a bi-log normal particle size distribution, in which the accumulation mode has a geometric mean of about 100 nm. The sampling probe could be rotated,

with a datum position when the probe faces into the flow. Once allowance is made for the inevitable variability of particulate matter (PM) measurements, the results show that probe orientation did not affect the sample for either the particle number (Pn) or particle mass (Pm). (There should of course be no difference between the measurements at, for example, 90° and 270°.)

It is also important, once a sample has been taken, that it is delivered to a measurement device without loss or change. Particle losses arise owing to inertial impaction, diffusion, gravitational settling, electrostatic attraction and thermophoresis. The magnitude of each depends on the sampling system, and the residence time. Change can occur as a result of nucleation and/or condensation onto existing particles and coagulation – even if particle mass is conserved the number distribution can change. All of these effects can be reduced by the use of dilution, often with heated air. Dilution:

(i) reduces the concentration of water and hydrocarbons in the sample, making nucleation and condensation less likely (even when diluting with cold gas, the reduction in partial pressure dominates the reduction in saturation pressure);

(ii) can be used to reduce the concentration into the measurable range of an instrument, or extend the interval between needing to clean systems;

(iii) reduces the PM concentration, so that the rate of coagulation is reduced (with sufficient dilution, the PM number concentration and size distribution can be considered to be 'frozen', so the effect of sample residence time is reduced).

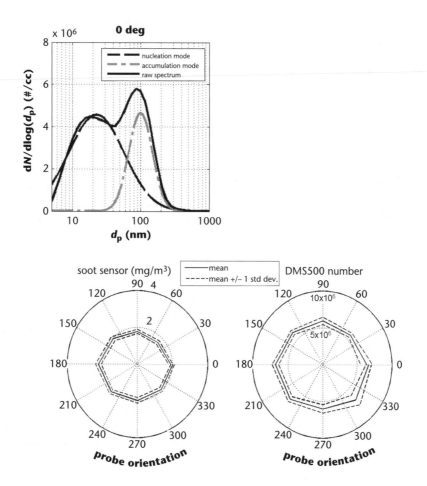

Figure 14.32

The effect of probe rotation (the probe facing into the flow is 0°) when isokinetically sampling the exhaust of an SGDI gasoline engine. The accumulation mode has a geometric mean size of about 100 nm (Braisher, 2010).

Dilution can be achieved by using a mass flow controller or some form of positive displacement pump, an ejector diluter or a rotating disc diluter. The ejector diluter (figure 14.33) uses high-pressure air and a critical flow orifice to entrain the exhaust sample, and this helps to reduce the effect of exhaust system pressure pulsations on the dilution ratio. The exhaust flow to atmosphere also ensures that the flow for analysis is at a constant pressure.

Figure 14.34 shows a rotating disc diluter for dilution ratios in the range 1:20 to 1:500. Rotating disc diluters should be used for secondary dilution, since otherwise there is a risk of condensation. The concentrated sample enters via the inlet, and passes through the rotating disc in one of the many axial holes in the disc. The sample is then filtered, to provide particle-free diluent gas, and the flow rate is measured using a mass flowmeter. Particle-free gas then passes through the rotating disc again, and into the main instrument. The plates on both sides of the rotating disc have ports that are wider than the hole diameter in the disc, so that there is continuous

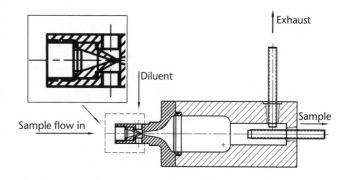

Figure 14.33

Ejector diluter that provides a diluted sample flow at ambient pressure for subsequent analysis.

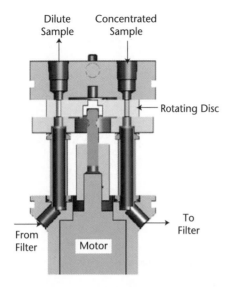

Figure 14.34
Cambustion rotating disc diluter for dilution ratios in the range 1:20 to 1:500, with acknowledgement to Cambustion Ltd.

flow. If the rotating disc is stationary, then all of the incoming sample is filtered, and no particles will reach the instrument. When the disc is rotated, the volume of gas (containing particles) that is trapped in the holes in the disc is introduced downstream of the filtered sample. The volume flow rate of sample gas which by-passes the filter is thus proportional to the disc rotation speed, the number of holes in the disc and the volume of each hole, all of which will be accurately known or measured.

Particle loss mechanisms

It can be inferred from figure 14.32 that exhaust particulates are small enough to move with the flow, so that particle loss by impaction is usually negligible. Hinds (1999) devotes a chapter to the way particles follow a flow, for which the Stokes number is central. The Stokes number is a measure of the ability of a particle to follow gas streamlines, and is the ratio of the so-called stopping distance of a particle and a characteristic system dimension (such as pipe diameter). For sampling velocities and particle sizes encountered with engines, it is usually safe to assume that the effect of impaction is small.

Thermophoresis is defined as the motion of a particle due to asymmetrical forces that arise in a temperature gradient. When a particle is located in a temperature gradient, the high-temperature side will be subject to more vigorous collisions from the gas molecules than is the cooler side, and the net effect will be for the particle to move down the temperature gradient. Temperature gradients almost inevitably exist in the sampling system piping.

Measurement techniques for particulate matter

There are a number of techniques for particle sizing and mass determination, such as Scanning Mobility Particle Sizing (SMPS), Differential Mobility Spectrometer (DMS) and Electrical Low Pressure Impactor (ELPI). The ELPI works by charging the particles, and then accelerating them through orifices of decreasing size. There is inertial separation onto the thirteen impactor stages that cover particles down to 0.030 μm. The SMPS and DMS can both measure smaller particles and are discussed in more detail.

Scanning Mobility Particle Sizing (SMPS) is capable of measuring particles in the size range 0.01–1 μm. An inertial impactor is used to remove particles above 1 μm, and part of the sample is then charged. The sample next flows through a Differential Mobility Analyser (DMA) which classifies particles on the basis of their electrical mobility. Electrical mobility is the ratio of the electrostatic force to the aerodynamic drag. The positively charged particles are attracted to the central rod, and the subsequent trajectories are determined by the size of the particles; the central rod voltage is varied in the range of 0–11 kV, so that a narrow electrical mobility range of particles is allowed to exit through a slit, where they enter a Condensation Particle Counter (CPC). Each of these will now be described in more detail.

Figure 14.35 shows the operating principle of a Differential Mobility Analyser (DMA). The charged and diluted exhaust gas (the polydisperse aerosol, meaning that all sizes of particle are present) enters adjacent to the outer electrode. There is a co-flow of filtered air (the sheath flow), and the velocities of the two flows are matched so as to minimise mixing; also to minimise mixing, the dividing wall between the two flows has

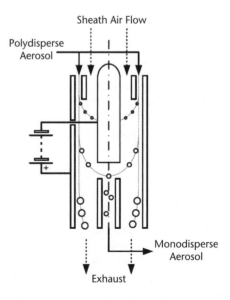

Figure 14.35
A Differential Mobility Analyser (DMA).

a knife-edge. Particles that are large will not be deflected much and will exit with the sheath flow (the exhaust flow in figure 14.35), while very small particles will be attracted onto the central electrode. However, particles of a specific size range (determined by the central electrode voltage) will exit through a slot that provides the monodisperse aerosol output. If a bipolar charger is used then negatively charged particles will be collected on the outer electrode.

Only particles above about 50 nm can be detected optically, but a Condensation Particle Counter (CPC) can be used to 'grow' the particles. Figure 14.36 shows how the particle-laden air becomes saturated with butanol vapour at about 25°C, before being cooled to about 8°C. When it is cooled, the vapour becomes supersaturated and it will condense onto the particles, since they act as nucleation sites. An alcohol is used in preference to water as the particles are hydrophobic (owing to the presence of condensed hydrocarbons). Particles as small as a few nm can be detected in a CPC if the level of supersaturation is high enough. The particles are counted as they pass through the measuring volume and scattered light is detected by the photodetector, and concentrations in the range 10^{-4}–$10^{7}/cm^3$ can be measured (Hinds, 1999).

The Differential Mobility Spectrometer (DMS) is a real-time instrument for measuring size-resolved particle number concentration, using the classifier column shown in figure 14.37; it is described in detail by Reavell *et al.* (2002). Like a DMA, it classifies particles on the basis of their electrical mobility, from which an equivalent diameter is inferred.

Sample from the engine exhaust is first mixed with a metered flow of filtered dilution air. This reduces the water content of the sample gas to prevent condensation inside the instrument, and reduces high concentrations to slow particle agglomeration processes in the heated sampling line. This dilution stage is typically a factor of 5:1. A diluted sample of aerosol first passes through a cyclone which acts as a cut-off filter to remove particles above 1000 nm. The sample gas passes through a critical flow orifice, which reduces the pressure from ambient to about 250 mbar. This reduced pressure helps to maintain a short transit time through the sample line, and reduce sample line losses. (This assists in maintaining a 10–90 per cent response time of 300 ms with a 5-metre heated line, or 200 ms without.) The

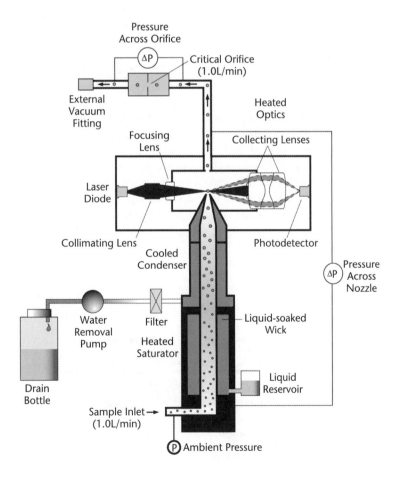

Figure 14.36
Schematic diagram of a Condensation Particle Counter (CPC), with acknowledgement to TSI, Inc.

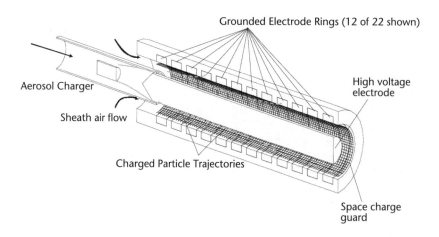

Figure 14.37
Classifier column of the Cambustion Differential Mobility Spectrometer, with acknowledgement to Cambustion Ltd.

sample line is fitted with an electrically conductive tube (to minimise losses of charged particles) and is electrically heated to reduce hydrocarbon condensation. Additional dilution can be provided by a rotating disc diluter (see figure 14.34).

The aerosol then enters an electrostatic precipitator (to remove small charged particles) and a unipolar charger producing positive ions that brings the particles to a known charge state. The charged particles enter the classifying column adjacent to the inner electrode, after which they are subjected to a radial electric field. The positively charged particles are attracted towards the concentric grounded outer electrodes comprising 22 electrometer rings. The current due to positively charged particles landing on the rings is measured and is proportional to the number of particles impacting upon each ring. The larger particles, with a lower electrical mobility, land further down the column, owing to the increased time taken to traverse the radial gap between the electrodes. From the current at each ring and its position, the inversion algorithm can provide size-resolved number concentration.

The Photo-Acoustic Soot Sensor (PASS) gives a real-time measurement of soot mass concentration in mg/m^3; a measuring cell is shown in figure 14.38. When light is absorbed by the soot, some of the energy is subsequently dissipated as heat that flows to the surrounding gas, causing a localised increase in pressure (Arnott *et al.*, 1999). By modulating the light source, the resulting modulated pressure changes, or sound wave, can be detected by a microphone. By modulating at such a frequency as to cause resonance, the acoustic signal can be amplified by the build-up of a standing wave (Arnott *et al.*, 1999). This microphone signal is linearly related to the soot mass concentration in the measuring cell and has a resolution of $50 \ \mu g/m^3$ – Schindler *et al.* (2004). The laser wavelength is chosen so that its energy is absorbed by the soot, but not by any possible gaseous species.

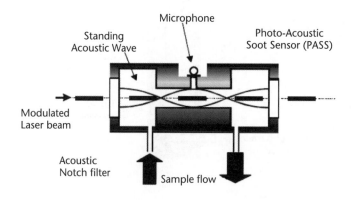

Figure 14.38
The measuring cell of the AVL Photo-Acoustic Soot Sensor (PASS), adapted from Schindler *et al.* (2004).

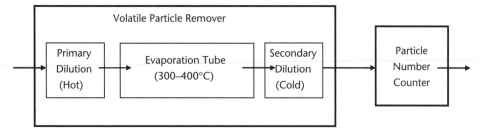

Figure 14.39
Volatile Particle Remover and Particle Number Counter as specified by the PMP protocol in UN-ECE Regulation No. 83.

Particle counting in emissions tests

A protocol for measuring particle number during drive cycles has been developed by a United Nations Economic Commission for Europe (UN-ECE) sponsored Particulate Measurement Programme (PMP) – Andersson *et al.* (2007). The motivation for this is the increasing difficulty with ever smaller mass measurements (see section 14.4.5), and the concerns over health effects associated with particle number. A schematic of the measurement system is shown in figure 14.39.

The sample conditioning is performed using a Volatile Particle Remover (VPR), which converts volatile material from the particulate phase into the gas phase, and is placed upstream of a Particle Number Counter (PNC). The VPR consists of an initial stage of hot dilution, an Evaporation Tube and a second stage of cold dilution. The initial hot dilution has to be capable of providing a dilution factor range of 10–200, using filtered air, with an exit temperature of 150±5°C. The secondary dilution has to be capable of maintaining a dilution factor in the range of 10–30, when supplied with filtered dilution air. The secondary dilution is selected so that the sample outlet temperature is below 35°C and the particle concentration is less than the upper threshold of the single particle count mode of the particle number counter. The evaporation tube (240 mm long and 6 mm diameter), located between the primary and secondary dilution stages, needs to have a wall temperature maintained at 300±20°C with a residence time of 0.2–0.5 s.

The VPR has to be capable of vaporising > 99 per cent of a concentration of $10^4/cm^3$ of tetracontane particles, 30 nm in diameter. Inevitably there are particle losses within the VPR and these are characterised and corrected for through Particle Concentration Reduction Factors (PCRFs). The PCRFs (defined as the ratio of upstream-to-downstream particle number concentrations) are evaluated using solid particles of 30, 50 and 100 nm diameter. The arithmetic mean of the PCRFs at 30, 50 and 100 nm diameters is calculated for each possible dilution ratio and is then applied to each particle number concentration result.

The Particle Number Counter (PNC) is based on a Condensation Particle Counter (CPC, see figure 14.36), and is required to have a counting accuracy of ±10 per cent from

Table 14.3 *Particle number counter (PNC) required counting efficiencies*

Particle size (nm)	Counting efficiency (per cent)
23	50 ± 12
41	> 90

$1/cm^3$ to the upper threshold of the single particle count mode. The required counting efficiencies are given in table 14.3.

If a volatile particle remover is used and the size distribution is known, then a 'digital filter' (f) can be created to achieve the counting efficiencies defined in table 14.3:

$$d_p \geq 14, f = 1 - \exp\left[-3.54\left(\frac{d_p - 14}{40}\right)^{1.09}\right]$$

$$d_p < 14, f = 0$$

It has also been shown that the particle number count in the Cambustion DMS 500 accumulation-mode fit gives a good approximation to the measurements obtained with a PMP compliant measurement system (Braisher *et al.*, 2010).

14.4.7 Determination of EGR and exhaust residual (ER) levels

The exhaust gas recirculation (EGR) level can be defined volumetrically, on a molar basis, or gravimetrically. The volumetric definition is simpler, as the EGR level is defined as the percentage reduction in volume flow rate of air at a fixed operating point:

$$\%\text{EGR} = \frac{\dot{V}_0 - \dot{V}_e}{\dot{V}_0} \times 100 \qquad (14.34)$$

where $\dot{V}_0$ = volume flow rate of air with no EGR
$\dot{V}_e$ = volume flow rate of air with EGR.

With a diesel engine the fixed operating point will correspond to the same speed, and either the same fuelling level or the same torque. With spark ignition engines the speed should be fixed, and either the torque is allowed to change or changes will have to be made to the throttling and/or fuelling level.

Alternatively the EGR level can be deduced, by comparing the concentration of a particular species in the exhaust with its concentration in the inlet manifold. It is usual to measure the carbon dioxide, as it is present in the most significant quantities, and it can be measured accurately. The mixture in the manifold is assumed to be well mixed, and a vacuum pump is needed to draw the sample from the inlet manifold (since spark ignition engine manifolds can be operating at less than 0.2 bar absolute). There are two definitions of EGR on the molar basis, with the denominator representing either the number of kilomoles of mixture being induced (14.35a), or the number of kilomoles of fuel and air being induced (14.35b):

$$\%\mathrm{EGR} = \frac{(CO_2)_m}{(CO_2)_e} \times 100 \qquad (14.35a)$$

$$\%\mathrm{EGR} = \frac{(CO_2)_m}{(CO_2)_e - (CO_2)_m} \times 100 \qquad (14.35b)$$

where　$(CO_2)_m$ = percentage of carbon dioxide in the inlet manifold

$(CO_2)_e$ = percentage of carbon dioxide in the exhaust manifold.

(*Note*: if there is a significant level of carbon dioxide in the air, then this should be subtracted from both values.)

If the temperature and composition of the exhaust just prior to mixing in the inlet manifold, and the temperature and composition of the air (and fuel) are both known, then it is possible to convert from equation (14.34) to (14.35), and vice versa. Such a calculation would assume isenthalpic mixing of the gas streams (that is, with no external heat transfer).

The volumetric definition of EGR is probably best suited to diesel engines, since at light loads the concentrations of carbon dioxide would be low and equation (14.35) would become ill-conditioned. An alternative approach that can be suitable for diesel engines is to measure the oxygen level in the inlet manifold and the exhaust manifold. In contrast, for spark ignition engines in which the throttle position might be varied and the levels of carbon dioxide in the exhaust will invariably be about 10 per cent, then equation (14.35) is more useful.

Since a gravimetric definition of EGR could also be used, it is essential to state or establish which method of measuring EGR has been used, and how it has been defined.

The exhaust residuals are those products of combustion that are not displaced from the cylinder during the gas exchange processes. They are a consequence of the clearance volume at tdc; the relative pressures in the inlet manifold, cylinder and exhaust manifold; and the valve timing. High levels of exhaust residuals will occur when there is a large clearance volume, or a wide valve overlap, or a low inlet manifold pressure (Toda *et al.*, 1976). This has also been referred to in chapter 7, section 7.4.1.

The exhaust residual fraction in the trapped charge can be deduced by assuming perfect mixing of the residuals with the incoming charge, and then extracting a sample during the compression process. Many researchers have developed their own high-speed gas sampling valve, and a typical example has been described by Yates (1988). Sampling valves are also available commercially, and the minimum sampling duration is about 1 ms.

In the absence of exhaust gas recirculation, the residual fraction on a molar basis (rf_M) is given by

$$rf_M = \frac{(CO_2)_c}{(CO_2)_e} \qquad (14.36)$$

where $(CO_2)_c$ = percentage of carbon dioxide from the sampling valve.

To convert equation (14.36) to a mass fraction, it is necessary to know the mean molecular mass of the cylinder contents and the exhaust gas. The mean molecular mass can be calculated once the composition is known, and this can be found either by measurement, or, if the air/fuel ratio is known, by assuming idealised combustion. Idealised combustion is illustrated in chapter 3 by examples 3.1, 3.4 and 3.6.

When EGR is being used, some of the carbon dioxide trapped in the cylinder prior to combustion will be a consequence of the exhaust gas recirculation. As the cylinder contents comprise both residuals and the gas drawn in from the inlet manifold, and the carbon dioxide level in each of these is known, it can be shown that

$$rf_M = \frac{(CO_2)_c - (CO_2)_m}{(CO_2)_e - (CO_2)_m} \qquad (14.37)$$

As with equation (14.36), so equation (14.37) could be converted to a mass fraction if the molar composition of the exhaust and induction gases are known.

Sampling valves can also be used to extract samples during or after combustion from discrete sites within the clearance volume. Such measurements are particularly useful in gaining insights into the diesel combustion process (Whitehouse, 1987).

14.4.8　Determination of the air/fuel ratio from exhaust emissions

The methods used in calculating the air/fuel ratio from the exhaust emissions have already been introduced in chapter 3 by example 3.2. The generalised combustion of an oxygenate fuel is given by

$$C_xH_yO_z + \frac{(x + y/4 - z/2)}{\phi}\left(O_2 + \frac{79}{21}N_2\right) \to aCO + bCO_2$$
$$+ cH_2O + dH_2 + eO_2 + fC_mH_n + gNO_q + hN_2$$

$$(14.38)$$

In general the emissions measurements do not include hydrogen or nitrogen, and are made on a dry basis. Equation (14.38) is most conveniently solved if it is assumed that there are sufficient reactants to produce 100 kmol of *dry products*, since *a* will correspond to the measured percentage of carbon monoxide and so on (that is why there are assumed to be *x* kmol of carbon atoms rather than 1 kmol).

When the fuel composition is known (as the ratio of *x:y* and *x:z*), then there are four atomic balances and four unknowns, *x* and ϕ, and the concentration of the hydrogen and water vapour in the products.

As the nitrogen is not usually measured but found by difference, there are only in fact three independent equations. However, as the ratio of fuel to air is being evaluated (the variable *x* could be eliminated in the ratio of fuel to air) there are only three unknowns: ϕ, *c* and *d*. Unfortunately, the solution of these equations is ill-conditioned, and the usual approach is to make assumptions about the amount of hydrogen present.

This can be illustrated by considering a carbon balance and the nitrogen balance:

C balance: $\quad x = a + b + f \times m \quad$ (14.39)

where $f \times m = $ (ppm HC) $\times 10^{-4}$.

N$_2$ balance: $\quad \dfrac{79}{21\phi}(x + y/4 - z/2) = h \quad$ (14.40)

where *a* is the percentage of carbon monoxide etc.

Equation (14.39) allows *x* to be evaluated, so this can be substituted into equation (14.40).

The number of kmols of nitrogen can be found by difference, if assumptions are made about the number of kmols of hydrogen (*c*). The options are

(i) To neglect the hydrogen – acceptable for weak mixtures.

(ii) To make assumptions about the relative quantities of CO and H$_2$ (for example, from data such as those in figure 3.17).

(iii) To consider the water–gas equilibrium.

The normal approach is to consider the water–gas equilibrium, and to assume a typical value. Spindt (1965) adopted a value of 3.5, and this has been used by others subsequently:

$$bCO_2 + dH_2 \rightleftharpoons aCO + cH_2O$$
$$3.5 = \frac{a \times c}{b \times d} \quad (14.41)$$

The other atomic balances (hydrogen, oxygen and nitrogen) can also be used to determine the equivalence ratio, and this provides a means of checking the quality of the emissions measurements, and the assumptions about the hydrogen level. The solution of these equations is straightforward, but

lengthy and tedious, so it is best suited to the use of charts or a computer program.

A series of charts for different composition hydrocarbons has been prepared by Eltinge (1968), and the chart for a fuel with a 1.9 H : C ratio (typical of gasoline) is shown in figure 14.40.

The use of the chart is illustrated by the inset in figure 14.40 which considers an exhaust comprising 1 per cent CO, 2 per cent O$_2$ and 13.0 per cent CO$_2$. If the emissions measurements were self-consistent (not necessarily the same as being correct) and the combustion was complete, then the triangle formed by the iso-composition lines would collapse into a point. In other words, the larger the triangle, the greater the inconsistency in the emissions measurements.

The parameter S_x is meant to indicate the level of fuel maldistribution in a multi-cylinder engine. Consider a two-cylinder engine which is operating on a stoichiometric mixture overall, but one cylinder is operating with an equivalence ratio of 0.9 and the other with 1.11. The result of averaging these two exhaust streams will be a higher level of carbon monoxide and oxygen (but a lower level of carbon dioxide) than if both cylinders were operating with a stoichiometric mixture. Examination of figure 14.40 shows that this will lead to a higher value of S_x.

The alternative to using charts is to employ a computer program, and this is particularly convenient when the exhaust gas analysers are linked directly to a computer to undertake both data analysis and logging. Such a program can also be used to conduct a sensitivity analysis, in which the effect of possible measurement errors on the computed air/fuel ratio can be analysed. The results of a sensitivity analysis are shown in figure 14.41 (Stone, 1989a).

Figure 14.41 has used hypothetical emissions data that have been evaluated for a hydrocarbon fuel (H:C ratio of 1.8) at a specified air/fuel ratio. The effect of a +1 percentage point error in each of the emissions (CO$_2$, CO and O$_2$) has been examined in turn, and the error in the computed air/fuel ratio has then been evaluated and plotted against the specified air/fuel ratio. AFR1 has been evaluated by the oxygen/hydrogen atomic balance, while AFR2 has been evaluated by the carbon balance (equations 14.39 and 14.40). Figure 14.41 shows that AFR2, the carbon balance, is more likely to give reliable results, and that the computed air/fuel ratio is much more sensitive to measurement errors when the air/fuel ratio is significantly weak – such as might occur with diesel engines. The data used in preparing figure 14.41 show that in the equivalence ratio range of $0.7 < \phi < 1.5$, a +1 percentage point error in any of the emissions measurements leads to less than a 1 per cent error in the equivalence ratio.

The computer program can also be used to investigate the significance of uncertainty in the fuel composition. Such an exercise has been summarised in table 14.4 for a fuel with a H:C ratio of 1.8 for a range of air/fuel ratios.

Inspection of the data in table 14.4 shows that in the region

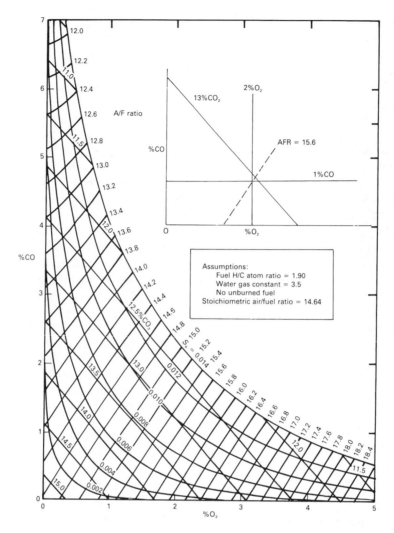

Figure 14.40
Exhaust emissions chart for evaluating the air/fuel ratio and maldistribution parameter (S_x) for a typical gasoline (adapted from Eltinge, 1968).

of stoichiometry (as would be encountered with homogeneous charge spark ignition engines), the effect of uncertainty in fuel composition on the equivalence ratio is negligible. For the H:C ratio range of 1.6–2.0, there would be an error of less than 1 per cent in the computed equivalence ratio. Thus, figure 14.40 can be used for a wide range of gasolines, and the results will be most accurate if expressed as an equivalence ratio.

For the weaker mixture, as might be encountered with a diesel engine, table 14.4 demonstrates that the results would be expressed more accurately as an air/fuel ratio when there

Table 14.4 *Sensitivity of computed air/fuel ratio and equivalence ratio (using the carbon balance) to uncertainty in the fuel composition (actual composition 1.8 H:C)*

AFR	ϕ	$AFR_{1.6}$ (assuming 1.6 H:C)	$\phi_{1.6}$	$AFR_{2.0}$ (assuming 2 H:C)	$\phi_{2.0}$
11.05	1.313	10.83	1.310	11.26	1.314
20.02	0.725	19.80	0.718	20.22	0.732
30.02	0.483	29.95	0.475	30.08	0.492
40.03	0.362	40.09	0.355	39.92	0.371
50.03	0.290	50.28	0.283	49.83	0.297
60.04	0.242	60.37	0.236	59.63	0.248
80.07	0.181	80.64	0.176	79.32	0.187

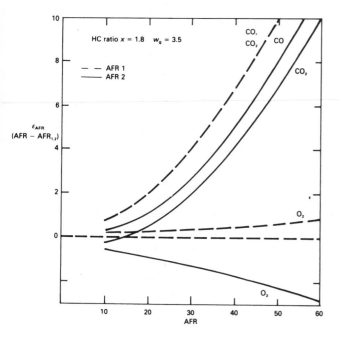

Figure 14.41
The errors in the computed air/fuel ratios caused by 0.01 errors on each of the principal emissions (CO_2, CO and O_2).

is uncertainty over the fuel composition. For air/fuel ratios in the range of 20–80, the uncertainty in the fuel composition led to an error of less than 1 per cent in the calculated air/fuel ratio. Since it is possible to calculate the air/fuel ratio by two independent ways when the fuel composition is known, this implies that if the fuel composition is unknown, then it is possible to calculate the air/fuel ratio and the fuel composition. Table 14.4 has shown that the computed air/fuel ratio is insensitive to the assumed composition of the fuel. An unfortunate corollary of this is that when emissions measurements are used to evaluate the air/fuel ratio and fuel composition, the equations for determining the fuel composition are ill-conditioned and the fuel composition result will be prone to large errors. If the fuel composition is not known, it is best to guess the composition and express the results as an equivalence ratio in the region of stoichiometry, or as an air/fuel ratio for weak mixtures.

A sensitivity analysis also shows that with weak mixtures the value of the water gas constant has little effect on the computed air/fuel ratio. However, for rich mixtures, the water–gas constant has a significant and almost equal effect on the air/fuel ratios calculated from the carbon balance and the hydrogen/oxygen balance. Thus a comparison of these two computed air/fuel ratios will not indicate whether the water–gas equilibrium constant has been assigned the correct value. However, examination of wet exhaust emissions data presented by Heywood (1988) shows that the water–gas constant remains within the range of 3–4, and this uncertainty would cause an error of less than ± 0.25 per cent ϕ.

14.5   Computer-based combustion analysis

14.5.1   Introduction

The increasing power and falling costs of personal computer systems have meant that very satisfactory systems can be bought for a modest amount. There is, of course, a need for a high-speed data acquisition card that allows for the conversion of analogue voltages to digital signals. The cost of such a computer-based data acquisition system is now less than the cost of a piezoelectric pressure transducer and its amplifier.

There are essentially two types of combustion analysis undertaken:

(i)   burn rate analysis – usually associated with spark ignition engines, and

(ii)   heat release analysis – usually associated with diesel engines.

These will be described shortly, but first it will be useful to review a computer-based data acquisition system. A possible specification is in table 14.5.

When selecting a data acquisition card, it is necessary to decide on:

(i)   the resolution and accuracy (for example, 10 bit)

(ii)   the number of channels

(iii)   the maximum sampling rate

(iv)   how much data are to be collected.

Firstly, the lower the resolution of the card, the lower the cost for a given sampling rate. An analogue-to-digital converter (ADC) with 8 bits might appear adequate, as this will give a resolution of 1 part in 256. In other words, the ADC will cause an uncertainty of about 0.4 per cent, and this is comparable with the accuracy of a piezoelectric pressure transducer. However, to achieve this ADC accuracy it is necessary to use the full dynamic voltage range of the input, and this may not be convenient. Furthermore, it is desirable to minimise any sources of additional errors. In practice, 16- or 12-bit resolution ADC cards are most likely to be used.

Secondly, it is necessary to identify how many channels might need to be logged. It is possible to superimpose a marker voltage (at say bdc) on the cylinder pressure signal,

Table 14.5   *The specification of a computer-based data acquisition system*

Channels	1, 2, 4, 8 or 16
Multiplexing overhead	1 μs
Resolution	12 bits
Input voltage range	± 10 V
Max. sampling rate	1 MHz

and then use software later to separate the marker voltage. However, it is simpler to use a separate channel for the reference flag. Other channels might be wanted for measuring fuel line pressure or injector needle lift. Furthermore, measurements might be wanted from one or more cylinders. For this type of application it is usual to have a single ADC, and to multiplex the inputs so that each is read in turn. With multiplexing, many systems introduce a multiplexing over-head (the time to switch from one channel to the next). Thirdly, the maximum sampling rate has to be identified. This is determined by the number of channels being sampled, and the frequency of the signals that are to be recorded. Sampling theory determines that the sampling rate should be twice the highest frequency that is to be recorded from the signal. Consider the cylinder pressure transducer which might have a response that is within 1 per cent of linear to a frequency of 10 kHz. This might suggest that the maximum useful sampling rate is 20 kHz. However, if a phenomenon such as knock is being investigated, then a higher sampling rate will be wanted, even though the non-linearity of the transducer is becoming significant. In general, the sampling will be controlled by a shaft encoder on the engine that presents a signal to the 'external clock' input of the data acquisition system. This arrangement has the advantage that the angular position of each reading will be known. Thus the sampling rate will be influenced by the engine speed and the crank angle resolution that is required. For example, the injector needle lift might be wanted with a $\frac{1}{4}°$ resolution to determine the start of injection. Thus

$$\text{Sampling rate (sample/s)} = 6 \times \text{engine speed (rpm)}$$
$$\times \text{ number of readings/degree}$$
$$\times \text{ number of channels enabled}$$
$$(14.42)$$

For example, with a diesel engine operating at 4000 rpm, with four channels enabled and readings taken every $\frac{1}{4}°$, there would be 384k sample/s.

Finally, the amount of data to be collected has to be identified, as this influences the maximum sampling rate. Large quantities of data need to be written to some form of disk, and this is slower than writing to RAM (random access memory). Consider measurements of cycle-by-cycle variation in a four-stroke spark ignition engine. The requirement for memory is

$$\text{Memory} = \text{number of cycles} \times \text{number of channels} \times$$
$$\text{number of readings/degrees} \times$$
$$(180 \times \text{number of strokes}) \times$$
$$\text{number of bytes/sample} \quad (14.43)$$

For example, if 300 cycles of a four-stroke engine are to be recorded from two channels, with readings every $\frac{1}{2}°$ and 2

bytes required for each reading (for a 12-bit reading), then about 3.4 MB of memory will be required.

There are other issues to be considered with computer-based data acquisition systems. Firstly, when channels are being multiplexed, the channels are not all being read at the same time. For example, table 14.5 shows that the system has a delay of 1 microsecond between reading successive channels. This becomes significant with high sampling rates, but it can be corrected for in software that either assigns the correct angle to the reading or interpolates between readings to give the value at a specific crank angle. Secondly, it is essential that the ADC is coupled to a sample/hold circuit. If this is not the case, slight changes in the signal during the analogue-to-digital conversion process can lead to large errors.

When the data are being analysed, it must be remembered that they were originally a continuous analogue signal that has been assigned to a progression of digital levels. For example, when such a signal is differentiated numerically by looking at the difference in successive values, the result is very noisy. A more satisfactory result can be obtained by using a higher-order finite difference approach and smoothing the signal.

Taylor series can be used to represent the signal, and the following example calculates the first derivative, but includes the terms up to the fourth derivative. Consider a function $f(a)$ in which the interval between values is h:

$$f(a_{n+2}) = f(a_n) + \frac{2hf'(a_n)}{1!} + \frac{2^2h^2f''(a_n)}{2!} + \frac{2^3h^3f'''(a_n)}{3!} + \frac{2^4h^4f''''(a_n)}{4!} + \ldots$$

$$f(a_{n+1}) = f(a_n) + \frac{hf'(a_n)}{1!} + \frac{h^2f''(a_n)}{2!} + \frac{h^3f'''(a_n)}{3!} + \frac{h^4f''''(a_n)}{4!} + \ldots$$

$$f(a_n) = f(a_n) \quad (14.44)$$

$$f(a_{n-1}) = f(a_n) - \frac{hf'(a_n)}{1!} + \frac{h^2f''(a_n)}{2!} - \frac{h^3f'''(a_n)}{3!} + \frac{h^4f''''(a_n)}{4!} - \ldots$$

$$f(a_{n-2}) = f(a_n) - \frac{2hf'(a_n)}{1!} + \frac{2^2h^2f''(a_n)}{2!} - \frac{2^3h^3f'''(a_n)}{3!} + \frac{2^4h^4f''''(a_n)}{4!} - \ldots$$

These equations can be combined to eliminate the second, third and fourth derivatives ($f''(a_n)$, $f'''(a_n)$ and $f''''(a_n)$), to give

$$f'(a_n) = (f(a_{n-2}) - 8f(a_{n-1}) + 8f(a_{n+1}) - f(a_{n+2}))/12h$$
$$(14.45)$$

However, figure 14.42 shows that the derivative will still be noisy, and it is advantageous to apply smoothing. The following smoothing algorithm for $(2b + 1)$ values is widely used:

$$a_n = \frac{1}{b^2}\left[a_{n-(b-1)} + 2a_{n-(b-2)} + 3a_{n-(b-3)} + \ldots + ba_n \right.$$
$$\left. + \ldots 3a_{n+(b-3)} + 2a_{n+(b-2)} + a_{n+(b-1)}\right] \quad (14.46)$$

Note: the terms in equation (14.46) are only evaluated when the part of the subscript in brackets is not negative.

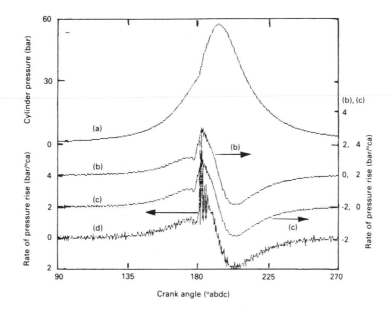

Figure 14.42
Cylinder pressure data recorded every 1/2°ca from a direct injection diesel engine at 1000 rpm with 60 per cent load. Trace (d) shows the first derivative evaluated by equation (14.45). Trace (b) illustrates the consequence of applying equation (14.47) twice to smooth trace (d). Trace (c) illustrates when equation (14.47) is used twice to smooth trace (a), prior to differentiation with equation (14.45).

This is illustrated by the simplest case when $b = 2$:

$$a_n = \frac{(a_{n-1}) + 2(a_n) + (a_{n+1})}{4} \tag{14.47}$$

The smoothing equation can be applied recursively (that is, more than once) and this is illustrated by figure 14.42. Figure 14.42 also shows that the order of smoothing and differentiation is not important, a result that can be shown algebraically. It is also possible to combine the smoothing and differentiation equations. The smoothing equation would be applied to each of the terms in equation (14.45) in turn, and the combined equation will reduce the amount of arithmetic to be undertaken by the computer. Temporary variables can also be used, to minimise the access of variables stored in the arrays. The theory behind these smoothing equations is beyond the scope of this book, but a thorough treatment is provided by Anon (1979), which also describes digital filtering techniques. However, it must be remembered that the smoothing process loses information; for example, any maxima will be reduced in magnitude. Care must be taken to avoid over-smoothing.

The use of the first and second pressure derivatives has already been illustrated in chapter 11, figure 11.1, in which the start of diesel combustion has been defined as the first minimum in the first derivative after the start of injection; this can be identified by the zero value in the second derivative.

14.5.2 Burn rate analysis

A burn rate analysis is usually applied to the combustion data

from spark ignition engines to calculate the mass fraction burnt (mfb). A widely used technique is the approach devised by Rassweiler and Withrow (1938), which is illustrated by figure 14.43. After the start of combustion, the pressure rise (Δp) during a crank angle interval ($\Delta\theta$) is assumed to be made of two parts (figure 14.43) – a pressure rise due to combustion (Δp_c) and a pressure change due to the volume change (Δp_v).

$$\Delta p = \Delta p_c + \Delta p_v \tag{14.48}$$

As the crank angle (θ_i) increments to its next value (θ_{i+1}) the volume changes from V_i to V_{i+1}, and the pressure changes from p_i to p_{i+1}. It is assumed that the pressure change due to the change in volume can be modelled by a polytropic process

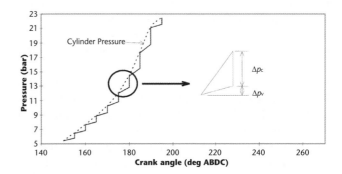

Figure 14.43
Separation of the pressure change (Δp) into a pressure rise due to the volume change (Δp_v) and combustion (Δp_c).

with an exponent k. Substituting for Δp_v, equation (14.48) becomes

$$p_{i+1} - p_i = \Delta p_c + p_i \left[\left(\frac{V_i}{V_{i+1}} \right)^k - 1 \right]$$

from which Δp_c can be evaluated:

$$\Delta p_c = p_{i+1} - p_i (V_i/V_{i+1})^k \tag{14.49}$$

The pressure rise due to combustion is not directly proportional to the mass of fuel burned, as the combustion process is not occurring at constant volume. The pressure rise due to combustion has to be referenced to a datum volume, for example the clearance volume at tdc, V_c:

$$\Delta p_c^* = \Delta p_c V_i/V_c \tag{14.50}$$

The end of combustion occurs after N increments, and is defined by the pressure rise due to combustion becoming zero. If it is assumed that the referenced pressure rise due to combustion is proportional to the mass fraction burned (mfb) then

$$\text{mfb} = \sum_0^i \Delta p_c^* / \sum_0^N \Delta p_c^* \tag{14.51}$$

The summation of the referenced pressure rise due to combustion is illustrated by figure 14.44 along with the normalisation of the data to give the mass fraction burned.

The summation of the pressure rise due to combustion (the denominator of equation 14.51) contains information about the completeness of combustion. By looking at the distribution of the pressure rise due to combustion, it is possible to set an upper bound for its value, and then to calculate a completeness of combustion. This has been done by Ball *et al.* (1998), who demonstrate how the average completeness of combustion can be correlated with the unburnt hydrocarbon emissions.

Since the volume change is small when the piston is in the region of tdc, the computed mass fraction burned is insensitive to slight errors in the positioning of tdc. However, the method does depend on using an appropriate value of the polytropic index, k. Rassweiler and Withrow (1938) evaluated the polytropic index for before and after combustion and used an appropriately averaged value during combustion. For the results shown here in figure 14.44, the polytropic index was only evaluated during compression. This leads to the fall in the referenced pressure due to combustion and the mass fraction burned in figure 14.44 after the end of combustion, as the polytropic index is lower during the expansion process than during compression. This is a consequence of the heat transfer and the presence of combustion products. Ball *et al.* (1998) show how a separate polytropic index can be used for the burnt gases, so as to avoid this 'droop'.

During compression the polytropic index is usually within the range of 1.2–1.3 for a spark ignition engine, and a suitable

value can be chosen by the user. Alternatively, the polytropic index can be evaluated from the compression process prior to ignition. By evaluating the logarithmic values of pressure and volume, a least squares straight line fit can be used to determine the polytropic index. However, care is needed because of two reasons:

1 there might be errors in the pressure datum

2 during the initial part of compression the pressure rise is small, and discretisation errors from the ADC are more significant.

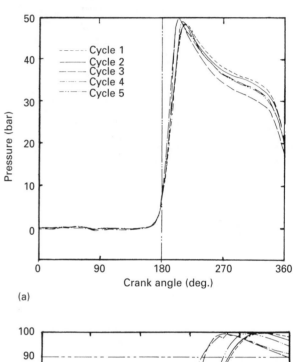

(a)

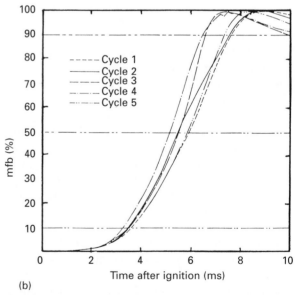

(b)

Figure 14.44

The referenced pressure rise attributable to combustion (a), and the mass fraction burnt (mfb) as a function of time (b), for the five successive cycles of figure 4.25.

Both effects are minimised if the initial part of the compression is ignored; for example, the polytropic index could be evaluated up to ignition, from half way between the inlet valve closure and ignition.

The Rassweiler and Withrow method contains several assumptions. It is assumed that the referenced pressure rise due to combustion is proportional to the mass fraction burned in each increment. There is no explicit allowance for heat transfer, dissociation, or change in composition of the gases; though to some extent an allowance is made, as the polytropic index is allowed to vary from the ratio of the gas specific heat capacities. These shortcomings have been investigated by Stone and Green-Armytage (1987), who used a thermodynamic analysis to make a direct comparison with the same data being analysed by the Rassweiler and Withrow method. The thermodynamic model divided the combustion chamber into two zones, and took into account dissociation and heat transfer within the combustion process. Notwithstanding the substantial differences between the two approaches, the results were in surprisingly close agreement. This was attributed to the temperature of the burnt gas being almost constant during combustion, so that the effects of dissociation and heat transfer had an almost uniform influence throughout combustion.

Since the Rassweiler and Withrow method is simple to calculate, it is an appropriate and popular method when cycle-by-cycle variations in combustion are to be analysed. It has already been used in chapter 4, section 4.4 to calculate the 0–10 per cent, 0–50 per cent and 0–90 per cent burn times in table 4.8.

14.5.3   Heat release analysis

A heat release analysis is normally applied to combustion in diesel engines, but there is no reason why it should not be applied to spark ignition engines. (Similarly, the mass fraction burnt analysis can be applied to diesel engines.) The heat release analyses compute how much heat would have to be added to the cylinder contents, in order to produce the observed pressure variations. The usual assumption is to have a single zone; this implies that the products and reactants are fully mixed.

If the First Law of Thermodynamics is applied to a control volume in which there is no mass transfer, then the heat released by combustion (δQ_{hr}) is given by:

$$\delta Q_{hr} = dU + \delta W + \delta Q_{ht} \tag{14.52}$$

where δQ_{ht} = the heat transfer with the chamber walls.

Equation (14.52) is a simplification of equation (11.1). Also implicit in equation (14.52) is that there is no allowance for differences in the properties of the reactants and products, and that there is a uniform temperature. Each of the terms in equation (14.52) has to be evaluated:

$$dU = mc_v dT \tag{14.53}$$

From the equation of state ($pV = mRT$)

$$m dT = \frac{1}{R}(p dV + V dp) \tag{14.54}$$

Substitution of equation (14.54) into equation (14.53) gives

$$dU = \frac{c_v}{R}(p dV + V dp) \tag{14.55}$$

Substituting equation (14.55) into equation (14.52) and noting that $\delta W = p dV$ gives, on an incremental angle basis:

$$\frac{dQ_{hr}}{d\theta} = \frac{c_v}{R}\left(p\frac{dV}{d\theta} + V\frac{dp}{d\theta}\right) + p\frac{dV}{d\theta} + \frac{dQ_{ht}}{d\theta} \tag{14.56}$$

The $dV/d\theta$ term is defined from the geometry of the engine (equation 11.6a), and the $dp/d\theta$ term has been recorded from the engine. However, c_v and R are functions of temperature, and the heat transfer cannot be readily evaluated. But, if semi-perfect gas behaviour is assumed (such that $c_p/c_v = \gamma$ and $R = c_p - c_v$), then equation (14.56) can be written as

$$\frac{dQ_{hr}}{d\theta} - \frac{dQ_{ht}}{d\theta} = \frac{1}{\gamma - 1}\left(p\frac{dV}{d\theta} + V\frac{dp}{d\theta}\right) + p\frac{dV}{d\theta}$$
$$\frac{dQ_n}{d\theta} = \frac{\gamma}{\gamma - 1}p\frac{dV}{d\theta} + \frac{1}{\gamma - 1}V\frac{dp}{d\theta} \tag{14.57}$$

where $dQ_n/d\theta$ = the net heat release.

The gas temperature can be found from the equation of state ($pV = mRT$), since the pressure and volume are known, and it has been assumed that the mass is constant. The gas properties vary with temperature, but as the variation is modest, it is acceptable in most cases to evaluate the properties at the gas temperature computed in the previous increment. Equations such as (11.8) and (11.9) can be used to evaluate u and R, from which γ can then be evaluated. Once the gas temperature has been evaluated, it is possible to estimate the heat transfer, by assuming a wall temperature and employing a heat transfer correlation (see chapter 11, section 11.2.4 Engine gas-side heat transfer). Sometimes the heat release analysed in this way is known as the gross heat release, to emphasise the difference from the net heat release.

Examples of heat release are provided here by the BICCAS (*BICERI Combustion Analysis Software*), which can be used in conjunction with a Computerscope data acquisition system (Whiteman, 1989).

The data used here are from a high compression ratio lean-burn gas engine, which has been described by Ladommatos and Stone (1991). Figure 14.45 shows the indicator diagram for operation at 1500 rpm with wide-open throttle and an equivalence ratio of 0.6, and ignition 18°btdc. Plotting the pressure/volume diagram on a logarithmic basis provides several insights:

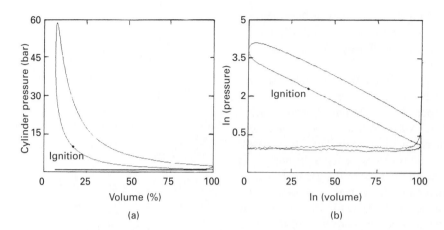

Figure 14.45
The indicator diagrams on a linear (a) and logarithmic (b) basis, for a gas engine operating at 1500 rpm with wide-open throttle and an equivalence ratio of 0.6 with ignition 28°btdc.

(a) The pumping loop can be seen more clearly (and this is an alternative to the linear magnification, for example, in figure 14.16).

(b) As the compression process is a straight line, this demonstrates that assuming a polytropic process is a good model.

(c) The departure from a straight line just before tdc indicates the start of combustion.

(d) The combustion is fairly symmetric about tdc, but the end of combustion is in fact much later and less clearly defined.

(e) The shallow curve of the expansion stroke is initially a consequence of the final stage of combustion, and later on, a consequence of heat transfer.

Although the engine being discussed here is a spark ignition engine, there is no reason why a heat release analysis cannot

be applied. Figure 14.46 shows both the net heat release rate, and the cumulative net heat release. The heat release analysis evaluates the data on a differential basis (equation 14.57), and this leads to noise in the computed result – especially at the lower pressures where the discretisation (ADC steps) are a larger proportion of the signal. The negative net heat release rate (figure 14.46a) implies that there is heat transfer to the cylinder surfaces, and ignition (28°btdc) should be close to the minimum net heat release rate. This is because the heat transfer rate increases as the temperature and density of the gas rise during combustion, but once combustion commences, the heat release rate will add to the (negative) heat transfer. The minimum net heat release rate is ill-defined in a spark ignition engine, because the initial rate of combustion is very low. In a diesel engine, the fuel vaporisation also contributes to the negative heat release rate prior to combustion. However, the initial combustion rate in a diesel engine is very rapid (because of the combustion of the pre-mixed

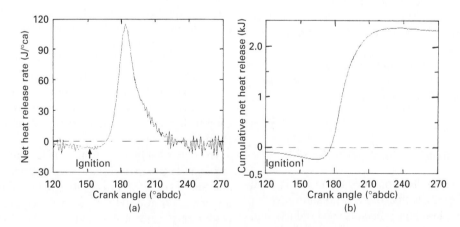

Figure 14.46
The pressure data for figure 14.45a have been analysed for: (a) the net heat release rate; (b) the cumulative net heat release.

reactants formed during the ignition delay period). The initial combustion is so rapid, in a diesel engine, that the start of combustion is often defined as when the net heat release rate becomes positive (see figure 11.2). Figure 14.46a illustrates that this would be an unsatisfactory way of defining the start of combustion in a spark ignition engine.

The cumulative net heat release in figure 14.46b is an integration of the results in figure 14.46a; the process of integration smooths the noise present in figure 14.46a. The minimum in the cumulative heat release corresponds to the zero-crossing of the net heat release rate. The zero-crossing of the cumulative net heat release occurs somewhat later (about 10°); this has no physical significance and should not be used to define the start of combustion – even with a diesel engine.

Near the end of combustion, the heat transfer to the combustion chamber becomes greater than the heat release from the combustion process. Thus, the net heat release rate becomes negative (in other words the cumulative heat release starts to fall) and this can be used to estimate the end of combustion.

Figure 14.47 shows the gas temperature (calculated by applying the equation of state: $pV = mRT$), and an estimate of the heat transfer from the Woschni correlation (chapter 11, equations 11.28 to 11.30). The gas temperature in figure 14.47a assumes a single zone for the combustion, and thus represents an estimate of the mean temperature. The heat flux calculation (figure 14.47b) requires an estimate of the combustion chamber surface temperature; this, and the choice of heat transfer correlation, influence the calculated heat flux. Because of these sources of uncertainty, the (gross) heat release and its rate have not been calculated here.

The final graph in this sequence (figure 14.48) is the mass fraction burnt calculated by the Rassweiler and Withrow method (equations 14.48–14.51). When figure 14.48 is compared with the cumulative net heat release (figure 14.46b) it can be seen that they have a very similar form.

14.6 | Advanced test systems

Engine test cells are becoming increasingly complex for several reasons. Additional instrumentation such as exhaust gas analysis has become necessary and, in the search for the smaller gains in fuel economy, greater accuracy is also necessary. Consequently the cost of engine test cells has escalated, but fortunately the cost of computing equipment has fallen.

Computers can be used for the control of a test and data acquisition, thus improving the efficiency of engine testing. The computer can also process all the data, carry out statistical analyses, and plot all the results. The design of the test facility and computer system will depend on its use – whether it is for research, development, endurance running or production testing. There are, of course, overlaps in these areas, but the facility described here will be for development work.

Descriptions of computer-controlled test facilities are given by Watson et al. (1981) and Donnelly et al. (1981). Complete facilities are marketed by firms such as Froude Hofmann and Schenck; a Schenck system is described here. A block diagram of a computerised test facility is given in figure 14.49. Separate microprocessors (micros) are used for data acquisition and test point control. A single microprocessor could be used but that would reduce the data-acquisition rate, and reduce the storage space for data and test cycles. The printer and monitors are linked to the micros so that the test pattern can be chosen and then monitored. The host computer can be linked to many such microsystems, and it provides a more powerful data-processing system with greater storage. Data from many tests can then be archived on disks or CDs. The host computer can also provide sophisticated graphics facilities and plotters that would be underused if dedicated to a single test cell.

A typical requirement is to produce an engine fuel-consumption map. This requires running the engine over a wide range of discrete test points. Each test point is specified

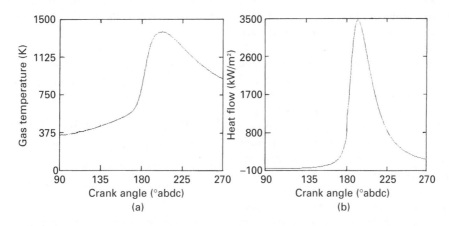

Figure 14.47
The calculated mean gas temperature (a) and the heat flux (b) calculated by the Woschni (1967) correlation.

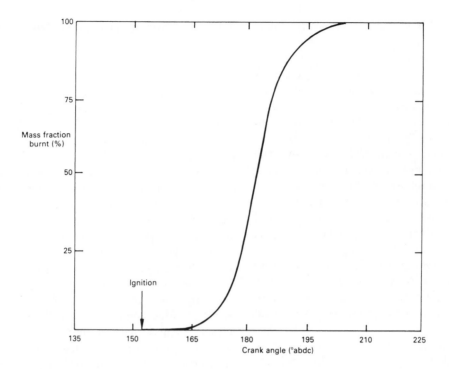

Figure 14.48
The mass fraction burnt calculated by the method of Rassweiler and Withrow (1938) for the pressure data in figure 14.45a.

in terms of speed and load (or throttle position), and the sequence and duration of each test point are stored in the micro. The software for setting up a test program is stored in an EPROM (erasable programmable read only memory) and is designed so that new or existing test programs can be readily used. At each test point the dynamometer controller and throttle controller work in conjunction to obtain the desired test condition. Additional parameters such as ignition timing and air/fuel mixture can also be controlled. At each test point a command is sent to the data-acquisition system to sweep all the channels (that is, to take measurements from all the specified transducers). At the end of the test the data can be sent for analysis on the host computer. At each test point, commands can be sent to instrumentation that does not take continuous readings; for instance, the initiation of a fuel-consumption measurement.

The data-acquisition system also monitors values from the instrumentation (such as engine speed, water temperature) and if the values fall outside a certain range then an alarm can be set. If the values fall outside a specified wider range, the engine can be stopped and a record kept of what caused the shutdown. The software also permits the separate calibration of transducer channels. The transducer inputs have to be linear and the calibration is defined in terms of specified inputs.

Other parameters such as oil and water temperature normally have separate control systems. These would apply closed-loop control to the flow through the appropriate heat exchanger.

When transient tests are required, a computer-controlled test facility is almost essential. Transient testing might be required for simulating an urban drive cycle, testing the load acceptance of an engine that is intended to drive a generating set, and measuring the smoke emissions from a turbocharged truck engine. With a turbocharged diesel engine, transient testing is particularly important, since under these conditions the emissions of smoke and noise are higher.

Most instrumentation in a test cell will have a suitable response for measuring transients, but the most notable exception is likely to be the torque measurement. If the torque measurement is from the torque reaction on the dynamometer case (T_d), this does not include the torque required to accelerate or decelerate the dynamometer:

$$T_e = T_d + I_d \dot{\omega} \qquad (14.58)$$

where T_e = engine torque output
I_d = inertia of the dynamometer and the coupling
$\dot{\omega}$ = angular acceleration.

The two options for measuring the torque output are:

(i) To measure the torque in the coupling.

(ii) If the inertia of the dynamometer is known, differentiate the speed, and add to the torque reaction on the dynamometer.

Unfortunately, both options have problems associated with them: there are torque fluctuations occurring in the coupling,

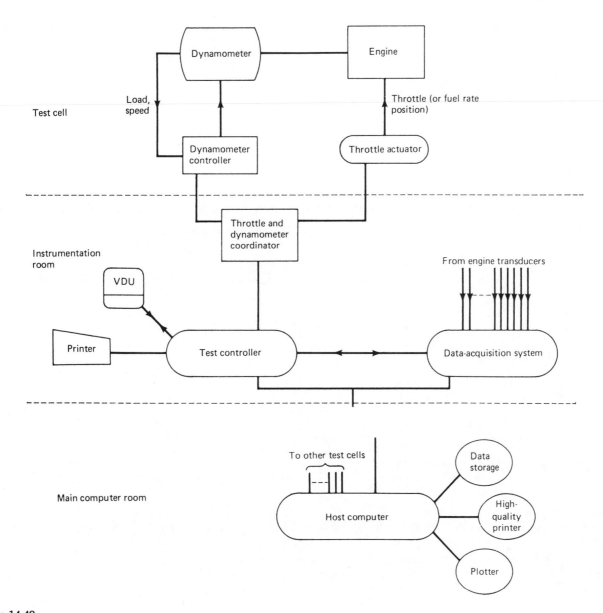

Figure 14.49
Computer-based test facility.

and there is noise associated with differentiating a speed signal.

In test cells for production testing it is usual to have a pallet-mounting system for the engine, in order to minimise the cell downtime. Such systems may also be justified in development testing where many different engines are being tested.

14.7 | Conclusions

Engine testing is an important aspect of internal combustion engines, as it leads to a better understanding of engine operation. This is true whether the engine is a teaching experiment or part of an engine-development programme. As in any experiment, it is important to consider the accuracy of the results. The first decision is the accuracy level that is required; too high a level is expensive in both time and equipment. The second decision is to assess the accuracy of a given test system; this is of ever-increasing difficulty owing to the rising sophistication of the test equipment.

The nature of development testing is also changing. The use of computer control and data acquisition has been complemented by increasing levels of engine instrumentation. This leads to large quantities of data, and a need for effective post-processing. Needless to say, such systems are expensive, but the need for their use is held back by two factors:

1 The number of different engines produced is reduced by manufacturers standardising on engine ranges, and by collaboration between companies.

2 The decreasing cost of computing time, and the use of increasingly powerful computer models, leads to greater optimisation prior to the start of development testing.

However, these factors are balanced by increasing restrictions on engine emissions, and the ever-rising difficulty of improving engine fuel economy.

14.8 | Question

14.1 Experiments are being conducted on a 3.9-litre four-stroke diesel engine, which is being operated with three different inlet conditions:

(i) naturally aspirated with air at an inlet temperature of 30°C;
(ii) naturally aspirated with air at an inlet temperature of 150°C;
(iii) naturally aspirated, but with a mixture of 30 per cent O_2 and 70 per cent CO_2 (molar basis) at an inlet temperature of 150°C.

At a speed of 2000 rpm in mode (iii), the following data were obtained:

Brake power	16.2	33.2 kW
bsfc	405.9	345.0 g/kWh
Oxygen level in the exhaust	18.0	7.8 per cent (dry gas)

At a bmep of 2.6 bar, the engine performance for cases (ii) and (iii) are:

	(ii)	(iii)
Pressure at the end of compression	38	31 bar
Effective phasing of combustion	0	12 °atdc

(a) Use the Willans' line construction to estimate the fmep of the engine.
(b) Stating clearly any assumptions, calculate the volumetric efficiency of the engine, and comment on the result.
   Assume that the effective compression ratio of the engine in 14:1 (instead of the nominal value of 15.5:1, to allow for the inlet valves closing slightly after bottom dead centre).
(c) If the pressure at the start of compression is 1 bar, calculate the value of the polytropic index, and compare this with the value for the ratio of the specific heat capacities (use the data from the tables in Appendix A).
(d) By assuming that the combustion can be treated as instantaneous at the 'effective phasing of combustion', use an Otto cycle analysis to apportion the reduction in cycle efficiency from case (ii) to case (iii), between the effects of the lower value of the ratio of the heat capacities and the increased ignition delay.

CHAPTER 15

Case studies

15.1 | Introduction

The three engines that have been chosen as case studies are the Rover K Series spark ignition engine, the Chrysler 2.2-litre spark ignition engine, and the Ford (high-speed) 2.5-litre DI diesel engine. These engines have been chosen because their characteristics are likely to be seen also in subsequent engines. The Rover engine uses a pent-roof combustion chamber; this permits the use of a high compression ratio and the combustion of lean mixtures, both of which lead to economical operation. The Chrysler 2.2-litre spark ignition engine is typical of practice in the USA and thus has low emissions of carbon monoxide, unburnt hydrocarbons and oxides of nitrogen.

The Ford compression ignition engine achieves short combustion times and thus high engine speeds by meticulous matching of the air motion and fuel injection. By utilising direct injection, as opposed to indirect injection into a pre-chamber, the pressure drop and heat transfer in the throat to the pre-chamber are eliminated. This immediately leads to a

10–15 per cent improvement in economy and better cold starting performance.

15.2 | The Rover K series engine

15.2.1 Introduction

The Rover K series engine has been selected as a case study for a number of reasons. Firstly, it has many innovative design features that are well documented, and the engine is constructed both as an in-line four-cylinder engine and as a V6. Secondly, there are a number of significant variations, including the Variable Valve Control (VVC) system (already discussed in chapter 7, section 7.4.2), and the variable-geometry induction system (already discussed in chapter 7, section 7.6.2, and used on the V6 engine). Thirdly, as well as being used by Rover, it has also been used by Lotus and Caterham Cars, and has been manufactured under licence by Kia.

Table 15.1 summarises the specifications of some of the K series engine variants.

Table 15.1 *Specification of K series engine variants*

	K8		K16				V6
	1100	1400	1400	1600	1800	1800 VVC	2500
Bore (mm)	75.0	75.0	75.0	80.0	80.0	80.0	80.0
Stroke (mm)	59.0	79.0	79.0	79.0	89.3	89.3	82.8
Connecting-rod length (mm)	141.7	131.5	131.5	131.5	141.7	141.7	
Compression ratio	10.5	10.5	10.5	10.5	10.5	10.5	10.5
Maximum power (kW/litre)	38	39	54	53	49.4	60	51.6
at speed (rpm)	5500	5500	6000	6000	5500	7000	6500
Maximum bmep (bar)	10.1	10.1	11.2	11.4	11.5	12.1	12.1
at speed (rpm)	4000	3500	3500	3000	3000	4500	4000
Valve timing							
inlet opens (°btdc)	12	12	12	12	12	9–46	2
inlet closes (°abdc)	52	52	52	52	52	31–69	54
exhaust opens (°bbdc)	52	52	52	52	52	52	50
exhaust closes (°atdc)	12	12	12	12	12	12	14
Inlet valve diameter (mm)			27.3	27.3	27.3	31.5	31.5
Exhaust valve diameter (mm)			24.0	24.0	24.0	27.3	27.3

The designation K8 indicates two valves/cylinder, while all the other variants have four valves/cylinder. Table 15.1 shows that all members of the K series engine family have a high specific output, and that there is significant commonality in components, through the use of the same bore and valve diameters. The VVC system used on the K16 1800 engine and the variable-geometry induction system used on the 2.5-litre V6 engine, both give a very high bmep (12.1 bar). The VVC system maintains a high bmep at high engine speeds, to give the highest specific output of 60 kW/litre. The 1.8-litre non-VVC engine uses the same cylinder head as the smaller derivatives, so the gas flow is comparatively restricted, and this accounts for the lowest specific output (49.4 kW/litre) of the four valves/cylinder derivatives.

15.2.2 The K16 base engine design

The design and development of the Rover K16 engine has been described in detail by Stone *et al.* (1990). The engine was originally introduced with a compression ratio of 9.5:1 but this has been raised to 10.5:1. The camshaft has a lift of 8.2 mm. Figure 15.1 is a sectioned view of the engine, and this shows some of the unusual design features. The ten long cylinder head bolts pass through holes in the block, and are secured into the main bearing ladder. The cylinder head loads are thus transferred directly to the main bearings. A precise clamping load is obtained by tightening the bolts to a specified torque, and then rotating the bolt through an additional angle (360° for the K series engine). However, the bolt is designed to yield on initial warm-up, so that the clamping load is only a function of the bolt material properties.

The engine is compact (600 mm long, 695 mm wide and 627 mm high) and is made up of comparatively few main components: cylinder head cover, camshaft upper bearing ladder, cylinder head, cylinder block, main bearing ladder and sump. In the case of the 1.8-litre engine a cast alloy sump is used to provide additional beam stiffness to the engine/transmission assembly. The oilways are cast into the camshaft and main bearing ladders, and there are oil drain holes and crankcase breathing holes cast into the cylinder block and the cylinder head.

The cylinder head and block are low-pressure sand castings, and the cylinder head bolts ensure that these components are always loaded in compression. The main bearing ladder is a high-pressure die-casting in LM 24 alloy; it is located by a pair of hollow dowels and is retained by bolts around its periphery. Figure 15.2 shows longitudinal and transverse sections of the engine. The cylinder block is cast in LM 25 alloy, and the top hung wet liners (centrifugally cast in BS 1452 G220) are sealed into the crankcase at their lower ends by a pair of O-rings. In the 1.6, 1.8 litre and V6 engines a 'damp' liner is used. For a fixed bore centre distance, the bore could only be increased by using a new type of cylinder liner. Instead of an O-ring seal, the lower seal is formed using Hylomar sealant.

The clamping system minimises the bore distortion, so as to ensure low friction levels, low wear rates and a low oil consumption. The block contains an integral housing for the water pump on its front face.

The cylinder head is cast in LM 25TF alloy, and the tappet bores (for hydraulic bucket tappets) and the camshaft bearings are an integral part of it. The cylinder head gasket is of stainless steel, with conventional flame sealing rings between the cylinder head and liners, but with elastomeric seals for all fluid seals. Extensive use is made of anaerobic jointing compounds and elastomeric seals elsewhere in the engine. The valve stems are of diameter 6 mm, and the garter-type stem seals are integral to the valve spring seats. Single valve springs have been used with a target float speed of 7200 rpm. The camshafts are driven by a 25 mm wide toothed belt, which also drives the water pump. The belt has a design life of 160 000 km.

The crankshaft is made from spheroidal graphite cast iron, and has five main bearings. It has 15.8 mm thick webs with 8 counter-balance masses. The connecting-rods are drop forgings (080X47 material), which have been heat treated to give an ultimate tensile stress of 775 MN/m². The pistons have two compression rings and an oil control ring. Their mass is 360 g, inclusive of the piston pin. The piston pin is of 30 mm diameter, with an interference fit into the connecting-rod. The little end of the connecting-rod is induction heated for assembly. Ultra-light-weight pistons have been adopted for the 1.6-litre engine, with a mass of 194 g. There is selective assembly of the main bearing and big-end bearing shells. There are three grades of bearing bore, and three grades of journal diameter. The appropriate bearing shells (with an aluminium–tin bearing surface) are selected automatically during assembly.

15.2.3 The Rover K series engine coolant and lubricant circuits

The lubrication system is shown in figure 15.3. The oil pump is a multi-lobe type, directly mounted on the pulley end of the crankshaft. The oil pressure relief valve vents into the pump inlet, so as to minimise aeration of the oil. The oil passes through a filter, and is then distributed through internal oilways. Most of the internal oilways are cast into the main bearing and camshaft bearing ladders; notable exceptions are drilled holes in the cylinder block and head. Otherwise the oil drillings are quite short, to supply the bearings and tappets.

Details of the cooling system are shown in figure 15.4. The outflow from the pump is fed by internal passageways to a tapered duct along the exhaust valve side of the cylinder head. The coolant is metered (by holes in the cylinder head gasket) into the cylinder block liner pockets, and thence through metering holes in the cylinder head gasket into a duct running along the inlet side of the cylinder head. The coolant flows transversely in the cylinder head from the ducts along

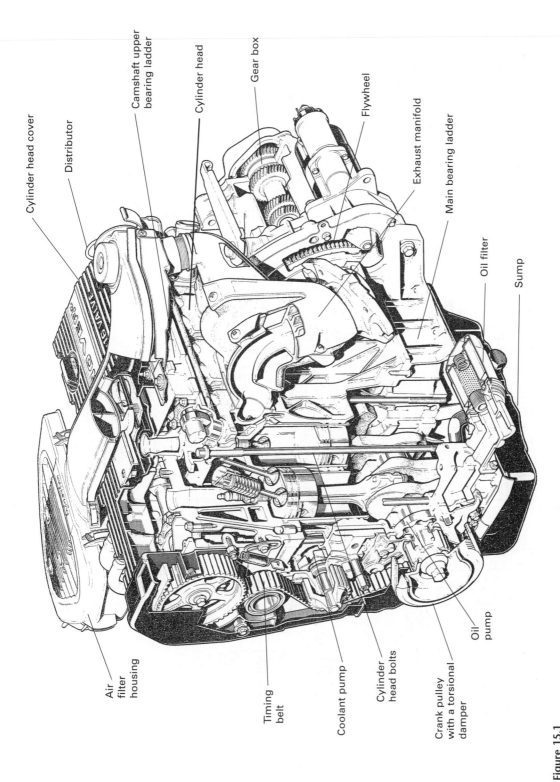

Cylinder head cover

Distributor

Camshaft upper bearing ladder

Cylinder head

Gear box

Flywheel

Exhaust manifold

Main bearing ladder

Oil filter

Sump

Air filter housing

Timing belt

Coolant pump

Cylinder head bolts

Crank pulley with a torsional damper

Oil pump

Figure 15.1
Sectioned view of the Rover K16 engine and gearbox.

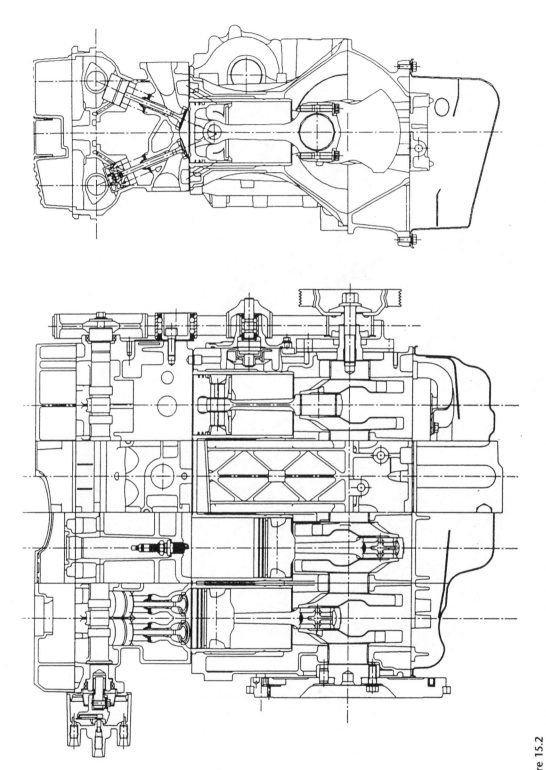

Figure 15.2
Longitudinal and transverse sections of the Rover K series engine (Stone *et al.*, 1990). © IMechE/Professional Engineering Publishing Limited.

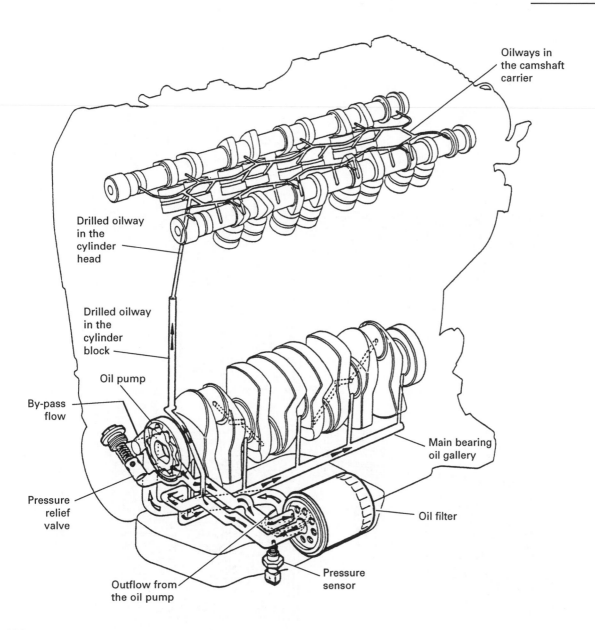

Figure 15.3
Details of the Rover K series engine lubrication system.

the inlet and exhaust sides into a central duct in which there is a longitudinal flow. The transverse flow in the cylinder head is in cast passages around the valves, and in drilled holes between the valves. The drillings ensure that critical areas are cooled by high-velocity flows. There are only 1.2 litres of coolant in the cylinder block and head, and this facilitates a rapid warm-up. The coolant flow patterns were optimised with the help of computer simulations, which were validated by the experimental tests.

15.2.4 The Rover K series engine manifolding, fuelling and ignition systems

Originally the K series engine was introduced with a single-point fuel injection system, but subsequent developments have led to the use of multi-point fuel injection. In the 1.4-litre engine, all the injectors are operated simultaneously, while for the 1.6-litre engine, the injectors are fired as two pairs. In the case of the V6 and VVC engines sequential fuel

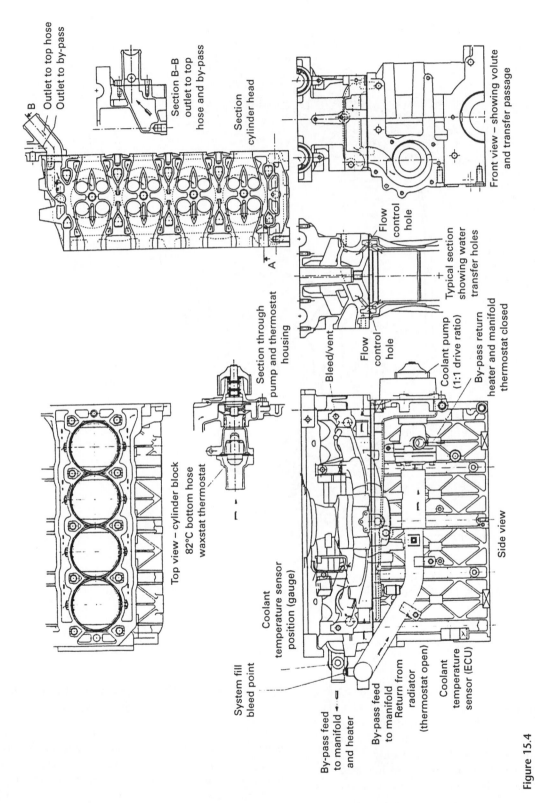

Figure 15.4

Details of the Rover K series engine cooling circuit (Stone *et al.*, 1990). © IMechE/Professional Engineering Publishing Limited.

injection is used, in which the fuel is injected into the inlet port of each cylinder at the same angle relative to the valve events.

With single-point fuel injection there is substantial wetting by fuel of the inlet manifold. Good cold-start and warm-up characteristics were achieved by using a positive temperature coefficient (PTC) heater immediately below the injector. The induction system had to be designed for good air and fuel flow. The inlet air temperature is controlled by a temperature-controlled flap valve, which can source the air from a heat exchanger around the exhaust manifold when needed. During development it was found that small leaks upstream of the air filter could lead to disproportionate resonances which were audible in the passenger compartment.

The ignition is fully mapped using the Rover MEMS (Modular Engine Management System), which also controls the fuelling. A single coil with an HT distributor is used on all but the 1.8-litre VVC and V6 engines. A distributor is not possible on the VVC or V6 engines, since the distributor is located at the end of the exhaust camshaft. In the VVC engine, the inlet camshaft needs to be driven from both ends, and this is achieved by a toothed belt drive between the flywheel ends of the camshafts. In the V6 engine, the timing belt from the crankshaft only drives the inlet camshafts. The exhaust camshafts are driven by a toothed belt drive from the inlet camshafts at the flywheel end of the engine. These engines use a distributorless ignition system with double-ended coils.

The engine speed and position are deduced from an inductive sensor operating with a reluctor ring that is mounted on the flywheel and has a missing tooth. The fuel injection timing map is shown in figure 15.5.

15.2.5 The Rover K series engine performance development

Steady-state flow tests were used to optimise the cylinder head design for the target values of Mach Index and barrel swirl. The port shape finally selected was one that was insensitive to the parameters that were most difficult to control during manufacture. The engine was developed with the help of LDA and hot wire anemometry measurements for flow field determination, and ionisation sensors to monitor flame growth.

The original engine was developed for lean-burn operation, but in the event emissions legislation has required a three-way catalyst to be used. However, the development of a fast, stable lean-burn combustion system in an engine with low friction led to fuel economy benefits of about 20 g/kWh in the range 3–7 bar bmep (and greater gains at lighter loads, compared to the previous engine). Figure 15.6 shows the fuel economy and emissions benefits associated with lean burn, for a constant-speed low-load (2 bar bmep) operating point. The fuel consumption reduces with the increasing air/fuel ratio at a fixed part load, since frictional losses are constant, and the weaker mixtures have a higher indicated efficiency, until the onset of large cycle-by-cycle variations in combus-

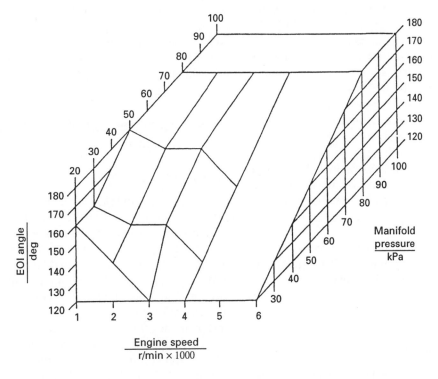

Figure 15.5
Fuel injection timing map for the Rover K series engine (Stone *et al.*, 1990). © IMechE/Professional Engineering Publishing Limited.

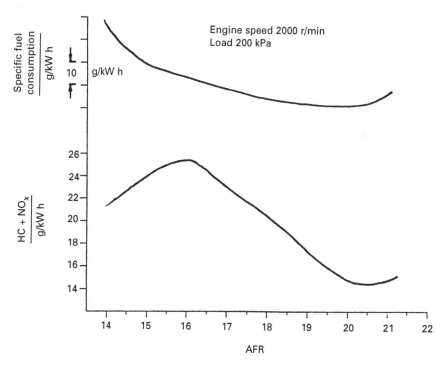

Figure 15.6
The dependence of emissions and fuel economy on the air/fuel ratio in the Rover K series engine (Stone *et al.*, 1990). © IMechE/Professional Engineering Publishing Limited.

tion. These cycle-by-cycle variations in combustion are also responsible for the rise in hydrocarbon emissions, which accounts for the upturn in the combined HC + NO$_x$ emissions at the weak air/fuel ratio. With weakening mixtures the NO$_x$ emissions fall, so operation is limited by the rise in hydrocarbon emissions. The NO$_x$ emissions are very sensitive to the ignition timing, with less advanced ignition timings giving lower NO$_x$ emissions. The hydrocarbon emissions are much less sensitive than the NO$_x$ emissions, and are likely to have the opposite trend. However, as shown in figure 15.7, the retarded ignition timings also increase the fuel consumption. Figure 15.7 also shows how a 5 per cent fuel-consumption penalty can be used to obtain a 30 per cent reduction in the HC + NO$_x$ emissions. In stoichiometrically operated engines, the lean-burn capability can be exploited as a high tolerance to EGR, which can be used to reduce the part-load pumping losses.

The variable-geometry induction system used on the V6 engine, and the Variable Valve Control (VVC) system, have already been discussed in chapter 7, and table 15.1 shows the significant improvements in output that these systems achieve.

15.2.6 Concluding remarks

The preceding sections have described the development of the K series engine family, but before its introduction there was substantial durability testing. The first phase of testing was on 150 prototype engines with over 55 000 hours of dynamometer testing. This was followed by a series of pre-production runs, which mimicked as closely as possible the engines that would be produced once full-scale manufacture commenced. These engines were also subject to extensive durability testing on dynamometers and vehicles.

15.3 Chrysler 2.2-litre spark ignition engine

15.3.1 Background

The increasing emphasis on fuel economy and reduction of polluting emissions, and a trend towards compact, front wheel drive (fwd) cars all favour the adoption of a four-cylinder in-line spark ignition engine. This particular engine was specifically designed for fwd applications, and attention was paid to performance, fuel economy, durability, cost, emissions, engine weight and size, serviceability and manufacturing. The philosophy, development and manufacture of this engine are described by Weertman and Dean (1981).

For fwd applications with the engine mounted transversely, a four-cylinder configuration provides a compact package. The overall length of the engine can be reduced by siamesing the bores and having an under-square cylinder

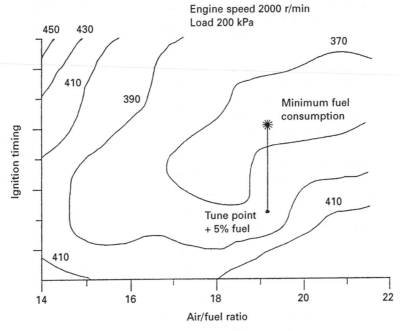

(a) Contours of specific fuel consumption (g/kW h)

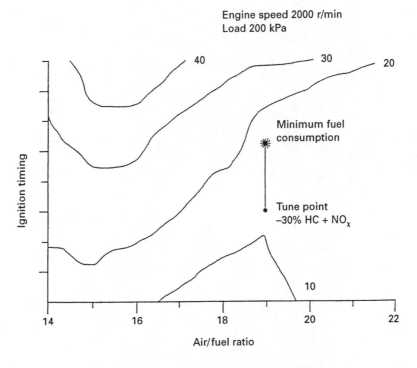

(b) Contours of specific HC + NO$_x$ (g/kW h)

Figure 15.7
The dependence of emissions and fuel economy on the air/fuel ratio and ignition timing in the Rover K series engine (Stone *et al.*, 1990).
© IMechE/Professional Engineering Publishing Limited.

(the stroke being greater than the bore). An isometric cut-away view of the engine is shown in figure 15.8; the bore is 87.5 mm and the stroke is 92.0 mm, giving a swept volume of 2.213 litres.

The single overhead camshaft is driven by a toothed belt, and operates the in-line valves via rocker arms. The cylinder head is alloy, and the cylinder block is cast iron; the induction and exhaust manifolds are on the same side of the engine to provide a large clear space for accessories. The timing belt also drives an accessory shaft, which drives the distributor and oil pump. Since these have been removed from figure 15.8 they are shown separately in figure 15.9. The engine has been used in a very wide range of Chrysler vehicles in the following forms: carburetted with normal and high compression ratios, electronic fuel injection with normal compression ratio, and as a turbocharged engine with multi-point electronic fuel injection.

15.3.2 The cylinder head

The cylinder head is cast from aluminium alloy, and the camshaft runs directly in the cylinder head without shell bearings. A detailed view of the cylinder head and combustion chamber is shown in figure 15.10.

The camshaft is made of a hardenable cast iron; the cam lobes are induction hardened and phosphate treated. The five camshaft journals are pressure lubricated, and 0.8 mm drilled holes also direct an oil jet onto the cam-lobe/rocker-arm contact area. The rocker arm is cast iron, and pivots on a hydraulic lash adjuster (hydraulic tappet). The tappet adjustment range is 4.5 mm, and the valve geometry is such that the valves can never hit the piston.

The valves and associated components have also been designed for long life. The valve spring rests on a steel washer to prevent the spring embedding in the cylinder head. The Viton (a heat-resistant synthetic rubber) seals on the valve

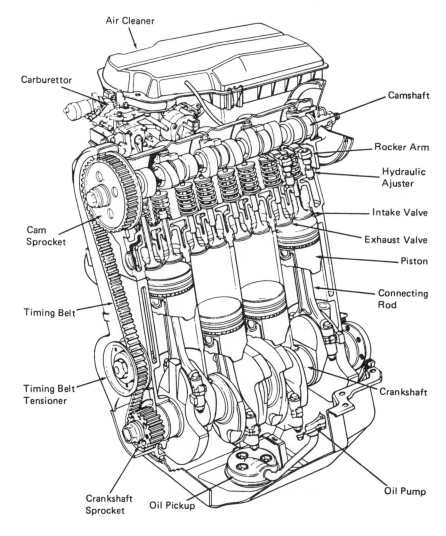

Figure 15.8
Chrysler 2.2-litre four-cylinder engine – longitudinal section (reprinted with permission © 1981 Society of Automotive Engineers, Inc.).

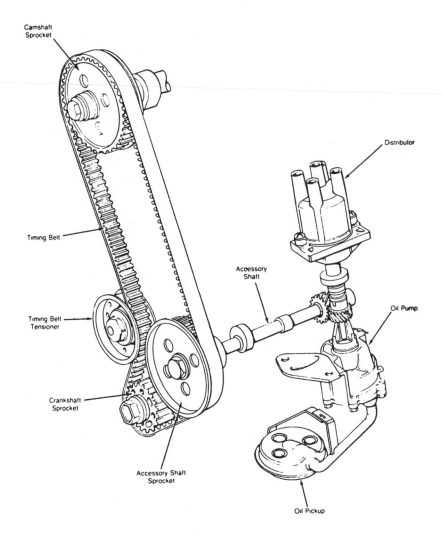

Camshaft
Sprocket

Distributor

Timing Belt

Accessory
Shaft

Oil Pump

Timing Belt
Tensioner

Crankshaft
Sprocket

Accessory Shaft
Sprocket

Oil Pickup

Figure 15.9

Chrysler 2.2-litre four-cylinder engine: timing belt drive–accessory shaft–distributor and oil pump (reprinted with permission © 1981 Society of Automotive Engineers, Inc.).

spindles prevent oil entering the combustion chamber through the valve guides; this could be a problem with the copious supply of lubricant to the overhead camshaft. Table 15.2 highlights some of the design and materials aspects of the valve gear.

The valve guides and seat inserts are all inserted after pre-cooling in liquid nitrogen.

The compact combustion chamber is as cast, and the piston is flat topped; there are two separate squish or quench areas. The spark plug is located very close to the centre of the combustion chamber.

To ensure predictable and reliable clamping of the cylinder head to the block, a 'joint control' system is used, in which carefully toleranced bolts are loaded to their yield point. The cylinder head gasket has minimal compression relaxation, and there are stainless steel bore flanges for the cylinder bore sealing.

15.3.3 The cylinder block and associated components

The cylinder block is cast iron, and the weight is kept to a minimum (35 kg), by having a nominal wall thickness of 4.5 mm and a skirt depth of only 3 mm below the crankshaft centre-line. There are five main bearings, with bearing caps of the same material – this enables bore circularity to be maintained without resorting to a honing operation. All the shell bearings are either a lead–aluminium alloy or tri-metal on a steel backing strip.

The crankshaft is of nodular cast iron, with main bearing journal diameters of 60 mm, and crank-pin or big-end journal diameters of 50 mm. All the crankshaft journals have under-cut radiused fillets; these are deep rolled to give a fatigue strength improvement of 35 per cent.

Table 15.2 *Chrysler 2.2-litre engine valve components*

	Inlet valve	Exhaust valve
Head diameter (mm)	40.6	35.4
Stem diameter (mm)	8	8 (tapered to allow for thermal expansion)
Valve material	SAE 1541 steel with hardened tip	Head: 21–2N steel Stem: SAE 4140 steel with hardened tip
Stem finish	Flash chrome	Heavy chrome
Valve guide	Low alloy cast iron	Hardenable iron
Valve seat	Sintered copper–iron alloy	Sintered cobalt–iron alloy
Valve lift (mm)	10.9	10.9
Valve timing: open	12° btdc	48° bbdc
close	52° abdc	16° atdc

The connecting-rod and cap are made from forged steel, with a centre distance of 151 mm. A squirt hole in the connecting-rod provides cylinder lubrication from the big-end.

The piston design is the result of finite element modelling and high-speed engine tests. The piston is made from aluminium alloy, and has cast-in-steel struts at the gudgeon-pin bosses to control the thermal expansion. The top ring is made from nodular cast iron with a molybdenum-filled, radiused face for wear and scuff resistance. In contrast the second ring has a phosphate-coated, tapered face; both rings are 1.5 mm thick. The oil control ring is of a three-piece construction with a stainless steel expander, and chromium-faced side rails.

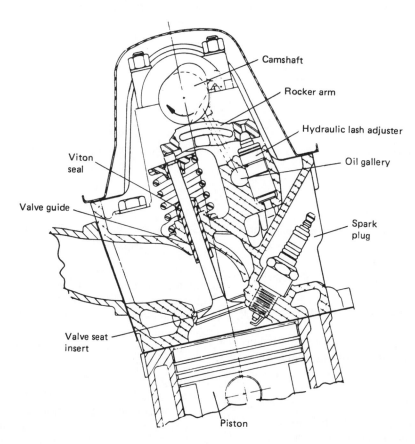

Figure 15.10

Chrysler 2.2-litre engine cylinder head section (reprinted with permission © 1981 Society of Automotive Engineers, Inc.).

Table 15.3 *Performance of the Chrysler 2.2-litre engine and its derivatives*

Engine description	Max. torque; speed		Max. power; speed	
	(N m)	(rpm)	(kW)	(rpm)
Standard, EDE	161	3200	72	5200
Turbo I	217	3200	106	5600
Turbo II	256	3200	130	4800
Lotus/Shelby	305	3200	169	6000

15.3.4 Combustion control

Combined electronic control is used for the ignition timing and the air/fuel mixture preparation. Signals from seven transducers are processed electronically:

(i)   ambient air temperature

(ii)   engine load

(iii)   carburettor throttle plate, open or closed

(iv)   engine speed

(v)   engine coolant temperature

(vi)   exhaust gas oxygen level

(vii) engine starting.

The air/fuel mixture is prepared in a twin choke carburettor with progressive choke operation. The carburettor has an electric choke, and an electronic feedback system for mixture strength control. The electronic engine-management system ensures optimum fuel-economy performance, and driveability, while still meeting stringent exhaust emission regulations.

The induction and exhaust passages were optimised by air flow testing with plastic models. The compact combustion chamber, in an aluminium alloy cylinder head, with significant squish or quench areas, permits a compression ratio of 8.9:1 to be used with 91 RON fuel; the final performance is shown in figure 15.11. The engines have a compression ratio of 9.6:1, or 9:1 if naturally aspirated, while the turbocharged engine has the compression ratio reduced to 8.1:1.

In all versions of the engine, emission control is by a combination of exhaust gas recirculation (EGR), air injection (exhaust manifold outlet cold; catalytic converter hot) and a catalytic converter below the exhaust manifold. The catalytic converter uses a three-way catalyst on a monolithic substrate.

15.3.5 Engine development

The output of the Chrysler 2.2-litre engine has been increased by turbocharging. The performances of the three different turbocharged derivatives are compared in table 15.3 with the baseline engine performance.

The Turbo II engine uses an air-to-air inter-cooler, and the maximum boost pressure has been increased from 0.7 bar to 0.83 bar (Anon, 1986). The Turbo II engine uses a forged steel crankshaft, and a tuned induction system that raises the torque curve by about 10 per cent (this is particularly useful for the low speeds at which the turbocharger performance is

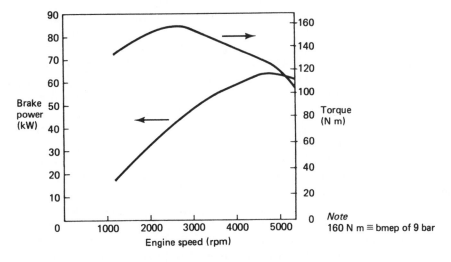

Figure 15.11
Chrysler 2.2-litre engine output (reprinted with permission © 1981 Society of Automotive Engineers, Inc.).

limited). The primary runners from each cylinder to the manifold are 370 mm long, and give a torque improvement in the range of 4800 rpm. The secondary runner (from the inlet manifold to the throttle body) is 300 mm long, and provides a torque improvement in the 2000 rpm range.

The Lotus/Shelby engine employs four valves per cylinder and a Garrett T3 turbocharger (Birch, 1988).

15.4 | Ford 2.5-litre DI diesel engine

15.4.1 Background

Compression ignition engines with direct fuel injection (DI) have typically a 10–15 per cent fuel economy benefit over indirect engines and are easier to start. However, indirect engines were used in light automotive applications, because the restricted speed range of direct injection engines (say up to 3000 rpm) led to a poor power-to-weight ratio, and necessitated a multi-ratio gearbox. Indirect injection engines achieved their greater speed range by injecting fuel into a pre-chamber where there is rapid air motion. The air motion is generated in a throat that connects the pre-chamber and the main chamber; but this also leads to pressure drops and the high heat transfer coefficients that account for the reduced efficiency.

The Ford high-speed direct injection engine achieves a high engine speed (4000 rpm) by meticulous attention to the fuel injection and air motion. This four-cylinder engine is naturally aspirated, slightly over-square, with a bore of 93.7 mm and a stroke of 90.5 mm. Since there is no need to pump air into and out of the pre-chamber the compression ratio can be lower, while still maintaining good starting performance. Consequently, the compression ratio is 19:1 while the compression ratio for an indirect injection engine is typically 22 : 1. In 1984 this engine was the first automotive high-speed direct injection (HSDI) diesel engine (Bird, 1985).

15.4.2 Description

An isometric sectioned drawing of the 1992 model year engine is shown in figure 15.12. A single toothed belt drives the camshaft and fuel injection pump, while poly-vee belt drives are used for the other engine auxiliaries. The inlet and exhaust valves are all in-line, but are offset slightly from the cylinder bore axes; the induction and exhaust manifolds are on opposite sides of the engine to provide a cross-flow system. The engine is inclined at an angle of $22\frac{1}{2}°$ to the vertical, and this is shown more clearly in figure 15.13, a sectioned view of the engine.

The camshaft is mounted at the side of the cylinder block and the cam followers operate the valves through short push rods and rocker arms. The valve clearance is adjusted by spherically ended screws in the rocker arms, which engage with the push rods. To ensure valve train rigidity, the rocker shaft is carried by five pedestal bearings. The single valve spring rests on a hardened steel washer (to reduce wear), and the spring retainer is connected to the valve stem by a multi-groove collet. The valves operate in valve guides inserted into the cast iron cylinder head; the valve seats are hardened by induction heating after machining. The valve stem seals restrict the flow of lubricating oil down the valve stem, in order to reduce the build-up of carbon deposits.

The cylinder head for a direct injection engine is simpler than a comparable indirect injection design since there is no pre-chamber or heater plug. The cast iron cylinder head contains a dual-acting thermostat, which restricts water circulation to within the engine and heater until a temperature of about 82°C is attained. The water flow to the cylinder head from the cylinder block is controlled by graduated holes in the cylinder head gasket. To ensure optimum clamping of the cylinder head to the block, a torque-to-yield system is used on the cylinder head bolts.

The cylinder bores are an integral part of the cast iron cylinder block. A finite element vibration analysis was used in the design of the cylinder block to help minimise the engine noise levels. The main and big-end bearing shells use an aluminium–tin alloy. The crankshaft is made from nodular cast iron, with induction hardened journals to reduce wear. All the crankshaft journals are under-cut and fillet-rolled to improve the fatigue life. The connecting-rods are made from an air hardening steel that is relatively easy to machine, despite strengths of up to 930 MN/m^2.

The pistons are made from an aluminium alloy and are expansion-controlled; this enables the bore clearance to be reduced to a minimum of 8 μm, thereby minimising piston slap and its ensuing noise. This close tolerance is obtained by selective assembly during manufacture with four grades of piston skirt diameter. The piston carries an oil control ring and two compression rings; the top ring is located in a cast iron insert to achieve long life.

The piston clearance at top dead centre is carefully controlled to ensure the correct power output and smoke conformity at the higher speeds. This clearance is controlled by selective assembly of the pistons and connecting-rods to match a given crankshaft and block. Four grades of connecting-rod (by length) are matched with five grades of piston height (gudgeon-pin centre to piston crown). This enables the compression ratio to be kept to within ± 0.7 of the nominal value.

To enable a high-speed direct injection engine to operate there has to be rapid, yet controlled, combustion. This is attained by meticulous attention to the fuel injection equipment and the in-cylinder air motion.

The fuel injector is inclined at an angle of 23° to the cylinder axis, and it is offset from the centre-line of the engine (away from the inlet and exhaust valves). The shallow toroidal bowl in the piston is also offset from the cylinder axis. The peak fuel

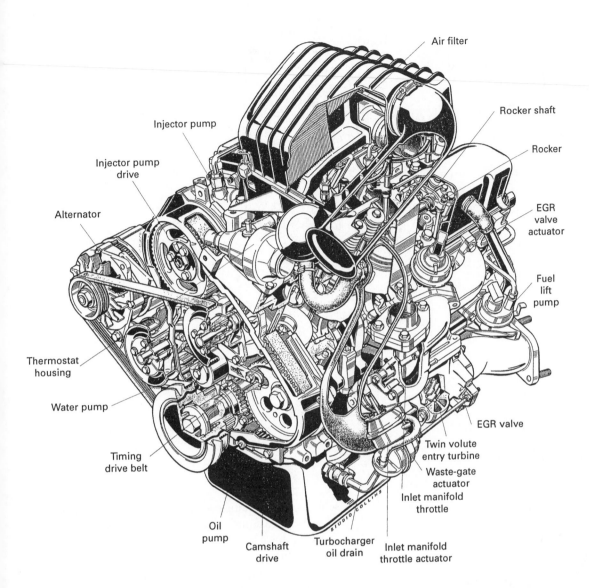

Air filter

Rocker shaft

Rocker

Injector pump

EGR
valve
actuator

Injector pump
drive

Fuel
lift
pump

Alternator

Thermostat
housing

Water pump

EGR valve

Twin volute
entry turbine

Timing
drive belt

Waste-gate
actuator

Inlet manifold
throttle

Oil
pump

Camshaft
drive

Turbocharger
oil drain

Inlet manifold
throttle actuator

Figure 15.12
Ford 2.5-litre turbocharged HSDI low emission engine (courtesy of Ford).

injection pressure is about 700 bar; this is twice the traditional value for a rotary type fuel pump.

Direct injection engines have inherently good cold starting characteristics, and the automatic provision of excess fuel gives quick-starting down to temperatures of −10°C. For temperatures down to −20°C an electrically operated flame heater is used. Also, for this lower temperature range an additional battery is fitted. For cold-starting and warming-up, the injection timing is automatically advanced to avoid the white smoke from unburnt fuel. Independently, a separate water-temperature sensor causes the engine idle speed to be raised until the engine reaches its normal operating temperature.

Air flow management is a crucial aspect in the development of the high-speed direct injection engine. The swirl is generated by the shape of the inlet port (figure 15.14), and the correct trade-off is needed between swirl and volumetric efficiency. To limit the smoke output a high air flow rate is needed at high speeds, while at low speeds a high swirl is needed. Unfortunately inducing more swirl reduces the volumetric efficiency. During production every cylinder head is checked for the correct air flow performance.

The aluminium induction manifold is the result of a CAD study, and the manufacture is tightly controlled to ensure accurate alignment with the ports in the cylinder head. Any misalignment between the passages in the inlet manifold and the cylinder head has a disproportionately serious effect on the flow efficiency. The consequences would be particularly

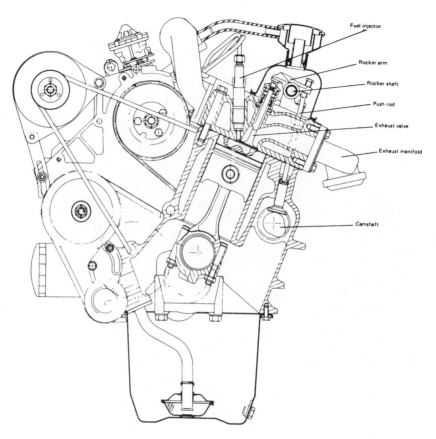

Figure 15.13
Cross-section of the Ford 2.5-litre naturally aspirated DI diesel (courtesy of Ford).

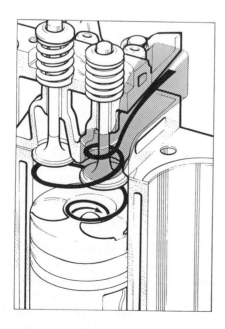

Figure 15.14
Swirl generation in the Ford 2.5-litre DI diesel (courtesy of Ford).

serious on a direct injection engine. To improve the air flow at the valve throat, the valve seat is angled at 30°, and the valve throat is angled at 45°. During development a decrease in valve diameter was found to improve the air flow – this is explained by the reduced interference between the inlet air flow and the cylinder wall.

Turbocharged versions of the engine have been produced using a Garrett T2 turbocharger and a KKK K04 turbocharger (Bostock and Cooper, 1992). Table 15.4 compares the performance of the naturally aspirated and turbocharged engines. The higher torque at a lower speed improves the driveability.

15.4.3 Combustion system

The combustion system has been developed since the initial launch in 1984, so as to meet emissions legislation. This has also led to a slight increase in power (52 kW at 4000 rpm), a higher torque that now occurs at a lower speed (146 N m at 2500 rpm) and a lower minimum specific fuel consumption of 220 g/kWh (Bird *et al.*, 1989). The performance of the engine is illustrated by the specific fuel consumption map shown in figure 15.15.

Table 15.4 *Comparison of the performance of the naturally aspirated and turbocharged Ford 2.5 DI engines*

Engine	Maximum torque; speed	Maximum power; speed
Naturally aspirated	146 N m; 2500 rpm	52 kW; 4000 rpm
Turbocharged (Garrett T2)	173 N m; 2300 rpm	60 kW; 3900 rpm
Turbocharged (KKK K04)	226 N m; 2100 rpm	75 kW; 4000 rpm
Turbocharged and inter-cooled	255 N m; 2100 rpm	86 kW; 4000 rpm

A key element in the refinement of the combustion system was the adoption of a Stanadyne Slim Tip Pencil Nozzle (STPN). This nozzle has a very long injector needle, with the spring and its adjustment remote from the nozzle tip. The reduction in the injector body diameter means that the Slim Tip Pencil Nozzle can be mounted so that its axis is only offset 5.5 mm from the cylinder axis; this compares with 9.5 mm for the original injector. Moving the injector closer to the cylinder axis led to a direct reduction in smoke at a given operating point, as a result of better fuel/air mixing and air utilisation. In parallel, a new design of re-entrant bowl with a higher compression ratio (20.7) had led to performance improvements with the original injector (designated 17/21). Thus tests were conducted with the new combustion chamber bowl, in order to make comparisons between the two injectors; the two engine builds are shown in figure 15.16.

The emissions legislation in Europe and the USA for cars and light trucks is in terms of engines being tested in vehicles over a drive cycle. This is not very convenient (or necessarily informative) when prototype engines are being evaluated. Instead, a widely adopted approach is to identify load/speed

operating points that are representative of the drive cycles. The load/speed points have to be identified by a statistical analysis that correlates steady-state engine/dynamometer tests with vehicle/chassis dynamometer drive cycle tests. For the Ford 2.5 DI diesel engine four test points were identified, and the emissions data were expressed in terms of a four-point average.

The emissions results for the two engine builds (figure 15.16) are shown in figure 15.17. Each build was tested with the standard injection timing, and with 2° advance and 2° retard. The Slim Tip injector gave lower gaseous emissions, and a smoke level that was always more than 1 Bosch Smoke Unit lower. The low emission target box in figure 15.17, which represents the emissions legislation requirements, assumed that EGR could be applied to reduce the NO_x emissions by 50 per cent. The EGR is drawn into the inlet manifold by throttling the air inlet and using a conventional diaphragm-operated EGR valve. The EGR system has been described by Brandstetter and Howard (1989). The EGR level (in terms of volume flow) rises from zero at full load, to 50 per cent at low load.

Further optimisation of the combustion system was undertaken, and this involved the use of 60 variations in the injector nozzle, covering:

orifice size, number (4, 5 or 6), length, and spray pattern
opening pressure, valve lift and spring rate sac volumes
valve to seat diameter ratios

The experimental engine programme was backed by rig testing of the injection equipment and computer modelling of the injection system. The result was a five-hole nozzle with a spray cone angle of 150° and a hole size of either 0.21 mm or 0.22 mm according to the selection of injector pump. The injector development included a novel extension to the injector needle that further reduced the sac volume in the injector tip. This reduction in sac volume reduced the unburnt hydrocarbon emissions.

Other refinements concerned the piston and ring pack. An AEconoguide low-friction piston was adopted, and the piston bowl was moved towards the cylinder axis – this led to a higher combustion efficiency and a potential for reduced noise with retarded injection. The oil control ring has been improved and the ensuing lower lubricant consumption also leads to lower emissions of particulates and unburned hydrocarbons. The piston ring pack has also been moved up, so the top ring is

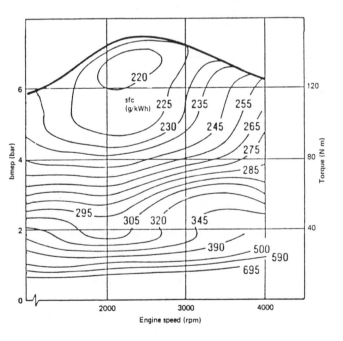

Figure 15.15
Specific fuel consumption contours (g/kWh) for the Ford 2.5-litre DI diesel engine (adapted from Brandstetter and Howard, 1989).

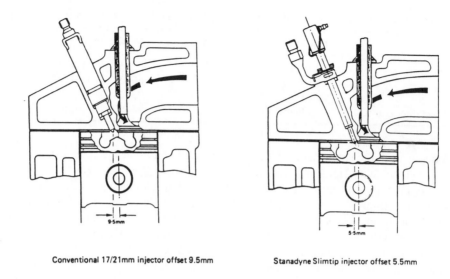

Conventional 17/21mm injector offset 9.5mm   Stanadyne Slimtip injector offset 5.5mm

Figure 15.16
Combustion system development for the Ford 2.5-litre DI diesel engine (Bird *et al.*, 1989).

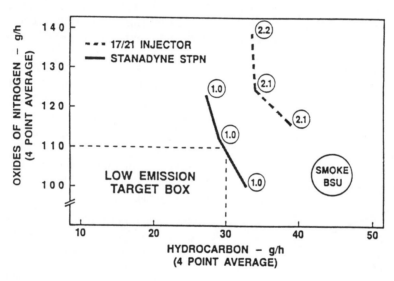

Figure 15.17
Comparison of gaseous emissions of conventional 17/21 mm and Stanadyne Slim Tip injectors (Bird *et al.*, 1989).

now 7 mm away from the piston crown. As with spark ignition engines, reductions in crevice volumes led to lower hydrocarbon emissions. The response of this engine to changes in the fuel cetane rating and changes in fuel volatility have been reported by Cooke *et al.* (1990). They conclude that there is limited potential for reducing exhaust emissions with conventional diesel fuel blending components.

15.4.4 Turbocharged engine development

Apart from the turbocharger, the main development associated with the KKK K04 turbocharged engine was the use of an electronic engine management system. The peak cylinder pressure was limited to 110 bar, to ensure an acceptable piston pin loading, and the exhaust temperatures were limited to 720°C to maximise the exhaust valve and turbine life. Figure 15.18 shows how the bmep has been increased significantly compared to the naturally aspirated engine, and despite more stringent emissions requirements the minimum specific fuel consumption has not deteriorated.

The turbocharged engine uses the Lucas EPIC system (see chapter 6, figure 6.29 in section 6.5.3), which is based around an electrically controlled rotary fuel pump (Bostock and Cooper, 1992). Sensors are connected to the electronic engine controller, in the manner shown in figure 15.19, with measurements of

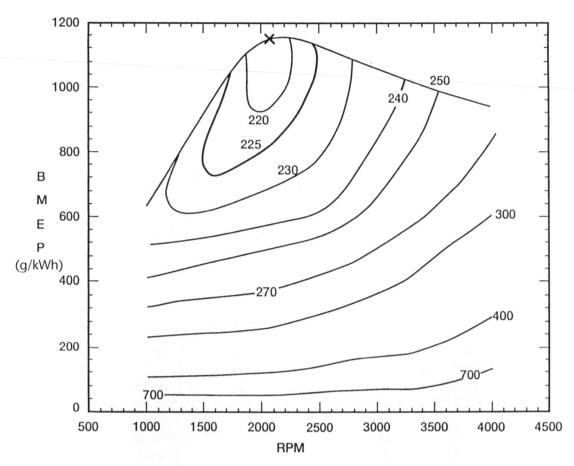

Figure 15.18
The specific fuel consumption map (g/kWh) for the Ford 2.5 HSDI turbocharged diesel engine (courtesy of Ford).

inlet manifold absolute pressure	inlet manifold temperature
coolant temperature	crankshaft speed and position
EGR valve position	EGR throttle position
driver ('throttle') demand	injector pump feedback signals

For every engine operating condition (starting, idling, limited by the governor, providing engine braking and so on) there are sets of rules that determine the required fuelling (quantity and timing) and the EGR level, as a function of load and speed.

Figure 15.19 shows that the EGR valve and the air throttle are operated by vacuum actuators, with a feedback signal telling the ECU its position. The vacuum is generated by the vacuum pump for the braking servo, and the vacuum signal level is adjusted by controlling the on-to-off ratio of a pulsed solenoid valve which switches between the vacuum and atmospheric pressure. This system ensures accurate positional control of the EGR valve. The EGR valve position correlates well with EGR level, so it is not necessary to have a direct measurement of the EGR level (for example, by measuring the oxygen level in the inlet manifold). However, it is

necessary to ensure that the EGR calibration does not drift with engine condition. The effect of faults such as a blocked air filter, wear on the engine reducing its volumetric efficiency, a restricted exhaust system and deposits in the EGR system all need to be reviewed. Such a study was undertaken by Lancefield *et al.* (1996) using a system modelling approach to model the dynamic response. Once the model had been verified with experimental data, it was used to investigate different engine management system strategies. The favoured engine management system strategies were then tested on the engine.

Figure 15.20 shows a typical start of injection timing map. In general the injection is more advanced as the speed and load are increased. However, there are additional rules, so that when the coolant is cold, the injection is more advanced. There are also five different maps for temperatures in the range -20 to $+60°C$.

In the normal driving mode, the fuel injected is the minimum of

1 that signalled by the driver demand,

2 the maximum allowed by the pre-set torque curve, or

Schematic diagram

1 Electronic control unit (ECU)
2 Fuel tank
3 Fuel lift pump
4 Fuel filter
5 Fuel injection pump
6 Injection nozzles
7 Air cleaner
8 Turbocharger
9 Throttle plate
10 Cross-over duct
11 Manifold absolute pressure (MAP) sensor
12 Air charge temperature (ACT) sensor
13 Engine coolant temperature (ECT) sensor
14 EGR vacuum flow regulator
15 Exhaust gas recirculation (EGR) valve
16 Engine speed sensor
17 Accelerator pedal position sensor
18 Power hold relay
19 Engine management system lamp
20 Self test connector
21 Self test connector (LUCAS)

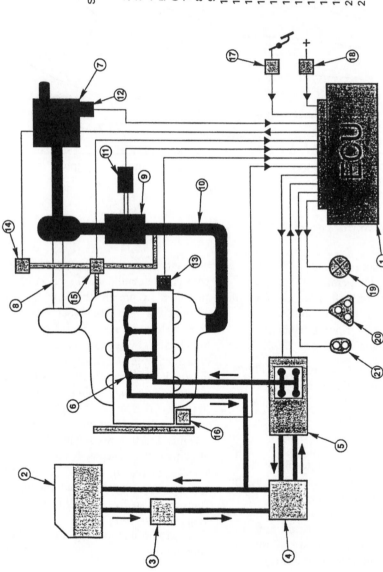

Figure 15.19
The Lucas EPIC system applied to the Ford 2.5 HSDI turbocharged diesel engine (courtesy of Ford).

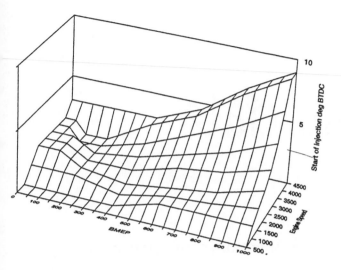

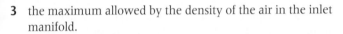

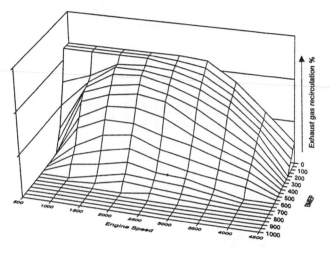

Figure 15.20
Typical start of injection timing map for the Ford 2.5 HSDI turbocharged diesel engine (courtesy of Ford).

Figure 15.21
EGR map for the Ford 2.5 HSDI turbocharged diesel engine (courtesy of Ford).

3 the maximum allowed by the density of the air in the inlet manifold.

The air density is calculated from the manifold absolute pressure and temperature, and was found always to be known to an accuracy of better than 3 per cent, thus not needing an air mass flowmeter. The fuelling is limited when there is a lower than expected boost pressure, such as high-altitude operation, a turbocharger fault, or rapid engine acceleration. If the waste gate should fail in a closed position, the maximum fuelling is limited by the pre-set torque curve, to avoid the engine combustion pressures rising too high.

EGR is a well-established method of controlling NO_x emissions, and the Ford HSDI engine was the first application in which cooled EGR was used. Figure 15.21 shows the EGR schedule, with up to half of the air being replaced by EGR at light loads. The intake throttle valve (after the compressor) is

needed in order to generate a sufficient pressure differential at low loads to achieve sufficient flow of the exhaust gases. At speeds above 1500 rpm with zero fuel delivery (corresponding to engine braking) the EGR level is set to zero. This fills the intake system with air, since engine braking is frequently followed by a need for acceleration, and if EGR had been applied then there is a possibility of overfuelling since it takes a finite time for the EGR to be eliminated from the inlet manifold.

Comparatively minor changes were needed to the base engine for turbocharging (Bostock and Cooper, 1992). There was some strengthening, and the coolant and oil-pump flow rates were increased, and an oil cooler was added. A squeeze cast piston was used for its higher strength (see chapter 12, section 12.4) and the timing belt drive was changed to a deeper tooth profile to carry the increased loading from the injection pump.

APPENDIX A

Thermodynamic data

A.1 | Introduction

The thermodynamic tables presented here for enthalpy and internal energy differ from those that are usually available, since they incorporate the enthalpy of formation. This means that there is no need for separate tabulations of calorific values, and it will be found that energy balances for combustion calculations are greatly simplified. The enthalpy of formation (ΔH_f^0) is perhaps more familiar to physical chemists than to engineers.

> The **enthalpy of formation** (ΔH_f^0) of a substance is the standard reaction enthalpy for the formation of the compound from its elements in their reference state.

> The **reference state** of an element is its most stable state (for example, carbon *atoms*, but oxygen *molecules*) at a specified temperature and pressure, usually 298.15 K and a pressure of 1 bar. In the case of atoms that can exist in different forms, it is necessary to specify their form, for example, carbon is as graphite, not diamond.

Combustion calculations are most readily undertaken by using absolute (sometimes known as sensible) internal energies or enthalpy. In steady-flow systems where there is displacement work, enthalpy should be used; this has been illustrated by figure 3.8. When there is no displacement work then internal energy should be used (figure 3.7). Consider now figure 3.8 in more detail. With an adiabatic combustion process from reactants (R) to products (P) the enthalpy is constant, but there is a substantial rise in temperature:

$$H_\mathrm{1R} = H_\mathrm{2P}, \quad T_2 > T_1 \tag{A.1}$$

In the case of the isothermal combustion process (IR → IP) the temperature (T) is obviously constant, and the difference in enthalpy corresponds to the isobaric calorific value $(-\Delta H_T^0)$:

$$\Delta H_T^0 = H_{\mathrm{R},T} - H_{\mathrm{P},T} \tag{A.2}$$

A.2 | Thermodynamic principles

In the following sections, it will be seen how the thermodynamic data for internal energy, enthalpy, entropy and Gibbs function can all be determined from measurements of heat capacity and phase change enthalpies (or internal energies). Furthermore, when the energy change associated with a chemical reaction is measured, the enthalpy of formation can be deduced. This in turn leads to 'absolute' values of internal energy, enthalpy, entropy and Gibbs energy, from which it is possible to derive the equilibrium constant for any reaction.

A.2.1 Determination of internal energy, enthalpy, entropy and Gibbs energy

Figure 3.8 shows that the enthalpies of both reactants and products are in general non-linear functions of temperature. The Absolute Molar Enthalpy tables presented here use a datum of *zero enthalpy for elements when they are in their standard state at a temperature of 25°C*. The enthalpy of any substance at 25°C will thus correspond to its enthalpy of formation, ΔH_f^0. The use of these tables will be illustrated, after a description of how they have been developed.

Tables are not very convenient for computational use, so instead molar enthalpies and other thermodynamic data are evaluated from analytical functions; a popular choice is a simple polynomial. For species i

$$H_i(T) = A_i + B_i T + C_i T^2 + D_i T^3 + E_i T^4 + F_i T^5 \tag{A.3}$$

from which it can be deduced (since $\mathrm{d}H = C_\mathrm{p}\mathrm{d}T$) that

$$C_{\mathrm{p},i}(T) = B_i + 2C_i T + 3D_i T^2 + 4E_i T^3 + 5F_i T^4 \tag{A.4}$$

As $U = H - RT$

$$U_i(T) = A_i + (B_i - R_0)T + C_i T^2 + D_i T^3 + E_i T^4 + F_i T^5 \tag{A.5}$$

and $\mathrm{d}U = C_\mathrm{v}\mathrm{d}T$, so

$$C_{\mathrm{v},i}(T) = (B_i - R_0) + 2C_i T + 3D_i T^2 + 4E_i T^3 + 5F_i T^4 \tag{A.6}$$

and as $\quad dH = C_p dT = TdS$

$\qquad dS = (C_p/T)dT$

integrating equation (A.4) gives

$$S_i^0 = B_i \ln(T) + 2C_i T + 3/2 D_i T^2 + 4/3 E_i T^3 + 5/4 F_i T^4 + G_i \tag{A.7}$$

where G_i is an integration constant that is used to set the zero datum, for example 0 K by Rogers and Mayhew (1988) – the same datum does not have to be used as for enthalpy.

A more common choice is to use a polynomial function to describe the specific heat capacity variation, and to divide through by the Molar Gas Constant (R_0). Equation (A.4) becomes

$$C_{p,i}(T)/R_0 = a_i + b_i T + c_i T^2 + d_i T^3 + e_i T^4 \tag{A.8}$$

where $a_i \equiv B_i/R_0, \quad b_i \equiv 2C_i/R_0, \quad c_i \equiv 3D_i/R_0$ etc. Thus

$$H_i(T)/R_0 = a_i T + b_i T^2/2 + c_i T^3/3 + d_i T^4/4 + e_i T^5/5 + f_i \tag{A.9}$$

and

$$S_i^0(T)/R = a_i \ln(T) + b_i T + c_i/2 T^2 + d_i/3 T^3 + e_i/4 T^4 + g_i \tag{A.10}$$

A polynomial fit will only be satisfactory over a small temperature range (300–1000 K or 1000–5000 K), and such polynomial equations ought never to be used outside their range. As the specific heat capacity variation with temperature has a 'knee' between 900 K and 2000 K, a single polynomial is never likely to be satisfactory. Instead, two polynomials can be used which give identical values of C_p at the transition between the low range and the high range (the transition temperature is usually chosen between 1000 and 2000 K). Examples of such polynomials are presented by Gordon and McBride (1971), and were used as the basis for constructing the tables used here. The coefficients for the evaluation of the thermodynamic tables are summarised in table A.1.

The tables at the end of this Appendix present the enthalpy (H), internal energy (U), entropy (S) and Gibbs energy (G) for

gaseous species (table A.4) and fuels (table A.5). The tables have been extrapolated below 300 K, and these data should be used with caution. The entropy datum of zero at 0 K cannot be illustrated by the evaluation of equation (A.10), since there is a singularity in this equation at 0 K, and in any case there will be phase changes. Phase changes will lead to isothermal changes in entropy, and each different phase will have a different temperature/entropy relationship. Instead, use is made of the values of entropy for substances at 25°C in their standard state at a pressure of 1 bar. (**WARNING** – many sources use a datum pressure of 1 atm.)

The entropy can be evaluated at other pressures from:

$$S = S^0 - R_0 \ln(p/p^0) \tag{A.11}$$

The internal energy (U) and Gibbs energy (G) are by definition

$$H = U + pV = U + R_0 T \quad \text{and} \quad G^0 = H^0 - TS^0 \tag{A.12}$$

with the superscript ^{0} referring to the datum pressure (p^0) of 1 bar. The Gibbs energy can be evaluated at other pressures in a similar way to the entropy (equation A.11), and through the use of this equation:

$$G = G^0 + (H - H^0) - T(S - S^0) = G^0 + R_0 T \ln(p/p^0) \tag{A.13}$$

The changes in S and G between state 1 (p_1, T_1) and state 2 (p_2, T_2), for a gas or vapour, are given by

$$S_2 - S_1 = (S_2 - S_2^0) + (S_2^0 - S_1^0) + (S_1^0 - S_1)$$
$$S_2 - S_1 = (S_2^0 - S_1^0) - R_0 \ln(p_2/p_1) \tag{A.14}$$

and $\quad G_2 - G_1 = (G_2 - G_2^0) + (G_2^0 - G_1^0) + (G_1^0 - G_1)$
$$G_2 - G_1 = (G_2^0 - G_1^0) + R_0 T_2 \ln(p_2/p^0) - R_0 T_1 \ln(p_1/p^0) \tag{A.15}$$

When the entropy and Gibbs energy of a mixture are being evaluated, the properties of the individual constituents are summed, but the pressures (p_1 and p_2) now refer to the partial pressures of each constituent.

The use of these tables in combustion calculations is best illustrated by an example.

EXAMPLE

A mixture of carbon monoxide and 10 per cent excess air at 25°C is burnt at constant pressure, and it is assumed that no carbon monoxide is present in the products. Treat air as 79 per cent nitrogen and 21 per cent oxygen, and estimate the adiabatic flame temperature.

Solution:
The stoichiometric equation is $CO + x(O_2 + 79/21\, N_2) \rightarrow CO_2 + 79x/21\, N_2$
Balancing of the oxygen *atoms* gives: $1 + 2x = 2$, or $x = 0.5$
With 10 per cent excess air the combustion equation is

$\quad CO + 0.55(O_2 + 79/21\, N_2) \rightarrow CO_2 + 0.05 O_2 + 2.07\, N_2$

At 25°C, the enthalpy of the reactants (H_R) is

$H_R = -110.525 + 0.55(0 + 0) = -110.525$ MJ/kmol CO

Since the flame is adiabatic, $H_P = H_R = -110.525$ MJ/kmol CO
Thus a temperature has to be found at which the enthalpy of the products will sum to -110.525 MJ/kmol.

1st guess 2000 K

$$H_{P,2000} = 1 \times -302.128 + 0.05 \times 59.171 + 2.07 \times 56.114 = -183.013$$

2nd guess 3000 K

$$H_{P,3000} = 1 \times -240.621 + 0.05 \times 98.116 + 2.07 \times 92.754 = -43.714$$

The *3rd guess* can be based on linear interpolation:

$T_3 = 2000 + 1000 \times (-110.525 + 183.013)/(-43.714 + 183.013) = 2520$ K

More accurate interpolation of the tables would lead to a temperature of 2523 K, but this is of limited purpose, since dissociation results in a temperature of about 2350 K.

A.2.2 Equilibrium constants

Chemical reactions move towards an equilibrium in which both the reactants and products are present. If the concentration of the products is much greater than that of the reactants, the reaction is said to be 'complete'. However, at the elevated temperatures associated with combustion there may be reactants, products and partial products of combustion (such as CO) present. Chemical reactions proceed in such a way as to minimise the Gibbs energy of the system, since this is the requirement for any system to be in equilibrium. This can be established by considering equation (2.11):

$$W_{REV} = G_{R0} - G_{P0} = -\Delta G_0 \tag{2.11}$$

If a system has the capability for doing work, the Gibbs energy will be reduced, thus when no more work can be done, the system will be in equilibrium and the Gibbs energy will be a minimum. For reacting mixtures it is helpful to introduce a parameter to define the extent of a reaction (ξ).

At a constant temperature and pressure, consider a reaction in which A is in equilibrium with B:

$$A \rightleftharpoons B$$

For an infinitesimal change $d\xi$ of A into B:

the change in amount of A present is $\quad dn_A = -d\xi$, and
the change in amount of B present is $\quad dn_B = +d\xi$

$$\tag{A.16}$$

The change in Gibbs energy at constant temperature and pressure, when the concentration of a species changes (with no other changes in composition of the mixture) is known as its chemical potential (μ). Thus

$$\mu = (\partial G/\partial n)_{T,p,n_j} \tag{A.17}$$

The subscript n_j indicates that there is no change in the amounts of any of the other species that might be present.

For our simple system with only substances A and B, the change in Gibbs energy is given by

$$dG = (\partial G/\partial n)_{T,p,n_B} dn_A + (\partial G/\partial n)_{T,p,n_A} dn_B = \mu_A dn_A + \mu_B dn_B$$
$$= -\mu_A d\xi + \mu_B d\xi \tag{A.18}$$

The change in Gibbs energy with a change in the extent of the reaction is

$$(\partial G/\partial \xi)_{T,p} = -\mu_A + \mu_B \tag{A.19}$$

and at equilibrium $\quad (\partial G/\partial \xi)_{T,p} = 0 \tag{A.20}$

Since the Gibbs energy is now a minimum, there can be no scope for the system to do any work, and the system will be at equilibrium.

This now needs to be extended to a multi-component reaction, in which a kmols of species A, b kmols of species B and so on react to produce c kmols of species C, d kmols of species D:

$$aA + bB \rightleftharpoons cC + dD \quad \text{or} \quad aA + bB - cC - dD = 0$$

This can be generalised as: $\quad \Sigma v_i A_i = 0 \tag{A.21}$

where v_i is the stoichiometric coefficient of species A_i.

When the extent of the reaction changes by $d\xi$, the amounts of the reactants and products change by:

$$dn_A = -ad\xi, \quad dn_B = -bd\xi, \quad dn_C = +cd\xi, \quad \text{and}$$
$$dn_D = +dd\xi \tag{A.22}$$

and in general: $\quad dn_i = -v_i d\xi \tag{A.23}$

So, at constant pressure and temperature, the change in Gibbs energy is

$$dG = \mu_A dn_A + \mu_B dn_B + \mu_C dn_C + \mu_D dn_D$$
$$= (-a\mu_A - b\mu_B + c\mu_C + d\mu_D)d\xi \tag{A.24}$$

Table A.1 Coefficients in equations (A.8), (A.9) and (A.10), for the evaluation of thermodynamic data from Gordon and McBride (1971), except argon, from Reid et al. (1987)

	N_2	O_2	H_2	CO	CO_2	H_2O	NO	Ar	OH	O	H
					1000–5000 K						
a_i	0.28963E1	0.36220E1	0.31002E1	0.29841E1	0.44608E1	0.27168E1	0.31890E1	2.5016	0.29106E1	0.25421E1	2.5
b_i	0.15155E-2	0.73618E-3	0.51119E-3	0.14891E-2	0.30982E-2	0.29451E-2	0.13382E-2	0	0.95932E-3	-0.27551E-4	0
c_i	-0.57235E-6	-0.19652E-6	0.52644E-7	-0.57900E-6	-0.12393E-5	-0.80224E-6	-0.52899E-6	0	-0.19442E-6	-0.31028E-8	0
d_i	0.99807E-10	0.36202E-10	-0.34910E-10	0.10365E-9	0.22741E-9	0.10227E-9	0.95919E-10	0	0.13757E-10	0.45511E-11	0
e_i	-0.65224E-14	-0.28946E-14	0.36946E-14	-0.69354E-14	-0.15526E-13	-0.48472E-14	-0.64848E-14		0.14225E-15	-0.43681E-15	
f_i	-0.90586E3	-0.12020E4	-0.87738E3	-0.14245E5	-0.48961E5	-0.29906E5	0.98283E4	-745.852	0.39354E4	0.29231E5	0.25472E5
g_i	0.61615E1	0.36151E1	-0.19629E1	0.63479E1	-0.98636E0	0.66306E1	0.67458E1	0.43529E1	0.54423E1	0.49203E1	-0.460120E0
					300–1000 K						
a_i	0.36748E1	0.36256E1	0.30574E1	0.37101E1	0.24008E1	0.40701E1		2.5016			
b_i	-0.12082E-2	-0.18782E-2	0.26765E-2	-0.16191E-2	0.87351E-2	-0.11084E-2		0			
c_i	0.23240E-5	0.70555E-5	-0.58099E-5	0.36924E-5	-0.66071E-5	0.41521E-5		0			
d_i	-0.63218E-9	-0.67635E-8	0.55210E-8	-0.20320E-8	0.20022E-8	-0.29637E-8		0			
e_i	-0.22577E-12	0.21556E-11	-0.18123E-11	0.23953E-12	0.63274E-15	0.80702E-12		0			
f_i	-0.10612E4	-0.10475E4	-0.98890E3	-0.14356E5	-0.48378E5	-0.30280E5		-745.852			
g_i	0.23580E1	0.43053E1	-0.22997E1	0.29555E1	0.96951E1	-0.32270E0		0.43529E1			

Table A.2 *Coefficients for the evaluation of $\log_{10} K_p$; pressure units – bar, temperature units – kK*

Constants Reaction	A	B	C	D	E
$\frac{1}{2}H_2 \rightleftharpoons H$	0.432168	–0.112464 E2	0.266983 E1	–0.745744 E-1	0.242484 E-2
$\frac{1}{2}O_2 \rightleftharpoons O$	0.310805	–0.129540 E2	0.3214932 E1	–0.738336 E-1	0.344645 E-2
$\frac{1}{2}N_2 \rightleftharpoons N$	0.389716	–0.245828 E2	0.3142192 E1	–0.963730 E-1	0.585643 E-2
$\frac{1}{2}O_2 + \frac{1}{2}H_2 \rightleftharpoons OH$	–0.141784	–0.213308 E1	0.853461	0.355015 E-1	–0.310227 E-2
$\frac{1}{2}N_2 + \frac{1}{2}O_2 \rightleftharpoons NO$	0.150879 E-1	–0.470959 E1	0.646096	0.272805 E-2	–0.154444 E-2
$H_2 + \frac{1}{2}O_2 \rightleftharpoons H_2O$	–0.752364	0.124210 E2	–0.262575 E1	0.259556	–0.162687 E-1
$CO + \frac{1}{2}O_2 \rightleftharpoons CO_2$	–0.415302 E-2	0.148627 E2	–0.475460 E1	0.124699	–900227 E-2

and in general: $dG = (\Sigma v_i \mu_i)d\xi$,   and $(\partial G/\partial \xi)_{T,p} = (\Sigma v_i \mu_i)$

$$(A.25)$$

and at equilibrium $(\partial G/\partial \xi)_{T,p} = 0$ so $\Sigma v_i \mu_i = 0$ (A.26)

For species i: $\mu_i = (\partial G/\partial n_i)_{T,p,n_j}$ (A.17)

For a pure substance, the chemical potential (μ_i) is simply the molar Gibbs energy G/n_i. For one mole of gas:

$$G = G^0 + R_0 T \ln(p/p^0)$$ (A.13)

so $\mu = \mu^0 + R_0 T \ln(p/p^0)$ (A.27)

where the superscript ^{0} refers to the use of a pressure datum of 1 bar.

For gaseous species i with ideal gas behaviour (enthalpy is not a function of pressure):

$$\mu_i = \mu_i^0 + R_0 T \ln(p_i/p^0)$$ (A.28)

or $\mu_i = \mu_i^0 + R_0 T \ln(p_i^*)$ (A.29)

where p_i^* represents the numerical value of the partial pressure of component i, when the pressure is expressed in units of **bar**.

For the mixture at equilibrium: $\Sigma v_i \mu_i = 0$ (A.26)

Combining equations (A.26) and (A.29) gives:

$$\Sigma v_i(\mu_i^0 + R_0 T \ln(p_i^*)) = \Sigma v_i(G_i^0 + R_0 T \ln(p_i^*)) = 0$$

or $\Delta G^0 = -R_0 T \Sigma(v_i \ln p_i^*) = -R_0 T \ln K_p$ (A.30)

where $\Delta G^0 = \Sigma v_i G_i^0$, the change in Gibbs (or free) energy for the reaction

and $\Sigma(v_i \ln p_i^*) = K_p$ (A.31)

where K_p is the equilibrium constant of the reaction.

This is more frequently expressed as $K_p = \Pi p_i^{*v_i}$, with Π denoting the product of the terms that follow. The equilibrium constant has a strong temperature dependency, so it is convenient to tabulate $\ln K_p$. Although it has no pressure dependency, it is essential to use the appropriate pressure units for the partial pressures unless $\Sigma v_i = 0$.

For the multi-component reaction $aA + bB \rightleftharpoons cC + dD$:

$$K_p = \Pi p_i^{*v_i} = (p_C^{*c} \times p_D^{*d})/(p_A^{*a} \times p_B^{*b})$$ (3.6)

remembering that p_i^* is the numerical value of the partial pressure of component i when the pressure is expressed in units of **bar**.

The equilibrium constant can be determined from the change in Gibbs energy of the reaction at the relevant temperature:

$$\ln K_p = -\Delta G^0/R_0 T \quad \text{where} \quad \Delta G^0 = \Sigma v_i G_i^0$$ (A.32)

Thus the equilibrium constants can be calculated from the Gibbs energy values in tables A.5 and A.6, and this is indeed how table A.6 has been produced.

Tabulations of the equilibrium constants can be found in many sources (such as Howatson *et al.*, 1991; Haywood, 1972; and Rogers and Mayhew, 1988). For calculations, an analytical expression is frequently more convenient, and an appropriate form can be found by dividing the equation for the Gibbs energy by $R_0 T$. Such an equation was used by Olikara and Borman (1975):

$$\log_{10} K_p = A \ln(T) + B/T + C + DT + ET^2$$ (A.33)

who evaluated these coefficients for a number of equilibria in the range 600–4000 K. Olikara and Borman used a temperature unit of kK and pressure units of atmospheres; in table A.2 the pressure units have been converted to bar.

Regardless of the source of the equilibrium constant data, it is essential to pay strict attention to

(a) the pressure units

(b) the form of the equation.

Consider now the equilibrium between H_2O, H and OH, which is not detailed in table A.2:

$$H_2O \rightleftharpoons H + OH$$

$$K_p = \frac{p_H p_{OH}}{p_{H_2O}}$$ (A.34)

However, it will be shown how this can be reformulated in terms of the following tabulated equilibria:

$$\frac{1}{2}H_2 \rightleftharpoons H \qquad K_{p,H} = p_H \big/ p_{H_2}^{\frac{1}{2}}$$

$$\frac{1}{2}O_2 + \frac{1}{2}H_2 \rightleftharpoons OH \qquad K_{p,OH} = p_{OH} \big/ \left(p_{H_2}^{\frac{1}{2}} \times p_{O_2}^{\frac{1}{2}} \right)$$

$$H_2 + \frac{1}{2}O_2 \rightleftharpoons H_2O \qquad K_{p,H_2O} = p_{H_2O} \big/ \left(p_{H_2} \times p_{O_2}^{\frac{1}{2}} \right)$$

Equation (A.34) can be rewritten as

$$K_p = \frac{p_H p_{OH}}{p_{H_2O}} \times \frac{p_{H_2}^{\frac{1}{2}} p_{O_2}^{\frac{1}{2}}}{p_{H_2} p_{O_2}^{\frac{1}{2}}} = \frac{\dfrac{p_H}{p_{H_2}^{\frac{1}{2}}} \times \dfrac{p_{OH}}{p_{H_2}^{\frac{1}{2}} \times p_{O_2}^{\frac{1}{2}}}}{\dfrac{p_{H_2O}}{p_{H_2} p_{O_2}^{\frac{1}{2}}}}$$

Thus
$$K_p = \frac{K_{p,H} \times K_{p,OH}}{K_{p,H_2O}} \qquad (A.35)$$

or
$$\log_{10} K_p = \log_{10} K_{p,H} + \log_{10} K_{p,OH} - \log_{10} K_{p,H_2O}$$

This result can be generalised and applied to the values of the equilibrium constants of formation of species (K_f) from the elements in their standard state. The JANAF Tables (1985) tabulate $\log_{10} K_f$ for numerous species. For elements in their standard state (such as N_2, O_2, He etc.) $\log_{10} K_f$ is zero:

$$\log_{10}(K_p)_{\text{reaction}} = \sum_i \nu_i \log_{10}(K_f)_i \qquad (A.36)$$

A.2.3 Reaction rates

The *Law of Mass Action* states that the rate of a chemical reaction is proportional to the production of the concentrations of the reactant species, with each species raised to the power of its stoichiometric coefficient. For the reaction:

$$\sum_{i=1}^{n} \nu_{R_i} M_{R_i} = \sum_{i=1}^{m} \nu_{P_i} M_{P_i} \qquad (A.37)$$

where ν_i is the stoichiometric coefficient of species M_i
there are n species in the reactants, R
there are m species in the products, P.

The *Law of Mass Action* can be written as

$$R^+ = k^+ \prod_{i=1}^{n} [M_{R_i}]^{\nu_{R_i}}$$

$$R^- = k^- \prod_{i=1}^{m} [M_{P_i}]^{\nu_{P_i}} \qquad (A.38)$$

where R^+, R^- are the forward and reverse reaction rates
k^+, k^- are the forward and reverse rate coefficients
and
$[M_i]$ is concentration of species i.

The net rate of formation of the product species M_{P_i} is the difference between the forward and reverse reaction rates:

$$\frac{d[M_{P_i}]}{dt} = \nu_{P_i}(R^+ - R^-) \qquad (A.39)$$

Similarly the net rate of formation of the reactant species M_{R_i} is

$$-\frac{d[M_{R_i}]}{dt} = \nu_{R_i}(R^+ - R^-) \qquad (A.40)$$

The rate 'constants' or coefficients (k^+, k^-) frequently have an Arrhenius form:

$$k = A \exp\left(-\frac{E_A}{RT}\right) \qquad (A.41)$$

where A is the pre-exponential factor (which can be a weak function of temperature) and
E_A is the activation energy.

The exponential term in equation (A.41) (the Boltzmann factor) establishes the fraction of all the collisions between species that have an energy greater than the activation energy (E_A).

At equilibrium the forward and reverse reaction rates (R^+, R^-) will be equal, so that rearranging equation (A.38) gives

$$\frac{k^+}{k^-} = \prod_{i=1}^{m} [M_{P_i}]^{\nu_{P_i}} \Big/ \prod_{i=1}^{n} [M_{R_i}]^{\nu_{R_i}} = K_C \qquad (A.42)$$

or
$$K_C = \prod_{i=1}^{n} [M_i]^{\nu_i} \qquad (A.43)$$

where K_C is the equilibrium constant based on concentrations
M_i refers to either products or reactants
and ν_i is negative for reactants.

It is now necessary to relate the equilibrium constant based on composition (K_C) to the equilibrium constant based on pressure (K_p), from equation (A.31):

$$K_p = \prod_{i=1}^{n} (p_i)^{\nu_i} \qquad (A.44)$$

If the system pressure is p, and the mole fraction of each species is x_i, then

$$K_p = \prod_{i=1}^{n} (p_i)^{\nu_i} = \prod_{i=1}^{n} (x_i p)^{\nu_i} = (p)^{\sum_i \nu_i} \prod_{i=1}^{n} x_i^{\nu_i} \qquad (A.45)$$

The concentration of species i ($[M_i]$) can be found from the equation of state:

$$[M_i] = p_i/R_0 T = x_i p/R_0 T \qquad (A.46)$$

Substituting equation (A.46) into equation (A.43) and rearranging gives:

$$K_c = \prod_{i=1}^{n} (x_i p/R_0 T)^{\nu_i} = \left(\frac{p}{R_0 T}\right)^{\sum_i \nu_i} \prod_{i=1}^{n} x_i^{\nu_i} \qquad (A.47)$$

A comparison of equations (A.47) and (A.45) shows that

$$K_p = K_C(R_0 T)^{\sum_i v_i} \tag{A.48}$$

Thus K_C and K_p are only equal when there are equal number of moles of reactants and products ($\Sigma_i v_i = 0$), and furthermore K_p will have units that are pressure raised to the power $-\Sigma v_i$. The numerical value of K_p will be independent of the system pressure, but not in general the pressure units.

In general, chemical reactions (including combustion) are not fully described by the simplest equation that can balance the reactants and products. The following simple stoichiometric equation describing the oxidation of methane can be used for thermostatic purposes (calculation of atom balances, energy, equilibrium etc.), but not the reaction rates:

$$CH_4 + 2O_2 \rightleftharpoons CO_2 + 2H_2O \tag{A.49}$$

The likelihood of such a three-body collision occurring with sufficient energy to react is so small that this mechanism can be considered impossible. Instead, the oxidation of a hydrocarbon will involve a complex oxidation mechanism, comprising sequential and parallel elementary reactions that include intermediate compounds. Thus the Law of Mass Action will not apply to the simple stoichiometric equation, but to the constituent elementary mechanism reactions. So far as stoichiometry is concerned, there are many equally valid ways of representing the same reaction. For example, the oxidation of nitric oxide to nitrogen dioxide:

$$NO + O \rightleftharpoons NO_2 \tag{A.50a}$$

or

$$2NO + O_2 \rightleftharpoons 2NO_2 \tag{A.50b}$$

Experiments in which the reaction rates have been measured show that the termolecular reaction (equation A.50b) is the relevant formulation, since

$$\frac{d[NO_2]}{dt} = k[NO]^2[O_2] \tag{A.51}$$

This introduces the need to define the order of a reaction, for example

Rate = $k[A]$	1st order
Rate = $k[A][B]$	2nd order, but 1st order in A
Rate = $k[A]^2$	2nd order

The order of a reaction cannot be determined from stoichiometry or equilibrium, but it can be determined experimentally. Not all reactions have such simple rate expressions; non-integer powers and more complex relations are possible, although these may be indications that the full mechanism has not been identified. For example

$$H_2 + Br_2 \rightleftharpoons 2HBr$$

$$\frac{d[HBr]}{dt} = \frac{k[H_2][Br_2]^{\frac{1}{2}}}{1 + k'[HBr]/[Br_2]} \tag{A.52}$$

But this can be explained by a combination of the following elementary reactions:

$$Br_2 \rightleftharpoons 2Br \tag{A.53a}$$

$$Br + H_2 \rightleftharpoons HBr + H \tag{A.53b}$$

$$H + Br_2 \rightleftharpoons HBr + Br \tag{A.53c}$$

Equation (A.53a) is chain initiating in the forward direction and chain termination in the reverse direction. Equations (A.53b) and (A.53c) are chain propagating in the forward direction and chain inhibiting in the reverse direction.

The derivation of equation (A.52) is on the premise that the reverse reaction of equation (A.53c) is negligible and the assumption that the concentrations of the atomic hydrogen and bromine quickly attain their equilibrium concentrations. Without this steady-state approximation the differential equations cannot be solved explicitly; the approximation is justifiable since the concentration of the atomic species is very low. If the concentration of a species is very low and almost constant, by definition the rate of change of its concentration will also be very low.

When the rate constant can be described by the Arrhenius equation, the activation energy and the pre-exponential term can be found by plotting the log of the rate constant against the reciprocal of the temperature. Equation (A.41) becomes:

$$\ln(k) = \ln A - (E/R_0)/T \tag{A.54}$$

The zero intercept determines the pre-exponential term, and the slope enables the actuation energy to be found.

The scatter in the data in figure A.1 gives scope for several lines to be fitted. It should also be noted (which is often the case) that there are no data for very high or very low temperatures. The high-temperature data are needed for combustion, yet they are difficult to obtain because of the speed of the reactions and the high temperatures. The low-temperature data might be needed for environmental modelling, and the difficulty here is the slow speed of the reactions.

A.2.4 Decomposition of hydrocarbons

The decomposition of many hydrocarbons and oxygenates is often described by a first-order equation (that is, the reaction rate is proportional to the concentration of the hydrocarbon or oxygenate). However, this does not mean these reactions are unimolecular, and spectrographic measurements can be used to show the presence of intermediate species (including radicals). This implies there are chain reactions, and Rice and Herzfeld identified possible mechanisms, which are named

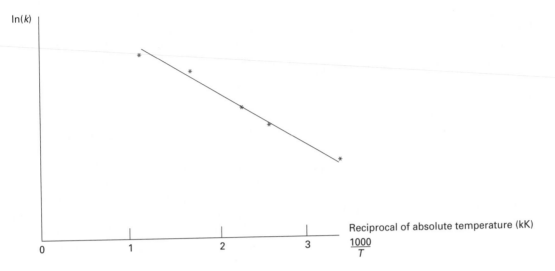

Figure A.1 Evaluation of rate data from experimental data.

after them. The mechanisms in general form (in which R refers to any radical and M any molecule), are:

(1) dissociation $M_1 \rightarrow R_1 + R_1'$ k_1

The radical R_1' is assumed to play no part in the reaction, although there are exceptions that need to be considered separately, for example the decomposition of ethane into two methyl radicals. The radical R_1 initiates the following chain reactions:

(2) propagation $R_1 + M_1 \rightarrow R_1H + R_2$ k_2
(3) dissociation $R_2 \rightarrow R_1 + M_2$ k_3
(4) termination $R_1 + R_2 \rightarrow M_3$ k_4
(5) termination $2R_1 \rightarrow M_4$ k_5
(6) termination $2R_2 \rightarrow M_5$ k_6

where only one of the three termination steps (4), (5) and (6) would be expected to be important in a given reaction.

Linear dependence

If reaction (4) is the relevant chain-termination step, the steady-state expressions for the concentration of R_1 and R_2 are

$$\frac{d[R_1]}{dt} = k_1[M_1] - k_2[R_1][M_1] + k_3[R_2] - k_4[R_1][R_2] = 0$$
(A.55)

and

$$\frac{d[R_2]}{dt} = k_2[R_1][M_1] - k_3[R_2] - k_4[R_1][R_2] = 0$$
(A.56)

Rearranging equation (A.55) gives

$$[M_1] = \frac{k_3[R_2] - k_4[R_1][R_2]}{k_2[R_1] - k_1}$$
(A.57)

Rearranging equation (A.56) gives

$$[M_1] = \frac{k_3[R_2] + k_4[R_1][R_2]}{k_2[R_1]}$$
(A.58)

Equating equations (A.57) and (A.58) to eliminate $[M_1]$, and then dividing both sides by $[R_2]$ gives

$$\frac{k_3[R_2] - k_4[R_1][R_2]}{k_2[R_1] - k_1} = \frac{k_3[R_2] + k_4[R_1][R_2]}{k_2[R_1]}$$

$$\frac{k_3 - k_4[R_1]}{k_2[R_1] - k_1} = \frac{k_3 + k_4[R_1]}{k_2[R_1]}$$
(A.59)

Multiplying by the denominators gives

$$k_2 k_3[R_1] - k_2 k_4[R_1]^2 = k_2 k_3[R_1] - k_1 k_4[R_1] - k_1 k_3 + k_2 k_4[R_1]^2$$

Rearrange and divide by $2k_2 k_4$:

$$[R_1]^2 + \left(\frac{k_1}{2k_2}\right)[R_1] - \frac{k_1 k_3}{2k_2 k_4} = 0$$
(A.60)

Since $[R_1]$ has to be positive, only one root of the quadratic equation is valid:

$$[R_1] = \frac{k_1}{4k_2} + \sqrt{\left(\frac{k_1}{4k_2}\right)^2 + \frac{k_1 k_3}{2k_2 k_4}}$$
(A.61)

The reaction rate k_1 is very small compared to k_2, since reaction (1) involves the disruption of a C–C bond. Equation (A.61) can thus be approximated as follows:

$$[R_1] \approx \left(\frac{k_1 k_3}{2 k_2 k_4}\right)^{\frac{1}{2}} \tag{A.62}$$

The overall rate for the decomposition of M_1 is given by

$$-\frac{d[M_1]}{dt} \approx k_1[M_1] + k_2[R_1][M_1] \tag{A.63}$$

Substituting from equation (A.62) gives

$$-\frac{d[M_1]}{dt} \approx \left[1 + \left(\frac{k_2 k_3}{2 k_1 k_4}\right)^{\frac{1}{2}}\right] k_1[M_1] \tag{A.64}$$

Since k_2 and k_3 will both be large compared to k_1 and k_4, equation (A.64) can be approximated as

$$-\frac{d[M_1]}{dt} \approx \left(\frac{k_1 k_2 k_3}{2 k_4}\right)^{\frac{1}{2}} [M_1] \tag{A.65}$$

which is a first-order reaction.

Half-power dependence

When reaction (6) is the chain-termination reaction:

$$\frac{d[R_1]}{dt} = k_1[M_1] - k_2[R_1][M_1] + k_3[R_2] = 0 \tag{A.66}$$

$$\frac{d[R_2]}{dt} = k_2[R_1][M_1] - k_3[R_2] - 2k_6[R_2]^2 = 0 \tag{A.67}$$

Note: For reaction (6) the concentration of R_2 is squared (in accordance with the Law of Mass Action) *and* multiplied by 2, since the reaction consumes two R_2 radicals.

After manipulation with similar assumptions to that for linear dependence, it can be shown that

$$-\frac{d[M_1]}{dt} = k_3 \sqrt{\frac{k_2}{2 k_6}} [M_1]^{\frac{1}{2}} \tag{A.68}$$

Three-halves power dependence

If reaction (5) is the chain-termination reaction, it can be shown that the overall rate of decomposition of M_1 is three-halves:

$$-\frac{d[M_1]}{dt} \approx k_2 \left(\frac{k_1}{2 k_5}\right)^{\frac{1}{2}} [M_1]^{\frac{3}{2}} \tag{A.69}$$

Specific examples of the thermal decomposition of many hydrocarbons and oxygenates can be found in many Physical Chemistry texts, for example, Atkins (1990). These texts also provide justification of the chosen kinetic scheme from physical evidence.

A.3 | Thermodynamic data

The following fuel properties (tables A.3, A.4, A.5 and A.6) have been obtained from the listings of *Physical and Thermodynamic Properties of Pure Chemicals* by Daubert and Danner (1989). This comprehensive compilation covers the properties of solids, liquids and gases, with analytical expressions and coefficients that enable the temperature dependency of the following properties to be determined:

Solid density
Liquid density
Vapour pressure
Enthalpy of vaporisation
Solid specific heat capacity
Liquid specific heat capacity
Ideal gas specific heat capacity
Second virial coefficient (polynomials used in the Equation of State)
Liquid viscosity
Vapour viscosity
Liquid thermal conductivity
Vapour thermal conductivity
Surface tension

Table A.3 *Boiling points, enthalpy of vaporisation, liquid density and specific heat capacity, molar masses, standard enthalpy of formation, standard state entropy and calorific values for fuels, derived from Daubert and Danner (1989)*

Fuel	Formula	Boiling point at 1 atm (°C)	Enthalpy of vaporisation[1] at 298.15 K (MJ/kmol)	Density[1] (kg/m³)	$C_{p,l}$ [1,2] (kJ/kmol K)
Methane	CH_4	−161.5	8.171	422.5	57[a]
Propane	C_3H_8	−42.0	18.743	582.5	106.3[b]
Benzene	C_6H_6	80.1	33.790	872.9	135.6
Toluene	C_7H_8	110.6	38.341	864.7	156.1
n-Heptane	C_7H_{16}	98.4	36.630	681.5	224.7
Iso-octane (2,2,4-Trimethylpentane)	C_8H_{18}	99.2	35.142	690.4	236.4
n-Hexadecane (Cetane)	$C_{16}H_{34}$	286.9	79.641	769.7	501.7
α-Methylnaphthalene	$C_{10}H_7CH_3$	244.7	59.387	1017.2	224.4[c]
Methanol	CH_3OH	64.7	38.012	789.6	81.6
Ethanol	C_2H_5OH	78.3	42.512	785.9	113.0
Nitromethane	CH_3NO_2	101.2	38.365	1112.7	106.3

Fuel	Formula	M (kg/kmol)	H_f (MJ/kmol)	S^0 [3] (kJ/kmol K)	Calorific values[4] at 298.15 K (MJ/kmol)	(MJ/kg)
Methane	CH_4	16.043	−74.52	186.27	802.64	50.031
Propane	C_3H_8	44.096	−104.68	270.2	2043.15	46.334
Benzene	C_6H_6	78.114	82.93	269.20	3169.47	40.575
Toluene	C_7H_8	92.141	50.00	319.74	3771.88	40.936
n-Heptane	C_7H_{16}	100.204	−187.65	427.98	4501.53	44.924
Iso-octane (2,2,4-Trimethylpentane)	C_8H_{18}	114.231	−224.01	422.96	5100.50	44.651
n-Hexadecane (Cetane)	$C_{16}H_{34}$	226.446	−374.17	781.02	10033.03	44.307
α-Methylnaphthalene	$C_{10}H_7CH_3$	142.2	116.86	377.44	5654.61	39.765
Methanol	CH_3OH	32.042	−200.94	239.88	676.22	21.104
Ethanol	C_2H_5OH	46.069	−234.95	280.64	1277.55	27.731
Nitromethane	CH_3NO_2	61.040	−74.73	275.01	681.52	11.165

Note 1:
Properties have been evaluated at 25°C, except when a substance is a gas at this temperature, in which case the evaluation refers to the normal boiling point.

Note 2:
These data were obtained from the *Handbook of Chemistry and Physics* (70th edn, CRC Press, 1990), with the exception of the following data:
a Y. S. Touloukian and T. Makita, *Specific Heat, nonmetallic liquids and gases* (Plenum, 1970);
b *International Critical Tables, Vol V* (McGraw-Hill, 1929);
c Daubert and Danner (1989).

Note 3:
The entropy values tabulated here for the standard state (S^0) refer to a pressure of 1 atm (1.01325 bar), while the entropy values evaluated in tables A.4 and A.5 use a pressure of 1 bar as the datum; this accounts for the slight differences in the numerical values for entropy at 298.15 K. The standard state values can refer to a hypothetical state, and this is indeed the case for many of these fuels, which cannot exist as a vapour at a pressure of 1 atm and a temperature of 298.15 K.

Note 4:
The calorific values have been determined from the difference in the enthalpies of formation of the fuel and products, with all reactants and products in the vapour phase; this is known as the Net or Lower Calorific Value (LCV). When the water vapour in the products of combustion has been condensed to its liquid state, the calorific value of the fuel is known as the Gross or Higher Calorific Value (HCV). Thus

$$HCV = LCV + (n \times H_{fg})_{H_2O}$$

where the enthalpy of condensation of the water vapour, $H_{fg} = 43.99$ MJ/kmol H_2O.

Table A.4 Thermodynamic data for gaseous species

'ABSOLUTE' MOLAR ENTHALPY (MJ/kmol)

T (K)	Air	N₂	O₂	H₂	CO	CO₂	H₂O	NO	Ar	OH	O	H	T (K)
0	-8.774	-8.823	-8.709	-8.222	-119.363	-402.239	-251.763	81.717	-6.201	32.721	243.041	211.787	0
100	-5.784	-5.812	-5.755	-5.584	-116.336	-399.897	-248.414	84.423	-4.121	35.180	245.154	213.866	100
200	-2.855	-2.864	-2.858	-2.804	-113.387	-396.934	-245.097	87.231	-2.041	37.716	247.264	215.944	200
298.15	**0.0**	**0.0**	**0.0**	**0.0**	**-110.525**	**-393.512**	**-241.824**	**90.080**	**0.000**	**40.277**	**249.333**	**217.984**	**298.15**
300	0.053	0.053	0.054	0.056	-110.471	-393.443	-241.761	90.135	0.038	40.326	249.372	218.023	300
400	2.973	2.969	3.028	2.959	-107.550	-389.509	-238.371	93.125	2.118	43.005	251.477	220.101	400
T (K)	Air	N₂	O₂	H₂	CO	CO₂	H₂O	NO	Ar	OH	O	H	T (K)
500	5.930	5.909	6.089	5.880	-104.594	-385.208	-234.899	96.195	4.198	45.752	253.580	222.180	500
600	8.942	8.893	9.247	8.811	-101.584	-380.605	-231.326	99.337	6.278	48.564	255.681	224.259	600
700	12.018	11.934	12.500	11.750	-98.505	-375.753	-227.639	102.547	8.358	51.437	257.780	226.337	700
800	15.164	15.042	15.837	14.704	-95.353	-370.699	-223.831	105.817	10.438	54.369	259.876	228.416	800
900	18.379	18.219	19.244	17.681	-92.128	-365.477	-219.899	109.143	12.518	57.358	261.971	230.495	900
T (K)	Air	N₂	O₂	H₂	CO	CO₂	H₂O	NO	Ar	OH	O	H	T (K)
1000	21.655	21.459	22.708	20.686	-88.838	-360.112	-215.841	112.518	14.598	60.399	264.063	232.573	1000
1100	24.984	24.756	26.213	23.729	-85.496	-354.624	-211.652	115.939	16.678	63.492	266.154	234.652	1100
1200	28.357	28.100	29.754	26.814	-82.107	-349.051	-207.337	119.400	18.758	66.633	268.243	236.730	1200
1300	31.772	31.489	33.328	29.942	-78.676	-343.395	-202.904	122.897	20.838	69.820	270.330	238.809	1300
1400	35.224	34.917	36.934	33.112	-75.208	-337.664	-198.359	126.427	22.918	73.050	272.416	240.888	1400
T (K)	Air	N₂	O₂	H₂	CO	CO₂	H₂O	NO	Ar	OH	O	H	T (K)
1500	38.711	38.380	40.571	36.323	-71.707	-331.867	-193.711	129.986	24.998	76.322	274.501	242.966	1500
1600	42.228	41.876	44.237	39.574	-68.176	-326.012	-188.966	133.572	27.078	79.634	276.585	245.045	1600
1700	45.774	45.400	47.931	42.865	-64.618	-320.105	-184.131	137.180	29.158	82.982	278.668	247.124	1700
1800	49.345	48.949	51.652	46.194	-61.037	-314.151	-179.212	140.808	31.238	86.365	280.751	249.202	1800
1900	52.940	52.522	55.399	49.560	-57.435	-308.158	-174.215	144.454	33.318	89.781	282.833	251.281	1900
T (K)	Air	N₂	O₂	H₂	CO	CO₂	H₂O	NO	Ar	OH	O	H	T (K)
2000	56.556	56.114	59.171	52.962	-53.815	-302.128	-169.146	148.116	35.398	93.228	284.915	253.359	2000
2100	60.190	59.725	62.967	56.399	-50.178	-296.068	-164.009	151.792	37.478	96.705	286.997	255.438	2100
2200	63.842	63.351	66.787	59.870	-46.527	-289.981	-158.810	155.480	39.558	100.208	289.079	257.517	2200
2300	67.510	66.991	70.630	63.373	-42.863	-283.870	-153.553	159.179	41.638	103.738	291.162	259.595	2300
2400	71.192	70.644	74.495	66.908	-39.189	-277.737	-148.243	162.886	43.718	107.292	293.246	261.674	2400
T (K)	Air	N₂	O₂	H₂	CO	CO₂	H₂O	NO	Ar	OH	O	H	T (K)
2500	74.887	74.308	78.381	70.474	-35.504	-271.587	-142.883	166.602	45.797	110.868	295.330	263.753	2500
2600	78.595	77.981	82.288	74.068	-31.811	-265.420	-137.478	170.326	47.877	114.465	297.416	265.831	2600
2700	82.313	81.664	86.216	77.691	-28.111	-259.239	-132.031	174.055	49.957	118.083	299.503	267.910	2700
2800	86.041	85.354	90.163	81.341	-24.403	-253.045	-126.545	177.790	52.037	121.718	301.592	269.988	2800
2900	89.779	89.051	94.130	85.016	-20.689	-246.838	-121.022	181.531	54.117	125.371	303.684	272.067	2900

(continued)

Table A.4 Thermodynamic data for gaseous species (continued)

'ABSOLUTE' MOLAR ENTHALPY (MJ/kmol)

T (K)	Air	N₂	O₂	H₂	CO	CO₂	H₂O	NO	Ar	OH	O	H	T (K)
3000	93.527	92.754	98.116	88.716	-16.969	-240.621	-115.467	185.276	56.197	129.040	305.777	274.146	3000
3100	97.282	96.464	102.120	92.439	-13.244	-234.393	-109.881	189.026	58.277	132.724	307.872	276.224	3100
3200	101.046	100.179	106.142	96.185	-9.514	-228.155	-104.266	192.780	60.357	136.421	309.971	278.303	3200
3300	104.818	103.900	110.182	99.953	-5.778	-221.907	-98.625	196.538	62.437	140.132	312.072	280.382	3300
3400	108.597	107.625	114.238	103.741	-2.039	-215.650	-92.959	200.300	64.517	143.855	314.177	282.460	3400
T (K)	Air	N₂	O₂	H₂	CO	CO₂	H₂O	NO	Ar	OH	O	H	T (K)
3500	112.383	111.356	118.312	107.549	1.706	-209.383	-87.271	204.066	66.597	147.589	316.284	284.539	3500
3600	116.177	115.091	122.402	111.377	5.455	-203.106	-81.561	207.836	68.677	151.334	318.396	286.617	3600
3700	119.978	118.832	126.507	115.223	9.209	-196.819	-75.832	211.610	70.757	155.089	320.511	288.696	3700
3800	123.786	122.577	130.627	119.086	12.968	-190.522	-70.083	215.389	72.837	158.853	322.630	290.775	3800
3900	127.600	126.326	134.762	122.968	16.731	-184.215	-64.317	219.171	74.917	162.627	324.754	292.853	3900
T (K)	Air	N₂	O₂	H₂	CO	CO₂	H₂O	NO	Ar	OH	O	H	T (K)
4000	131.421	130.081	138.911	126.866	20.499	-177.898	-58.533	222.958	76.997	166.409	326.881	294.932	4000
4100	135.249	133.840	143.074	130.781	24.272	-171.570	-52.733	226.748	79.077	170.200	329.014	297.011	4100
4200	139.083	137.604	147.249	134.713	28.049	-165.232	-46.918	230.543	81.157	173.998	331.150	299.089	4200
4300	142.923	141.373	151.436	138.662	31.831	-158.884	-41.088	234.342	83.237	177.805	333.292	301.168	4300
4400	146.769	145.147	155.634	142.627	35.617	-152.526	-35.242	238.145	85.317	181.620	335.439	303.246	4400
T (K)	Air	N₂	O₂	H₂	CO	CO₂	H₂O	NO	Ar	OH	O	H	T (K)
4500	150.621	148.925	159.842	146.610	39.408	-146.159	-29.383	241.951	87.397	185.443	337.590	305.325	4500
4600	154.478	152.707	164.060	150.610	43.202	-139.783	-23.509	245.761	89.477	189.273	339.747	307.404	4600
4700	158.340	156.493	168.286	154.628	47.000	-133.400	-17.621	249.574	91.557	193.112	341.909	309.482	4700
4800	162.206	160.282	172.519	158.665	50.801	-127.012	-11.719	253.390	93.636	196.960	344.076	311.561	4800
4900	166.075	164.074	176.758	162.722	54.604	-120.619	-5.803	257.207	95.716	200.816	346.249	313.640	4900
5000	169.947	167.868	181.001	166.800	58.409	-114.224	0.127	261.025	97.796	204.682	348.427	315.718	5000
T (K)	Air	N₂	O₂	H₂	CO	CO₂	H₂O	NO	Ar	OH	O	H	T (K)

Note:
The term 'ABSOLUTE' Molar Enthalpy adopted in this table uses a datum of **zero enthalpy for elements when they are in their standard state at a temperature of 25° C**. The enthalpy of any molecule at 25°C will thus correspond to its enthalpy of formation, ΔH_f°. The tables have been extrapolated below 300 K, and these data should be used with caution. The enthalpies have been evaluated by the integration of a polynomial function that describes the molar specific heat capacity (C_p) variation with temperature. The difference in Enthalpy of Reactants and Products at 25°C will thus correspond to the Constant Pressure Calorific Value of the reaction.

(continued)

Table A.4 Thermodynamic data for gaseous species (continued)

'ABSOLUTE' MOLAR INTERNAL ENERGY (MJ/kmol)

T (K)	Air	N_2	O_2	H_2	CO	CO_2	H_2O	NO	Ar	OH	O	H	T (K)
0	-8.774	-8.823	-8.709	-8.222	-119.363	-402.239	-251.763	81.717	-6.201	32.721	243.041	211.787	0
100	-6.615	-6.643	-6.586	-6.415	-117.167	-400.729	-249.246	83.592	-4.953	34.349	244.322	213.034	100
200	-4.518	-4.527	-4.521	-4.467	-115.050	-398.597	-246.759	85.569	-3.704	36.053	245.601	214.281	200
298.15	**-2.480**	**-2.480**	**-2.479**	**-2.476**	**-113.004**	**-395.991**	**-244.303**	**87.601**	**-2.479**	**37.798**	**246.854**	**215.505**	**298.15**
300	-2.441	-2.441	-2.440	-2.438	-112.966	-395.937	-244.256	87.640	-2.456	37.831	246.877	215.528	300
400	-0.352	-0.356	-0.298	-0.367	-110.876	-392.835	-241.697	89.799	-1.207	39.680	248.151	216.776	400
T (K)	Air	N_2	O_2	H_2	CO	CO_2	H_2O	NO	Ar	OH	O	H	T (K)
500	1.773	1.752	1.931	1.723	-108.751	-389.366	-239.057	92.037	0.041	41.595	249.423	218.023	500
600	3.953	3.904	4.258	3.822	-106.572	-385.594	-236.314	94.349	1.290	43.575	250.692	219.270	600
700	6.198	6.114	6.680	5.930	-104.325	-381.574	-233.459	96.727	2.538	45.617	251.959	220.517	700
800	8.513	8.390	9.186	8.053	-102.004	-377.351	-230.483	99.166	3.787	47.718	253.225	221.764	800
900	10.896	10.735	11.761	10.198	-99.611	-372.960	-227.382	101.660	5.035	49.874	254.487	223.012	900
T (K)	Air	N_2	O_2	H_2	CO	CO_2	H_2O	NO	Ar	OH	O	H	T (K)
1000	13.341	13.145	14.393	12.371	-97.153	-368.427	-224.155	104.204	6.284	52.085	255.749	224.259	1000
1100	15.838	15.610	17.067	14.583	-94.642	-363.770	-220.798	106.793	7.532	54.346	257.008	225.506	1100
1200	18.380	18.123	19.777	16.837	-92.084	-359.028	-217.315	109.422	8.781	56.655	258.265	226.753	1200
1300	20.963	20.680	22.519	19.133	-89.485	-354.204	-213.713	112.088	10.029	59.011	259.521	228.000	1300
1400	23.584	23.276	25.294	21.472	-86.849	-349.304	-210.000	114.787	11.278	61.410	260.776	229.247	1400
T (K)	Air	N_2	O_2	H_2	CO	CO_2	H_2O	NO	Ar	OH	O	H	T (K)
1500	26.239	25.908	28.099	23.851	-84.179	-344.339	-206.183	117.515	12.526	63.851	262.029	230.495	1500
1600	28.925	28.572	30.934	26.271	-81.479	-339.315	-202.270	120.268	13.775	66.330	263.282	231.742	1600
1700	31.639	31.265	33.796	28.730	-78.753	-334.239	-198.266	123.045	15.023	68.847	264.534	232.989	1700
1800	34.379	33.983	36.686	31.228	-76.003	-329.117	-194.178	125.842	16.272	71.399	265.785	234.236	1800
1900	37.143	36.724	39.601	33.762	-73.233	-323.955	-190.013	128.657	17.520	73.983	267.035	235.483	1900
T (K)	Air	N_2	O_2	H_2	CO	CO_2	H_2O	NO	Ar	OH	O	H	T (K)
2000	39.927	39.485	42.542	36.333	-70.444	-318.757	-185.775	131.487	18.769	76.599	268.286	236.730	2000
2100	42.730	42.264	45.507	38.938	-67.638	-313.529	-181.469	134.332	20.017	79.244	269.536	237.978	2100
2200	45.551	45.059	48.495	41.578	-64.819	-308.273	-177.102	137.188	21.266	81.916	270.787	239.225	2200
2300	48.387	47.868	51.507	44.250	-61.987	-302.993	-172.676	140.055	22.514	84.615	272.039	240.472	2300
2400	51.237	50.689	54.540	46.954	-59.144	-297.692	-168.198	142.932	23.763	87.337	273.291	241.719	2400
T (K)	Air	N_2	O_2	H_2	CO	CO_2	H_2O	NO	Ar	OH	O	H	T (K)
2500	54.101	53.521	57.595	49.688	-56.291	-292.373	-163.670	145.816	25.011	90.082	274.544	242.966	2500
2600	56.977	56.364	60.670	52.451	-53.429	-287.038	-159.096	148.708	26.260	92.848	275.798	244.213	2600
2700	59.864	59.214	63.767	55.242	-50.560	-281.688	-154.480	151.606	27.508	95.633	277.054	245.461	2700
2800	62.761	62.073	66.883	58.060	-47.683	-276.325	-149.825	154.510	28.757	98.438	278.312	246.708	2800
2900	65.667	64.939	70.018	60.904	-44.801	-270.950	-145.134	157.419	30.005	101.259	279.571	247.955	2900

(continued)

Table A.4 Thermodynamic data for gaseous species (continued)

'ABSOLUTE' MOLAR INTERNAL ENERGY (MJ/kmol)

T (K)	Air	N_2	O_2	H_2	CO	CO_2	H_2O	NO	Ar	OH	O	H	T (K)
3000	68.583	67.811	73.172	63.772	-41.913	-265.564	-140.410	160.333	31.254	104.096	280.833	249.202	3000
3100	71.507	70.689	76.345	66.664	-39.019	-260.168	-135.656	163.251	32.502	106.949	282.098	250.449	3100
3200	74.440	73.573	79.535	69.579	-36.120	-254.762	-130.872	166.173	33.751	109.815	283.364	251.697	3200
3300	77.380	76.462	82.744	72.515	-33.216	-249.345	-126.063	169.100	34.999	112.694	284.634	252.944	3300
3400	80.328	79.356	85.969	75.472	-30.308	-243.919	-121.229	172.031	36.248	115.585	285.907	254.191	3400
T (K)	Air	N_2	O_2	H_2	CO	CO_2	H_2O	NO	Ar	OH	O	H	T (K)
3500	83.283	82.255	89.211	78.448	-27.395	-238.484	-116.372	174.965	37.496	118.488	287.184	255.438	3500
3600	86.245	85.159	92.469	81.444	-24.477	-233.038	-111.494	177.904	38.745	121.402	288.464	256.685	3600
3700	89.214	88.068	95.743	84.459	-21.555	-227.583	-106.595	180.847	39.993	124.325	289.747	257.932	3700
3800	92.191	90.981	99.032	87.491	-18.627	-222.117	-101.678	183.794	41.242	127.258	291.035	259.180	3800
3900	95.174	93.900	102.336	90.541	-15.696	-216.642	-96.743	186.744	42.490	130.200	292.327	260.427	3900
T (K)	Air	N_2	O_2	H_2	CO	CO_2	H_2O	NO	Ar	OH	O	H	T (K)
4000	98.163	96.823	105.653	93.608	-12.759	-211.156	-91.791	189.700	43.739	133.151	293.623	261.674	4000
4100	101.160	99.751	108.984	96.692	-9.818	-205.659	-86.823	192.659	44.987	136.110	294.924	262.921	4100
4200	104.162	102.684	112.328	99.792	-6.872	-200.153	-81.839	195.622	46.236	139.077	296.229	264.168	4200
4300	107.171	105.621	115.684	102.909	-3.922	-194.636	-76.840	198.590	47.484	142.053	297.540	265.415	4300
4400	110.186	108.563	119.050	106.043	-0.967	-189.109	-71.826	201.561	48.733	145.036	298.855	266.663	4400
T (K)	Air	N_2	O_2	H_2	CO	CO_2	H_2O	NO	Ar	OH	O	H	T (K)
4500	113.206	111.509	122.427	109.195	1.993	-183.574	-66.798	204.536	49.981	148.027	300.175	267.910	4500
4600	116.232	114.460	125.813	112.363	4.956	-178.030	-61.756	207.515	51.230	151.027	301.500	269.157	4600
4700	119.262	117.414	129.208	115.550	7.922	-172.479	-56.699	210.496	52.478	154.034	302.831	270.404	4700
4800	122.296	120.372	132.609	118.756	10.892	-166.921	-51.629	213.480	53.727	157.050	304.167	271.651	4800
4900	125.334	123.333	136.017	121.981	13.863	-161.360	-46.544	216.465	54.975	160.075	305.508	272.898	4900
5000	128.374	126.296	139.429	125.228	16.836	-155.797	-41.446	219.452	56.224	163.110	306.854	274.146	5000
T (K)	Air	N_2	O_2	H_2	CO	CO_2	H_2O	NO	Ar	OH	O	H	T (K)

Note:
The term 'ABSOLUTE' Molar Internal Energy adopted in the table uses the same datum as the enthalpy table, namely a datum of **zero enthalpy for elements when they are in their standard state at a temperature of 25°C**. The tables have been extrapolated below 300 K, and these data should be used with caution.
The difference in Internal Energy of Reactants and Products at 25°C will thus correspond to the Constant Volume Calorific Value of the reaction. When there is a difference in the number of kmols of gaseous reactants and products, the Constant Pressure Calorific Value (the difference in Enthalpy of Reactants and Products at 25°C) will differ from the Constant Volume Calorific Value.

(continued)

Table A.4 *Thermodynamic data for gaseous species (continued)*

'ABSOLUTE' MOLAR ENTROPY (kJ/kmolK)

T (K)	Air	N₂	O₂	H₂	CO	CO₂	H₂O	NO	Ar	OH	O	H	T (K)
0	0.0	0.0	0.0	0.0	0.0	0.0	0.0	0.0	0.0	0.0	0.0	0.0	0
100	162.167	159.512	173.442	100.054	165.543	179.639	152.512	179.394	132.087	157.595	138.332	92.008	100
200	182.484	179.963	193.531	119.277	186.001	199.953	175.510	198.821	146.504	175.143	152.960	106.416	200
298.15	**194.096**	**191.614**	**205.150**	**130.689**	**197.646**	**213.811**	**188.820**	**210.398**	**154.809**	**185.550**	**161.376**	**114.715**	**298.15**
300	194.276	191.794	205.332	130.868	197.826	214.041	189.028	210.580	154.938	185.713	161.507	114.844	300
400	202.675	200.182	213.880	139.216	206.229	225.329	198.775	219.176	160.921	193.418	167.564	120.824	400
T (K)	Air	N₂	O₂	H₂	CO	CO₂	H₂O	NO	Ar	OH	O	H	T (K)
500	209.272	206.741	220.707	145.734	212.823	234.913	206.519	226.023	165.563	199.545	172.257	125.462	500
600	214.760	212.179	226.463	151.077	218.310	243.299	213.032	231.751	169.355	204.669	176.087	129.252	600
700	219.502	216.866	231.476	155.608	223.054	250.773	218.713	236.698	172.561	209.097	179.322	132.456	700
800	223.701	221.015	235.931	159.553	227.262	257.519	223.795	241.064	175.339	213.012	182.122	135.232	800
900	227.487	224.756	239.943	163.058	231.060	263.668	228.425	244.980	177.788	216.531	184.589	137.680	900
T (K)	Air	N₂	O₂	H₂	CO	CO₂	H₂O	NO	Ar	OH	O	H	T (K)
1000	230.938	228.170	243.591	166.224	234.525	269.319	232.700	248.536	179.980	219.735	186.793	139.870	1000
1100	234.111	231.312	246.935	169.127	237.712	274.541	236.693	251.796	181.962	222.682	188.786	141.851	1100
1200	237.046	234.222	250.015	171.811	240.661	279.390	240.447	254.807	183.772	225.415	190.604	143.660	1200
1300	239.779	236.934	252.876	174.315	243.407	283.917	243.995	257.606	185.437	227.965	192.275	145.324	1300
1400	242.337	239.474	255.548	176.664	245.976	288.163	247.362	260.222	186.978	230.359	193.821	146.864	1400
T (K)	Air	N₂	O₂	H₂	CO	CO₂	H₂O	NO	Ar	OH	O	H	T (K)
1500	244.743	241.863	258.057	178.879	248.392	292.162	250.569	262.678	188.413	232.617	195.259	148.298	1500
1600	247.013	244.119	260.423	180.977	250.671	295.941	253.631	264.991	189.756	234.754	196.604	149.640	1600
1700	249.162	246.255	262.662	182.972	252.827	299.522	256.562	267.179	191.017	236.783	197.867	150.900	1700
1800	251.204	248.284	264.789	184.875	254.874	302.925	259.373	269.253	192.206	238.717	199.057	152.088	1800
1900	253.147	250.216	266.815	186.694	256.822	306.166	262.075	271.224	193.330	240.564	200.183	153.212	1900
T (K)	Air	N₂	O₂	H₂	CO	CO₂	H₂O	NO	Ar	OH	O	H	T (K)
2000	255.002	252.058	268.750	188.439	258.679	309.258	264.675	273.102	194.397	242.332	201.251	154.278	2000
2100	256.775	253.820	270.602	190.116	260.453	312.215	267.181	274.896	195.412	244.028	202.267	155.292	2100
2200	258.474	255.507	272.379	191.731	262.152	315.047	269.600	276.611	196.379	245.658	203.235	156.259	2200
2300	260.104	257.125	274.087	193.288	263.780	317.763	271.936	278.255	197.304	247.227	204.161	157.183	2300
2400	261.671	258.680	275.732	194.793	265.344	320.373	274.196	279.833	198.189	248.739	205.048	158.068	2400
T (K)	Air	N₂	O₂	H₂	CO	CO₂	H₂O	NO	Ar	OH	O	H	T (K)
2500	263.180	260.175	277.318	196.248	266.848	322.884	276.384	281.350	199.038	250.199	205.899	158.916	2500
2600	264.634	261.616	278.851	197.658	268.296	325.302	278.504	282.810	199.854	251.610	206.717	159.732	2600
2700	266.037	263.006	280.333	199.025	269.693	327.635	280.560	284.218	200.639	252.975	207.505	160.516	2700
2800	267.393	264.348	281.768	200.352	271.041	329.888	282.555	285.576	201.395	254.297	208.264	161.272	2800
2900	268.705	265.645	283.160	201.642	272.345	332.066	284.493	286.889	202.125	255.579	208.998	162.001	2900

(continued)

Table A.4 *Thermodynamic data for gaseous species (continued)*

'ABSOLUTE' MOLAR ENTROPY (kJ/kmol K)

T (K)	Air	N₂	O₂	H₂	CO	CO₂	H₂O	NO	Ar	OH	O	H	T (K)
3000	269.975	266.901	284.512	202.896	273.606	334.173	286.376	288.159	202.830	256.823	209.708	162.706	3000
3100	271.206	268.117	285.825	204.117	274.827	336.216	288.208	289.388	203.513	258.031	210.395	163.388	3100
3200	272.401	269.296	287.102	205.306	276.012	338.196	289.990	290.580	204.173	259.205	211.061	164.048	3200
3300	273.562	270.441	288.345	206.466	277.161	340.118	291.726	291.736	204.813	260.347	211.708	164.687	3300
3400	274.690	271.554	289.556	207.597	278.277	341.987	293.417	292.860	205.434	261.458	212.336	165.308	3400
T (K)	Air	N₂	O₂	H₂	CO	CO₂	H₂O	NO	Ar	OH	O	H	T (K)
3500	275.788	272.635	290.736	208.701	279.363	343.803	295.066	293.951	206.037	262.540	212.947	165.910	3500
3600	276.857	273.687	291.889	209.779	280.419	345.571	296.675	295.013	206.623	263.595	213.542	166.496	3600
3700	277.898	274.712	293.013	210.832	281.448	347.294	298.245	296.047	207.193	264.624	214.121	167.065	3700
3800	278.913	275.711	294.112	211.863	282.450	348.973	299.778	297.055	207.747	265.628	214.687	167.620	3800
3900	279.904	276.685	295.186	212.871	283.427	350.612	301.275	298.037	208.288	266.608	215.238	168.160	3900
T (K)	Air	N₂	O₂	H₂	CO	CO₂	H₂O	NO	Ar	OH	O	H	T (K)
4000	280.872	277.635	296.237	213.858	284.381	352.211	302.740	298.996	208.814	267.566	215.777	168.686	4000
4100	281.817	278.564	297.264	214.825	285.313	353.773	304.172	299.932	209.328	268.502	216.303	169.199	4100
4200	282.741	279.471	298.271	215.772	286.223	355.301	305.573	300.847	209.829	269.417	216.818	169.700	4200
4300	283.644	280.357	299.256	216.701	287.113	356.795	306.945	301.741	210.318	270.313	217.322	170.189	4300
4400	284.529	281.225	300.221	217.613	287.984	358.256	308.289	302.615	210.797	271.190	217.816	170.667	4400
T (K)	Air	N₂	O₂	H₂	CO	CO₂	H₂O	NO	Ar	OH	O	H	T (K)
4500	285.394	282.074	301.167	218.508	288.836	359.687	309.606	303.470	211.264	272.049	218.299	171.134	4500
4600	286.242	282.905	302.094	219.387	289.669	361.088	310.897	304.308	211.721	272.891	218.773	171.591	4600
4700	287.072	283.719	303.002	220.251	290.486	362.461	312.163	305.128	212.168	273.717	219.238	172.038	4700
4800	287.886	284.517	303.894	221.101	291.286	363.806	313.406	305.931	212.606	274.527	219.695	172.476	4800
4900	288.684	285.299	304.768	221.938	292.071	365.124	314.625	306.718	213.035	275.322	220.142	172.904	4900
5000	289.466	286.066	305.625	222.762	292.839	366.416	315.823	307.489	213.455	276.103	220.582	173.324	5000
T (K)	Air	N₂	O₂	H₂	CO	CO₂	H₂O	NO	Ar	OH	O	H	T (K)

Note:
The entropy is evaluated by the integration of $dS = (C_p/t)dT$.
When C_p is described by a polynomial function there is a singularity at 0 K. The datum is provided by the values of entropy for substances at 25°C in their standard state at a pressure of 1 bar. (**WARNING** – Many sources use a datum pressure of 1 atm.) The entropy can be evaluated at other pressures from:

$$S = S^0 - R \ln (p/p^0)$$

with the superscript ^{0} referring to the datum pressure (p^0) of 1 bar. When the entropy of a mixture is being evaluated, the properties of the individual constituents are summed, but the pressure (p) now refers to the partial pressure of each constituent.

(continued)

Table A.4 Thermodynamic data for gaseous species (continued)

'ABSOLUTE' MOLAR GIBBS ENERGY (MJ/kmol)

T (K)	Air	N₂	O₂	H₂	CO	CO₂	H₂O	NO	Ar	OH	O	H	T (K)
0	−8.77	−8.82	−8.70	−8.22	−119.36	−402.23	−251.76	81.71	−6.20	32.72	243.04	211.78	0
100	−22.00	−21.76	−23.09	−15.58	−132.89	−417.86	−263.66	66.48	−17.33	19.42	231.32	204.66	100
200	−39.35	−38.85	−41.56	−26.66	−150.58	−436.92	−280.19	47.46	−31.34	2.68	216.67	194.66	200
298.15	**−57.87**	**−57.13**	**−61.16**	**−38.96**	**−169.45**	**−457.25**	**−298.12**	**27.35**	**−46.15**	**−15.04**	**201.21**	**183.78**	**298.15**
300	−58.23	−57.48	−61.54	−39.20	−169.81	−457.65	−298.47	26.96	−46.44	−15.38	200.92	183.57	300
400	−78.09	−77.10	−82.52	−52.72	−190.04	−479.64	−317.88	5.45	−62.25	−34.36	184.45	171.77	400
T (K)	Air	N₂	O₂	H₂	CO	CO₂	H₂O	NO	Ar	OH	O	H	T (K)
500	−98.70	−97.46	−104.26	−66.98	−211.00	−502.66	−338.15	−16.81	−78.58	−54.02	167.45	159.44	500
600	−119.91	−118.41	−126.63	−81.83	−232.57	−526.58	−359.14	−39.71	−95.33	−74.23	150.02	146.70	600
700	−141.63	−139.87	−149.53	−97.17	−254.64	−551.29	−380.73	−63.14	−112.43	−94.93	132.25	133.61	700
800	−163.79	−161.77	−172.90	−112.93	−277.16	−576.71	−402.86	−87.03	−129.83	−116.04	114.17	120.23	800
900	−186.35	−184.06	−196.70	−129.07	−300.08	−602.77	−425.48	−111.33	−147.49	−137.52	95.84	106.58	900
T (K)	Air	N₂	O₂	H₂	CO	CO₂	H₂O	NO	Ar	OH	O	H	T (K)
1000	−209.28	−206.71	−220.88	−145.53	−323.36	−629.43	−448.54	−136.01	−165.38	−159.33	77.27	92.70	1000
1100	−232.53	−229.68	−245.41	−162.31	−346.97	−656.61	−472.01	−161.03	−183.48	−181.45	58.48	78.61	1100
1200	−256.09	−252.96	−270.26	−179.35	−370.90	−684.31	−495.87	−186.36	−201.76	−203.86	39.51	64.33	1200
1300	−279.94	−276.52	−295.41	−196.66	−395.10	−712.48	−520.09	−211.99	−220.23	−226.53	20.37	49.88	1300
1400	−304.04	−300.34	−320.83	−214.21	−419.57	−741.09	−544.66	−237.88	−238.85	−249.45	1.06	35.27	1400
T (K)	Air	N₂	O₂	H₂	CO	CO₂	H₂O	NO	Ar	OH	O	H	T (K)
1500	−328.40	−324.41	−346.51	−231.99	−444.29	−770.11	−569.56	−264.03	−257.62	−272.60	−18.38	20.51	1500
1600	−352.99	−348.71	−372.44	−249.98	−469.24	−799.51	−594.77	−290.41	−276.53	−295.97	−37.98	5.62	1600
1700	−377.80	−373.23	−398.59	−268.18	−494.42	−829.29	−620.28	−317.02	−295.57	−319.55	−57.70	−9.40	1700
1800	−402.82	−397.96	−424.96	−286.58	−519.81	−859.41	−646.08	−343.84	−314.73	−343.32	−77.55	−24.55	1800
1900	−428.03	−422.88	−451.54	−305.16	−545.39	−889.87	−672.15	−370.87	−334.00	−367.29	−97.51	−39.82	1900
T (K)	Air	N₂	O₂	H₂	CO	CO₂	H₂O	NO	Ar	OH	O	H	T (K)
2000	−453.44	−448.00	−478.32	−323.91	−571.17	−920.64	−698.49	−398.08	−353.39	−391.43	−117.58	−55.19	2000
2100	−479.03	−473.29	−505.29	−342.84	−597.13	−951.71	−725.08	−425.48	−372.88	−415.75	−137.76	−70.67	2100
2200	−504.80	−498.76	−532.44	−361.93	−623.26	−983.08	−751.92	−453.06	−392.47	−440.23	−158.03	−86.25	2200
2300	−530.73	−524.39	−559.77	−381.18	−649.55	−1014.72	−779.00	−480.80	−412.16	−464.88	−178.40	−101.92	2300
2400	−556.81	−550.18	−587.26	−400.59	−676.01	−1046.63	−806.31	−508.71	−431.93	−489.68	−198.86	−117.68	2400
T (K)	Air	N₂	O₂	H₂	CO	CO₂	H₂O	NO	Ar	OH	O	H	T (K)
2500	−583.06	−576.13	−614.91	−420.14	−702.62	−1078.79	−833.84	−536.77	−451.79	−514.63	−219.41	−133.53	2500
2600	−609.45	−602.22	−642.72	−439.84	−729.38	−1111.20	−861.58	−564.98	−471.74	−539.72	−240.04	−149.47	2600
2700	−635.98	−628.45	−670.68	−459.67	−756.28	−1143.85	−889.54	−593.33	−491.76	−564.95	−260.75	−165.48	2700
2800	−662.65	−654.82	−698.78	−479.64	−783.31	−1176.73	−917.69	−621.82	−511.87	−590.31	−281.54	−181.57	2800
2900	−689.46	−681.32	−727.03	−499.74	−810.48	−1209.82	−946.05	−650.44	−532.04	−615.80	−302.41	−197.73	2900

(continued)

Table A.4 *Thermodynamic data for gaseous species (continued)*

'ABSOLUTE' MOLAR GIBBS ENERGY (MJ/kmol)

T (K)	Air	N_2	O_2	H_2	CO	CO_2	H_2O	NO	Ar	OH	O	H	T (K)
3000	−716.39	−707.94	−755.41	−519.97	−837.78	−1243.14	−974.59	−679.20	−552.29	−641.42	−323.34	−213.97	3000
3100	−743.45	−734.69	−783.93	−540.32	−865.20	−1276.66	−1003.32	−708.07	−572.61	−667.17	−344.35	−230.27	3100
3200	−770.63	−761.56	−812.58	−560.79	−892.75	−1310.38	−1032.23	−737.07	−592.99	−693.03	−365.42	−246.65	3200
3300	−797.93	−788.55	−841.35	−581.38	−920.41	−1344.29	−1061.32	−766.19	−613.44	−719.01	−386.56	−263.08	3300
3400	−825.35	−815.65	−870.25	−602.08	−948.18	−1378.40	−1090.57	−795.42	−633.95	−745.10	−407.76	−279.58	3400

T (K)	Air	N_2	O_2	H_2	CO	CO_2	H_2O	NO	Ar	OH	O	H	T (K)
3500	−852.87	−842.86	−899.26	−622.90	−976.06	−1412.69	−1120.00	−824.76	−654.53	−771.30	−429.03	−296.14	3500
3600	−880.50	−870.18	−928.39	−643.82	−1004.05	−1447.16	−1149.59	−854.21	−675.16	−797.60	−450.35	−312.76	3600
3700	−908.24	−897.60	−957.64	−664.85	−1032.14	−1481.80	−1179.33	−883.76	−695.85	−824.02	−471.73	−329.44	3700
3800	−936.08	−925.12	−986.99	−685.99	−1060.34	−1516.62	−1209.23	−913.42	−716.60	−850.53	−493.17	−346.18	3800
3900	−964.02	−952.74	−1016.46	−707.22	−1088.63	−1551.60	−1239.29	−943.17	−737.40	−877.14	−514.67	−362.97	3900

T (K)	Air	N_2	O_2	H_2	CO	CO_2	H_2O	NO	Ar	OH	O	H	T (K)
4000	−992.06	−980.46	−1046.03	−728.56	−1117.02	−1586.74	−1269.49	−973.02	−758.26	−903.85	−536.22	−379.81	4000
4100	−1020.20	−1008.27	−1075.71	−750.00	−1145.51	−1622.04	−1299.83	−1002.97	−779.16	−930.65	−557.83	−396.70	4100
4200	−1048.42	−1036.17	−1105.48	−771.53	−1174.08	−1657.49	−1330.32	−1033.01	−800.12	−957.55	−579.48	−413.65	4200
4300	−1076.74	−1064.16	−1135.36	−793.15	−1202.75	−1693.10	−1360.95	−1063.14	−821.13	−984.54	−601.19	−430.64	4300
4400	−1105.15	−1092.24	−1165.33	−814.87	−1231.51	−1728.85	−1391.71	−1093.36	−842.18	−1011.61	−622.95	−447.68	4400

T (K)	Air	N_2	O_2	H_2	CO	CO_2	H_2O	NO	Ar	OH	O	H	T (K)
4500	−1133.65	−1120.40	−1195.40	−836.67	−1260.35	−1764.75	−1422.60	−1123.66	−863.29	−1038.77	−644.75	−464.77	4500
4600	−1162.23	−1148.65	−1225.57	−858.57	−1289.27	−1800.78	−1453.63	−1154.05	−884.44	−1066.02	−666.61	−481.91	4600
4700	−1190.90	−1176.98	−1255.82	−880.55	−1318.28	−1836.96	−1484.78	−1184.52	−905.63	−1093.35	−688.51	−499.09	4700
4800	−1219.64	−1205.40	−1286.17	−902.62	−1347.37	−1873.28	−1516.06	−1215.07	−926.87	−1120.76	−710.45	−516.32	4800
4900	−1248.47	−1233.89	−1316.60	−924.77	−1376.54	−1909.72	−1547.46	−1245.71	−948.15	−1148.26	−732.44	−533.59	4900
5000	−1277.38	−1262.46	−1347.12	−947.00	−1405.78	−1946.30	−1578.99	−1276.42	−969.48	−1175.83	−754.48	−550.90	5000

Note:

The Gibbs function (G) is by definition:

$$G^0 = H - TS^0$$

with the superscript ^{0} referring to the datum pressure (p^0) of 1 bar. The Gibbs function can be evaluated at other pressures in a similar way to the entropy through the use of:

$$G = G^0 + RT \ln(p/p^0)$$

When the Gibbs function of a mixture is being evaluated, the properties of the individual constituents are summed, but the pressure (p) now refers to the partial pressure of each constituent.

Table A.5 *Thermodynamic data for fuels*

ABSOLUTE MOLAR ENTHALPY (MJ/kmol)

T (K)	CH$_4$	C$_3$H$_8$	C$_6$H$_6$	C$_7$H$_8$	C$_7$H$_{16}$	C$_8$H$_{18}$	C$_{16}$H$_{34}$	C$_{10}$H$_7$CH$_3$	CH$_3$OH	C$_2$H$_5$OH	CH$_3$NO$_2$	T (K)
0	−82.660	−115.697	74.264	36.827	−213.096	−252.754	−431.071	100.199	−210.113	−246.415	−84.821	0
100	−80.470	−114.640	73.170	36.863	−210.350	−249.752	−425.008	98.184	−207.653	−244.471	−83.125	100
200	−77.740	−110.826	76.290	41.489	−201.545	−239.795	−405.297	104.238	−204.545	−240.553	−79.660	200
298.15	**−74.520**	**−104.680**	**82.930**	**50.000**	**−187.650**	**−224.01**	**−374.170**	**116.860**	**−200.940**	**−234.950**	**−74.730**	**298.15**
300	−74.454	−104.543	83.086	50.195	−187.342	−223.66	−373.480	117.155	−200.867	−234.829	−74.624	300
400	−70.606	−96.060	93.061	62.502	−168.357	−202.065	−330.987	135.847	−196.714	−227.463	−68.200	400
500	−66.197	−85.627	105.760	77.961	−145.160	−175.677	−279.136	159.351	−192.197	−218.620	−60.563	500
600	−61.231	−73.475	120.774	96.150	−118.274	−145.103	−219.135	186.822	−187.444	−208.465	−51.871	600
700	−55.724	−59.814	137.734	116.679	−88.176	−110.9	−152.081	217.537	−182.602	−197.161	−42.272	700
800	−49.694	−44.836	156.315	139.186	−55.298	−73.567	−78.961	250.894	−177.834	−184.870	−31.901	800
900	−43.170	−28.713	176.235	163.340	−20.025	−33.548	−0.648	286.411	−173.320	−171.754	−20.880	900
1000	−36.185	−11.600	197.253	188.838	17.303	8.767	82.093	323.729	−169.256	−157.974	−9.319	1000
1100	−28.779	6.370	219.172	215.406	56.392	53.044	168.609	362.607	−165.858	−143.690	2.685	1100
1200	−21.000	25.083	241.839	242.801	96.994	99.002	258.358	402.929	−163.357	−129.061	15.046	1200
1300	−12.902	44.443	265.142	270.810	138.908	146.418	350.910	444.696	−162.000	−114.245	27.692	1300
1400	−4.546	64.374	289.012	299.247	181.978	195.122	445.945	488.032	−162.055	−99.399	40.564	1400
1500	4.001	84.819	313.424	327.957	226.094	245	543.253	533.181	−163.803	−84.681	53.614	1500

(continued)

Table A.5 Thermodynamic data for fuels (*continued*)

ABSOLUTE MOLAR INTERNAL ENERGY (MJ/kmol)

T (K)	CH_4	C_3H_8	C_6H_6	C_7H_8	C_7H_{16}	C_8H_{18}	$C_{16}H_{34}$	$C_{10}H_7CH_3$	CH_3OH	C_2H_5OH	CH_3NO_2	T (K)
0	−82.660	−115.697	74.264	36.827	−213.096	−252.754	−431.071	100.199	−210.113	−246.415	−84.821	0
100	−81.302	−115.471	72.339	36.032	−211.181	−250.583	−425.839	97.352	−208.484	−245.303	−83.957	100
200	−79.403	−112.489	74.628	39.826	−203.207	−241.458	−406.960	102.575	−206.208	−242.216	−81.323	200
298.15	**−76.999**	**−107.159**	**80.451**	**47.521**	**−190.129**	**−226.489**	**−376.649**	**114.381**	**−203.419**	**−237.429**	**−77.209**	**298.15**
300	−76.948	−107.037	80.592	47.701	−189.836	−226.154	−375.975	114.660	−203.362	−237.323	−77.118	300
400	−73.932	−99.386	89.735	59.177	−171.683	−205.391	−334.313	132.521	−200.040	−230.788	−71.526	400
500	−70.354	−89.785	101.603	73.804	−149.318	−179.834	−283.293	155.194	−196.354	−222.777	−64.720	500
600	−66.220	−78.464	115.785	91.161	−123.263	−150.092	−224.124	181.833	−192.433	−213.454	−56.859	600
700	−61.544	−65.634	131.914	110.859	−93.996	−116.72	−157.901	211.717	−188.423	−202.981	−48.092	700
800	−56.346	−51.487	149.664	132.535	−61.949	−80.218	−85.612	244.242	−184.486	−191.522	−38.552	800
900	−50.653	−36.197	168.752	155.857	−27.508	−41.031	−8.131	278.928	−180.803	−179.237	−28.363	900
1000	−44.499	−19.915	188.938	180.523	8.988	0.453	73.778	315.414	−177.571	−166.289	−17.633	1000
1100	−37.925	−2.776	210.026	206.260	47.246	43.898	159.463	353.461	−175.004	−152.836	−6.461	1100
1200	−30.977	15.106	231.862	232.824	87.016	89.025	248.381	392.951	−173.334	−139.038	5.068	1200
1300	−23.711	33.634	254.333	260.001	128.099	135.609	340.101	433.887	−172.809	−125.054	16.883	1300
1400	−16.186	52.734	277.372	287.606	170.338	183.482	434.304	476.391	−173.695	−111.040	28.924	1400
1500	−8.471	72.348	300.952	315.485	213.622	232.528	530.781	520.709	−176.274	−97.153	41.143	1500

(continued)

Table A.5 *Thermodynamic data for fuels (continued)*

ABSOLUTE MOLAR ENTROPY (kJ/kmol)

T (K)	CH₄	C₃H₈	C₆H₆	C₇H₈	C₇H₁₆	C₈H₁₈	C₁₆H₃₄	C₁₀H₇CH₃	CH₃OH	C₂H₅OH	CH₃NO₂	T (K)
0	0.000	0.000	0.000	0.000	0.000	0.000	0.000	0.000	0.000	0.000	0.000	0
100	154.754	220.323	222.796	255.497	313.582	293.32	524.594	288.137	204.204	231.886	232.013	100
200	173.463	245.721	242.855	285.836	372.338	359.722	656.095	327.134	225.502	258.289	255.365	200
298.15	**186.270**	**270.200**	**269.200**	**319.740**	**427.980**	**422.96**	**781.020**	**377.440**	**239.880**	**280.640**	**275.010**	**298.15**
300	186.709	270.877	269.941	320.612	429.229	424.35	783.545	378.644	240.342	281.265	275.584	300
400	197.739	295.137	298.432	355.785	483.534	486.12	905.102	432.056	252.261	302.348	293.974	400
500	207.556	318.341	326.667	390.157	535.134	544.819	1020.442	484.322	262.329	322.022	310.969	500
600	216.594	340.454	353.982	423.250	584.058	600.453	1129.626	534.308	270.989	340.503	326.790	600
700	225.073	361.484	380.091	454.853	630.395	653.11	1232.860	581.597	278.452	357.907	341.570	700
800	233.117	381.467	404.880	484.880	674.259	702.919	1330.416	626.103	284.821	374.305	355.408	800
900	240.796	400.444	428.328	513.312	715.779	750.026	1422.599	667.914	290.143	389.745	368.382	900
1000	248.152	418.466	450.463	540.165	755.090	794.59	1509.738	707.217	294.429	404.258	380.558	1000
1100	255.207	435.587	471.349	565.480	792.333	836.776	1592.170	744.261	297.674	417.869	391.996	1100
1200	261.974	451.865	491.067	589.313	827.653	876.755	1670.243	779.336	299.857	430.597	402.749	1200
1300	268.454	467.358	509.716	611.729	861.196	914.701	1744.310	812.760	300.949	442.455	412.870	1300
1400	274.646	482.126	527.403	632.801	893.109	950.789	1814.728	844.868	300.916	453.457	422.409	1400
1500	280.541	496.230	544.243	652.608	923.542	985.196	1881.854	876.010	299.718	463.612	431.412	1500

(continued)

Table A.5 *Thermodynamic data for fuels (continued)*

ABSOLUTE MOLAR GIBBS ENERGY (MJ/kmol)

T (K)	CH$_4$	C$_3$H$_8$	C$_6$H$_6$	C$_7$H$_8$	C$_7$H$_{16}$	C$_8$H$_{18}$	C$_{16}$H$_{34}$	C$_{10}$H$_7$CH$_3$	CH$_3$OH	C$_2$H$_5$OH	CH$_3$NO$_2$	T (K)
0	−82.660	−115.697	74.264	36.827	−213.096	−252.754	−431.071	100.199	−210.113	−246.415	−84.821	0
100	−95.946	−136.672	50.891	11.314	−241.708	−279.034	−477.467	69.370	−228.073	−267.660	−106.326	100
200	−112.432	−159.970	27.719	−15.678	−276.012	−311.739	−536.516	38.811	−249.645	−292.211	−130.733	200
298.15	**−130.056**	**−185.240**	**2.668**	**−45.330**	**−315.252**	**−350.116**	**−607.031**	**4.326**	**−272.460**	**−318.623**	**−156.724**	**298.15**
300	−130.467	−185.806	2.104	−45.988	−316.111	−350.965	−608.544	3.561	−272.970	−319.208	−157.299	300
400	−149.702	−214.115	−26.312	−79.811	−361.771	−396.513	−693.028	−36.975	−297.618	−348.402	−185.790	400
500	−169.975	−244.798	−57.573	−117.118	−412.727	−448.086	−789.357	−82.810	−323.361	−379.631	−216.047	500
600	−191.188	−277.747	−91.615	−157.800	−468.709	−505.375	−896.911	−133.763	−350.037	−412.767	−247.945	600
700	−213.275	−312.853	−128.329	−201.718	−529.452	−568.077	−1015.083	−189.581	−377.519	−447.696	−281.371	700
800	−236.188	−350.009	−167.589	−248.718	−594.705	−635.902	−1143.293	−249.989	−405.691	−484.314	−316.227	800
900	−259.886	−389.113	−209.260	−298.641	−664.226	−708.571	−1280.988	−314.712	−434.448	−522.525	−352.424	900
1000	−284.336	−430.066	−253.211	−351.328	−737.787	−785.823	−1427.645	−383.488	−463.686	−562.232	−389.877	1000
1100	−309.507	−472.776	−299.311	−406.623	−815.175	−867.41	−1582.778	−456.080	−493.299	−603.346	−428.511	1100
1200	−335.368	−517.155	−347.441	−464.374	−896.190	−953.104	−1745.934	−532.275	−523.185	−645.777	−468.253	1200
1300	−361.892	−563.123	−397.489	−524.438	−980.647	−1042.693	−1916.693	−611.892	−553.234	−689.437	−509.039	1300
1400	−389.049	−610.603	−449.353	−586.675	−1068.375	−1135.982	−2094.674	−694.783	−583.337	−734.239	−550.808	1400
1500	−416.811	−659.526	−502.941	−650.956	−1159.219	−1232.795	−2279.529	−780.833	−613.379	−780.100	−593.503	1500

Table A.6 *Equilibrium constants*

At a given temperature, the standard (referring here to a pressure of 1 bar) free enthalpy of reaction or Gibbs energy change (ΔG^0) is related to the equilibrium constant (K_p) by

$$\Delta G_T^0 = -R_0 T \ln K_p$$

and the following values of the equilibrium constants have been calculated from the Gibbs energy tabulations in tables A.4 and A.5. The chemical reactions considered here are presented in the form

$$\Sigma \nu_i A_i = 0$$

where ν_i is the stoichiometric coefficient of the substance A_i.
The partial pressures of the species in equilibrium are found from

$$\ln K_p = \Sigma \nu_i \ln p_i^*$$

where the dimensionless quantity p_i^* is numerically equal to the partial pressure of substance A_i, in units of **bar**.

Reaction number	Valid range: 300–5000 K
1	$-2H + H_2 = 0$
3	$-2NO + N_2 + O_2 = 0$
5	$-\frac{1}{2}H_2 - OH + H_2O = 0$
7	$-CO - H_2O + CO_2 + H_2 = 0$

Reaction number	Valid range: 300–5000 K
2	$-2O + O_2 = 0$
4	$-H_2 - \frac{1}{2}O_2 + H_2O = 0$
6	$-CO - \frac{1}{2}O_2 + CO_2 = 0$

$\ln K_p$ – bar

				Reaction numbers				
T (K)	1	2	3	4	5	6	7	T (K)
100	511.058	584.208	213.879	284.475	331.098	328.849	44.375	100
200	250.154	285.590	105.451	139.970	162.100	159.694	19.723	200
298.15	**163.990**	**187.014**	**69.785**	**92.206**	**106.332**	**103.762**	**11.556**	**298.15**
300	162.906	185.774	69.337	91.604	105.630	103.058	11.454	300
400	119.151	135.735	51.277	67.320	77.322	74.670	7.350	400
500	92.822	105.639	40.433	52.689	60.291	57.617	4.928	500
600	75.220	85.531	33.199	42.896	48.908	46.244	3.349	600
700	62.612	71.139	28.027	35.875	40.758	38.124	2.249	700
800	53.130	60.326	24.146	30.591	34.632	32.037	1.447	800
900	45.735	51.902	21.126	26.468	29.858	27.308	0.840	900
1000	39.803	45.153	18.709	23.160	26.031	23.528	0.369	1000
1100	34.938	39.623	16.732	20.446	22.895	20.439	-0.007	1100
1200	30.873	35.009	15.083	18.179	20.279	17.869	-0.310	1200
1300	27.426	31.100	13.688	16.258	18.062	15.698	-0.560	1300
1400	24.464	27.746	12.492	14.607	16.160	13.840	-0.767	1400
1500	21.892	24.835	11.455	13.175	14.510	12.232	-0.942	1500
1600	19.637	22.286	10.548	11.919	13.065	10.828	-1.091	1600
1700	17.643	20.035	9.748	10.810	11.790	9.591	-1.219	1700
1800	15.867	18.032	9.036	9.823	10.655	8.494	-1.330	1800
1900	14.275	16.238	8.400	8.940	9.640	7.514	-1.426	1900

(continued)

Table A.6 *Equilibrium constants (continued)*

T (K)	1	2	3	4	5	6	7
2000	12.840	14.622	7.827	8.143	8.726	6.633	-1.510
2100	11.540	13.160	7.309	7.422	7.899	5.838	-1.584
2200	10.356	11.829	6.838	6.766	7.146	5.117	-1.649
2300	9.273	10.613	6.408	6.167	6.460	4.460	-1.707
2400	8.280	9.498	6.015	5.617	5.830	3.858	-1.759
2500	7.364	8.471	5.653	5.111	5.251	3.306	-1.805
2600	6.518	7.523	5.319	4.644	4.716	2.797	-1.847
2700	5.733	6.645	5.010	4.211	4.221	2.327	-1.884
2800	5.004	5.829	4.723	3.808	3.761	1.891	-1.918
2900	4.324	5.069	4.457	3.433	3.333	1.486	-1.948
3000	3.689	4.359	4.208	3.083	2.934	1.108	-1.975
3100	3.095	3.695	3.976	2.756	2.560	0.756	-2.000
3200	2.537	3.072	3.758	2.449	2.210	0.426	-2.022
3300	2.012	2.487	3.554	2.160	1.881	0.117	-2.043
3400	1.518	1.936	3.363	1.888	1.572	-0.173	-2.061
3500	1.052	1.416	3.182	1.631	1.280	-0.447	-2.078
3600	0.611	0.925	3.012	1.389	1.005	-0.705	-2.093
3700	0.194	0.460	2.851	1.159	0.744	-0.948	-2.107
3800	-0.202	0.020	2.699	0.941	0.497	-1.178	-2.120
3900	-0.577	-0.397	2.555	0.735	0.263	-1.396	-2.131
4000	-0.934	-0.794	2.419	0.538	0.041	-1.603	-2.141
4100	-1.273	-1.172	2.289	0.351	-0.171	-1.799	-2.150
4200	-1.597	-1.532	2.166	0.173	-0.372	-1.986	-2.159
4300	-1.906	-1.875	2.049	0.003	-0.564	-2.163	-2.166
4400	-2.201	-2.202	1.937	-0.159	-0.747	-2.332	-2.173
4500	-2.482	-2.515	1.830	-0.315	-0.922	-2.494	-2.179
4600	-2.752	-2.815	1.729	-0.463	-1.090	-2.648	-2.185
4700	-3.010	-3.101	1.632	-0.606	-1.250	-2.795	-2.189
4800	-3.258	-3.376	1.539	-0.743	-1.403	-2.936	-2.193
4900	-3.496	-3.640	1.450	-0.874	-1.551	-3.071	-2.197
5000	-3.724	-3.893	1.365	-1.000	-1.692	-3.200	-2.200
5500	-4.740	-5.022	0.988	-1.564	-2.322	-3.773	-2.209
6000	-5.586	-5.964	0.678	-2.037	-2.847	-4.247	-2.210
T (K)	1	2	3	4	5	6	7

Answers to numerical problems

2.3 6 bar, 32 per cent, 22.3.

2.4 6.91 bar, 20.7 per cent, 72.0 per cent.

2.5 (a) 74.1 kW, 174 N m; (b) 3000 rpm: 9.9 bar, 11.0 bar; 5500 rpm: 7.3 bar, 8.1 bar; (c) 25.3 per cent, 58.3 per cent; (d) 91 per cent, 12.1.

2.8 (a) 84 g/MJ; (b) 2.35 litres, 91 mm, 8.5 bar.

2.9 (a) 66 per cent; (b) 57 per cent.

2.10 660 K, 2590 K, 1529 K, 42 per cent.

2.12 0.21.

2.13 $mc_v T_1(1 - r_v^{\gamma-1})$, $mc_v T_3(1 - 1/r_v^{\gamma-1})$, $mc_v \phi(1 - 1/r_v^{\gamma-1})$, $\phi/(\phi + T_1 r_v^{\gamma-1})$.

3.2 C: 0.849, H: 0.151; 13.6.

3.3 C: 0.846, H: 0.154; 0.835; 15.0.

3.4 721 K, 21.3 bar; 4098 K, 89 bar.

3.5 P_{CO_2} 6.76 bar, P_{CO} 2.16 bar, P_{O_2} 1.08 bar; P_{CO_2} 3.80 bar, P_{CO} 4.13 bar, P_{O_2} 2.07 bar.

3.6 51.

3.9 (i) 8.9 per cent CO_2, 13.8 per cent H_2O, 3.7 per cent O_2, 73.5 per cent N_2; (ii) 29.1 per cent; (iii) 1.23 kg CH_4 : 1 kg $C_{16}H_{34}$.

3.10 (i) 2.6 per cent H_2; (ii) 26 per cent.

3.11 Assume the fuel is a hydrocarbon with only CO_2, O_2, H_2O and N_2 in the exhaust. (i) 21.7; (ii) 1.81 H : 1C, 1 kg H to 6.63 kg C; (iii) 14.5; (iv) 0.67.

3.12 (i) 255 MJ/kmol fuel, 31.7 per cent; (ii) 39 kW.

3.13 (i) 13.2; (ii) 85.2 per cent C, 14.8 per cent H, or 5.76 kg C/kg H; (iii) 14.9 : 1.

3.14 $k = 2159$.

3.16 (a) 19.1 bar.

3.17 (c) 39.6 MJ/kmol fuel.

3.18 0.24 per cent, 7.6 MJ/kg fuel.

3.19 0.9; 2871 K.

3.20 0.268 CO, 0.732 CO_2, 1.87 H_2O, 0.13 H_2, 6.77 N_2.

3.23 1.52 CO, 6.48 CO_2, 8.24 H_2O, 0.76 H_2, 42.7 N_2; 27.4 per cent, 11.9 per cent; 1.4 per cent.

3.24 (a) $C_x H_{1.53x}$; (b) 11.4; (c) 14.1, 1.23; (d) 1.7.

3.25 (i) 12.8; (ii) 5.45 kg C/kg H; (iii) 15.1.

3.26 (a) 0.91; (b) 2499 K; (c) 806 K.

3.27 (a) 1965 K; (b) 2875 K; (c) 513 K; (d) 2133 K.

4.7 (i) 4.51 bar; (ii) 28.1 per cent; (iii) 8.62 litres/100 km; (iv) 61.3 per cent.

4.8 (i) 45 K; (ii) 12.7 K; (iii) CO_2 12.1 per cent, O_2 2.9 per cent, N_2 84.9 per cent; 0.805.

4.9 (i) 0.59; (ii) 0.31 per cent; (iii) 11.2 bar.

5.6 (ii) 0.54, 0.48.

5.7 (i) bsp = bsfc $\times \bar{p}_b \times \bar{v}_p/4$;
(ii) AFR = $(\rho \times \eta_{vol})/(\bar{p}_b \times$ bsfc);
(iii) bsac = bsfc $\times$ AFR.

5.8 (i) 56.5 kW, 19.3; (ii) 6.8, 7.3 bar; (iii) 0.34.

6.1 29 per cent.

7.1 62 kW, 5.7 bar, 29 per cent, 26 per cent.

9.1 54°C, 671°C, 92 per cent.

9.4 $[p_2/p_1]^{(\gamma_a-1)/\gamma_a} - 1 =$
$\eta_c \eta_{mech} \eta_t (c_{p,ex}/c_{p,a})(T_3/T_1)(1 - [p_4/p_3]^{(\gamma_{ex}-1)/\gamma_{ex}}) (1 + 1/\text{AFR})$

9.8 (a) 0.94, 8.63 kW; (b) 0.61 kW.

9.9 2.2, 80 000 rpm, 414 K, 6.1 kW.

9.10 (a) 628 kW; (b) 0.404; (c) 0.57; (d) 0.78.

9.11 22.4, 89 600 rpm, 38.6 kW.

9.12 (1) 70 per cent; (2) 19.7 bar; (3) 85 per cent; (4) 555 kW, 32 per cent.

9.14 (a) 47 kW, 0.77, 0.83.

9.15 $\dfrac{p_2}{p_1} = \left[\eta_c \eta_m \eta_t \left(\dfrac{T_3}{T_1}\right)\left(\dfrac{1+\text{AFR}}{\text{AFR}}\right) \right]^{\frac{\gamma}{\gamma-1}}$

9.16 (a) LP: 106 kW, 0.68; MP: 113 kW, 0.66; (c) 0.67, 0.85; (d) 0.97; (e) LP: 1.20 kJ/kg K; MP: 2.00 kJ/kg K.

9.17 1.44 kg/m^3, 0.89, 6 per cent increase in bsfc.

9.18 16.3 kW, 91°C, 90 per cent.
11.1 34 kg mm, 2.4 kg mm.
11.2 4.5 C/D kg mm.
13.1 (a) 1.05 bar; (b) (ii) 0.82, (iii) 0.80; (c) (ii) 1.38,
(iii) 1.30, 1.36, 1.24;
(d) 14.4 percentage points for change in heat capacities,
and 1.8 percentage points by the delayed combustion.

The use of SI units

SI (Système International) Units are widely used, and adopt prefixes in multiple powers of one-thousand to establish the size ranges. Using the watt (W) as an example of a base unit:

picowatt	(pW)	10^{-12}	W
nanowatt	(nW)	10^{-9}	W
microwatt	(μW)	10^{-6}	W
milliwatt	(mW)	10^{-3}	W
watt	(W)	1	W
kilowatt	(kW)	10^{3}	W
megawatt	(MW)	10^{6}	W
gigawatt	(GW)	10^{9}	W
terawatt	(TW)	10^{12}	W

It is unusual for any single unit to have such a size range, nor are the prefixes nano (10^{-9}) and giga (10^{9}) very commonly used.

An exception to the prefix rule is the base unit for mass – the kilogram. Quantities of 1000 kg and over commonly use the tonne (t) as the base unit (1 tonne (t) = 1000 kg).

Sometimes a size range using the preferred prefixes is inconvenient. A notable example is volume; here there is a difference of 10^{9} between mm^3 and m^3. Consequently it is very convenient to make use of additional metric units:

$$1 \text{ cm} = 10^{-2} \text{ m}$$

thus

$$1 \text{ cm}^3 = 10^3 \text{ mm}^3 = 10^{-6} \text{ m}^3$$
$$1 \text{ litre (L)} = 1000 \text{ cm}^3 = 10^6 \text{ mm}^3 = 10^{-3} \text{ m}^3$$

Pressure in SI units is the unit of force per unit area (N/m^2), and this is sometimes denoted by the Pascal (Pa). A widely used unit is the bar (1 bar = 10^5 N/m^2), since this is nearly equal to the standard atmosphere:

1 standard atmosphere (atm) = 1.01325 bar

A unit commonly used for low pressures is the torr:

$$1 \text{ torr} = \frac{1}{760} \text{ atm}$$

In an earlier metric system (cgs), 1 torr = 1 mm Hg.

The unit for thermodynamic temperature (T) is the kelvin with the symbol K (*not* °K). Through long established habit a truncated thermodynamic temperature is used, called the Celsius temperature (t). This is defined by

$$t = (T - 273.15)°\text{C}$$

Note that (strictly) temperature differences should always be expressed in terms of kelvins.

Some additional metric (non-SI) units include:

Length 1 micron = 10^{-6} m
1 angstrom (Å) = 10^{-10} m
Force 1 dyne (dyn) = 10^{-5} N
Energy 1 erg = 10^{-7} N m = 10^{-7} J
1 calorie (cal) = 4.1868 J
Dynamic viscosity 1 poise (P) = 1 g/cm s = 0.1 N s/m²
Kinematic viscosity 1 stokes (St) = 1 cm²/s = 10^{-4} m²/s

A very thorough and complete set of definitions for SI units, with conversions to other unit systems, is given by Haywood (1972).

Conversion factors for non-SI units

Exact definitions of some basic units:

Length 1 yard (yd) = 0.9144 m

Mass 1 pound (lb) = 0.453 592 37 kg

Force 1 pound force (lbf) $= \dfrac{9.806\,65}{0.3048}$ pdl

(1 poundal (pdl) = 1 lb ft/s^2)

Most of the following conversions are approximations:

Length 1 inch (in) = 25.4 mm
1 foot (ft) = 0.3048 m
1 mile (mile) $\approx$ 1.61 km

Area 1 square inch (sq. in) = 645.16 mm^2
1 square foot (sq. ft) $\approx$ 0.0929 m^2

Volume 1 cubic inch (cu. in) $\approx$ 16.39 cm^3
1 gallon (gal) $\approx$ 4.546 litres
1 US gallon $\approx$ 3.785 litres

Mass 1 ounce (oz) $\approx$ 28.35 g
1 pound (lb) $\approx$ 0.4536 kg
1 ton (ton) $\approx$ 1016 kg
1 US short ton $\approx$ 907 kg

Density 1 lb/ft^3 $\approx$ 16.02 kg/m^3

Force 1 pound force (lbf) $\approx$ 4.45 N

Pressure 1 lbf/in^2 $\approx$ 6.895 kN/m^2
1 in Hg $\approx$ 3.39 kN/m^2
1 in H$_2$O $\approx$ 0.249 kN/m^2

Dynamic viscosity 1 lb/ft s $\approx$ 1.488 kg/m s
N s/m^2

Kinematic viscosity 1 ft^2/s $\approx$ 0.0929 m^2/s

Energy 1 ft lbf $\approx$ 1.356 J

Power 1 horse power (hp) $\approx$ 745.7 W

Specific fuel 1 lb/hp h $\approx$ 0.608 kg/kW
consumption $\approx$ 0.169 kg/MJ

Torque 1 ft lbf $\approx$ 1.356 N m

Energy 1 therm (= 10^5 Btu) $\approx$ 105.5 MJ

Temperature 1 rankine (r) $= \dfrac{1}{1.8}$ K

$$\left\{ \begin{array}{l} t_F = (T_R - 459.67)^\circ F \\ \text{thus } t_F + 40 = 1.8(t_C + 40) \end{array} \right\}$$

Specific heat capacity ⎫
Specific entropy   ⎬ 1 Btu/lb R = 4.1868 kJ/kg K
Specific energy   ⎭ 1 Btu/lb = 2.326 kJ/kg

Bibliography

The most prolific source of published material on internal combustion engines is the Society of Automotive Engineers (SAE) of America. Some of the individual papers are selected for inclusion in the annual *SAE Transactions*. Other SAE publications include *Progress in Technology* (PT) and *Specialist Publications* (SP), in which appropriate papers are grouped together. Examples are

SP-532 *Aspects of Internal Combustion Engine Design*
PT-24 *Passenger Car Diesels*

The SAE also organise a wide range of meetings and conferences, and publish the magazine *Automotive Engineering*.

In the United Kingdom the Institution of Mechanical Engineers (I. Mech. E.) publish *Proceedings* and hold conferences, some of which relate to internal combustion engines. The Automobile Division also publishes the bi-monthly *Automotive Engineer*.

The other main organisers of European conferences include:

CIMAC   Conseil International des Machines à Combustion
FISITA   Fédération International des Sociétés d'Ingénieur et de Techniciens de l'Automobile
IAVD   International Association for Vehicle Design
ISATA   International Symposium on Automotive Technology and Automation.

Many books are published on internal combustion engines, and this can be seen in the list of references. However, since books can become dated, care and discretion are necessary in the use of old material.

The following books are useful texts:

Rowland S. Benson, *The Thermodynamics and Gas Dynamics of Internal Combustion Engines*, Vol. I (eds J. H. Horlock and D. E. Winterbone), Oxford University Press, 1982.
Rowland S. Benson, *The Thermodynamics and Gas Dynamics of Internal Combustion Engines*, Vol. II (eds J. H. Horlock and D. E. Winterbone), Oxford University Press, 1986.
Rowland S. Benson and N. D. Whitehouse, *Internal Combustion Engines*, Vol. 1, Pergamon, Oxford, 1979.
Rowland S. Benson and N. D. Whitehouse, *Internal Combustion Engines*, Vol. 2, Pergamon, Oxford, 1979.
G. P. Blair, *The Basic Design of Two-Stroke Engines*, SAE, Warrendale, Pennsylvania, USA, 1990.
Colin R. Ferguson, *Internal Combustion Engines – Applied Thermosciences*, Wiley, New York, 1986.
E. M. Goodger, *Hydrocarbon Fuels – Production, Properties and Performance of Liquids and Gases*, Macmillan, London, 1975.
John B. Heywood, *Internal Combustion Engine Fundamentals*, McGraw-Hill, New York, 1988.
L. R. C. Lilly (ed.), *Diesel Engine Reference Book*, Butterworth, London, 1984.
Charles Fayette Taylor, *The Internal-Combustion Engine in Theory and Practice, Vol. I: Thermodynamics, Fluid Flow, Performance*, 2nd edn (revised), MIT Press, Cambridge, Massachusetts, 1985.
Charles Fayette Taylor, *The Internal-Combustion Engine in Theory and Practice, Vol. II: Combustion, Fuels, Materials, Design*, Revised Edition, MIT Press, Cambridge, Massachusetts, 1985.
N. Watson and M. S. Janota, *Turbocharging the Internal Combustion Engine*, Macmillan, London, 1982.
J. H. Weaving (ed.), *Internal Combustion Engineering*, Elsevier Applied Science, London and New York, 1990.

Finally, *Engines – the search for power* by John Day (published by Hamlyn, London, 1980) is a copiously illustrated book describing the development of all types of engine.

References

Achleitner E., Backer H. and Funaioli A. (2007). 'Direct injection systems for Otto engines', *SAE Paper* 2007-01-1416

Adams L. F. (1975). *Engineering Measurements and Instrumentation*, EUP, London

Adrian R. J. (1991). 'Particle imaging techniques for experimental fluid mechanics', *Annual Review of Fluid Mechanics*, Vol. 23, pp. 261–304

Ahmad T. and Theobald M. A. (1989). 'A survey of variable valve actuation technology', *SAE Paper* 891674

Aikawa K., Sakurai T. and Jetter J. J. (2010). 'Development of a predictive model for gasoline vehicle particulate matter emissions', *SAE International Journal of Fuels and Lubricants*, Vol. 3, No. 2, pp. 610–22

Alcock J. F., Robson F. V. B. and Mash C. (1958). 'Distribution of heat flow in high duty internal combustion engines', *CIMAC*, pp. 723–49

Aleiferis P. G., Serras-Pereira J., van Romunde Z., Caine J. and Wirth M. (2010). 'Mechanisms of spray formation and combustion from a multi-hole injector with E85 and gasoline', *Combustion and Flame*, Vol. 157, pp. 735–56

Allard A. (1982). *Turbocharging and Supercharging*, Patrick Stephens, Cambridge

Anderson J. E., Kramer U., Mueller S. A. and Wallington T. J. (2010). 'Octane numbers of ethanol– and methanol–gasoline blends estimated from molar concentrations', *Energy and Fuels*, Vol. 24, pp. 6576–85. DOI: 10.1021/ef101125c

Andersson J., Giechaskiel B., Muñoz-Bueno R., Sandbach E. and Dilara, P. (2007). 'Particle Measurement Programme (PMP) Light-duty Inter-laboratory Correlation Exercise (ILCE_LD) Final Report', *European Commission Joint Research Centre*, Informal document No. GRPE-54-08-Rev.1

Ando H. (1997). 'Combustion control strategies for gasoline engines', Paper S433/001/96, *Lean Burn Combustion Engines*, pp. 3–17, I. Mech. E. Seminar Publication, MEP, London

Andrae T., Brinck G. and Kalghatgi G. (2008). 'HCCI experiments with toluene reference fuels modelled by a semidetailed chemical kinetic model', *Combustion and Flame*, Vol. 155, No. 4, pp. 696–712

Annand W. J. D. (1963). 'Heat transfer in the cylinders of reciprocating internal combustion engines', *Proc. I. Mech. E.*, Vol. 177, No. 36, pp. 973–90

Annand W. J. D. (1986). 'Heat transfer in the cylinder and porting', in Horlock J. H. and Winterbone D. E. (eds), *The Thermodynamics and Gas Dynamics of Internal Combustion Engines*, Vol. II, Oxford University Press (*see also* Benson, 1982)

Annand W. J. D. and Ma T. (1971). 'Instantaneous heat transfer rates to the cylinder head surface of a small compression ignition engine', *Proc. I. Mech. E.*, Vol. 185, No. 72, pp. 976–87

Annand W. J. D. and Roe G. E. (1974). *Gas Flow in the Internal Combustion Engine*, Foulis, Yeovil

Annual Book of ASTM Standards, Vol. 05.04 – Test Methods for Rating Motor, Diesel and Aviation Fuels, ASTM, Philadelphia, Pennsylvania

Anon (1979). *Programs for Digital Signal Processing*, IEEE/Wiley, Chichester

Anon (1984a). 'Catalytic exhaust-purification for Europe?', *Automotive Engineer*, Vol. 9, No. 1

Anon (1984b). 'Variable inlet valve timing aids fuel economy', *CME*, Vol. 31, No. 4, p. 19

Anon (1986). 'Chrysler, Turbo II', *Automotive Engineering*, Vol. 94, No. 10, pp. 56–7

Anon (1990). 'Isuzu reveals ''bolt-on'' compact engine', *Vehicle Engineering and Design*, p. 3, Design Council, London

Aoyagi Y. (2007). 'HCCI combustion with early and multiple injections in the heavy-duty diesel engine', in Zhao H. (ed.), *HCCI and CAI for the Automotive Industry*, CRC, New York

Arai M. and Miyashita S. (1990). 'Particulate regeneration improvement on actual vehicle under various conditions', Paper C394/012, *Automotive Power Systems – Environment and Conservation*, I. Mech. E. Conf. Proc., MEP, London

Araya K. and Tsunematsu S. (1987). 'Single droplet combustion of sunflower oil', *SAE Paper* 870590

Arcoumanis C. (1997). Private communication, Imperial College, London

Arcoumanis C., Cutter P. A., Foulkes D. and Tabaczynski R. (1997). 'Spray, combustion, and heat transfer studies in a Ricardo hydra direct-injection diesel engine, Paper S433/001/96, *Lean Burn Combustion Engines*, pp. 201–8, I. Mech. E. Seminar Publication, MEP, London

Arnott P. W., Moosmüller H., Rogers F. C., Jin T. and Bruch R. (1999). 'Photo-acoustic spectrometer for measuring light absorption by aerosol: instrument description', *Atmospheric Environment*, Vol. 33, No. 17, pp. 2845–52

Asmus T. W. (1984). 'Effects of valve events on engine operation', in Hilliard J. C. and Springer G. S. (eds), *Fuel Economy in Road Vehicles Powered by Spark Ignition Engines*, Plenum, New York

Assanis D. N. and Heywood J. B. (1986). 'Development and use of computer simulation of the turbocompounded diesel system for engine performance and components heat transfer studies', *SAE Paper* 860329

ASTM Procedure D323, *Annual Book of ASTM Standards, Vol. 05.01*, ASTM, Philadelphia, Pennsylvania

Atkins P. W. (1994). *Physical Chemistry*, 5th edn, Oxford University Press

Bachalo W. D., Brena de la Rosa A. and Sankar S. V. (1991). 'Diagnostics for fuel spray characterization', in Chigier N. (ed.), *Combustion Measurements*, Hemisphere Pub. Corp.

Baker A. (1979). *The Component Contribution*, Hutchinson Benham, London

Ball J., Raine R. and Stone R. (1998). 'Combustion analysis and cycle-by-cycle variations in spark ignition engine combustion, Parts I and II', *Proc. I. Mech. E., Part D*, Vol. 212, pp. 381–99 and pp. 507–24, J. Automotive Engineering, London

Ball J. K., Stone C. R. and Collings N. (1999). 'Cycle-by-cycle modelling of NO formation and comparison with experimental data', *Proc. I. Mech. E., Part D*, Vol. 213, J. Automotive Engineering, London

Baranescu R. and Challen B. (1999). *Diesel Engine Reference Book* (2nd edn), Butterworth–Heinemann

Barnes-Moss H. W. (1975). 'A designer's viewpoint', in *Passenger Car Engines*, I. Mech. E. Conf. Proc., pp. 133–47, MEP, London

Baron P. A. and Willeke K. (eds) (2001). *Aerosol Measurement. Principles, Techniques and Applications*, John Wiley and Sons, Inc.

Baruah P. C. (1986). 'Combustion and cycle calculations in spark ignition engines', in Horlock J. H. and Winterbone D. E. (eds), *The Thermodynamics and Gas Dynamics of Internal Combustion Engines*, Vol. II, Oxford University Press (*see also* Benson, 1982)

Batt R. J., McMillan J. A. and Bradbury I. P. (1996). 'Lubricity additives – performance and non-harm effects in low sulphur fuels', *SAE Paper 961943*

Bazari Z., Smith L. A., Banisoleiman K. and French B. A. (1996). 'An engineering building block approach to engine simulation with special reference to new application areas', Paper C499/017, *Computers in Reciprocating Engines*, pp. 203–17, I. Mech. E. Conf. Publication, MEP, London

Beard C. A. (1958). 'Some aspects of valve gear design', *Proc. I. Mech. E.*, Vol. 2, pp. 49–62

Beard C. A. (1984). 'Inlet and exhaust systems', in Lilly L. R. C. (ed.), *Diesel Engine Reference Book*, Butterworth, London

Beckwith P., Denham M. J., Lang G. J. and Palmer F. H. (1986). 'The effects of hydrocarbon in methanol automotive fuels', *7th International Symposium on Alcohol Fuels*, Paris

Belardini P., Bertoli C., Corcoine F. and Police G. (1983). 'Ignition delay measurement in a direct injection diesel engine', Paper C86/83, *Int. Conf. on Combustion in Engineering*, Vol. II, I. Mech. E. Conf. Proc., pp. 1–8, MEP, London

Benjamin S. F. (1988). 'The development of the GTL ''barrel swirl'' combustion system with application to four-valve spark ignition engines', Paper C54/88, *Int. Conf. Combustion in Engines – Technology and Applications*, I. Mech. E. Conf. Proc., MEP, London

Benson R. S. (1960). 'Experiments on a piston controlled port', *The Engineer*, Vol. 210, pp. 875–80

Benson R. S. (1977). 'A new dynamic model for the gas exchange process in two-stroke loop and cross scavenged engines', *Int. J. Mech. Sci.*, Vol. 19, pp. 693–711

Benson R. S. (1982). *The Thermodynamics and Gas Dynamics of Internal Combustion Engines*, Vol. I (eds Horlock J. H. and Winterbone D. E.), Oxford University Press, p. 9 (*see also* Horlock and Winterbone, 1986)

Benson R. S. and Brandham P. T. (1969). 'A method for obtaining a quantitative assessment of the influence of charging efficiency on two-stroke engine performance', *Int. J. Mech. Sci.*, Vol. 11, pp. 303–12

Benson R. S. and Whitehouse N. D. (1979). *Internal Combustion Engines*, Vols 1 and 2, Pergamon, Oxford

Benson R. S., Garg R. D. and Woollatt D. (1964). 'A numerical solution of unsteady flow problems', *Int. J. Mech. Sci.*, Vol. 6, pp. 117–44

Beretta G. P., Rashidi M. and Keck J. C. (1983). 'Turbulent flame propagation and combustion in spark ignition engines', *Combustion and Flame*, Vol. 52, pp. 217–45

Bevc F. (1997). Advances in solid oxide fuel cells and integrated power plants', *Proc. Inst. Mech. Engrs, Part A*, Vol. 211, pp. 359–66

Biddulph T. W. and Lyn W. T. (1966). 'Unaided starting of diesel engines', *Proc. I. Mech. E.*, Vol. 181, Part 2A

Birch S. (1988). 'Lotus/Shelby', *Automotive Engineering*, Vol. 96, No. 2, p. 172

Bird G. L. (1985). 'The Ford 2.5 litre direct injection naturally aspirated diesel engine', *Proc. I. Mech. E.*, Vol. 199, No. D2

Bird G. L., Duffy K. A. and Tolan L. E. (1989). 'Development and application of the Stanadyne new slip tip pencil injector', *Diesel Injection Systems*, I. Mech. E. Seminar, MEP, London

Blackmore D. R. and Thomas A. (1977). *Fuel Economy of the Gasoline Engine*, Macmillan, London

Blair G. P. (1990). *The Basic Design of Two-Stroke Engines*, SAE, Warrendale, Pennsylvania

Blair G. P. (1996). *Design and Simulation of Two-Stroke Engines*, SAE, Warrendale, Pennsylvania

Blundell D. W., Turner J. W. G., Pearson R. J., Patel R. and Young J. (2010). 'The omnivore wide-range auto-ignition engine: results to date using 98RON unleaded gasoline and E85 fuels', *SAE Paper 2010-01-0846*

Boehman A. L., Song J. and Alam M. (2005). 'Impact of biodiesel blending on diesel soot and the regeneration of particulate filters', *Energy & Fuels*, Vol. 19, pp. 1857–64.

de Boer C. D., Johns R. J. R., Grigg D. W., Train B. M., Denbratt I. P. and Linna J-R. (1990). 'Refinement with performance and economy for four-valve automotive engines', Paper C394/053, *Automotive Power Systems*, pp. 147–56, Inst. Mech. Engrs Conf. Proc., MEP, London

Borgnakke C. (1984). 'Flame propagation and heat-transfer effects in spark-ignition engines', in Hilliard J. C. and Springer G. S. (eds), *Fuel Economy of Road Vehicles Powered by Spark Ignition Engines*, Plenum, New York

Bostock P. G. and Cooper L. (1992). 'Turbocharging the Ford 2.5 HSDI diesel engine', I. Mech. E., *Seminar Diesel Fuel Injection Systems*, 14–15 April 1992, MEP, London

Boucher R. F. and Kitsios E. E. (1986). 'Simulation of fluid network dynamics by transmission line modelling', *Proc. I. Mech. E.*, Vol. 200, No. C1, pp. 21–9

Bradley D., Lau A. K. C. and Lawes M. (1992). 'Flame stretch rate as a determinant of turbulent burning velocity', *Phil. Trans. R. Soc. Lond.*, Vol. A338, pp. 359–87

Braisher M. H. (2010). 'Particulate matter emissions from engines and vehicles', *DPhil Thesis*, University of Oxford

Braisher M., Stone R. and Price P. (2010). 'Particle number emissions from a range of European vehicles', *SAE Paper 2010-01-0786*

Brandstetter W. and Howard J. (1989). 'The second generation of the Ford 2.5 litre direct injection diesel engine', *ATZ*, Vol. 91, No. 6, pp. 327–9

Brogan M. S., Will N. S., Twigg M. V., Wilkins A. J. J., Jordan K. and Brisley R. J. (1998). 'Advances in DENOx catalyst technology for European Stage IV emissions levels', *Future Engine and Systems Technologies*, I. Mech. E. Seminar Publication, Professional Engineering Publications, London

Brown C. N. (1987). 'An investigation into the performance of a Waukesha VRG220 SI engine, fuelled by CH_4–CO_2 mixtures', *Final Year Project Report*, Brunel University, London

Brown A. G. (1991). *PhD Thesis*, Brunel University, London

Brown C. N. and Ladommatos N. (1991). 'The effects of mixture preparation and trapped residuals on the performance of a port-injected spark-ignition engine at low load and low speed', *Proc. I. Mech. E.*, Vol. 205

Brown A. G., Stone C. R. and Beckwith P. (1996). 'Cycle-by-cycle variations in spark ignition engine combustion, Part I Flame speed and combustion measurements, and a simplified turbulent combustion model', *SAE Paper 960612. Also in Advances in Engine Combustion and Flow Diagnostics* (SP-1157)

BS 7800: 1992 *Specification for high octane (super) unleaded petrol (gasoline) for motor vehicles*, BSI, London

BS EN 228: 1993 *Specification for unleaded petrol (gasoline) for motor vehicles*, BSI, London

BS EN 590: 1997 *Specification for automotive diesel fuel*, BSI, London

BS EN 25163: 1994 *Methods of test for petroleum and its products. Motor and aviation-type fuels. Determination of knock characteristics motor method*, BSI, London

BS EN 25164: 1994 *Methods of test for petroleum and its products. Motor and aviation-type fuels. Determination of knock characteristics research method*, BSI, London

Campbell C. (1978). *The Sports Car*, 4th edn, Chapman & Hall, London

Carabateas N. E., Taylor A. M. K. P., Whitelaw J. H., Ishii K. and Miyano H. (1996). 'Droplet velocities within the pent-roof of a Honda Vtec-E cylinder head', *Lean Burn Combustion Engines*, I. Mech. E. Seminar Publication, MEP, London

Caris D. F. and Nelson E. E. (1958). 'A new look at high compression engines', *SAE Paper 61A*

Caton J. A. and Heywood J. B. (1981). 'An experimental and analytical study of heat transfer in an engine exhaust port', *Int. J. Heat Mass Transfer*, Vol. 24, No. 4, pp. 581–95

Charlton S. J. (1986). *SPICE User Manual*, School of Engineering, University of Bath

Charlton S. J. (1998). 'Control technologies of compression-ignition engines', in Sher E. (ed.), *Handbook of Air Pollution from Internal Combustion Engines*, Academic Press, Boston, Massachusetts

Charlton, S. J., Keane A. J., Leonard H. J. and Stone C. R. (1990). 'Application of variable valve timing to a highly turbocharged diesel engine', *Turbochargers and Turbocharging*, I. Mech. E. Conf. Proc., MEP, London

Charlton S. J., Stone C. R., Leonard H. J., Elliott C. and Newman M. (1991). 'Transient simulation of a highly turbocharged diesel engine with variable valve timing', *2nd Int. Conf. Computers in Engine Technology*, I. Mech. E., London

Chen J. C. (1966). 'A correlation for boiling heat transfer to saturated fluids in convective flow', *Industrial and Engineering Chemistry – Process Design and Development*, Vol. 5, No. 3, pp. 322–9

Chen S. K. and Flynn P. (1965). 'Development of a compression ignition research engine', *SAE Paper 650733*

Chen A., Lee K. C. and Yianneskis M. (1995). 'Velocity characteristics of steady flow through a straight generic inlet port', *Int. J. Numerical Methods in Fluids*, Vol. 21, pp. 571–90

Chen X., Fu H., Smith S. and Sandford M. (2009). 'Investigation of combustion robustness in catalyst heating operation on a spray guided DISI engine,' *SAE Paper 2009-01-1489*

Chen L., Xu F., Stone R. and Richardson D. (2011). 'Spray imaging, mixture preparation and particulate matter emissions using a GDI engine fuelled with stoichiometric gasoline/ethanol blends', *Internal Combustion Engines: Improving Performance, Fuel Economy and Emissions*, I. Mech. E. Conf., MEP, London, November 2011

Cheng W. K., Galliott F., Chen T., Sztenderowiicz M. and Collings N. (1990). 'In cylinder measurements of residual gas concentration in spark ignition engine', *SAE Paper 900485*

Cheng W. K., Hamrin D., Heywood J. B., Hochgreb S., Min K. and Norris M. G. (1993). 'An overview of hydrocarbon emissions mechanisms in spark ignition engines', *SAE Paper 932708*

Cheng W. K., Summers T. and Collings N. (1998). 'The fast-response flame ionization detector', *Progress in Energy and Combustion Science*, Vol. 24, No. 2, pp. 89–124

Cockshott C. P., Vernon J. P. and Chambers P. (1983). 'An air mass flowmeter for test cell instrumentation', *4th Inst. Conf. on Automotive Electronics*, pp. 20–6, I. Mech. E. Conf. Proc., MEP, London

Cohen H., Rogers G. F. C. and Saravanamuttoo H. I. H. (1972). *Gas Turbine Theory*, 2nd edn., Longman, London

Collings N. and Willey J. (1987). 'Cyclically resolved emissions from a spark ignition engine', *SAE Paper 871691*

Collins D. and Stokes J. (1983). 'Gasoline combustion chambers – compact or open?' *SAE Paper 830866*

Collis D. C. and Williams M. J. (1959). 'Two-dimensional convection from heated wires at low Reynolds numbers', *Journal of Fluid Mechanics*, Vol. 6, pp. 357–84

CONCAWE (2007). *Well-to-Wheels Analysis of Future Automotive Fuels and Powertrains in the European Context*, Version 2c, March 2007, http://ies.jrc.ec.europa.eu/WTW

Cooke J. A., Roberts D. D., Horrocks R. and Ketcher D. A. (1990). 'Automotive diesel emissions: a fuel appetite study on two light duty direct injection diesel engines', Paper C394/058, *Automotive Power Systems – Environment and Conservation*, I. Mech. E. Conf. Proc., MEP, London

Correa S. M. (1992). 'A review of NO_x formulation under gas-turbine combustion conditions', *Combust. Sci. Tech.*, Vol. 87, pp. 329–62

CSTI (Council of Science and Technology Institutes) (1992). 'The greenhouse effect: fact or fiction?', *Environmental Information Paper 1* [available from the UK Institute of Physics]

Cuttler D. H., Girgis N. S. and Wright C. C. (1987). 'Reduction and analysis of combustion data using linear and non-linear regression techniques', Paper C17187, *I. Mech. E. Conf. Proc.*, MEP, London

Dale A., Cole A. C. and Watson N. (1988). 'The development of a turbocharger turbine test facility', in *Experimental Methods in Engine Research and Development*, I. Mech. E. Seminar, MEP, London

Daneshyar H. (1976). *One-Dimensional Compressible Flow*, Pergamon, Oxford

Daniels J. (1996). 'General Motors' new DI diesel engine', *Automotive Engineer*, Vol. 21, No. 5, pp. 24–8

Daubert T. E. and Danner R. P. (1989). *Physical and Thermodynamic Properties of Pure Chemicals*, Taylor and Francis, Washington DC

Davies G. O. (1983). 'The preparation and combustion characteristics of coal derived transport fuels', Paper C85/83, *Int. Conf. on Combustion in Engineering*, Vol. II, I. Mech. E. Conf. Proc., MEP, London

Davies P. O. A. L. and Fisher M. J. (1964). 'Heat transfer from electrically heated cylinders', *Proc. Royal Soc.*, Vol. A280, pp. 486–527

DoEn (1991). *The Ozone Layer*, Department of the Environment, London

Donnelly M. J., Junday J. and Tidmarsh D. H. (1981). 'Computerised data acquisition and processing system for engine test beds', *3rd Int. Conf. on Automotive Electronics*, MEP, London

Downs D. and Wheeler R. W. (1951–52). 'Recent developments in knock research', *Proc. I. Mech. E. (AD)*, Pt III, p. 89

Downs D., Griffiths S. T. and Wheeler R. W. (1961). 'The part played by the preparational stage in determining lead anti-knock effectiveness', *J. Inst. Petrol.*, Vol. 47, p. 1

Duclos J. M., Bruneaux G. and Baritaud T. A. (1996). '3D modelling of combustion and pollutants in a 4-valve SI engine; effect of fuel and residuals distribution and spark location', *SAE Paper 961964*

Dudley W. M. (1948). 'New methods in valve cam design', *SAE Quarterly Transactions*, Vol. 2, No. 1, pp. 19–33

Dunn J. (1985). 'Top-hat piston engine set to make a comeback', *The Engineer*, 12 December 1985, p. 38

Eade D., Hurley R. G., Rutter B., Inman G. and Bakshi R. (1996). 'Exhaust gas ignition', *Automotive Engineering*, Vol. 104, No. 4, pp. 70–3

Eastham D. R., Parker D. D. and Summerton K. (1995). 'Developments in tri-metal bearings', Paper 2, *T&N Symposium*

Eastwood P. (2008). *Particulate Emissions from Motor Vehicles*, Wiley

Echekki T. (2009). 'Topical review: multiscale methods in turbulent combustion: strategies and computational challenges', *Computational Science and Discovery* 2 013001. DOI: 10.1088/1749-4699/2/1/013001

Eggers M., Becher S., Eggert T., Druhe M-A., Kreh A. and Weissar H-G. (2004). 'Emission control legislation and measuring techniques', in *Bosch Automotive Handbook*, 6th edn, Robert Bosch

Eichelberg G. (1939). 'Some new investigations on old combustion-engine problems' (in four parts), *Engineering*, Vol. 149 [original work published in German, *ZDVI*, Vol. 67, in 1923]

Eltinge L. (1968). 'Fuel–air ratio and distribution from exhaust gas composition', *SAE Paper 680114*. Also in *SAE Transactions*, Vol. 77

Enga B. E., Buchman M. F. and Lichtenstein I. E. (1982). 'Catalytic control of diesel particulates', *SAE Paper 820184*. Also in *SAE P-107*

Ernest R. P. (1977). 'A unique cooling approach makes aluminium alloy cylinder heads cost effective', *SAE Paper 770832*

Felton G. N. (1996). 'Compact rotary spill pump with electronic control for high speed direct injection engines', *SAE Paper 960865*

Ferguson C. R. (1986). *Internal Combustion Engines – Applied Thermosciences*, Wiley, New York

Finlay I. C., Harris D., Boam D. J. and Parks B. I. (1985). 'Factors influencing combustion chamber wall temperatures in a liquid-cooled automotive, spark-ignition engine', *Proc. I. Mech. E.*, Vol. 199, No. D3, pp. 207–14

Finlay I. C., Boyle R. J., Pirault J-P. and Biddulph T. (1987a). 'Nucleate and film boiling of engine coolants, flowing in a uniformly heated duct of small cross section', *SAE Paper 870032*

Finlay I. C., Tugwell W., Pirault J-P. and Biddulph T. (1987b). 'The influence of coolant temperature on the performance of a 1100cc engine employing a dual circuit cooling system', *XIXth International Symposium, International Centre for Heat and Mass Transfer, Heat and Mass Transfer in Gasoline and Diesel Engines*, Dubrovnik, August 1987

Finlay I. C., Boam D. J., Bingham J. F. and Clark T. (1990). 'Fast response FID measurements of unburned hydrocarbons in the exhaust port of a firing gasoline engine', *SAE Paper 902165*

Fisher F. B. (1986). 'Development of vaned diffuser compressors for heavy duty diesel engine turbochargers', Paper C108/86, *Turbochargers and Turbocharging*, I. Mech. E. Conf. Proc., MEP, London

Fisher E. H. (1989). 'Means of improving the efficiency of automotive cooling systems', Paper C372/001, *Small Engines Conference*, I. Mech. E. Conf. Proc., pp. 71–6, MEP, London

Fitzen M., Hatz W., Eiser A., Heiduk T. and Riegner, J. (2008). 'Der Audi 3,0l TFSI – die neue V6 Spitzenmotorisierung', 'The Audi 3.0 litre TFSI – the new top-of-the-range V6 engine', *29th Internationales Wiener Motorensymposium 2008*

Fleisch T. H. and Meurer P. C. (1995). 'DME, the diesel fuel for the 21st century?', *AVL Conference 'Engine and Environment'*, Graz, Austria

Ford (1982). *Ford Energy Report*, Interscience Enterprises, Channel Islands, UK

Forlani E. and Ferrati E. (1987). 'Microelectronics in electronic ignition – status and evolution', Paper 87002, *16th ISATA Proceedings*

Fortnagel M. (1990). 'The Mercedes–Benz diesel engine range for passenger cars and light duty trucks as viewed from emissions aspects', *Automotive Power Systems – Environment and Conservation*, I. Mech. E. Conf. Publication 1990–9, MEP, London

Fosberry R. A. C. and Gee D. E. (1961). 'Some experiments on the measurement of exhaust smoke emissions from diesel engines', *MIRA Report 1961/5*

Fox J. W., Cheng W. K. and Heywood J. B. (1993). 'A model for predicting residual gas fraction in a spark-ignition engine', *SAE Paper 931025*

Fraidl G. K. (1987). 'Spray quality of mixture preparation systems', *EAEC Int. Conf. on New Developments in Powertrain and Chassis Engineering*, Strasbourg, Vol. 1, pp. 232–46

Fraidl G. K., Mikulic L. A. and Quissek F. (1990). 'Development strategies for low emission high performance four-valve engines', *Proc. Instr. Mech. Engrs*, Vol. 204D, pp. 59–65

Francis R. J. and Woollacott P. N. (1981). 'Prospects for improved fuel economy and fuel flexibility in road vehicles', *Energy Paper No. 45*, Department of Energy, HMSO, London

Frank W., Hüthwohl G. and Maurer B. (2003). 'SCR systems for heavy duty trucks: progress towards meeting Euro 4 emission standards in 2005', *9th DEER Conference*, August 2003, Newport RI

Frankl G., Barker B. G. and Timms C. T. (1989), 'Electronic unit injectors for direct injection engines', *Diesel Fuel Injection Systems*, I. Mech. E. Seminar, MEP, London

Fraser N., Blaxill H., Lumsden G. and Bassett M. (2009). 'Challenges for increased efficiency through gasoline engine downsizing', *SAE Paper 2009-01-1053*

French C. C. J. (1970). 'Problems arising from the water cooling of engine components', *Proc. I. Mech. E.*, Vol. 184, Pt 1, No. 29, pp. 507–42

French C. C. J. (1984). 'Thermal loading', in Lilly L. R. C. (ed.), *Diesel Engine Reference Book*, Butterworth, London

French C. C. J. and Atkins K. A. (1973). 'Thermal loading of a petrol engine', *Proc. Inst. Mech. Engrs*, Vol. 187, pp. 561–73

Freudenschuss O. (1988). 'A new high speed light duty DI diesel engine', *SAE Paper 881209*

Fujita T., Kiga S. and Tsuruta S. (2008). 'The innovative variable valve event and lift system (VVEL) for the new NISSAN V6 and V8 engine', *29th Internationales Wiener Motorensymposium 2008*

Gardner F. J. (1997). Thermodynamic processes in solid oxide and other fuel cells', *Proc. Inst. Mech. Engrs, Part A*, Vol. 211, pp. 367–80

Garrett K. (1990). 'Fuel quality diesel emissions and the City Filter', *Automotive Engineer*, Vol. 15, No. 5, pp. 51–5

Gatowski J. A., Smith M. K. and Alkidas A. C. (1989). 'An experimental investigation of surface thermometry and heat flux', *J. Exp. Thermal Fluid Science*, Vol. 2, No. 3, pp. 280–92

Gaydon A. G. and Wolfhard H. G. (1979). *Flames, their Structure, Radiation and Temperature*, 4th edn, Chapman & Hall, London

Gilchrist J. M. (1947). 'Chart for the investigation of thermodynamic cycles in internal combustion engines and turbines', *Proc. I. Mech. E.*, Vol. 156, pp. 335–48

Glikin P. E. (1985). 'Fuel injection in diesel engines', *Proc. I. Mech. E.*, Vol. 199, No. 78, pp. 1–14

Goodger E. M. (1975). *Hydrocarbon Fuels*, Macmillan, London

Goodger E. M. (1979). *Combustion Calculations*, Macmillan, London

Gordon S. and McBride B. J. (1971). 'Computer program for calculation of complex chemical equilibrium compositions, rocket performance, incident and reflected shocks, and Chapman–Jouget detonations', *NASA SP-273 Interim Report*

Gordon S. and McBride B. J. (1976). 'Computer program for calculation of complex chemical equilibrium compositions, rocket performance, incident and reflected shocks and Chapman–Jouget detonations', *NASA SP-273 Interim Revision N78-17724*

Gosman A. D. (1986). 'Flow processes in cylinders', in Horlock J. H. and Winterbone D. E. (eds), *The Thermodynamics and Gas Dynamics of Internal Combustion Engines*, Vol. II, Oxford University Press

Greene A. B. and Lucas G. G. (1969). *The Testing of Internal Combustion Engines*, EUP, London

Greenhalgh D. A. (1983). 'Gas phase temperature and concentration diagnostics with lasers', *Int. Conf. on Combustion in Engineering*, Vol. I, I. Mech. E. Conf. Publication 1983–3, MEP, London

Greeves G. and Wang C. H. T. (1981). 'Origins of diesel particulate mass emission', *SAE Paper 810260*

Griffiths J. F. and Barnard J. A. (1995). *Flame and Combustion*, Blackie Academic and Professional, London

Gruden D. and Kuper P. F. (1987). 'Heat balance of modern passenger car SI engines', *XIXth International Symposium, International Centre for Heat and Mass Transfer, Heat and Mass Transfer in Gasoline and Diesel Engines*, Dubrovnik, August

Guerrassi N. and Dupraz P. (1998). 'A common rail injection system for high speed direct injection diesel engines', *SAE Paper 980803*

Gunther D. (1988). 'Exhaust emissions engineering', in Adler U. and Bauer H. (eds), *Automotive Electric/Electronic Systems*, Robert Bosch, Stuttgart

Hahn H. W. (1986). 'Improving the overall efficiency of trucks and buses. *Proc. I. Mech. E.*, Vol. 200, No. D1, pp. 1–13

Hall M. J. and Bracco F. V. (1986). 'Cycle resolved velocity and turbulence measurements near the cylinder wall of a firing S.I. engine', *SAE Paper 861530*

de Haller R. (1945). 'The application of a graphic method to some dynamic problems in gases', *Sulzer Technical Review*, Vol. 1, No. 6

Hancock D., Fraser N., Jeremy M., Sykes R. and Blaxill H., (2008). 'A new 3 cylinder 1.2 l advanced downsizing technology demonstrator engine', *SAE Paper 2008-01-0611*

Hanson R. K. and Salimian S. (1984). 'Survey of rate constants in the N/H/O system', in Gardiner W. C. (ed.), *Combustion Chemistry*, Chapter 6, Springer-Verlag

Hara S., Nakajima Y. and Nagumo S. (1985). 'Effects of intake valve closing timing on SI engine combustion', *SAE Paper 850074*

Harada J., Yamada T. and Watanabe K. (2008). 'The new L4 gasoline engines with VALVEMATIC system', *29th Internationales Wiener Motorensymposium 2008*

Hardenberg H. O. and Hase F. W. (1979). 'An empirical formula for computing the pressure rise delay of a fuel from its cetane number and from the relevant parameters of direct injection diesel engines', *SAE Paper 790493, Also in SAE Trans.*, Vol. 88

Harman R. T. C. (1981). *Gas Turbine Engineering*, Macmillan, London

Hartmann V. and Mallog J. (1988). 'In-cylinder flow analysis as a practical aid in the development of internal combustion engines', in *Int. Conf. Combustion in Engines – Technology and Applications*, I. Mech. E. Conf. Proc., MEP, London

Hawker P. N. (1995). 'Diesel emission control technology', *Platinum Metal Review*, Vol. 39, No. 1, pp. 2–8

Hawley G. J., Brace C. J., Wallace F. J. and Horrocks R. W. (1998). 'Combustion related emissions in CI engines', in Sher E. (ed.), *Handbook of Air Pollution from Internal Combustion Engines*, Academic Press, Boston, Massachusetts

Hay N., Watt P. M., Ormerod M. J., Burnett G. P., Beesley P. W. and French B. A. (1986). 'Design study for a low heat loss version of the Dover engine', *Proc. I. Mech. E.*, Vol. 200, No. DI, pp. 53–60

Haywood R. W. (1972). *Thermodynamic Tables in SI (Metric) Units*, 2nd edn, Cambridge University Press

Haywood R. W. (1980). *Analysis of Engineering Cycles*, 3rd edn, Pergamon International Library, Oxford

Heimrich M. J., Albu S. and Osborn J. (1991). 'Electrically-heated catalyst system conversions on two current-technology vehicles', *SAE Paper 910612*

Heisler H. (1995). *Advanced Engine Technology*, Edward Arnold, London

Herzog P. (1998). 'HSDI diesel engine developments towards Euro IV', *Future Engine and System Technologies*, I. Mech. E. Seminar Publication, Professional Engineering Publications, London

Heywood J. B. (1988). *Internal Combustion Engine Fundamentals*, McGraw-Hill, New York

Heywood J. B., Higgins J. M., Watts P. A. and Tabaczynski R. J. (1979). 'Development and use of a cycle simulation to predict SI engine efficiency and NO_x emissions', *SAE Paper 790291*

Hicks R. A., Lawes M., Sheppard C. G. W. and Whitaker B. J. (1994). 'Multiple laser sheet imaging investigation of turbulent flame structure in a spark ignition engine', *SAE Paper 941992*

Hinds W. C. (1999). *Aerosol Technology. Properties, Behaviour, and Measurement of Airborne Particles*, John Wiley and Sons, Inc.

Hiroyasu H., Arai M. and Tabata M. (1989). 'Empirical equations for the Sauter mean diameter of a diesel spray', *SAE Paper 890464*

Hoag K. L. (2006). *Vehicular Engine Design*, SAE International

Hoag K. L., Brands M. C. and Bryzik W. (1985). 'Cummins/TACOM adiabatic engine program', *SAE Paper 850356. Also in SAE SP-610*

Hochgreb S. (1998). 'Combustion-related emissions in SI engines', in Sher E. (ed.), *Handbook of Air Pollution from Internal Combustion Engines*, Academic Press, Boston, Massachusetts

Hohenberg G. F. (1979). 'Advanced approaches for heat transfer calculations', *SAE Paper 790825. Also in SAE Trans.*, Vol. 88

Hollingworth P. and Hodges R. A. (1991). 'The history and mathematical development of cam profile design in Rover', Paper C427/1/107, *Autotech Seminar Papers, Seminar 1, Mechanical Components and Systems*, I. Mech. E., London

Holmes M., Willcocks D. A. R. and Bridgers B. J. (1988). 'Adaptive ignition and knock control', *SAE Paper 885065*

Hooper P. R. (2005). 'Stepped piston engines for multifuel UAV application', *Proc. I. Mech. E. Conf. on Propulsion Systems for Unmanned Aircraft*, Bristol, UK, 14 April 2005, pp. 83–95 (Professional Engineering Publishing Limited, London)

Hooper P. R., Al-Shemmeri T. and Goodwin M. J. (2011). 'Advanced modern low-emission two-stroke cycle engines', *Proc. I. Mech. E., Part D: J. Automobile Engineering*, Vol. 225, published online 4 August 2011. DOI: 10.1177/0954407011408649

Horie K. and Nishizawa K. (1992). 'Development of a high fuel economy and high performance four-valve lean burn engine', Paper C448/014, *Combustion in Engines*, Inst. Mech. Engrs Conf. Proc., pp. 137–43, MEP, London

Horlock J. H. and Winterbone D. E. (eds) (1986). *The Thermodynamics and Gas Dynamics of Internal Combustion Engines*, Vol. II, Oxford University Press (*see also* Benson, 1982)

Howarth M. H. (1966). *The Design of High Speed Diesel Engines*, Constable, London

Howatson A. M., Lund P. G. and Todd J. D. (1991). *Engineering Tables and Data*, Chapman & Hall, London

Hudson C., Ladommatos N., Schmid F. and Stone R. (1988). 'Digital instrumentation for combustion study and monitoring in internal combustion engines', *ISATA Paper 88095*

Hurn R. W. and Smith H. M. (1951). 'Hydrocarbons in the diesel boiling range', *Industrial Engineering and Chemistry*, Vol. 43, p. 2788

Ikoma T., Abe S., Sonoda Y., Suzuki H., Suzuki Y. and Basaki M. (2006). 'Development of V-6 3.5-liter engine adopting new direct injection system,' *SAE Paper 2006-01-1259*. DOI: 10.4271/2006-01-1259

Isaji H., Satou T., Kobayashi A., Oookouchi M., Yoichi M., Shimasaki H. and Inoue K. (2011). 'Development of new I3 1.2L supercharged gasoline engine', *JSAE Paper 20115705*

Ives A. P. and Trenne M. V. (1981). 'Closed loop electronic control of diesel fuel injection', *3rd Int. Conf. on Automotive Electronics*, I. Mech. E. Conf. Publication 1981–10, MEP, London

Iwamoto J. and Deckker B. E. L. (1985). 'Application of random-choice method to the calculation of unsteady flow in pipes', *SAE Paper 851561*

Iwamoto Y., Noma K., Nakayama O., Yamauchi T. and Ando, H. (1997). 'Development of gasoline direct injection engine', *SAE Technical Paper, 970541*

Jackson N. S., Stokes J., Whitaker P. A. and Lake T. H. (1997). 'Stratified and homogeneous charge operation for the DISI engine – high power with low fuel consumption and emissions', *SAE Technical Paper 970543*

James E. H. (1990). 'Further aspects of combustion modelling in spark ignition engines', *SAE Paper 900684*

JANAF (1985). *Thermochemical Tables*, 3rd edn (eds Chase *et al.*), J. Phys. Chem. Ref. Data, Vol. 14, Suppl. 1

Jante A. (1968). 'Scavenging and other problems of two-stroke cycle spark-ignition engines', *SAE Paper 680468*

Jenny E. (1950). 'Unidimensional transient flow with consideration of friction, heat transfer, and change of section', *Brown Boveri Review*, Vol. 37, No. 11, p. 447

Jochheim J., Hesse D., Duesterdick T., Engeler W., Neyer D., Warren J. P., Wilkins A. J. J. and Twigg M. V. (1996). 'A study of the catalytic reduction of NO_x in diesel exhaust', *SAE Paper 962042*

Johnson T. (2008). 'Diesel engine emissions and their control – an overview', *Platinum Metals Rev.*, Vol. 52, No. 1, pp. 23–37

Joyce M. (1994). 'Jaguar's supercharged 6-cylinder engine', *Turbochargers and Turbocharging*, I. Mech. E. Conf. Proc. 1994–6, MEP, London

Judge A. W. (1967). *High Speed Diesel Engines*, 6th edn, Chapman & Hall, London

Judge A. W. (1970). *Motor Manuals 2: Carburettors and Fuel Injection Systems*, 8th edn, Chapman & Hall, London

Kalghatgi, G. (2005). 'Auto-ignition quality of practical fuels and implications for fuel requirements of future SI and HCCI engines', *SAE Technical Paper* 2005-01-0239, 2005. DOI: 10.4271/2005-01-0239

Kalghatgi G. (2007). 'Fuel effects in CAI gasoline engines', in Zhao H. (ed.), *HCCI and CAI for the Automotive Industry*, CRC, New York

Kalghatgi G. (2009). 'Is gasoline the best fuel for advanced diesel engines? – Fuel effects in "premixed enough" compression ignition (CI) engines', *Towards Clean Diesel Engines*, TDCE2009

Kalghatgi G. T. and Bradley D. (2012). 'Pre-ignition and "super-knock" in turbo-charged spark-ignition (SI) engines', *Int. J. Engine Research*, Vol. 13, No. 1. DOI: 10.1177/1468087411431890

Kalghatgi G. T., Bradley D., Andrae J. and Harrison A. J. (2009). 'The nature of "superknock" and its origins in SI engines', *I. Mech. E. Conf. on Internal Combustion Engines: Performance, Fuel Economy and Emissions*, London, December 8–9, 2009, Chandos Publishing, 2009, ISBN 978 1 843 34607 4

Katsoulakos P. S. (1983). 'Effectiveness of the combustion of emulsified fuels in diesel engines', *Int. Conf. on Combustion in Engineering*, Vol. II, I. Mech. E. Conf. Publication 1983–3, MEP, London

Katys G. P. (1964). *Continuous Measurement of Unsteady Flow* (ed. Walker G. E.) (translated by Barrett D. P.), Pergamon, Oxford

Kavalov B. and Peteves S. D. (2005). *Status and Perspectives of Biomass-to Liquid Fuels in the European Union*, Institute for Energy – JRC EUR 21745 EN, ISBN 92 894 9784 X

Kaye G. C. (1989). 'The evolution of the Sabre "Marathon" diesel engine for offshore powerboat racing', *Sprint Rated Diesel Engines*, I. Mech. E. Seminar, pp. 1–12, I. Mech. E., London

Keck J. C. (1982). 'Turbulent flame structure and speed in spark ignition engines', *Proc. 19th Int. Symp. on Combustion*, The Combustion Institute, pp. 1451–66

Keck J. C., Heywood J. B. and Noske G. (1987). 'Easy flame development and burning rates in spark-ignition engines', *SAE Paper 870104*

Khalighi B. (1990). 'Intake generated swirl and tumble motions in a four valve engine with various intake configurations', *SAE Paper 90059*

King L. V. (1914). 'On the convection of heat from small cylinders in a stream of fluid: determination of the convective constants of small platinum wires with application to hot wire anemometry', *Proc. Royal Soc.*, Vol. 214A, No. 14, p. 373

Kittelson D. B. and Collings N. (1987). 'Origin of the response of electrostatic particle probes', *SAE Paper 870476*

Knight B. E. (1960–61). 'Fuel injection system calculations', *Proc. I. Mech. E.*, No. 1

Knoll R. (1996). 'The two-stroke list diesel – a new approach to motorize small comfort cars', *FISITA Congress*, Prague, 16–23 June

Kobayashi H., Yoshimura K. and Hirayama T. (1984). 'A study on dual circuit cooling for higher compression ratio', *SAE Paper 841294*

Kojima K., Hirota T., Inoue T. and Yakushiji K. (1992). 'Effects of exhaust emission control devices and fuel composition on speciated emissions of SI engines', Paper C448/015, *Combustion in Engines*, Inst. Mech. Engrs Conf. Proc., pp. 145–9, MEP, London

Körfer T., Lamping M., Kolbeck A., Genz T., Pischinger S., Busch H. and Adolph D. (2007). 'The potential of downsizing of diesel engines considering performance and emissions challenges', *Internal Combustion Engines: Performance, Fuel Economy and Emissions*, I. Mech. E. Conf., 11–12 December, 2007

Krieger R. B. and Borman G. L. (1986). 'The computation of apparent heat release for internal combustion engines', *ASME Paper 66-WA/DGP-4*

Kume T., Iwamoto Y., Iida K., Murakami N., Akishino K. and Ando H. (1996). 'Combustion control technologies for direct injection SI engines', *SAE Paper* 960600

Kyriakides S. C. and Glover A. R. (1988). 'A study of the correlation between in-cylinder air motion and combustion in gasoline engines', in *Int. Conf. Combustion in Engines – Technology and Applications*, I. Mech. E. Conf. Proc., MEP, London

Ladommatos N. and Stone R. (1991). 'Conversion of a diesel engine for gaseous fuel operation at high compression ratio', *SAE Paper* 910849

Ladommatos N., Abdelhalim S. M., Zhao H. and Hu Z. (1998). 'The effects of carbon dioxide in exhaust gas recirculation on diesel engine emissions', *Proc. I. Mech. E.*, Vol. 212, Pt D, pp. 25–42

Lai M-C. D. and Dingle P. J. (2005). *Diesel Common Rail and Advanced Fuel Injection Systems*, SAE International

Lancaster D. R. (1976). 'Effects of engine variables on turbulence in a spark ignition engine', *SAE Paper* 760159

Lancefield T., Cooper L. and French B. (1996). 'Design the control and simulation of EGR', *Automotive Engineer*, Vol. 21, No. 1, pp. 28–31

Lavoie G. A., Heywood J. B. and Keck J. C. (1970). 'Experimental and theoretical study of nitric oxide formation in internal combustion engines', *Comb. Sci. Technology*, Vol. 1, pp. 313–26

Law C. K. (2006). *Combustion Physics*, CUP, Cambridge, ISBN-13: 978 0 521 87052 8

Lawes K. T. and Mason B. R. (2011). 'The RCV Engine. Technical Analysis and Real World Benefits', *Internal Combustion Engines: Improving Performance, Fuel Economy and Emissions*, I. Mech. E. Conf., Woodhead Publishing

Lee J. and Min K. (2005). 'Effects of needle response on spray characteristics in high pressure injector driven by piezo actuator for common-rail injection system', *Journal of Mechanical Science and Technology*, Vol. 19, No. 5, pp. 1194–1205

Lehrle R. S., West H. and Wyszynski M. L. (1995). 'On-line mass-spectrometric characterization of hydrocarbons in engine exhaust gases', *Proc. Inst. Mech. Engrs*, Vol. 209, Pt D, pp. 307–24

Leshner M. D. (1983). 'Evaporative engine cooling for fuel economy – 1983', *SAE Paper* 831261

Lewis B. and von Elbe G. (1961). *Combustion Flames and Explosions of Gases*, 2nd edn, Academic Press, New York

Light J. T. (1989), 'Advanced anhydrous engine cooling systems', *SAE Paper* 891635

Liou T-M., Hall M., Santavicca D. A. and Bracco F. V. (1984). 'Laser doppler velocimetry measurements in valved and ported engines', *SAE Paper* 840375

Livengood J. C. and Wu P. C. (1955). 'Correlation of autoignition phenomenon in internal combustion engines and rapid compression machines', *Proc 5th International Symposium on Combustion*, pp. 347–56, ISSN 0082 0784

Lovell W. G. (1948). 'Knocking characteristics of hydrocarbons', *Industrial and Engineering Chemistry*, Vol. 40, No. 12, pp. 2388–438

Lumsden G., Oude Nijeweme D., Fraser N. and Blaxill H. (2009). 'Development of a turbocharged direct injection downsizing demonstrator engine', *SAE Paper* 2009-01-1503

Lyon D. (1986). 'Knock and cyclic dispersion in a spark ignition engine', *Petroleum Based Fuels and Automotive Applications*, I. Mech. E. Conf. Proc., MEP, London

Ma T. (1988). 'Effect of variable valve timing on fuel economy', *SAE Paper* 880390

Maekawa M. (1957). *Text of Course*, JSME No. G36, p. 23

Maly R. R. (1984). 'Spark ignition: its physics and effect on the internal combustion engine', in Hilliard J. C. and Springer C. S. (eds), *Fuel Economy of Road Vehicles Powered by Spark Ignition Engines*, Plenum, New York

Maly R. R. (1998). 'Progress in combustion research', *I. Mech. E. Combustion Engines Group Prestige Lecture*, 8 October, London

Maly R. R. and Vogel M. (1978). 'Initiation and propagation of flame fronts in lean CH_4–air mixtures by the three modes of the ignition spark', *17th Int. Conf. on Combustion*, pp. 821–31, The Combustion Institute

Mark P. E. and Jetten W. (1986). 'Propylene glycol: a new base fluid for automotive coolants', in Beal R. E. (ed.), *Engine Coolant Testing: Second Symposium*, ASTM STP 87, pp. 61–77

Mattavi J. N. and Amann S. A. (1980). *Combustion Modelling in Reciprocating Engines*, Plenum, New York

Mauss F. and Peters N. (1993). 'Reduced kinetic mechanisms for pre-mixed methane–air flames', in Peters N. and Rogg B. (eds), *Reduced Kinetic Mechanisms for Applications in Combustion Systems*, Chapter 5, Springer-Verlag

May M. (1979). 'The high compression lean burn spark ignited 4-stroke engine', *I. Mech. E. Conf. Publication 1979–9*, MEP, London

McAllister M. and Buckley D. (2009). 'Future gasoline engine downsizing technologies – CO_2 improvements and engine design considerations', *I. Mech. E. Internal Combustion Engines: Performance, Fuel and Emissions Conference*, London

McMahon P. J. (1971). *Aircraft Propulsion*, Pitman, London

Mellor A. M. (1972). 'Current kinetic modelling techniques for continuous flow combustion systems', in Cornelius W. G. and Agnew W. G. (eds), *Emissions from Continuous Combustion Systems*, Plenum, New York

Mercer A. D. (1980a). 'Experience of the British Standards Institution in the field of engine coolants', in Ailor W. H. (ed.), *Engine Coolant Testing: State of its Art*, ASTM STP 705, pp. 24–41

Mercer A. D. (1980b). 'Laboratory research in the development and testing of inhibited coolant in boiling heat-transfer conditions', in Ailor W. H. (ed.), *Engine Coolant Testing: State of its Art*, ASTM STP 705, pp. 53–80

Merchant R., Bradbury I. P., Ashton S. and Vincent M. W. (1997). 'Effect on vehicle performance of changes in automotive diesel fuel composition', *Automotive Fuels for the 21st Century*, I. Mech. E. Conf. Publication, MEP, London

Merdjani S. and Sheppard C. G. W. (1993). 'Gasoline engine cycle simulation using the Leeds turbulent burning velocity correlations', *SAE Paper* 932640

Metghalchi M. and Keck J. C. (1982). 'Burning velocities of air with methanol, isooctane and indolene at high pressure and temperature', *Combustion and Flame*, Vol. 48, pp. 191–210

Meyer R. and Heywood J. B. (1997). 'Liquid fuel transport mechanisms into the cylinder of a firing port-injected SI engine during warm-up', *SAE Paper* 970865

Meyer E. W., Green R. and Cops M. H. (1984). 'Austin–Rover Montego programmed ignition system', Paper C446/84, *VECON '84 Fuel Efficient Power Trains and Vehicles*, I. Mech. E. Conf. Publication 1984–14, MEP, London

Mikulic L. A., Quissek F. and Fraidl G. K. (1990). 'Development of low emission high performance four valve engines', *SAE Paper* 90027

Miller J. A. and Bowman C. T. (1989) 'Mechanism and modelling of nitrogen chemistry in combustion', *Progress in Energy and Combustion Science*, Vol. 15, pp. 287–338

Millington B. W. and Hartles E. R. (1968). 'Frictional losses in diesel engines', *SAE Paper 680590*

Milton B. E. (1998). 'Control technologies in spark-ignition engines', in Sher E. (ed), *Handbook of Air Pollution from Internal Combustion Engines*, Academic Press, Boston, Massachussetts

Mittelbach M. and Remschmidt C. (2006). *Biodiesel – the Comprehensive Handbook*, 3rd edn, Selbstverl, Graz, Austria, ISBN 3 200 00249 2

Mock R. and Lubitz K. (2008). 'Piezoelectric injection systems', chapter 13 in Heywang W., Lubitz K. and Wersing W. (eds), *Piezoelectricity – Evolution and Future of a Technology*, Springer Series in Materials Science

Monaghan M. L. (1990). 'The diesel passenger car in a green world', *SIA Diesel Congress*, Lyon, France

Moore C. H. and Hoehne J. L. (1986). 'Combustion chamber insulation effect on the performance of a low heat rejection Cummins V-903 engine', *SAE Paper 860317*

Mullins P. (1995). 'A look at the future of diesel piston assemblies', *High Speed Diesels and Drives*, June, pp. 61–3

Mundy H. (1972). 'Jaguar V12 engine: its design and development history', *Proc. I. Mech. E.*, Vol. 186, paper 34/72, pp. 463–77

Muranaka S., Takagi Y. and Tshida T. (1987). 'Factors limiting the improvement in thermal efficiency of SI engine at higher compression ratio', *SAE Paper 870548*

Nahum A., Foster-Pegg R. W. and Birch D. (1994). '*The Rolls-Royce Crecy*', Rolls-Royce Heritage Trust, Derby, England

Nakajima Y., Sugihara K. and Takagi Y. (1979). 'Lean mixture or EGR – which is better for fuel economy and NO$_x$ reduction?', *Proceedings of Conference on Fuel Economy and Emissions of Lean Burn Engines*, I. Mech. E. Conf. Proc., MEP, London

Nandi M. K. and Jacobs D. C. (1995). 'Cetane response of di-tertiary-butyl peroxide in different diesel fuels', *SAE Paper 952368*

Narasimhan S. L. and Larson J. M. (1985). 'Valve gear wear and materials', *SAE Paper 851497*

Needham J. R. and Doyle D. M. (1985). 'The combustion and ignition quality of alternative fuels in light duty diesels', *SAE Paper 852101*

Newton K., Steeds W. and Garrett T. K. (1983). *The Motor Vehicle*, 10th edn, Butterworth, London

Ng K., Yianneskis M., Foster D. P. E. and Ganti G. (1996). 'A numerical investigation of inlet charge direction on the scavenging behaviour of a two-stroke engine', *Proc. 3rd Int. Conf. on Computers in Reciprocating Engines and Gas Turbines*, I. Mech. E. Conf. Trans., C499/056/96, pp. 269–80

Obert E. F. (1973). *Internal Combustion Engines and Air Pollution*, Harper & Row, New York

O'Callaghan T. M. (1974). 'Factors affecting the coolant heat transfer surfaces in an IC engine', *General Engineer*, October 1974, pp. 220–30

Oh S. H. (1988). 'Thermal response of monolithic catalytic converters during sustained engine misfiring: a computational study', *SAE Paper 881591*

Ohata A. and Ishida Y. (1982). 'Dynamic inlet pressure and volumetric efficiency of four-cycle four cylinder engine, *SAE Paper 820407*

Olikara C. and Borman G. L. (1975). 'A computer program for calculating properties of equilibrium combustion products with some applications to I.C. engines', *SAE Paper 750468*

Onion G. and Bodo L. B. (1983). 'Oxygenate fuels for diesel engines: a survey of worldwide activity', *Biomass*, Vol. 3, pp. 77–133

Osborne A. G. (1985). 'Low cost FM transducer for diesel injector needle lift', *Developments in Measurements and Instrumentation in Engineering*, I. Mech. E. Conf. Proc., MEP, London

Osborne R., Stokes J., Ceccarini D., Jackson N., Lake T., Joyce M., Visser S., Miche N., Begg S. M., Heikal M. R., Kalian N., Zhao H. and Ma T. (2008). 'The 2/4SIGHT project – development of a multi-cylinder two-stroke/four-stroke switching gasoline engine', *JSAE Technical Paper 20085400*, 79-08

Otte R., Müller J., Weberbaurer F. and Guggenberger B. (2007). 'Homogeneous diesel combustion challenges under vehicle compliant operating conditions', *Internal Combustion Engines: Performance, Fuel Economy and Emissions*, I. Mech. E. Conf., 11–12 December, 2007

Owen K. and Coley T. (1995). *Automotive Fuels Reference Book*, 2nd edn, SAE, Warrendale, Pennsylvania

Palmer F. H. (1986). 'Vehicle performance of gasoline containing oxygenates', Paper C319/86, *Int. Conf. Petroleum Based Fuels and Automotive Applications*, pp. 36–46, I. Mech. E. Conf. Publication 1986–11, MEP, London

Palmer F. H. and Smith A. M. (1985). 'The performance and specification of gasoline', in Hancock E. G. (ed.), *Technology of Gasoline*, Blackwell Scientific Publications, Oxford

Partridge I. M. and Greeves G. (1998). 'Interpreting diesel combustion with a fuel spray computer model', *Combustion Engines and Hybrid Vehicles*, I. Mech. E. Conf. Proc., C529/025/98, London

Pashley N. C. (1994). The Destruction of Waste Volatile Organic Compounds Using a Spark Ignition Lean Burn Gas Engine, *MSc Thesis*, UMIST, Manchester

Peters N. (1993). 'Flame calculations with reduced mechanisms', in Peters N. and Rogg B. (eds), *Reduced Kinetic Mechanisms for Applications in Combustion Systems*, Chapter 1, Springer-Verlag

Peters N. and Rogg B. (eds) (1993). *Reduced Kinetic Mechanisms for Applications in Combustion Systems*, Springer-Verlag

Pfluger F. (1997). 'Two stage turbocharging for commercial diesel engines', *Diesel Progress*, Jan.–Feb., pp. 18–20

Phillips I. (1994). 'Telling little green lies?' *The Times*, 10 Oct., London

Piccone A. and Rinolfi R. (1998). 'Fiat third generation DI diesel engines', *Future Engine and System Technologies*, I. Mech. E. Seminar Publication, Professional Engineering Publications, London

Pischinger F. F. (1998). 'Compression-ignition engines – introduction', in Sher E. (ed.), *Handbook of Air Pollution from Internal Combustion Engines*, Academic Press, Boston, Massachusetts

Pischinger R. and Cartellieri W. (1972). 'Combustion system parameters and their effect upon diesel engine exhaust emissions', *SAE Paper 720756. Also in SAE Trans.*, Vol. 81

Plint M. A. and Martyr A. (1995). *Engine Testing: Theory and Practice*, Butterworth-Heinemann, London

Polach W. and Leonard R. (1994). 'Exhaust gas treatment', in Adler U., Bauer H. and Beer A. (eds), *Diesel Fuel Injection*, Robert Bosch, Stuttgart

Poloni M., Winterbone D. E. and Nichols J. R. (1987). 'Calculation of pressure and temperature discontinuity in a pipe by the method of characteristics and the two step Lax–Wendroff method', *ASME Conference*, Boston, 14–16 December 1987

Pouille J-P., Lauga V., Schonfeld S., Strobel M. and Stommel P. (1998). 'Application strategies of lean NO$_x$ catalyst systems to diesel passenger cars', *Future Engine and Systems Technologies*, I. Mech. E. Seminar Publication, Professional Engineering Publications, London

Poulos S. G. and Heywood J. B. (1983). 'The effect of chamber geometry on spark-ignition engine combustion', *SAE Paper 830334. Also in SAE Trans.*, Vol. 92

Price P. (2009). 'Direct injection gasoline engine particulate emissions', *DPhil Thesis*, University of Oxford

Price P., Stone R., Collier T. and Davies M. (2006). 'Particulate matter and hydrocarbon emissions measurements: comparing first and second generation DISI with PFI in single cylinder optical engines', *SAE Paper* 2006-01-1263.

Price P., Stone R., Misztal J., Xu H., Wyszynski M., Wilson T. and Qiao J. (2007). 'Particulate emissions from a gasoline homogeneous charge compression ignition engine', *SAE Paper* 2007-01-0209

Priede T. and Anderton D. (1974). 'Likely advances in mechanics, cooling, vibration and noise of automotive engines', *Proc. I. Mech. E.*, Vol. 198D, No. 7, pp. 95–106

Radermacher K. (1982). 'The BMW eta engine concept', *Proc. I. Mech. E.*, Vol. 196

Raine R. R., Stone C. R. and Gould J. (1995). 'Modelling of nitric oxide formation in spark ignition engines with a multi-zone burned gas', *Combustion and Flame*, Vol. 102, pp. 241–55

Raj A., Celnik M. S., Shirley R., Sander M., Patterson R. I. A., West R. H. and Kraft M. (2009). 'A statistical approach to develop a detailed soot growth model using PAH characteristics', *Combustion and Flame*, Vol. 156, pp. 896–913

Rao K. K., Winterbone D. E. and Clough E. (1992). 'Combustion and emission studies in high-speed DI diesel engines', Paper C448/070, *Combustion in Engines*, Inst. Mech. Engrs. Conf. Proc., pp. 137–43, MEP, London

Rask R. B. (1981). 'Comparison of window, smoothed ensemble, and cycle-by-cycle data reduction techniques for laser Doppler anemometer measurements of in-cylinder velocity', in Morel T., Lohnmann R. P. and Rackley J. M. (eds), *Fluid Mechanics of Combustion Systems*, pp. 11–20, ASME

Rassweiler G. M. and Withrow L. (1938). 'Motion pictures of engine flame correlated with pressure cards', *SAE Paper* 800131 (originally presented in January 1938)

Reavell K. StJ., Collings N., Peckham M. and Hands T. (1997). 'Simultaneous fast response NO and HC measurements from a spark ignition engine', *SAE Paper* 971610

Reavell K., Hands T. and Collings N. (2002). 'A fast response particulate spectrometer for combustion aerosols', *SAE Paper* 2002-01-2714

Reeves M., Garner C. P., Dent J. C. and Halliwell N. A. (1996). 'Particle image velocimetry of in-cylinder flow in a multi-valve internal combustion engine', *Proc. I. Mech. E.*, Pt 210D, pp. 63–70

Reid R. C., Prausnitz J. M. and Poling B. E. (1987). *The Properties of Gases and Liquids*, 4th edn, McGraw-Hill, New York

Reynolds W. C. (1992). *STANJAN User Manual*, Stanford University

Rhodes D. B. and Keck J. C. (1985). 'Laminar burning speed measurements of indolene–air–diluent mixtures at high pressures and temperature', *SAE Paper* 850047

Ricardo H. R. and Hempson J. G. G. (1968). *The High Speed Internal Combustion Engine*, 5th edn, Blackie, London

Rieck J. S., Collins N. R. and Moore J. S. (1998). 'OBD-II performance of three-watt catalysts', *Automotive Engineering*, Vol. 106, No. 7, pp. 33–5

Roark R. J. and Young W. C. (1976). *Formulas for Stress and Strain*, 5th edn, McGraw-Hill, New York

Rogers G. F. C. and Mayhew Y. R. (1980a). *Engineering Thermodynamics, Work and Heat Transfer*, 3rd edn, Longman, London

Rogers G. F. C. and Mayhew Y. R. (1980b). *Thermodynamic and Transport Properties of Fluids, SI Units*, 3rd edn, Blackwell, Oxford

Rogers G. F. C. and Mayhew Y. R. (1988). *Thermodynamic and Transport Properties of Fluids*, 4th edn, Blackwell, Oxford

Rose D., Ladommatos N. and Stone R. (1994). 'In-cylinder mixture excursions in a port-injected engine during fast throttle-openings', SAE Congress 1994, *SAE Paper* 940382. *Also in* SP-1029 *Electronic Engine Controls*, and *SAE Transactions*, Vol. 103

Rowe L. C. (1980). 'Automotive engine coolants: a review of their requirement and methods of evaluation', in Ailor W. H. (ed.), *Engine Coolant Testing: State of its Art*, ASTM STP 705, pp. 3–23

Russell M. F. (1998). 'The dependence of diesel combustion on injection rate', *Future Engine and System Technologies*, I. Mech. E. Seminar Publication, Professional Engineering Publications, London

Sadler M., Stokes J., Edwards S. P., Zhao H. and Ladommatos N. (1998). 'Optimization of the combustion system for a direct injection gasoline engine, using a high speed in-cylinder sampling valve', *Future Engine and System Technologies*, I. Mech. E. Seminar Publication, Professional Engineering Publications, London

SAE (1987). 'SAE recommended practice, engine terminology and nomenclature', *SAE Handbook*, p. 24.01, SAE, Warrendale, Pennsylvania

Sandford M. H., Allen J. and Tudor R. (1998). 'Reduced fuel consumption and emissions through cylinder deactivation', *Future Engine and System Technologies*, I. Mech. E. Seminar Publication, Professional Engineering Publications, London

Sandford M., Page G. and Crawford P. (2009). 'The All New AJV8', *SAE Technical Paper* 2009-01-1060

Schenk P. M., Thomas-Hall S. R., Stephens E., Marx U. C., Mussgnug J. H., Posten C., Kruse O. and Hankamer B. (2008). 'Second generation biofuels: high-efficiency microalgae for biodiesel production', *Bio-Energy Research*, Vol. 1, pp. 20–43

Scherenberg H. (1955). 'Rückblick über 25 Jahre Benzin-Einspritzung in Deutschland', *MTZ*, Vol. 9, No. 16

Schindler W., Haisch C., Beck H. A., Niessner R., Jacob E. and Rothe E. (2004). 'A photo-acoustic sensor system for time resolved quantification of diesel soot emissions', *SAE Paper* 2004-01-0968

Schönborn A., Ladommatos N., Allan R., Williams J. and Rogerson J. (2008). 'Effect of the molecular structure of individual fatty acid alcohol esters (biodiesel) on the formation of NO_x and particulate matter in the diesel combustion process', *SAE Paper* 2008-01-1578

Schwarz C. H., Schunemann E., Durst B., Fischer J. and Witt A. (2006). 'Potentials of the spray-guided BMW DI combustion system', *SAE Paper* 2006-01-1265

Scott D. (1996). 'Piston technology trends', *Automotive Engineer*, Vol. 21, No. 1, p. 68.S

Seizinger D. E., Marshall W. F. and Brooks A. L. (1985). 'Fuel influences on diesel particulates', *SAE Paper* 850546

Sheehan J., Dunahay T., Benemann J. and Roessler P. (1998). 'A look back at the U.S. Department of Energy's aquatic species program – biodiesel from algae', *National Renewable Energy Lab. Report* NREL/TP-580-24190

Sheppard C. G. W. (1998). Private communication, University of Leeds

Sher E. (1990). 'Scavenging the two-stroke engine', *Progress in Energy and Combustion Science*, Vol. 16, pp. 95–124

Sher E. (1998a). 'Environmental aspects of air pollution', in Sher E. (ed.), *Handbook of Air Pollution from Internal Combustion Engines*, Academic Press, Boston, Massachusetts

Sher E. (ed.) (1998b). *Handbook of Air Pollution from Internal Combustion Engines*, Academic Press, Boston, Massachusetts

Shillington S. A. C. (1998). 'The unstoppable versus the immovable – a personal view of the future facing gasoline direct injection in the face of Euro IV', *Future Engine and System Technologies*, I. Mech. E. Seminar Publication, Professional Engineering Publications, London

Sihling K. and Woschni G. (1979). 'Experimental investigation of the instantaneous heat transfer in the cylinder of a high speed diesel engine', *SAE Paper* 790833

Simonini S. (1988). Combustion Analysis of IC Engines Using a CCD Camera and Image Processing, *MSc Thesis*, University of Oxford

Smallbone A., Bhave A., Morgan N., Kraft M., Cracknell R. and Kalghatgi G. (2010). 'Simulating combustion of practical fuels and blends for modern engine applications using detailed chemical kinetics', *SAE Technical Paper* 2010-01-0572

Smith P. H. (1967). *Valve Mechanisms for High Speed Engines*, Foulis, Yeovil

Smith P. H. (1968). *Scientific Design of Exhaust and Intake Systems*, Foulis, Yeovil

Smith G. D. (1985). *Numerical Solution of Partial Differential Equations – Finite Difference Methods*, Oxford University Press

de Soete G. C. (1983). 'Propagation behaviour of spark ignited flames in the early stages', *Int. Conf. on Combustion in Engineering*, Vol. 1, Mech. E. Conf. Proc., MEP, London

Soltau J. P. (1960). 'Cylinder pressure variations in petrol engines', *I. Mech. E. Conf. Proc.*, July 1960, pp. 96–116

Solvat O., Marez P. and Belot G. (2000). 'Passenger car serial application of a particulate filter system on a common-rail, direct-injection diesel engine', *SAE Paper* 2000-01-0473

Sorrell A. J. (1989). Spark Ignition Engine Performance During Warm-up, *PhD Thesis*, Brunel University, London

Soteriou C., Andrews R. and Smith M. (1998). 'Diesel injection – laser light sheet illumination of development of cavitation in orifices', *Combustion Engines and Hybrid Vehicles*, I. Mech. E. Conf. Proc., London

Spicher U. and Heindenreich T. (2010). 'Stratified-charge combustion in direct injection gasoline engines', in Zhao H. (ed.), *Advanced Direct Injection Combustion Engine Technologies and Development: Gasoline and Gas Engines*, Woodhead Publishing Ltd and CRC Press, ISBN 1 84569 389 2

Spindt R. S. (1965). 'Air–fuel ratios from exhaust gas analysis', *SAE Paper* 650507

Stevens J. L. and Tawil N. (1989). 'A new range of industrial diesel engines with emphasis on cost reduction techniques', Paper C372/028, *The Small Internal Combustion Engine*, pp. 137–44, I. Mech. E. Conf. Proc., MEP, London

Stokes J., Hundleby G. E., Lake T. H. and Christie M. J. (1992). 'Development experience of a poppet-valved two-stroke flagship engine', *SAE Paper* 920778

Stone C. R. (1988). 'The efficiency of Roots compressors, and compressors with fixed internal compression', *Proc. I. Mech. E.*, Vol. 202, No. A3

Stone C. R. (1989a). 'Air flow measurement in internal combustion engines', *SAE Paper* 890242

Stone R. (1989b). *Motor Vehicle Fuel Economy*, Macmillan, London

Stone R. (1989c). *Review of Induction System Design*, Notes for Cranfield MSc Course in Automotive Product Engineering

Stone C. R. and Green-Armytage D. I. (1987). 'Comparison of methods for the calculation of mass fraction burnt from engine pressure–time diagrams', *Proc. I. Mech. E.*, Vol. 201, No. D1

Stone R. and Kwan E. (1989). 'Variable valve actuation mechanisms and the potential for their application', *SAE Paper* 890673. *Also in SAE Trans.*, Vol. 98

Stone R. and Ladommatos N. (1992a). 'Combustion analysis of sunflower oil in a diesel engine and its impact on lubricant quality', *SAE Paper* 921631

Stone C. R. and Ladommatos N. (1992b). 'The measurement and analysis of swirl in steady flow', *SAE Off-Highway Congress*, Milwaukee, *SAE Paper* 921642. *Also in SAE* SP-931, *Diesel Combustion, Emissions and Exhaust After Treatment* (ed. Ziejewski M.)

Stone C. R. and Steele A. B. (1989). 'Measurement and modelling of ignition system energy and its effect on engine performance', *Proc. I. Mech. E.*, Vol. 203, Pt D, No. 4, pp. 277–86

Stone C. R. and Wright S. D. (1994). 'Non-linear and unsteady flow analysis of flow in a viscous flowmeter', *Trans. Inst. Meas. Control*, Vol. 16, No. 3, pp. 128–42

Stone R. D., Crabb D., Richardson P. and Draper A. (1990). 'The design and development of the Rover K16 engine', *Proc. I. Mech. E.*, Vol. 204, Pt D4, pp. 221–36

Stone C. R., Carden T. R. and Podmore T. (1993). 'Analysis of the effect of swirl on combustion and emissions in a spark ignition engine', *Proc. I. Mech. E.*, Vol. 207, Pt D4, pp. 295–305

Suzuki T. (1997). 'Development and perspective of the diesel combustion system for commercial vehicles', *Combustion Engine Group Prestige Lecture*, Institution of Mechanical Engineers, 22 May 1997, London

Swartzlander M. (2006). 'Eaton TVS supercharger – performance improvement of the roots supercharger through optimization of rotor helix angle and port geometry', *11th Supercharging Conference*, Dresden

Swartzlander M. (2010). 'Performance range adaptation in modern mechanical TVS® superchargers',*15th Supercharging Conference*, Dresden

Szybist J. P., Song J., Alam M. and Boehman A. L. (2007). 'Biodiesel combustion, emissions and emission control', *Fuel Processing Technology*, Vol. 88, No. 7, pp. 679–91

Tabaczynski R. J. (1976). 'Turbulent combustion in spark-ignition engines', *Prog. Energy Combustion Sciences*, Vol. 2, pp. 143–65

Tabaczynski R. J. (1983). 'Turbulence measurements and modelling in reciprocating engines – an overview', *Int. Conf. on Combustion in Engineering*, Vol. I, I. Mech. E. Conf. Publication 1983–3, MEP, London

Tabaczynski R. J., Ferguson C. R. and Radhakrishnan K. (1977). 'A turbulent entrainment model for spark-ignition engine combustion', *SAE Paper* 770647. *Also in SAE Trans.*, Vol. 86

Tabaczynski R. J., Trinker F. H. and Shannon B. A. S. (1980). 'Further refinement and validation of a turbulent flame propagation model for spark ignition engines', *Combustion and Flame*, Vol. 39, pp. 111–21

Tagawa M. and Ohta Y. (1997). 'Two-thermocouple probe for fluctuating temperature measurement in combustion – rational estimation of mean and fluctuating time constants', *Combustion and Flame*, Vol. 109, pp. 549–60

Tawfig M., Jager D. and Charlton S. J. (1990). 'In-cylinder measurement of mixture strength in a divided-chamber natural gas engine', *Fast Response FID Workshop*, September 1990, Brunel University, London

Taylor C. F. (1985a). *The Internal-Combustion Engine in Theory and Practice*, Vol. I, MIT Press, Cambridge, Massachusetts

Taylor C. F. (1985b). *The Internal-Combustion Engine in Theory and Practice*, Vol. II, MIT Press, Cambridge, Massachusetts

Thompson A. A., Lambert S. W. and Mulqueen S. (1997). 'Prediction and precision of cetane number improver response equations', *SAE Paper 972901*

Thornton N. and Blohm T. J. (2002). 'Milestone completion report for system emission reduction (SER) analysis', NREL/MP-540-33044, *National Renewable Energy Laboratory*, Colorado

Thring R. H. (1989). 'Homogeneous charge compression ignition (Hcci) engines', *SAE Paper 892068*

Timoney D. J. (1985). 'A simple technique for predicting optimum fuel air mixing conditions in a direct injection diesel engine with swirl', *SAE Paper 851543*

Toda T., Nohira H. and Kobashi K. (1976). 'Evaluation of burned gas ratio (BGR) as a predominant factor to NO_x', *SAE Paper 760765. Also in SAE Trans.*, Vol. 85

Tomita E. and Hamamoto Y. (1988). 'The effect of turbulence on combustion in cylinder of spark ignition engine – evaluation of entrainment model', *SAE Paper 880128*

Trier C. J., Rhead M. M. and Fussey D. E. (1990). 'Evidence for the pyrosynthesis of parent polycyclic aromatic compounds in the combustion chamber of a diesel engine', Paper C394/003, *Automotive Power Systems – Environment and Conservation*, I. Mech. E. Conf. Proc., MEP, London

Tsai S-F. (1994). An Improved Simulation of Two-Stroke Engines, *PhD Thesis*, UMIST

Tsuchiya K. and Hiramo S. (1975). 'Characteristics of 2-stroke motorcycle exhaust HC emission and effects of air–fuel ratio and ignition timing', *SAE Paper 750908*

Tullis S. and Greeves G. (1996). 'Improving NO_x versus BSFC and EUI 200 using EGR and pilot injection for heavy-duty diesel engines', *SAE Paper 960843*

Turner J. W. G. and Pearson R. J. (2010). 'The turbocharged direct injection spark-ignition engine', in Zhao H. (ed.), *Advanced Direct Injection Combustion Engine Technologies and Development: Gasoline and Gas Engines*, Woodhead Publishing Ltd and CRC Press, ISBN 1 84569 389 2

Turner J. W. G., Blundell D. W., Pearson R. J., Patel R., Larkman D. B., Burke P., Richardson S., Green N. M., Brewster S., Kenny R. G. and Kee R. J. (2010). 'Project Omnivore: a variable compression ratio ATAC 2-stroke engine for ultra-wide-range HCCI operation on a variety of fuels', *SAE Paper 2010-01-1249*

Twigg M. V. and Phillips P. R. (2009). 'Cleaning the air we breathe – controlling diesel particulate emissions from passenger cars', *Platinum Metals Rev.*, Vol. 53, No. 1, pp. 27–34

Twiney B., Stone R., Chen X. and Edmunds G. (2010a). 'Investigation of combustion robustness in catalyst heating operation on a spray guided DISI engine, Part 1 – measurements of spark parameters and combustion', *SAE Paper 2010-01-0593*

Twiney B., Stone R., Chen X. and Edmunds G. (2010b). 'Investigation of combustion robustness in catalyst heating operation on a spray guided DISI engine, Part II – measurements of spray development, combustion imaging and emissions', *SAE Paper 2010-01-0603*

Uchiyama H., Chiku T. and Sayo S. (1977). 'Emission control of two-stroke automobile engine', *SAE Paper 770766. Also in SAE Trans.*, Vol. 86

Ueda K., Kaihara K., Kurose K. and Ando H. (2001). 'Idling stop system coupled with quick start features of gasoline direct injection', *SAE Paper 2001-01-0545*

Unger H., Schneider J., Schwarz C. and Koch K-F. (2008). 'Die VALVETRONIC–Erfahrungen aus 7 Jahren Großserie und Ausblick in die Zukunft', *29th Internationales Wiener Motorensymposium 2008*

Veynante D. and Vervisch L. (2002). 'Turbulent combustion modeling', *Progress in Energy and Combustion Science*, Vol. 28, Issue 3, pp. 193–266, ISSN 0 360 1285, 10.1016/S0360-1285(01)00017-X

Vincent M. W. (1991). *Valve Seat Recession: Use of Unleaded Gasoline in Older Engines*, Federation of British Historic Vehicle Clubs

Wade W. R., Havstad P. H., Ounsted E. J., Trinkler F. H. and Garwin I. J. (1984). 'Fuel economy opportunities with an uncooled DI diesel engine', *SAE Paper 841286. Also in SAE SP-610*

Wakeman A. C., Ironside J. M., Holmes M., Edwards S. I. and Nutton D. (1987). 'Adaptive engine controls for fuel consumption and emission reduction', *SAE Paper 870083*

Walker J. (1997). 'Caterpillar presents new engines, plans for growth', *Diesel Progress*, Sept.–Oct., pp. 66–9

Walker J. (1998). 'Low sulfur fuel stimulates particulate trap sales', *Diesel Progress*, May–June, pp. 78–9

Wallace W. B. (1968). 'High-output medium-speed diesel engine air and exhaust system flow losses', *Proc. I. Mech. E.*, Vol. 182, Pt 3D, pp. 134–44

Wallace F. J. (1986a). 'Engine simulation models with filling-and-emptying methods', in Horlock J. H. and Winterbone D. E. (eds), *The Thermodynamics and Gas Dynamics of Internal Combustion Engines*, Vol. II, Oxford University Press (*see also* Benson, 1982)

Wallace F. J. (1986b). 'Filling and emptying methods', in Horlock J. H. and Winterbone D. E. (eds), *The Thermodynamics and Gas Dynamics of Internal Combustion Engines*, Vol. II, Oxford University Press (*see also* Benson, 1982)

Wallace F. J., Tarabad M. and Howard D. (1983). 'The differential compound engine – a new integrated engine transmission system for heavy vehicles', *Proc. I. Mech. E.*, Vol. 197A

Walzer P. (1990). 'Automotive power systems for future environmental problems', Paper C394/036, *Automotive Power Systems – Environment and Conservation*, I. Mech. E. Conf. Proc., MEP, London

Walzer P., Heirrich H. and Langer M. (1985). 'Ceramic components in passenger-car diesel engines', *SAE Paper 850567*

Wang X., Stone R., Buttsworth D., Stevens R. and Arita A. (2006). 'Finite element analysis of eroding type surface thermocouple with application to engine heat flux measurement', *SAE Paper 2006-01-1045*

Warga J. (1994). 'Nozzles and nozzle holders', in Adler U., Bauer H. and Beer A. (eds), *Diesel Fuel Injection*, Robert Bosch, Stuttgart

Watanabe Y., Ishikawa H. and Miyahara M. (1987). 'An application study of evaporative cooling to heavy duty diesel engines', *SAE Paper 870023*

Waters M. H. L. (1992). 'Road vehicle fuel economy', *State of the Art Review 3*, HMSO, London

Watkins L. H. (1991). 'Air pollution from road vehicles', *State of the Art Review 1*, HMSO, London

Watson N. (1979). *Combustion Gas Properties*, Internal Report, Dept of Mechanical Engineering, Imperial College of Science and Technology, London

Watson N. and Janota M. S. (1982). *Turbocharging the Internal Combustion Engine*, Macmillan, London

Watson N., Wijeyakumar S. and Roberts G. L. (1981). 'A microprocessor controlled test facility for transient vehicle engine system development', *3rd Int. Conf. on Automotive Electronics*, MEP, London

Weertman W. L. and Dean J. W. (1981). 'Chrysler Corporation's new 2.2 liter 4 cylinder engine', *SAE Paper 810007*

Wentworth J. T. (1971). 'Effect of combustion chamber surface temperature on exhaust hydrocarbon concentration', *SAE Paper 710587*

Whitehouse N. D. (1987). 'An estimate of local instantaneous condition in a diesel engine', *1st Int. Conf. on Heat and Mass Transfer in Gasoline and Diesel Engines*, Dubrovnik

Whitehouse J. A. and Metcalfe J. A. (1956). 'The influence of lubricating oil on the power output and fuel consumption of modern petrol and compression ignition engines', *MIRA Report* 1956/2

Whitehouse N. D. and Way R. J. B. (1968). 'Rate of heat release in diesel engines and its correlation with fuel injection data', *Proc. I. Mech. E.*, Vol. 177, No. 2, pp. 43–63

Whiteley F. (1995). 'VVC for Rover K series', *Automotive Engineer*, Vol. 20, No. 2, pp. 52–3

Whiteley F. (1996). 'Rover KV6 engine', *Automotive Engineer*, Vol. 21, No. 2, p. 56

Whiteman P. R. (1989). 'PC based engine combustion analysis', *Session 24: Computer Aided Engineering for Powertrains*, Autotech, I. Mech. E., London

Wiebe I. (1967). 'Halbempirische Formel für die Verbrennungsgeschwindigkeit', *Verlag der Akademie der Wissenschaften der UdSSR*, Moscow

Wiedenmann H. M., Raff L. and Noack R. (1984). 'Heated zirconia oxygen sensor for stoichiometric and lean air–fuel ratios', *SAE Paper* 840141

Willumeit H. P. and Steinberg P. (1986). 'The heat transfer within the combustion engine', *MTZ*, Vol. 47, No. 1, pp. 9–14

Willumeit H. P., Steinberg P., Otting H., Scheibner B. and Lee W. (1984). 'New temperature control criteria for more efficient gasoline engines', *SAE Paper* 841292

Wilson C. E., Sadler J. P. and Michels W. J. (1983). *Kinematics and Dynamics of Machinery*, Harper & Row, New York

Winterbone D. E. (1986). 'Transient performance', in Horlock J. H. and Winterbone D. E. (eds), *The Thermodynamics and Gas Dynamics of Internal Combustion Engines*, Vol. II, Oxford University Press (*see also* Benson, 1982)

Winterbone D. E. (1990a). 'The theory of wave action approaches applied to reciprocating engines', in Weaving J. H. (ed.), *Internal Combustion Engineering*, Elsevier Applied Science, London and New York

Winterbone D. E. (1990b). 'Application of wave action techniques', in Weaving J. H. (ed.), *Internal Combustion Engineering*, Elsevier Applied Science, London and New York

Winterbone D. E. and Pearson R. J. (1999). *Design Techniques for Engine Manifolds: Wave Action Methods for IC Engines*, Professional Engineering Publications

Winterbone D. E. and Pearson R. J., (2000). *Theory of Engine Manifold Design: Wave Action Methods for IC Engines*, Professional Engineering Publications

Winterbone D. E., Yates D. A. and Clough E. (1997). 'Combustion processes in engines', in Ray S. (ed.), *High Speed Photography and Photonics*, pp. 362–92, Focal Press, London

Wirth M., Zimmermann D., Friedfeldt R., Caine J., Schamel A., Davies M., Peirce G., Storch A., Ries-Muller K., Gansert K. P., Pilgram G., Ortmann R., Wurfel G. and Gerhardt J. (2004). 'A cost optimised gasoline spray guided direct injection system for improved fuel economy', in *I. Mech. E. Fuel Economy and Engine Downsizing*

Witze P. O. (1980). 'A critical comparison of hot-wire anemometry and laser Doppler velocimetry for IC engine applications', *SAE Paper* 800132

Witze P. O. and Green R. M. (1997). 'LIF and flame-emission imaging of liquid fuel films and pool fires in an SI engine during a simulated cold start', *SAE Paper* 970866

Wolf G. (1982). 'The large bore diesel engine', *Sulzer Technical Review*, Vol. 3

Worthen R. P. and Rauen D. G. (1986). 'Valve parameters measured in firing engine', *Automotive Engineering*, Vol. 94, No. 6, pp. 40–7

Woschni G. (1967). 'A universally applicable equation for the instantaneous heat transfer coefficient in the internal combustion engine', *SAE Paper* 670931. *Also in SAE Trans.*, Vol. 76, p. 3065

Wyszynski L. P., Stone C. R. and Kalghatgi G. T. (2002). 'The volumetric efficiency of direct and port injection gasoline engines with different fuels', *SAE Paper* 2002-01-0839

Yamaguchi J. (1986). 'Optical sensor feeds back diesel ignition timing', *Automotive Engineering*, Vol. 94, No. 4, pp. 84–5

Yamaguchi J. (1997). 'Toyota readies gasoline/electric hybrid system', *Automotive Engineering*, Vol. 105, No. 7, pp. 55–8

Yamamoto H., Kato F., Kitagawa J. and Machida M. (1991). 'Warm-up characteristics of thin wall honeycomb catalysts', *SAE Paper* 910611

Yates D. (1988). 'The compact high-speed gas sampling valve for an internal combustion engine', *Experimental Methods in Engine Research and Development*, I. Mech. E. Seminar, MEP, London

Yokota K., Hattori H., Shimizu M. and Furukawa H. (1986). 'A high BMEP diesel engine with variable geometry turbocharger', *Turbochargers and Turbocharging*, I. Mech. Conf. Proc. C119/86

Zambare V. V. (1998). Study of Combustion and Emissions in Direct-Injection Diesel Engines Using the Two-Colour Method, *PhD Thesis*, UMIST, Manchester

Zeldovich Ya. B. (1946). 'The oxidation of nitrogen in combustion and explosions', *Acta Physiochim. U.R.S.S.*, Vol. 21, pp. 577–628

Zhao H. (2007). *HCCI and CAI engines for the automotive industry*, Woodhead Publishing

Zhao H. (ed) (2010). *Advanced Direct Injection Combustion Engine Technologies and Development: Gasoline and Gas Engines*, Woodhead Publishing Ltd and CRC Press, ISBN 1 84569 389 2

Zhao H. and Ladommatos N. (2001). *Engine Combustion Instrumentation and Diagnostics*, SAE, Warrendale, Pennsylvania

Zhao H., Collings N. and Ma T. (1994). 'Two-dimensional temperature distributions of combustion chamber surfaces in a firing spark ignition engine', *Proc. I. Mech. E.*, Part 208D, pp. 99–108

Zhao F., Lai M. C. and Harrington D. L. (1999). 'Automotive spark-ignited direct-injection gasoline engines', *Prog. Energy Comb. Sci.*, Vol. 25. pp. 437–562

Ziejewski M. (1983). 'Vegetable oils as a potential alternate fuel in direct injection diesel engines', *SAE Paper* 831359

Index